# Introduction to Hydrology

Series in Civil Engineering

*Series Editor*
Russell C. Brinker
*New Mexico State University*

| | |
|---|---|
| Brinker and Wolf | ELEMENTARY SURVEYING, *6th edition* |
| Clark, Viessman, and Hammer | WATER SUPPLY AND POLLUTION CONTROL, *3rd edition* |
| Jumikis | FOUNDATION ENGINEERING |
| McCormac | STRUCTURAL ANALYSIS, *3rd edition* |
| McCormac | STRUCTURAL STEEL DESIGN, *2nd edition* |
| Meyer | ROUTE SURVEYING AND DESIGN, *4th edition* |
| Mikhail | OBSERVATIONS AND LEAST SQUARES |
| Moffitt | PHOTOGRAMMETRY, *2nd edition* |
| Moffitt and Bouchard | SURVEYING, *6th edition* |
| Salmon and Johnson | STEEL STRUCTURES: DESIGN AND BEHAVIOR |
| Spangler and Handy | SOIL ENGINEERING, *3rd edition* |
| Viessman, Knapp, Lewis, and Harbaugh | INTRODUCTION TO HYDROLOGY, *2nd edition* |
| Wang and Salmon | REINFORCED CONCRETE DESIGN, *3rd edition* |

SECOND EDITION

# Introduction to Hydrology

Warren Viessman, Jr.
*Environment and Natural Resources Policy Division*
*Library of Congress*

John W. Knapp
*Virginia Military Institute*

Gary L. Lewis
*University of Nebraska*

*The Late*
Terence E. Harbaugh

IEP—Dun—Donnelley
HARPER & ROW, PUBLISHERS
New York Hagerstown San Francisco London

INTRODUCTION TO HYDROLOGY, Second Edition

 For information address Harper & Row, Publishers, Inc., 10 East 53rd Street, New York, N.Y. 10022.
Published simultaneously in Canada by Fitzhenry & Whiteside, Ltd., Toronto.

Library of Congress Cataloging in Publication Data
Main entry under title:

Introduction to hydrology.
(IEP series in civil engineering)
First ed. by W. Viessman, Jr., T. E. Harbaugh, and J. W. Knapp.
Includes bibliographical references and index.
1. Hydrology. I. Viessman, Warren. Introduction to hydrology.
GB661.V53 1977 551.4′8 77-2190
ISBN 0-7002-2497-1

*To*
*Jeanne,*
*Peggy,*
*Gayle,*
*and*
*Karen*

# Contents

# Preface

A desire to minimize the ecologic impact of the development and use of natural resources is now firmly established. It is, therefore, vitally important for engineers and scientists to thoroughly understand the nature of a particular environment before attempting to manipulate it or govern its uses. This book deals with one crucial resource—water.

There is sufficient material for either a one- or a two-semester, beginning course in hydrology. Although primarily intended for undergraduates, the content is sufficiently rigorous for a first graduate course as well. And because the book emphasizes hydrologic principles and their practical application to realistic problems, practitioners should find the material useful as a reference.

The second edition has been strengthened and expanded significantly in the areas of simulation and the application of hydrologic techniques to problems encountered on both large and small watersheds. Urban hydrology has been given much greater emphasis. And the organization of the book has been revised so that one moves from background material into rational models for hydrologic analysis and water resources planning. A chapter on water quality models has been added to introduce the important linkages between water quality and water quantity.

Frequent citations are made to computer applications and several programs included. Many solved examples illustrate the principles developed. Statistical and numerical methods are widely employed.

Many sources have supplied subject matter for the book. The authors hope that suitable acknowledgment has been given to them and wish to thank their colleagues and students for helpful comments and review.

# 1

# Introduction

## 1-1 Hydrology Defined

Hydrology is an earth science. It encompasses the occurrence, distribution, movement, and properties of the waters of the earth and their environmental relationships. Closely allied fields include geology, climatology, meteorology, and oceanography. The emphasis of this book is on engineering hydrology.

## 1-2 A Brief History

Ancient philosophers focused their attention on the nature of processes involved in the production of surface flows and other phenomena related to the origin and occurrence of water in various stages of the perpetual cycle of water being conveyed from the sea to the atmosphere to the land and back again to the sea. Unfortunately, early speculation was often faulty.[1,4–6*] For example, Homer believed in the existence of large subterranean reservoirs that supplied rivers, seas, springs, and deep wells. It is interesting to note, however, that Homer understood the dependence of flow in the Greek aqueducts on both conveyance cross section and velocity. This knowledge was lost to the Romans and the proper relationship between area, velocity,

*Superior numbers indicate references at the end of the chapter.

and rate of flow remained unknown until Leonardo da Vinci discovered it during the Italian Renaissance.

During the first century B.C. Marcus Vitruvius, in Volume 8 of his treatise *De Architectura Libri Decem* (the engineer's chief handbook during the Middle Ages) set forth a theory generally considered to be the predecessor of modern notions of the hydrologic cycle. He hypothesized that rain and snow falling in mountainous areas infiltrated the earth's surface and later appeared in the lowlands as streams and springs.

In spite of the inaccurate theories proposed in ancient times, it is only fair to state that practical application of various hydrologic principles was often carried out with considerable success. For example, about 4000 B.C. a dam was constructed across the Nile to permit reclamation of previously barren lands for agricultural production. Several thousand years later a canal to convey fresh water from Cairo to Suez was built. Mesopotamian towns were protected against floods by high earthen walls. The Greek and Roman aqueducts and early Chinese irrigation and flood control works are also significant projects.

Near the end of the fifteenth century the trend toward a more scientific approach to hydrology based on the observation of hydrologic phenomena became evident. Leonardo da Vinci and Bernard Palissy independently reached an accurate understanding of the water cycle. They apparently based their theories more on observation than on purely philosophical reasoning. Nevertheless, until the seventeenth century it seems evident that little if any effort was directed toward obtaining quantitative measurements of hydrologic variables.

The advent of what might be called the "modern" science of hydrology is usually considered to begin with the studies of such pioneers as Perrault, Mariotte, and Halley in the seventeenth century.[1,4] Perrault obtained measurements of rainfall in the Seine River drainage basin over a period of three years. Using these and measurements of runoff, and knowing the drainage area size, he showed that rainfall was adequate in quantity to account for river flows. He also made measurements of evaporation and capillarity. Mariotte gauged the velocity of flow of the River Seine. Recorded velocities were translated into terms of discharge by introducing measurements of the river cross section. The English astronomer Halley measured the rate of evaporation of the Mediterranean Sea and concluded that the amount of water evaporated was sufficient to account for the outflow of rivers tributary to the sea. Measurements such as these, although crude, permitted reliable conclusions to be drawn regarding the hydrologic phenomena being studied.

The eighteenth century brought forth numerous advances in hydraulic theory and instrumentation. The Bernoulli piezometer, the

Pitot tube, Bernoulli's theorem, and the Chézy formula are some examples.[8]

During the nineteenth century experimental hydrology flourished. Significant advances were made in groundwater hydrology and in the measurement of surface water. Such significant contributions as Hagen-Poiseuille's capillary flow equation, Darcy's law of flow in porous media, and the Dupuit-Thiem well formula were evolved.[9–11] The beginning of systematic stream gauging can also be traced to this period. Although the basis for modern hydrology was well established in the nineteenth century, much of the effort was empirical in nature. The fundamentals of physical hydrology had not yet been well established or widely recognized. In the early years of the twentieth century the inadequacies of many earlier empirical formulations became well known. As a result, interested governmental agencies began to develop their own programs of hydrologic research. From about 1930 to 1950, rational analyses began to replace empiricism.[3] Sherman's unit hydrograph, Horton's infiltration theory, and Theis's nonequilibrium approach to well hydraulics are outstanding examples of the great progress made.[12–14]

Since 1950 a theoretical approach to hydrologic problems has largely replaced less sophisticated methods of the past. Advances in scientific knowledge permit a better understanding of the physical basis of hydrologic relationships, and the advent and continued development of high speed digital computers have made possible, in both a practical and an economic sense, extensive mathematical manipulations that would have been overwhelming in the past. The effective utilization of "men, models, methods, and machines" appears to be at hand.[7]

For a more comprehensive historical treatment, the reader is referred to the works of Meinzer, Jones, Biswas, et al.[1,2,4,5,15]

## 1-3 The Hydrologic Cycle

The hydrologic cycle is a continuous process by which water is transported from the oceans to the atmosphere to the land and back to the sea. Many subcycles exist. The evaporation of inland water and its subsequent precipitation over land before returning to the ocean is one example. The driving force for the global water transport system is provided by the sun, which furnishes the energy required for evaporation. Note that the water quality also changes during passage through the cycle; for example, sea water is converted to fresh water through evaporation.

The complete water cycle is global in nature. World water problems require studies on regional, national, international, continental, and global scales.[17] Practical significance of the fact that the total supply of fresh water available to the earth is limited and very small compared with the salt water content of the oceans has received little attention. Thus waters flowing in one country cannot be available at the same time for use in other regions of the world. Raymond L. Nace of the U.S. Geological Survey has aptly stated that "water resources are a global problem with local roots."[17] Modern hydrologists are obligated to cope with problems requiring definition in varying scales of significant order of magnitude difference. In addition, developing techniques to control weather must receive careful attention, since climatological changes in one area can profoundly affect the hydrology and therefore the water resources of other regions.

## 1-4 The Hydrologic Budget

Because the total quantity of water available to the earth is finite and indestructible, the global hydrologic system may be looked upon as closed. Open hydrologic subsystems are abundant, however, and these are usually the type analyzed by the engineering hydrologist. For any system, a water budget can be developed to account for the hydrologic components.

To illustrate, assume the simple and highly restricted system of Fig. 1-1. Consider a completely impervious inclined plane surface (water cannot be transmitted through the surface), confined on all four sides with an outlet at corner A. Since this surface is assumed to be a perfect plane, there are no depressions in which water can be stored. If a rainstorm input is applied, an output or outflow, designated as surface runoff, will be developed at A. The hydrologic water budget for this system can be represented by the following differential equation:

$$I - Q = \frac{dS}{dt} \tag{1-1}$$

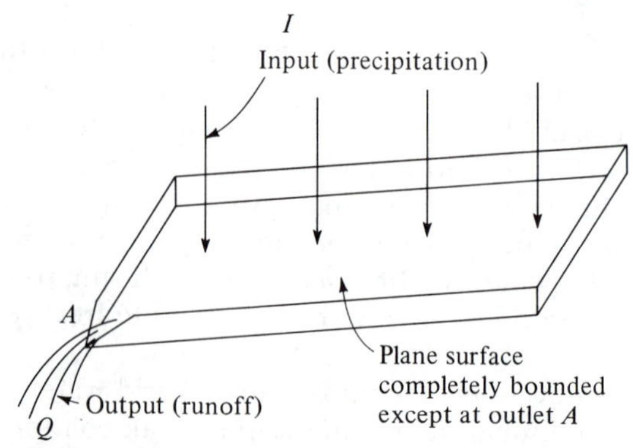


**Fig. 1-1.** *Simple hydrologic system model.*

where

$I$ = inflow per unit time
$Q$ = outflow per unit time
$dS/dt$ = the change in storage within the system per unit time

Until a minimal depth is accumulated on the surface, outflow cannot occur, but, as a storm intensifies, the depth retained on the surface (surface detention) increases. At the cessation of input, water held on the surface becomes the outflow from the system. For the example cited, all input eventually becomes outflow, neglecting the small quantity electrically bound to the surface and any evaporation during the period of input (a reasonable assumption for the system depicted). This elementary illustration should suggest that any hydrologic system can be described by a hydrologic budget which accounts for the disposition of inputs to the system and changes in storage. The simplicity of the budget equation is often misleading, however, since, as will be seen later, the terms in the equation cannot always be adequately or easily quantified.

A more generalized version of the hydrologic budget will explain the various components of a hydrologic cycle and provide an insight to problem-solving techniques for composite hydrologic regions. Such regions may be topographically defined, politically limited, or arbitrarily specified. A watershed or drainage basin is a topographically defined area drained by a river/stream or system of connecting rivers/streams such that all outflow is discharged through a single outlet. Past water resources studies have often been conducted on watersheds or drainage basins because these areas simplify application of the water budget. Theoretically, an accounting is possible for any kind of region, although data availability and the degree of refinement of analytical methods will determine the feasibility of this in a practical sense.

To demonstrate the nature of a generalized hydrologic budget, use is made of Figs. 1-2, 1-3, and 1-4. Figure 1-2 is a conceptual model of the hydrologic cycle. Precipitation in the form of rain, snow, hail, and so forth, comes from atmospheric water vapor and constitutes the primary input. Some rainfall may be intercepted by trees, grass, other vegetation, and structural objects, and will eventually return to the atmosphere by evaporation. Once precipitation reaches the ground, some of it may fill depressions (become depression storage), part may penetrate the ground (infiltrate) to replenish soil moisture and groundwater reservoirs, and some may become surface runoff—that is, flow over the earth's surface to a defined channel such as a stream. The flow diagrams of Fig. 1-3 show the disposition of infiltration, depression storage, and surface runoff.

Water entering the ground may take several paths. Some may be

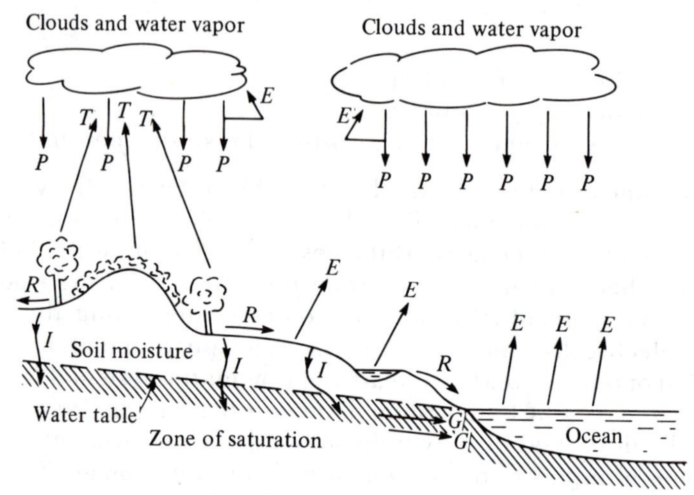


***Fig. 1-2.*** *The hydrologic cycle. Legend: T, transpiration; E, evaporation; P, precipitation; R, surface runoff; G, groundwater flow; and I, infiltration.*

directly evaporated if adequate transfer from the soil to the surface is maintained. This can easily occur where a high groundwater table (free water surface) is within the limits of capillary transport to the ground surface. Vegetation using soil moisture or ground water directly can also transmit infiltrated water to the atmosphere by a process known as *transpiration.* Infiltrated water may likewise replenish soil moisture deficiencies and enter storage provided in groundwater reservoirs which in turn maintain dry weather streamflow. Important bodies of groundwater are usually flowing so that infiltrated water reaching the saturated zone may be transported for considerable distances before it is discharged. Groundwater movement is subject, of course, to physical and geological constraints.

Water stored in depressions will eventually evaporate or infiltrate the ground surface. Surface runoff ultimately reaches minor channels (gullies, rivulets, and the like), flows to major streams and rivers, and finally reaches an ocean. Along the course of a stream, evaporation and infiltration can also occur.

The foregoing discussion suggests that the hydrologic cycle, while simple in concept, is actually exceedingly complex. Paths taken by particles of water precipitated in any area are numerous and varied before the sea is reached. The time scale is on the order of seconds, minutes, days, or years.

A general hydrologic equation can be developed based on the concepts illustrated in Figs. 1-2 and 1-3. Figure 1-4, a more abstract version of Fig. 1-2, represents schematically the hydrologic cycle of a region and it serves a useful purpose, since it is readily translated into mathematical terms. Hydrologic variables $P$, $E$, $T$, $R$, $G$, and $I$ are as defined in Fig. 1-2. Subscripts $s$ and $g$ are introduced to denote vectors originating above and below the earth's surface respectively. For example, $R_g$ signifies groundwater flow that is effluent to a surface stream, and $E_s$ represents evaporation from surface water bodies or other surface storage areas. Letter $S$ stands for storage. The region under consideration specified as $A$ has a lower boundary below which water will not be found. The upper boundary is the earth's surface. Vertical bounds are arbitrarily set as projections of the periphery of the region. Remember from Eq. 1-1 that the water budget is a balance between inflows, outflows, and changes in storage. Then Fig. 1-4 can be translated into the following mathematical statements, where all values are given in units of volume per unit time:

1. hydrologic budget above the surface

$$P + R_1 - R_2 + R_g - E_s - T_s - I = \Delta S_s \tag{1-2}$$

2. hydrologic budget below the surface

$$I + G_1 - G_2 - R_g - E_g - T_g = \Delta S_g \tag{1-3}$$

3. hydrologic budget for the earth (sum of Eqs. 1-2 and 1-3)

$$P - (R_2 - R_1) - (E_s + E_g) - (T_s + T_g) - (G_2 - G_1) = \Delta(S_s + S_g) \tag{1-4}$$

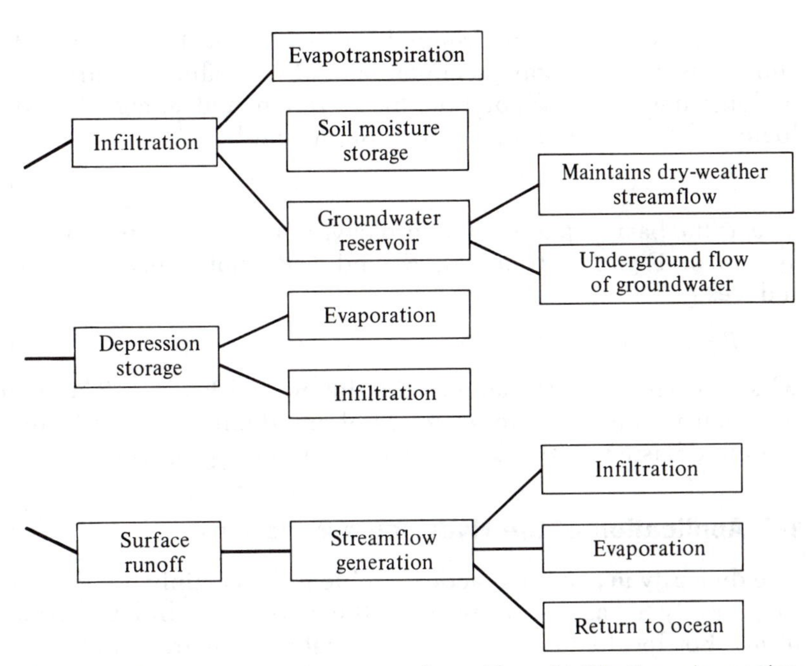


***Fig. 1-3.*** *Flow diagrams indicating disposition of infiltration, depression storage, and surface runoff.*

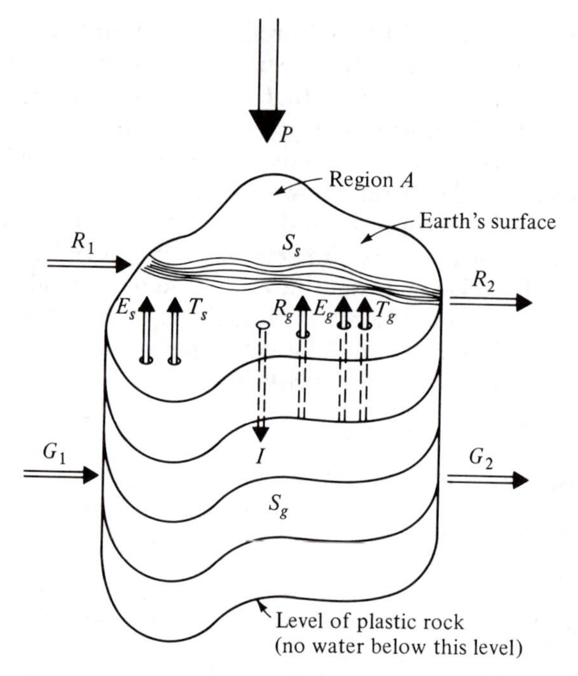


***Fig. 1-4.*** *Schematic diagram of a hydrologic cycle for a region.*

If the subscripts are dropped from Eq. 1-4 so that letters without subscripts refer to total precipitation and net values of surface flow, underground flow, evaporation, transpiration, and storage, the hydrologic budget for a region can be written simply as

$$P - R - G - E - T = \Delta S \tag{1-5}$$

This is the basic equation of hydrology. For the simplified hydrologic system of Fig. 1-1, terms $G$, $E$, and $T$ do not apply and Eq. 1-5 reduces to

$$P - R = \Delta S \tag{1-6}$$

which is basically the same as the differential Eq. 1-1. The general equation is applicable to exercises of any degree of complexity and therefore basic to the solution of all hydrologic problems.

## 1-5 Application of the Hydrologic Equation

The difficulty in solving practical problems lies mainly in the inability to properly measure or estimate the various hydrologic equation terms. For local studies, reliable estimates often are made, but on a global scale quantification is usually crude. Precipitation is measured by rain or snow gauges located throughout an area. Surface flows can be measured using various devices such as weirs, flumes, velocity meters, and depth gauges located in the rivers and streams of the area. Under good conditions these measurements are frequently 95% or more accurate, but large floods cannot be directly measured by current methods and data on such events are sorely needed. Soil moisture can be measured using neutron probes and gravimetric methods; infiltration determined locally by infiltrometers or estimated through the use of precipitation runoff data. Areal estimates of soil moisture and infiltration are generally very crude, however. The extent and rate of movement of groundwater are usually exceedingly difficult to determine, and adequate data on quantities of groundwater are not always available. Knowledge of the geology of a region is essential for groundwater estimates if they are to be more than just rough guides. The determination of the quantities of water evaporated and transpired is also extremely difficult under the present state of development of the science. Most estimates of evapotranspiration are obtained by using evaporation pans, energy budgets, mass transfer methods, or empirical relationships. A predicament inherent in the analysis of large drainage basins is the fact that rates of evaporation, transpiration, and groundwater movement are often highly heterogeneous.

The hydrologic equation is a useful tool; the reader should understand that it can be employed in various ways to estimate the magnitude and time distribution of hydrologic variables. An introductory example is given here, and others will be found throughout the book.

***Example 1-1*** In a given year, a 10,000-mi² watershed received 20 in. of precipitation. The annual rate of flow measured in the river draining the area was found to be 6000 cfs (cubic feet per second). Make a rough estimate of the combined amounts of water evaporated and transpired from the region during the year of record.

Beginning with the basic hydrologic equation

$$P - R - G - E - T = \Delta S \tag{1-5}$$

and since evaporation and transpiration can be combined,

$$ET = P - R - G - \Delta S \tag{1-7}$$

The term $ET$ is the unknown to be evaluated and $P$ and $R$ are specified. The equation thus has five variables and three unknowns and cannot be solved without additional information.

In order to get a solution, two assumptions will be made. First, since the drainage area is quite large (measured in hundreds of square miles), a presumption that the groundwater divide (boundary) follows the surface divide will probably be reasonable. In this case the $G$ component may be considered zero. The vector $R_g$ exists but is included in $R$. The foregoing assumption is usually not valid for small areas and must therefore be used carefully.

It will also be presupposed that $\Delta S = 0$, thus implying that the groundwater reservoir volume has not changed during the year. For such short periods this assumption can be very inaccurate, even for well-watered regions with balanced withdrawals and good recharge potentials. In arid areas where groundwater is being mined ($\Delta S$ consistently negative), it would be an unreasonable supposition in many cases. Nevertheless, the assumption will be made here for illustrative purposes and qualified by saying that past records of water levels in the area have revealed an approximate constancy in groundwater storage. Hydrology is not an exact science, and reasonable well-founded assumptions are required if practical problems are to be solved.

Using the simplifications just outlined, the working relationship reduces to

$$ET = P - R$$

which can be solved directly. First change $R$ into inches per year so that the units are compatible:

$$\frac{\text{ft}^3}{\text{sec}} \times \frac{1}{\text{area} - (\text{ft}^2)} \times \frac{\text{sec}}{\text{yr}} \times \frac{\text{in.}}{\text{ft}} = R, \text{ in.}$$

$$R = \frac{6000 \times 86{,}400 \times 365 \times 12}{10^4 \times (5280)^2} = 8.1 \text{ in.}$$

Therefore, $ET = 20 - 8.1 = 11.9$ in./yr

The amount of evapotranspiration for the year in question is estimated to be 11.9 in. This is admittedly a crude approximation but could serve as a useful guide for water resources planning.

## 1-6 Hydrologic Data

Hydrologic data are needed to describe precipitation; streamflows; evaporation; soil moisture; snow fields; sedimentation; transpiration; infiltration; water quality; air, soil, and water temperatures; and other variables or components of hydrologic systems. Sources of data are numerous, with the U.S. Geological Survey being the primary one for streamflow and groundwater facts. The National Weather Service (NOAA or National Oceanic and Atmospheric Administration) is the major collector of meteorologic data. Many other federal, state, and local agencies and other organizations also compile hydrologic data. For a complete listing of these organizations see References 3 and 16.

## 1-7 Common Units of Measurements

Stream and river flows are usually recorded as cubic meters per second ($\text{m}^3$/sec), cubic feet per second (cfs) or second-feet (sec-ft); groundwater flows and water supply flows are commonly measured in gallons per minute (gal/min), hour (hr), or day (gpm, gph, gpd), or millions of gallons per day (mgd); flows used in agriculture or related to water storage are often expressed as acre-feet (acre-ft), acre-feet per unit time, inches (in.) or centimeters (cm) depth per unit time or acre-inches per hour (acre-in./hr).

Volumes are often given as gallons, cubic feet, cubic meters, acre-feet, second-foot-days, and inches or centimeters. An acre-foot is equivalent to a volume of water 1 foot deep over 1 acre of land (43,560 $ft^3$). A second-foot-day (cfs-day, sfd) is the accumulated volume produced by a flow of 1 cfs in a 24-hr period. Inches or centimeters of depth relate to a volume equivalent to that many inches or centimeters of water over the area of concern.

Rainfall depths are usually recorded in inches or centimeters whereas rainfall rates are given in inches or centimeters per hour. Evaporation, transpiration, and infiltration rates are usually given as inches or centimeters depth per unit time. Some useful constants and tabulated values of several of the physical properties of water are given in Appendix B at the end of the book.

## 1-8 Application of Hydrology to Environmental Problems

The fact that man cannot exist without water is fundamental. It is also true that at times water becomes a threat to the health and welfare of individuals, communities, and even regions. Because water is such an indispensable and yet potentially damaging resource, its effective management has become an important goal of society.

Everyone uses water in a variety of ways: for drinking, cooking meals, and bathing. Water is also used in industry and agriculture and in many other important applications. The 1976 average per capita rate of consumption of water in American cities is about 165 gpd, triple the amount used 100 years ago. In future years it appears that the average rate will continue to increase although at a pace likely to be tempered by technological changes and improved management policies. A little reflection on the nature of the trend in population shift from rural to urban centers underlines the increasing burden being placed on the water resources of heavily populated regions.

Although the use of water for domestic activities is vital, its use in industry, commerce, agriculture, recreation, and esthetics is also basic to the national welfare. Water for recreation has become the most important single use in some multiple-purpose, water resource development projects, and this shift is expected to accelerate.

Our national concern over environmental quality is tied in part to the country's water resource. It is clear that improved techniques for planning, managing, and developing the environment are high on the priority list. A greater emphasis will have to be placed on humanistic and social values. At the same time, effective planning and management will result only if we are able clearly to understand the physical systems with which we must deal. This is the challenge to the hydrologist. Reliable mathematical models are needed to evaluate and predict the performance of hydrologic systems under expected or known conditions of stress. The effects of physical works such as dams, drainage systems, channelization, and the like must be understood before construction begins. The hydrologist has to play a fundamental role in our quest for a better managed and understood environment. His competence will have a tremendous impact on the success of all water resources activities, regardless of scale.

It is the intent of this book to provide the reader with a basic understanding of the physical aspects of hydrologic systems and the manner in which hydrologic components can be synthesized for effective planning and preparation of design models.

## Problems

1-1. The storage in a river reach at a given time is 16 acre-ft. At the same instant, the inflow is 500 cfs and the outflow is 700 cfs. One hour later the inflow rate is 700 cfs and the outflow is 740 cfs. Determine the change in storage in the reach that occurred during the hour. Is the storage at the end of the hour greater or less than the original value?

1-2. The storage in a reach of a river is 16.0 acre-ft at a given time. Determine the storage (acre-ft) 1 hr later if the average rates of inflow and outflow during the hour are 720 and 600 cfs, respectively.

1-3. Twelve inches of water evaporated from a 500-acre vertical-walled reservoir during 24 hr. Storm water was added to the reservoir at a constant rate of 1008 cfs during this period. Determine the volume (acre-ft) of water released during the day (through the bottom of the reservoir) if the water level in the reservoir was the same at the beginning and end of the day.

1-4. The annual evaporation from a lake (with a surface area of 3650 acres) is 120 in. Determine the average daily evaporation rate (acre-ft/day) during the year.

1-5. The evaporation rate from the surface of a 3650-acre lake is 100 acre-ft per day. Determine the depth change (ft) in the lake during a 365-day year if the inflow to the lake is 25.2 cfs. Is the change in lake depth an increase or a decrease?

1-6. One and one-half inches of runoff are equivalent to how many acre-feet if the drainage area is 25 $mi^2$ (1 acre = 43,560 $ft^2$)?

1-7. One-half inch of rain per day is equivalent to an average rate of how many cubic feet per second if the area is 500 acres?

1-8. A flow of 12 cfs enters a 600-acre reservoir. By first converting to inches per hour, determine the time (in hours) required to raise the water level 6.0 in.

1-9. Rain falls at an average intensity of 0.4 in./hr over a 600-acre area for 3 days. (a) Determine the average rate of rainfall in cfs; (b) determine the 3-day volume of rainfall in acre-ft; and (c) determine the 3-day volume of rainfall in inches of equivalent depth over the 600-acre area.

1-10. Three inches of runoff volume from a 120-mi$^2$ area are equivalent to how many acre-feet?

1-11. Twelve cubic feet of water per second are added to a vertical-walled reservoir with a surface area of 600 acres. How many hours will it take to raise the water level in the reservoir 1 ft?

## References


1. Jones, P. B., G. D. Walker, R. W. Harden, and L. L. McDaniels, "The Development of the Science of Hydrology," Circular No. 63-03, Texas Water Commission, Apr. 1963.
2. Mead, W. D., *Notes on Hydrology.* (Chicago: D. W. Mead, 1904).
3. Chow, Ven Te (ed.), *Handbook of Applied Hydrology.* (New York: McGraw-Hill Book Company, 1964).
4. Meinzer, O.E., *Hydrology,* Vol. 9 of *Physics of the Earth.* (New York: McGraw-Hill Book Company, 1942). Reprinted by Dover Publications, Inc., New York, 1949.
5. Krynine, P.D., "On the Antiquity of Sedimentation and Hydrology," *Bull. Geol. Soc. Amer.,* 70 (1960).
6. Kazmann, Raphael G., *Modern Hydrology.* (New York: Harper & Row, Publishers, 1965).
7. Snyder, Willard M., and John B. Stall, "Men, Models, Methods, and Machines in Hydrologic Analysis," *Proc. ASCE, J. Hyd. Div.,* 91, No. HY2 (Mar. 1965).
8. Rouse, Hunter, and Simon Ince, *History of Hydraulics,* Iowa Institute of Hydraulic Research, State University of Iowa, 1957.
9. Hagen, G. H. L., "Ueber die Bewegung des Wassers in engen cylindrischen Rohren," *Poggendorfs Ann. Physik. Chem.,* 16 (1839).
10. Darcy, Henri, *Les fontaines publiques de la ville de Dijon.* (Paris: V. Dalmont, 1856).
11. Dupuit, A. J., *Études théoriques et practiques sur le mouvement des eaux dans les canauxs découverts et à travers les terrains perméables,* 2nd ed. (Paris: Dunod, 1863).
12. Sherman, L. K., "Stream Flow from Rainfall by the Unit-Graph Method," *Eng. News-Rec.,* 108 (1932).
13. Horton, R. E., "The Role of Infiltration in the Hydrologic Cycle," *Trans. Amer. Geophys. Union,* 14 (1933).
14. Theis, C. V., "The Relation Between the Lowering of the Piezometric Surface and the Rate and Duration of a Well Using Ground Water Recharge," *Trans. Amer. Geophys. Union,* 16 (1935).
15. Biswas, Asit K., "Hydrologic Engineering Prior to 600 B.C.," *Proc. ASCE, J. Hyd. Div.,* Proc. Paper 5431, Vol. 93, No. HY5 (Sept. 1967).
16. Todd, D. K. (ed.), *The Water Encyclopedia.* (New York: Water Information Center, Inc., 1970).
17. Nace, Raymond L., "Water Resources: A Global Problem with Local Roots," *Environmental Science and Technology,* 1, No. 7 (July 1967).

# 2

# Precipitation

Streamflow is a function of hydrologic input and physical, vegetative, and climatic characteristics. Water derived from precipitation does not all appear as streamflow. Various fractions of it are diverted into paths that do not terminate in the regional surface transport system. The geologic, topographic, vegetative, and man-made features of the earth all are important modifiers of any area's streamflow characteristics.

Once precipitation strikes the ground, it may go into storage on the surface or in the soil and groundwater reservoirs beneath the surface. Character of soil and rocks largely determines which storage system the precipitated water will enter. Opportunity for evaporation and transpiration is also affected by geologic and topographic features.

## 2-1 Water Vapor

The fraction of water vapor in the atmosphere is very small compared to quantities of other gases present, but it is exceedingly important and is largely responsible for prevailing weather conditions. Precipitation is derived from this atmospheric water. The moisture content of the air is also a significant factor in local evaporation processes. Thus it is necessary for a hydrologist to be acquainted with ways for evaluating the atmospheric water vapor content and to understand the thermodynamic effects of atmospheric moisture.

Under most conditions of practical interest (modest ranges of pressure and temperature, provided that the condensation point is excluded) water vapor essentially obeys the gas laws. Atmospheric moisture is derived from evaporation and transpiration, the principal source being evaporation from the oceans. Precipitation over the United States comes largely from oceanic evaporation, the water vapor being transported over the continent by the primary atmospheric circulation system.

Measures of water vapor or atmospheric humidity are related basically to conditions of evaporation and condensation occurring over a level surface of pure water. Consider a closed system containing approximately equal volumes of water and air maintained at the same temperature. If the initial condition of the air is dry, evaporation takes place and the quantity of water vapor in the air increases. A measurement of pressure in the air space will reveal that as evaporation proceeds, pressure in the air space increases because of an increase in partial pressure of the water vapor (vapor pressure). Evaporation continues until vapor pressure of the overlying air equals the surface vapor pressure [a measure of the excess of water molecules leaving (evaporating from) the water surface over those returning]. At this point evaporation ceases, and if the temperatures of the air space and water are equal, the air space is said to be *saturated.* If the container had been open instead of closed, the equilibrium would not have been reached, and all of the water would eventually have evaporated.

### Measures of Atmospheric Moisture

Some commonly used measures of atmospheric moisture or humidity are vapor pressure, absolute humidity, specific humidity, mixing ratio, relative humidity, and dew point temperature.

Partial pressure is the pressure exerted on the surface of a container by an individual component of a gaseous mixture. In accordance with Dalton's law for gaseous mixtures, each gas exerts a pressure approximately proportional to the product of the total pressure of the mixture and the volume percentage of the gas. The partial pressure exerted by water vapor is called *vapor pressure.* Saturation vapor pressure is the partial pressure of water vapor at a specific temperature in saturated air.

Absolute humidity is the mass of water vapor contained in a unit volume of space. Units of measurement are normally grams per cubic meter ($g/m^3$). It can be determined by using the equation of state for an ideal gas. In meteorological applications the equation of state is generally written in a form slightly different from that most commonly used. If we consider that $M$ is the mass of some volume $V$ which has a gram-molecular weight $m$, then the number of gram molecules $n$ contained in volume $V$ will be determined by the relation

$$n = \frac{M}{m} \tag{2-1}$$

The usual form of the equation of state is

$$pV = nRT \tag{2-2}$$

where

$R$ = the universal gas constant
$T$ = the absolute temperature

When the value for $n$ from Eq. 2-1 is inserted into Eq. 2-2, the following expression results:

$$pV = \frac{M}{m} RT \tag{2-3}$$

By definition, $V/M = 1/\rho = v$ where $v$ is the specific volume ($cm^3/g$). Introducing this in Eq. 2-3 yields

$$pv = \frac{R}{m} T \tag{2-4}$$

or

$$p = \rho \frac{R}{m} T \tag{2-5}$$

Equations 2-4 and 2-5 are commonly used in meteorologic work. Occasionally, the term $R/m$ is called $R'$ or $B$ and is termed an *individual gas constant* which depends upon the gas.

It follows that the equation of state for water vapor may be written as

$$e = \rho_w \frac{R}{m_w} T \tag{2-6}$$

where

$\rho_w$ = the vapor density or absolute humidity
$e$ = the partial pressure of the water vapor

Specific humidity is the mass of water vapor contained within a unit mass of moist space. It is normally expressed as grams per gram (g/g) or grams per kilogram (g/kg) and may be calculated using

$$q = \frac{0.622e}{p - 0.378e} \tag{2-7}$$

where

$q$ = the value of specific humidity in grams per gram
$e$ = the vapor pressure in millibars
$p$ = the total pressure of the moist space in millibars (mb)

The definition of mixing ratio is similar to that for specific humidity with the exception that it is the ratio of the mass of water vapor to the mass of dry air rather than to the mass of moist space. For most practical purposes the specific humidity and mixing ratio can be considered equal.

The dew point temperature is that temperature at which an air mass just becomes saturated if cooled at constant pressure and if moisture is neither added nor removed.

Relative humidity is the gravimetric ratio of water vapor in unit space to the water vapor moist space can hold at that temperature. For all practical purposes it can be considered as the ratio of the water vapor pressure existing to that which would prevail under saturated conditions. It may be stated in percent in the following form:

$$H = \frac{100e}{e_s} \tag{2-8}$$

where $e_s$ = the saturated vapor pressure. It should be noted that the ratio of vapor densities could also be used. The relative humidity is a term widely used in weather analyses, since most of the atmosphere is unsaturated and it is useful to determine the degree of saturation that exists. Figure 2-1 gives saturation and 50% relative humidity curves.

### Amount of Precipitable Water

Estimates of the amount of precipitation that might occur over a given region with favorable conditions are often useful. These may be obtained by calculating the amount of water contained in a column of atmosphere extending up from the earth's surface. This quantity is known as the *precipitable water* W, although it cannot all be removed from the atmosphere by natural processes. Precipitable water is usually expressed in centimeters or inches.

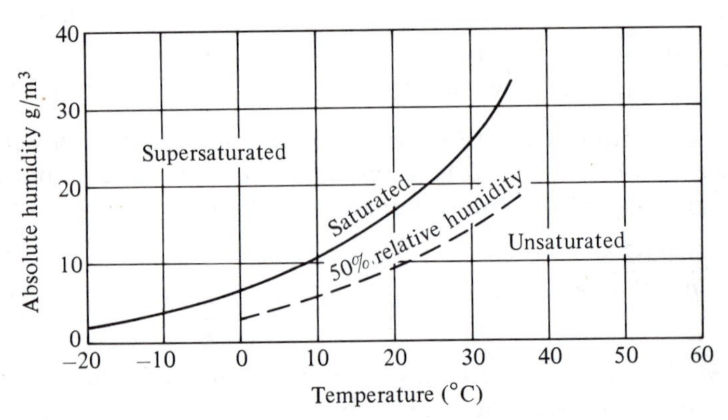


**Fig. 2-1.** *Saturation and 50% relative humidity curves.*

An equation for computing the amount of precipitable water in the atmosphere can be derived as follows. Consider a column of air having a square base 1 cm on a side. The total water mass contained in this column between elevation zero and some height $z$ would be

$$W = \int_0^z \rho_w \, dz \tag{2-9}$$

where $\rho_w$ = the absolute humidity. The hydrostatic equation may be written as

$$dp = -\rho g dz \tag{2-10}$$

where $\rho$ = the total air density and is thus the sum of the vapor density and the total density of the dry gases $\rho_d$, or $\rho = (\rho_w + \rho_d)$. The hydrostatic equation may be solved for $dz$, yielding

$$dz = -\frac{dp}{\rho g} \tag{2-11}$$

Substituting the above expressions for $\rho$ and $dz$ in Eq. 2-9, we obtain

$$W = \frac{1}{g} \int_p^{p_0} \frac{\rho_w}{(\rho_w + \rho_d)} \, dp \tag{2-12}$$

The ratio of densities shown above is actually the specific humidity $q$, and inasmuch as $q$ is approximately equal to $0.622e/p$, Eq. 2-12 may be written as

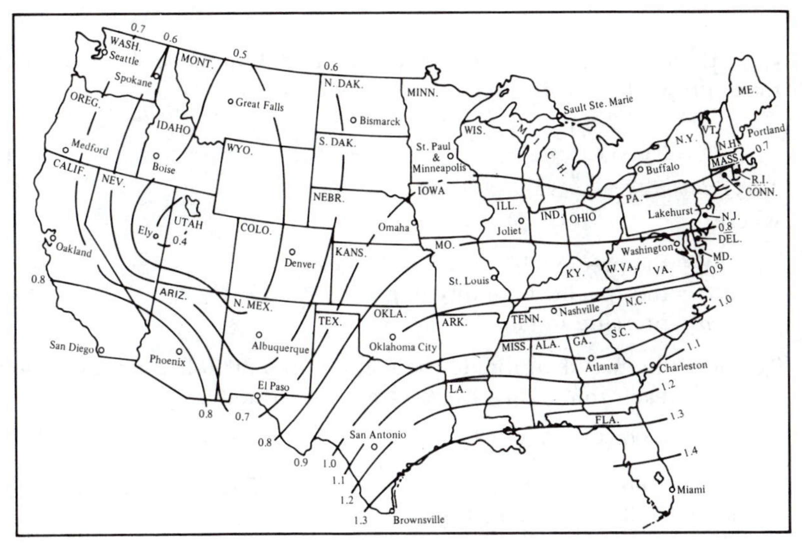


**Fig. 2-2.** *Mean precipitable water, in inches, to an elevation of 8 km. (U.S. Weather Bureau.)*

$$W = \frac{0.622}{g} \int_{p}^{p_0} e \frac{dp}{p} \tag{2-13}$$

with little loss of accuracy, where $W$ is given in (g/cm²). Considering the density of the precipitable water equal to 1, and since integration is performed over a unit area, we find that the value of $W$ is also the depth of precipitable water in centimeters. The integral can be evaluated graphically or by dividing the atmosphere into layers of approximately uniform specific humidities and solving for these individually and then summing. Figure 2-2 illustrates the average amount of precipitable water for the continental United States up to an elevation of 8 km.[1] Example 2-1 illustrates a computational procedure.

***Example 2-1*** Given the data in Table 2-1 on elevation, temperature, and vapor pressure, compute the depth of precipitable water in inches for a column of atmosphere 18,000 ft in height.

***Table 2-1***

| | | | | | | | | | | |
|---|---|---|---|---|---|---|---|---|---|---|
| Elevation, 1000s ft | 0 | 2 | 4 | 6 | 8 | 10 | 12 | 14 | 16 | 18 |
| Temperature, °F | 59 | 52 | 45 | 38 | 30 | 23 | 16 | 9 | 2 | −5 |
| Pressure, mb | 1013 | 942 | 875 | 812 | 753 | 697 | 644 | 595 | 550 | 500 |
| Vapor pressure, mb | 7.0 | 5.0 | 3.8 | 3.2 | 2.0 | 1.6 | 1.1 | 0.8 | 0.6 | 0.4 |

Equation 2-13 can be converted to yield $W$ in inches by multiplying by the appropriate conversion factor. This yields

$$W = 0.0004 \int_{p}^{p_0} q\, dp \tag{2-14}$$

where

$W$ = the precipitable water in inches
$q$ = the specific humidity in g/kg
$p$ = pressure in mb

To solve this, we shall divide the column into 2000-ft zones and use mean values for each zone. Thus Eq. 2-14 can be written as

$$W = 0.0004 \sum \bar{q}\, \Delta p \tag{2-15}$$

where $\bar{q}$ is the mean value of $q$ for each zone. The computations are shown in Table 2-2. The final result is

$$W = 0.0004 \times 1013 = 0.41 \text{ in.}$$

**Table 2-2** *Calculations for Example 2-1*

| Elevation (1000s ft) | Specific humidity (g/kg = 622$e/p$) | Mean specific humidity, $\bar{q}$ | $\Delta p$ | $\bar{q}\,\Delta p$ |
|---|---|---|---|---|
| 0 | 4.32 | 3.81 | 71 | 271 |
| 2 | 3.30 | 2.98 | 67 | 199 |
| 4 | 2.66 | 2.55 | 63 | 161 |
| 6 | 2.45 | 2.05 | 59 | 121 |
| 8 | 1.65 | 1.53 | 56 | 86 |
| 10 | 1.41 | 1.23 | 53 | 65 |
| 12 | 1.06 | 0.95 | 49 | 47 |
| 14 | 0.84 | 0.76 | 45 | 34 |
| 16 | 0.68 | 0.59 | 50 | 29 |
| 18 | 0.50 | | | |
| | | | | $\sum = 1013$ |

### Geographic and Temporal Variations

The quantity of atmospheric water vapor varies with location and time. These variations may be attributed mainly to temperature and source of supply considerations. The greatest concentrations can be found near the ocean surface in the tropics, the concentrations generally decreasing with latitude, altitude, and distance inland from coastal areas.

About half the atmospheric moisture can be found within the first mile above the earth's surface. This is because the vertical transport of vapor is mainly through convective action, which is slight at higher altitudes. It is also of interest that there is not necessarily any relation between the amount of atmospheric water vapor over a region and the resulting precipitation. The amount of water vapor contained over dry areas of the Southwest, for example, at times exceeds that over considerably more humid northern regions, even though the latter areas experience precipitation while the former do not.

## 2-2 Precipitation

Precipitation is the primary input vector of the hydrologic cycle. Its primary forms are rain, snow, and hail and several variations of these forms such as drizzle and sleet. Precipitation is derived from atmospheric water, its form and quantity thus being influenced by the action of other climatic factors such as wind, temperature, and atmospheric pressure. Atmospheric moisture is a necessary but not sufficient condition for precipitation. Continental air masses are usually very dry so that most precipitation is derived from moist maritime air that originates over the oceans. In North America about 50% of the evaporated water is taken up by continental air and moves back again to the sea. This is generally contradictory to some proposed theories which indicate a relationship between the amount of water evaporated in a region and its precipitation.

Because considerable water may accumulate in clouds without any precipitation resulting, condensation can occur independently of precipitation. It is therefore useful to consider each of these processes individually so that a better understanding of the occurrence of precipitation can be had.

### Condensation

Condensation may be attributed to one or more of the following causes: (1) dynamic or adiabatic cooling, (2) mixing of air masses having different temperatures, (3) contact cooling, and (4) cooling by radiation. However, the most important cause is dynamic cooling, and it is this process which produces nearly all precipitation. Condensation rates associated with other causes are usually small, and these rarely produce appreciable precipitation. Dew, frost, and fog are condensation forms commonly associated with radiational and contact cooling.

The condensation of water vapor into cloud droplets (average size $\frac{1}{2500}$-in. diameter) occurs on small hygroscopic particles, which are termed *condensation nuclei.* Particles of ocean salt and combustion products containing sulfurous and nitrous acids constitute the most active nuclei. Such nuclei are usually less than 1 micron ($\mu$) in diameter, but very large nuclei on the order of 5 $\mu$ are known to be present occasionally. The number of salt nuclei present range from about 10 to 1000 per $cm^3$, while the number of combustion nuclei is directly related to the extent and nature of regional industrial operations.[2] Under conditions usually encountered, it takes anywhere from several seconds to produce a droplet of about 10 $\mu$ to approximately a day for a small raindrop (about 3 mm) to grow on a condensation nuclei. This natural growth process is extremely slow and at best produces only very small cloud droplets. As a result, the condensation mechanism cannot alone be expected to produce any significant form of precipitation.

### Formation of Precipitation

Two processes are considered to be capable of supporting the growth of droplets of sufficient mass (droplets of from about 500 to 4000 $\mu$ in diameter) to overcome air resistance and consequently fall to the earth as precipitation. These are known as the *ice crystal process* and the *coalescence process.*

The coalescence process is one by which the small cloud droplets increase their size due to contact with other droplets through collision. Water droplets may be considered as falling bodies that are subjected to both gravitational and air resistance effects. Fall velocities at equilibrium (terminal velocities) are proportional to the square of the radius of the droplet; thus the larger droplets will descend more quickly than the smaller ones. As a result, smaller droplets are often overtaken by larger droplets, and the resulting collisions tend to unite the drops, producing increasingly larger particles. Very large drops (order of 7 mm in diameter) break up into small droplets that repeat the coalescence process and produce somewhat of a chain effect. In this manner, sufficient large raindrops may be produced to generate significant precipitation. This process is considered to be particularly important in tropical regions or in warm clouds.

An important type of growth is known to occur if ice crystals and water droplets are found to exist together at subfreezing temperatures down to about −40°C. Under these conditions, certain particles of clay minerals and organic and ordinary ocean salts serve as freezing nuclei so that ice crystals are formed. The vapor pressure under these conditions is higher over the water droplets than over the ice crystals, and thus condensation occurs on the surface of the crystals. The ice crystals grow in size, and uneven particle size distributions develop, which further favor growth through contact with other particles. This is considered to be a very important precipitation-producing mechanism.

The artificial inducement of precipitation has been studied extensively in recent years, and these studies are continuing. It has been demonstrated that condensation nuclei supplied to clouds can induce precipitation. The ability of man to insure the production of precipitation or to control its geographic location or timing has not yet been attained. Nevertheless, the future of this phase of weather control is promising.

Many legal as well as technological problems are associated with the prospects of "rainmaking" processes. Of interest here is the impact on hydrologic estimates that uncontrolled or only partially controlled artificial precipitation might have. Many naturally occurring hydrologic variables are considered as statistical variates that are either randomly distributed or are distributed with a random component. If the distribution or time series of the variable can be modeled, an inference as to the frequency of occurrence of significant hydrologic events of a given magnitude (such as precipitation) can be made. If, however, artificial controls are used and if the effects of these cannot be reliably predicted, frequency analyses may prove to be totally unreliable tools.

### Precipitation Types

Dynamic or adiabatic cooling is the primary cause of condensation and is responsible for most rainfall. Thus it can be seen that vertical transport of air masses is a requirement for precipitation. Precipitation may be classified according to the conditions that generate vertical air motion. In this respect, the three major categories of precipitation type are *convective, orographic,* and *cyclonic.*

**Convective Precipitation** Convective precipitation is typical of the tropics and is brought about by heating of the air at the interface with the ground. This heated air expands with a resultant reduction in weight. During this period, increasing quantities of water vapor are taken up; the warm moisture-laden air becomes unstable; and pronounced vertical currents are developed. Dynamic cooling takes place, causing condensation and precipitation. Convective precipitation may be in the form of light showers or storms of extremely high intensity (thunderstorms are a typical example).

**Orographic Precipitation** Orographic precipitation results from the mechanical lifting of moist horizontal air currents over natural barriers such as mountain ranges. The rainfalls of the Pacific Northwest are typical examples.

**Cyclonic Precipitation** Cyclonic precipitation is associated with the movement of air masses from high pressure regions to low pressure regions. These pressure differences are created by the unequal heating of the earth's surface.

Cyclonic precipitation may be classified as frontal or nonfrontal. Any barometric low can produce nonfrontal precipitation as air is lifted through horizontal convergence of the inflow into a low pressure area. Frontal precipitation results from the lifting of warm air over cold air at the contact zone between air masses having different characteristics. If the air masses are moving so that warm air replaces colder air, the front is a known as a *warm front*; if, on the other hand, cold air displaces warm air, the front is said to be *cold.* If the front is not in motion, it is said to be a *stationary front.* Figure 2-3 illustrates a vertical section through a frontal surface.

### Thunderstorms

Many areas of the United States are subjected to severe convective storms, which are generally identified as thunderstorms because of their electrical nature. These storms, although usually very local in nature, are often productive of very intense rainfalls that are highly significant when local and urban drainage works are considered.

Thunderstorm cells develop from vertical air movements associ-

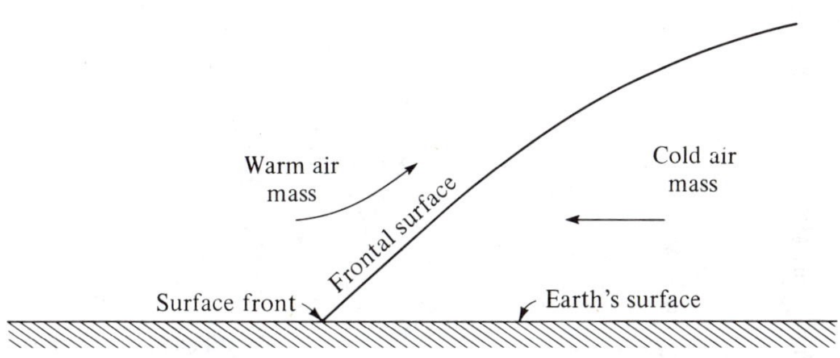


***Fig. 2-3.*** *Vertical cross section through a frontal surface.*

ated with intense surface heating or orographic effects. There are three primary stages in the life history of a thunderstorm. These are the *cumulus stage,* the *mature stage,* and the *dissipating stage.* Figure 2-4 illustrates each of these stages.

All thunderstorms begin as cumulus clouds, although few such clouds ever reach the stage of development needed to produce such a storm. The cumulus stage is characterized by strong updrafts that often reach altitudes of over 25,000 ft. Vertical wind speeds at upper levels are often as great as 35 mph. As indicated in Fig. 2-4a, there is considerable horizontal inflow of air (entrainment) during the cumulus stage. This is an important element in the development of the storm, as additional moisture is provided. Air temperatures inside the cell are greater than those outside, as indicated by the convexity of the isotherms viewed from above. The number and size of the water droplets increase as the stage progresses. The duration of the cumulus stage is approximately 10 to 15 min.

The strong updrafts and entrainment support increased condensation and the development of water droplets and ice crystals. Finally, when the particles increase in size and number so that surface precipitation occurs, the storm is said to be in the mature stage. In this stage strong downdrafts are created as falling rain and ice crystals cool the air below. Updraft velocities at the higher altitudes reach up to 70 mph in the early periods of the mature stage. Downdraft speeds of over 20 mph are usual above about 5000 ft in elevation. At lower levels, frictional resistance tends to decrease the downdraft velocity. Gusty surface winds move outward from the region of rainfall. Heavy precipitation is often derived during this period which is usually on the order of 15 to 30 min.

In the final or dissipating stage, the downdraft becomes predominant until all of the air within the cell is descending and being dynamically heated. Since the updraft ceases, the mechanism for condensation ends and precipitation tails off and ends.

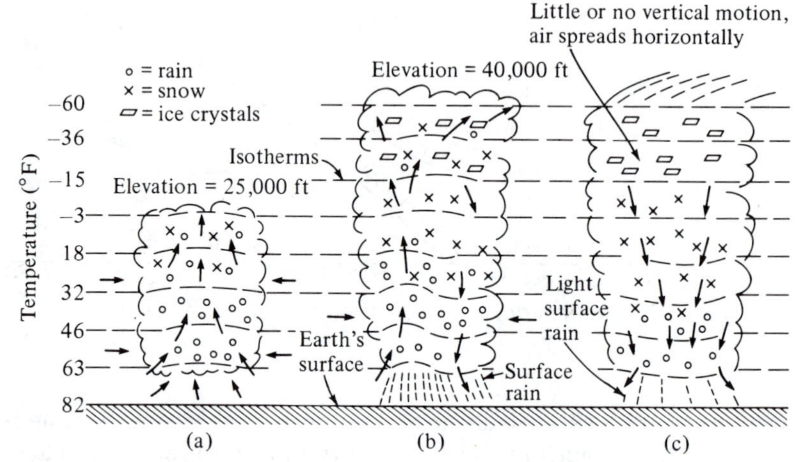


***Fig. 2-4.*** *Cumulus, mature, and dissipating stages of a thunderstorm cell. Legend: o, rain; x, snow; and ▱ ice crystals. (Department of the Army.)*

## Precipitation Data

Considerable data on precipitation are available in publications of the National Weather Service.[3,4] Other sources include various state and federal agencies engaged in water resources work. For critical regional studies it is recommended that all possible data be compiled; often the establishment of a gauging network will be necessary.

## Precipitation Variability

Precipitation varies geographically, temporally, and seasonally. Figure 2-5 indicates the mean annual precipitation for the continental United States, while Fig. 2-6 gives an example of seasonal differences. It should be understood that both regional and temporal variations in precipitation are very important in water resources planning and hydrologic studies. For example, it may be very important to know that the cycle of minimum precipitation coincides with the peak growing season in a particular area, or that the period of heaviest rainfall should be avoided in scheduling certain construction activities.

Precipitation amounts sometimes vary considerably within short distances. Records have shown differences of 20% or more in the catch of rain gauges less than 20 ft apart. Precipitation is usually measured with a rain gauge placed in the open so that no obstacle projects within the inverted conical surface having the top of the gauge as its apex and a slope of 45°. The catch of a gauge is influenced by the wind which usually causes low readings. Various devices such as Nipher and Alter shields have been designed to minimize this error in measurement. Precipitation gauges may be of the recording or nonrecording type. The former are required if the time distribution of pre-

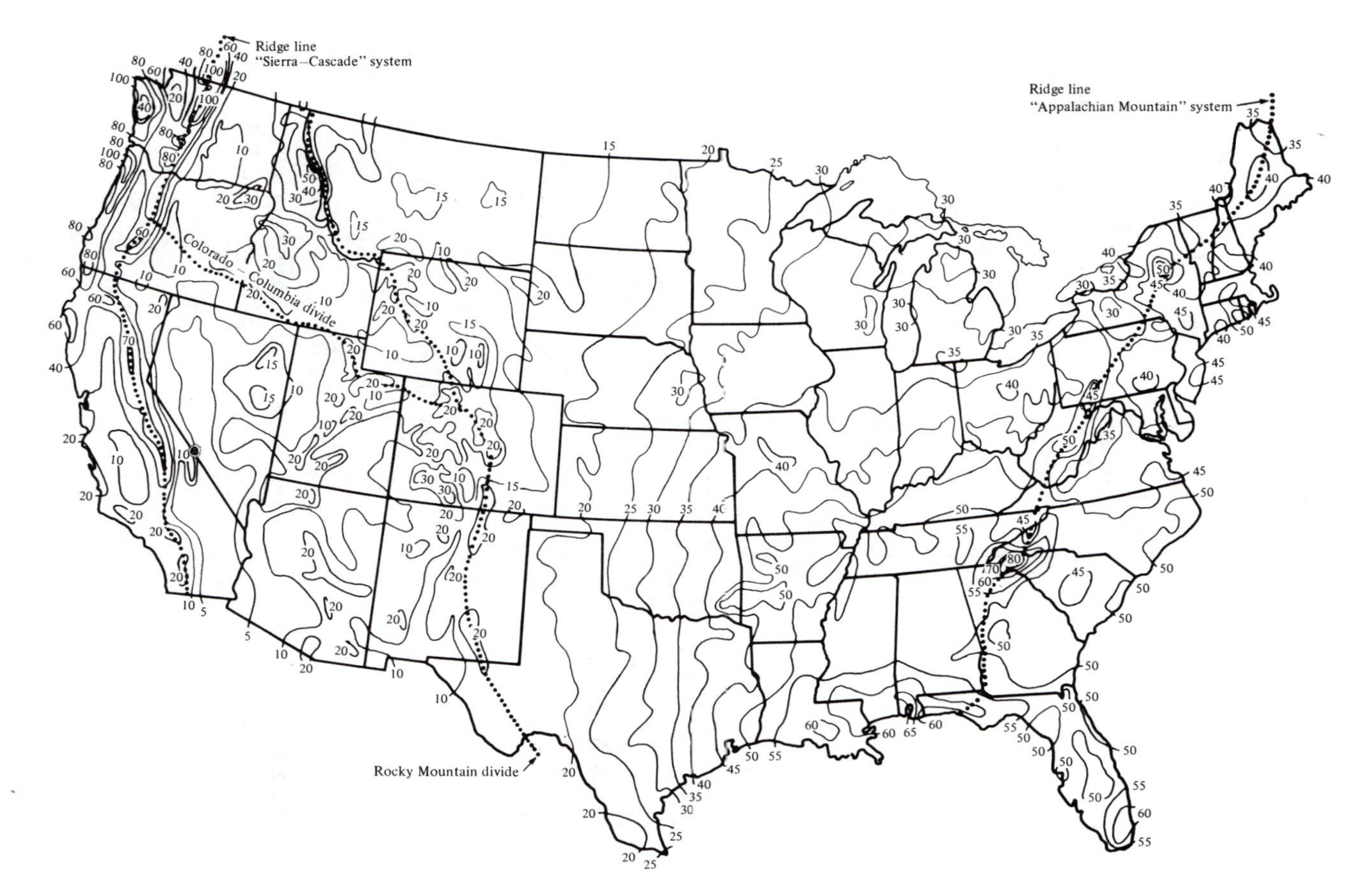


**Fig. 2-5.** *Mean annual precipitation in inches, 1899–1938. (U.S. Department of Agriculture, Soil Conservation Service.)*

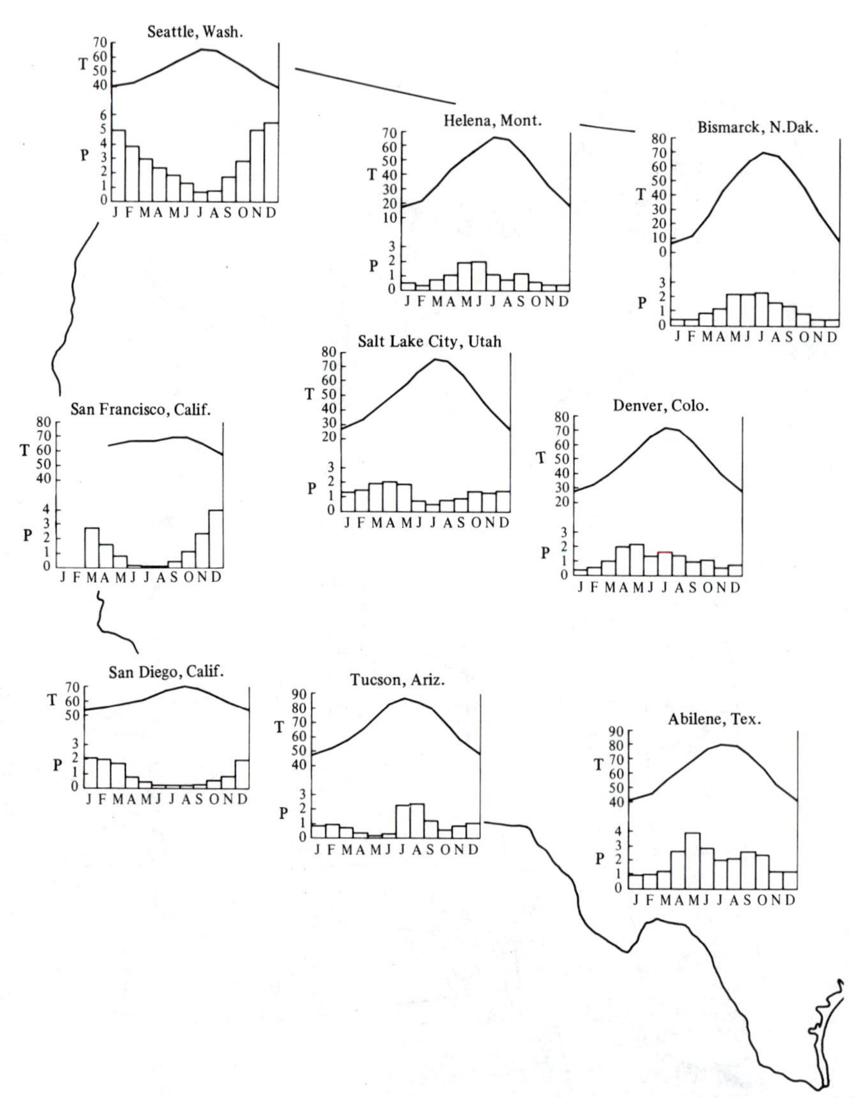


***Fig. 2-6.*** *Precipitation and temperature distributions. Legend: T, mean monthly temperature (°F); P, mean monthly precipitation (in.). (U.S. Department of Agriculture, Soil Conservation Service.)*

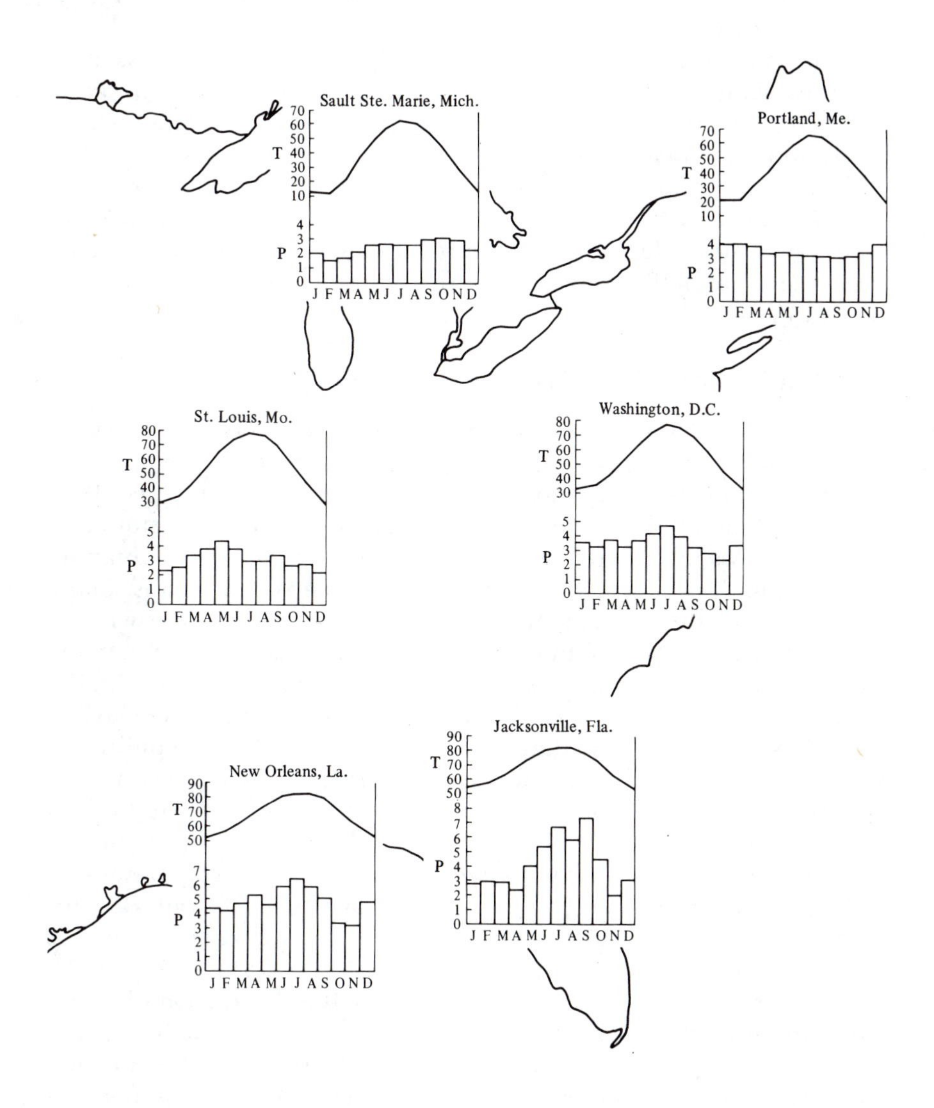
Sault Ste. Marie, Mich.
Portland, Me.
St. Louis, Mo.
Washington, D.C.
Jacksonville, Fla.
New Orleans, La.
T
P
J F M A M J J A S O N D

cipitation is to be known. Information about the features of gauges is readily available.[2]

Because precipitation varies spatially, it is usually necessary to use the data from several gauges to estimate the average precipitation for an area and to evaluate its reliability (see Chapter 6). This is especially important in forested areas where the variation tends to be large.

Time variations in rainfall intensity are extremely important in the rainfall-runoff process, particularly in urban areas (see Fig. 2-7a). The areal distribution is also significant and highly correlated with the time history of outflow (see Fig. 2-7b). These considerations are discussed in greater detail in following chapters.

## 2-3 Distribution of the Precipitation Input

Total precipitation is distributed in numerous ways. That intercepted by vegetation and trees may be equivalent to the total precipitation input for relatively small storms. Once interception storage is filled, raindrops begin falling from leaves and grass, where water stored on these surfaces eventually becomes depleted through evaporation. Precipitation that reaches the ground may take several paths. Some water will fill depressions and eventually evaporate; some will infiltrate the soil. Part of the infiltrated water may strike relatively impervious strata near the soil surface and flow approximately parallel to it as interflow until an outlet is reached. Other portions may replenish soil moisture in the upper soil zone, and some infiltrated water may reach the groundwater reservoir which sustains dry weather streamflow. The component of the precipitation input that exceeds the local infiltration rate will develop a film of water on the surface (surface detention) until overland flow commences. Detention depths varying from $\frac{1}{8}$ to $1\frac{1}{2}$ in. for various conditions of slope and surface type have been reported.[2] Overland flow ultimately reaches defined channels and becomes streamflow.

Figure 2-8 illustrates in a general way the disposition of a uniform storm input to a natural drainage basin. Although such an input is not to be expected in nature, the indicated relationships are representative of actual conditions. Modifications resulting from nonuniform storms will be discussed as they arise.

In Fig. 2-8a note that the storm input is distributed uniformly over time $t_a$ at a rate equal to $i$ (dimensionally equal to $LT^{-1}$). This input is dissected into components $i_1$ through $i_4$, the sum of which is equal to $i$ at any time $t$. Figure 2-8b illustrates the manner in which infiltrated water is further subdivided into interflow, groundwater, and soil moisture. Figure 2-8c shows the transition from overland flow supply into streamflow. The mechanics of these processes will be treated in detail in later sections. The nature of the curves presented

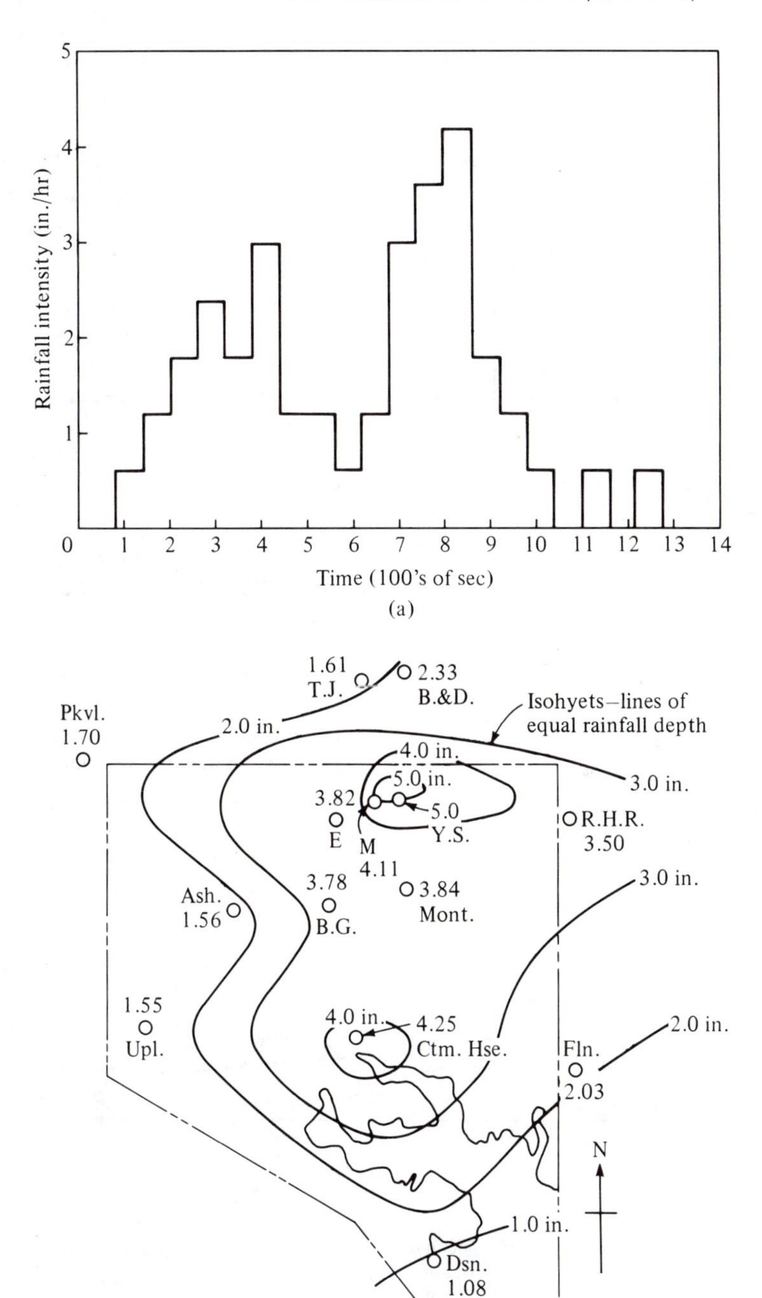


***Fig. 2-7.*** *(a) Rainfall distribution in a convective storm June, 1960, Baltimore, Maryland. (b) Isohyetal pattern, storm of September 10, 1957, Baltimore, Maryland. Key: O, recording rain gauge.*

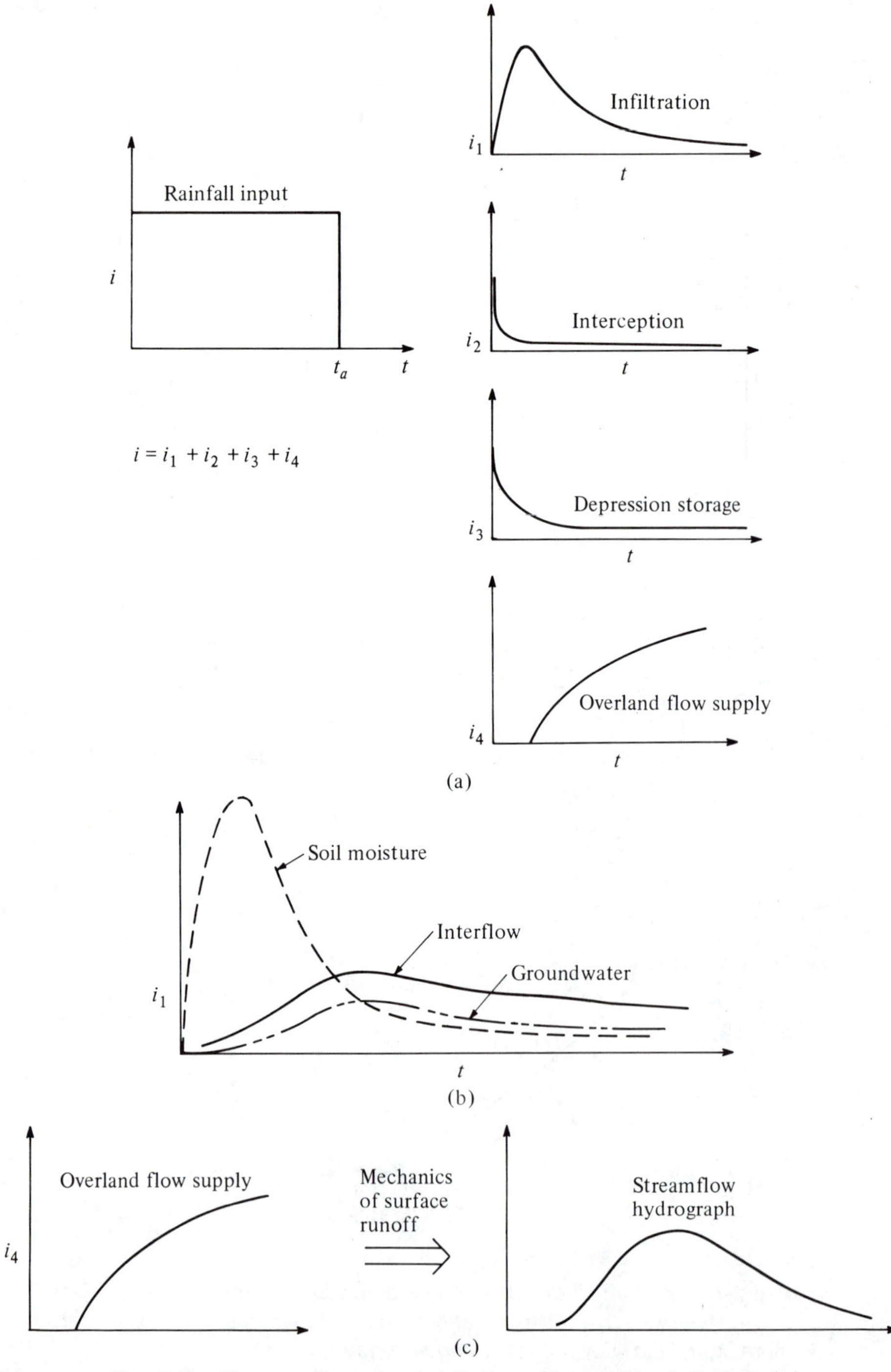


***Fig. 2-8.*** *The runoff process: (a) disposition of precipitation, (b) components of infiltration, and (c) disposition of overland flow supply.*

depicts the general runoff process. It should be realized, however, that actual graphs of infiltration and/or other factors versus time might appear quite different in form and relative magnitude when compared with these illustrations because of the effects of nonuniform storm patterns, antecedent conditions, and other conditions.

The rate and areal distribution of runoff from a drainage basin are determined by a combination of physiographic and climatic factors. Important climatic factors include the form of precipitation (rain, snow, hail), the type of precipitation (convective, orographic, cyclonic), the quantity and time distribution of the precipitation, the character of the regional vegetative cover, prevailing evapotranspiration characteristics, and the status of the soil moisture reservoir. Physiographic factors of significance include geometric properties of the drainage basin, land-use characteristics, soil type, geologic structure, and characteristics of drainage channels (geometry, slope, roughness, and storage capacity).

Large drainage basins often react differently from smaller ones when subjected to a precipitation input. This can be explained in part by such factors as geologic age, relative impact of land-use practices, size differential, variations in storage characteristics, and other causes. Chow defines a small watershed as a drainage basin whose characteristics do not filter out (1) fluctuations characteristic of high intensity, short-duration storms; or (2) the effects of land management practices.[5] On this basis, small basins may vary from less than an acre up to 100 $mi^2$. A large basin is one in which channel storage effectively filters out the high frequencies of imposed precipitation and effects of land-use practices.

## 2-4 Point Precipitation

Precipitation events are recorded by gauges at specific locations. The resulting data permit determination of the frequency and character of precipitation events in the vicinity of the site. Point precipitation data are used collectively to estimate areal variability of rain and snow and are also used individually for developing design storm characteristics for small urban or other watersheds. Design storms are discussed in detail in Chapter 12.

Point rainfall data are used to derive intensity-duration-frequency curves such as those shown in Fig. 2-9. Such curves are used in the Rational Method for urban storm drainage design (Chapter 11); their construction is discussed in Chapter 5. In applying the Rational Method, a rainfall intensity is used which represents the average intensity of a storm of given frequency for a selected duration. The frequency chosen should reflect the economics of flood damage reduction. Frequencies of 1 to 10 yr are commonly used where residential areas are to be protected. For higher-value districts, 10- to 20-yr or

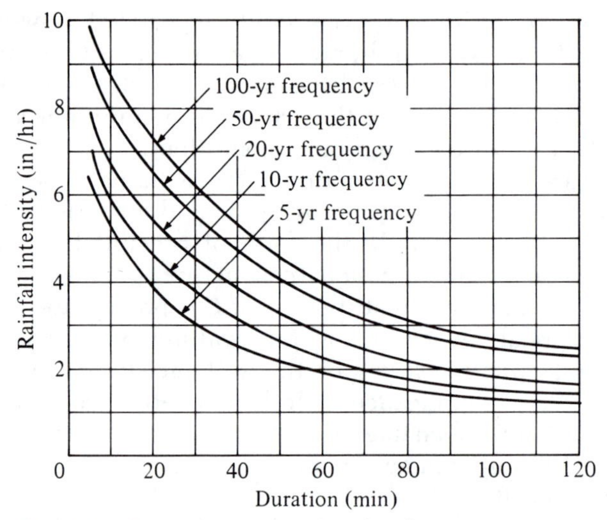


***Fig. 2-9.*** *Typical intensity-duration-frequency curves.*

higher return periods are often selected. Local conditions and practice normally dictate the selection of these design criteria.

It is occasionally necessary to estimate point rainfall at a given location from recorded values at surrounding sites. This can be done to complete missing records or to determine a representative precipitation to be used at the point of interest. The National Weather Service has developed a procedure for this which has been verified on both theoretical and empirical bases.[6]

Consider that rainfall is to be calculated for point $A$ in Figure 2-10. Establish a set of axes running through $A$ and determine the absolute coordinates of the nearest surrounding points $B$, $C$, $D$, $E$, and $F$. These are recorded in columns 3 and 4 of Table 2-3. The estimated precipitation at $A$ is determined as a weighted average of the other five points. The weights are reciprocals of the sums of the squares of $\Delta X$ and $\Delta Y$; that is, $D^2 = \Delta X^2 + \Delta Y^2$, and $W = 1/D^2$. The estimated rainfall at the point of interest is given by $\sum(P \times W)/\sum W$. In the special case where rainfall is known in only two adjacent quadrants (I and II, for example), the estimate is given as $\sum(P \times W)$. This has the effect of reducing estimates to zero as the points move from an area of precipitation to one with no records. This is considered to be the most logical procedure for handling this unusual case.[6] The estimated result will always be less than the greatest and greater than the smallest surrounding precipitation. For special effects such as mountain influences, an adjustment procedure can be applied.

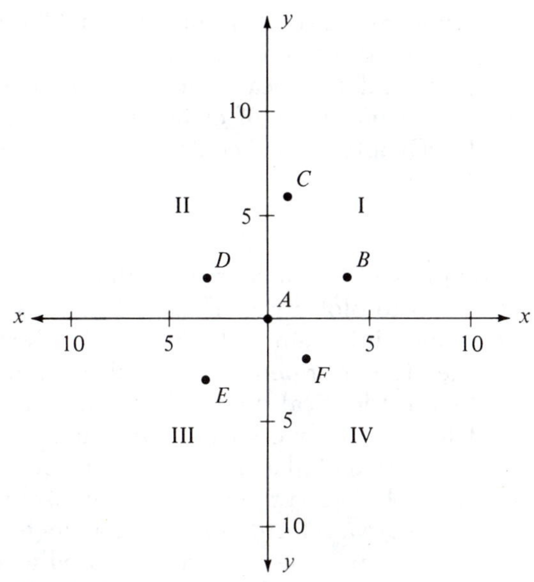


*Fig. 2-10.* Four quadrants surrounding precipitation station A.

## 2-5 Areal Precipitation

For most hydrologic analyses, it is important to know the areal distribution of precipitation. Usually, average depths for representative portions of the watershed are determined and used for this purpose. The most direct approach is to use the arithmetic average of gauged quantities. This procedure is satisfactory if gauges are uniformly distributed and the topography is flat. Other commonly used methods are the isohyetal method and the Thiessen method. The reliability of rainfall measured at one gauge in representing the average depth over a surrounding area is a function of (1) the distance from the gauge to

**Table 2-3** *Determination of Point Rainfall from Data at Nearby Gauges*

| (1) | (2) | (3) | (4) | (5) | (6) | (7) |
|---|---|---|---|---|---|---|
| Point | Rainfall (in.) | $\Delta X$ | $\Delta Y$ | $D^2$ | $W \times 10^3$ | $P \times W \times 10^3$ |
| A | — | — | — | — | — | — |
| B | 1.60 | 4 | 2 | 20 | 50 | 80.0 |
| C | 1.80 | 1 | 6 | 37 | 27.0 | 48.6 |
| D | 1.50 | 3 | 2 | 13 | 76.9 | 115.4 |
| E | 2.00 | 3 | 3 | 18 | 55.6 | 111.2 |
| F | 1.70 | 2 | 2 | 8 | 125.0 | 212.5 |
| Sums | — | — | — | — | 334.5 | 567.7 |

*Note:* Estimated precipitation ($P$) at A = 567.7/334.5; $P$ = 1.70 in.

the center of the representative area, (2) the size of the area, (3) topography, (4) the nature of the rainfall of concern (storm event versus mean monthly, for example), and (5) local storm pattern characteristics.[7] For more information on errors of estimation, the reader should consult Refs. 6 and 7. Chapter 6 also contains a discussion of areal variability of precipitation.

### Isohyetal Method

Isohyetal maps can be prepared for any area to interpolate areal rainfall values. The first step is to plot the rain gauge locations on a suitable map and to record the rainfall amounts (Fig. 2-11). Next, an interpolation between gauges is performed and rainfall amounts at selected increments are plotted. Identical depths from each interpolation are then connected to form isohyets (lines of equal rainfall depth). The areal average is the weighted average of depths between isohyets; that is, the mean value between the isohyets. The isohyetal method is the most accurate approach for determining average precipitation over an area, but its proper use requires a skilled analyst and careful attention to topographic and other factors that impact on areal variability.

### Thiessen Method

Another method of calculating areal rainfall averages is the Thiessen method. In this procedure the area is subdivided into polygonal subareas using rain gauges as centers. The subareas are used as weights in estimating the watershed average depth. Thiessen diagrams are constructed as shown in Fig. 2-12. This procedure is not suitable for mountainous areas because of orographic influences. The Thiessen network is fixed for a given gauge configuration, and polygons must be reconstructed if any gauges are relocated.

## 2-6 Probable Maximum Precipitation

The probable maximum precipitation (PMP) is the critical depth-duration-area rainfall relationship for a given area during the seasons of the year which would result from a storm containing the most critical meteorological conditions considered probable of occurrence.[8] Such storm events are used in flood flow estimates by the U.S. Corps of Engineers and other water resources agencies. The critical meteorological conditions are based on analyses of air-mass properties (effective precipitable water, depth of inflow layer, wind, temperature, and other factors), synoptic situations during recorded storms in the region, topography, season, and location of the area. The rainfall derived is termed probable maximum precipitation, since it is subject to limitations of meteorological theory and data and is based on the

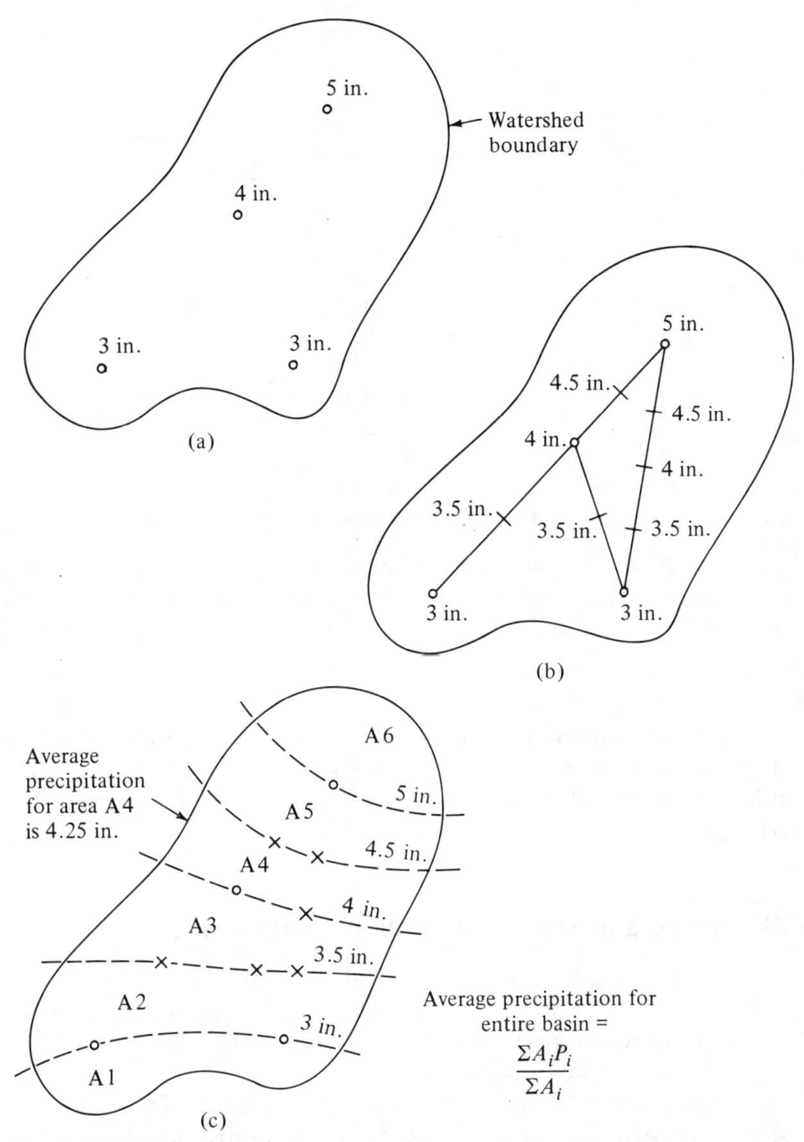


***Fig. 2-11.*** *Construction of an isohyetal map. (a) Locate rain gauges and plot values; (b) interpolate between rain gauges; and (c) plot isohyetals.*

most effective combination of factors controlling rainfall intensity.[8] An earlier designation of "maximum possible precipitation" is synonymous. Additional information on PMP is given in Chapter 12.

The seasonal variation of PMP is important in the design and operation of multipurpose structures and in flooding considerations that may occur in combination with snowmelt. In both of these cases,

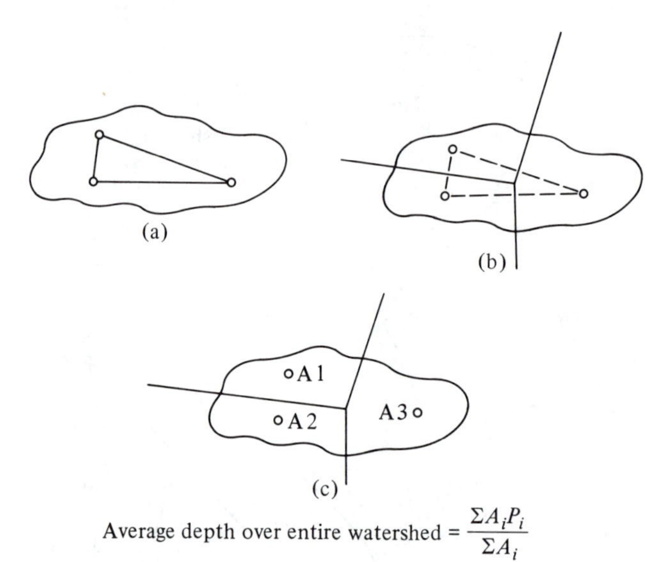


***Fig. 2-12.*** *Construction of a Thiessen diagram. (a) Connect rain gauge locations; (b) draw perpendicular bisectors; and (c) calculate Thiessen weights ($A_1$, $A_2$, $A_3$).*

annual probable maximums might be less important than seasonal maximums. Figures 2-13 and 2-14 display 24-hr PMP for the eastern half of the United States for 200-mi² watersheds during the month of August.

## 2-7 Gross and Net Precipitation Inputs

In the runoff process, the final outcome of a storm event—the outflow hydrograph—is of vital concern. This involves first a knowledge of the causative storm and, second, a means to relate this storm input to the outflow hydrograph.

A rain gauge network can provide a measure of the time distribution of gross precipitation. Such data can be displayed graphically as shown in Fig. 2-7, placed on magnetic tape or punched cards for computer use, or stored in some other form. The net (excess) precipitation that contributes directly to the surface runoff supply is equivalent to the gross precipitation minus losses to interception, storm period evaporation, depression storage, and infiltration. The relationship between excess precipitation $P_e$ and gross precipitation $P$ is thus

$$P_e = P - \sum \text{losses} \tag{2-16}$$

where the losses include all deductions from the gross storm input.

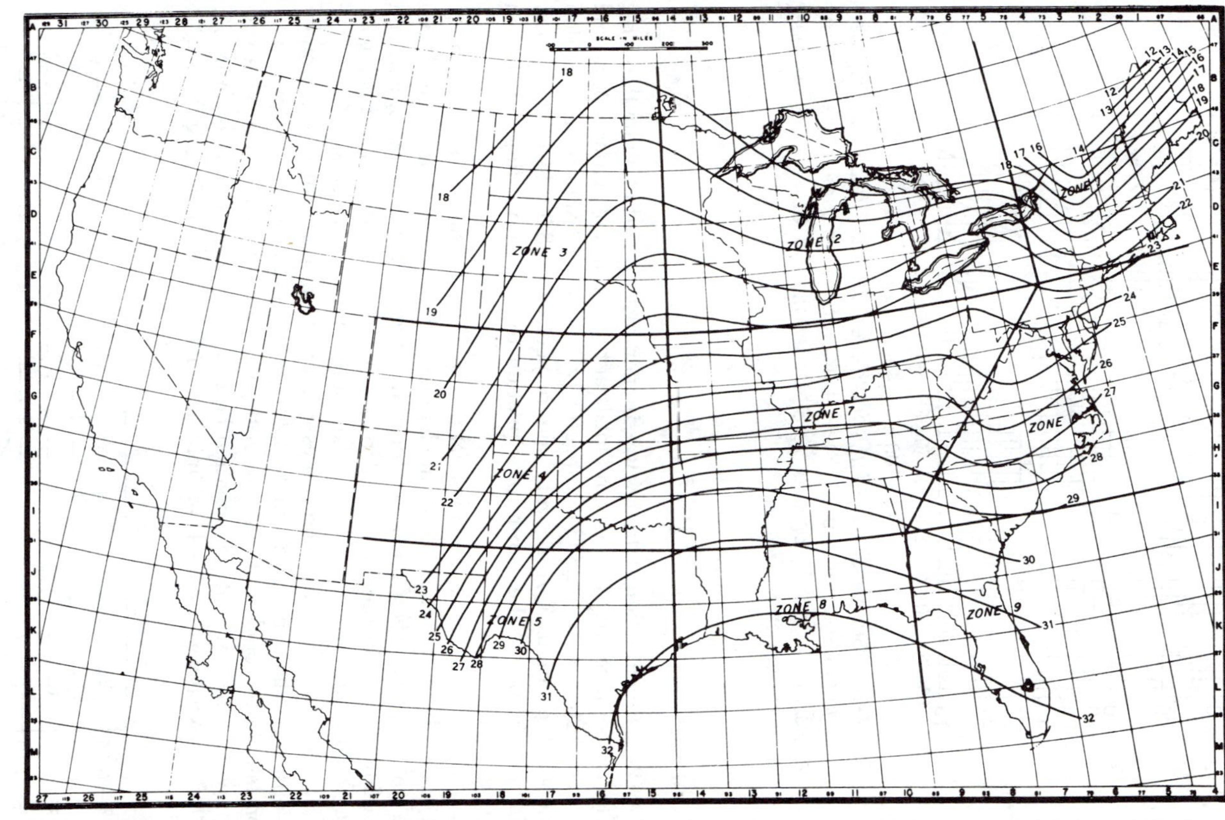


**Fig. 2-13.** *Probable maximum precipitation for 200 mi² in 24 hr (in.) (U.S. Department of Commerce, National Weather Service.)*

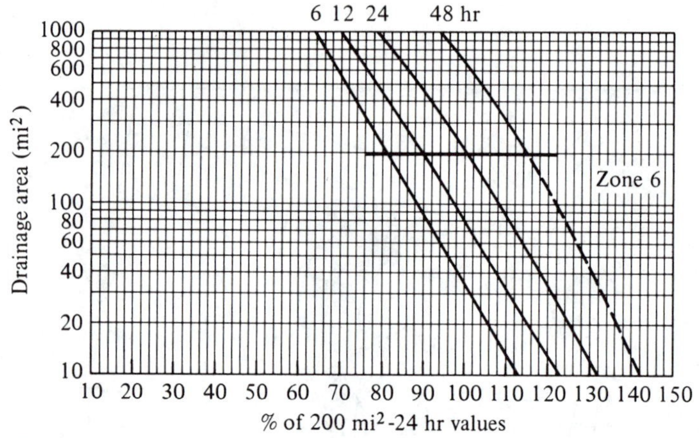


***Fig. 2-14.*** *Seasonal variation, depth-area-duration relationships; percentage to be applied to 200 mi²-24 hr probable maximum precipitation values for August in Zone 6. (U.S. Department of Commerce, National Weather Service.)*

## Problems

2-1. Precipitation station $X$ was inoperative for part of a month during which a storm occurred. The storm totals at three adjacent stations, $A$, $B$, and $C$, were 4.20, 3.50, and 4.80 in. respectively. The normal annual precipitation amounts for stations $X$, $A$, $B$, and $C$ are, respectively, 38.50, 44.10, 36.80, and 47.20 in. Estimate the storm precipitation for station $X$ using a weighted average.

2-2. Compute the mean annual precipitation for the river basin (see p. 41) using the arithmetic average method, the Thiessen Polygon Method, and an isohyetal analysis. Construct Thiessen polygons and isohyets of constant rainfall using all 11 precipitation stations. Areas inside isohyets and polygons may be determined by geometry, planimeter, or a grid overlay.

2-3. The following chart represents a portion of a chart sent to you by an observer who operates a weighing-type precipitation guage. As precipitation falls in the bucket, the height of the pen point increases. (a) What was the average intensity of precipitation between 6 A.M. and noon on Aug. 10, 1971? (b) What was the total precipitation on (1) Aug. 10? (2) Aug. 11?

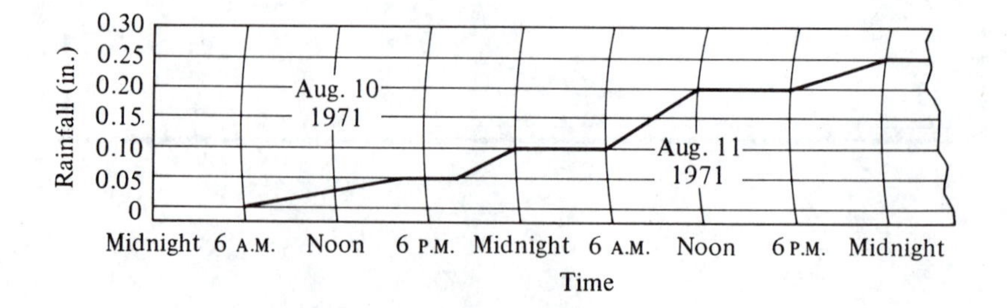

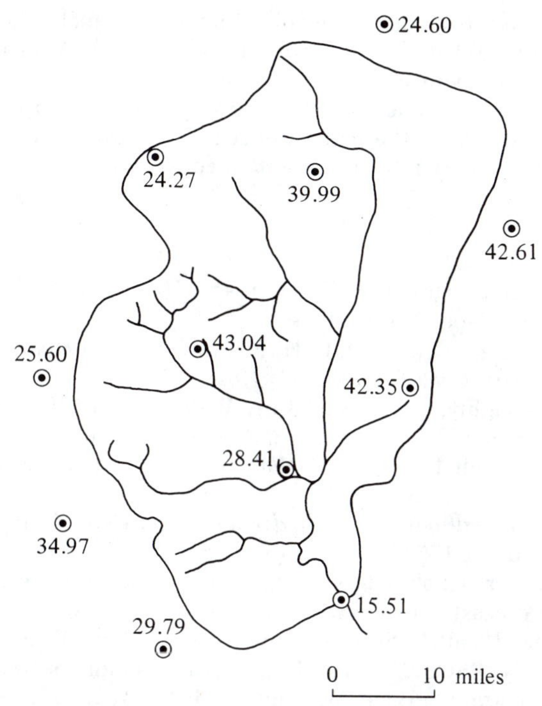


*Network of precipitation stations in the Kings River Watershed above Piedra, California. Depths are mean annual values in inches (for Prob. 2-2).*

2-4. Drainage areas within each of the isohyetal lines for a storm are tabulated for a 5500-acre basin. Use the isohyetal method to determine the average precipitation depth within the basin for the storm. Make a conceptual sketch.

| Isohyetal Interval (in.) | Area (acres) |
|---|---|
| 0–2 | 3000 |
| 2–4 | 1500 |
| 4–6 | 1000 |
| 6–8 | 0 |

2-5. A mean draft of 120 mgd is to be developed from a 200-mi² catchment area. At the flow line the reservoir is estimated to be 4000 acres. Annual rainfall is 35 in.; the mean annual runoff is 10 in.; and the mean annual evaporation is 40 in. Find the net gain or loss in storage. Compute the volume of water evaporated. State this figure in a form such as the number of years the volume could supply a given community.

2-6. Discuss how you would go about collecting data for analysis of the water budget of a region. What agencies would you contact? What other sources of information would you seek out?

2-7. For an area of your choice, plot the mean monthly precipitation versus time. Explain how this fits the pattern of seasonal water uses for the area. Will the form of precipitation be an important consideration?

## References

1. Shands, A. L., "Mean Precipitable Water in the United States," U.S. Weather Bureau, Tech. Paper No. 10, 1949.
2. Linsley, R. K., Jr., M. A. Kohler, and J. L. H. Paulhus, *Applied Hydrology.* (New York, McGraw-Hill Book Company, 1949).
3. Miller, D. W., J. J. Geraghty, and R. S. Collins, *Water Atlas of the United States.* (Port Washington, N.Y.: Water Information Center, Inc., 1963).
4. U.S. Weather Bureau, Tech. Papers 1–33 (Washington, D.C.: Government Printing Office).
5. Chow, Ven Te (ed.), *Handbook of Applied Hydrology.* (New York, McGraw-Hill Book Company, 1964).
6. Staff, Hydrologic Research Laboratory, "National Weather Service River Forecast System Forecast Procedures," NOAA Techn. Mem. NWS HYDRO 14, National Weather Service, Silver Spring, Md., Dec. 1972.
7. Mockus, V., Section 4, "SCS National Engineering Handbook on Hydrology," Soil Conservation Service, Washington, D.C., Aug. 1972.
8. Riedel, J. T., J. F. Appleby, and R. W. Schloemer, "Seasonal Variation of the Probable Maximum Precipitation East of the 105th Meridian for Areas from 10 to 1000 Square Miles and Durations of 6, 12, 24, and 48 Hours," Hydrometeorological Rept. No. 33, U.S. Weather Bureau, Washington, D.C., 1967.

# 3

# Hydrologic Abstractions

Problems of hydrologic design usually require a modeling of the precipitation runoff process. To determine runoff as a residual of the gross precipitation input, it is necessary to calculate those fractions of the precipitation input which are lost to the runoff process through various natural or artificial mechanisms.

## 3-1 Evaporation

Evaporation is the process by which water is transferred from land and water masses of the earth to the atmosphere. Because there is a continuous exchange of water molecules between an evaporating surface and its overlying atmosphere, it is common in hydrologic practice to define evaporation as the *net rate of vapor transfer.* It is a function of solar radiation, differences in vapor pressure between a water surface and the overlying air, temperature, wind, atmospheric pressure, and the quality of evaporating water. Conversion of snow or ice into water vapor is in reality sublimation rather than evaporation, since water molecules do not pass through a liquid phase. Otherwise, the effects of these two processes are the same.

The role of evaporation in the hydrologic cycle has been discussed. The practical significance of evaporation losses to a particular region will be illustrated.

Mean annual lake evaporation for the continental United States varies from a low of about 20 in. in the northeast to a high of about 86 in. in southern California. The mean rate of lake evaporation in arid and semiarid regions is often in excess of the local precipitation depth for that area. As a result, significant quantities of water are lost to those areas for beneficial uses.

About 75 in. of water are estimated to be evaporated annually from the pool of Elephant Butte Reservoir in New Mexico, a storage site located along the Rio Grande River, which furnishes irrigation water to much of southern New Mexico. The average surface area of the pool is roughly 13,760 acres so that approximately 86,000 acre-ft of water are evaporated annually. For a normal annual irrigation requirement of 3 acre-ft per acre, a loss in irrigation potential of about 29,000 acres is indicated. This same amount of water could meet the water supply needs of New York City (about 1 billion gpd) for almost 1 month. Evaporation loss from a larger reservoir such as Lake Mead (mean surface area of 124,200 acres and mean annual evaporation of 849,000 acre-ft) could supply the city of New York for approximately 1 yr.[14]

These examples illustrate the importance of evaporation in water development planning. About 30 in. of water are received annually as precipitation in the United States, with perhaps 70% of this amount returned to the atmosphere through evapotranspiration, and the remaining 30% appearing as runoff to the oceans. Figure 3-1 shows the mean annual evaporation from shallow lakes and reservoirs in the United States.

The determination of evaporation is generally based on the water budget, energy budget, mass transfer techniques, or pan-evaporation data. Direct measurements of evaporation are not easily obtained for large bodies of water because of the extensive surfaces involved. In fact, of all variables included in the general hydrologic equation, surface runoff is the only one that readily permits direct evaluation, since it is confined within well-defined geometric boundaries that permit determination of both rate and cross-sectional area of flow. The choice of method used to determine evaporation depends on the required accuracy of results and the type of instrumentation available. Accuracy is related to the varying degree of reliability with which the method's parameters can be determined.

Evaporation is a crucial consideration in water resources planning and management programs, in evaluating the potential for water resources development, and in water supply studies. During storm periods, vapor pressure gradients are reduced and evaporation is usually not significant. In very dry regions, for example, the mean hourly evaporation rate during precipitation-free periods would be on the order of 0.008 in./hr. Comparing this with a relatively small storm having an average intensity of 0.5 in./hr, it is easily seen that during

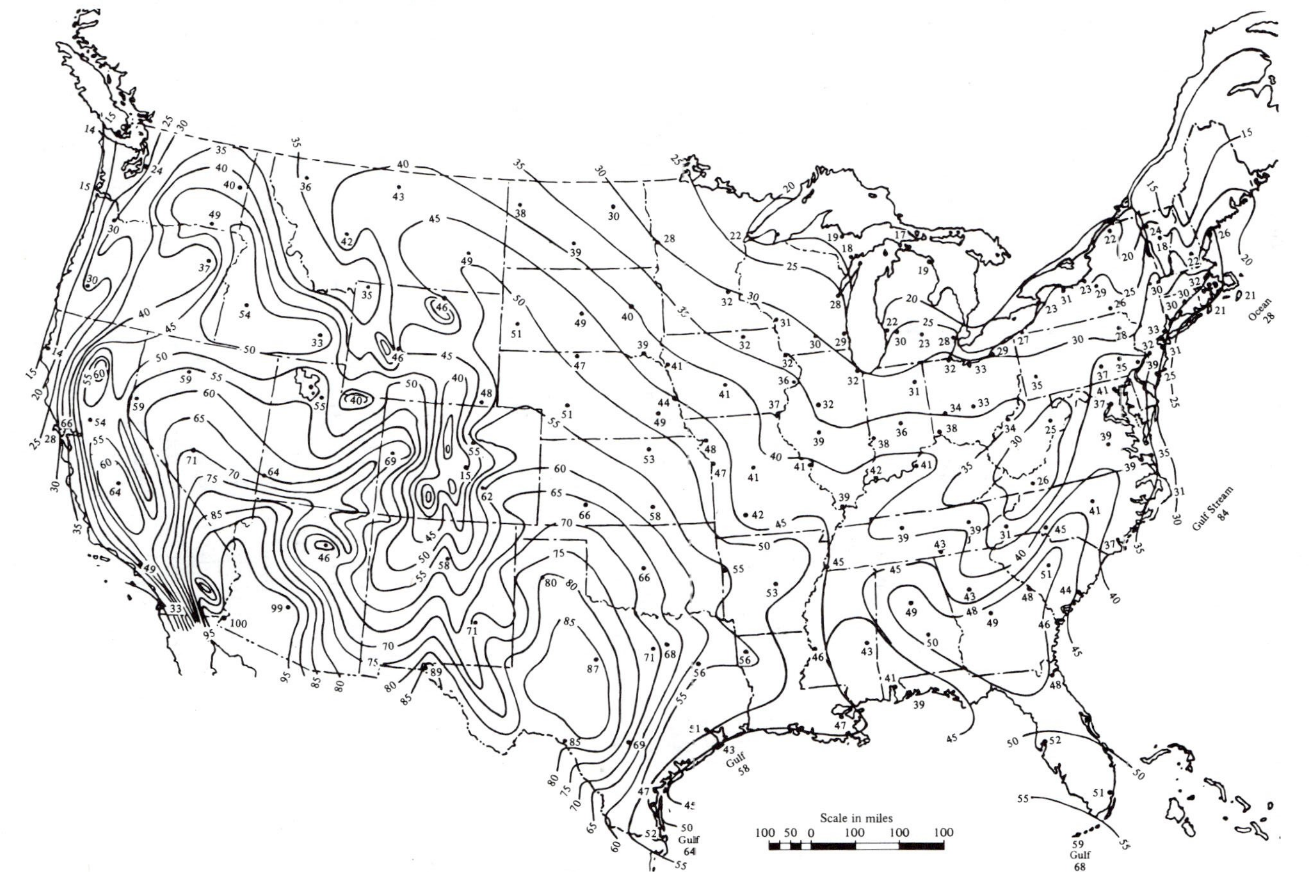


**Fig. 3-1.** *Mean annual evaporation from shallow lakes and reservoirs, in inches. Note: Evaporation from large deep lakes and reservoirs, particularly in arid regions, will be substantially less in spring and summer, greater in fall and winter, and less for the year than the values here shown. Evaporation from the surfaces of soil and vegetation immediately after rains or irrigation will begin at greater rates and diminish rapidly with the supply of available moisture. Great local differences in topography and climate in mountainous regions cause large local differences in evaporation not adequately shown here, particularly in the western states. (U.S. Department of Agriculture, Soil Conservation Service.)*

important storm periods evaporation can be safely disregarded with respect to flow estimation. As a result it is common practice to make no deduction from gross storm precipitation for evapotranspiration.

**Water Budget** The water budget method for determining reservoir evaporation has already been mentioned in discussing the use of the hydrologic equation. Basically, it is a very simple procedure, but it seldom produces reliable results. In this method, reservoir (lake, pool) evaporation $E_s$ can be computed by rearranging Eq. 1-2.

$$E_s = P + R_1 - R_2 + R_g - T_s - I - \Delta S_s \tag{3-1}$$

It is useful to deal with the net transfer of seepage through the ground $(R_g - I) = O_s$ and consider that the transpiration term $T_s$ equals zero. With these few modifications, Eq. 3-1 becomes

$$E_s = P + R_1 - R_2 + O_s - \Delta S_s \tag{3-2}$$

All the terms are in volume units for a time period of interest, and $\Delta t$ should be at least a week. In general, however, the method would more likely be used to estimate monthly or annual evaporation from a particular reservoir. Note that all errors in measuring inflow, precipitation, net seepage, and change in storage are reflected in the final estimate of evaporation. Precipitation, runoff, and changes in storage can often be determined within reasonable limits of accuracy, but evaluation of the net seepage $O_s$ is frequently subject to appreciable errors; if the magnitude of $O_s$ is on the order of $E_s$, very large errors are possible. Seepage estimates usually come from measurements of groundwater levels and/or soil permeability. The water budget is usable on a continuous basis if a stage-seepage relationship for the lake can be established.

In cases where the water budget for a lake is defined by only two unknowns, net seepage and evaporation, these losses can be separated and evaluated in a relatively simple manner by assuming that evaporation is proportional to the product $u(e_0 - e_a)$. This is the mass transfer product described later in the section on mass transfer techniques. The variables are wind velocity $u$, saturation vapor pressure $e_0$ (related to water surface temperature), and vapor pressure of the air $e_a$. When the product $u(e_0 - e_a)$ is zero, evaporation may be neglected.

Periods of no surface inflow or outflow are desirable for net seepage determination, since during such intervals the only losses are evaporation and seepage. Under these conditions, whenever the mass transfer product is equal to zero, the change in water elevation is considered equivalent to the net seepage loss. Normally, a daily plot of change in elevation versus $u(e_0 - e_a)$ is obtained and a best fitting line constructed. The intercept of this line on the change in stage axis is the net seepage rate. Values of net seepage estimated in this manner can be used on a long-term basis in the water budget equation if the net seepage does not change appreciably over extended periods. Unfortunately, this condition is rarely representative, since net seepage is a function of reservoir stage and season of the year in many cases. Unless these effects can be calculated, net seepage values determined from limited data have little utility.

Good estimates of evaporation using the water budget equation have been obtained, as exemplified by research conducted on Lake Hefner, Oklahoma.[12] Under optimal conditions, the order of accuracy of the method is about 10%.

Example 1-1 illustrated the use of the water budget in resolving basin evapotranspiration. For such estimates, reliable values can be expected if the period of time chosen is 1 yr or longer. Short period values may also be obtained if observations are adequate. Mean annual evapotranspiration is successfully judged by using long-time averages of precipitation and surface flows, and competent information on the fluctuation of storage. Adequate short-period estimates are also possible if variables in the budget equation can be satisfactorily quantified on a short-term basis.

**Energy Budget Method** The energy budget method illustrates an application of the continuity equation written in terms of energy. It has been employed to compute the evaporation from oceans and lakes, that is, for Lake Hefner in Oklahoma and at Elephant Butte Reservoir in New Mexico.[4,12] The equation accounts for incoming and outgoing energy balanced by the amount of energy stored in the system. The accuracy of estimates of evaporation using the energy budget is highly dependent upon the reliability and preciseness of measurement data. Under good conditions, average errors of perhaps 10% for summer periods and 20% for winter months can be expected.

The energy budget equation for a lake may be written as

$$Q_0 = Q_s - Q_r + Q_a - Q_{ar} + Q_v - Q_{bs} - Q_e - Q_h - Q_w \tag{3-3}$$

where

$Q_0$ = increase in stored energy by the water
$Q_s$ = solar radiation incident at the water surface
$Q_r$ = reflected solar radiation
$Q_a$ = incoming long-wave radiation from the atmosphere
$Q_v$ = net energy advected (net energy content of incoming and outgoing water) into the water body
$Q_{ar}$ = reflected long-wave radiation
$Q_{bs}$ = long-wave radiation emitted by the water
$Q_e$ = energy used in evaporation
$Q_h$ = energy conducted from water mass as sensible heat
$Q_w$ = energy advected by evaporated water

All the terms are in calories per square centimeter per day (cal/cm²-day). Heating brought about by chemical changes and biological proccesses is neglected as is the energy transfer that occurs at the water-ground interface. The transformation of kinetic energy into thermal energy is also excluded. These factors are usually very small, in a quantitative sense, when compared with other terms in the budget if large reservoirs are considered. As a result, their omission has little effect on the reliability of results.

During winter months when ice cover is partial or complete, the energy budget only occasionally yields adequate results because it is difficult to measure reflected solar radiation, ice-surface temperature, and the areal extent of the ice cover. Daily evaporation estimates based on the energy budget are not feasible in most cases because reliable determination of changes in stored energy for such short periods is impractical. Periods of a week or longer are more likely to provide satisfactory measurements.

In using the energy budget approach, it has been demonstrated that the required accuracy of measurement is not the same for all variables.[4] For example, errors in measurement of incoming long-wave radiation as small as 2% can introduce errors of 3 to 15% in estimates of monthly evaporation, while errors on the order of 10% in measurements of reflected solar energy may cause errors of only 1 to 5% in calculated monthly evaporation.

To permit the determination of evaporation by Eq. 3-3, it is common to use the following relationships:

$$B = \frac{Q_h}{Q_e} \tag{3-4}$$

where $B$ is known as Bowen's ratio,[13] and

$$Q_w = \frac{c_p Q_e (T_e - T_b)}{L} \tag{3-5}$$

where

$c_p$ = the specific heat of water (cal/g-°C)
$T_e$ = the temperature of evaporated water (°C)
$T_b$ = the temperature of an arbitrary datum usually taken as 0°C
$L$ = the latent heat of vaporization (cal/g)

Introducing these expressions in Eq. 3-3 and solving for $Q_e$, we obtain

$$Q_e = \frac{Q_s - Q_r + Q_a - Q_{ar} - Q_{bs} - Q_0 + Q_v}{1 + B + c_p(T_e - T_b)/L} \tag{3-6}$$

To determine the depth of water evaporated per unit time, the expression

$$E = \frac{Q_e}{\rho L} \tag{3-7}$$

may be used where

$E$ = evaporation ($cm^3/cm^2$-day)

$\rho$ = the mass density of evaporated water ($g/cm^3$)

The energy budget equation thus becomes

$$E = \frac{Q_s - Q_r + Q_a - Q_{ar} - Q_{bs} - Q_0 + Q_v}{\rho[L(1 + B) + c_p(T_e - T_b)]} \tag{3-8}$$

The Bowen ratio can be computed using

$$B = 0.61 \frac{p}{1000} \frac{(T_0 - T_a)}{(e_0 - e_a)} \tag{3-9}$$

where

$p$ = the atmospheric pressure (mb)

$T_0$ = the water surface temperature (°C)

$T_a$ = the air temperature (°C)

$e_0$ = the saturation vapor pressure at the water surface temperature (mb)

$e_a$ = the vapor pressure of the air (mb)

This expression circumvents the problem of evaluating the sensible heat term, which does not lend itself to direct measurement.

### Mass Transfer Techniques

Mass transfer equations are based primarily on the concept of the turbulent transfer of water vapor (by eddy motion) from an evaporating surface to the atmosphere. Many equations, both theoretical and empirical, have been developed. Most are similar in form to a relationship between evaporation and vapor pressure first recognized by Dalton,[14]

$$E = \kappa(e_0 - e_a) \tag{3-10}$$

where

$E$ = the direct evaporation

$\kappa$ = a coefficient dependent upon the wind velocity, atmospheric pressure, and other factors

$e_0$ and $e_a$ = the saturation vapor pressure at water surface temperature and the vapor pressure of air, respectively

Theoretical mass transfer equations are based on the concepts of discontinuous and continuous mixing at the air-liquid interface.

Empirical approaches often require exacting and costly instrumentation and observations, so their general utility is limited. The complexity of the equations varies from simple expressions such as Eq. 3-10 to complex relationships like Sutton's equation[15] for a circular lake of radius $r$.

$$E = \frac{0.623}{p} G' \rho u^{(2-n)/(2+n)} r^{(4+n)/(2+n)} (e_0 - e_a) \tag{3-11}$$

where

$E$ = the evaporation (cm/day)
$\rho$ = the mass density of the air (g/cm³)
$u$ = the average wind velocity (cm/sec)
$r$ = the radius of circular lake (cm)
$p$ = the atmospheric pressure (mb)
$e_0$ and $e_a$ = as previously defined
$n$ = an empirical constant
$G'$ = a complex function

A commonly used empirical equation has been developed by Meyer.[16] This equation takes the form

$$E = C(e_0 - e_a)\left(1 + \frac{W}{10}\right) \tag{3-12}$$

where

$E$ = the daily evaporation in inches depth
$e_0$ and $e_a$ = as previously defined but in units of inches of Hg
$W$ = the wind velocity in mph measured about 25 ft above the water surface
$C$ = an empirical coefficient

For daily data on an ordinary lake, $C$ will be about 0.36. For wet soil surfaces, small puddles, and shallow pans, the value of $C$ is approximately 0.50.

***Example 3-1*** Find the daily evaporating $E$ from a lake for a day during which the following mean values were obtained: air temperature 87°F, water temperature 63°F, wind speed 10 mph, and relative humidity 20%.

Interpolating from Table 3-1, we find that

$$e_0 = 0.58 \text{ in. Hg}$$
$$e_a = 1.29 \times 0.20 = 0.26 \text{ in. Hg}$$

Using Eq. 3-12, we obtain

$$\begin{aligned} E &= 0.36\,(0.58 - 0.26)\left(1 + \frac{10}{10}\right) \\ &= 0.36 \times 0.32 \times 2 \\ &= 0.23 \text{ in./day} \end{aligned}$$

***Table 3-1*** *Water Vapor Pressure at Various Temperatures*

| Temp (°F) | Vapor Pressure | | |
|---|---|---|---|
| | *Millibars* | *Psi* | *(in. Hg)* |
| 32 | 6.11 | 0.09 | 0.18 |
| 40 | 8.36 | 0.12 | 0.25 |
| 50 | 12.19 | 0.18 | 0.36 |
| 60 | 17.51 | 0.26 | 0.52 |
| 70 | 24.79 | 0.36 | 0.74 |
| 80 | 34.61 | 0.51 | 1.03 |
| 90 | 47.68 | 0.70 | 1.42 |
| 100 | 64.88 | 0.96 | 1.94 |

Investigations of the utility of mass transfer equations conducted at Lake Hefner indicated that a simple equation using wind speed and vapor pressure differences yielded as good results as any that were tested.[12] The equation was

$$E = Nu(e_0 - e_a) \tag{3-13}$$

where

$E$ = the evaporation (cm/day)
$N$ = a coefficient
$u$ = the wind velocity at 2 m above the water surface (m/sec)
$e_0$ and $e_a$ = as previously defined (mb)

The value of $N$ can be determined by comparative studies between mass transfer and energy budget methods. This is the preferred approach. If such an evaluation is not available, it can be approximated using,

$$N = \frac{0.0291}{A^{0.05}} \tag{3-14}$$

where $A$ = the surface area of the water surface in square meters. For values of $A$ less that about $4 \times 10^6$ m$^2$, variations in wind exposure may become important and Eq. 3-14 should be used with caution. When $N$ is based on comparative studies using the energy budget, average errors in evaporation estimates of about 15% can be expected, while errors of roughly 30% would likely obtain if Eq. 3-14 were employed.

### Evaporation Determination Using Pans

The most widely used method of finding reservoir evaporation is by means of evaporation pans. The standard weather bureau Class A pan, built of unpainted galvanized iron, is currently the most popular. It is 4 ft in diameter, 10 in. deep, and is mounted 12 in. above the ground on a wooden frame. Relationships developed between pan and actual evaporation from large bodies of water such as lakes indicate multiplying the former by a factor of 0.70 to 0.75 (pan coefficient) gives the equivalent lake evaporation. Ratios of annual reservoir evaporation to pan evaporation are consistent from year to year and region to region, while monthly ratios often show considerable variation.

Estimates of reservoir evaporation based on short-period pan observations (less than 1 yr) may be seriously in error. The use of a pan coefficient to estimate evaporation from an ungauged location should reflect the geographic variability in heat transfer through the sides of the class A pan. For lakes subjected to significant amounts of advected energy, local pan-lake relationships should be established.

Available data indicate that the annual ratio of lake evaporation to class A pan evaporation is essentially 0.70, provided that net advection is balanced by the change in energy stored, conduction through the pan is negligible, and the pan is located so that its exposure conditions are representative of the body of water being considered.

### Selection of a Method for Determining Evaporation

Any of the methods described can be employed to determine evaporation. Usually, instrumentation for energy budget and mass transfer methods is quite expensive and the cost to maintain observations is substantial. For these reasons, the water budget method and use of evaporation pans are more common. The pan method is the least expensive and will frequently provide good estimates of annual evaporation. Any approach selected is dependent, however, upon the degree of accuracy required. As our ability to evaluate the terms in the water budget and energy budget improves, so will the resulting estimates of evaporation.

### The Nature of Evaporating Surface

Evaporation from a particular surface is directly related to the opportunity for evaporation (availability of water) provided by that surface. For open bodies of water, evaporation opportunity is 100%, while for soils it varies from a high of 100% when the soil is highly saturated—for example, during storm periods—to essentially zero at stages of very low moisture content. Other types of surfaces provide diverse degrees of evaporation opportunity and, except in rare cases, these will almost always vary widely with time.

### Evaporation Reduction

Evaporation losses can be greatly significant at any location. Consequently, the concept of evaporation reduction is receiving widespread attention. Evaporation losses from soils can be controlled by employing various types of mulches or by chemical alteration. They may be reduced from open waters by (1) storing water in covered reservoirs, (2) making increased use of underground storage, (3) controlling aquatic growths, (4) building storage reservoirs with minimal surface area, and (5) through the use of chemicals. Some of these approaches may be impractical (covering large reservoirs) or uneconomical (large-scale vegetation control). All have potential advantages, however, under the proper circumstances.

The first four approaches need no explanation of the mechanism expected to control evaporation. The fifth method, chemical means, requires further comment. Research has shown that certain types of organic compounds such as hexadecanol and octadecanol form monomolecular films that are effective as evaporation inhibitors.[4,17] Studies by the Bureau of Reclamation indicate that evaporation may be suppressed by as much as 64% with hexadecanol films in 4-ft diameter pans under controlled conditions. Actual reductions on large bodies of water would be significantly less than this, however, due to problems in maintaining films against wind and wave action. Evaporation reduction in the range of 22 to 35% has been observed for some studies on small lakes of roughly 100 acres in size with reductions of 9 to 14% reported for Lake Hefner in Oklahoma (2500 acres).[18] Wind was a major problem at Lake Hefner, however.

In Australia, extended tests on medium-sized lakes (less than 2500 acres) have indicated savings of 30 to 50%, although adverse winds were generally not encountered.[18] Although considerable research and development work remains, the use of monomolecular films to control evaporation appears promising. Franzini has indicated that the cost per acre-foot of water saved by evaporation suppression is, in fact, competitive with various alternate means of increasing local water supplies, and these costs will likely decrease with advances in research.[17]

## 3-2 Transpiration

Root systems of plants absorb water in varying quantities. Most of this water is transmitted through the plant and escapes through pores in the leaf system. This is known as *stomatal transpiration*. Plants also lose water by other mechanisms, but usually this is negligible compared with that lost through the microscopic leaf apertures. Transpiration is basically a process by which water is evaporated from the airspaces in plant leaves. Therefore, it is controlled essentially by the same factors that dominate evaporation, namely, solar radiation, temperature, wind velocity, and vapor pressure gradients. In addition, transpiration is affected to some degree by the character of the plant and plant density.

Soil moisture content, when reduced to the wilting point (stage at which plants wilt and do not recover in a humid atmosphere) also affects transpiration. The effects of decreased soil moisture above the wilting point are not clearly established and are somewhat controversial. Nevertheless, it appears that as long as soil moisture lies between the limits of the wilting point and field capacity (the amount of water retained in a soil against gravity after percolation ceases) transpiration is not materially affected. Saturated soils can sometimes adversely affect plant life.

Diffusion of water vapor from plant leaves to the atmosphere is proportional to the vapor pressure gradient at the leaf-atmosphere interface. Upon absorbing solar radiation, leaves tend to become warmer than the surrounding air (often by as much as 5 to 10°F). The amount of water vapor held by the air at the leaf-air interface thus increases; more rapid water losses are favored; and transpiration follows a diurnal cycle which is approximately that of light intensity. It has also been demonstrated that transpiration and the rate of plant growth are related. Below a temperature of about 40°F the amount of water transpired is considered negligible.

Different species and types of plants often display considerably different demands on soil moisture even if the same environmental conditions prevail. For example, an oak tree may transpire as much as 170 qt of water a day, whereas a corn plant will transpire only about 2 qt. The area covered by the two root systems is, of course, significantly different. Various species also indicate different patterns in seasonal demands for water. Agricultural products obviously have their periods of greatest transpiration at the peak of the growing season.

Precise values for quantities of water transpired are difficult to acquire, since many variables are active in the process and these range widely from one region to another. Available estimates should be used with caution, and the conditions under which they were obtained should be determined before applying the data. Adequate relationships between climatic factors and transpiration become prerequisites if data derived in one climatic region are to have general utility.

Transpiration may be measured in the laboratory by using tanks wherein evaporation is eliminated and water losses are found by weighing. Coefficients must be derived before such data can be applied to field conditions and, even then the observations usually provide little more than an index to field water use. Large-scale field measurements of transpiration are virtually impossible under prevailing field conditions so it is common to find measures of consumptive use (combined evaporation plus transpiration) more widely adopted and of greater value to the practicing hydrologist. Most field observations are made by using lysimeters (grass or crop-covered containers for which a water budget is maintained). Table 3-2 gives some values of consumptive use for several crops in the Montrose area of Colorado.[19] These values are presented only to indicate their order of magnitude in this area during the growing or irrigation season. More complete information on consumptive use by various crops can be found elsewhere.[1,19,20]

For many small local projects it is not possible to carry out detailed field studies to determine the consumptive use of crops. In such cases it is common to use either the Blaney-Criddle or Penman method for estimating seasonal consumptive use.[21,22] The Blaney-Criddle method will be briefly described here.

The seasonal consumptive use for a particular crop can be computed using the relation

$$U = k_s B \tag{3-15}$$

where

$U$ = the consumptive use of water during the growing season (in.)
$k_s$ = a seasonal consumptive use coefficient applicable to a particular crop, empirically derived
$B$ = the summation of monthly consumptive use factors for a given season

The term $B$ can be expressed as

$$B = \sum \left( \frac{tp}{100} \right) \tag{3-16}$$

where

$t$ = the mean monthly temperature (°F)
$p$ = the monthly daytime hours given as percent of the year

**Table 3-2** *Consumptive Use for Crops in the Montrose, Colorado, Area During the Irrigation or Growing Season*

| Crop | Consumptive Use (in.) |
|---|---|
| Alfalfa | 26.5 |
| Corn | 19.7 |
| Small grain | 14.9 |
| Grass hay | 23.3 |
| Natural vegetation | 37.3 |

*Source:* H. F. Blaney, "Water and Our Crops," *Water, The Yearbook of Agriculture* (Washington, D.C.: U.S. Department of Agriculture, 1955).

If monthly values for the consumptive use coefficient $k$ are available, monthly consumptive use can be found by using

$$u = \frac{ktp}{100} \tag{3-17}$$

where $u$ is the monthly consumptive use (in.) and the other terms are as previously described. Selected values of $p$ and $k$ are available in the literature.[20,21] An example will illustrate the use of this equation.

***Example 3-2*** Determine the monthly consumptive use of an alfalfa crop grown in southern California for the month of July if the average monthly temperature is 72°F, average daytime hours in percent of the year is 9.88, and the mean monthly consumptive use coefficient for alfalfa is 0.85.

Using Eq. 3-17 we find that,

$$\begin{aligned} u &= \frac{ktp}{100} \\ &= 0.85 \times 72 \times \frac{9.88}{100} \\ &= 6.05 \text{ in. of water} \end{aligned}$$

### Transpiration Reduction

Water conservation through transpiration reduction is being seriously studied, and certain preventative practices are presently in use. Methods of control include the use of chemicals to inhibit water consumption (analogous to the use of films to control surface evaporation except that chemicals are applied in the root zone), harvesting of plants, improved irrigation practices, and actual removal or destruction of certain vegetative types.[27]

In arid regions of the Southwest, certain plants known as *phreatophytes* (plants capable of tapping the water table or capillary fringe) transpire enormous quantities of water each year without providing any particular apparent benefit (although this statement is open to question). Many of these plants, such as the salt cedar, grow in stream channels and tend to create flood control problems by restricting channels in addition to using valuable underground water supplies. In New Mexico there have been as many as 43,000 acres of salt cedar along the Pecos River alone. Control of these phreatophytes could result in estimated savings of over 200,000 acre-ft of water in a critically water-short region of the United States.[23] Conservation through transpiration control may be important, but the ecologic consequences of such control practices should be given careful consideration.

## 3-3 Evapotranspiration

In most cases of practical interest to the hydrologist, only total evaporation from an area—combined evaporation plus transpiration (consumptive use)—is of real interest. Various methods for determining evapotranspiration have been proposed, but there is no one system generally acceptable under all conditions. Basically, there are three major approaches:

1. Theoretical, based on physics of the process.
2. Analytical, based on energy or water budgets.
3. Empirical.

The water budget method was illustrated in Example 1-1. Its adequacy is dependent upon the accuracy with which the several terms in the budget equation can be evaluated. The energy budget can also be used to calculate field evapotranspiration in a manner similar to that previously described for lakes. For this application, however, the soil's thermal properties must be known, and temperature and vapor pressure gradients measured at two levels above the ground are needed in Bowen's ratio. For field plots the amount of energy advected usually can be neglected.

Mass transfer equations of the form previously discussed can also be used to estimate evapotranspiration. The Thornthwaite-Holzman equation is a good example of a mass transfer equation that has often been employed for this purpose. However, Linsley et al. indicate that there is some question as to the adequate verification of this model to estimate evapotranspiration losses.[24] The equation is expressed as

$$E = \frac{833\,\kappa^2\,(e_1 - e_2)\,(V_2 - V_1)}{(T + 459.4)\,\log_e\,(z_2/z_1)^2} \tag{3-18}$$

where it is assumed that the atmosphere is adiabatic and the wind speed and moisture are distributed logarithmically in a vertical direction.

$E$ = evaporation (in./hr)
$\kappa$ = von Kármán's constant (0.4)
$e_1$ and $e_2$ = vapor pressures (in. Hg)
$V_1$ and $V_2$ = wind speeds (mph)
$T$ = the mean temperature (°F) of the layer between the lower level $z_1$ and the upper level $z_2$

In view of the small differences between wind and vapor pressure to be expected at two levels so closely spaced, and since these gradients are directly related to the sought-after evaporation, highly exacting instrumentation is required to get reliable results.

A number of specific evapotranspiration equations are available, the Blaney-Criddle equation (Eq. 3-15) being representative.[28]

## 3-4 Simulation of Evapotranspiration

The volume of water that leaves a watershed through evapotranspiration is very significant over time, and continuous modeling processes require that estimates of this be incorporated. In most cases, potential evapotranspiration and soil moisture conditions are the predominant factors used in calculating actual evapotranspiration.

### Potential Evapotranspiration

Thornthwaite defined potential evapotranspiration as "the water loss which will occur if at no time there is a deficiency of water in the soil for the use of vegetation." In a practical sense, however, most investigators have assumed that potential evapotranspiration is equal to lake evaporation as determined from National Weather Service Class A pan records. This is not theoretically correct because the albedo (amount of incoming radiation reflected back to the atmosphere) of vegetated areas and soils ranges as high as 45%.[35] As a result, potential evapotranspiration should be somewhat less than free water surface evaporation. Errors in estimating free water evapotranspiration from pan records are such, however, as to make an adjustment for potential evapotranspiration of questionable value.

An equation for estimating potential evapotranspiration developed by the Agricultural Research Service (ARS) illustrates efforts to include vegetal characteristics and soil moisture in such a calculation. The evapotranspiration potential for any given day is determined as follows:[36]

$$ET = GI \cdot k \cdot E_p \cdot \left( \frac{S - SA}{AWC} \right)^x \qquad \textbf{(3-19)}$$

where

$ET$ = evapotranspiration potential (in./day)
$GI$ = growth index of crop in % of maturity
$k$ = ratio of $GI$ to pan evaporation, usually 1.0–1.2 for short grasses, 1.2–1.6 for crops up to shoulder height, and 1.6–2.0 for forest
$E_p$ = pan evaporation (in./day)
$S$ = total porosity
$SA$ = available porosity (unfilled by water)
$AWC$ = porosity drainable only by evapotranspirations
$x = AWC/G$ ($G$ = moisture freely drained by gravity)

The *GI* curves have been developed by expressing experimental data on daily evapotranspiration for several crops (Fig. 3-2) as a percentage of the annual maximal daily rate (Fig. 3-3). Equation 3-19 is used by the Agricultural Research Service in its USDAHL-74 model of watershed hydrology in combination with *GI* curves to calculate daily evapotranspiration. Representative values for *S*, *G*, and *AWC* are given in Table 3-3.

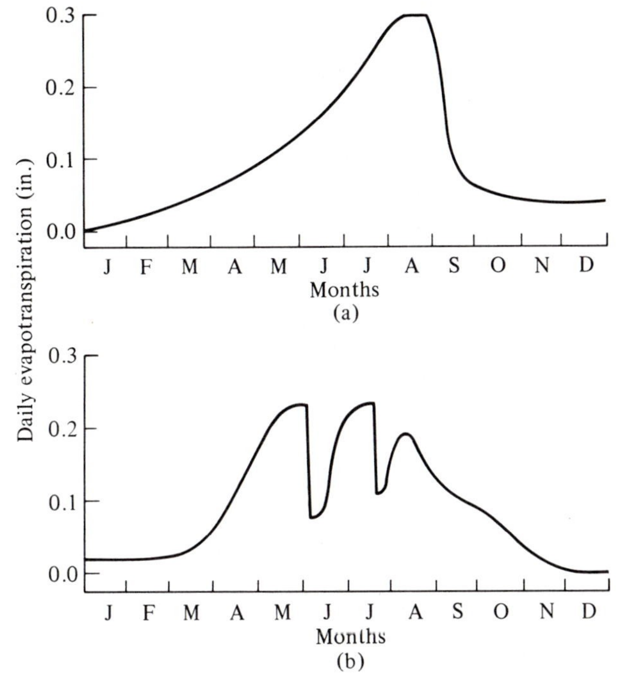


***Fig. 3-2.*** *Average daily consumption of water: (a), for year 1953 by corn, followed by winter wheat under irrigation; and (b), for year 1955, with irrigated first-year meadow of alfalfa, red clover, and timothy. Both measurements taken on lysimeter Y 102 C at the Soil and Water Conservation Research Station, Coshocton, Ohio. (After H. N. Holtan and N. C. Lopez, "USDAHL-74 Revised Watershed Hydrology," U.S. Department of Agriculture, ARS Tech. Bulletin No. 1518, Washington, D.C., 1975.)*

### Stanford Watershed Model

The Stanford Watershed Model (see Chapter 10 for complete discussion) is the basis for many continuous simulation models. In it, evapotranspiration is considered to occur from interception storage and upper zone storage at the potential rate. The upper zone simulates storage in depressions and highly permeable surface soils while the lower zone is linkage to groundwater storage. Evapotranspiration opportunity is considered to control evapotranspiration from the lower zone. Evapotranspiration opportunity is defined as the maximal amount of water available for evapotranspiration at a particular location during a prescribed time interval.[26] Figure 3-4 illustrates the nature of the relationship presumed in the Stanford Model between the evapotranspiration potential and the watershed area. In the modeling process, the evapotranspiration potential is filled if possible from interception storage and the upper zone in that order. If the potential cannot be satisfied from these sources, the remaining potential enters

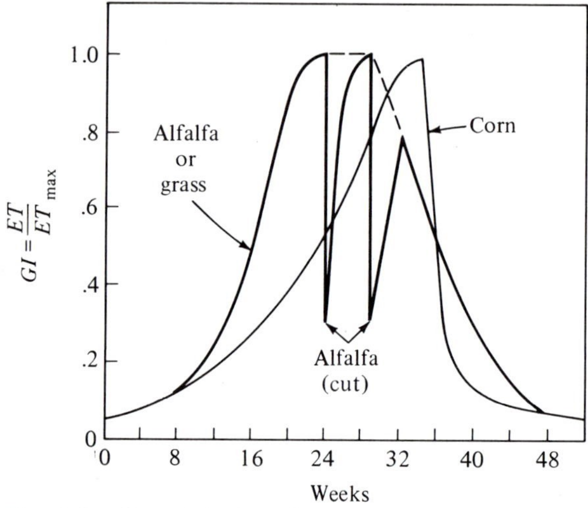


***Fig. 3-3.*** *Growth index, $GI = ET/ET_{max}$ from lysimeter records, irrigated corn, and hay for 1955, Coshocton, Ohio. (After H. N. Holtan and N. C. Lopez, "USDAHL-74 Revised Watershed Hydrology," U.S. Department of Agriculture, ARS Tech. Bulletin No. 1518, Washington, D.C., 1975.)*

***Table 3-3*** *Hydrologic Capacities of Soil Texture Classes*

| Texture Class | *S* (%) | *G* (%) | *AWC* (%) | *x* *AWC/G* |
|---|---|---|---|---|
| Coarse sand | 24.4 | 17.7 | 6.7 | 0.38 |
| Coarse sandy loam | 24.5 | 15.8 | 8.7 | 0.55 |
| Sand | 32.3 | 19.0 | 13.3 | 0.70 |
| Loamy sand | 37.0 | 26.9 | 10.1 | 0.38 |
| Loamy fine sand | 32.6 | 27.2 | 5.4 | 0.20 |
| Sandy loam | 30.9 | 18.6 | 12.3 | 0.66 |
| Fine sandy loam | 36.6 | 23.5 | 13.1 | 0.56 |
| Very fine sandy loam | 32.7 | 21.0 | 11.7 | 0.56 |
| Loam | 30.0 | 14.4 | 15.6 | 1.08 |
| Silt loam | 31.3 | 11.4 | 19.9 | 1.74 |
| Sandy clay loam | 25.3 | 13.4 | 11.9 | 0.89 |
| Clay loam | 25.7 | 13.0 | 12.7 | 0.98 |
| Silty clay loam | 23.3 | 8.4 | 14.9 | 1.77 |
| Sandy clay | 19.4 | 11.6 | 7.8 | 0.67 |
| Silty clay | 21.4 | 9.1 | 12.3 | 1.34 |
| Clay | 18.8 | 7.3 | 11.5 | 1.58 |

$S$ = total porosity − 15 bar moisture %, $G$ = total porosity − 0.3 bar moisture %, and $AWC = S$ minus $G$.

*Source:* Adapted from C. B. England, "Land Capability: A Hydrologic Response Unit in Agricultural Watersheds," U.S. Department of Agriculture, ARS 41-172, Sept. 1970. After H. N. Holtan, et al., "USDAHL-74 Revised Model of Watershed Hydrology," U.S. Department of Agriculture, ARS Tech. Bulletin No. 1518, Washington, D.C., 1975.

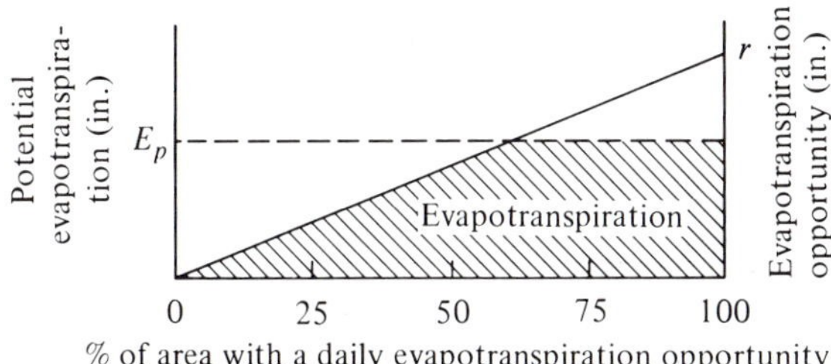


***Fig. 3-4.*** *Evapotranspiration relationship used in the Stanford Watershed Model. (After N. H. Crawford and R. K. Linsley, Jr., "Digital Simulation in Hydrology: Stanford Watershed Model IV," Tech. Rept. 39, Department of Civil Engineering, Stanford University, Stanford, Calif., July 1966.)*

as $E_p$ in Fig. 3-4 and the soil moisture is depleted accordingly. Evapotranspiration from stream surfaces is calculated at the potential rate and evapotranspiration from groundwater storage is determined by a parameter that represents the fraction of the total watershed area in which evapotranspiration from groundwater storage is assumed to occur at the potential rate.

## 3-5 Interception

Part of the storm precipitation that occurs is intercepted by vegetation and other forms of cover on the drainage area. Interception can be defined as that segment of the gross precipitation input which wets and adheres to above ground objects until it is returned to the atmosphere through evaporation. Precipitation striking vegetation may be retained on leaves or blades of grass, flow down the stems of plants and become stemflow, or fall off the leaves to become part of the throughfall. The modifying effect that a forest canopy can have on rainfall intensity at the ground (the throughfall) can be put to practical use in watershed management schemes.

The amount of water intercepted is a function of (1) the storm character, (2) the species, age, and density of prevailing plants and trees, and (3) the season of the year. Usually about 10 to 20% of the precipitation that falls during the growing season is intercepted and returned to the hydrologic cycle by evaporation. Water losses by interception are especially pronounced under dense closed forest stands—as much as 25% of the total annual precipitation. Schomaker has reported that the average annual interception loss by Douglas fir stands in western Oregon and Washington is about 24%.[3] A 10-yr-old loblolly pine plantation in the South showed losses on a yearly basis of approximately 14%, while Ponderosa pine forests in California were found to intercept about 12% of the annual precipitation. Mean interception losses of approximately 13% of gross summer rainfall were reported for hardwood stands in the White Mountains of New Hampshire. Additional information given in Table 3-4 includes some data on interception measurements obtained in Maine from a mature spruce-fir stand, a moderately well-stocked white and gray birch stand, and an improved pasture.[3]

Lull indicates that oak or aspen leaves may retain as much as 100 drops of water.[1] For a well-developed tree, interception storage on the order of 0.06 in. of precipitation could therefore be expected on the basis of an average retention of about 20 drops per leaf. For light showers (where gross precipitation $P < 0.01$ in.) 100% interception

***Table 3-4*** *Weekly Average Precipitation Catch of Standard U.S. Weather Bureau-Type Rain Gauges Located in a Spruce-Fir Stand, a Hardwood Stand, and a Pasture During the Winter of 1965–1966*

| | Weekly Average Precipitation Catch (in. of equivalent rain) | | | Precent Interception by Forest Cover | |
|---|---|---|---|---|---|
| *Measuring Date*[a] | *Spruce-Fir* | *Birch* | *Pasture* | *Spruce-Fir* | *Birch* |
| 11/ 9/65 | 0.24 | 0.33 | 0.39 | 38 | 15 |
| 11/16/65 | 1.01 | 1.25 | 1.45 | 30 | 14 |
| 11/23/65 | 1.01 | 1.23 | 1.36 | 26 | 10 |
| 12/10/65[b] | 1.41 | 1.65 | 1.79 | 21 | 8 |
| 12/17/65 | 0.55 | 0.81 | 0.87 | 37 | 7 |
| 12/30/65 | 0.66 | 0.95 | 1.08 | 39 | 12 |
| 1/ 4/66 | 0.20 | 0.25 | 0.26 | 23 | 4 |
| 1/12/66 | 0.36 | 0.55 | 0.61 | 41 | 10 |
| 1/18/66 | Trace | Trace | Trace | — | — |
| 1/25/66 | 0.25 | 0.58 | 0.59 | 58 | 2 |
| 2/ 1/66 | 1.38 | 1.91 | 1.96 | 30 | 3 |
| 2/ 8/66 | 0.05 | 0.07 | 0.06 | 17 | 16 |
| 2/11/66 | 0.29[c] | 0.02 | Trace | — | — |
| 2/15/66 | 0.76 | 0.81 | 0.98 | 22 | 17 |
| 2/21/66 | 0.17 | 0.22 | 0.22 | 23 | 0 |
| 3/ 2/66 | 0.86 | 1.23 | 1.45 | 41 | 16 |
| 3/ 7/66 | 0.76 | 0.84 | 0.97 | 22 | 13 |
| 3/15/66 | 0 | 0 | 0 | — | — |
| 3/29/66 | 0.73 | 1.13 | 1.27 | 43 | 11 |
| Total | 10.69 | 13.83 | 15.31 | 30.2 | 9.5 |

[a] The period between measuring dates is 7 days, except when precipitation occurred on the seventh day. In this event, measurement was postponed until precipitation ceased.
[b] Measurements were delayed until a method was devised to melt frozen precipitation on the site.
[c] This measurement in the spruce stand was the result of foliage drip during a thaw from previously intercepted snow.

*Source:* After C. E. Schomaker, "The Effect of Forest and Pasture on the Disposition of Precipitation," *Maine Farm Research* (July 1966).

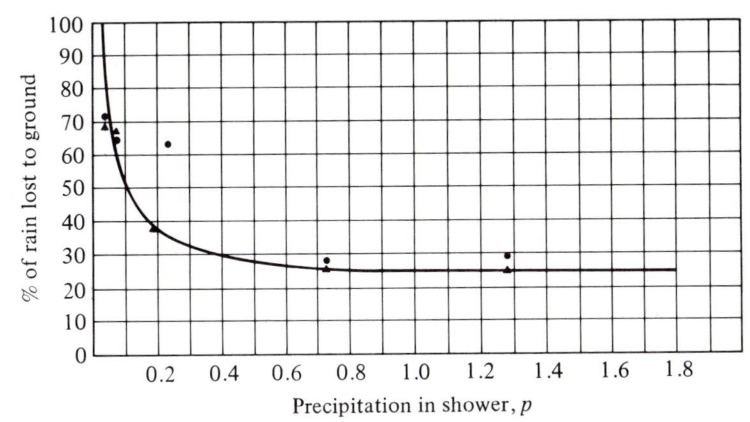


***Fig. 3-5.*** *Mean curve showing total percent of precipitation in a shower intercepted by various trees in 1917–1918. Note: The curve is drawn with respect to triangular points.* ● *Represents the mean of all the interceptometers including some uncorrected ones for trunk water and peripheral pans.* ▲ *Represents the mean of all interceptometers excluding nos. 4, 7, 8, 14, 15, 16, 17 which are peripherial pans or are uncorrected for trunk water. (After R. E. Horton, "Rainfall Interception,"* Monthly Weather Rev., *47 (1919): 603–623.)*

might occur, whereas for showers where $P > 0.04$ in., losses in the range of 10 to 40% are realistic.[5]

Figure 2-8 illustrates the general time distribution pattern of interception loss intensity. Most interception loss develops during the initial storm period and the rate of interception rapidly approaches zero thereafter. Total storm interception can be estimated by using[1,5]

$$L_i = S_i + KEt \tag{3-20}$$

where

$L_i$ = the volume of water intercepted (in.)

$S_i$ = the interception storage that will be retained on the foliage against the forces of wind and gravity (usually varies between 0.01 and 0.05 in.)

$K$ = the ratio of surface area of intercepting leaves to horizontal projection of this area

$E$ = the amount of water evaporated per hour during the precipitation period

$t$ = time (hr)

Total interception by an individual plant is directly related to the amount of foliage and its character and orientation. Figure 3-5 illustrates the mean percentage precipitation loss for various trees for various quantities of total precipitation.

Equation 3-20 can be used to estimate total interception losses. For analyses of individual storms it is also necessary to know or assume distribution of this abstraction. A general equation for estimating such losses is not available, since most studies have been related to particular species or experimental plots strongly associated with a given locality. In addition, the loss function varies with the storm's character. If adequate experimental data are available, the nature of the variance of interception versus time might be inferred. Otherwise, common practice is to deduct the estimated volume entirely from the initial period of the storm ("initial abstraction").

***Example 3-3*** Using the curve shown in Fig. 3-5 and the following equations derived by Horton[41] for interception by ash and oak trees, estimate the interception loss beneath trees during a storm having a total precipitation of 1.5 in.

1. For ash trees,

$$L_i = 0.015 + 0.23\,P$$
$$L_i = 0.015 + 0.23\,(1.5) = 0.36 \text{ in.}$$

2. For oak trees,

$$L_i = 0.03 + 0.22\,P$$
$$L_i = 0.03 + 0.22\,(1.5) = 0.36 \text{ in.}$$

3. Using the curve for mean percentage interception by trees, the percentage interception = 24.

$$0.24 \times 1.5 = 0.36 \text{ in.}$$

Thus, for this example, the storm loss beneath the trees resulting from interception would be about 0.36 in.

## 3-6 Depression Storage

Precipitation that reaches the ground may infiltrate, flow over the surface, or become trapped in numerous small depressions from which the only escape is evaporation or infiltration. The nature of depressions, as well as their size, is largely a function of the original land form and local land use practices. Because of extreme variability in the nature of depressions and the paucity of sufficient measurements, no generalized relationship with enough specified parameters for all cases is feasible. A rational model can be suggested, however, and information on the range of depression storage losses reported in the literature will be offered.

According to Linsley et al.[29] the volume of water stored by surface depressions at any given time can be approximated using

$$V = S_d(1 - e^{-kP_e}) \tag{3-21}$$

where

$V$ = the volume actually in storage at some time of interest
$S_d$ = the maximum storage capacity of the depressions
$P_e$ = the rainfall excess (gross rainfall minus evaporation, interception, and infiltration)
$k$ = a constant equivalent to $1/S_d$

The value of the constant can be determined by considering that if $P_e \approx 0$, essentially all of the water will fill depressions and $dV/dP_e$ will equal 1. This requires that $k = 1/S_d$. Estimates of $S_d$ may be secured by making sample field measurements of the area under study. Combining such data with estimates of $P_e$ permits a determination of $V$. The manner in which $V$ varies with time must still be estimated if depression storage losses are to be abstracted from the gross rainfall input.

One assumption regarding $dV/dt$ is that all depressions must be full before overland flow supply begins. Actually, this would not agree with reality unless the location of depressions were graded with the largest ones occurring downstream. If the depression storage were abstracted in this manner, the total volume would be deducted from the initial storm period such as shown by the shaded area in Fig. 3-6. Such postulates have been used with satisfactory results under special circumstances.[6]

Depression storage intensity can also be estimated using Eq. 3-21. If the overland flow supply rate $\sigma$ plus depression storage intensity equal $(i - f)$

where

$i$ = the rainfall intensity reaching the ground
$f$ = the infiltration rate

then the ratio of overland flow supply to overland flow plus depression storage supply can be proved equal to

$$\frac{\sigma}{(i - f)} = 1 - e^{-kP_e} \tag{3-22}$$

This expression can be derived by adjudging

$$\frac{\sigma}{(i - f)} = \frac{i - f - v}{i - f} \tag{3-23}$$

and noting that $v$ is equal to the derivative of Eq. 3-21 with respect to time. Then

$$v = \frac{d}{dt} S_d(1 - e^{-kP_e}) \tag{3-24}$$

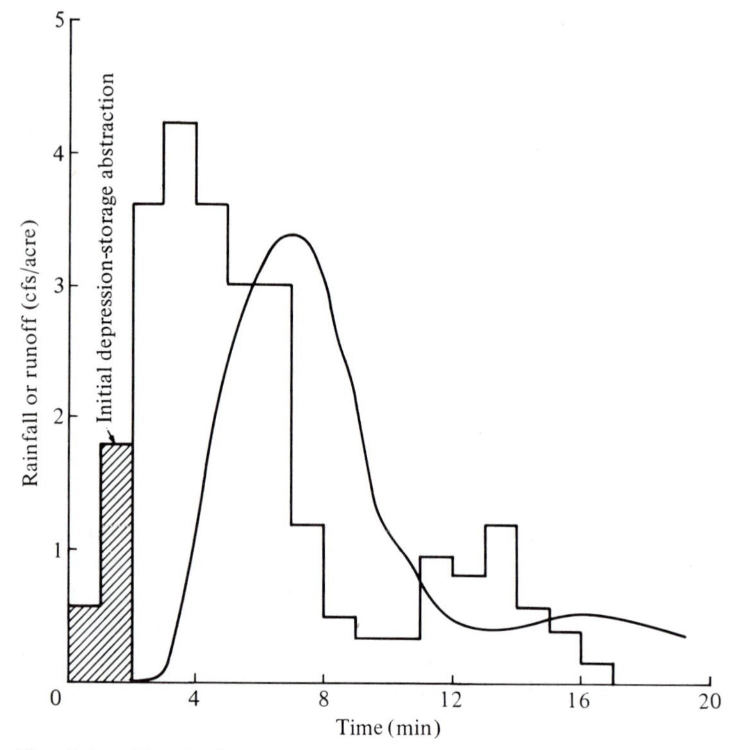


***Fig. 3-6.*** *Simple depression storage abstraction scheme.*

$$v = (S_d k e^{-kP_e}) \frac{dP_e}{dt} \tag{3-25}$$

It was shown that $k = 1/S_d$ so that

$$v = e^{-kP_e} \frac{dP_e}{dt} \tag{3-26}$$

The excess precipitation $P_e$ equals the gross rainfall minus infiltrated water and since the derivative with respect to time can be replaced by the equivalent intensity $(i - f)$, the intensity of depression storage becomes

$$v = e^{-kP_e}(i - f) \tag{3-27}$$

Inserting in Eq. 3-23, we obtain

$$\frac{\sigma}{(i - f)} = \frac{(i - f) - (i - f)e^{-kP_e}}{i - f} \tag{3-28}$$

and

$$\frac{\sigma}{(i - f)} = \frac{(i - f)(1 - e^{-kP_e})}{(i - f)} \tag{3-29}$$

or

$$\frac{\sigma}{(i-f)} = (1 - e^{-kP_e}) \tag{3-30}$$

Figure 3-7 illustrates a plot of this function versus the mass overland flow and depression storage supply $(P - F)$, where

$F$ = the accumulated mass infiltration[7]
$P$ = the gross precipitation

In the plot a mean depth of 0.25 in. for turf and 0.0625 in. for pavements was assumed. Maximum depths were 0.50 in. and 0.125 in., respectively.

The figure also depicts the effect on estimated overland flow supply rate which is derived from the choice of the depression storage model. For example, if it were assumed that all depressions must be full before overland flow begins, the graph shows that for a mean depth $\leq 0.25$ in. for turf areas, $\sigma = 0$, whereas at mean depths $\geq 0.25$ in., $\sigma = (i - f)$. For the exponential model, $\sigma$ always will be greater than zero. Tholin and Kiefer have recommended that a relationship somewhat intermediate between those previously mentioned is likely more representative of fully developed urban areas.[7] A cumulative normal probability curve was selected for this representation and is also described in Fig. 3-7. This approach is used in their model. The use of Fig. 3-7 is illustrated in example 3-4 at the end of the chapter.

Some values for depression storage losses from intense storms reported by Hicks are 0.20 in. for sand, 0.15 in. for loam, and 0.10 in. for clay.[8] Tholin and Kiefer have used values of 0.25 in. in pervious urban areas and 0.0625 in. for pavements.[7] Studies of four small impervious drainage areas by Viessman yielded the information shown in Fig. 3-8, where mean depression storage loss is highly correlated with slope. This is easily understood, since a given depression will hold its maximum volume if horizontally oriented. Using very limited data from a small, paved-street section, Turner devised the curves shown in Fig. 3-9.[9] Other sources of data related to surface storage are available in the literature.[1,2,10]

## 3-7 Infiltration

The role played by infiltration in the distribution of precipitation input is an exceedingly important one. It can affect not only the timing, but also the distribution and magnitude of surface runoff. For this reason reliable estimates of infiltration must be incorporated in any watershed model.[26]

Infiltration is the flow of water into the ground through the earth's surface. The rate $f$ at which it occurs is influenced by such fac-

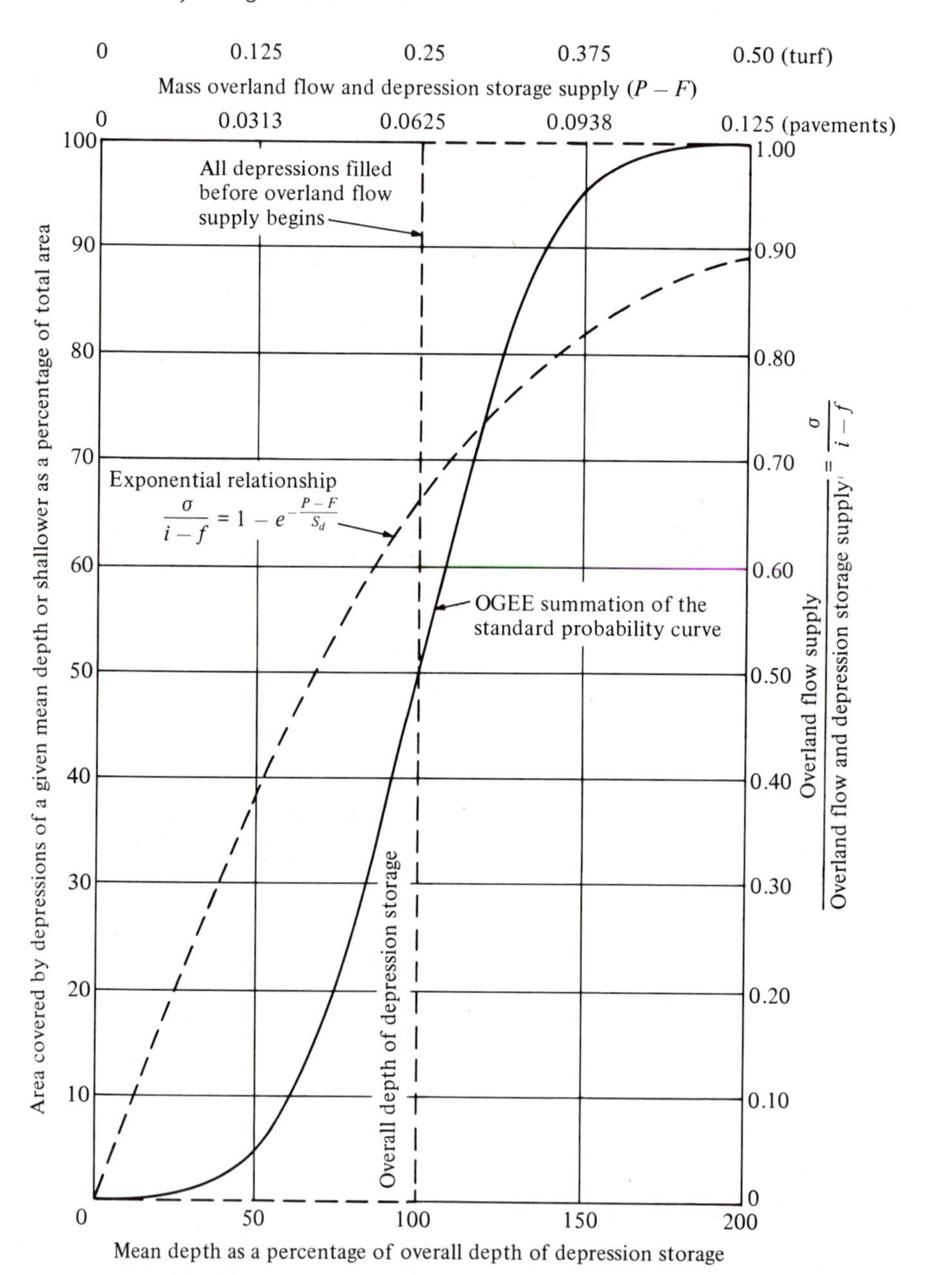


***Fig. 3.7.*** *Depth distribution curve of depression storage. Enter graph from top, read down to selected curve, and project right or left as desired. (After A. L. Tholin and C. J. Kiefer, "The Hydrology of Urban Runoff,"* Proc. ASCE, J. Sanitary Engineering Div., *85, No. 5A2 (Mar. 1959), Part 1, Proc. paper 198: 47–106.)*

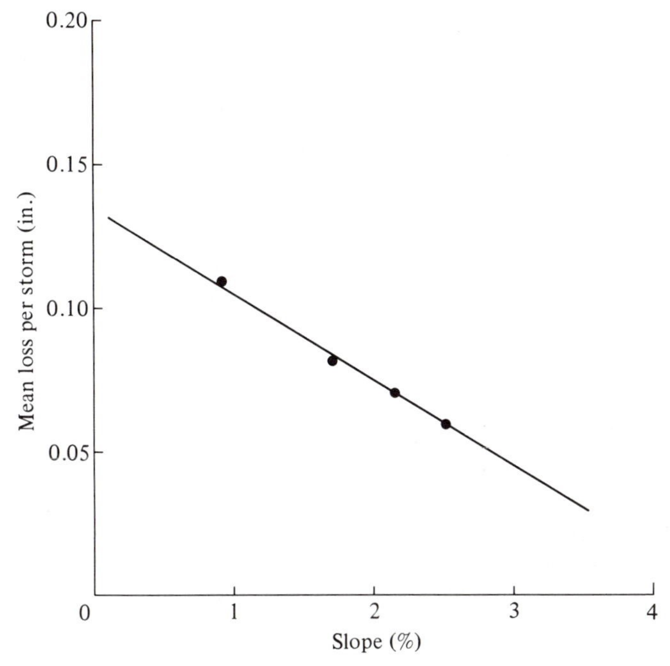


***Fig. 3-8.*** *Depression-storage loss versus slope for four impervious drainage areas. (After Warren Viessman, Jr., "A Linear Model for Synthesizing Hydrographs for Small Drainage Areas," paper presented at the Forty-eighth Annual Meeting of the American Geophysical Union, Washington, D.C., Apr. 1967.)*

tors as the type and extent of vegetal cover, the condition of the surface crust, temperature, rainfall intensity, physical properties of the soil, and water quality.

The rate at which water is transmitted through the surface layer is highly dependent upon the condition of the surface. For example, inwash of fine materials may seal the surface so that infiltration rates are low even when the underlying soils are highly permeable. After water crosses the surface interface, its rate of downward movement will be controlled by the transmission characteristics of the underlying soil profile. The volume of storage available below ground is also a factor affecting infiltration rates.

Considerable research on infiltration has taken place, but considering the infinite combinations of soil and other factors existing in nature, it is not hard to recognize that no adequately quantified general relationship exists.

### Measuring Infiltration

Commonly used methods for determining infiltration capacity are hydrograph analyses and infiltrometer studies. Infiltrometers are

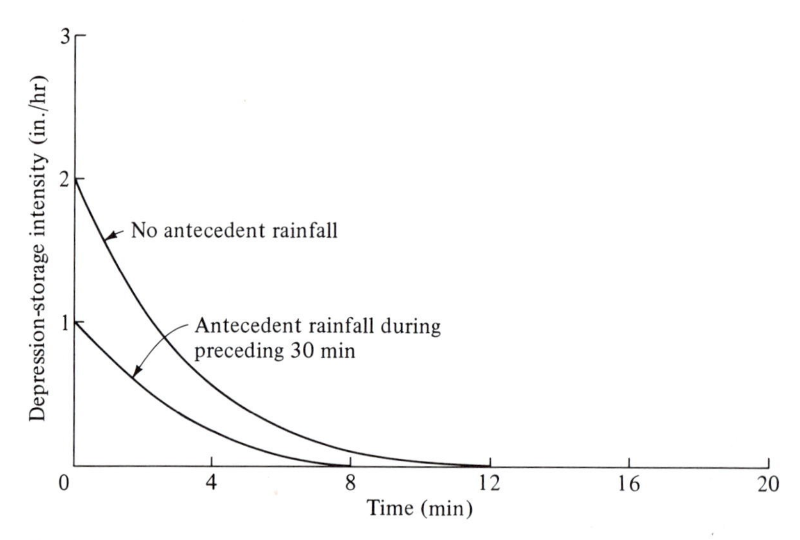


***Fig. 3-9.*** *Depression-storage intensity versus time for an impervious area. (After L. B. Turner, "Abstraction of Depression Storage from Storms and Small Impervious Areas," Master's thesis, University of Maine, Orono, Me., Aug. 1967.)*

usually classified as rainfall simulators or flooding devices. In the former, artificial rainfall is simulated over a small test plot and the infiltration calculated from observations of rainfall and runoff, with consideration given to depression storage and surface detention.[1] Flooding infiltrometers are usually rings or tubes inserted in the ground. Water is applied and maintained at a constant level and observations made of the rate of replenishment required.

Estimates of infiltration based on hydrograph analyses have the advantage over infiltrometers of relating more directly to prevailing conditions of precipitation and field. However, they are no better than the precision with which rainfall and runoff are measured. Of particular importance in such studies is the areal variability of rainfall. Several methods have been developed and are in use. Reference 1 gives a good description of these methods.

### Calculation of Infiltration

Infiltration calculations vary in sophistication from the application of reported average rates for specific soil types and vegetal covers to the use of conceptually sound differential equations governing the flow of water in unsaturated porous media. For small urban areas that respond rapidly to storm input, more precise methods are sometimes warranted. On large watersheds subject to peak flow production from prolonged storms, average or representative values may be adequate.

Past research efforts were generally directed toward (1) the development of empirical equations based on field observation, and (2) the solution of equations based on the mechanics of saturated flow in porous media.[10] In the final analysis, more basic approaches will probably yield the most reliable methods, but these efforts have not been satisfactorily translated into practical hydrologic tools.

The infiltration process was thoroughly studied by Horton in the early 1930s.[2] An outgrowth of his work, shown graphically in Fig. 3-10, was the following relationship for determining infiltration capacity:

$$f = f_c + (f_0 - f_c)e^{-kt} \tag{3-31}$$

where

$f$ = the infiltration capacity (depth/time) at some time $t$
$k$ = a constant representing the rate of decrease in $f$ capacity
$f_c$ = a final or equilibrium capacity
$f_0$ = the initial infiltration capacity

It indicates that if the rainfall supply exceeds the infiltration capacity, infiltration tends to decrease in an exponential manner. Although simple in form, difficulties in determining useful values for $f_0$ and $k$ restrict the use of this equation. The area under the curve for any time interval represents the depth of water infiltrated during that interval. The infiltration rate is usually given in in./hr and the time $t$ in min, although other time increments are used and the coefficient $k$ is determined accordingly.

By observing the variation of infiltration with time and develop-

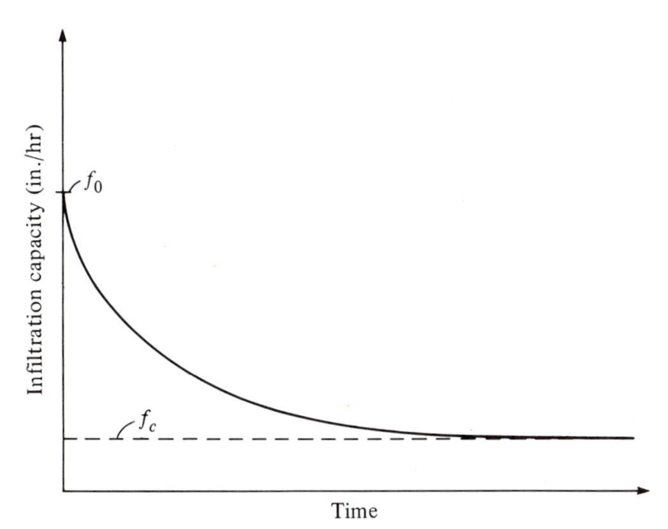


**Fig. 3-10.** *An infiltration capacity function.*

**Table 3-5** *Typical $f_1$ Values*

| Soil Group | $f_1$ (in./hr) |
|---|---|
| High (sandy soils) | 0.50–1.00 |
| Intermediate (loams, clay, silt) | 0.10–0.50 |
| Low (clays, clay loam) | 0.01–0.10 |

*Source:* After *ASCE Manual of Engineering Practice*, No. 28.

ing plots of $f$ versus $t$ as shown in Fig. 3-10, it is possible to estimate $f_0$ and $k$. Two sets of $f$ and $t$ are selected from the curve and entered in Eq. 3-31. Two equations having two unknowns are thus obtained; they can be solved by successive approximations for $f_0$ and $k$.

Typical infiltration rates at the end of 1 hr ($f_1$) are shown in Table 3-5. A typical relationship between $f_1$ and the infiltration rate throughout a rainfall period is shown graphically in Fig. 3-11a; Fig. 3-11b shows an infiltration capacity curve for normal antecedent conditions on turf. The data given in Table 3-5 are for a turf area and must be multiplied by a suitable cover factor for other types of cover complexes. A range of cover factors is listed in Table 3-6.

Total volumes of infiltration and other abstractions from a given recorded rainfall are obtainable from a discharge hydrograph (plot of the streamflow rate versus time) if one is available. Separation of the base flow (dry weather flow) from the discharge hydrograph results in a direct runoff hydrograph (DRH), which accounts for the direct surface runoff, that is, rainfall less abstractions. Direct surface runoff or precipitation excess in inches uniformly distributed over a watershed

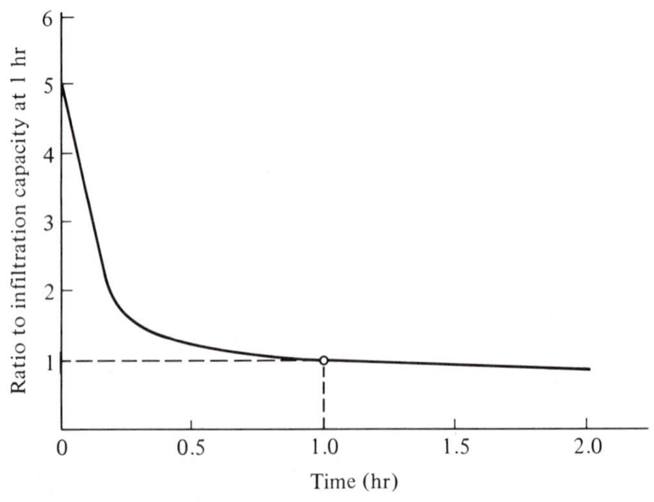


**Fig. 3-11a.** *Typical infiltration curve.*

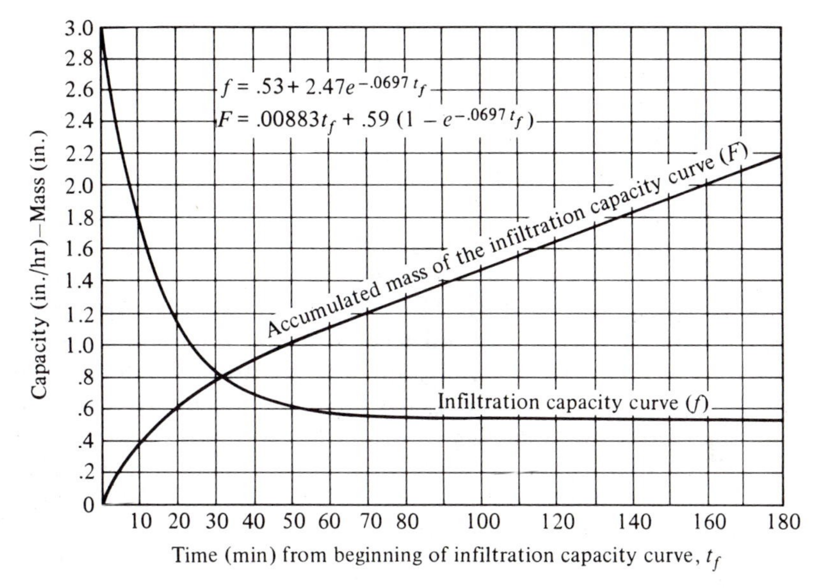


**Fig. 3-11b.** *Infiltration capacity and mass curves for normal antecedent conditions of turf areas. (After A. L. Tholin and Clint J. Kiefer, "The Hydrology of Urban Runoff,"* Proc. ASCE, J. Sanitary Engineering Div., *84, No. SA2 (Mar. 1959), Proc. paper 1984: 56.)*

:an readily be calculated by picking values of DRH discharge at equal ime increments through the hydrograph and applying the formula[13]

$$P_e = \frac{(0.03719)(\sum q_i)}{(A)(n_d)} \tag{3-32}$$

***Table 3-6*** *Cover Factors*

| | Cover | Cover factor |
|---|---|---|
| Permanent forest and grass | Good (1 in. humus) | 3.0–7.5 |
| | Medium ($\frac{1}{4}$–1 in. humus) | 2.0–3.0 |
| | Poor (<$\frac{1}{4}$ in. humus) | 1.2–1.4 |
| Close-growing crops | Good | 2.5–3.0 |
| | Medium | 1.6–2.0 |
| | Poor | 1.1–1.3 |
| Row crops | Good | 1.3–1.5 |
| | Medium | 1.1–1.3 |
| | Poor | 1.0–1.1 |

*Source:* After *ASCE Manual of Engineering Practice,* No. 28.

where

$P_e$ = precipitation excess (in.)
$q_i$ = DRH ordinates at equal time intervals (cfs)
A = drainage area ($mi^2$)
$n_d$ = number of time intervals/24-hr period

For most cases the difference between the original rainfall and the direct runoff can be considered as infiltrated water. Exceptions may occur in areas of excessive subsurface drainage or tracts of intensive interception potential. The calculated value of infiltration can then be assumed as distributed according to an equation of the form of Eq. 3-31 or it may be uniformly spread over the storm period. Choice of the method employed depends upon the accuracy requirements and size of the watershed.

Following Horton's work, numerous equations for infiltration rate have been proposed by Holtan and others.[11] In 1954 Philip developed an equation of the form[31]

$$\frac{dF}{dt} = \frac{k}{\mu}\left(1 + \left[\frac{(m - m_0)(P + H)}{F}\right]\right) \tag{3-33}$$

where

$\mu$ = fluid viscosity
$k$ = saturated permeability
$F$ = total volume of infiltrated water
$H$ = depth of the water on the soil surface
$m_0$ = initial moisture content of the soil
$P$ = capillary potential at the wetting front
$m$ = average moisture content to depth of the wetting front at time $t$

The use of this equation or equations such as Eq. 3-31 is hampered because of the independent variable, time. As long as the rate of precipitation exceeds infiltration capacity, no problem arises. When the supply rates drop below the infiltration capacity, difficulty occurs. Several investigators have circumvented this problem by introducing soil moisture as the dependent variable.[10,11,30] The following equation proposed by Huggins and Monke is an example,[10]

$$f = f_c + A\left(\frac{S - F}{T_p}\right)^P \tag{3-34}$$

where

A and $P$ = coefficients
S = the storage potential of a soil overlying the impeding layer ($T_p$ minus antecedent moisture)

$F$ = the total volume of water that infiltrates
$T_p$ = the total porosity of soil lying over the impeding stratum

The coefficients are determined using data from sprinkling infiltrometer studies. The variable $F$ must be calculated for each time increment in the iteration process. At the beginning of a storm $F = 0$ and $f$ is therefore known. In essence the continuity equation is solved for a block of soil with an inflow rate $f$ (or smaller if the rainfall is less) and an outflow determined according to Eq. 3-35. Expression $dS/dt \cdot \Delta t$ then gives the change in storage of the soil. When added to the storage at the beginning of the time increment, the total storage is obtained. Equation 3-34 is a modification of one originated by Holtan and Overton[11,30] and appears to have merit over the form of Eq. 3-31 if the rate of infiltration supply is less than infiltration capacity.

In order to use this relationship when the water supply rate only intermittently exceeds the infiltration capacity, the rate at which water drains from the "control zone," which determines the soil moisture content $(S - F)$, must be found. It is evaluated as follows:[10] (1) where the moisture content of the control zone is less than the field capacity (amount of water held in the soil after excess gravitational water has drained), the drainage rate is considered zero; (2) the drainage rate is assumed equal to the infiltration rate when the soil is saturated and the infiltration rate becomes constant; and (3) if the water content is between the field capacity and saturation, the drainage rate is computed as

$$\text{drainage rate} = f_c \left(1 - \frac{P_u}{G}\right)^3 \tag{3-35}$$

where

$P_u$ = the unsaturated pore volume
$G$ = maximum gravitational water, that is the total porosity minus the field capacity

Data from sprinkling infiltrometer studies of various watersheds of interest are used to estimate the coefficients in Eq. 3-34.[10]

Another equation for infiltration capacity has been developed by Holtan.[39,40]

$$f = a \cdot S_a^{1.4} + f_c \tag{3-36}$$

where

$f$ = the infiltration capacity (in./hr)
$a$ = the infiltration capacity in in.-hr/(in.)$^{1.4}$ of the available storage (index of surface-connected porosity)

$S_a$ = available storage in the surface layer (A-horizon in agricultural soils; that is, about first 6 in.) in inches of water equivalent

$f_c$ = the constant rate of infiltration after long wetting (in./hr)

This equation has been modified somewhat for use in the USDAHL-70 Watershed Model.[36]

$$f = GI \cdot a \cdot S_a^{1.4} + f_c \tag{3-37}$$

where $a$ is a vegetation parameter and $GI$ is the growth index of Eq. 3-19. Information about $a$ is given in Table 3-7.

In Eq. 3-37, it is assumed that the portion of the available storage connected to the surface is a function of the density of plant roots. This is given by the vegetation parameter $a$ which has been determined at plant maturity as the percentage of the ground surface area occupied by plant stems or root crowns. In this manner, the fraction of porosity in the agricultural A horizon that is surface-connected by mature plant roots to form conduits for air or water is represented.

### Stanford Watershed Model of Infiltration

In the Stanford Model, infiltration is calculated continuously in terms of two components: direct infiltration into the soil profile and increases in temporary storages such as surface depressions that delay infiltration.[26]

The moisture available is subjected to operations that govern direct flows into long-term lower zone and groundwater storages. That

**Table 3-7** *Tentative Estimates of the Vegetation Parameter a in Infiltration Equation* $f = GI \cdot a \cdot S_a^{1.4} + f_c$

| | Basal Area Rating[a] | |
|---|---|---|
| *Land use or cover* | *Poor condition* | *Good condition* |
| Fallow[b] | 0.10 | 0.30 |
| Row crops | 0.10 | 0.20 |
| Small grains | 0.20 | 0.30 |
| Hay (legumes) | 0.20 | 0.40 |
| Hay (sod) | 0.40 | 0.60 |
| Pasture (bunchgrass) | 0.20 | 0.40 |
| Temporary pasture (sod) | 0.40 | 0.60 |
| Permanent pasture (sod) | 0.80 | 1.00 |
| Woods and forests | 0.80 | 1.00 |

[a] Adjustments needed for "weeds" and "grazing."
[b] For fallow land only, "poor condition" means "after row crop," and "good condition" means "after sod."

*Source:* U.S. Department of Agriculture, Agricultural Research Service, 1975.

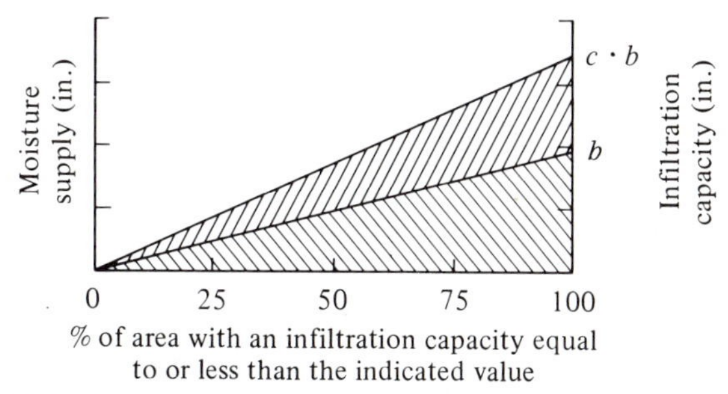


***Fig. 3-12.*** *Infiltration capacity relationship. (After N. H. Crawford and R. K. Linsley, Jr., "Digital Simulation in Hydrology: Stanford Watershed Model IV," Tech. Rept. 39, Department of Civil Engineering, Stanford University, Stanford, Calif., July 1966.)*

fraction of water determined to be remaining in surface detention after calculation of direct infiltration is disposed of according to operations of upper zone storages. In the Stanford Model the cumulative frequency distribution of infiltration capacity is assumed to take the linear form shown in Figure 3-12. Infiltration capacity is broken into two regions, one for lower zone and groundwater storage, the other for interflow (infiltrated water, which moves laterally through the upper soil layer until it enters a channel). Parameters for the curves of Fig. 3-12 are found by computer trials and are reported by Crawford, Linsley, and others.[26,35,37,38]

The overall Stanford Watershed Model and its operation are discussed in greater detail in Chapter 10.

### Temporal and Spatial Variability of Infiltration Capacity

The infiltration capacity generally varies both in space and time within a given drainage basin. Spatial variations occur because of differences in soil types and vegetation. The usual procedure used to accommodate this type of variation is to subdivide the total region into components having approximately uniform soil and vegetal cover properties.

The infiltration capacity at a given location in a watershed varies with time as shown in Fig. 3-10. The initial infiltration capacity is a function of antecedent conditions and can be estimated from a knowledge of the area's soil moisture or from an antecedent precipitation index. If precipitation occurs at a rate less than the $f$ capacity rate, the change in $f$ capacity with time will not be that given by the $f$ capacity curve; during periods of no precipitation, the infiltration capacity will recover. Example 3-4 illustrates these concepts.

***Example 3-4*** Given the rainfall pattern of Fig. 3-13 and the infiltration capacity curve of Fig. 3-14a, determine the overland flow supply rate $\sigma$.

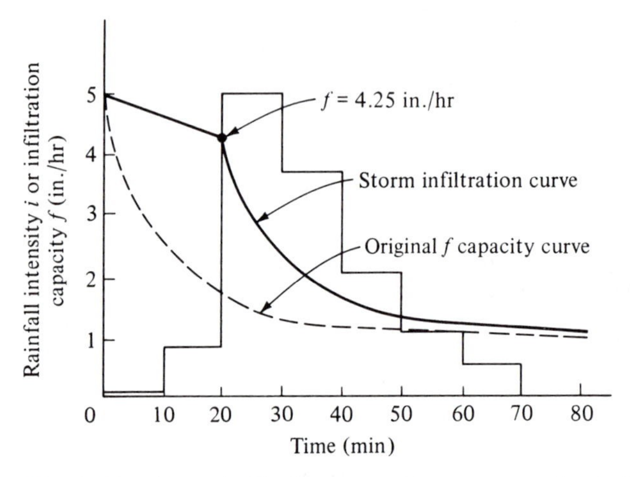


***Fig. 3-13.*** *Storm infiltration capacity curve constructed from original f capacity curve.*

Assume a turf cover and that the OGEE curve of Fig. 3-7 governs. Neglect interception losses.

In order to solve the problem, it is necessary to determine $P, F, i$, and $f$.

1. Construct a curve of mass infiltration $F$ versus $f$ capacity. This is done by calculating the areas under the curve in Fig. 3-14a at given times and plotting them versus $f$ capacity as shown in Fig. 3-14b. Calculations to determine cumulative infiltration are shown in Table 3-8.

***Table 3-8***

| Time Increment (min) | Average Height of Ordinate (in./hr) | Cumulative Infiltration ($F$-in.) |
|---|---|---|
| 0–5 | 4.00 | $4 \times 5/60 = 0.33$ |
| 5–15 | 2.50 | $2.5 \times 10/60 + 0.33 = 0.75$ |
| 15–30 | 1.50 | $1.5 \times 15/60 + 0.75 = 1.13$ |
| 30–60 | 1.15 | $1.15 \times 30/60 + 1.13 = 1.7$ |

Note that the $F$ values are plotted versus $f$ capacity at the end of the corresponding time interval. For example, the first value, 0.33, is plotted versus $f = 3.4$ which occurs at the end of the 5-min interval.

2. Determine the storm period infiltration

The storm pattern and original $f$ capacity curve are plotted as shown in Figure 3-13.

*a.* In the first 20 min $f > i$; therefore, all of the rainfall is infiltrated.

$$F = 0.1 \times \tfrac{1}{6} + 0.8 \times \tfrac{1}{6} = 0.15 \text{ in.}$$

*b*. From $F$ versus $f$ curve (Fig. 3-14b), for $F = 0.15$, $f = 4.25$ in./hr.
*c*. Use this as the initial value of $f$ at $t = 20$ min and shift the original $f$ capacity curve to the right to obtain the storm infiltration curve (Fig. 3-13). Note that this would not have been done if all rainfall intensities had exceeded the original $f$ capacity curve ordinates. Since at the end of 20 min, some $f$ capacity remained unfilled, the curve shift is carried out to accommodate this.

3. Having plots for the storm period infiltration and the rainfall versus time, values of $P$, $F$, $i$ and $f$ can be determined.
   *a*. Calculations for $P$ and $F$ are listed in Table 3-9.

   Note that the curve of $F$ versus $f$ (Fig. 3-14b) relates to the original $f$ capacity curve and is used to aid in constructing the storm $f$ curve while the values of $F$ calculated above are related to actual storm conditions. Rainfall intensities ($i$) are taken from the hyetograph to Fig. 3-14a.

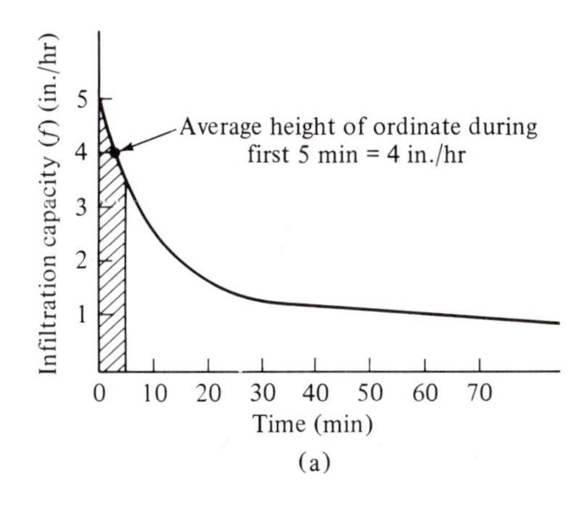


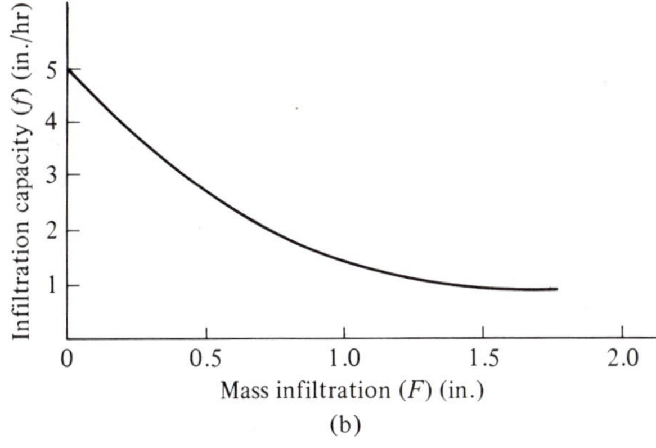


**Fig. 3-14.** *Infiltration capacity curve (a) and mass infiltration versus f curve (b) for Example 3-4.*

**Table 3-9**

| Time (min) | $i$ (in./hr) | Cumulative Precipitation, $P$ (in.) | $f$ (in./hr) | Cumulative Infiltration, $F$ (in.) |
|---|---|---|---|---|
| 10 | 0.10 | $0.10 \times 10/60 = 0.02$ | 0.10 | $0.10 \times 10/60 = 0.02$ |
| 20 | 0.80 | $0.80 \times 10/60 + 0.02 = 0.15$ | 0.80 | $0.8 \times 10/60 + 0.02 = 0.15$ |
| 30 | 5.00 | $5 \times 10/60 + 0.15 = 0.98$ | 2.90 | $2.9 \times 10/60 + 0.15 = 0.63$ |
| 40 | 3.70 | $3.7 \times 10/60 + 0.98 = 1.60$ | 1.80 | $1.8 \times 10/60 + 0.63 = 0.93$ |
| 50 | 2.00 | $2 \times 10/60 + 1.60 = 1.93$ | 1.40 | $1.4 \times 10/60 + 0.93 = 1.17$ |
| 60 | 1.10 | $1.1 \times 10/60 + 1.93 = 2.12$ | 1.10 | $1.1 \times 10/60 + 1.17 = 1.35$ |
| 70 | 0.50 | $0.5 \times 10/60 + 2.12 = 2.10$ | 0.50 | $0.5 \times 10/60 + 1.35 = 1.43$ |

Having determined $F$, $P$, $i$, and $f$, it is now possible to enter Fig. 3-7 using calculated $(P - F)$ values, and determine the ratio of overland flow supply $\sigma$ to $(i - f)$. Using this ratio and the calculated values of $(i - f)$ permits the determination of $\sigma$. These operations are tabulated in Table 3-10.

**Table 3-10** *Determination of the Overland Flow Supply Rate*

| Time (min) | $P$ (in.) | $F$ (in.) | $P - F$ (in.) | $i$ (in./hr) | $f$ (in./hr) | $(i - f)$ (in./hr) | $\frac{\sigma^a}{(i - f)}$ | $\sigma$ (in./hr) |
|---|---|---|---|---|---|---|---|---|
| 10 | 0.02 | 0.02 | 0 | 0.1 | 0.1 | — | 0 | 0 |
| 20 | 0.15 | 0.15 | 0 | 0.8 | 0.8 | — | 0 | 0 |
| 30 | 0.98 | 0.63 | 0.35 | 5.0 | 2.9 | 2.1 | 0.91 | 1.9 |
| 40 | 1.60 | 0.93 | 0.67 | 3.7 | 1.8 | 1.9 | 1.0 | 1.9 |
| 50 | 1.93 | 1.17 | 0.76 | 2.0 | 1.4 | 0.6 | 1.0 | 0.6 |
| 60 | 2.12 | 1.35 | 0.77 | 1.1 | 1.1 | — | 1.0 | 0 |
| 70 | 2.20 | 1.43 | 0.77 | 0.5 | 0.5 | — | 1.0 | 0 |

[a] From Fig. 3-7.

### Indexes

Infiltration indexes generally assume that infiltration occurs at some constant or average rate throughout a storm. Consequently, initial rates are underestimated and final rates are overstated if an entire storm sequence with little antecedent moisture is considered. The best application is to large storms on wet soils or storms where infiltration rates may be assumed to be relatively uniform.[1]

The most common index is termed the $\phi$ *index* for which the total volume of the storm period loss is estimated and distributed uniformly across the storm pattern. Then the volume of precipitation above the index line is equivalent to the runoff (Fig. 3-15). A variation is the $W$ index, which excludes surface storage and retention. Initial

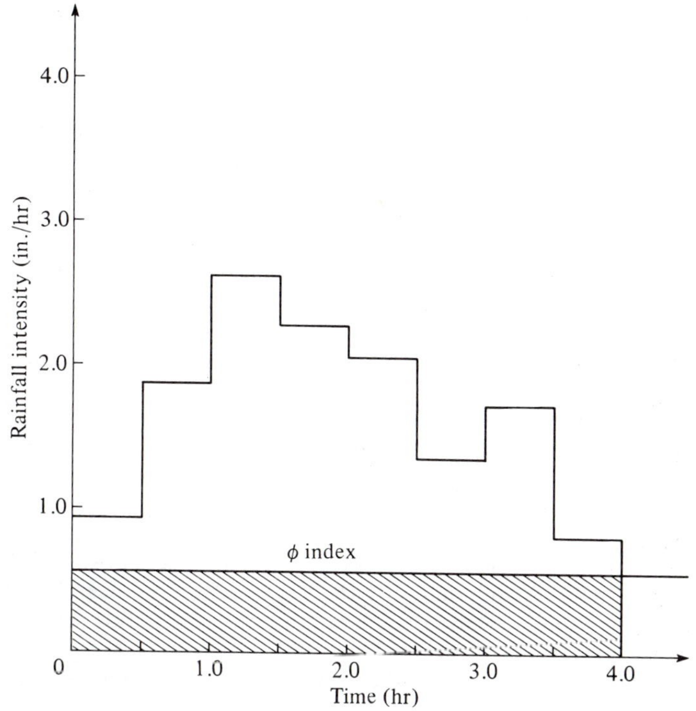


***Fig. 3-15.*** *Representation of a $\phi$ index.*

abstractions are often deducted from the early storm period to exclude initial depression storage and wetting.

To determine the $\phi$ index for a given storm, the amount of observed runoff is determined from the hydrograph and the difference between this quantity and the total gauged precipitation is then calculated. The volume of loss (including the effects of interception, depression storage, and infiltration) is distributed uniformly across the storm pattern as shown in Fig. 3-15.

Use of the $\phi$ index for determining the amount of direct runoff from a given storm pattern is essentially the reverse of this procedure. Unfortunately, the $\phi$ index determined from a single storm is not generally applicable to other storms, and unless it is correlated with basin parameters other than runoff, it is of little value.

## Problems

3-1. An 8000-mi² watershed received 20 in. of precipitation in a 1-yr period. The annual streamflow was recorded as 5000 cfs. Roughly estimate the combined amounts of water evaporated and transpired. Qualify your answer.

3-2. Find the daily evaporation from a lake during which the following data were obtained: air temperature 90°F, water temperature 60°F, wind speed 20 mph, and relative humidity 30%.

3-3. Find the monthly consumptive use of an alfalfa crop for a month when the mean temperature is 70°F, the average percentage of daytime hours for the year is 10, and the monthly consumptive use coefficient is 0.87.

3-4. Estimate the possible interception of a prevalent deciduous tree species in your area.

3-5. Use the precipitation pattern of Fig. 3-6 to estimate maximum cumulative depression storage of a paved parking lot 100 × 300 ft.

3-6. Show that $k = (f_0 - f_c)$ divided by the total volume of infiltration above the $f_c$ line of Fig. 3-10.

3-7. Refer to Fig. 3-13. Assume a 15% runoff. What will be the value of the $\phi$ index?

3-8. Consider an urban area of 60 acres, mostly residential. A rainfall occurs during a 30-min interval. If the total precipitation during this period is (a) 0.05 in., (b) 0.10 in., (c) 0.30 in., and (d) 1.3 in., during which of these storms will the estimation of runoff be most affected by the estimation of losses? Why?

3-9. During a given month a lake having a surface area of 350 acres has an inflow of 20 cfs, an outflow of 18 cfs, and a total seepage loss of 1 in. The total monthly precipitation is 1.5 in. and the evaporation loss is 4.0 in. Estimate the change in storage.

3-10. What are two methods that might be used to reduce evaporation from a small pond?

3-11. Compute the daily evaporation from a Class A pan if the amounts of water required to bring the level to the fixed point are as follows:

| Day | 1 | 2 | 3 | 4 | 5 |
|---|---|---|---|---|---|
| Rainfall (in.) | 0 | 0.65 | 0.12 | 0 | 0.01 |
| Water added (in.) | 0.29 | 0.55 | 0.07 | 0.28 | 0.10 |
| Evaporation | | | | | |

3-12. For Prob. 3-11, the pan coefficient is 0.70. What is the lake evaporation (in inches) for the 5-day period for a lake with a 250-acre surface area?

3-13. The pan coefficient for a Class A evaporation pan located near a lake is 0.7. A total of 0.50 in. of rain fell during a given day. Determine the depth of evaporation from the lake during the same day if 0.3 in. of water had to be added to the pan at the end of the day in order to restore the water level to its original value at the beginning of the day.

3-14. The following table lists the storm rainfall data and infiltration capacity data for a 24-hr storm beginning at midnight on Apr. 14 of the current year. (a) Plot the rainfall hyetograph and the $f$-capacity curve on rectangular coordinate paper; (b) determine the total storm precipitation in inches; and (c) by counting squares or by planimeter, determine the net storm rain by the $f$-capacity method.

*Rainfall Data for a Hypothetical Storm on Apr. 15 of the Current Year (beginning at midnight on Apr. 14)*

| Hour | Rainfall Intensity (in./hr) | Infiltration Capacity at Beginning of Hour (in./hr) | Hourly Deduction for Depression Storage (in./hr) |
|---|---|---|---|
| 1st | 0.41 | 0.200 | 0.20 |
| 2 | 0.49 | 0.160 | 0.14 |
| 3 | 0.32 | 0.125 | 0.04 |
| 4 | 0.31 | 0.100 | 0.02 |
| 5 | 0.22 | 0.085 | 0.00 |
| 6 | 0.08 | 0.070 | 0.00 |
| 7 | 0.07 | 0.065 | |
| 8 | 0.09 | 0.057 | |
| 9 | 0.08 | 0.052 | |
| 10 | 0.06 | 0.047 | |
| 11 | 0.11 | 0.044 | |
| 12 | 0.12 | 0.040 | |
| 13 | 0.15 | 0.037 | |
| 14 | 0.23 | 0.036 | |
| 15 | 0.28 | 0.035 | |
| 16 | 0.26 | 0.034 | |
| 17 | 0.21 | 0.033 | |
| 18 | 0.09 | 0.033 | |
| 19 | 0.07 | 0.033 | |
| 20 | 0.06 | 0.032 | |
| 21 | 0.03 | 0.032 | |
| 22 | 0.02 | 0.032 | |
| 23 | 0.01 | 0.031 | |
| 24 | 0.01 | 0.031 | |

3-15. Gross rain intensities during each hour of a 5-hr storm over a 1000-acre basin were 5, 4, 1, 3, and 2 in./hr, respectively. The direct surface runoff from the basin was 375 acre-ft. Determine the basin $\phi$ index.

3-16. The infiltration rate for excess rain on a small area was observed to be 4.5 in./hr at the beginning of rain, and it decreased exponentially to an equilibrium of 0.5 in./hr after 10 hr. A total of 30 in. of water infiltrated during the 10-hr interval. Determine the value of $k$ in Horton's equation $f = f_c + (f_0 - f_c)e^{-kt}$.

3-17. Tabulated below are total rainfall intensities during each hour of a frontal storm over a drainage basin.
(a) Plot the rainfall hyetograph (intensity versus time). (b) Determine the total storm precipitation amount in inches. (c) If the net storm rain is 2.00 in., determine the exact $\phi$ index (in./hr) for the drainage basin. (Note that by definition the area under the hyetograph above the $\phi$ index line must be 2.00 in.). (d) Determine the area of the drainage basin (acres) if the net rain is

2.00 in. and the measured volume of direct surface runoff is 2015 cfs-hr. (e) Using the $\phi$ index calculated in Part (c), determine the volume of direct surface runoff (acre-ft) that would result from the following storm:

| Hour | 1 | 2 | 3 | 4 |
|---|---|---|---|---|
| Intensity | 0.40 | 0.05 | 0.30 | 0.20 in./hr |

| Hour | Rainfall Intensity (in./hr) | Hour (cont.) | Rainfall Intensity (in./hr) |
|---|---|---|---|
| 1st | 0.41 | 13 | 0.15 |
| 2 | 0.49 | 14 | 0.23 |
| 3 | 0.22 | 15 | 0.28 |
| 4 | 0.31 | 16 | 0.26 |
| 5 | 0.22 | 17 | 0.21 |
| 6 | 0.08 | 18 | 0.09 |
| 7 | 0.07 | 19 | 0.07 |
| 8 | 0.09 | 20 | 0.06 |
| 9 | 0.08 | 21 | 0.03 |
| 10 | 0.06 | 22 | 0.02 |
| 11 | 0.11 | 23 | 0.01 |
| 12 | 0.12 | 24 | 0.01 |

3-18. Precipitation falls on a 500-acre drainage basin according to the following schedule:

| 30-min period | 1 | 2 | 3 | 4 |
|---|---|---|---|---|
| Intensity (in./hr) | 4.0 | 2.0 | 6.0 | 5.0 |

(a) Determine the total storm rainfall (in inches), and (b) determine the $\phi$ index for the basin if the net storm rain is 3.0 in.

3-19. The direct surface runoff volume from a 4.40-mi$^2$ drainage basin is determined by planimeter from the area under the hydrograph to be 10,080 cfs-hr. The hydrograph was produced by a 1.71 in./hr rain storm with a duration of 5 hr. Determine (a) the net rain, and (b) the $\phi$ index.

3-20. Infiltration capacities can be computed by measuring runoff from isolated plots subjected to an artificial rain by sprinkling. In an experiment a plot, 4 m wide and 12.5 m long, was sprinkled until the runoff became fairly constant at 0.5 liter/sec. The intensity of sprinkling amounted to 50 mm/hr. (a) What is the runoff in mm/hr? (b) What is the ultimate infiltration capacity $f_c$ in mm/hr? (c) What is the detention storage $D_a$ (expressed in mm depth) at the moment of cessation of rainfall if the runoff decreases as follows and $q/f$ stays constant after rain stops:

| Time | Runoff in liters/sec | |
|---|---|---|
| 0 cessation of rainfall | 0.50 | assume all runoff derived from detention storage |
| 5 min later | 0.25 | |
| 10 min later | 0.13 | |
| 15 min later | 0.05 | |
| 20 min later | 0 | |

## References

1. Chow, Ven Te (ed.), *Handbook of Applied Hydrology.* (New York: McGraw-Hill Book Company, 1964).
2. Horton, R. E., *Surface Runoff Phenomena: Part I, Analysis of the Hydrograph.* Horton Hydrol. Lab. Pub. 101. (Ann Arbor, Mich.: Edwards Bros., Inc., 1935).
3. Schomaker, C. E., "The Effect of Forest and Pasture on the Disposition of Precipitation," *Maine Farm Research* (July 1966).
4. Gunaji, N. N., et al., "Evaporation Reduction Investigation—Elephant Butte Reservoir, New Mexico," Engr. Exp. Sta., Tech. Rept. 25, Las Cruces, New Mexico State University, 1965.
5. Kittredge, Joseph, *Forest Influences.* (New York: McGraw-Hill Book Company, 1948).
6. Viessman, Warren, Jr., "A Linear Model for Synthesizing Hydrographs for Small Drainage Areas," paper presented at the Forty-eighth Annual Meeting American Geophysical Union, Washington, D.C., Apr. 1967.
7. Tholin, A. L., and C. J. Kiefer, "The Hydrology of Urban Runoff," *Trans. ASCE,* 125 (1960).
8. Hicks, W. I., "A Method of Computing Urban Runoff," *Trans. ASCE,* 109 (1944).
9. Turner, L. B., "Abstraction of Depression Storage from Storms on Small Impervious Areas," Master's thesis, University of Maine, Orono, Me., Aug. 1967.
10. Huggins, L. F., and E. J. Monke, "The Mathematical Simulation of the Hydrology of Small Watersheds," Tech. Rept. 1, Purdue University Water Resources Center, Lafayette, Ind., Aug. 1966.
11. Holtan, H. N., "A Concept for Infiltration Estimates in Watershed Engineering," U.S. Department of Agriculture, Agricultural Research Service, 1961, pp. 41–51.
12. "Water-Loss Investigations," Vol. 1, Lake Hefner Studies, U.S. Geologic Survey Professional Paper No. 269 (1954). (Reprint of U.S. Geological Survey Circ. 229, 1952.)
13. Bowen, I. S., "The Ratio of Heat Losses by Conduction and by Evaporation from Any Water Surface," *Phys. Rev.,* 27 (1926).
14. Anderson, E. R., L. J. Anderson, and J. J. Marciano, "A Review of Evaporation Theory and Development of Instrumentation," Lake Mead Water Loss Investigation; Interim Report, Navy Electronics Lab. Rept. No. 159 (Feb. 1950).

15. Sutton, O. G., "The Application to Micrometerology of the Theory of Turbulent Flow over Rough Surfaces," Royal Meteorological Society, *Quart. J.*, 75, No. 236 (Oct. 1949).
16. Meyer, A. F., "Evaporation from Lakes and Reservoirs," Minnesota Resources Commission, St. Paul, Minn., June 1944.
17. Franzini, J. B., "Evaporation Suppression Research," *Water and Sewage Works,* May 1961.
18. La Mer, Victor K., "The Case for Evaporation Suppression," *Chem. Eng.*, June 10, 1963.
19. Blaney, H. F., "Water and Our Crops," *Water, the Yearbook of Agriculture.* (Washington, D.C.: U.S. Department of Agriculture, 1955.)
20. Blaney, H. F., "Monthly Consumptive Use Requirements for Irrigated Crops," *Proc. ASCE, J. Irrigation and Drainage Div.*, 85, No. IRI (Mar. 1959).
21. Criddle, W. D., "Methods of Computing Consumptive Use of Water," *Proc. ASCE, J. Irrigation and Drainage Div.*, 84, No. IRI (Jan. 1958).
22. Penman, H. L., "Estimating Evaporation," *Trans. Amer. Geophys. Union,* 37, No. 1 (1956).
23. Hughes, E. H., "Research on Control of Phreatophytes," Proc. Ninth Ann. Water Conf., New Mexico State University, Las Cruces, Mar. 1964.
24. Linsley, R. K., Jr., M. A. Kohler, and J. L. H. Paulhus, *Hydrology for Engineers.* (New York: McGraw-Hill Book Company, 1958.)
25. Meier, W. L., Jr., "Analysis of Unit Hydrographs for Small Watersheds in Texas," Texas Water Commission, Bull. 6414, Austin, 1964.
26. Crawford, N. H., and R. K. Linsley, Jr., "Digital Simulation in Hydrology: Stanford Watershed Model IV," Tech. Rept. 39, Department of Civil Engineering, Stanford University, July 1966.
27. Roberts, N. J., "Reduction of Transpiration," *J. Geophy. Res.*, 66, No. 10 (Oct. 1961).
28. Rosenberg, N. J., H. E. Hart, and K. W. Brown, "Evapotranspiration—Review of Research," MP 20, Agricultural Experiment Station, University of Nebraska, Lincoln, 1968.
29. Linsley, R. K., Jr., M. A. Kohler, and J. L. H. Paulhus, *Applied Hydrology.* (New York: McGraw-Hill Book Company, 1949).
30. Overton, D. C., "Mathematical Refinement of an Infiltration Equation for Watershed Engineering." Washington, D.C.: U.S. Department of Agriculture, ARS 41–99, 1964.
31. Philip, J. R., "An Infiltration Equation with Physical Significance," *Soil Science,* 77 (1954).
32. Shands, A. L., "Mean Precipitable Water in the United States," U.S. Weather Bureau Tech. Papers No. 10 (1949).
33. Miller, D. W., J. J. Geraghty, and R. S. Collins, *Water Atlas of the United States.* (Port Washington, N.Y.: Water Information Center, Inc., 1963).
34. U.S. Weather Bureau, Tech. Papers 1–33 (Washington, D.C.: Government Printing Office).
35. Staff, Hydrologic Research Laboratory, "National Weather Service River Forecast System Forecast Procedures," U.S. Department of Commerce, NWS HYDRO 14, Washington, D.C., Dec. 1972.

36. Holtan, H. N., G. J. Stiltner, W. H. Henson, and N. C. Lopez, "USDAHL-74 Revised Model of Watershed Hydrology," U.S. Department of Agriculture, ARS Tech. Bulletin No. 1518, Washington, D.C., 1975.
37. Lumb, A. M., et al., "GTWS: Georgia Tech Watershed Simulation Model," School of Civil Engineering, ERC-0175, Georgia Institute of Technology, Atlanta, Ga., 1975.
38. Mitchell, W. D., "Unit Hydrographs in Illinois," Illinois Waterways Division, 1968.
39. Holtan, H. N., "A Concept for Infiltration Estimates in Watershed Engineering," U.S. Department of Agriculture, Agricultural Research Service, ARS 41-51, 1961.
40. Holtan, H. N., "A Model for Computing Watershed Retention from Soil Parameters," *Journal Soil and Water Conservation,* 20 (3) (1965): 91–94.
41. Horton, R. E., "Rainfall Interception," *Monthly Weather Rev.,* 47 (1919): 603–623.

# 4

# Streamflow

Surface water hydrology deals with the transfer of water along the earth's surface. Satisfactory surface water flow rate and quality are highly important to such fields as municipal and industrial water supply, flood control, streamflow forecasting, reservoir design, navigation, irrigation, drainage, water quality control, water-based recreation, and fish and wildlife management.

The relationship between precipitation and runoff is usually complex and is influenced by various storm pattern, antecedent, and basin characteristics. Because of these complexities and the frequent paucity of adequate data, many approximate formulas have been developed to relate rainfall and runoff. The earliest of these were usually crude empirical statements, whereas the trend now is to develop sound rational equations based on physical principles.

Streamflow is determined in the field by the use of channel geometry combined with measurements of velocity or by the use of special measuring devices such as weirs or flumes. Rating curves, which correlate flow depth with discharge, are widely employed. The literature on measuring techniques is extensive.[8,9,21,27]

## 4-1 Basin Characteristics Affecting Runoff

The nature of streamflow in a region is a function of the hydrologic input to that region and the physical, vegetative, and climatic characteristics of that region. As indicated by the hydrologic equation, all of the water that occurs in an area as a result of precipitation does not appear as streamflow. Fractions of the gross precipitation are diverted into paths that do not terminate in the regional surface transport system. Precipitation striking the ground can go into storage on the surface or in the soil and groundwater reservoirs beneath the surface. The character of the soil and rocks determines to a large extent the storage system into which precipitated water will enter. Opportunity for evaporation and transpiration will also be affected by the geologic and topographic nature of the area.

### Geologic Considerations

The principal geologic factors that affect surface waters are classified as lithologic and structural. Lithologic effects are associated with the composition, texture, and sequence of the rocks whereas structural effects relate mainly to discontinuities such as faults and folds. A fault is a fracture that results in the relative displacement of rock that was previously continuous. Folds are geologic strata that are contorted or bent. Variations in the erodibility of the different strata can easily lead to the evolvement of distinctive forms of drainage systems.

Both large-scale and local effects on the storage and movement of surface waters exist because of geologic activity and structure.[19] For example, drainage patterns are determined to a large extent by the nature of land forms. On the other hand, flowing surface waters also affect the surface geometry through the process of erosion. Thus the evolution of significant land forms resulting from volcanic activity, folding, and faulting affects drainage, whereas drainage patterns, having been generated, can also modify the land forms by creating valleys, deltas, and other geomorphic features.

Streams are classified as being young, mature, or old on the basis of their ability to erode channel materials. Young streams are highly active and usually flow rapidly so that they are continually cutting their channels. The sediment load imposed on these streams by their tributaries is transported without deposition. Mature streams are those in which the channel slope has been reduced to the point where flow velocities are just able to transport incoming sediment and where the channel depth is no longer being modified by erosion. A stream is classified as old when the channels in its system have become graded. The flow velocities of old streams are low due to gentle slopes that prevail. Wide meander belts, broad flood plains, and delta formation are also characteristic of old streams. The lower reaches of the Mississippi, Rhine, and Nile are examples. Flows in young river basins are often "flashy," whereas sluggish flows are common to older streams.

### Geomorphology of Drainage Basins

The description of a drainage basin in quantitative terms was an important forward step in hydrology and can be traced back in large part to the efforts of Robert E. Horton.[3] Langbein, Strahler, and others have expanded Horton's original work.[1,3,4]

To quantify the geometry of a basin, the fundamental dimensions of length, time, and mass are used. Many drainage basin features that are important to the hydrologist can be quantified in terms of length, length squared, or length cubed. Examples are elevation, stream length, basin perimeter, drainage area, and volume. The concept of geometric similarity can be applied to drainage basins just as it is to many other systems.[1] Most readers will be aware of model-prototype studies of aircraft, dams, and turbomachinery. Such studies involve considerations of geometric as well as dynamic similarity. In the same manner that inferences as to the operation of a prototype can sometimes be drawn from a geometrically similar model, inferences can also be drawn about the operation of one drainage area on the basis of information obtained from a similar one. Perfect similarity will never be realized if natural drainage systems are compared, but striking similarities have been observed which can often be put to practical use.

### Linear Measures

Important linear measures of drainage area characteristics include overland flow lengths and stream lengths. The concept of stream order is used with the linear dimension of stream length.

If the stream system in a drainage basin is clearly defined on a topographic map, the smallest tributaries are classified as order 1.[5] This is illustrated in Fig. 4-1. The point at which two first order streams join the channel is the beginning of a second order segment. Third order segments initiate where second order streams join the channel, and so on. The main stream channel that carries the flow from the entire tributary area upstream of a point of interest will necessarily be the highest order stream in that system.

The practical utility of the stream order system is based on the hypothesis that the site of the watershed, its channel dimensions, and streamflow are all proportional to the number of stream order, provided that a large enough sample is investigated. The order number permits comparisons of drainage systems that are quite different in size because the number is a dimensionless quantity. Such comparisons should be made at locations in the two systems that have a similar geometry; that is, second order streams, third order streams, and so forth.

Stream lengths are determined by the measurement of their

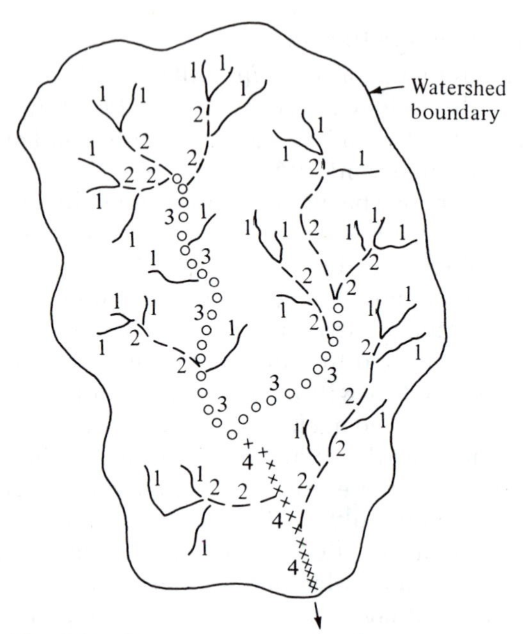


*Fig. 4-1.* Sketch indicating definition of stream order.

vertical projections onto a horizontal plane. Topographic maps are useful for obtaining such measurements. If the mean length of a stream segment $L_u$ of order $u$ is defined as $\bar{L}_u$, then it is possible to determine $\bar{L}_u$ using,[1]

$$\bar{L}_u = \frac{\sum_{i=1}^{N} L_{ui}}{N_u} \tag{4-1}$$

where $N_u$ = the number of stream segments of stream order $u$.

Another linear measure related to stream length is the distance $L_{ca}$ from a point of interest or observation on the main stream to a point on the primary channel that is nearest the center of gravity of the drainage area (center of gravity of the plane area of one drainage basin). Studies of basin lag (time between the centers of mass of effective storm input and the resulting output) have made use of this dimension.

Of particular significance in the physiographic development of a drainage basin is the overland flow length $L_g$. This is essentially the distance from the ridge line or drainage divide, measured along the path of surface flow which is not confined in any defined channel, to the intersection of this flow path with an established flow channel. If a drainage basin of the first order is the basic element of a larger drainage system, then a representative overland flow length can be determined for these first order basins. One approach is to measure a number of possible flow paths from a map of the area and to take the average of these. In some cases (for use with the "Rational Method," Chapter 11), the use of the longest overland flow length measured from the upstream end of the first order stream to the most remote point of flow that will terminate at this point is prescribed.

### Areal Measurements

Just as linear measures relate to many factors of hydrologic interest such as time of flow, so do areal measures. For example, the quantity of discharge from any drainage basin is obviously a function of the areal extent of that basin.

Relationships have been observed between the average area of basins of order $u$, ($\bar{A}_u$), and the average length of stream segments, $\bar{L}_u$. These variables are related by an exponential function. Studies of seven streams in the Maryland-Virginia area by Hack have produced the relationship[6]

$$L = 1.4A^{0.6} \tag{4-2}$$

where

$L$ = the stream length measured in miles to the drainage divide
$A$ = the drainage area in $mi^2$

Hack's observations indicate that as the drainage basin increases in size, it becomes longer and narrower; thus precise geometric similarity is not preserved.

Drainage area has long been used as a parameter in precipitation runoff equations or in simple equations indexing streamflow to area and occasionally other parameters. Many early empirical equations are of the form[1]

$$Q = cA^m \tag{4-3}$$

where

$Q$ = a measure of flow such as mean annual runoff
$A$ = the size of the contributing drainage area

Here $c$ and $m$ are determined by regression analysis (see Chapters 5 and 6); Fig. 4-2 illustrates a relationship of this form.

Other areal measures include definitions of the basin shape and the density of the drainage network or drainage density which is defined as the ratio of total channel segment lengths cumulated for all stream orders within a basin to the basin area. The stream frequency is defined as the summation of all segments in a drainage basin (total number of segments of all orders) divided by the drainage area.

### Basin Gradient

The slope of a drainage basin and its channels has a very profound effect on the surface runoff process of that region. Most stream channel profiles exhibit the characteristic of decreasing slope proceeding in a downstream direction. Figure 4-3 illustrates this particular trait. Also illustrated in the figure are the gross slope, which is the total elevation drop divided by the channel length, and the mean slope which is determined such that the areas between the average slope line and the stream profile are equal; that is, $A_1 = A_2$ in the figure. The gross slope and the mean slope are not very useful as parameters to describe drainage character, however, due to their generality; Fig. 4-3 should make this clear. Some mathematical functions used to describe stream profiles are linear relationships, and exponential, logarithmic, and power forms. A single numerical value to represent the primary channel profile has been used by Taylor and Schwartz.[7] This factor, known as the equivalent main stream slope $S_{st}$, is the slope of a uniform channel that is equivalent in length to the longest water course and has the same travel time. This factor has been found to be related to unit hydrograph lag (time from the center of mass of rainfall excess to the peak rate of runoff) and maximum discharge.

In addition to the slope of the stream channel, the overall land slope of the basin is an important topographic factor. A quantitative relationship between valley-wall slopes and stream channel slopes has been derived by Strahler.[1] A method of determining the slopes of

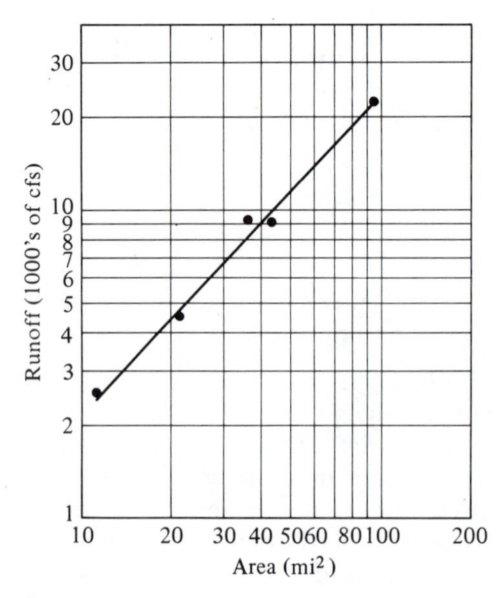


*Fig. 4-2. Runoff drainage area correlation for five Maryland streams (1933 storm data).*

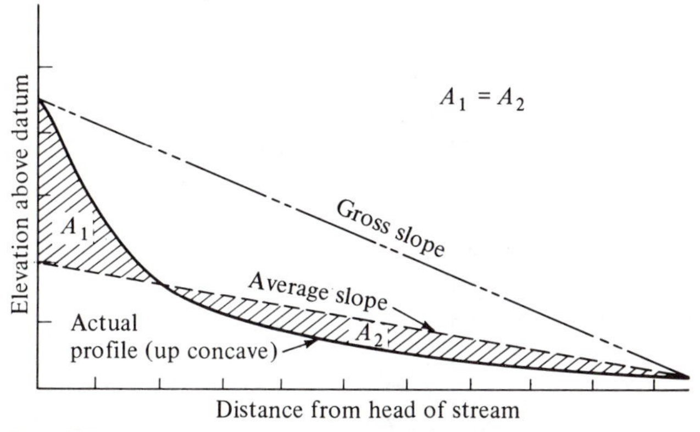


***Fig. 4-3.*** *Typical stream profile.*

a basin has been presented by Horton.[2] Briefly, the method involves superimposing a transparent grid over a topographic map of the drainage area in question. Each grid line is measured between its intersections with the drainage divide; the number of intersections of each grid line with a contour line is also needed. A determination of the land slope can then be made using

$$S = \frac{n \sec \theta}{l} h \tag{4-4}$$

where

$n$ = the total number of contour intersections by the horizontal and vertical grid lines
$l$ = the total length of grid line segments (horizontal and vertical)
$h$ = the contour interval
$\theta$ = the angle measured between a normal to the contours and the grid

Because $\theta$ is very difficult to measure, it is often neglected and separate values of average slope for the horizontal and vertical are computed and then averaged to obtain an estimate of the mean land slope. This procedure is illustrated in Fig. 4-4.

### Area Elevation Relationship

How the area within a drainage basin is distributed between contours is of interest for comparing drainage basins and gaining insight into the storage and flow characteristics of the basin (Fig. 4-5). For such studies, an area distribution curve such as that shown in Fig. 4-6 is used. The curve can be obtained by planimetering the areas between adjacent contours or by using a grid as in Fig. 4-4 and forming the ratio of the number of squares between contours to the total

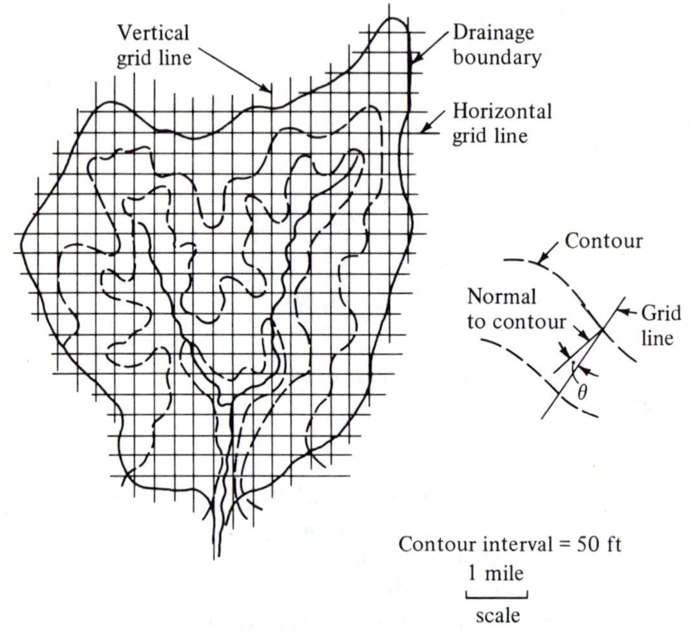


*Fig. 4-4.* *Determination of mean land slope.*

*Number of vertical intersections = 72.*
*Number of horizontal intersections = 120.*
*Total length of vertical grid segments = 103,900 ft.*
*Total length of horizontal grid segments = 101,200 ft.*

$$S_v = \frac{72 \times 50}{103{,}900} = 0.035 \text{ ft/ft} \qquad S_H = \frac{120 \times 50}{101{,}200} = 0.059 \text{ ft/ft}$$

$$\text{mean slope} = \frac{S_v + S_H}{2} = \frac{0.035 + 0.059}{2} = 0.047 \text{ ft/ft}$$

number of squares contained within the drainage boundaries. The mean elevation is determined as the weighted average of elevations between adjacent contours. The median elevation can be determined from the area elevation curves as the elevation at which the percent area is 50%.

### Drainage Basin Dynamics

Geomorphology like hydrology was largely qualitative in nature in its more formative years. With the passing of time and the greater need for reliable quantitative information, the science has progressed to the point where rational relationships between variables are being developed. These relationships are usually intended to quantify the interactions between the factors that modify the land form and the land form itself. In addition, equations relating the geomorphic prop-

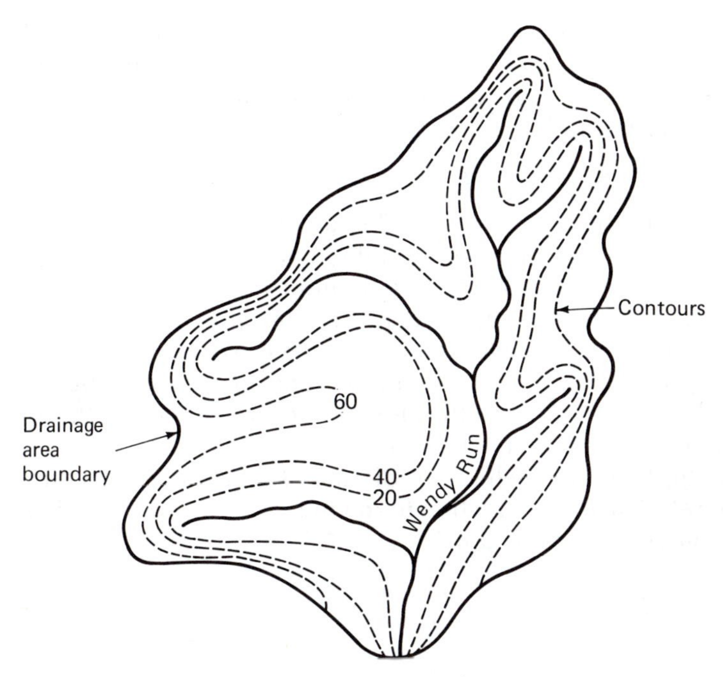


***Fig. 4-5.*** *Topographic map of Wendy Run Drainage Area showing 20-, 40-, and 60-ft contour lines.*

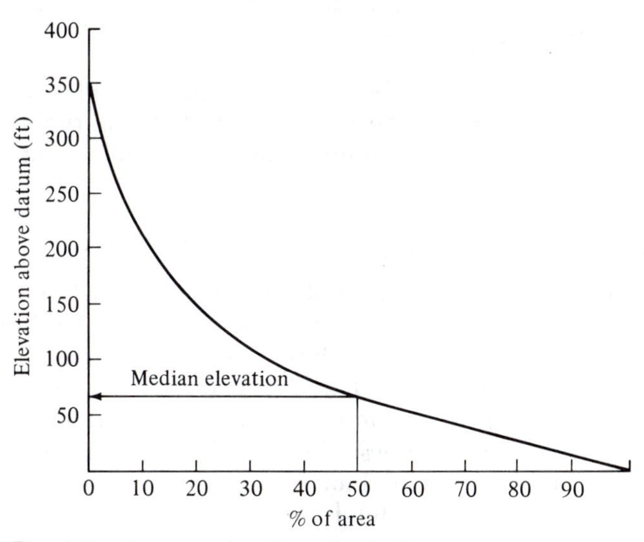


***Fig. 4-6.*** *An area-elevation distribution curve.*

erties to hydrologic, climatologic, or vegetative factors are being sought. Some of the functional relationships of particular significance to the hydrologist will be discussed in the following chapters.

Strahler has indicated that a relationship of the form

$$D = f(Q_r, K, H, \rho, \mu, g) \tag{4-5}$$

can be used to relate the horizontal scale of land form units to several controlling variables.[1] The terms in the functional relationship are the drainage density $D(L^{-1})$, the runoff intensity $Q_r(LT^{-1})$ [essentially equivalent to excess precipitation], the erosion proportionality factor $K$ [ratio of erosion intensity to eroding force $(L^{-1}T)$], the relief $H$ (L) which represents the variability of the land form in a vertical sense, the fluid density $\rho(ML^{-3})$, the fluid viscosity $\mu$ $(ML^{-1}T^{-1})$, and the gravitational acceleration $g$ $(LT^{-2})$.

By applying the techniques of dimensional analysis to the variables of relationship (4-5), the following expression involving four dimensionless groups obtains[1]

$$D = \frac{1}{H} f\left(Q_r K, \frac{Q_r \rho H}{\mu}, \frac{Q_r^2}{Hg}\right) \tag{4-6}$$

where

the product $DH$ = the ruggedness number, which reflects the combined attributes of steepness of slopes and length

the term $Q_r K$ = the Horton number and indicates the relative intensity of erosional processes

$\frac{Q_r \rho H}{\mu}$ and $\frac{Q_r^2}{Hg}$ = forms of the Reynolds and Froude numbers, respectively

The form of Eq. 4-6 embodies the dynamic aspects of geomorphic processes and has utility in the design of controlled experiments. It also provides a dimensionally sound basis for model-prototype studies.

### Artificial Control of Land Forms

If one considers the importance of physiographic factors to the hydrology of a basin, it follows that man-made changes of land forms might have profound effects as well. These effects, if controlled carefully, can serve useful purposes such as the retardation of erosion, or they might result in the destruction of valuable lands, structures, or personnel. Land use practices should therefore be carefully designed before being put into operation. Land use impacts on hydrologic processes are discussed in more detail in Chapter 11.

## 4-2 Some Elementary Precipitation-Streamflow Relationships

A common approach in connecting precipitation and runoff has been to plot precipitation versus runoff, find the slope of the trend line, and estimate the percentage of rainfall appearing as runoff. Runoff quantities determined this way are crude and subject to large errors. The degree of reliability is higher for drainage areas whose properties are least subject to seasonal or other types of variations, that is, an impervious area. Figure 4-7 illustrates the procedure. The resulting equation takes the form

$$Q = \left(\frac{1}{s}\right)(P - P_b) \tag{4-7}$$

where

$s$ = the slope of the line ($\Delta P/\Delta Q$)
$P_b$ = a base precipitation value below which $Q$ is zero

From Fig. 4-7 the relationship would be

$$Q = 0.57\,(P - 24)$$

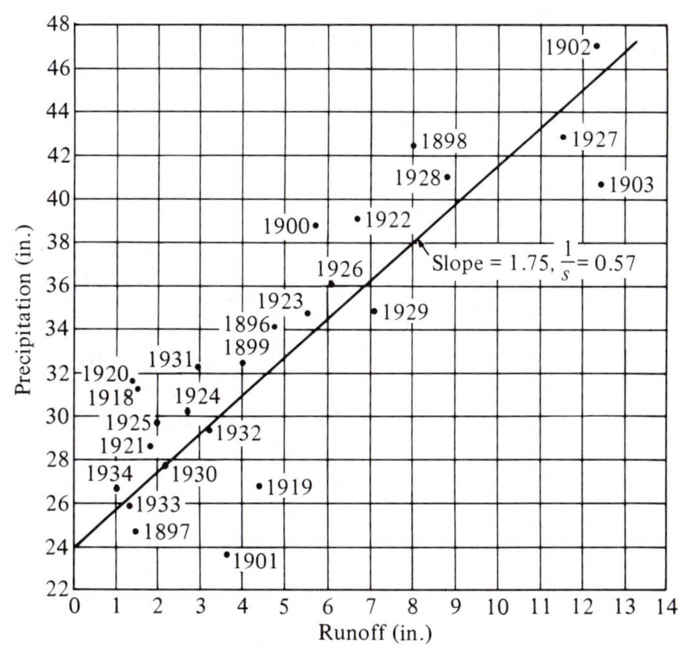


**Fig. 4-7.** *Annual precipitation and annual runoff in the Neosho River Basin above Iola, Kansas. (U.S. Geological Survey Data.)*

where $Q$ and $P$ are the annual runoff and precipitation, respectively, in inches. Considerable scatter of several data points from the assumed relationship indicates that this type of computation should be used with care. For rough approximations in preliminary planning studies, such methods are frequently helpful, however. Equations resembling 4-7 are improved if other parameters such as antecedent precipitation, soil moisture, season, and storm characteristics are included. Such relationships can be described using multiple regression techniques or graphical methods. Linsley et al. present a very complete treatment of methods for developing coaxial correlations involving several variables.[8]

Soil moisture relationships normally have a soil moisture index as the variable, since direct measurements of actual antecedent soil moisture are not generally practical. Indexes that have been inserted are groundwater flow at the beginning of the storm, antecedent precipitation, and basin evaporation.[9] Groundwater values should be weighted to reflect the effects of precipitation occurring within a few days of the storm because soil moisture changes from previous rains may be important. Pan-evaporation measurements can be employed to form soil moisture figures, since evaporation is related to soil moisture depletion.[10] Antecedent precipitation indexes (API) have probably received the widest use because precipitation is readily measured and relates directly to moisture deficiency of the basin.

A typical antecedent precipitation index is

$$P_a = aP_0 + bP_1 + cP_2 \tag{4-8}$$

where

$P_a$ = the antecedent precipitation index (in.)
$P_0, P_1$, and $P_2$ = the amounts of annual rainfall during the present year and for two years preceding the year in question

This index links annual rainfall and runoff values.[11] Coefficients $a$, $b$, and $c$ are found by trial and error or other fitting techniques to produce the best correlation between the runoff and the antecedent precipitation index. The sum of the coefficients must be 1.

Kohler and Linsley[12] have proposed the following API for use with individual storms:

$$P_a = b_1P_1 + b_2P_2 + \cdots + b_tP_t \tag{4-9}$$

where the subscripts on $P$ refer to precipitation which occurred that many days prior to the given storm, and the constants $b$ (less than unity) are assumed to be a function of $t$. Values for the coefficients can be determined by correlation techniques. In daily evaluation of the index, $b_t$ is considered to be related to $t$ by

$$b_t = K^t \tag{4-10}$$

where $K$ is a recession constant normally reported in the range 0.85 to 0.98. The initial value of the API ($P_{a0}$) is coupled to the API $t$ days later ($P_{at}$) by

$$P_{at} = P_{a0}K^t \tag{4-11}$$

To evaluate the index for a particular day based on that of the preceding one, Eq. 4-11 becomes simply

$$P_{a1} = KP_{a0} \tag{4-12}$$

because $t = 1$.

Various empirical relationships have been proposed. Most are based on correlating two or three variables and at best yield only rough approximations. In many cases these bonds were developed without considering physical principles or dimensional homogeneity. An added shortcoming is that many formulas fit only a specific watershed and have little general utility. Empirical equations demand great caution and an understanding of their origin.

***Example 4-1*** Precipitation depths $P_i$ for a 14-day period are listed in Table 4-1. The API on April 1 is 0.00. Use $K = 0.9$ and determine the API for each successive day.

*Solution*

Equation 4-9 reduces to

$$\text{API}_i = K(\text{API}_{i-1}) + P_i$$

**Table 4-1**

| Date ($i$) | Precipitation ($P_i$) | $\text{API}_i$ |
|---|---|---|
| Apr. 1 | 0.0 | 0.00 |
| 2 | 0.0 | 0.00 |
| 3 | 0.5 | 0.50 |
| 4 | 0.7 | 1.15 |
| 5 | 0.2 | 1.24 |
| 6 | 0.1 | 1.22 |
| 7 | 0.0 | 1.10 |
| 8 | 0.1 | 1.09 |
| 9 | 0.3 | 1.28 |
| 10 | 0.0 | 1.15 |
| 11 | 0.0 | 1.04 |
| 12 | 0.6 | 1.54 |
| 13 | 0.0 | 1.39 |
| 14 | 0.0 | 1.25 |

## 4-3 The Hydrograph

A knowledge of the size and time distributionof streamflows is essential to many aspects of water management and environmental planning.

A hydrograph is a continuous graph showing the properties of streamflow with respect to time, normally obtained by means of a continuous strip recorder which indicates stage versus time (stage hydrograph), and is then transformed to a discharge hydrograph by application of a rating curve. The term *hydrograph* will generally be taken herein to indicate a discharge hydrograph.

The hydrograph is the result of the physiographic and hydrometeorological effects of the watershed. Typically, it is a complex multiple peak curve.[21,36]

## 4-4 Components of the Hydrograph

A hydrograph has four component elements: (1) direct surface runoff, (2) interflow, (3) groundwater or base flow, and (4) channel precipitation. The rising portion of a hydrograph is known as the *concentration curve*; the region in the vicinity of the peak is called the *crest segment*; and the falling portion is the *recession.*[8] The shape of a hydrograph depends upon precipitation pattern characteristics and basin properties. Figure 4-8 illustrates the definitions presented.

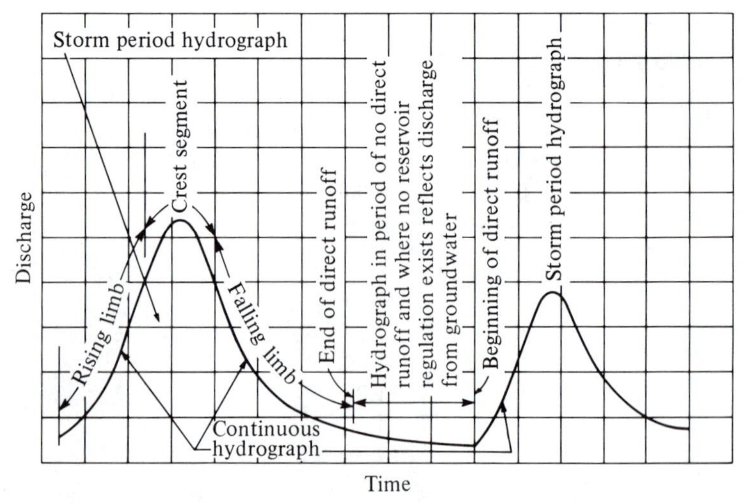


**Fig. 4-8.** *Hydrograph definition.*

## 4-5 Surface Flow Phenomenon within a Watershed

During a given rainfall, water is continually being abstracted to saturate the upper levels of the soil surface; however, this saturation or infiltration is only one of many continuous abstractions.[13,18,38] Rainfall is also intercepted by trees, plants, and roof surfaces, and at the same time is evaporated. Once rain falls and fulfills initial requirements of infiltration, natural depressions collect falling rain to form small puddles, creating *depression storage.* In addition, minute depths of water forming *detention storage* build up on permeable and impermeable surfaces within the watershed. This stored water gathers in small rivulets which carry the water originating as *overland flow* into small channels, then into larger channels, and finally as *channel flow* to the watershed outlet. Figure 4-9a illustrates the distribution of a prolonged uniform rainfall. Although such an event is not the norm, the concept is useful for showing the manner in which detention and depression storage would be distributed.

In general, the channel of a watershed possesses a certain amount of *base flow* during most of the year. This flow comes from groundwater or spring contributions and may be considered as the normal day-to-day flow. Discharge from precipitation excess—that is, after abstractions deducted from the original rainfall—constitutes the direct runoff hydrograph (DRH). Arrival of direct runoff at the outlet accounts for an initial rise in the DRH. As precipitation excess continues, enough time elapses for progressively distant areas to add to the outlet flow. Consequently, the duration of rainfall dictates the proportionate area of the watershed amplifying the peak, and the intensity of rainfall during this period of time determines the resulting highest discharge.

If the rainfall maintains a constant intensity for a long enough period of time, a state of equilibrium discharge is reached, as depicted by curve *A* in Fig. 4-9b. The inflection point on curve *A* indicates the time at which the entire drainage area contributes to the flow. At this time maximum storage of the watershed is only partially complete. As rainfall continues, maximum storage capacity is attained and equilibrium [inflow (rainfall) equals outflow (runoff)] reached. The condition of maximum storage and equilibrium is seldom if ever attained in nature. Extended rainfall may occur, but variations in intensity throughout its duration negate any possibility of a DRH of the theoretical shape for constant rainfall intensity.

A normal single-peak DRH generally possesses the shape shown by curve *B* in Fig. 4-9. The time to peak magnitude of this hydrograph depends upon the intensity and duration of the rainfall, and the size,

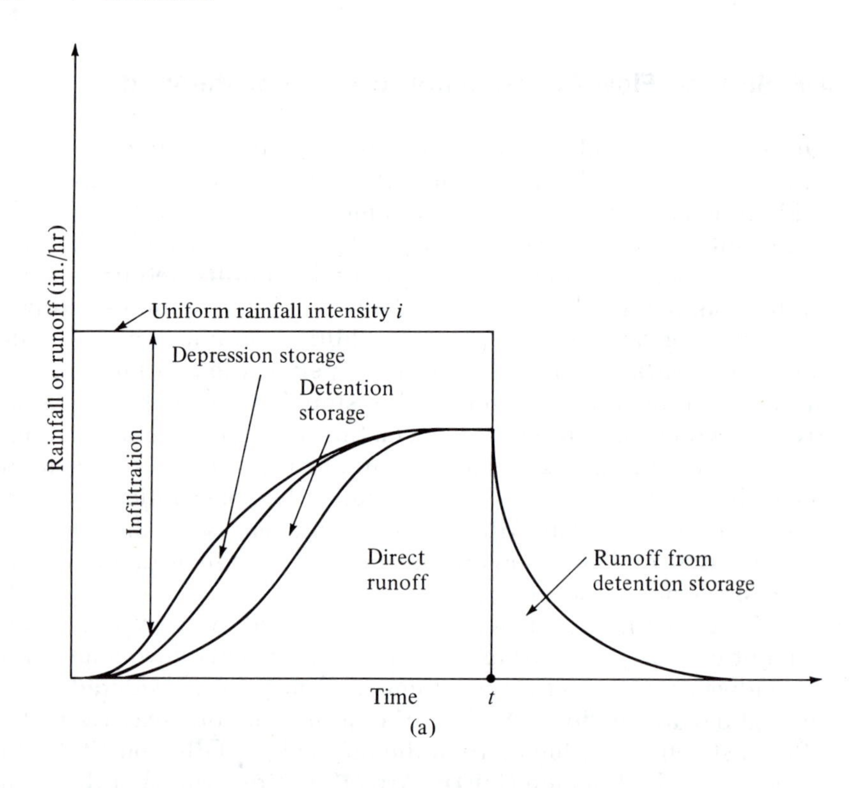


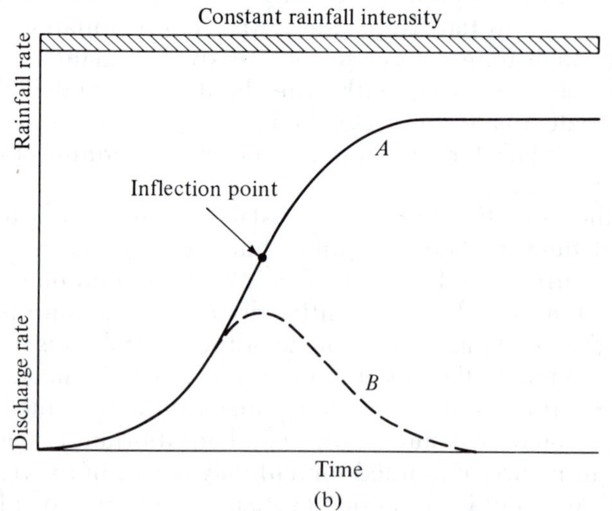


***Fig. 4-9.*** *(a) Distribution of a uniform storm rainfall for condition of no interception loss. Note that all water stored in depressions is ultimately evaporated or infiltrated while some detention storage is also subjected to these losses. (b) Equilibrium discharge hydrograph.*

slope, shape, and storage capacity of the watershed. Once peak flow has been reached for a given isolated rainstorm, the DRH begins to descend, its source of supply coming largely from water accumulated within the watershed such as detention and channel storage.

Processes involved in forming the DRH can be better understood by visualizing the precipitation excess as partially disposed of immediately by surface runoff while a portion remains held within the watershed boundaries and is released later from storage. Thus the shape and timing of the DRH are integrated effects of the duration and intensity of rainfall hydrometeorological factors as well as the effect of the physiographic factors of the watershed upon the storage capacity.

## 4-6 Hydrograph Components

It is important to understand how the hydrograph can be subdivided into its component parts, and helpful to look at the effect on hydrograph shape of precipitation and watershed features. Figure 4-10 is used in the discussion.

Figure 4-10a illustrates the hydrograph of a permanent stream during a period between precipitation events, known as a *base flow hydrograph* because groundwater sustains the flow. Four general conditions cause modification of the base flow hydrograph shape. They are described by Horton[41] using the following sets of inequalities:

| | | | |
|---|---|---|---|
| Set 1 | $i < f$ <br> $F < S_D$ | Set 3 | $i > f$ <br> $F < S_D$ |
| Set 2 | $i < f$ <br> $F > S_D$ | Set 4 | $i > f$ <br> $F > S_D$ |

where Set 1 would produce a hydrograph similar to that shown in Fig. 4-10a except for a very small rise due to direct channel precipitation. No overland flow occurs since $i < f$, and since the soil's field capacity $F$ is not reached, no interflow or added groundwater component develops. The entire effect of the storm would be to slightly reduce the soil moisture deficiency $S_D$. The field capacity is the amount of water held in the soil after excess gravitational water has drained.

The conditions described by Set 2 still do not produce direct surface runoff, although the components of interflow and groundwater flow are added to channel precipitation. The initial hydrograph would be modified, since the field capacity of the soil is exceeded. Figure 4-10b illustrates this condition. Note that deviation of the hydrograph from the original base flow curve is likely to be very small under these conditions.

Figure 4-10c illustrates a case where surface runoff becomes a

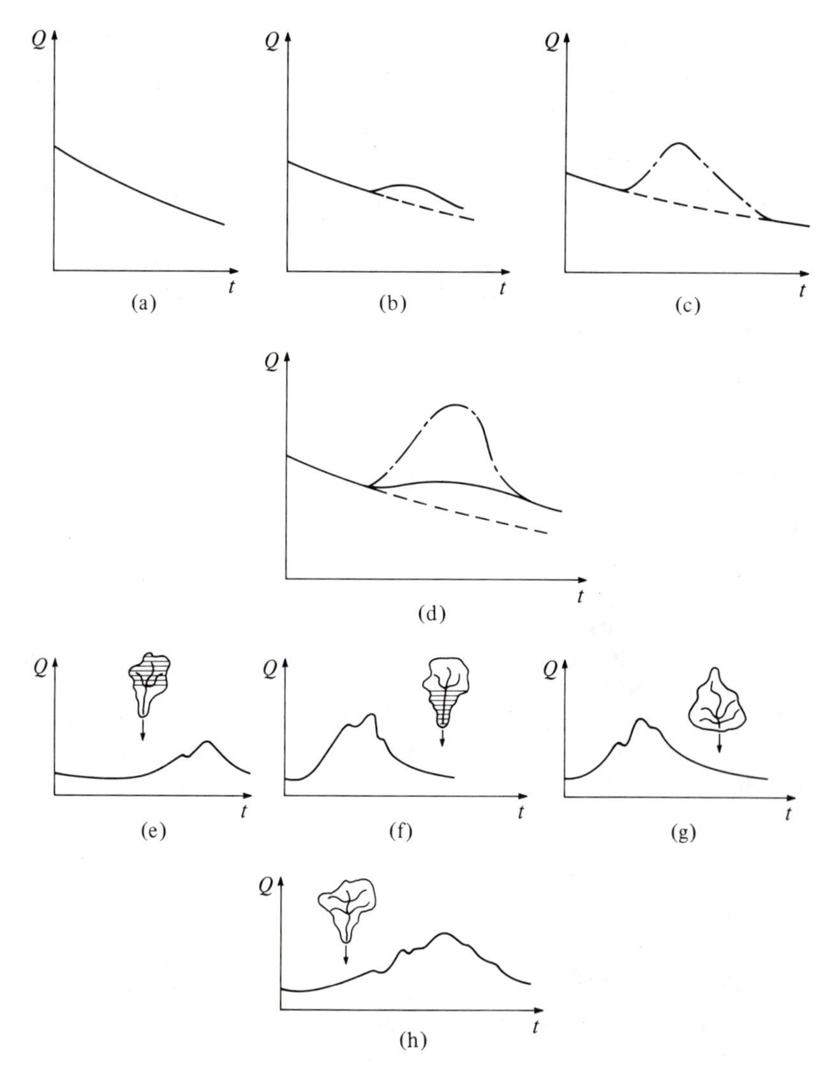


***Fig. 4-10.*** *Effects of storm and basin characteristics on hydrograph shape.*

component of flow because $i > f$. In this example, interflow and groundwater flow are zero, as soil moisture deficiency still exists although at a reduced level. Channel precipitation likewise constitutes a component.

In the final set, all four components exist with rainfall intensity exceeding infiltration rate and the field capacity of the soil is reached. This case would be typical of a large storm event.

Figures 4-10e–h illustrate how hydrograph shape can be modified by areal variations in rainfall and rainfall intensity and by watershed configuration.[14] Minor fluctuations shown in these hydrographs are linked to variations in storm intensity. In Fig. 4-10e only the delaying effects pertinent to a storm over the upstream section of the area are indicated. Figure 4-10f shows the reverse of this condition. Figures 4-10g–h depict the comparative effects of basin geometry.

In most hydrograph analyses, interflow and channel precipitation are grouped with surface runoff rather than treated independently. Channel precipitation begins with inception of rainfall and ends with the storm. Its distribution with respect to time is highly correlated with the storm pattern. The relative volume contribution tends to increase somewhat as the storm proceeds, since stream levels rise and the water surface area tends to increase. The fraction of watershed area occupied by streams and lakes is generally small, usually on the order of 5% or less, so the percentage of runoff related to channel precipitation is usually minor during important storms.

Interflow is that part of the subsurface flow which moves at shallow depths and reaches the surface channels in a relatively short period of time and therefore is commonly considered part of the direct surface runoff. Its distribution is commonly characterized by a slowly increasing rate up to the end of the storm period, followed by a gradual recession which terminates at the intersection of the surface flow hydrograph and base flow hydrograph. Figure 4-11 illustrates the approximate nature of the components of channel precipitation and interflow.

The base flow component is composed of the water that percolates downward until it reaches the groundwater reservoir and then

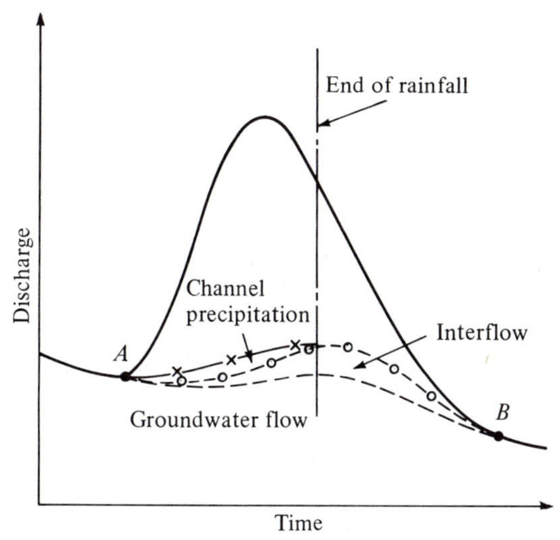


***Fig. 4-11.*** *Components of the hydrograph.*

flows to surface streams as groundwater discharge. The groundwater hydrograph may or may not show an increase during the actual storm period. Groundwater accretion resulting from a particular storm is normally released over an extended period measured in days for small watersheds and often in months or years for large drainage areas.

The surface runoff component consists of water which flows overland until a stream channel is reached. During large storms it is the most significant hydrograph component. Figure 4-11 illustrates the surface runoff and groundwater components of a hydrograph. As pointed out in Fig. 4-10, the relative magnitude of each component for a given storm is determined by a combination of many factors. Hydrographs are analyzed to provide knowledge of the way precipitation and watershed characteristics interact to form them. The degree of hydrograph separation required depends upon the objective of the study. For most practical work, surface runoff and groundwater components only are required. Research projects or more sophisticated analyses may dictate consideration of all components. When multiple storms occur within short periods, it is sometimes necessary to separate the overlapping parts of consecutive surface runoff hydrographs.

## 4-7 Groundwater Recession

Several techniques are used to separate surface and groundwater flows. Most of them are based on analyses of groundwater recession or depletion curves. If there is no added inflow to the groundwater reservoir, and if all groundwater discharge from the upstream area is intercepted at the stream-gauging point of interest, then groundwater discharge can be described by an equation of the form[15,16]

$$q_t = q_0 K^t \tag{4-13}$$

where

$q_0$ = a specified initial discharge
$q_t$ = the discharge at any time $t$ after flow $q_0$
$K$ = a recession constant

Time units frequently used are days for large watersheds, hours or minutes for small basins. A plot of Eq. 4-13 yields a straight line on semilogarithmic paper by plotting $t$ on the linear scale.

For most watersheds, groundwater depletion characteristics are approximately stable, since they closely fit watershed geology. Nevertheless, the recession constant varies with seasonal effects such as evaporation and freezing cycles and other factors. If we note that $q_t\,dt$ is equivalent to $-dS$, where $S$ is the quantity of water obtained from storage, integration of Eq. 4-13 produces

$$S = \frac{(q_t - q_0)}{\log_e K} \tag{4-14}$$

This equation determines the quantity of water released from groundwater storage between the times of occurrence of the two discharges of interest, or it can be used to calculate the volume of water still in storage at a time some chosen value of flow occurs. To get the latter, $q_t$ is set equal to zero and $q_0$ becomes the reference discharge. Figure 4-12 is a plot of Eqs. 4-13 and 4-14 and provides additional definition.

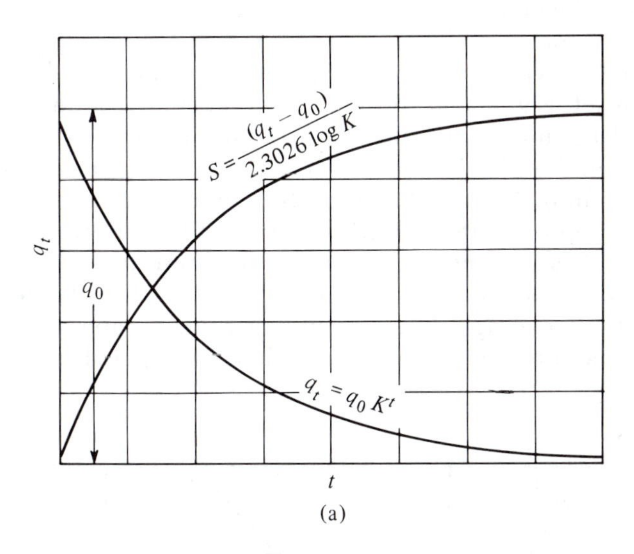

(a)

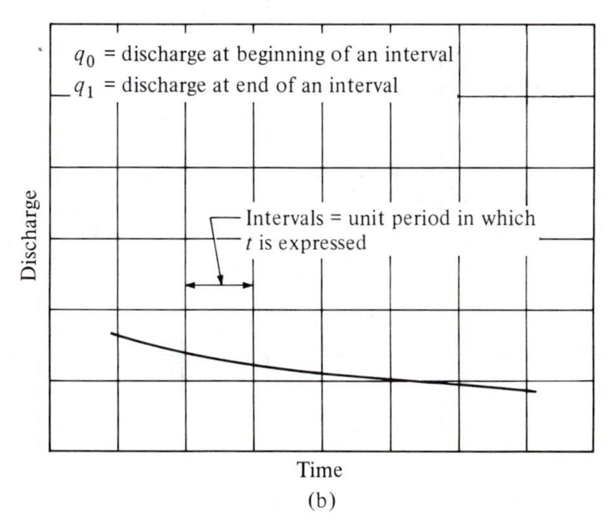

(b)

***Fig. 4-12.*** *Base flow model.*

Groundwater depletion curves can be analyzed by various methods to evaluate the recession constant $K$. One of these will be described. Data from a stream-gauging station are a prerequisite, and should reflect rainless periods with no upstream regulation such as a reservoir to affect flow at the gauging point. Otherwise an adjustment with its own errors is introduced.

From the streamflow data, plot a hydrograph (Fig. 4-12) to find values of discharge at the beginning and end of selected time intervals. Flows at the beginning of each interval are analogous to $q_0$ whereas those at the end are analogous to $q_1$. Next, select several time intervals and plot corresponding $q_0$'s versus $q_1$'s as shown in Fig. 4-13. The time period between consecutive values of $q$ should be identical for each datum set. Figures taken from recession curves of times that still reflect surface runoff will usually fall below and to the right of a 45° line drawn on the plot. These values will also be associated with larger numbers for $q$. Points taken from true groundwater recession periods should approximately describe a straight line. Because $q_1 = q_0 = 0$ when $q_0 = 0$, a straight line can be fitted graphically to the data points. The slope of this line is $q_1/q_0 = K$. Using this value, the depletion curve plots as a straight line on semilogarithmic paper ($t$ is the linear scale variable) or as a curve on arithmetic paper, Fig. 4-12a.

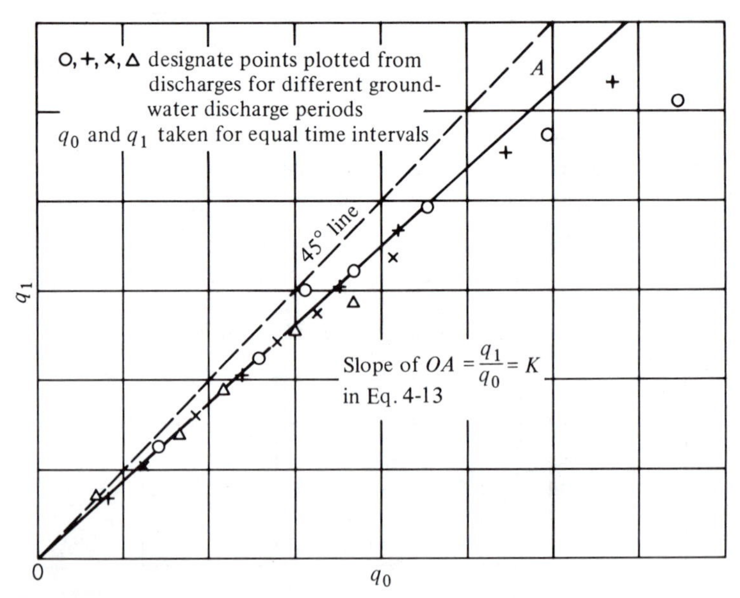


**Fig. 4-13.** *Graphical method for determining recession constant K. (U.S. Department of Agriculture, Soil Conservation Service.)*

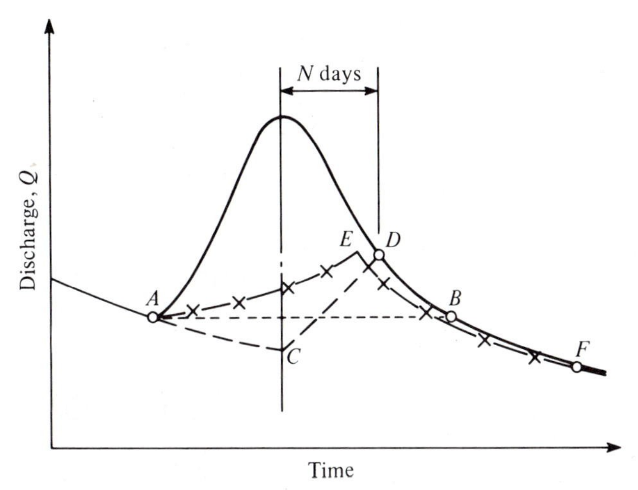


**Fig. 4-14.** *Illustration of some hydrograph separation techniques.*

## 4-8 Separation Techniques

Several methods for base flow separation are used when the actual amount of base flow is unknown. During large storms, the maximum rate of discharge is just slightly affected by base flow, and inaccuracies in separation are fortunately not important.

The simplest base flow separation technique is to draw a horizontal line from the point at which surface runoff begins, point $A$ in Fig. 4-14, to an intersection with the hydrograph recession as indicated by point $B$. A second method projects the initial recession curve downward from $A$ to $C$ which lies directly below the peak rate of flow. Then, point $D$ on the hydrograph, representing $N$ days after the peak, is connected to point $C$ by a straight line defining the groundwater component. The choice of $N$ is based on the formula[8]

$$N = A^{0.2} \tag{4-15}$$

where

$N$ = the time in days
$A$ = the drainage area in square miles

A third procedure employs a base flow recession curve that is fitted to the hydrograph and then computed in a time-decreasing direction. Point $F$, where the computed curve begins to deviate from the actual hydrograph, marks the end of direct runoff. The curve is projected backward arbitrarily to some point $E$ below the inflection point and its shape from $A$ to $E$ is arbitrarily assigned. A fourth widely used method is to draw a line between $A$ and $F$. All of these methods are

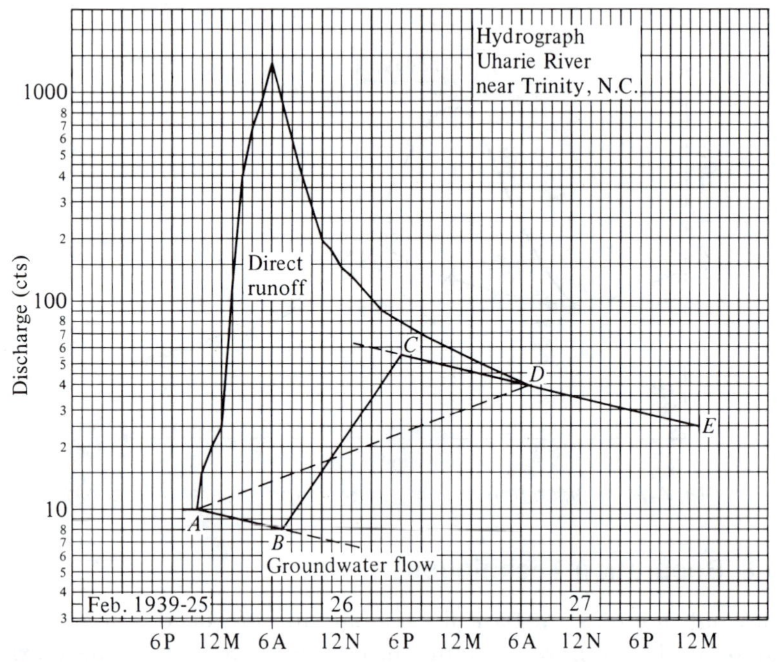


**Fig. 4-15.** *Illustration of base flow separation. (U.S. Department of Agriculture, Soil Conservation Service.)*

approximate since the separation of hydrographs is partly a subjective procedure. Figure 4-15 illustrates two separation techniques to determine surface runoff and groundwater flow components. Line *AD* represents the simple procedure of connecting the point of the beginning of direct runoff with an arbitrary point on the recession curve (an advantage over the horizontal line technique because the time base of direct runoff is much shorter). Curve *ABCD* is constructed from the extension of the base flow recession curve.

## 4-9 Time Base of the Hydrograph

The time base of a hydrograph is considered to be the time from which the concentration curve begins until the direct-runoff component essentially reaches zero. An equation for time base may take the form[7]

$$T_b = t_s + t_c \tag{4-16}$$

where

$T_b$ = the time base of the hydrograph
$t_s$ = the duration of runoff-producing rain
$t_c$ = the time of concentration

The definition of the time of concentration used herein is the flow time from the most remote point in the drainage area to the outlet of interest. It can be estimated for channels by the Chézy or Manning formulas. The hydrograph time base, related to both storm and basin characteristics, is a basic element of unit hydrograph theory.[17]

## 4-10 Basin Lag and Time of Concentration

The relative timing of hydrologic events must be known if drainage areas having subbasins are to be modeled or if continuous simulation is desired. A basic measure of timing is lag time or basin lag which locates the hydrograph's position relative to the causative storm pattern. It is that property of a drainage area which is defined as the difference in time between the center of mass of effective rainfall and the center of mass of runoff produced. Other definitions are also used mainly for ease of determination. Two of these are (1) the time interval from the maximum rainfall rate to the peak rate of runoff, and (2) the time from the center of mass of effective rainfall to the peak rate of flow. Time lag is characterized by the ratio of a flow length to a mean velocity of flow and is thus a property that is influenced by the shape of the drainage area, the slope of the main channel, channel geometry, and the storm pattern.

Various studies have been conducted for the purpose of developing relationships descriptive of time lag. Most prominent of these was the work by Snyder on large natural watersheds.[24] His original equation has been widely used and modified in various ways by other investigators. Eagleson has proposed an equation for lag time on sewered drainage areas having a minimum size of 141 acres.[42] An early investigation (1936) on small drainage areas (2 to 4 acres) was conducted by Horner in his classical work on urban drainage in St. Louis, Missouri.[43] Horner's work was inconclusive in that it did not yield a defined procedure, but he did conclude that the comparatively wide range in the lag at each location led to the inference that the lag was a variable, its value being determined more by rainfall characteristics than by characteristics of the drainage area.

Snyder's study based on data from the Appalachian Mountain Region produced the following equation for lag time:[1]

$$t_1 = C_t(L_{ca}L)^{0.3} \tag{4-17}$$

in which

$t_1$ = the lag time (hr) between the center of mass of the rainfall excess for a specified type of storm to the peak rate of flow

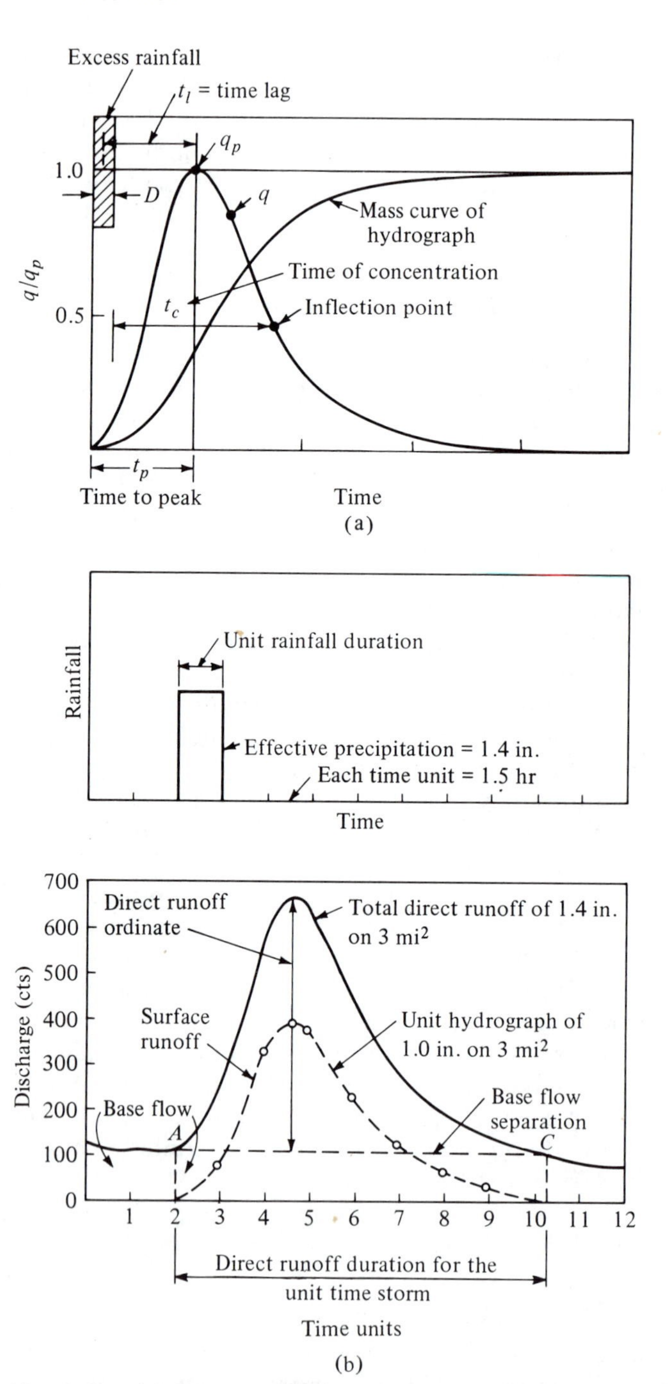


***Fig. 4-16.*** *(a) Dimensionless unit hydrograph; (b) illustration of the derivation of a unit hydrograph from an isolated storm.*

$L_{ca}$ = the distance along the main stream from the base gauge to a point opposite the center of gravity of the basin (mi)
$L$ = the maximum travel distance along the main stream (mi)
$C_t$ = coefficient depending upon the basin properties

For the area studied, the constant $C_t$ was found to vary from 1.8 to 2.2, with somewhat lower values for basins with steeper slopes. The constant is considered to include the effects of slope and storage. The value of $t_1$ is assumed to be constant for a given drainage area, but allowance is made for the use of different values of lag for different types of storms. The relationship is considered applicable to drainage areas ranging in size from 10 to 10,000 mi².

In a study of sewered areas ranging in size from 0.22 to 7.51 mi², Eagleson found the Snyder constant to vary from 0.21 to 0.32, with the estimated value for a 0.22-mi², 83% impervious area being 0.22.[42]

Eagleson, on the basis of his data, developed the equation

$$t_1 = \frac{\bar{L}}{(1.5/n)\,\bar{R}^{2/3}\,\bar{S}^{1/2}} \tag{4-18}$$

in which

$t_1$ = lag time, the center of mass of rainfall excess to the peak discharge (sec)
$\bar{L}$ = the mean travel distance (ft) which is equal to the length of that portion of the sewer which flows full
$n$ = the weighted Manning coefficient for the main sewer
$\bar{R}$ = the weighted hydraulic radius of the main sewer flowing full
$\bar{S}$ = the weighted physical slope of the main sewer

Eagleson's equation directly includes the effects of channel geometry and slope as well as basin shape, and thus represents a refinement of the Snyder approach. It also indirectly includes the important effect of the storm pattern.

Linsley gives examples of application of the following modified form of Snyder's equation:[10]

$$t_1 = C_t \frac{(LL_{ca})^a}{\sqrt{s}} \tag{4-19}$$

where $s$ is a weighted slope of the channel and the other variables are as defined previously.

Other investigators have represented time lag by equations of the form

$$t_1 = K \frac{L}{\sqrt{s}} \tag{4-20}$$

The SCS equations for time lag given in Sec. 4-17 can also be used.

Another important time parameter used in hydrologic analyses is the time of concentration ($t_c$). This is the time required for runoff to travel from the hydraulically most distant part of the storm area to the watershed outlet or some other point of reference downstream.[15] In hydrograph analysis, the time of concentration is the time from the end of excessive rainfall to the point on the falling limb of the hydrograph (inflection point) where the recession curve begins (where direct runoff ceases. See Fig. 4-16a).

## 4-11 Unit Hydrograph Development

Ways to predict flood peak discharges and discharge hydrographs from rainfall events have been studied intensively since the early 1930s. One method receiving considerable use and still employed to estimate peak discharge rates is called the *unit hydrograph method.* It was evolved from a method of unit graphs employed by Sherman[20,22] in 1932 and originally assumed that for a given duration rainfall, the hydrograph time base should remain constant.

The unit graph is defined as follows:[22]

> If a given one-day rainfall produces a 1-in. depth of runoff over the given drainage area, the hydrograph showing the rates at which the runoff occurred can be considered a unit graph for that watershed.

Application of a unit graph to design rainfall excess amounts other than 1 in. is accomplished simply by multiplying the rainfall excess amount by the unit graph ordinates, since the runoff ordinates for a given duration are assumed to be directly proportional to rainfall excess. Time periods other than 1 day are also widely used.

Implicit in deriving the unit graph or unit hydrograph is the assumption that rainfall is distributed in the same temporal and spatial pattern for all storms. This is generally not true; consequently variations in ordinates for different storms can be expected.

The construction of unit hydrographs for other than the derived duration is facilitated by a method known as the *S-hydrograph* developed by Morgan and Hulinghorst.[23] The procedure, as explained in Sec. 4-13, employs a unit hydrograph to form an S-hydrograph resulting from a continuous applied rainfall. The application of unit hydrograph theory to ungauged watersheds received early attention by Snyder[24] and also by Taylor and Schwartz,[25] who tried to relate features of the unit hydrograph to watershed characteristics. As a result of the attempted synthesis of data, these approaches are referred to as *synthetic unit-hydrograph methods.* The need to alter duration of a unit hydrograph encouraged studies to define the shortest possible storm duration—that is, an instantaneous unit rainfall. The concept of *instantaneous unit hydrographs* can be traced to Clark[26] in 1945.

Methods of deriving unit hydrographs vary and are subject to engineering judgment. The level of sophistication employed to unravel the problem depends largely upon the kind of issue in question. Several methods useful in the determination of unit hydrographs will be discussed. They are subdivided into (1) starting with unit hydrographs obtained from field data and manipulating them by S-hydrograph methods, and (2) constructing synthetic unit hydrographs.

## 4-12 Derivation of Unit Hydrographs

Data collection preparatory to deriving a unit hydrograph for a gauged watershed can be extremely time-consuming. Fortunately, many watersheds have available records of streamflow and rainfall, and these can be supplemented with office records of the Water Resources Division of the U.S. Geological Survey.[27] Rainfall records may be secured from "Climatological Data"[28] published for each state in the United States by the National Oceanic and Atmospheric Administration (NOAA). Sample data are given in Tables 4-2 and 4-3 and indicate only a portion of published information offered for selected watersheds. Hourly rainfall records for recording rainfall stations are published as a *Summary of Hourly Observations* for the location. Summaries are listed for approximately 300 first order stations in the United States.

To develop a unit hydrograph, it is desirable to get as many rainfall records as possible within the study area to insure that the amount and distribution of rainfall over the watershed is accurately known. Preliminary selection of storms to use in deriving a unit hydrograph for a watershed should be restricted to

1. Storms occurring individually, that is, simple storm structure.
2. Storms having uniform distribution of rainfall throughout the period of rainfall excess.
3. Storms having uniform spatial distribution over the entire watershed.

These restrictions place both upper and lower limits on size of the watershed to be employed. An upper limit of watershed size of approximately 1000 mi$^2$ is overcautious, although general storms over such areas are not unrealistic and some studies of areas up to 2000 mi$^2$ have used the unit hydrograph technique. The lower limit of watershed extent depends upon numerous other factors and cannot be precisely defined. A general rule of thumb is to assume about 1000 acres. Fortunately, other hydrologic techniques help resolve unit hydrographs for watersheds outside of this range. (See Chapter 7.)

The preliminary screening of suitable storms for unit hydrograph formation must meet more restrictive criteria before further analysis:

1. Duration of rainfall event should be approximately 10 to 30% of the drainage area lag time.

**Table 4-2** *Local Climatological Data*

Latitude 40° 49′ N    Longitude 96° 42′ W    Elevation (ground) 1150 ft    Standard Time Used: Central    MBAN #14971

| | Temperature °F | | | | | | | Weather Types on Dates of Occurrence | Snow, Ice Pellets or Ice on Ground at 06 A.M. (in.) | Precipitation | | Avg. Station Pressure (in.) Elev. 1189 (ft) m.s.l. | Wind | | | | | Sunshine | | Sky Cover (tenths) | | |
|---|---|---|---|---|---|---|---|---|---|---|---|---|---|---|---|---|---|---|---|---|---|---|
| | | | | | | Degree Days Base 65° | | 1 Fog<br>2 Heavy fog X<br>3 Thunderstorm<br>4 Ice pellets<br>5 Hail<br>6 Glaze<br>7 Duststorm<br>8 Smoke, haze<br>9 Blowing snow | | Water Equivalent (in.) | Snow, Ice Pellets (in.) | | | | | Fastest Mile | | | | | | |
| Date | Maximum | Minimum | Average | Departure from Normal | Average Dew Point | Heating | Cooling | | | | | | Resultant Direction | Resultant Speed (mph) | Average Speed (mph) | Speed (mph) | Direction | Hours and Tenths | % of Possible | Sunrise to Sunset | Midnight to Midnight | Date |
| 1 | 2 | 3 | 4 | 5 | 6 | 7A | 7B | 8 | 9 | 10 | 11 | 12 | 13 | 14 | 15 | 16 | 17 | 18 | 19 | 20 | 21 | 22 |
| 1 | 45 | 29 | 37 | −8 | | 28 | 0 | | T | 0.04 | .4 | | | | | 31 | NW | 4.4 | 34 | 8 | | 1 |
| 2 | 50 | 31 | 41 | −4 | | 24 | 0 | | 0 | 0 | 0 | | | | | 32 | NW | 12.2 | 96 | 0 | | 2 |
| 3 | 49 | 29 | 39 | −7 | | 26 | 0 | | 0 | 0 | 0 | | | | | 15 | NE | 12.7 | 99 | 5 | | 3 |
| 4 | 47 | 29 | 38 | −8 | | 27 | 0 | | 0 | 0 | 0 | | | | | 12 | NE | 5.5 | 43 | 7 | | 4 |
| 5 | 47 | 31 | 39 | −8 | | 26 | 0 | | 0 | 0 | 0 | | | | | 17 | NE | 8.3 | 64 | 8 | | 5 |
| 6 | 62 | 26 | 44 | −3 | | 21 | 0 | | 0 | 0 | 0 | | | | | 10 | S | 12.9 | 100 | 0 | | 6 |
| 7 | 76 | 39 | 58 | 10 | | 7 | 0 | | 0 | 0 | 0 | | | | | 28 | SW | 13.0 | 100 | | | 7 |
| 8 | 74 | 51 | 63 | 14 | | 2 | 0 | | 0 | 0 | 0 | | | | | 36 | NW | 7.8 | 61 | 7 | | 8 |
| 9 | 65 | 41 | 53 | 4 | | 12 | 0 | | 0 | 0 | 0 | | | | | 16 | N | 13.0 | 100 | 0 | | 9 |
| 10 | 77 | 39 | 58 | 8 | | 7 | 0 | Data | 0 | 0 | 0 | | | | | 33 | S | 12.3 | 94 | 1 | | 10 |
| 11 | 73 | 58 | 66 | 16 | | 0 | 1 | are not | 0 | 0 | 0 | | | | | 27 | S | 12.0 | 91 | 6 | | 11 |
| 12 | 73 | 48 | 61 | 10 | | 4 | 0 | entered | 0 | 0 | 0 | | | | | 26 | NE | 10.6 | 80 | 1 | | 12 |
| 13 | 64 | 42 | 53 | 2 | | 12 | 0 | in this | 0 | 0 | 0 | | | | | 27 | NE | 12.1 | 92 | 2 | | 13 |
| 14 | 70 | 35 | 53 | 1 | | 12 | 0 | column | 0 | 0 | 0 | | | | | 23 | SW | 12.8 | 96 | 5 | | 14 |
| 15 | 83 | 50 | 67 | 15 | | 0 | 2 | because | 0 | 0 | 0 | | | | | 33 | S | 12.5 | 94 | 6 | | 15 |
| 16 | 69 | 54 | 62 | 9 | | 3 | 0 | records | 0 | 0.05 | 0 | | | | | 19 | S | 0.5 | 4 | 10 | | 16 |

| | | | | | | | | | | | | | | | | | | | | | | |
|---|---|---|---|---|---|---|---|---|---|---|---|---|---|---|---|---|---|---|---|---|---|---|
| 17 | 74 | 50 | 62 | 9 | | 3 | 0 | of | 0 | 0.04 | 0 | | | | | 29 | S | 4.6 | 34 | 7 | | 17 |
| 18 | 72 | 57 | 65 | 12 | | 0 | 0 | weather | 0 | T | 0 | | | | | 30 | S | 1.6 | 12 | 10 | | 18 |
| 19 | 80 | 59 | 70 | 16 | | 0 | 5 | types | 0 | 0 | 0 | | | | | 30 | S | 6.2 | 46 | 8 | | 19 |
| 20 | 69 | 50 | 60 | 6 | | 5 | 0 | may be | 0 | 0.22 | 0 | | | | | 23 | S | 1.4 | 10 | 9 | | 20 |
| 21 | 73 | 50 | 62 | 7 | | 3 | 0 | incomplete | 0 | T | 0 | | | | | 20 | SE | 7.7 | 57 | 6 | | 21 |
| 22 | 60 | 45 | 53 | −2 | | 12 | 0 | | 0 | 0.14 | 0 | | | | | 25 | NE | 3.1 | 23 | 10 | | 22 |
| 23 | 67 | 42 | 55 | 0 | | 10 | 0 | | 0 | 0 | 0 | | | | | 17 | NW | 11.7 | 86 | 3 | | 23 |
| 24 | 72 | 44 | 58 | 2 | | 7 | 0 | | 0 | T | 0 | | | | | 14 | NE | 9.4 | 69 | 7 | | 24 |
| 25 | 63 | 50 | 57 | 1 | | 8 | 0 | | 0 | T | 0 | | | | | 19 | NE | 2.8 | 21 | 10 | | 25 |
| 26 | 62 | 47 | 55 | −2 | | 10 | 0 | | 0 | 0.18 | 0 | | | | | 22 | NE | 1.0 | 7 | 10 | | 26 |
| 27 | 59 | 42 | 51 | −6 | | 14 | 0 | | 0 | 0 | 0 | | | | | 26 | NW | 9.6 | 69 | 5 | | 27 |
| 28 | 60 | 36 | 48 | −9 | | 17 | 0 | | 0 | 0 | 0 | | | | | 20 | NW | 13.3 | 96 | 1 | | 28 |
| 29 | 64 | 35 | 50 | −8 | | 15 | 0 | | 0 | 0 | 0 | | | | | 17 | S | 10.5 | 75 | 8 | | 29 |
| 30 | 71 | 46 | 59 | 1 | | 6 | 0 | | 0 | 0.11 | 0 | | | | | 17 | SW | 7.0 | 50 | 7 | | 30 |
| | Sum | Sum | | | | Total | Total | | | Total | Total | | For the month: | | | | | Total | | Sum | Sum | |
| | 1970 | 1285 | | | | 321 | 8 | Number of days | | 0.78 | 0.4 | | | | | 36 | NW | 252.5 | % for month | 167 | | |
| | Avg. | Avg. | Avg. | Dep. | Avg. | Dep. | Dep. | Precipitation ≤0.01 in. 7 | | Dep. | | | | | | Date: | 08 | Possible | | Avg. | Avg. | |
| | 65.7 | 42.8 | 54.3 | 2.5 | | −81 | | | | −1.67 | | | | | | | | 399.2 | 63 | 5.6 | | |
| | | | | | | Season to date | | Snow, ice pellets ≤1.0 in. | | | | | | | | | | | | | | |
| | Number of days | | | | | Total | Total | | | Greatest in 24 hr and dates | | | | | | Greatest depth on ground of snow, ice pellets or ice and date | | | | | | |
| | Maximum temp. | | Minimum temp. | | | 5985 | 8 | Thunderstorms | | Precipitation | | Snow, ice pellets | | | | | | | | | | |
| | ≥90° | ≤32° | ≤32° | | ≤0° | Dep. | Dep. | Heavy fog X | | 0.22 | | 20 | 0.4 | 1 | | | | T | 1 | | | |
| | 0 | 0 | 6 | | 0 | 322 | | Clear 9 | | Partly cloudy 11 | | Cloudy 10 | | | | | | | | | | |

*Source:* U.S. Department of Commerce, National Oceanic and Atmospheric Administration, Environmental Data Service, Lincoln, Neb., 901 North 17th Street, Apr. 1971.

**Table 4-3** *Typical Small Watershed Data,* Kickapoo Creek near Kickapoo, Ill. *(5-5630), Latitude 40°48′00″, Longitude 89°48′00″; Drainage Area, 120 mi²; Mean Annual Precipitation, 34.02 in.*

| Water Year | Date | Peak Discharge (cfs) | Antecedent Base Discharge (cfs) | Direct Runoff, (in.) | Associated precipitation (in.) | Storm Duration (days) | Antecedent Precipitation (in.) | | Annual Runoff (in.) |
|---|---|---|---|---|---|---|---|---|---|
| | | | | | | | 5-day | 30-day | |
| 1945 | June 16 | 3,240 | 74 | 0.58 | 2.17 | 2 | 0.17 | 2.48 | 7.82 |
| 1946 | Jan. 5 | 5,840 | 17 | 1.90[a] | 1.64 | 2 | 0.29 | 1.82 | 7.82 |
| 1947 | June 17 | 3,800 | 39 | 0.56 | 1.54 | 2 | 0.41 | 5.80 | 8.42 |
| 1948 | Mar. 19 | 9,340 | 182 | 1.12 | 1.98 | 1 | 0.44 | 2.65 | 6.74 |
| 1949 | Feb. 13 | 6,150 | 24 | 1.42[a] | 0.77 | 1 | 0.00 | 2.95 | 5.84 |
| 1950 | July 17 | 7,300 | 10 | 1.15 | 3.89 | 1 | 0.07 | 3.61 | 10.97 |
| 1951 | July 22 | 18,400 | 34 | 1.39 | 3.59 | 1 | 0.45 | 6.98 | 12.99 |
| 1952 | June 14 | 3,800 | 62 | 0.62 | 2.46 | 2 | 1.28 | 4.90 | 10.89 |
| 1953 | July 21 | 2,420 | 20 | 0.17 | 0.72 | 2 | 0.68 | 5.01 | 5.66 |
| 1954 | June 1 | 3,950 | 167 | 0.55 | 1.79 | 1 | 2.43 | 2.72 | 5.31 |
| 1955 | Apr. 24 | 8,590 | 120 | 0.90 | 1.89 | 1 | 1.72 | 3.10 | 11.86 |
| 1956 | May 27 | 1,300 | 2 | 0.14 | 3.52 | 3 | 0.06 | 2.25 | 1.23 |
| 1957 | June 14 | 7,310 | 52 | 0.70 | 1.25 | 2 | 1.08 | 4.75 | 4.14 |
| 1958 | July 2 | 12,100 | 29 | 1.38 | 3.54 | 2 | 0.01 | 6.58 | 6.73 |
| 1959 | Apr. 26 | 6,000 | 26 | 0.49 | 1.85 | 2 | 0.01 | 1.97 | 4.81 |
| 1960 | Mar. 30 | 10,000[b] | 28 | 1.97 | 2.04 | 1 | 0.01 | 1.28 | 12.34 |

*a* Runoff increased by snowmelt.
*b* Approximate value.

*Source:* U.S. Geological Survey, Water Resources Division.

2. Direct runoff for the selected storm should range from 0.5 to 1.75 in.
3. A suitable number of storms should be analyzed to obtain an average of the ordinates for a selected unit hydrograph duration (approximately five events). Modifications may be made to adjust unit hydrograph durations by means of S-hydrographs or lagging procedures.
4. Direct runoff ordinates for each storm should be reduced so that each event represents 1 in. of direct runoff.
5. The final unit hydrograph of a specific duration for the watershed is obtained by averaging ordinates of selected events and adjusting the result to obtain 1 in. of direct runoff.

Constructing the unit hydrograph in this way produces the integrated effect of runoff resulting from a representative set of equal duration storms. Extreme rainfall intensity is not reflected in the determination. If intense storms are needed, a study of records should be made to ascertain their influence upon the discharge hydrograph by comparing peaks obtained utilizing the derived unit hydrograph and actual hydrographs from intense storms. If large discrepancies appear, a safety factor must be introduced.

Essential steps in developing a unit hydrograph for an isolated storm follow (see Table 4-4 and Figure 4-16b).

1. Analyze the streamflow hydrograph to permit separation of surface

***Table 4-4*** *Determination of a Unit Hydrograph from an Isolated Storm*

| 1 | 2 | 3 | 4 | 5 |
|---|---|---|---|---|
| *Time Unit* | *Total Runoff (cfs)* | *Base Flow (cfs)* | *Total Direct Runoff, (2) − (3) (cfs)* | *Unit Hydrograph Ordinate, (4) ÷ 1.4 (cfs)* |
| 1 | 110 | 110 | 0 | 0 |
| 2 | 122 | 122 | 0 | 0 |
| 3 | 230 | 120 | 110 | 78.7 |
| 4 | 578 | 118 | 460 | 328 |
| 4.7 | 666 | 116 | 550 | 393 |
| 5 | 645 | 115 | 530 | 379 |
| 6 | 434 | 114 | 320 | 229 |
| 7 | 293 | 113 | 180 | 129 |
| 8 | 202 | 112 | 90 | 64.2 |
| 9 | 160 | 110 | 50 | 35.7 |
| 10 | 117 | 105 | 12 | 8.6 |
| 10.5 | 105 | 105 | 0 | 0 |
| 11 | 90 | 90 | 0 | 0 |
| 12 | 80 | 80 | 0 | 0 |

runoff from groundwater flow, accomplished by the methods developed in Sec. 4-8.

2. Measure the total volume of surface runoff (direct runoff) from the storm producing the original hydrograph equal to the area under the hydrograph after groundwater base flow has been removed.
3. Divide the ordinates of direct runoff hydrograph by total direct runoff volume in inches and plot these results versus time as a unit graph for the basin.
4. Finally, the effective duration of the runoff-producing rain for this unit graph must be found from the hyetograph (time history of rainfall intensity) of the storm used.

Other procedures than those listed are required for complex storms or in developing synthetic unit graphs when few data are available. Unit hydrographs can also be transposed from one basin to another under certain circumstances. An example will illustrate the derivation and application of a unit hydrograph.

***Example 4-2*** Using the hydrograph given in Fig. 4-16b (p. 114), derive a unit hydrograph for the 3-mi² drainage area. Then from this unit hydrograph derive a hydrograph of direct runoff for the rainfall sequence given in Table 4-5.

**Table 4-5** *Unit Hydrograph Application*

| 1 | 2 | 3 | 4[a] | | | 5 |
|---|---|---|---|---|---|---|
| Time Unit Sequence | Rain Unit Number | Effective Rainfall (in.) | Hydrograph Ordinates for Rainfall Units 1–3 | | | Total Outflow Hydrograph Ordinates |
| | | | 1 | 2 | 3 | |
| 1 | 1 | 0.7 | 55.1 | — | — | 55.1 |
| 2 | 2 | 1.7 | 229 | 134 | — | 363 |
| 2.7 | 3 | 1.2 | 275 | — | — | — |
| 3 | — | — | 265 | 558 | 94.5 | 917.5 |
| 3.7 | — | — | — | 668 | — | — |
| 4 | — | — | 161 | 644 | 393 | 1218 |
| 4.7 | — | — | — | — | 472 | — |
| 5 | — | — | 90.5 | 389 | 455 | 934.5 |
| 6 | — | — | 44.9 | 219 | 275 | 538.9 |
| 7 | — | — | 25 | 109 | 155 | 289 |
| 8 | — | — | 6 | 60.7 | 77 | 143.7 |
| 9 | — | — | — | 14.6 | 42.8 | 57.4 |
| 10 | — | — | — | — | 10.3 | 10.3 |

[a] Values in column 4 obtained by multiplying the values in column 3 by unit hydrograph ordinates.

*Solution*

1. Separate the base or groundwater flow to get the total direct runoff hydrograph. A common method is to draw a straight line $AC$ which begins when the hydrograph starts an appreciable rise and ends where the recession curve intersects the base flow curve. The important point here is to be consistent in methodology from storm to storm.

2. Determine the duration of effective rainfall (rainfall that actually produces surface runoff). Effective rainfall volume must be equivalent to the volume of the direct surface runoff. Usually the unit time of effective rainfall will be 1 day, 1 hr, 12 hr, or some other interval appropriate for the size of drainage area studied. As stated previously, the unit storm duration should not exceed about 25% of the drainage area lag time. The effective portion of the rainstorm for this example is given in Fig. 4-16b along with its duration. The effective volume is 1.4 in.

The depth of direct runoff over the watershed is calculated using

$$\frac{\sum(DR \times \Delta t)}{\text{area}} \tag{4-21}$$

where $DR$ is the average height of the direct runoff ordinate during a chosen time period $\Delta t$ (in this case $\Delta t = 1.5$ hr). The values of $DR$ determined from Figure 4-16 are listed in Table 4-6.

**Table 4-6**

| Interval | $DR$ (cfs) | $\Delta t$ (hrs) | Direct Runoff (ft³) $DR \times \Delta t \times 3600$ |
|---|---|---|---|
| 2–3 | 60 | 1.5 | 324,000 |
| 3–4 | 290 | 1.5 | 1,566,000 |
| 4–5 | 550 | 1.5 | 2,970,000 |
| 5–6 | 435 | 1.5 | 2,349,000 |
| 6–7 | 255 | 1.5 | 1,377,000 |
| 7–8 | 120 | 1.5 | 648,000 |
| 8–9 | 70 | 1.5 | 378,000 |
| 9–10 | 20 | 1.5 | 108,000 |

$$\sum(DR \times \Delta t) = 9{,}720{,}000 \text{ ft}^3$$

$$\text{Total direct runoff} = \frac{9{,}720{,}000}{3 \times (5280)^2} = 1.4 \text{ in.}$$

3. Project the unit hydrograph base length to the abscissa to get the base flow separation line $AC$. Unit hydrograph theory assumes that for all storms of equal duration, regardless of intensity, the period of surface runoff is the same.

4. Tabulate the ordinates of direct runoff at peak rate of flow and at sufficient other positions to determine the hydrograph shape. Note that the direct runoff ordinate is the one above the base flow separation line.

5. Compute ordinates of the unit hydrograph by using

$$\frac{Q_s}{V_s} = \frac{Q_u}{1} \tag{4-22}$$

where

$Q_s$ = the magnitude of a hydrograph ordinate of direct runoff having a volume equal to $V_s$ (in.) at some instant of time after start of runoff
$Q_u$ = the ordinate of the unit hydrograph having a volume of 1 in. at same instant of time

In this example the values are obtained by dividing the direct runoff ordinates by 1.4. Table 4-4 outlines the computation of the unit hydrograph ordinates.

6. Using the values from Table 4-4, plot the unit hydrograph shown in Fig. 4-16b.

7. Utilizing the derived ordinates of the unit hydrograph, determine the ordinates of the hydrographs for each consecutive unit rainfall period from Table 4-5.

8. Find the synthesized hydrograph for unit storms 1–3 by plotting the three hydrographs and summing ordinates as indicated in Fig. 4-17.

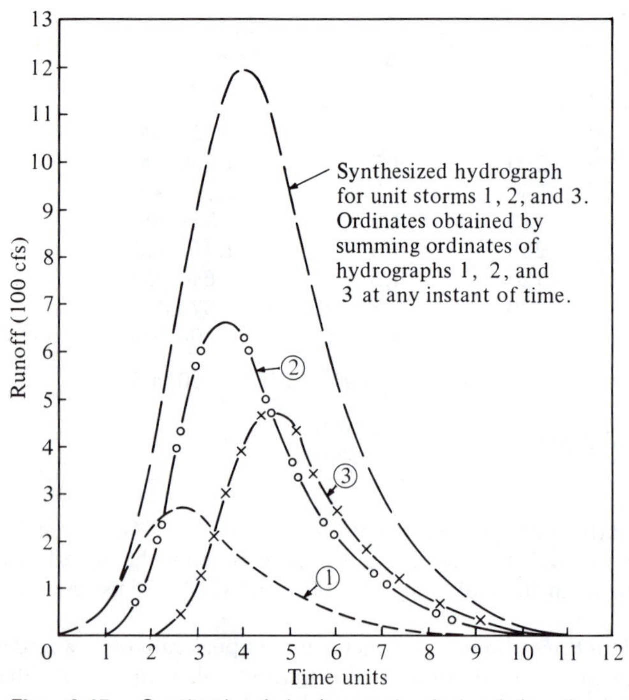


*Fig. 4-17. Synthesized hydrograph derived by the unit hydrograph method.*

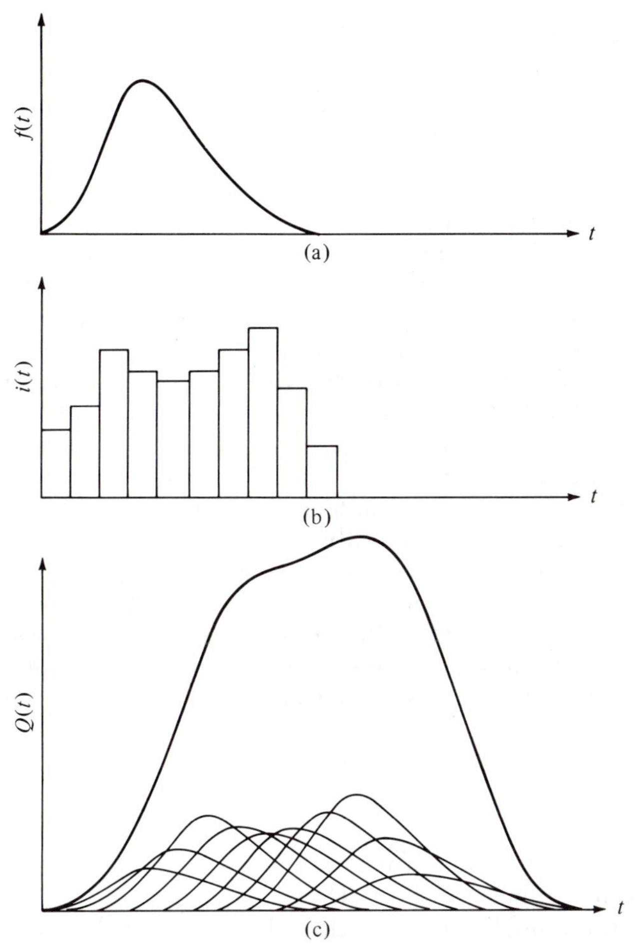


***Fig. 4-18.*** *Unit hydrograph description of the runoff process. (a) Unit hydrograph; (b) a sequence of 1-min storms; (c) superposition of runoff hydrographs for each of the 1-min storms. (After John C. Schaake, Jr., "Synthesis of the Inlet Hydrograph," Tech. Rept. No. 3, Department of Sanitary Engineering and Water Resources, Baltimore, Md., 1965.)*

## The Instantaneous Unit Hydrograph

The unit hydrograph method of estimating a runoff hydrograph can be used for storms of short duration. For example, if the duration of a storm is 1 min and a unit volume of surface runoff occurs, the resulting hydrograph is called the 1-min unit hydrograph. The hydrograph of runoff for any 1-min storm of constant intensity can be computed from the 1-min unit hydrograph by multiplying the ordinates of the 1-min unit hydrograph by the appropriate coefficient. A storm lasting for many minutes can be described as a sequence of 1-min storms (Fig. 4-18). The runoff hydrograph from each 1-min storm in this sequence can be obtained as in the preceding example. By superimposing the runoff hydrograph from each of the 1-min storms, the runoff hydrograph for the complete storm can be obtained.

From the unit hydrograph for any duration of uniform rain, the unit hydrograph for any other duration can be obtained. As the duration becomes shorter, the resulting unit hydrograph approaches an instantaneous unit hydrograph. The instantaneous unit hydrograph (IUH) is the hydrograph of runoff that would result if an inch of water were spread uniformly over an area and then allowed to run off.[44]

The IUH also has a significant meaning from a mathematical point of view. The ordinates of the IUH represent the relative effect of antecedent rainfall intensities on the runoff rate at any instant of time. By plotting the IUH with time increasing to the left rather than to the right (see Fig. 4-19), and then superimposing this plot over the rainfall hyetograph (plotted with time increasing to the right as in Fig. 4-19), the relative weight given to antecedent rainfall intensities (as a function of time into the past) is easily observed. In other words, the runoff rate at any time is computed as a weighted average of the previous rainfall intensities. Therefore, the computed runoff hydrograph is the weighted, moving average of the rainfall pattern and the weighting function is the time-reversed image of the unit hydrograph.[44]

Stated mathematically, the runoff rate at any time is given by

$$Q(t) = \int_0^t f(\tau) \quad i(t - \tau)\, d\tau$$

where

$Q(t)$ = the surface runoff rate at time $t$
$f(\tau)$ = the ordinate of the IUH at time $\tau$
$i(t - \tau)$ = the rainfall intensity (after abstraction of the appropriate infiltration losses, and so on) at time $t - \tau$

The variable $\tau$ represents time into the past so that time $t - \tau$ occurs before time $t$. The limits on the integral allow $\tau$ to vary between the present time (that is, $\tau = 0, t - \tau = 0$). The integral gives a continuous weighting of previous rainfall intensities by the ordinates of the IUH.

## 4-13 S-Hydrographs and Lagging Methods

Unit hydrographs developed according to procedures outlined in the previous section represent 1 in. of direct runoff resulting from a rainfall excess occurring in a specified period of time. The application to storms of larger or smaller durations will often require a different method to define additional unit hydrographs. The method of "lagging" is one possibility. It is based on the assumption that linear response of the watershed is not influenced by previous storms—that

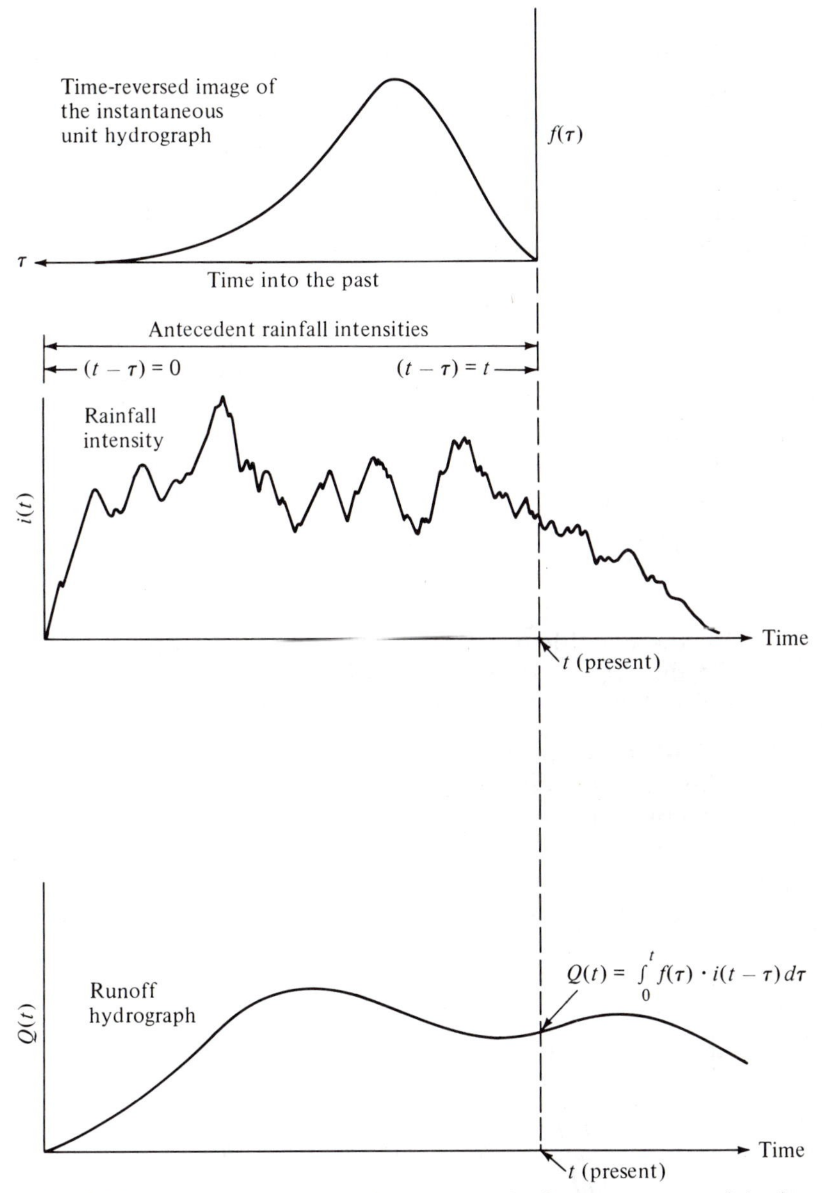


***Fig. 4-19.*** *Calculation of runoff rates with the instantaneous unit hydrograph. The runoff rate at any time is a weighted average of the antecedent rainfall intensities. The time-reversed image of the instantaneous unit hydrograph represents the weighting function. (After John C. Schaake, Jr., "Synthesis of the Inlet Hydrograph," Tech. Rept. No. 3, Department of Sanitary Engineering and Water Resources, Baltimore, Md., 1965.)*

is, one can superimpose hydrographs offset in time and the flows are directly additive. For example, if a 1-hr unit hydrograph is available for a given watershed, a unit hydrograph resulting from a 2-hr unit storm is obtained by plotting two 1-hr unit hydrographs, with the second unit hydrograph lagged 1 hr, adding ordinates, and dividing by 2. This is demonstrated in Fig. 4-20 where the dashed line represents the resulting 2-hr unit hydrograph. Thus the 1 in. of rainfall contained in the original 1-hr duration has been distributed over a 2-hr period.

Modifications of the original unit hydrograph duration can be made so that two 1-hr unit hydrographs are used to form a 2-hr unit hydrograph; two 2-hr unit hydrographs result in a 4-hr diagram, and so on. Care must be taken not to mix durations in the lagging procedure, since errors are introduced; hence a 1-hr and a 2-hr unit hydrograph do not represent a 3-hr unit hydrograph. Lagging procedure is therefore restricted to multiples of the original duration according to the expression

$$D^1 = nD \tag{4-23}$$

where

$D^1$ = the possible durations of the unit hydrograph
$D$ = the original duration of one unit hydrograph
$n$ = 1, 2, 3, . . .

The S-hydrograph method overcomes restrictions imposed by the lagging plan and allows construction of any duration-unit hydrograph. By observing the lagging system just described, it is apparent that for a 1-hr unit hydrograph, the 1-in. rainfall excess has an intensity

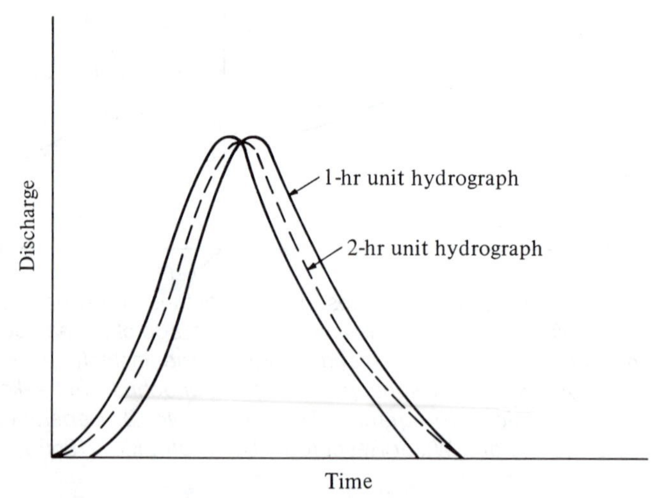


**Fig. 4-20.** *Unit hydrograph lagging procedure.*

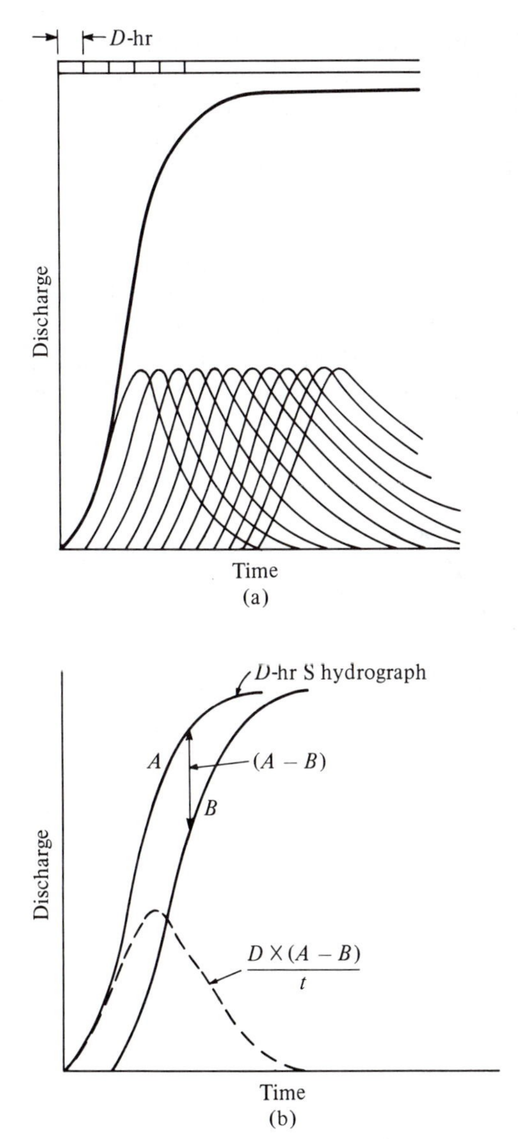


***Fig. 4-21.*** *S-hydrograph method.*

of 1 in./hr, whereas the 2-hr unit hydrograph is produced by a rainfall intensity of 0.5 in./hr. Continuous lagging of either one of these unit hydrographs is comparable to a continuously applied rainfall at either 0.5 in./hr or 1 in./hr intensity, depending upon which unit hydrograph is chosen.

As an example, using the 1-hr unit hydrograph, continuous lagging represents the direct runoff from a constant rainfall of 1 in./hr as shown in Fig. 4-21a. The cumulative addition of the initial unit hydrograph ordinates at time intervals equal to the unit storm duration results in an S-hydrograph (see Fig. 4-22). Graphically, construction of an S-hydrograph is readily accomplished with a pair of dividers. The maximum discharge of the S-hydrograph occurs at a time equal to $D$ hours less than the time base of the initial unit hydrograph as shown in Fig. 4-21a.

To pictorially construct a 2-hr unit hydrograph, simply lag the first S-hydrograph by a second S-hydrograph a time interval equal to the desired duration. The difference in S-hydrograph ordinates must then be divided by 2. Any duration-$t$ unit hydrograph may be obtained in the same manner once another duration $D$ unit hydrograph is known. Simply form a $D$-hr S-hydrograph; lag this S-hydrograph $t$ hr

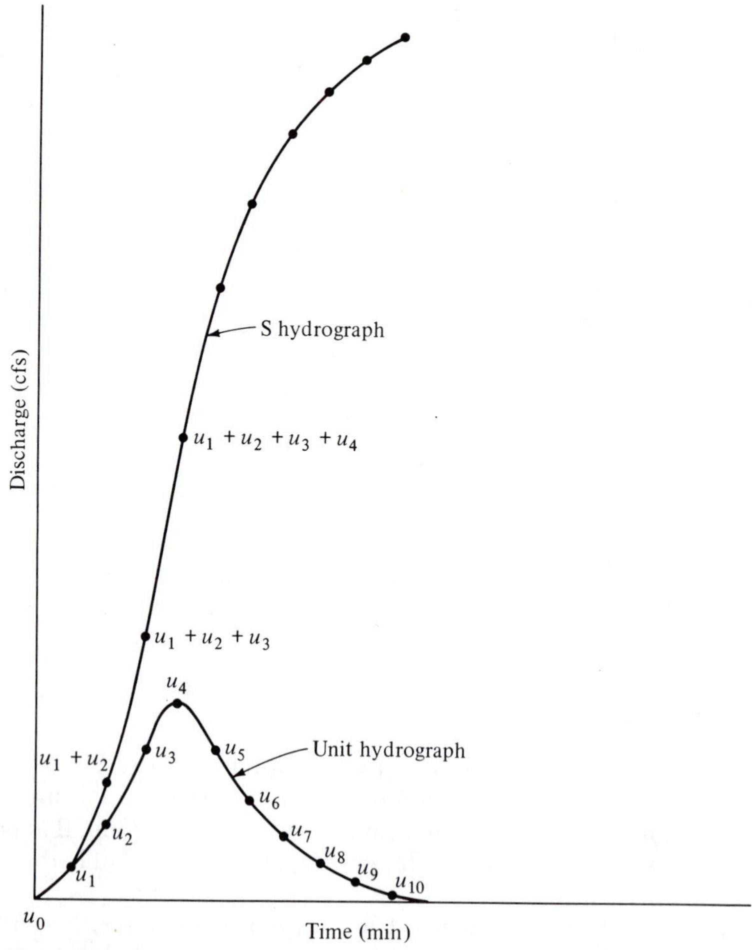


**Fig. 4-22.** *S-hydrograph.*

and multiply the difference in S-hydrograph ordinates by $D/t$. Accuracy of the graphical procedure depends upon the scales chosen to plot the hydrographs. Tabular solution of the S-hydrograph method is also employed, but hydrograph tabulations must be at intervals of the original unit hydrograph duration.

## 4-14 Matrix Methods of Unit Hydrographs*

Methods used to construct design storm hydrographs from known unit hydrographs require multiplying the design rainfall excess of each time period by the proper unit hydrograph ordinate. Lagging the hydrographs for each time period and adding the ordinates permits construction of the required storm hydrograph. A design storm hydrograph does not include the base flow at this stage of development. The base flow may be added to procure a more realistic design storm by studying base flow behavior and applying a representative value or values of discharge throughout the storm period.

The process of getting a design storm hydrograph with base flow excluded is graphically demonstrated for a three-period storm in Fig. 4-23. Equations representing the design storm hydrograph of Fig. 4-23c are given as Eqs. 4-24.

$$\begin{aligned}
p_1u_1 &= q_1 \\
p_2u_1 + p_1u_2 &= q_2 \\
p_3u_1 + p_2u_2 + p_1u_3 &= q_3 \\
p_3u_2 + p_2u_3 + p_1u_4 &= q_4 \\
p_3u_3 + p_2u_4 &= q_5 \\
p_3u_4 &= q_6
\end{aligned} \tag{4-24}$$

The relationship between the original unit hydrograph, the design rainfall, and the storm hydrograph time periods follows the form

$$n = j + i - 1 \tag{4-25}$$

where

$n$ = the number of storm hydrograph ordinates
$j$ = the number of unit hydrograph ordinates
$i$ = the number of periods of rainfall excess

In matrix notation Eqs. 4-24 can be written more concisely as

$$PU = Q \tag{4-26}$$

*See Appendix A for a brief review of matrix algebra.

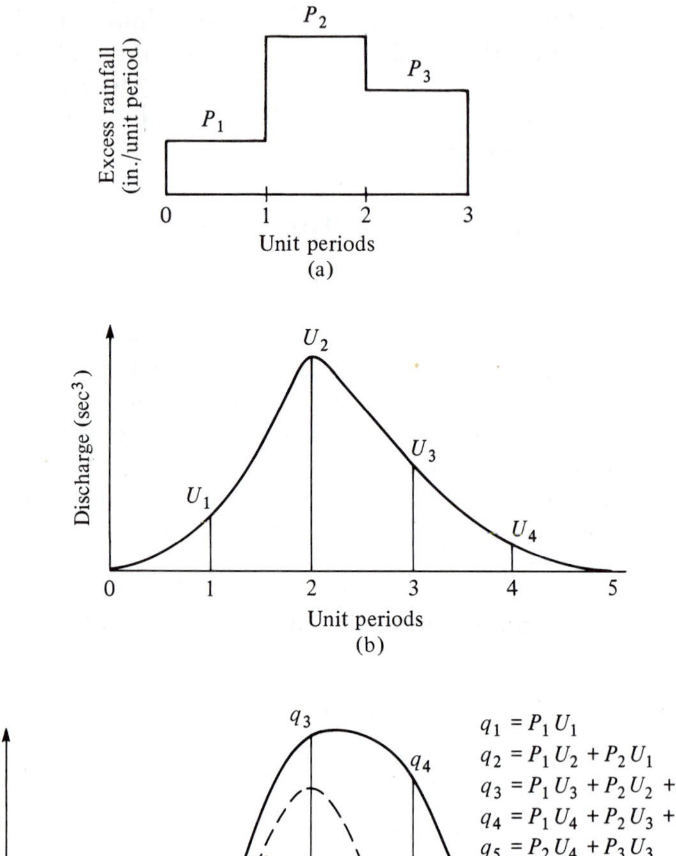


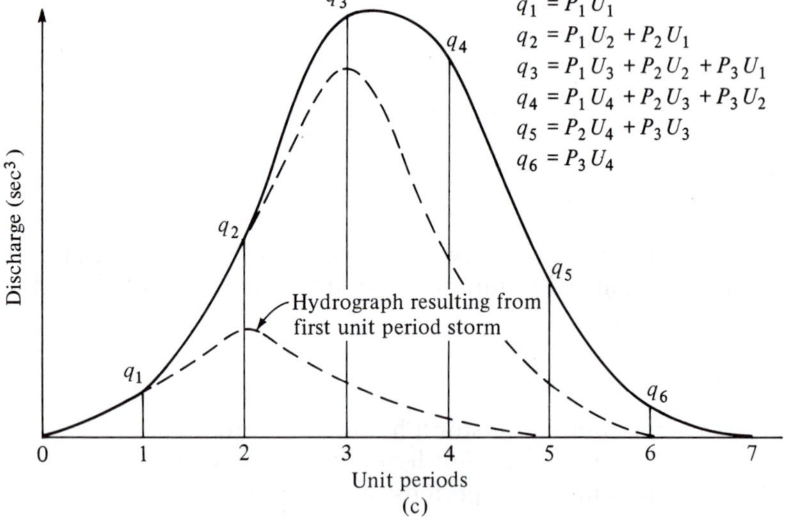


***Fig. 4-23.*** *Determination of a design storm hydrograph: (a) excess rainfall, (b) unit graph, and (c) surface runoff hydrograph.*

where

$$P = \begin{bmatrix} p_1 & 0 & 0 & 0 & 0 & 0 \\ p_2 & p_1 & 0 & 0 & 0 & 0 \\ p_3 & p_2 & p_1 & 0 & 0 & 0 \\ 0 & p_3 & p_2 & p_1 & 0 & 0 \\ 0 & 0 & p_3 & p_2 & 0 & 0 \\ 0 & 0 & 0 & p_3 & 0 & 0 \end{bmatrix}$$

$$U = \begin{bmatrix} u_1 \\ u_2 \\ u_3 \\ u_4 \\ 0 \\ 0 \end{bmatrix}$$

$$Q = \begin{bmatrix} q_1 \\ q_2 \\ q_3 \\ q_4 \\ q_5 \\ q_6 \end{bmatrix}$$

The reverse of the process shown in Fig. 4-23 is the basis of the matrix method. Thus having an existing storm hydrograph $Q$ less base flow, with known rainfall excess $P$, allows computation of unit hydrograph $U$ subject to restrictions imposed by matrix algebra. A solution of Eq. 4-26 can be of the form

$$P^{-1}PU = P^{-1}Q \qquad \textbf{(4-27)}$$

or

$$U = P^{-1}Q \qquad \textbf{(4-28)}$$

The answer is valid if, and only if, an inverse can be found for the precipitation matrix, but an inverse exists only if it is a square matrix with a nonvanishing determinant. As can be seen from Eq. 4-26, the precipitation matrix does not fulfill this criteria. However, by use of the transpose of the precipitation matrix, a square matrix is formed as follows

$$P^T PU = P^T Q \tag{4-29}$$

and the solution for the unit hydrograph matrix becomes

$$U = (P^T P)^{-1} P^T Q \tag{4-30}$$

Resolving Eq. 4-30 produces a trial unit hydrograph with duration equal to the selected time interval of precipitation elements. Reproduction of the initial storm hydrograph based upon this unit hydrograph, however, will generally not be exact. To minimize the error, adjustments can be made to unit hydrograph ordinates to obtain a "best" solution. This process is usually performed by reducing the square of the error or difference between the original and the derived storm hydrograph ordinates. Further refinements in this method are presented by Newton and Vineyard.[29] Generally, solution of the unit hydrograph by matrix methods is performed on a digital computer. Most computer installations have packaged programs for work with large systems and for determining the inverse and transpose of matrices.

## 4-15 Synthetic Methods of Unit Hydrographs

As previously noted, the linear characteristics exhibited by unit hydrographs for a watershed are a distinct advantage in constructing more complex storm discharge hydrographs. Generally, however, basic streamflow and rainfall data are not available to allow construction of a unit hydrograph except for relatively few watersheds; therefore, techniques have evolved that allow generation of "synthetic unit hydrographs."

## 4-16 Snyder's Method

One technique employed by the Corps of Engineers[30] is based on methods developed by Snyder[24] and expanded by Taylor and Schwarz.[25] Unfortunately, it does not provide a simple method of constructing the entire time distribution of the discharge hydrograph but allows computation of lag time, unit hydrograph duration, peak discharge, and hydrograph time widths at 50 and 75% of peak flow. Using these points, a sketch of the unit hydrograph is obtained and checked by Eq. 3-32 to see if it contains 1 in. of direct runoff (see Fig. 4-24).

### Time To Peak

Snyder's method of synthetic unit hydrographs relies upon correlation of the dependent variables of lag time and peak discharge with

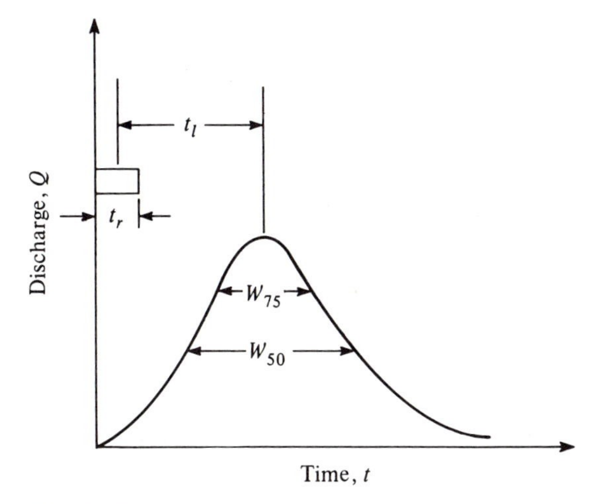


***Fig. 4-24.*** *Snyder's synthetic unit hydrograph.*

various physiographic watershed characteristics. A lag-time relationship derived by Snyder for watersheds from 10 to 10,000 mi² located in the Appalachian Highlands is

$$t_1 = C_t(LL_{ca})^{0.3} \tag{4-31}$$

where

$t_1$ = the lag time (hr). See Fig. 4-24 for definition.
$C_t$ = the coefficient representing variations of watershed slopes and storage
$L$ = length of the main stream channel (mi) from the outlet to the divide
$L_{ca}$ = length along the main channel to a point nearest the watershed centroid (mi)

It is assumed that lag time is a constant for a particular watershed —that is, uninfluenced by variations in rainfall intensities or similar factors. The use of $L_{ca}$ accounts for the watershed shape, and $C_t$ takes care of wide variations in topography, from plains to mountainous regions. Values of $C_t$ for Snyder's study ranged from 1.8 to 2.2 with an average of about 2.0. The coefficient $C_t$ accounts for variations in slopes and storage and did not vary greatly for study areas. Steeper slopes tend to generate lower values of $C_t$, with extremes of 0.4 noted in Southern California and 8.0 along the Gulf of Mexico. When snow pack accumulations influence peak discharge, values of $C_t$ will be between one-sixth to one-third of Snyder's values.

### Duration

The duration of rainfall excess for Snyder's synthetic unit hydrograph development is a function of lag time as shown by Eq. 4-32.

$$t_r = \frac{t_1}{5.5} \qquad \textbf{(4-32)}$$

where

$t_r$ = duration of the unit rainfall excess (hr)
$t_1$ = the lag time from the centroid of unit rainfall excess to the peak of the unit hydrograph

This synthetic technique always results in an initial unit hydrograph duration equal to $t_1/5.5$. However, since changes in lag time do occur with changes in duration of the unit hydrograph, the following equation was developed to allow lag time and peak discharge adjustments for other unit hydrograph durations.

$$t_{1R} = t_1 + 0.25\,(t_R - t_r) \qquad \textbf{(4-33)}$$

where

$t_{1R}$ = the adjusted lag time (hr)
$t_1$ = the original lag time (hr)
$t_R$ = the desired unit hydrograph duration (hr)
$t_r$ = the original unit hydrograph duration = $t_1/5.5$ (hr)

### Peak Discharge

If one assumes a given duration rainfall produces 1 in. of direct runoff, the outflow volume is some relatively constant percentage of inflow volume, and a simplified approximation of outflow volume is $t_1 \times Q_P$, then the equation for peak discharge can be written

$$Q_P = \frac{640 C_P A}{t_1} \qquad \textbf{(4-34)}$$

where

$Q_P$ = peak discharge (cfs)
$C_P$ = the coefficient accounting for flood wave and storage conditions. It is a function of lag time, duration of runoff producing rain, effective area contributing to peak flow and drainage area
$A$ = watershed size ($mi^2$)
$t_1$ = the lag time (hr)

Thus peak discharge can be calculated given lag time and coefficient of peak discharge $C_P$. Values for $C_P$ range from 0.4 to 0.8 and generally indicate retention or storage capacity of the watershed. Larger values of $C_P$ are generally associated with smaller values of $C_t$.

### Time Base

The time base of a synthetic unit hydrograph by Snyder's method is:

$$T = 3 + \frac{t_1}{8} \tag{4-35}$$

where

$T$ = the base time of the synthetic unit hydrograph (days)
$t_1$ = the lag time (hr)

Equation 4-35 gives reasonable estimates for large watersheds but will produce excessively large values for smaller areas. A general rule of thumb for small areas is to use three to five times the time to peak as a base value when sketching a unit hydrograph.

### Hydrograph Construction

From Eqs. 3-32, 4-31, 4-32, 4-34, and 4-35 plot critical points for the unit hydrograph and sketch a synthetic unit hydrograph, remembering that total direct runoff amounts to 1 in. An analysis by the Corps of Engineers (see Fig. 4-25) gives additional assistance in plotting time widths for points on the hydrograph located at 50 and 75% of

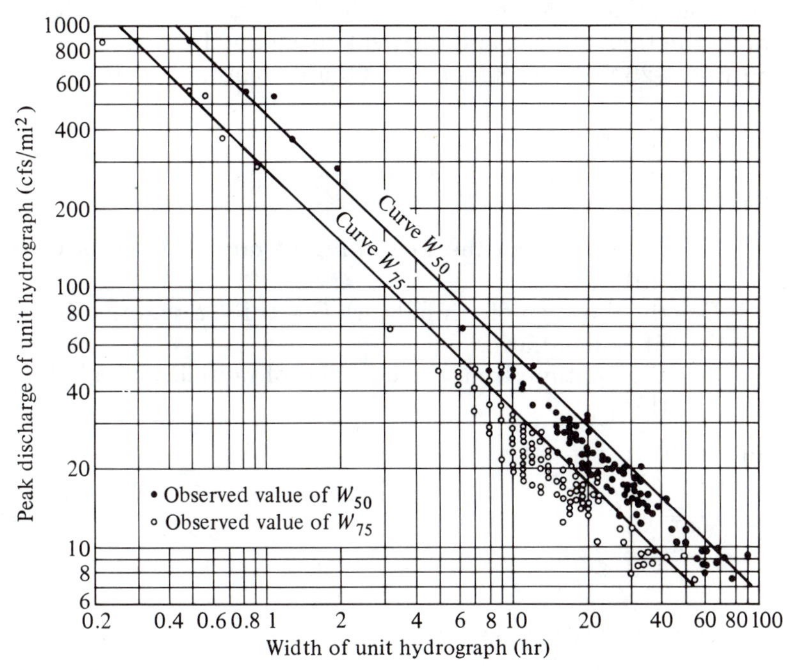


**Fig. 4-25.** *Unit hydrograph width at 50 and 75% of peak flow. Key: •, observed value of $W_{50}$; ∘, observed value of $W_{75}$.*

peak discharge. As a general rule of thumb, the time width at $W_{50}$ and $W_{75}$ ordinates should be proportioned each side of the peak in a ratio of 1:2 with the short time side on the left of the synthetic unit hydrograph peak. As noted earlier, for smaller watersheds, Eq. 4-35 gives unrealistic values for the base time. If this occurs, a value can be estimated by multiplying total time to the peak by a value of from 3 to 5. This ratio can be modified based upon the amount and time rate of depletion of storage water within the watershed boundaries.

The application of Snyder's synthetic unit hydrograph method to areas other than the original study area should be preceded by a reevaluation of coefficients $C_t$ and $C_P$ in Eqs. 4-31 and 4-34. This analysis can be accomplished by the use of unit hydrographs in the region under study which have the proper lag-time-rainfall-duration ratio, that is, $t_r = t_1/5.5$. If another rainfall duration is selected, variations of $C_t$ and $C_P$ can be expected.

## 4-17 SCS Method

A method developed by the Soil Conservation Service for constructing synthetic unit hydrographs is based on a dimensionless hydrograph (Fig. 4-26). This dimensionless graph is the result of an analysis of a large number of natural unit hydrographs from a wide range in size and geographic locations. The method requires only the determination of the time to peak and the peak discharge by Eqs. 4-36 and 4-37 and Fig. 4-26.[40] Parameters $t_p$ and $Q_P$ are computed as follows:

$$t_P = \frac{D}{2} + t_1 \tag{4-36}$$

where

$t_p$ = the time from the beginning of rainfall to peak discharge (hr)
$D$ = the duration of rainfall (hr). $D = 0.133\, t_c$ where $t_c$ is the time of concentration.
$t_1$ = the lag time from the centroid of rainfall to peak discharge (hr)

and

$$Q_P = \frac{484A}{t_p} \tag{4-37}$$

where

$Q_P$ = peak discharge (cfs)
$A$ = drainage area ($mi^2$)
$t_p$ = the time to peak (hr)

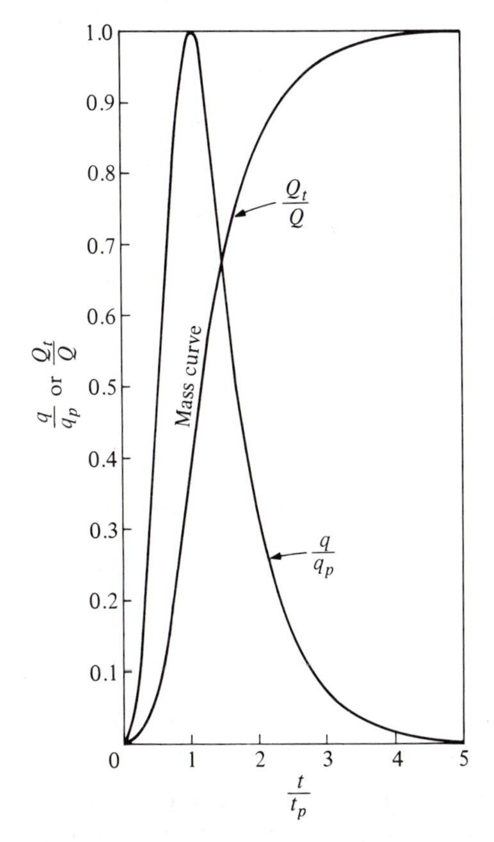


***Fig. 4-26.*** *Dimensionless unit hydrograph and mass curve. (After V. Mockus, "Use of Storm and Watershed Characteristics in Synthetic Hydrograph Analysis and Application," U.S. Soil Conservation Service, 1957.)*

A relationship of $t_1$ to size of watershed can be used to estimate lag time. Relationships from two geographic regions are

$$t_1 = 1.44A^{0.6} \quad \text{Texas} \tag{4-38a}$$

$$t_1 = 0.54A^{0.6} \quad \text{Ohio} \tag{4-38b}$$

By finding a value of $t_1$, a synthetic unit hydrograph of chosen duration $D$ is obtained from Fig. 4-26.

Another equation used by the SCS is

$$t_l = \frac{l^{0.8}(S+1)^{0.7}}{1900\,Y^{0.5}} \tag{4-39}$$

where

$t_l$ = the lag time (hr)
$l$ = length to divide in feet
$Y$ = average watershed slope in percent

$S$ = the potential maximum retention (in.) = $(1000/CN) - 10$, where $CN$ is a curve number described in Chapter 12

***Example 4-3*** For a drainage area of 70 mi$^2$ having a lag time of $7\frac{1}{2}$ hr, derive a unit hydrograph of duration 4 hr. Use the SCS dimensionless unit hydrograph.

1. Using Eq. 4-36, we obtain

$$t_P = \tfrac{4}{2} + 7\tfrac{1}{2} = 9\tfrac{1}{2} \text{ hr}$$

2. From Eq. 4-37

$$Qp = \frac{484 \times 70}{9.5}$$

$$Qp = 3560 \text{ cfs occurring at } t = 9\tfrac{1}{2} \text{ hr}$$

3. Using Fig. 4-26, we find that
   a. The peak flow occurs at $t/t_p = 1$ or at $t = 9\frac{1}{2}$ hr.
   b. The time base of the hydrograph = $5\,t_p$ or 47.5 hr.
   c. The hydrograph ordinates are
      (1) At $t/t_p = 0.5$, $q/q_p = 0.45$; thus at $t = 4.75$ hr, $q = 1600$ cfs.
      (2) At $t/t_p = 2$, $q/q_p = 0.28$; thus at $t = 19$ hr, $q = 995$ cfs.
      (3) At $t/t_p = 3$, $q/q_p = 0.07$; thus at $t = 28.5$ hr, $q = 249$ cfs.

## 4-18 Gray's Method

A method of generating synthetic unit hydrographs for midwestern watersheds has been developed by Gray.[31] An approximate upper limit of watershed size for application of this method to the geographic areas of Central Iowa, Missouri, Illinois, and Wisconsin is 94 mi$^2$. The method is based on dimensionalizing the incomplete gamma distribution and results in a dimensionless graph of the form

$$Q_{t/P_R} = \frac{25.0(\gamma')^q}{\Gamma(q)}\left(e^{-\gamma' t/P_R}\right)\left(\frac{t}{P_R}\right)^{q-1} \tag{4-40}$$

where

$Q_{t/P_R}$ = % flow/0.25 $P_R$ at any given $t/P_R$ value
The factors $q$ and $\gamma$ = shape and scale parameters, respectively
$\Gamma$ = the gamma function of $q$ which is equal to $(q-1)!$*
$e$ = the base of the natural logarithms
$P_R$ = the period of rise (min)
$t$ = time (min)

The relationship for $\gamma'$ is defined as $\gamma' = \gamma P_R$ and $q = 1 + \gamma'$.

* If $q$ is not an integer $\Gamma(q) = \Gamma(N + Z) = (N - 1 + Z)(N - 2 + Z) \ldots (1 + Z)\,\Gamma(1 + Z)$, where $N$ equals the integer.

This form of the dimensionless unit hydrograph (Fig. 4-27) allows computation of the discharge ordinates for the unit hydrograph at times equal to $\frac{1}{4}$ intervals of the period of rise $P_R$, that is, the time from the beginning of rainfall to the time of peak discharge of the unit hydrograph.

Correlations with physiograph characteristics of the watershed can be developed to get the values of both $P_R$ and $\gamma'$.

As an example, the storage factor $P_R/\gamma'$ has been linked with watershed parameter $L/\sqrt{S_c}$, where $L$ is the length of the main channel of the watershed in miles measured from the outlet to the uppermost part of the watershed (Fig. 4-28); $S_c$ is defined as an average slope in percent obtained by plotting the main channel profile and drawing a straight line through the outlet elevation such that the positive and negative areas between the stream profile and the straight line are equal. The storage factor $P_R/\gamma'$ can also be correlated with the period of rise $P_R$ as shown in Fig. 4-29. These two correlations allow solution of Eq. 4-40 and produce a synthetic unit hydrograph of duration $P_R/4$ for an ungauged area.

The solution proceeds as follows:

1. Determine $L$, $S_c$, and $A$ for the ungauged watershed.
2. Determine parameters $P_R$, $\gamma'$, and $q$.
   a. With $L/\sqrt{S_c}$, use Fig. 4-28 to select $P_R/\gamma'$.
   b. With $P_R/\gamma'$, use Fig. 4-29 to obtain $P_R$. Compute $\gamma'$ as the ratio $P_R/P_R/\gamma'$.

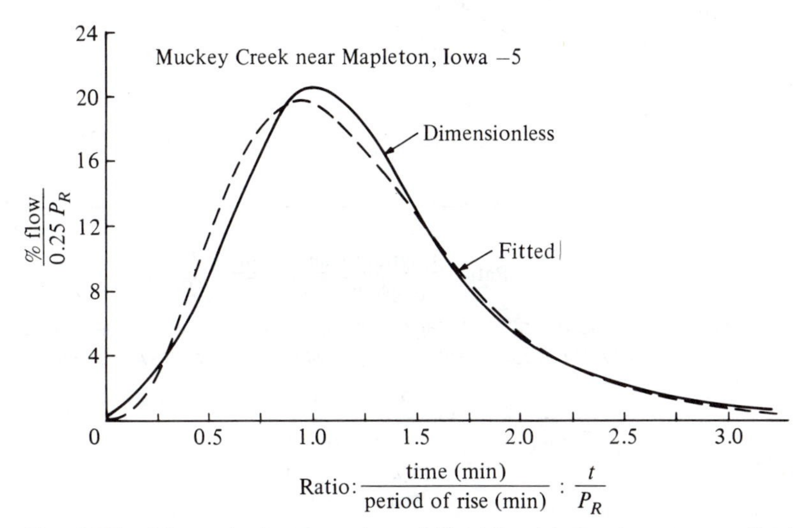


**Fig. 4-27.** *Dimensionless graph and fitted two-parameter gamma distribution for watershed 5. (After D. M. Gray, "Synthetic Unit Hydrographs for Small Drainage Areas,"* Proc. ASCE, J. Hyd. Div., *87, No. HY4 (July 1961): 41.)*

c. Substitute $\gamma'$ obtained in step 2b into the equation, $q = 1 + \gamma'$, and solve for $q$.
3. Compute the ordinates for the dimensionless graph using Eq. 4-40. Compute the % flow/0.25$P_R$ for values of $t/P_R = 0.125, 0.375, 0.625$ . . . and every succeeding increment of $t/P_R = 0.250$ until the sum of the percentage flows approximates 100% (column 4, Table 4-7). Also compute the peak percentage by substituting $t/P_R = 1$.
4. Compute the unit hydrograph.
   a. Compute the necessary conversion factor to convert the volume of the direct runoff under the dimensionless graph to 1 in. of precipitation excess over the entire watershed.
   (1) The volume of the unit hydrograph = V

$$V = 1 \text{ in.} \times A \text{ mi}^2 \times 640 \frac{\text{acre}}{\text{mi}^2} \times \frac{1}{12 \text{ in./ft}} \times 43{,}560 \frac{\text{ft}^2}{\text{acre}}$$

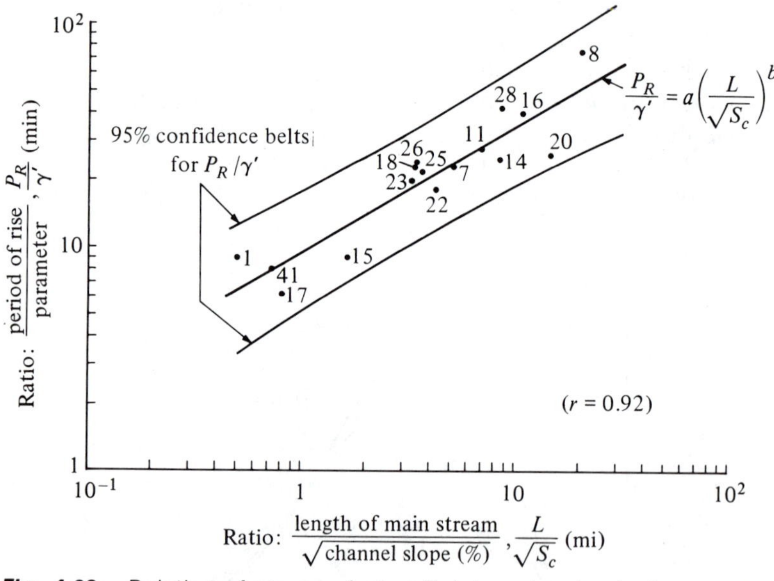


***Fig. 4-28.*** *Relation of storage factor, $P_R/\gamma'$, and watershed parameter, $L\sqrt{S_c}$, for watersheds in central Iowa-Missouri-Illinois-Wisconsin.*

| *Area* | a | b |
|---|---|---|
| Ill., Iowa, Mo., Wisc., | 9.27 | 0.562 |
| Ohio | 11.4 | 0.531 |
| Neb. and W. Iowa | 7.4 | 0.498 |

*(After D. M. Gray, "Synthetic Unit Hydrographs for Small Drainage Areas,"* Proc. ASCE, J. Hyd. Div., *87, No. HY4 (July 1961): 45.)*

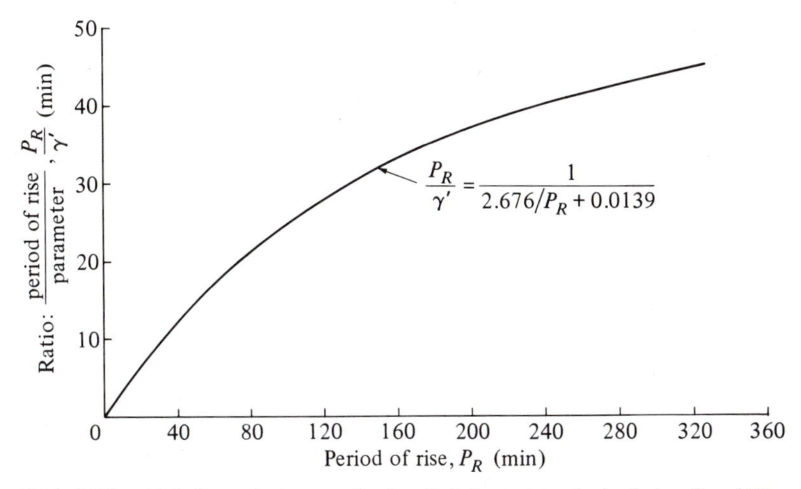


***Fig. 4-29.*** *Relation of storage factor $P_R/\gamma'$, and period of rise $P_R$. (After D. M. Gray, "Synthetic Unit Hydrographs for Small Drainage Areas,"* Proc. ASCE, J. Hyd. Div., *87, No. HY4 (July 1961): 48.)*

(2) The volume of the dimensionless graph = $V_D$

$$V_D = \sum \text{cfs} \times 0.25 \times P_R \times 60 \frac{\text{sec}}{\text{min}}$$

(3) Solve for $\sum$ cfs by equating $V$ and $V_D$, since they must be equal.

b. Convert the dimensionless graph ordinates to the unit hydrograph ordinates

$$Q_u = \frac{\%\ \text{flow}/0.25\, P_R}{100} \sum \text{cfs}$$

c. Translate time base of dimensionless graph to absolute time units by multiplying $t/P_R \times P_R$ for each computed point. Remember that runoff does not commence until the centroid of rainfall, or at a time $P_R/8$.

An example problem demonstrates the solution of Gray's method for a Missouri watershed. A plot representing the derived hydrograph is shown in Fig. 4-30. Note again that direct runoff commences at the centroid of rainfall. Thus it is necessary to add $D/2$ or $P_R/8$ to column 2, Table 4-7, to obtain the proper times of unit hydrograph ordinates.

***Example 4-4*** For the given data, use Gray's method to construct a unit hydrograph.

Green Acre Branch, drainage area = 0.62 mi², length = 0.98 mi, $S_c$ = 1.45%.

*Procedure*

1. a. Fig. 4-28; $L/\sqrt{S} = 0.813$ mi; $P_R/\gamma' = 8.25$ min.
   b. Fig. 4-29; $P_R/\gamma' = 8.25$ min; $P_R = 24.9$ min.
   c. $q = 1 + \gamma' = 4.02$; $\gamma' = 3.02$.
2. a.

**Table 4-7** *Tabulation for Example 4-2*

| (1) $t/P_R$ | (2) Time (min) | (3) % flow/ $0.25\ P_R$ | (4) Cumulated Flow | (5) UHG (cfs) |
|---|---|---|---|---|
| 0.000 | 0 | 0 | 0 | 0 |
| 0.125 | 3.11 | 0.488 | 0.448 | 17.3 |
| 0.375 | 9.33 | 5.800 | 6.248 | 224 |
| 0.625 | 15.55 | 12.700 | 18.948 | 490 |
| 0.875 | 21.77 | 16.350 | 35.298 | 631 |
| 1.000 | 24.90 | 16.850 | — | 651 |
| 1.125 | 27.99 | 16.250 | 51.548 | 628 |
| 1.375 | 34.21 | 14.200 | 65.748 | 548 |
| 1.625 | 40.43 | 11.100 | 76.848 | 428 |
| 1.875 | 46.65 | 7.970 | 84.818 | 308 |
| 2.125 | 52.87 | 5.550 | 90.368 | 214 |
| 2.375 | 59.29 | 3.560 | 93.928 | 138 |
| 2.625 | 65.51 | 2.28 | 96.208 | 88.0 |
| 2.875 | 71.73 | 1.410 | 97.618 | 54.4 |
| 3.125 | 77.95 | 0.864 | 98.482 | 33.3 |
| 3.375 | 84.17 | 0.500 | 99.982 | 19.3 |

3. a. (1) $V = 1 \times 0.62 \times 640 \times 43{,}560/12 = 14.4 \times 10^5$ ft³.
   (2) $V_D = 0.25 \times 24.9 \times 60 \times \sum$ cfs $= 373.5 \sum$ cfs.
   (3) $\sum$ cfs $= 3860$.
   b. Column 5 is tabulated by multiplying 3860 times values in column 3 divided by 100.
   c. Column 2 is secured by multiplying 24.9 times values in column 1.
   d. Column 3 comes from solution of Eq. 4-40.

A sample Fortran IV program is shown in Fig. 4-31 demonstrating the solution of Gray's method for a drainage area of 14.22 mi². Results of the physiographic correlations for this area produced values of $P_R = 128$ min and $\gamma' = 4.4139$, which results in $q = 5.4139$, and $\Gamma(q) = 47$. These values are used as inputs to the computer program although routines could be developed to allow internal computation of all values by simply inputting physiographic parameters of the watershed.

## 4-19 Simulation Methods in Hydrology

Hydrologic problems are often complex and/or unique. In such cases, physical models can sometimes be built and operated to determine

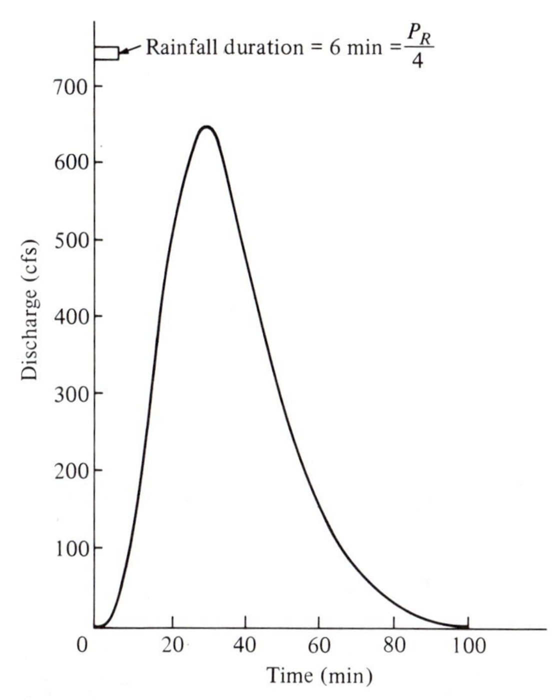


***Fig. 4-30.*** *Derived hydrograph of Example 4-4 using Gray's method. Note that Gray's method results in a unit hydrograph for a $P_R$/4-hr storm.*

characteristics of the system.[32] Such models are occasionally effective but are usually costly and lack versatility.

All systems are analyzed with the aid of mathematical methods. These fall into two categories: (1) the analytic or direct approach, and (2) the trial-and-error or nonanalytic approach. Analytic solutions are the most desirable, since they yield the wanted result directly. Unfortunately, equations describing hydrologic systems are often so complex and numerous that direct solutions are impossible and nonanalytic or trial-and-error solutions must be resorted to. The principal advantage of the trial-and-error method is that relatively simple equations are usually involved. However, a separate trial is required for each combination of system parameters.

Crawford and Linsley note that simulation is an indirect approach to the study of the behavior or response of a system.[33] Systems may be simulated by physical models, analog models, or digital models. Digital simulation is used to analyze large and complex systems.[33,34] It has the important advantage of high speed and does not require the extensive "hardware" often needed for physical or analog models. The digital program becomes the model and its parameters can be changed to allow for experiment or to represent any particular

```
FORTRAN IV G LEVEL  21                    MAIN

0001              TPR = 0.125
0002              QT = 0.
0003              T = 16.
0004              GAMMA = 47.
0005              V = 1.*14.22*640.*43560./12.
0006              VD = .25*128.*60.
0007              CFS = V/VD
0008              WRITE(6,11)
0009           11 FORMAT (8X,'TPR',8X,'ACCUM TIME',8X,'Q',8X,'QT',8X,'UNIT GRAPH')
0010              A = 4.41379
0011              B = 5.41379
0012              DO 10 I = 1,50
0013              Q = ((25.0*(A**B)*(TPR**(B-1.)))/(GAMMA*2.71828**(A*TPR)))
0014              QT = QT+Q
0015              IF (100.-QT) 20,20,21
0016           21 UH = CFS*Q/100.
0017              WRITE(6,25) TPR,T,Q,QT,UH
0018           25 FORMAT(6X,F6.3,9X,F6.1,8X,F6.3,6X,F5.1,7X,F8.0)
0019              TPR = TPR + 0.250
0020           10 T = T + 32.
0021           20 CONTINUE
0022              U = ((25.0*(A**B))/(GAMMA*2.71828**A))
0023              WRITE(6,6) U
0024              U = U*CFS/100.
0025              WRITE(6,7) U
0026            6 FORMAT(6X,'U = ',F8.1)
0027            7 FORMAT(6X,'U = ',F10.1)
0028              STOP
0029              END
```

| TPR | ACCUM TIME | Q | QT | UNIT GRAPH |
|---|---|---|---|---|
| 0.125 | 16.0 | 0.098 | 0.1 | 17. |
| 0.375 | 48.0 | 4.147 | 4.2 | 714. |
| 0.625 | 80.0 | 13.114 | 17.4 | 2256. |
| 0.875 | 112.0 | 19.208 | 36.6 | 3305. |
| 1.125 | 144.0 | 19.320 | 55.9 | 3324. |
| 1.375 | 176.0 | 15.540 | 71.4 | 2674. |
| 1.625 | 208.0 | 10.776 | 82.2 | 1854. |
| 1.875 | 240.0 | 6.722 | 88.9 | 1157. |
| 2.125 | 272.0 | 3.875 | 92.8 | 667. |
| 2.375 | 304.0 | 2.100 | 94.9 | 361. |

```
 2.625      336.0     1.084     96.0      186.
 2.875      368.0     0.537     96.5       92.
 3.125      400.0     0.257     96.8       44.
 3.375      432.0     0.120     96.9       21.
 3.625      464.0     0.055     97.0        9.
 3.875      496.0     0.024     97.0        4.
 4.125      528.0     0.011     97.0        2.
 4.375      560.0     0.005     97.0        1.
 4.625      592.0     0.002     97.0        0.
 4.875      624.0     0.001     97.0        0.
 5.125      656.0     0.000     97.0        0.
 5.375      688.0     0.000     97.0        0.
 5.625      720.0     0.000     97.0        0.
 5.875      752.0     0.000     97.0        0.
 6.125      784.0     0.000     97.0        0.
 6.375      816.0     0.000     97.0        0.
 6.625      848.0     0.000     97.0        0.
 6.875      880.0     0.000     97.0        0.
 7.125      912.0     0.000     97.0        0.
 7.375      944.0     0.000     97.0        0.
 7.625      976.0     0.000     97.0        0.
 7.875     1008.0     0.000     97.0        0.
 8.125     1040.0     0.000     97.0        0.
 8.375     1072.0     0.000     97.0        0.
 8.625     1104.0     0.000     97.0        0.
 8.875     1136.0     0.000     97.0        0.
 9.125     1168.0     0.000     97.0        0.
 9.375     1200.0     0.000     97.0        0.
 9.625     1232.0     0.000     97.0        0.
 9.875     1264.0     0.000     97.0        0.
10.125     1296.0     0.000     97.0        0.
10.375     1328.0     0.000     97.0        0.
10.625     1360.0     0.000     97.0        0.
10.875     1392.0     0.000     97.0        0.
11.125     1424.0     0.000     97.0        0.
11.375     1456.0     0.000     97.0        0.
11.625     1488.0     0.000     97.0        0.
11.875     1520.0     0.000     97.0        0.
12.125     1552.0     0.000     97.0        0.
12.375     1584.0     0.000     97.0        0.
U =      19.9
U =      3431.8
```

**Fig. 4-31.** *Sample FORTRAN IV program for Gray's method.*

condition of interest. Simulation of a hydrologic system through use of models has many virtues. The model can be much more easily operated and observed than the real system. Another important advantage is that it is possible to compress real-time scales on the order of years to time scales on the order of minutes. As a result, long periods of real time can be successfully studied.

Digital simulation expresses physical systems in mathematical terms that involve various parameters. These mathematical models are then improved and verified by simulating the systems reaction to known inputs and outputs, continuing until the model is considered to be a reliable representation of the prototype system. The procedure is analogous to that of verifying a physical model. Once this has been accomplished, the model can be used for a variety of purposes such as project planning and project operation where model parameters and system input are varied to determine the effects of changes on the system output.[34] The variations may demonstrate seasonal effects, impact of land management practices, and many other situations of practical interest.

According to Crawford and Linsley the goal of digital simulation of a hydrologic system is to develop a practical hydrologic model that is a skeleton of the hypothetical "absolute knowledge" model.[33] A prime example of its application is development of the Stanford Watershed Models.[9,33] These models use various time durations of rainfall input to generate outflow hydrographs for a watershed. A general digital computer model builds the values of model parameters by adjusting such components as evapotranspiration losses, infiltration capacities, and groundwater recession rates to represent a particular basin. Chapter 10 treats simulation methods in greater detail.

## 4-20 Statistical and Probability Approaches to Runoff Determination

Hydrologists are concerned with estimating streamflows. Two approaches are available. The first is an indirect approach in which runoff is computed on the basis of observed or expected precipitation. The second is founded on direct analyses of runoff records without resort to precipitation data. Such investigations usually include frequency studies to indicate the likelihood of certain runoff events taking place. A knowledge of the frequency of runoff events is helpful in determining risks associated with proposed designs or anticipated operating schemes. Frequency analyses are usually directed toward studies of maximum (flood) and minimum (drought) flows.[35,37] Figure 4-32 illustrates a typical drought frequency analysis. Unfortunately, many existing runoff records are short-term; as a result they limit utility for reliable frequency analyses. Few adequate records are available earlier than about 1900. In some cases, sequential gen-

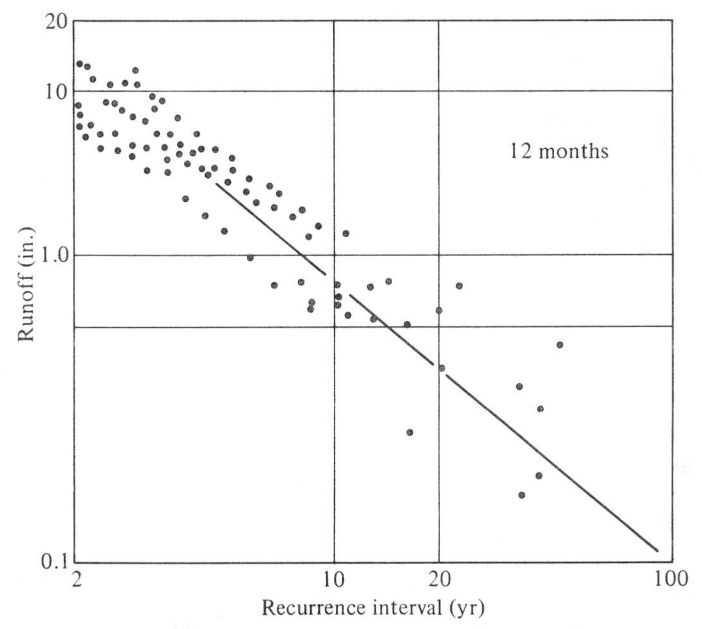


***Fig. 4-32.*** *Low-flow frequency data consolidated for Five Rivers. (After William Whipple, Jr., "Regional Drought Frequency Analysis,"* Proc. ASCE, J. Irrigation and Drainage Div. *92, No. IR2, Paper No. 4837 (June 1966): 11–31.)*

erating techniques can be used to develop synthetic records of any desired length (operational hydrology, Chapter 10).

Time-series analyses are particularly pertinent to the problem of estimating trends, cycles, and fluctuations in hydrologic data. They also permit derivation of generating processes by which synthetic records of runoff can be developed. Chapters 5 and 6 are devoted to statistical methods in hydrology.

## 4-21 Streamflow Forecasting

Surface water hydrology is basic to the design of many engineering works, and important in water quality and other management schemes. In addition, the ability to reliably forecast flows for short periods into the future is of signal value in operating storage and other works and in planning proper actions during times of flood.[8,39] A good example is the operation of a reservoir with an uncontrolled inflow but having a means of regulating the outflow. If information on the nature of the inflow is determinable in advance, then the reservoir can be operated by some decision rule to minimize downstream flood damage. Such operations can be computerized to continually improve estimates based on incoming data and thus offer direction on the nature of the releases to be made. For river forecasts to be reliable, adequate, dependable data on various watershed and meteorologic conditions are needed on a continuing basis. Modern monitoring stations capable of telemetering data to computer control centers provide an important support function for forecasting. The methods used to forecast flows are basically the same ones employed in design, namely, precipitation runoff equations, unit hydrographs, watershed models, and flow-routing techniques.

## Problems

4-1. Given the following storm pattern, unit storm, and unit hydrograph, determine the composite hydrograph.

| Unit Storm = 1 Unit of Rainfall for 1 Unit of Time | | | | | |
|---|---|---|---|---|---|
| Actual storm | Time units | 1 | 2 | 3 | 4 |
| Pattern | Rainfall units | 1 | 1 | 4 | 2 |

Unit hydrograph (triangular) with base length = 6 time units; time of rise = 2 time units, and maximum ordinate = $\frac{1}{2}$ ranfall unit height.

4-2. Given a unit rainfall duration of 1 time unit, an effective precipitation of 1.5 in., and the following hydrograph and storm sequence. Determine (a) the unit hydrograph, and (b) the hydrograph for the given storm sequence.

Storm Sequence

| Time units | 1 | 2 | 3 | 4 |
|---|---|---|---|---|
| Precipitation (in.) | 0.4 | 1.1 | 2.0 | 1.5 |

Hydrograph

| Time units | 1 | 2 | 3 | 4 | 4.5 | 5 | 6 | 7 | 8 | 9 | 10 | 11 | 12 | 13 |
|---|---|---|---|---|---|---|---|---|---|---|---|---|---|---|
| Flow (cfs) | 100 | 98 | 220 | 512 | 620 | 585 | 460 | 330 | 210 | 150 | 105 | 75 | 60 | 54 |

4-3 Do Prob. 4-2 if the storm sequence is as follows:

Storm Sequence

| Time units | 1 | 2 | 3 |
|---|---|---|---|
| Precipitation, in. | 0.3 | 1.4 | 0.9 |

4-4. For a drainage basin of your choice, plot the annual precipitation in inches versus the runoff in inches. Does the relationship appear to be strong? Under what conditions might you use this?

4-5. Select a raingauge record of interest. Use the annual values as data to calculate the coefficients of an antecedent precipitation index of the form of Eq. 4-8.

4-6. Given an initial infiltration capacity rate of 5.0 in./hr, a final rate of 0.30 in./hr, and a time constant $k$ of 0.35, find the infiltration capacity at 10 min, 30 min, 1 hr, 2 hr, and 6 hr. Plot these values.

4-7. Refer to Fig. 4-8. Replot this hydrograph and use two different techniques to separate the base flow.

4-8. Obtain streamflow data for a water course of interest. Plot the hydrograph for a major runoff event and separate the base flow.

4-9. For the event of Prob. 4-8, tabulate the precipitation causing the surface runoff and determine the duration of runoff producing rain. Estimate the time of concentration and use Eq. 4-16 to estimate the time base of the hydrograph. Compare this with the time base computed from the hydrograph.

4-10. Using U.S. Geological Survey records, or other data, select a streamflow hydrograph for a large, preferably single-peaked runoff event. Separate the base flow and determine a unit hydrograph for the area.

4-11. For the unit hydrograph of Prob. 4-1, construct an S-hydrograph.

4-12. For the unit hydrograph computed in Prob. 4-2, construct an S-hydrograph.

4-13. Use the S-hydrograph of Prob. 4-12 to find a 3 time-unit unit hydrograph.

4-14. Given a watershed of 100 mi², assume that $C_t = 1.8$, the length of main stream channel is 18 mi, and the length to a point nearest the centroid is 10 mi. Use Snyder's method to find (a) the time lag, (b) the duration of the synthetic unit hydrograph, and (c) the peak discharge.

4-15. Apply Snyder's method to the determination of a synthetic hydrograph for a drainage area of your choice.

4-16. Use Fig. 4-26 to determine a unit hydrograph for a rainfall duration of 2 hr if the drainage area is 60 mi² and Eq. 4-38a is applicable.

4-17. Solve Prob. 4-16 using Eq. 4-38b.

4-18. For the given data, use Gray's method to determine a unit hydrograph: drainage area = 1.0 mi², length = 0.6 mi, $S_c$ = 1.3%.

4-19. A drainage area contains 30 mi². The length of the main channel is 10 mi and the representative watershed slope is 2.5%. Use Gray's method to determine a unit hydrograph.

4-20. Tabulated below are total hourly discharge rates at a particular cross section of a stream. The drainage area above the section is 1.0 acre. (a) Plot the hydrograph on rectangular coordinate paper and label the rising limb (concentration curve), the crest segment, and the recession limb. (b) Determine the hour of cessation of the direct runoff using a semilog plot at $Q$ versus time. (c) Use the base flow portion of your semilog plot to determine the groundwater recession constant, $K$. (d) Carefully construct and label base flow separation curves on the graph of part (a), using two different methods.

| Time (hr) | Q (cfs) | Time (hr) | Q (cfs) |
|---|---|---|---|
| 0 | 102 | 8 | 210 |
| 1 | 100 | 9 | 150 |
| 2 | 98 | 10 | 105 |
| 3 | 220 | 11 | 75 |
| 4 | 512 | 12 | 60 |
| 5 | 630 | 13 | 54 |
| 6 | 460 | 14 | 48.5 |
| 7 | 330 | 15 | 43.5 |

4-21. What does the area under an outflow hydrograph represent?

4-22. Actual discharge rates for a flood hydrograph passing the point of concentration for a 600-acre drainage basin are given in the table. The flood was produced by a uniform rainfall rate of 2.75 in./hr which started at 9 A.M., abrutly ended at 11 A.M. and resulted in 5.00 in. of direct surface runoff. The base flow (derived from influent seepage) prior to, during, and after the storm was 100 cfs.

| Time | 8 A.M. | 9 | 10 | 11 | 12 | 1 P.M. | 2 | 3 | 4 | 5 | 6 | 7 |
|---|---|---|---|---|---|---|---|---|---|---|---|---|
| Measured Discharge | 100 | 100 | 300 | 500 | 700 | 800 | 600 | 400 | 300 | 200 | 100 | 100 |

(a) At what times did direct surface runoff begin and cease? (b) Determine the $\phi$ index (in./hr) for the basin. (c) Derive the 2-hr unit hydrograph ordinates (cfs) for each time listed. (d) Determine the time of concentration for the basin. (e) At what time would direct surface runoff cease if the rainfall of 2.75 in./hr had begun at 9 A.M. and had lasted for 8 hr rather than 2? (f) Determine the peak discharge rate (cfs) and the direct surface runoff (in.) for an actual rainfall of 2.75 in./hr and a duration of 8 hr.

4-23. Determine the drainage density of the basin shown. Area = 6400 acres. Lengths are in miles.

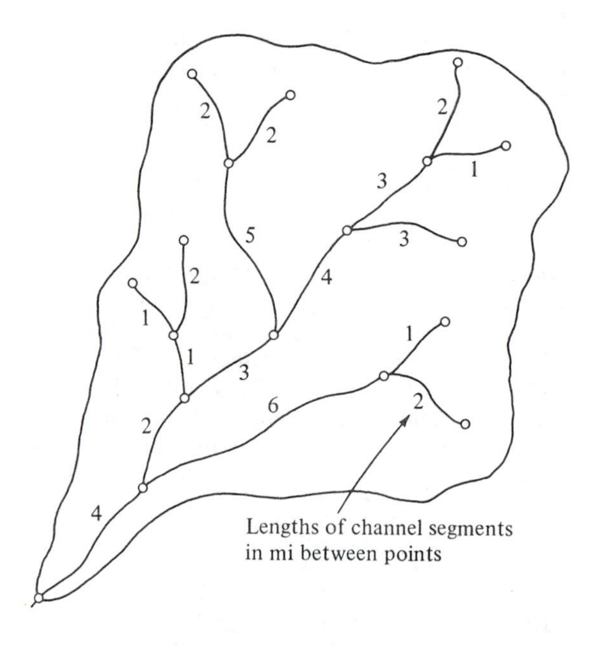


4-24. Measured total hourly discharge rates (cfs) from a 3.10 $mi^2$ drainage basin are listed in the accompanying table. The hydrograph was produced by a rainstorm having a uniform intensity of 2.60 in./hr starting at 9 A.M. and abruptly ending at 11 A.M. The base flow from 8 A.M. to 3 P.M. was a constant 100 cfs. The volume of direct surface runoff, determined as the area under the direct surface runoff hydrograph, is 1000 cfs-hr.

| Time | 8 A.M. | 9 | 10 | 11 | 12 | 1 P.M. | 2 | 3 |
|---|---|---|---|---|---|---|---|---|
| Measured Discharge | 100 | 100 | 300 | 600 | 400 | 200 | 100 | 100 |

(a) At what time did the direct surface runoff begin? (b) Determine the net rain (in.) corresponding to the volume of the direct surface runoff of 1000 cfs-hr. (c) Determine the $\phi$ index for the basin. (d) Derive a 2-hr unit hydrograph for the basin by tabulating time in hours and discharge in cfs. (e) What is the time of concentration of the basin? (f) For the same basin, use the derived 2-hr unit hydrograph to determine the direct surface runoff *rate* (cfs) at 4 P.M. on a day when excess (net) rainfall began at 1 P.M. and continued at a net intensity of 2 in./hr for 4 hr, ceasing abruptly at 5 P.M.

4-25. A 5-hr unit hydrograph for a 4250-acre basin is shown in the accompanying sketch. The given hydrograph actually appeared as a direct surface runoff hydrograph from the basin, caused by rain falling at an intensity of 0.30 in./hr for a duration of 5 hr, beginning at $t = 0$. (a) Determine the time of concentration of the basin. (b) Determine the $\phi$ index for the basin. (c) What percentage of the drainage basin was contributing to direct surface runoff 4 hr after rain began ($t = 4$)? Why? (d) Use your response to part (c) to determine $Q_p$, as shown in the sketch. Do not scale $Q_p$ from the drawing. (e) Note that rain continued to fall between $t = 3$ and $t = 5$. Why did the hydrograph form a plateau between $t = 3$ and $t = 5$, rather than continue to rise during those 2 hr? (f) Use the given 5-hr unit hydrograph to determine the direct surface runoff rate (cfs) at 7 P.M. on a day when rain fell at an intensity of 0.60 in./hr from 1 P.M. to 11 P.M.

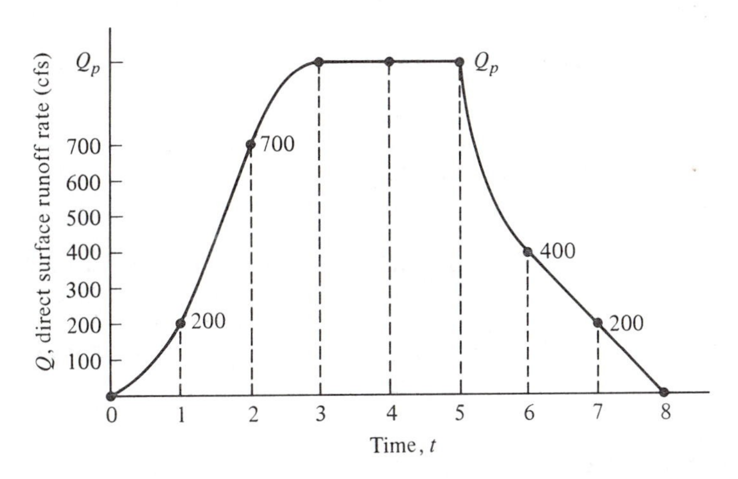


4-26. The 2-hr unit hydrograph for a basin is given by the following table.

| Time (hr) | 0 | 1 | 2 | 3 | 4 | 5 | 6 | 7 | 8 | 9 |
|---|---|---|---|---|---|---|---|---|---|---|
| Q (cfs) | 0 | 60 | 200 | 300 | 200 | 120 | 60 | 30 | 10 | 0 |

(a) Determine the hourly discharge values (cfs) from the basin for a net rain of 5 in./hr and a rainfall duration of 2 hr. (b) Determine the direct surface runoff (in.) for the storm of part (a). What is the direct surface runoff for a net rain of 0.5 in./hr and a duration of 2 hr? (c) Rain falls on the basin at a rate of 4.5 in./hr for a 2-hr period and abruptly increases to a rate of 6.5 in./hr for a second 2-hr period. Convert these actual intensities to net rain intensities using a $\phi$ index of 0.5 in./hr. Construct a table that properly lags and amplifies the 2-hr unit hydrograph, and determine the hourly ordinates (cfs) of direct surface runoff for the storm. The derived direct surface runoff hydrograph should begin and end with zero discharge values.

4-27. Given the following 2-hr unit hydrograph for a drainage basin, determine hourly ordinates of the 4-hr unit hydrograph:

| Time (hr) | 0 | 1 | 2 | 3 | 4 | 5 | 6 |
|---|---|---|---|---|---|---|---|
| $Q$ (cfs) | 0 | 50 | 300 | 400 | 200 | 50 | 0 |

4-28. Use the following 4-hr unit hydrograph for a basin to determine the peak discharge rate (cfs) resulting from a net rain of 3.0 in./hr for a 4-hr duration followed immediately by 2.0 in./hr for a 4-hr duration.

| Time (hr) | 0 | 2 | 4 | 6 | 8 | 10 |
|---|---|---|---|---|---|---|
| $Q$ (cfs) | 0 | 200 | 300 | 100 | 50 | 0 |

## References


1. Strahler, A. N., "Geology—Part II," *Handbook of Applied Hydrology* (New York: McGraw-Hill Book Company, 1964).
2. Horton, R. E., "Discussion of Paper, Flood Flow Characteristics by C. S. Jarvis," *Trans. ASCE,* 89 (1926).
3. Horton, R. E., "Drainage Basin Characteristics," *Trans. Amer. Geophys. Union,* 13 (1932).
4. Langbein, W. B. et al., "Topographic Characteristics of Drainage Basins," U.S. Geological Survey, Water Supply Paper, 968-c (1947).
5. Horton, R. E., "Erosional Development of Streams and Their Drainage Basins: Hydrophysical Approach to Qualitative Morphology," *Bull. Geol. Soc. Amer.,* 56 (1945).
6. Hack, J. T., "Studies of Longitudinal Stream Profiles of Small Watersheds," Tech. Rept. 18, Columbia University, Department of Geology, New York, 1959.
7. Taylor, A. B. and Schwartz, H. E., "Unit Hydrograph Lag and Peak Flow Related to Basin Characteristics," *Trans. Amer. Geophys. Union,* 33 (1952).
8. Linsley, R. K., Jr., M. A. Kohler, and J. L. H. Paulhus, *Applied Hydrology,* (New York: McGraw-Hill Book Company, 1949).
9. Chow, Ven Te (ed.), *Handbook of Applied Hydrology* (New York: McGraw-Hill Book Company, 1964).
10. Linsley, R. K., Jr., and W. C. Ackerman, "Method of Predicting the Runoff from Rainfall," *Trans. ASCE,* 107 (1942).
11. Butler, S. S., *Engineering Hydrology* (Englewood Cliffs, N.J.: Prentice-Hall, Inc., 1957).

12. Kohler, M. A., and R. K. Linsley, Jr., "Predicting the Runoff from Storm Rainfall," U.S. Weather Bureau, Research Paper 34, 1951.
13. Horton, R. E., "Erosional Development of Streams and Their Drainage Basins: Hydrophysical Approach to Quantitative Morphology," *Bull. Geol. Soc. Amer.*, 56 (1945).
14. DeWiest, R. J. M., *Geohydrology* (New York: John Wiley & Sons, Inc., 1965).
15. "Hydrology," Engineering Handbook Section 4, U.S. Department of Agriculture, Soil Conservation Service, 1972.
16. Barnes, B. S., "Discussion of Analysis of Runoff Characteristics by O. H. Meyer," *Trans. ASCE*, 105 (1940).
17. Sherman, L. K., "Streamflow from Rainfall by the Unit-Graph Method," *Eng. News-Rec.*, 108 (1932).
18. Horton, R. E., "An Approach Toward a Physical Interpretation of Infiltration Capacity," *Proc. Soil Science Soc. Amer.*, 5 (1940): 399–417.
19. Meinzer, O. E., *Hydrology* (New York: Dover Publications, Inc., 1942).
20. Mitchell, W. D., "Unit Hydrographs in Illinois," Illinois Waterways Div., 1948.
21. American Society of Civil Engineers, "Hydrology Handbook," Manuals of Engineering Practice, No. 28, New York, ASCE, 1957.
22. Sherman, L. K., "Stream-Flow from Rainfall by the Unit-Graph Method," *Eng. News-Rec.*, 108 (Apr. 1932): 501–505.
23. Morgan, Rand, and D.W. Hulinghorst, "Unit Hydrographs for Gauged and Ungauged Watersheds," U.S. Engineers Office, Binghamton, N.Y., July 1939.
24. Snyder, F. F., "Synthetic Unit Graphs," *Trans. Amer. Geophys. Union*, 19 (1938): 447–454.
25. Taylor, A. B., and H. E. Schwartz, "Unit Hydrograph Lag and Peak Flow Related to Basin Characteristics," *Trans. Am. Geophys. Union*, 33 (1952): 235–246.
26. Clark, C. O., "Storage and the Unit Hydrograph," *ASCE Trans.*, 110 (1945): 1419–1446.
27. "Water Supply Papers," U.S. Geological Survey, Water Resources Division (Washington, D.C.: Government Printing Office, 1966–1970).
28. "Hourly Precipitation Data," National Oceanic and Atmospheric Administration (Washington, D.C.: Government Printing Office, 1971).
29. Newton, D. W., and J. W. Vineyard, "Computer-Determined Unit Hydrograph from Floods," *Proc. ASCE, J. Hyd. Div.*, 93, HY 5 (1967): 219–236.
30. "Flood-Hydrograph Analysis and Computations," U.S. Army Corps of Engineers, Engineering and Design Manuals, Em1110-2-1405 (Washington, D.C.: Government Printing Office, Aug. 1959).
31. Gray, D. M., "Synthetic Unit Hydrographs for Small Drainage Areas," *Proc. ASCE, J. Hyd. Div.*, 87, No. HY4 (July 1961).
32. Grace, R. A., and P. S. Eagleson, "Scale Model of Urban Runoff from Storm Rainfall," *Proc. ASCE, J. Hyd. Div.*, HY 3, Paper 5249, May 1967.
33. Crawford, N. H., and R. K. Linsley, Jr., "Digital Simulation in Hydrology, Stanford Watershed Model IV," *Tech. Rept.* 39, Department of Civil Engineering, Stanford University, Stanford, Calif., July 1966.

34. Beard, Leo R., "Hydrologic Simulation to Water-Yield Analysis," *Proc. ASCE, J. Irrigation and Drainage Div.*, 93, No. IR1 (Mar. 1967).
35. Beard, Leo R., "Statistical Methods in Hydrology," U.S. Army Engineer District, Sacramento, Calif., 1962.
36. DeCoursey, Donn G., "A Runoff Hydrograph Equation," U.S. Department of Agriculture, Agricultural Research Service, Feb. 1966, pp. 41–116.
37. Whipple, William W., Jr., "Regional Drought Frequency Analysis," *Proc. ASCE, J. Irrigation and Drainage Div.*, 92, No. IR2 (1966).
38. Philip, J. R., "An Infiltration Equation with Physical Significance," *Soil Science*, 77 (1954).
39. Quick, Michael C., "River Flood Flows: Forecasts and Probabilities," *Proc. ASCE, J. Hyd. Div.*, 91, HY 3 (May 1965).
40. Mockus, V., "Use of Storm and Watershed Characteristics in Synthetic Hydrograph Analysis and Application," U.S. Department of Argiculture, Soil Conservation Service, 1957.
41. Horton, R. E., *Surface Runoff Phenomena* (Ann Arbor, Mich.: Edwards Brothers, Inc., 1935).
42. Eagleson, Peter S., "Characteristics of Unit Hydrographs for Sewered Areas," paper presented before the ASCE, Los Angeles, Calif., 1959, unpublished.
43. Horner, W. W., and Flynt, F. L., "Relation Between Rainfall and Runoff from Small Urban Areas," *Trans. ASCE*, 62, No. 101 (Oct. 1956): 140–205.
44. Schaake, John C., Jr., "Synthesis of the Inlet Hydrograph," Tech. Rept. No. 3, Department of Sanitary Engineering and Water Resources, The Johns Hopkins University, Baltimore, Md., 1965.

# 5 Point-Frequency Analysis

Many hydrologic processes are so complex that they can be interpreted and explained only in a *probabilistic* sense. Hydrologic events appear as uncertainties of nature and are the result, it must be assumed, of an underlying process with random or *stochastic* components. The information to investigate these processes is contained in records of hydrologic observations. Methods of statistical analysis provide ways to reduce and summarize observed data, to present information in precise and meaningful form, to determine the underlying characteristics of the observed phenomena, and to make predictions concerning future behavior.

Statistical analysis involves two basic sets of problems, one *descriptive*, the other *inferential*. The former is a straightforward application of statistical methods, requiring few decisions and representing little risk. The inferential problem, however, entails decisions bearing some risk, and requires an understanding of the methods employed and the dangers involved in predicting and estimating. The most common inferential problem is to describe the whole class of possible occurrences when only a portion of them has been observed. The whole class is the *population* and the portion observed is the *sample*.

The variables in the process under study are *continuous* if they may take on all values in the range of occurrence, including figures differing only by an infinitesimal amount; they are *discrete* if they are restricted to specific, incremental values. Distribution of the variables over the range of occurrence is defined in terms of the frequency or probability with which different values have occurred or might occur. Historical sequences of floods, droughts, streamflow, and rainfall are assumed to result from underlying physical processes that are *stationary*—that is, their characteristics remain unchanged by long-term trends or other effects. Usually, the length of record as well as the design life for an engineering project are relatively short compared with geologic history and tend to temper, if not to justify, the assumption of stationarity.[1,2]

## 5-1 Concepts of Probability

The laws of probability underlie any study of the statistical nature of repeated observations or trials. The probability of a single event, say $E_1$, is defined as the relative number of occurrences of the event after a long series of trials. Thus $P(E_1)$, the probability of event $E_1$, is $n_1/N$ for $n_1$ occurrences of the same event in $N$ trials if $N$ is sufficiently large. The number of occurrences $n_1$ is the *frequency*, and $n_1/N$ the *relative frequency*.

Often the probabilities and the rules governing their manipulation are known intuitively or from experience. In the familiar coin-tossing experiment, $P(\text{heads}) = P(\text{tails}) = \frac{1}{2}$. Each outcome of a single toss (a trial) has a finite probability, and the sum of the probabilities of all possible outcomes is 1. Also, the outcomes are *mutually exclusive*; that is, if one occurs, say a head, then a tail cannot occur. In two successive tests, there are four possible outcomes—HH, TT, HT, TH—each with a probability of $\frac{1}{4}$. In this case, because each trial is independent of the other one, probabilities for each outcome are found by $P(\text{first trial}) \times P(\text{second trial}) = \frac{1}{2} \times \frac{1}{2} = \frac{1}{4}$. Again, the sum of the probabilities of the possible outcomes is 1. Note that the probability of getting exactly one head and one tail during the experiment (without regard to the order) is $P(\text{HT}) + P(\text{TH}) = \frac{1}{2}$.

Summarizing the rules of probability indicated by coin tossing, we find:

1. The probability of an event is nonnegative and never exceeds one.

$$0 \leq P(E_i) \leq 1 \tag{5-1}$$

2. The sum of the probabilities of all possible outcomes in a single trial is 1.

$$\sum_i P(E_i) = 1 \tag{5-2}$$

3. The probability of a number of *independent* and *mutually exclusive* events is the sum of the probabilities of the separate events.

$$P(E_1 \cup E_2) = P(E_1) + P(E_2) \tag{5-3}$$

The probability statement, $P(E_1 \cup E_2)$, signifies a union of probabilities and is read "the probability of $E_1$ or $E_2$."

4. The probability of two *independent* events' occurring simultaneously or in succession is the product of the individual probabilities.

$$P(E_1 \cap E_2) = P(E_1) \times P(E_2) \tag{5-4}$$

$P(E_1 \cap E_2)$ is called the *intersection* or *joint probability* and is read "the probability of $E_1$ and $E_2$."

Consider the following example of events that are not independent or mutually exclusive: An urban drainage canal reaches flood stage each summer with relative frequency of .10; power failures in industries along the canal occur with probability of .20; experience shows that when there is a flood the chances of a power failure are raised to .40. The probability statements are

$$P(\text{flood}) - P(F) = .10 \qquad P(\text{power failure}) = P(P) = .20$$
$$P(\text{no flood}) = P(\bar{F}) = .90 \qquad P(\text{no power failure}) = P(\bar{P}) = .80$$
$$P(\text{power failure given that a flood occurs}) = .40$$

The last statement is called a *conditional probability.* It signifies the joint occurrence of events and is usually written $P(P \mid F)$. Rules 3 and 4 no longer are strictly applicable. If rule 3 applied, $P(F \cup P) = P(F) + P(P) = .3$. If the events remained independent, the joint probabilities would be

$$P(F \cap P) = .1 \times .2 = .02$$
$$P(F \cap \bar{P}) = .1 \times .8 = .08$$
$$P(\bar{F} \cap P) = .9 \times .2 = .18$$
$$P(\bar{F} \cap \bar{P}) = .9 \times .8 = .72$$

The probability of a flood or a power failure during the summer would be the sum of the first three joint probabilities above.

$$P(F \cup P) = P(F \cap P) + P(F \cap \bar{P}) + P(\bar{F} \cap P) = .28$$

The events are dependent, however, from the statement of conditional probability: When a flood occurs with $P(F) = .1$, a power failure will occur with probability .4, and true joint probability is $P(F) \times P(P \mid F) = .1 \times .4 = .04 = P(F \cap P)$. The union probability is then $P(F \cup P) = P(F) + P(P) - P(F \cap P) = .1 + .2 - .04 = .26$. Note the contrast

$P(F \cup P) = .30$ for mutually exclusive events
$P(F \cup P) = .28$ for joint but independent events
$P(F \cup P) = .26$ otherwise

The new, more general rule for the union of probabilities is

$$5.\ P(E_1 \cup E_2) = P(E_1) + P(E_2) - P(E_1 \cap E_2) \tag{5-5}$$

and a sixth rule should be added for conditional probabilities:

$$6.\ P(E_1 \mid E_2) = \frac{P(E_1 \cap E_2)}{P(E_2)} \tag{5-6}$$

A very important concept of independence is expressed in a variation of rule 6, namely, $P(E_1 \mid E_2) = P(E_1)$ if events $E_1$ and $E_2$ are independent. This further explains rule 3 that $P(E_1) \times P(E_2) = P(E_1 \cap E_2)$ for independent events.

The example of flooding can be extended to show some interesting features about probabilities and risks associated with hydrologic phenomena: $P(F) = .10$ implies a 10% chance each year for the flood level to be reached or exceeded. In the long run, the level would be reached on the average once in 10 yr. Thus the average *return period** $T$ in years is defined as

$$T = \frac{1}{P(F)} = \frac{1}{1 - P(\bar{F})} \tag{5-7}$$

and the following general probability relationships hold:

1. The probability that $F$ will occur in any year

$$P(F) = \frac{1}{T} \tag{5-8}$$

2. The probability that $F$ will not occur in any year

$$P(\bar{F}) = 1 - P(F) = 1 - \frac{1}{T} \tag{5-9}$$

3. The probability that $F$ will not occur for $n$ successive years

$$P_1(\bar{F}) \times P_2(\bar{F}) \cdots P_n(\bar{F}) = P(\bar{F})^n = \left(1 - \frac{1}{T}\right)^n \tag{5-10}$$

4. The probability $R$, called *risk,* that $F$ will occur at least once in $n$ successive years

$$R = 1 - \left(1 - \frac{1}{T}\right)^n = 1 - \{P(\bar{F})\}^n \tag{5-11}$$

* The terms *return period* and *recurrence interval* are used interchangeably to denote the reciprocal of the annual probability of exceedence.

*Table 5-1* Return Periods Associated with Various Degrees of Risk and Expected Design Life

| Risk (%) | Expected Design Life (years) | | | | | | | |
|---|---|---|---|---|---|---|---|---|
| | 2 | 5 | 10 | 15 | 20 | 25 | 50 | 100 |
| 75 | 2.00 | 4.02 | 6.69 | 11.0 | 14.9 | 18.0 | 35.6 | 72.7 |
| 50 | 3.43 | 7.74 | 14.9 | 22.1 | 29.4 | 36.6 | 72.6 | 144.8 |
| 40 | 4.44 | 10.3 | 20.1 | 29.9 | 39.7 | 49.5 | 98.4 | 196.3 |
| 30 | 6.12 | 14.5 | 28.5 | 42.6 | 56.5 | 70.6 | 140.7 | 281 |
| 25 | 7.46 | 17.9 | 35.3 | 52.6 | 70.0 | 87.4 | 174.3 | 348 |
| 20 | 9.47 | 22.9 | 45.3 | 67.7 | 90.1 | 112.5 | 224.6 | 449 |
| 15 | 12.8 | 31.3 | 62.0 | 90.8 | 123.6 | 154.3 | 308 | 616 |
| 10 | 19.5 | 48.1 | 95.4 | 142.9 | 190.3 | 238 | 475 | 950 |
| 5 | 39.5 | 98.0 | 195.5 | 292.9 | 390 | 488 | 976 | 1949 |
| 2 | 99.5 | 248 | 496 | 743 | 990 | 1238 | 2475 | 4950 |
| 1 | 198.4 | 498 | 996 | 1492 | 1992 | 2488 | 4975 | 9953 |

Table 5-1 shows return periods associated with various levels of risk.

***Example 5-1*** What return period must a highway engineer use in his design of a critical underpass drain if he is willing to accept only a 10% risk that flooding will occur in the next 5 years?

*Solution*

$$R = 1 - \left(1 - \frac{1}{T}\right)^n$$

$$.10 = 1 - \left(1 - \frac{1}{T}\right)^5$$

$$T = 48.1 \text{ years}$$

## 5-2 Probability Distributions

Random variables, either discrete or continuous, are characterized by the distribution of probabilities attached to the specific values that the variable may assume. A random variable throughout its range of occurrence is generally designated by a capital letter, and a specific value or outcome of the random process is designated by a small letter. For example, $P(X = x_1)$ is the probability that random variable $X$ takes on the value $x_1$. A shorter version is $P(x_1)$. Figure 5-1 shows the probability distribution of the number of cloudy days in a week. It is a discrete distribution because the number of days is exact; in the record from which the relative frequencies were taken, a day had to be described as

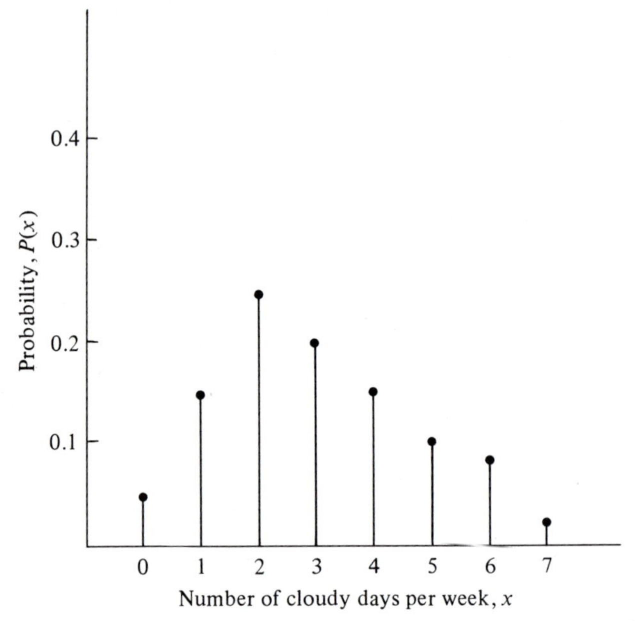


***Fig. 5-1.*** *Probability distribution of cloudy days per week.*

$P(0) = 0.05$ $P(4) = 0.15$
$P(1) = 0.15$ $P(5) = 0.10$
$P(2) = 0.25$ $P(6) = 0.08$
$P(3) = 0.20$ $P(7) = 0.02$

cloudy or not. Observe that each of the seven events has a finite probability and the sum is 1, that is,

$$\sum_i P(x_i) = 1$$

Another convenient description of probabilities is the *cumulative distribution function*, CDF, defined as the probability that any outcome in $X$ is less than or equal to a stated, limiting value $x$. The cumulative distribution function is denoted $F(x)$. Thus

$$F(x) = P(X \leqslant x) \tag{5-12}$$

and the function increases monotonically from a lower limit of zero to an upper bound of unity. Figure 5-2 is the CDF of the number of cloudy days in a week derived from Fig. 5-1 by taking cumulative probabilities. The function shows that the probability is 90% that the number of cloudy days in the week will be 5 or less. Conversely, there is a 10% probability that it will be cloudy for 6 or 7 days. This complementary cumulative probability is sometimes called $G(x)$, where[3]

$$G(x) = 1 - F(x) = P(X \geqslant x) \tag{5-13}$$

Continuous variables present a slightly different picture. Figure 5-3 is the *histogram* of an 85-yr record of annual stream-flows. The observations were grouped into nine intervals ranging from 0 to 900 cfs and the number falling in each interval was plotted as frequency on the left ordinate. A convenient alternative is to plot the relative frequency as shown by the right ordinate. The CDF for the streamflow record is shown in Fig. 5-4.

As the number of observations increase, the continuous distribution will be developed by reducing the size of the intervals. In the limit, the broken curves of Figs. 5-3 and 5-4 will appear as those in Fig. 5-5. There is a subtle difference between the ordinates of Figs. 5-3 and 5-5a. Since relative frequency is synonymous with probability, it is convenient to reconstitute the histogram so that the area in each interval

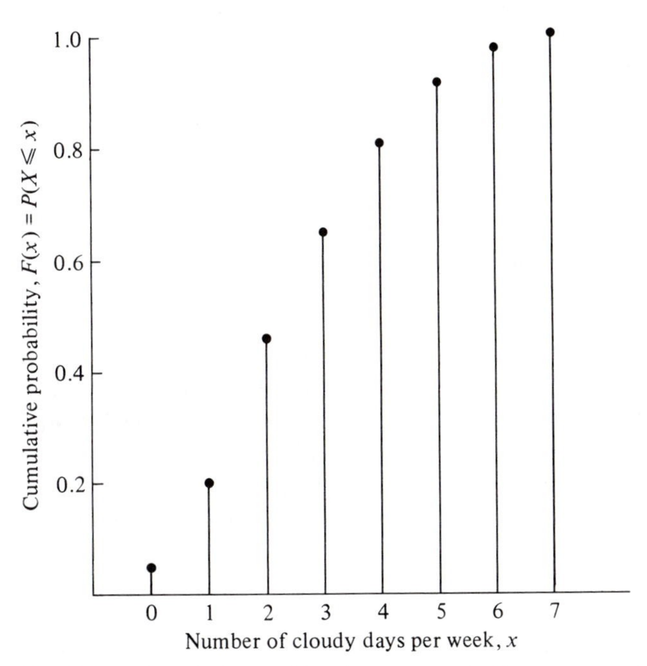


**Fig. 5-2.** *Cumulative distribution of cloudy days per week.*

| $x$ | $P(x)$ | $F(x)$ |
|---|---|---|
| 0 | 0.05 | 0.05 |
| 1 | 0.15 | 0.20 |
| 2 | 0.25 | 0.45 |
| 3 | 0.20 | 0.65 |
| 4 | 0.15 | 0.80 |
| 5 | 0.10 | 0.90 |
| 6 | 0.08 | 0.98 |
| 7 | 0.02 | 1.00 |

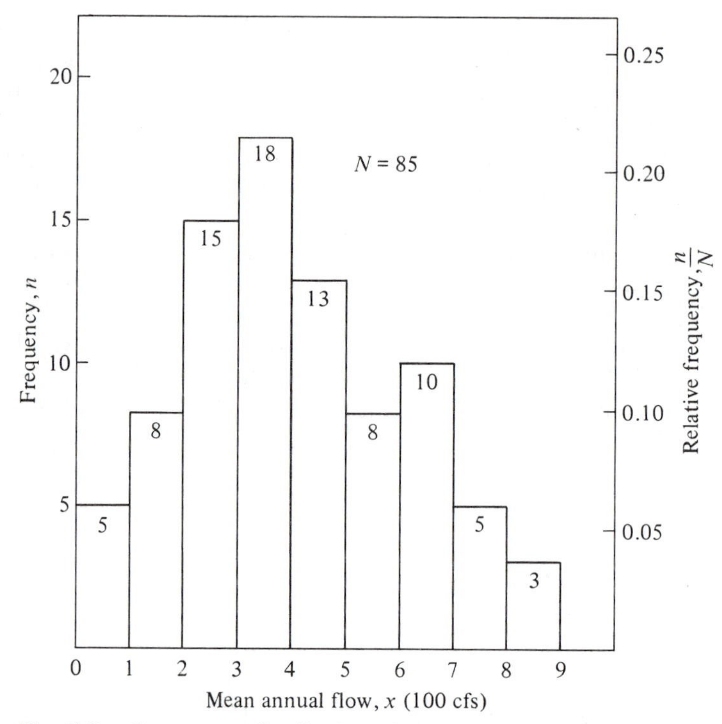


*Fig. 5-3.* *Frequency distribution of mean annual flows.*

represents probability; the total area contained is thus unity. To do this, the ordinate in each interval, say $n/N$ for relative frequency or probability, is divided by the interval width, $\Delta x$. The ratio $n/N\,\Delta x$ is literally the probability per unit length in the interval and therefore represents the average density of probability. The probability $n/N$ in the interval is represented on the CDF (before the limiting process as $\Delta F(x)$, or $F(x + \Delta x/2) - F(x - \Delta x/2)$. We then can define

$$f(x) = \lim_{\Delta x \to 0} \frac{\Delta F(x)}{\Delta x} = \frac{dF(x)}{dx} \tag{5-14}$$

which is called the *probability density function*, PDF.[3] This function is the density (or intensity) of probability at any point; $f(x)\,dx$ is described as the differential probability.

For continuous variables, $f(x) \geq 0$, since negative probabilities have no meaning. Also, the function has the property that

$$\int_{-\infty}^{\infty} f(x)\,dx = 1 \tag{5-15}$$

which again is the requirement that the probabilities of all outcomes sum to 1. Further, the probability that $x$ will fall between the limits $a$ and $b$ is written

$$P(a \leqslant X \leqslant b) = \int_a^b f(x)\,dx \tag{5-16}$$

Note that the probability that $x$ takes on a particular value, say $a$, is zero, that is,

$$\int_a^a f(x)\,dx = 0$$

which emphasizes that finite probabilities are defined only as areas under the PDF between finite limits.

The CDF can now be defined in terms of the PDF as

$$P(-\infty \leqslant X \leqslant x) = P(X \leqslant x) = F(x) = \int_{-\infty}^{x} f(u)\,du \tag{5-17}$$

where $u$ is used as a dummy variable to avoid confusion with the limit of integration. The area under the CDF has no meaning, only the ordinates, or the difference in ordinates. For example, $P(x_1 \leqslant X \leqslant x_2)$, which is equivalent to Eq. 5-16, can be evaluated as $F(x_2) - F(x_1)$.

For discrete distributions that cannot be summarized in integral form, there are analogous arithmetic statements corresponding to the

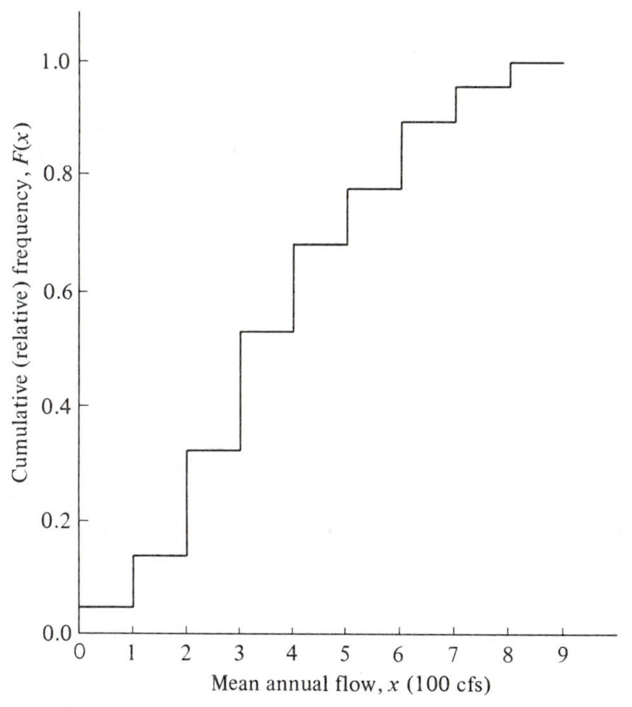


***Fig. 5-4.*** *Cumulative frequency distribution of mean annual flows.*

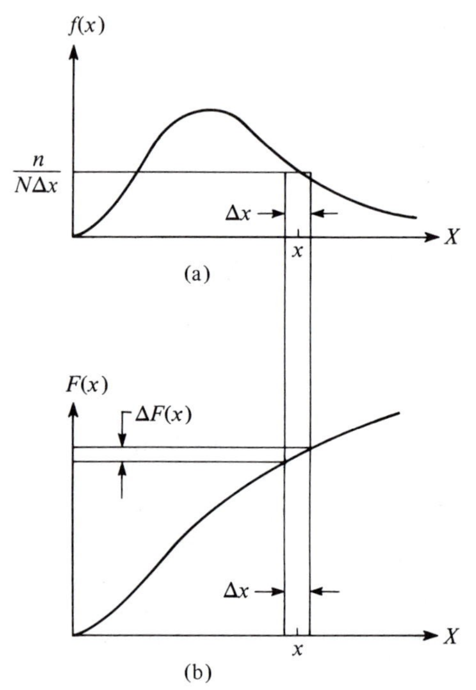


***Fig. 5-5.*** *Continuous probability distributions: (a) probability density function, and (b) cumulative distribution function.*

properties given in Eqs. 5-15, 5-16, and 5-17. In particular, the distribution of sampled data taken from a continuous distribution is a special case of discrete distributions and can be given in the form of arithmetic summations.[3] Thus

$$\sum_i f(x_i) = 1 \tag{5-18}$$

$$P(a \leqslant X \leqslant b) = \sum_{x_i \geqslant a}^{x_i \leqslant b} f(x_i) \tag{5-19}$$

$$P(X \leqslant x_k) = \sum_{i=1}^{k} f(x_i) \tag{5-20}$$

For a finite number of observations in the sample, $f(x_i)$ is the probability of $x_i$ for each outcome in the sample space and therefore $P(x_i) = P(x_1) = P(x_2) = \cdots = 1/N$. Hence $f(x_i)$ can be replaced with $P(x_i)$ in Eqs. 5-18, 5-19, and 5-20.

## 5-3 Moments of Distributions

The mathematical properties of statistical distributions can be defined completely in terms of the moments of the distribution. Fortunately, the moments represent parameters that usually have physical or geometrical significance. Readers will recognize the analogy between statistical moments and the moments of areas studied in solid mechanics.

The $r$th moment about the origin is defined as[4]

$$\mu_r' = \int_{-\infty}^{+\infty} x^r f(x)\, dx \tag{5-21}$$

or

$$\mu_r' = \sum_{i=1}^{n} x_i^r f(x_i) = \frac{1}{n} \sum_{i=1}^{n} x_i^r \tag{5-22}$$

The first moment about the origin is the *mean*, or as it commonly is known, the average. It determines the distance from the origin to the centroid of the distribution frequency function; its importance as a parameter is so complete that it is written simply $\mu$ instead of $\mu_1'$. The prime is used to signify moments taken about the origin.

Moments can be defined about other axes than the origin; the axis used extensively in defining higher moments is the *mean* or, as given above, the first moment about the origin. Thus

$$\mu_r = \int_{-\infty}^{\infty} (x - \mu)^r f(x)\, dx \tag{5-23}$$

or

$$\mu_r = \frac{1}{n} \sum_{i=1}^{n} (x_i - \mu)^r \tag{5-24}$$

Whenever $\mu_r'$ or $\mu_r$ are defined for $r = 1 \ldots$, the distribution $f(x)$ is completely defined. It seldom is necessary to compute more than the first three moments; several important distributions require only two. The moments are used to specify the parameters and descriptive characteristics of distributions that follow in the next article. Because various characteristics of distributions are described by combinations of the moments about the mean and origin, the following relationships are occasionally helpful:[1,4]

$$\mu_1 = 0 \tag{5-25}$$

$$\mu_2 = \mu_2' - \mu^2 \tag{5-26}$$

$$\mu_3 = \mu_3' - 3\mu_2'\mu + 2\mu^3 \tag{5-27}$$

## 5-4 Distribution Characteristics

Characteristics of statistical distributions are described by the parameters of probability functions which, in turn, are expressed in terms of the moments. The principal characteristics are *central tendency*, the grouping of observations or probability about a central value; *variability*, the dispersion of the variate or observations; and, *skewness*, the degree of asymmetry of the distribution. The theoretical functions shown in Fig. 5-6 exhibit approximately the same grouping about a central value, but $f_2$ has much greater variability than $f_1$, and $f_2$ possesses a pronounced right-skew while $f_1$ is symmetrical.

In introducing the parameters of distributions, the usual sequence of statistical problems will be followed—that is, we shall derive the parameters from the distribution of sample data and use these as estimates of the parameters of the population distribution. Thus the summation forms are used to compute moments. For example, the mean of sample data is designated $\bar{x}$ and it is used as the best estimate of the population mean. By convention, Greek letters are used to denote population parameters.

### Central Tendency

The familiar arithmetic average, the *mean*, is the most reliable measure of central tendency. It is the first moment about the origin and is designated

$$\bar{x} = \frac{1}{n} \sum_{i=1}^{n} x_i \tag{5-28}$$

The statistic, $\bar{x}$, is the best estimate of the population mean, $\mu$.

Means other than the arithmetic mean—for example, the geometric mean or harmonic mean—are needed in some instances, but the one defined in Eq. 5-28 is sufficient for most purposes. Two additional measures of central tendency are the *median*, which is the middle value of the observed data and divides the distribution into equal parts, and the *mode*, which in discrete variables is the value occurring most frequently and in continuous variables is the peak value of probability density.

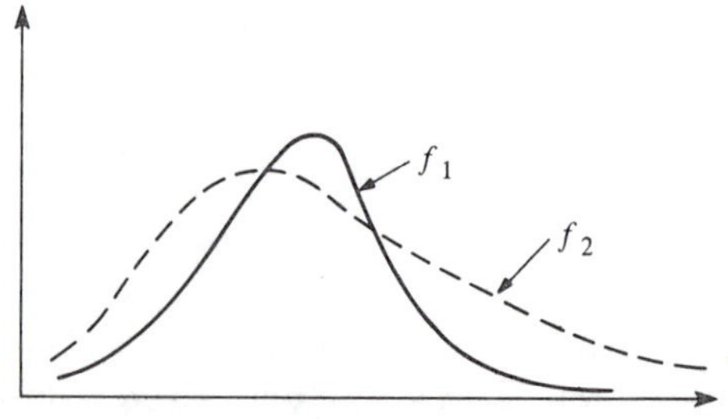


**Fig. 5-6.** *Symmetrical and skewed probability distributions.*

### Variability

Dispersion can be represented by the total range of values or by the average deviation about the mean; however, the parameter of statistical importance is the mean squared deviation as measured by the second moment about the mean. The parameter is termed the variance and is designated by

$$\sigma^2 = \frac{1}{n}\sum_{i=1}^{n}(x_i - \mu)^2 \tag{5-29}$$

But the population mean is not known precisely and therefore it is necessary to compute instead

$$s^2 = \frac{1}{n-1}\sum_{i=1}^{n}(x_i - \bar{x})^2 \tag{5-30}$$

As the best estimate of $\sigma^2$, the quantity $s^2$ is found using $(n - 1)$ in place of $n$ in Eq. 5-29. The reasoning for this substitution involves the loss of a degree of freedom by using $\bar{x}$ instead of $\mu$, but a proof is beyond the scope of this text.

The square root of the variance is a statistic known as the *standard deviation* ($\sigma$ or $s$), in which form variability is measured in the same units as the variate and the mean, and hence is easier to interpret and manipulate. The *coefficient of variation* ($C_v$), defined as $\sigma/\mu$ or $s/\bar{x}$, is an expression useful in comparing relative variability.

### Skewness

A fully symmetrical distribution would exhibit the property that all odd moments equal zero. A skewed distribution, however, would have excessive weight to either side of the center and the odd moments would exist. The third moment is used to define skewness:

$$\alpha = \frac{1}{n}\sum_{i=1}^{n}(x_i - \mu)^3 \tag{5-31}$$

The best estimate of the population skewness must be computed by

$$a = \frac{n}{(n-1)(n-2)}\sum_{i=1}^{n}(x_i - \bar{x})^3 \tag{5-32}$$

A *coefficient of skewness* ($C_s$) is represented by $\alpha/\sigma^3$ or $a/s^3$. For symmetrical distributions, the third moment is zero and $C_s = 0$, for right skewness (that is, the long tail to the right side) $C_s > 0$, and for left skewness $C_s < 0$. The property of skewness is of questionable statistical value when it must be estimated from less than 50 sample data points.

***Example 5-2*** Determine the parameters and compare the distributions of annual rainfall for the records shown in Table 5-2.

**Table 5-2** *Annual Rainfall for Selected Cities*

| | Annual rainfall (in.) | | |
|---|---|---|---|
| Year | *Anniston, Ala.* | *Los Angeles, Calif.* | *Richmond, Va.* |
| 1928 | 48 | 9 | 43 |
| 1927 | 49 | 19 | 44 |
| 1926 | 55 | 19 | 38 |
| 1925 | 98 | 9 | 31 |
| 1924 | 43 | 8 | 47 |
| 1923 | 53 | 6 | 49 |
| 1922 | 56 | 15 | 52 |
| 1921 | 47 | 20 | 31 |
| 1920 | 69 | 11 | 51 |
| 1919 | 57 | 9 | 40 |
| 1918 | 61 | 18 | 41 |
| 1917 | 64 | 8 | 43 |
| 1916 | 99 | 23 | 37 |
| 1915 | 54 | 17 | 36 |
| 1914 | 40 | 23 | 34 |
| 1913 | 47 | 17 | 38 |
| 1912 | 58 | 10 | 36 |
| 1911 | 44 | 18 | 37 |
| 1910 | 44 | 5 | 43 |
| 1909 | 64 | 24 | 34 |
| 1908 | 44 | 19 | 53 |
| 1907 | 51 | 15 | 49 |
| 1906 | 71 | 21 | 47 |

*Solution*

| Parameter | Anniston | Los Angeles | Richmond |
|---|---|---|---|
| Mean, $\bar{x}$ | 57.2 in. | 14.9 in. | 41.5 in. |
| Standard deviation, $s$ | 15.5 in. | 5.9 in. | 6.7 in. |
| Coefficient of variance, $C_v = s/\bar{x}$ | 0.27 | 0.40 | 0.16 |
| Coefficient of skewness, $C_s = a/s^3$ | 1.69 | −0.15 | 0.16 |

***Comments:*** (1) Anniston's record shows a high annual average and a fairly large variability. In particular, Anniston's distribution has a pronounced right skew, caused principally by two very large observed values in this short period of record. (2) Los Angeles has a small annual average but a very large variability and a slightly negative skewness. (3) Richmond has the most uniform distribution: a relatively small variability and only a slight positive skewness.

### A Note on Computation

In the past, special forms of the moment equations were developed as an aid for computation by hand or with a desk-top calculator. These had easier expressions for summing and raising to powers, or short-cuts that relied on the frequency of points falling within class intervals. Since the advent of modern electronic computing facilities, however, the need for these forms has largely disappeared. Standard library programs are available for nearly all statistical computations. An example is STIL (Statistical Interpretive Language) an IBM package with extremely simple input. Electronic computation will prove easier and freer of mistakes than laborious handwork.

## 5-5 Types of Probability Distributions

Many well-defined theoretical probability distributions have been used to describe hydrologic processes, and the common ones have been applied successfully to other groups of statistical phenomena. It should be emphasized, however, that any theoretical distribution is not an exact representation of the natural process but only a description which approximates the underlying phenomenon and has proved useful in describing the observed data. Table 5-3 summarizes the common distributions, giving the PDF, mean, and variance of the functions. The distributions presented in the table have enjoyed wide application and have been derived and discussed in many standard textbooks on statistics. In the material to follow, only special aspects of the most used distributions are given. In describing the relationship of distribution parameters to statistical moments, the scheme known as the *method of moments* is followed in fitting theoretical distributions to sample data; that is, the best estimates of population moments are calculated from the sample data and are assumed to be equivalent to the moments of the theoretical distribution. In this way a general solution is obtained for the fitted distribution.

The use of discrete probability distributions is restricted generally to those random events in which the outcome can be described either as a success or failure. Furthermore, the successive trials are independent and the probability of success remains constant from trial to trial.[3,4] In a sense, the common discrete distributions are counting or enumerating techniques.

The binomial distribution is frequently used to approximate other distributions, and vice versa. For example, with discrete values, when $n$ is large and $p$ small (such that $np < 5$ preferably), the binomial approaches the Poisson distribution. This is a single-parameter distribution ($\lambda = np$), and is very useful in describing arrivals in queueing theory. When $p$ approaches one-half and $n$ grows large, the

**Table 5-3** *Table of Common Distributions*

| Distribution of Random Variable $X$ | Probability Distribution Function | Range | Mean $\bar{x}$ or $\mu$ | Variance $s^2$ or $\sigma^2$ |
|---|---|---|---|---|
| Binomial | $P(x) = \frac{n!}{x!(n-x)!} p^x(1-p)^{n-x}$ | $0 \leq x < n$ | $np$ | $np(1-p)$ |
| Poisson | $P(x) = \frac{\lambda^x e^{-\lambda}}{x!}$ | $0 \leq x \leq \cdots$ | $\lambda$ | $\lambda$ |
| Uniform | $f(x) = \frac{1}{b-a}$ | $a \leq x \leq b$ | $(b+a)/2$ | $\frac{(b-a)^2}{12}$ |
| Exponential | $f(x) = \frac{1}{a} e^{-x/a}$ | $0 \leq x \leq \infty$ | $a$ | $a^2$ |
| Normal | $f(x) = \frac{1}{\sigma\sqrt{2\pi}} e^{-(x-\mu)^2 2\sigma^2}$ | $-\infty \leq x \leq \infty$ | $\mu$ | $\sigma^2$ |
| Lognormal ($y = \ln x$) | $f(y) = \frac{1}{\sigma_y\sqrt{2\pi}} \exp[-(y-\mu_y)^2/2\sigma_y^2]$ | $-\infty \leq y \leq \infty$ ($0 \leq x \leq \infty$) | $\mu_y$ | $\sigma_y^2$ |
| Gamma | $f(x) = \frac{x^\alpha e^{-x/\beta}}{\beta^{\alpha+1}\Gamma(\alpha+1)}$ | $0 \leq x \leq \infty$ | $\beta(\alpha+1)$ | $\beta^2(\alpha+1)$ |
| Extreme value | $f(x) = \alpha \exp\{-\alpha(x-u) - e^{-\alpha(x-u)}\}$ | $-\infty \leq x \leq \infty$ | $u + \frac{0.5772}{\alpha}$ | $\frac{\pi^2}{6\alpha^2}$ |

binomial becomes indistinguishable from the normal distribution described in the next section.

## 5-6 Continuous Frequency Distributions

Most hydrologic variables are assumed to come from a continuous random process, and the common continuous distributions are used to fit historical sequences, as in frequency analysis, for example. Other applications are important, too, for continuous distributions. The elementary uniform distribution is the basis for computing random numbers so important in simulation studies. The whole body of material in the area of testing and estimating depends on derived distributions like Student's "*t*," chi-squared, and the *F* distribution. The explanations that follow concern the more common distributions applied in fitting hydrologic sequences. The reader is referred to standard texts for more detailed treatment.[3–6]

### Normal Distribution

The normal distribution is a symmetrical, bell-shaped frequency function, also known as the Gaussian distribution or the natural law of errors. It describes many processes that are subject to random and independent variations. The whole basis for a large body of statistics involving testing and quality control is the normal distribution. Although it often does not perfectly fit sequences of hydrologic data, it has wide application, for example, in dealing with transformed data that subsequently follow the normal distribution and in estimating sample reliability by virtue of the central limit theorem.

The normal distribution has two parameters, the mean, $\mu$, and the standard deviation, $\sigma$, for which $\bar{x}$ and $s$, derived from sample data, are substituted. By a simple transformation, the distribution can be written as a single-parameter function only. Thus when $z = (x - \mu)/\sigma$, $dx = \sigma\, dz$, the PDF becomes

$$f(z) = \frac{1}{\sqrt{2\pi}} e^{-z^2/2} \tag{5-33}$$

and the CDF

$$F(z) = \frac{1}{\sqrt{2\pi}} \int_{-\infty}^{z} e^{-u^2/2}\, du \tag{5-34}$$

The variable $z$ is called the standard unit; it is normally distributed with zero mean and unit standard deviation. Tables of areas under the standard normal curve, as given in Appendix C, Table 1, serve all normal distributions after standardization of the variables.

Given a cumulative probability, for example, the deviate $z$ is found in the table of areas and $x$ is found from the inverse transform:

$$x = \mu + z\sigma \qquad \text{or} \qquad x = \bar{x} + zs \tag{5-35}$$

### Lognormal Distribution

Many hydrologic variables exhibit a marked right skewness, partly due to the influence of natural phenomena having values greater than zero, or some other lower limit, and being unconstrained, theoretically, in the upper range. In such cases, frequencies will not follow the normal distribution, but, fortunately, the variables often are functionally normal and their logarithms follow a normal distribution.[7] Thus the PDF shown in Table 5-3 for the lognormal comes from substituting $y = \ln x$ in the normal. With $\mu_y$ and $\sigma_y$ as the mean and standard deviation, respectively, the following relationships have been found to hold between the characteristics of the untransformed variate $x$ and the transformed variate $y$:[1,7]

$$\mu = \exp\,(\mu_y + \sigma_y^2/2) \tag{5-36}$$

$$\sigma^2 = \mu^2[\exp\,(\sigma_y^2) - 1] \tag{5-37}$$

$$\alpha = [\exp\,(3\sigma_y^2) - 3\,\exp\,(\sigma_y^2) + 2]C_v^3 \tag{5-38}$$

$$C_v = [\exp\,(\sigma_y^2) - 1]^{1/2} \tag{5-39}$$

$$C_s = 3C_v + C_v^3 \tag{5-40}$$

Also $\mu_y = \ln M$, where $M$ is the median value *and* the geometric mean of the $x$'s.

The lognormal is especially useful because the transformation opens the extensive body of theoretical and applied uses of the normal distribution. Since both the normal and lognormal are two-parameter distributions, it is necessary only to compute the mean and variance of the untransformed variate $x$ and solve Eqs. 5-36 and 5-37 simultaneously. Information on three-parameter or truncated lognormal distributions can be found in the literature.[1,7]

### Gamma (and Pearson Type III)

The gamma distribution has wide application in mathematical statistics and has been used increasingly in hydrologic studies now that computing facilities make it easy to evaluate the *gamma function* instead of relying on the painstaking method of using tables of the Incomplete Gamma Function that lead to the CDF, $P(X < x)$. In greater use is a special case of gamma: the *Pearson Type III*. This distribution has been widely adopted as the standard method for flood frequency analysis in a form known as the *Log Pearson III* in which the transform $y = \log x$ is used to reduce skewness.[8–10] Although all three moments are required to fit the distribution, it is extremely flexible in that a zero skew will reduce the Log Pearson III distribution to a lognormal and the Pearson Type III to a normal. Tables of the cumulative function are available and will be explained in a later section.[10,11] A very important property of gamma variates as well as normal variates (including transformed normals) is that the sum of two such variables retains the same distribution. This feature is important in generating synthetic hydrologic sequences.[12,13]

### Gumbel's Extremal Distribution

The theory of extreme values considers the distribution of the largest (or smallest) observations occurring in each group of repeated samples. The distribution of the $n_1$ extreme values taken from $n_1$ samples, with each sample having $n_2$ observations, depends on the distribution of the $n_1 n_2$ total observations. Gumbel was the first to employ extreme value theory for analysis of flood frequencies.[14] Chow has shown that the Gumbel distribution is essentially a lognormal with constant skewness.[15] The CDF of the density function given in Table 5-3 is

$$P(X \leq x) = F(x) = \exp \{- \exp [-\alpha(x - u)]\} \quad \textbf{(5-41)}$$

a convenient form to evaluate the function. Parameters $\alpha$ and $u$ are given as functions of the mean and standard deviation in Table 5-3. Tables of the double exponential are usually in terms of the reduced variate, $y = \alpha(x - u)$.[16] Gumbel also has proposed another extreme value distribution which appears better to fit instantaneous (minimum annual) drought flows.[17, 18]

## 5-7 Frequency Analysis

The statistical methods presented in the earlier sections are used most frequently in describing hydrologic data such as rainfall depths and intensities, peak annual discharge, flood flows, low-flow durations, and the like. In the remainder of this chapter, the fitting of distributions will be restricted to point data, that is, observations at a single station of measurement. Time series, time-dependent information, and correlated variables for regional frequency studies are treated in the next chapter.

Two methods of fitting distributions and using them for frequency analysis will be described: One is a straightforward plotting technique for the cumulative distribution and the other considers the use of frequency factors. The cumulative distribution provides a rapid means of determining the probability of the event equal to or less than some specified quantity. This attribute is used to obtain recurrence intervals for observed data. As a general rule, frequency analysis should be avoided when working with records shorter than 10 yr, and in estimating frequencies of expected hydrologic events greater than twice the record length.

## 5-8 Plotting Distributions

Summarizing cumulative distribution functions is simplified by employing probability paper, so constructed that the CDF for certain well-balanced distributions plots as a straight line. Normal probability paper, for example, has an origin at the midpoint, or the line of 50% probability, then distorts the scales equally and increasingly on either side to infinity. Thus a symmetrical distribution whose CDF is a perfect S-curve when plotted on cartesian coordinates becomes a straight line on probability paper. The other scale of the special paper is kept in arithmetic units. Similarly, the lognormal distribution becomes a straight-line CDF by using the same probability scale with the other axis in logarithmic values. In both cases the mean of the normal variate occurs at the 50% point, and variations at equal deviations above and below the mean will plot as a straight line. A special paper for the Gumbel extreme value distribution uses recurrence interval rather than its reciprocal, the exceedence probability.

The probability of the event can be obtained by use of a "plotting position." When annual maximum values are being analyzed, for example, the recurrence interval is defined as the mean time in years, with $N$ future trials, for the $m$th largest value among annual maxima to be exceeded once on the average. The mean number of exceedences for this condition can be shown to be

$$\bar{x} = N \frac{m}{n+1} \tag{5-42}$$

where

$\bar{x}$ = the mean number of exceedences
$N$ = the number of future trials
$n$ = the number of values
$m$ = the rank of descending values, with largest equal to 1

If the mean number of exceedences $\bar{x} = 1$, $N = T$, and

$$T = \frac{n+1}{m} \tag{5-43}$$

indicating that the recurrence interval is equal to the number of years of record plus 1, divided by the rank of the event.

Several plotting position formulas are available[19] and are noted in Table 5-4. The range in recurrence intervals obtained for 10 years of record can be observed in the right-hand column. These formulas

**Table 5-4** *Plotting Position Formulas*

| Method | Solve for $P(X > x)$ | For $m = 1$ and $n = 10$ | |
|---|---|---|---|
| | | $P$ | $T$ |
| California | $\frac{m}{n}$ | .10 | 10 |
| Hazen | $(2m - 1)/2n$ | .05 | 20 |
| Beard | $1 - (0.5)^{1/n}$ | .067 | 14.9 |
| Weibull | $\frac{m}{n+1}$ | .091 | 11 |
| Chegadayev | $\frac{m - 0.3}{n + 0.4}$ | .067 | 14.9 |
| Blom | $\frac{m - 3/8}{n + 1/4}$ | .061 | 16.4 |
| Tukey | $\frac{3m - 1}{3n + 1}$ | .065 | 15.5 |

do not account for the sample size, that is, the length of record. A general formula that does account for sample size was given by Gringorten[20] and has the general form

$$T = \frac{n + 1 - 2a}{m - a} \tag{5-44}$$

where

$n$ = the number of years of record
$m$ = the rank
$a = 0 < a < 1$, depending upon $n$

| $n$ | 10 | 20 | 30 | 40 | 50 |
|---|---|---|---|---|---|
| $a$ | .448 | .443 | .442 | .441 | .440 |
| $n$ | 60 | 70 | 80 | 90 | 100 |
| $a$ | .440 | .440 | .440 | .439 | .439 |

The technique in all cases is to arrange the data in increasing or decreasing order of magnitude and to assign order number $m$ to the ranked values. The most efficient formula for computing plotting positions for unspecified distributions,[19] and the one now commonly used for most sample data, is

$$P = \frac{m}{n + 1} \tag{5-45}$$

When $m$ is ranked from lowest to highest, $P$ is an estimate of the probability of values being equal to or less than the ranked value, that is, $P(X \leq x)$; when the rank is from highest to lowest, $P$ is $P(X \geq x)$. For probabilities expressed in percentages, the value is $100m/(n + 1)$.

## 5-9 Frequency Factors

Chow[21] has proposed use of

$$x = \bar{x} + Ks \tag{5-46}$$

as the general equation for hydrologic frequency analysis, where $K$ is the *frequency factor.* It literally is the number of standard deviations above and below the mean to attain the probability point of interest. For two-parameter distributions it varies only with probability, or the equivalent in frequency analysis, the recurrence interval, $T$. It varies with the coefficient of skewness in skewed distributions and can be affected greatly by the number of years of record. For the normal distribution, and for the transformed variate in a log-normal distribution, the deviate $z$ given in standard normal tables is synonymous with the frequency factor.

### Log Pearson III

Frequency factors for the Pearson Type III (logarithmic or arithmetic) are shown in Appendix C for various recurrence intervals (or exceedence probabilities) and skew coefficients. As outlined by the Water Resources Council,[10] the fitting technique involves transforming annual floods to logarithmic values ($y_i = \log x_i$) and finding the mean, standard deviation, and skew coefficients of the logarithms. Flood magnitudes ($Q$) are estimated from

$$\log Q = \bar{y} + Ks_y \tag{5-47}$$

which is the same form as Eq. 5-46. Note that $K = \phi(T, C_s)$, a function of both recurrence interval and skewness. Because the skewness coefficient has a much greater variability than the mean or standard deviation, Beard[9] has recommended that only average regional coefficients of skew be employed in flood analysis for a single station unless the record exceeds 100 years. In practice this may be impractical to attain, and it will be best to compute all parameters and compare results with any other records, experience, or regional studies. The use of logarithms to reduce the skewness of an already skewed distribution helps. Hazen recommended that the computed skewness for Pearson III analysis be multiplied by a factor of $(1 + 8.5/n)$ to obtain an adjusted skewness when dealing with small samples.[22] Chow has developed $K$ versus $T$ curves for the distribution.[1]

***Example 5-3*** For rainfall data developed in Example 5-2, fit distribution functions to the records of Richmond, Virginia, and Los Angeles, California.

*Solution*

Both distributions exhibit small skewness and are approximately normal. For the purposes of illustration, the Richmond data are fitted with the normal and the Los Angeles data with a Pearson Type III.

1. The data are arrayed for plotting in Table 5-5.

***Table 5-5***

| $m$ | Richmond | Los Angeles | $100m/(n+1)$ |
|---|---|---|---|
| 1 | 53 | 24 | 4.2 |
| 2 | 52 | 23 | 8.3 |
| 3 | 51 | 23 | 12.5 |
| 4 | 49 | 21 | 16.7 |
| 5 | 49 | 20 | 20.8 |
| 6 | 47 | 19 | 25.0 |
| 7 | 47 | 19 | 29.2 |
| 8 | 44 | 19 | 33.3 |
| 9 | 43 | 18 | 37.5 |
| 10 | 43 | 18 | 41.7 |
| 11 | 43 | 17 | 45.8 |
| 12 | 41 | 17 | 50.0 |
| 13 | 40 | 15 | 54.7 |
| 14 | 38 | 15 | 58.3 |
| 15 | 38 | 11 | 62.5 |
| 16 | 37 | 10 | 66.7 |
| 17 | 37 | 9 | 70.8 |
| 18 | 36 | 9 | 75.0 |
| 19 | 36 | 9 | 79.7 |
| 20 | 34 | 8 | 83.3 |
| 21 | 34 | 8 | 87.5 |
| 22 | 31 | 6 | 91.7 |
| 23 | 31 | 5 | 95.8 |

The points are plotted in Figs. 5-7 and 5-8 as exceedence probability (left-hand sale) versus inches of rainfall.

2. The theoretical normal of best fit is a straight line through $(\bar{x} - s)$ at 15.9%, $\bar{x}$ at 50%, and $(\bar{x} + s)$ at 84.1%. Thus for Richmond,

| $X$ | Plotting Position (right-hand scale) |
|---|---|
| $\bar{x} - s = 41.5 - 6.7 = 34.8$ | 15.9 |
| $\bar{x} = 41.5$ | 50.0 |
| $\bar{x} + s = 41.5 + 6.8 = 48.3$ | 84.1 |

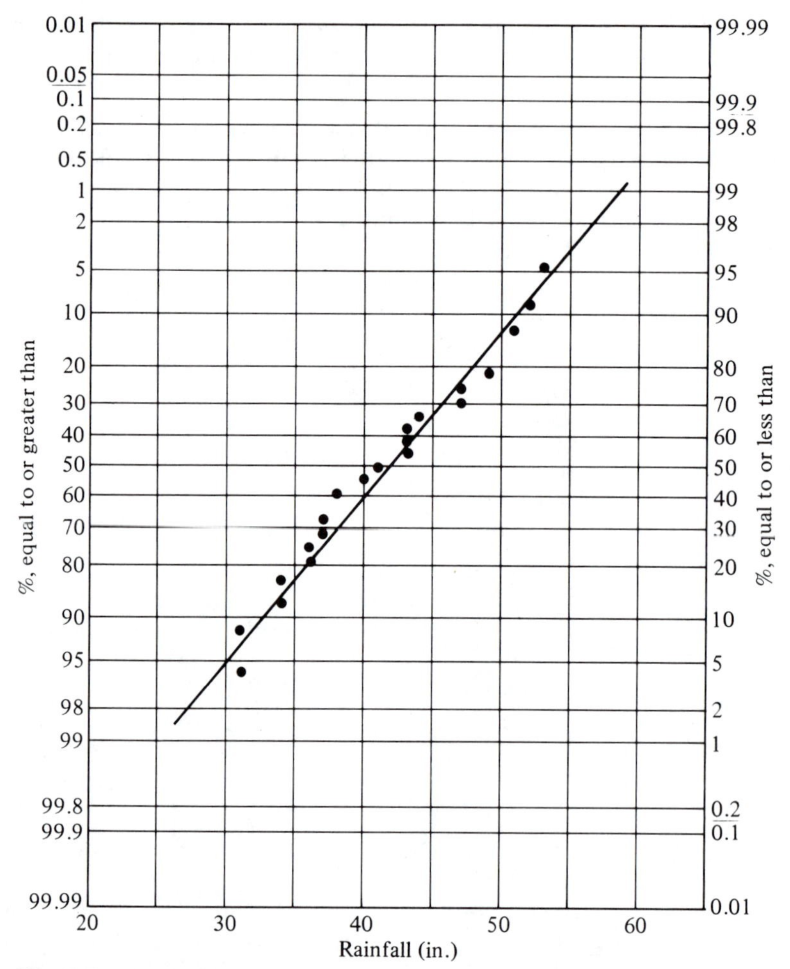


***Fig. 5-7.*** *Annual rainfall for Richmond, Virginia, 1906–1928, plotted on normal probability paper.*

3. The plotting positions (read as % chance) come from Table II, Appendix C, for the computed skewness. Thus for Los Angeles,

| % chance | 99 | 95 | 90 | 80 | 50 | 20 | 10 | 4 | 2 | 1 | 0.5 |
|---|---|---|---|---|---|---|---|---|---|---|---|
| $K(C_s = -0.16)$ | −2.44 | −1.69 | −1.30 | −0.83 | 0.03 | 0.85 | 1.26 | 1.69 | 1.97 | 2.21 | 2.43 |
| $x = 14.9 + K(5.9)$ | 0.5 | 4.9 | 7.2 | 10.0 | 15.1 | 19.9 | 22.3 | 24.9 | 26.5 | 27.9 | 29.2 |

## Gumbel's Extreme Value

Equation 5-41 can be solved for the recurrence interval $T$ and for the variate $x$, as follows:

$$\frac{1}{T} = 1 - F(x) = 1 - \exp\{-\exp[-\alpha(x - u)]\} \tag{5-48}$$

$$x = u - \frac{1}{\alpha} \ln [\ln T - \ln (T - 1)] \tag{5-49}$$

On substituting into Eq. 5-46, the general frequency equation, with $u$ and $\alpha$ as defined in Table 5-3, the frequency factor for the extreme value distribution becomes

$$K = -\frac{\sqrt{6}}{\pi}\left(0.5772 + \ln \ln \frac{T}{T - 1}\right) \tag{5-50}$$

It should be noted that this expression for $K$ is valid only in the limit, that is, as $n$ approaches infinity. For a finite sample, $K$ varies

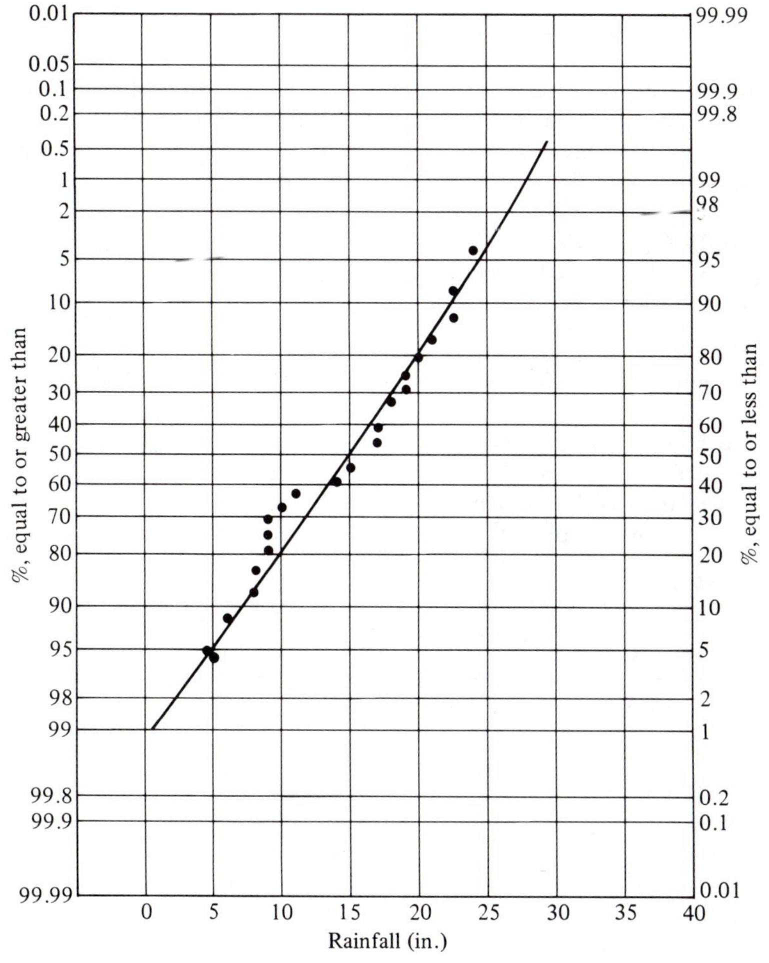


**Fig. 5-8.** *Annual rainfall for Los Angeles, California, 1906–1928.*

**Table 5-6** *Extreme-Value Frequency Factors*

| Sample Size | Recurrence Interval | | | | | | |
|---|---|---|---|---|---|---|---|
| | *10* | *20* | *25* | *50* | *75* | *100* | *1000* |
| 15 | 1.703 | 2.410 | 2.632 | 3.321 | 3.721 | 4.005 | 6.265 |
| 20 | 1.625 | 2.302 | 2.517 | 3.179 | 3.563 | 3.836 | 6.006 |
| 25 | 1.575 | 2.235 | 2.444 | 3.088 | 3.463 | 3.729 | 5.842 |
| 30 | 1.541 | 2.188 | 2.393 | 3.026 | 3.393 | 3.653 | 5.727 |
| 40 | 1.495 | 2.126 | 2.326 | 2.943 | 3.301 | 3.554 | 5.476 |
| 50 | 1.466 | 2.086 | 2.283 | 2.889 | 3.241 | 3.491 | 5.478 |
| 60 | 1.446 | 2.059 | 2.253 | 2.852 | 3.200 | 3.446 | |
| 70 | 1.430 | 2.038 | 2.230 | 2.824 | 3.169 | 3.413 | 5.359 |
| 75 | 1.423 | 2.029 | 2.220 | 2.812 | 3.155 | 3.400 | |
| 100 | 1.401 | 1.998 | 2.187 | 2.770 | 3.109 | 3.349 | 5.261 |

with the sample or length of record as shown in Table 5-6. $K$ versus $T$ curves have also been developed.[1] In Eq. 5-50, when $K = 0$, $T = 2.33$ yr; thus in flood frequency analysis the recurrence interval of the mean annual flood is commonly designated the 2.33-yr event.

***Example 5-4*** The mean of the annual maximum discharges at a streamflow site with 25 yr of record is 1000 cfs. The standard deviation is 400 cfs. Estimate the magnitude of the 50-yr flood.

*Solution*

From Table 5-6, $K = 3.088$; $x = \bar{x} + Ks = 1000 + 3.088(400) = 2235$ cfs.

***Example 5-5*** The annual maximum discharge data in Table 5-7 have been obtained from the U.S. Geological Survey Water Resources Division for a small stream in Missouri. Rank the data and plot on extreme-value probability paper. The data are plotted in Fig. 5-9.

**Table 5-7**

| Water Year | Annual Maximum Discharge (cfs) | Rank | $(n + 1)/m$ |
|---|---|---|---|
| 1967 | 2510 | 8 | 1.375 |
| 1966 | 4150 | 1 | 11.0 |
| 1965 | 2990 | 5 | 2.2 |
| 1964 | 2120 | 10 | 1.1 |
| 1963 | 3555 | 2 | 5.5 |
| 1962 | 2380 | 9 | 1.22 |
| 1961 | 2550 | 7 | 1.57 |
| 1960 | 2800 | 6 | 1.83 |
| 1959 | 3300 | 3 | 3.67 |
| 1958 | 3150 | 4 | 2.75 |

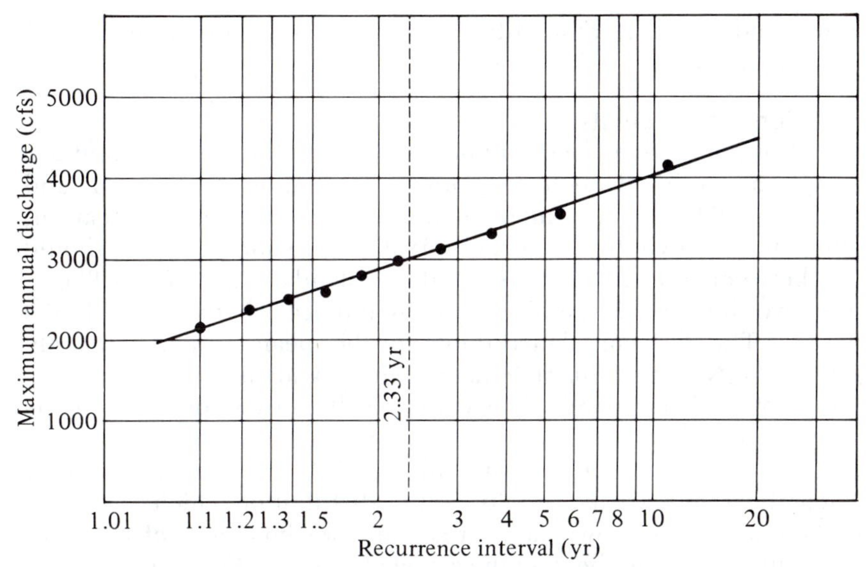


***Fig. 5-9.*** *Annual floods on a small Missouri stream.*

## 5-10 Reliability of Statistical and Frequency Studies

A truly remarkable development of theoretical statistics is the *central limit theorem.* As a consequence of the law of large numbers, the central limit theorem states that for a population with finite variance $\sigma^2$ and a mean $\mu$, the distribution of sample means—that is, a number of equally good means from repeated samples—will be distributed themselves as a normal distribution with mean $\mu$ and a variance equal to $\sigma^2/n$, where $\sigma$ is the population standard deviation. This theorem does not limit the type of underlying population distribution but says that the distribution of the sample means will approach a normal distribution as the sample size increases. The statistic $\sigma/\sqrt{n}$ is the standard deviation of the distribution of means and is called the *standard error* of the mean. Listed in Table 5-8 are several parameters of distributions and their standard errors. It is apparent that standard

***Table 5-8***

| Measure | Standard Error |
|---|---|
| Mean | $\sigma/\sqrt{n}$ |
| Standard deviation | $\sigma/\sqrt{2n}$ |
| Coefficient of variation | $C_v\sqrt{1 + 2C_v^2}/\sqrt{2n}$ |
| Coefficient of skewness | $\sqrt{6n(n-1)/(n+1)(n-2)(n+3)}$ |

errors, and therefore reliability, are almost completely a function of the sample size.

### Confidence Limits

It is possible to place confidence limits on the measurement of a sample mean based upon the normal distribution of all means and regardless of the underlying population. Hence, as mentioned earlier, approximately two-thirds of the observations of a normal variate should fall between the limits of +1 and −1 standard deviation. Therefore, two-thirds of all sample means should occur between the limits $\pm \sigma/\sqrt{n}$. The 95% confidence limits for the mean are approximately $\mu \pm 2\sigma/\sqrt{n}$. Note that this statement requires knowledge of the underlying population variance. However, usually only $s^2$ instead of $\sigma^2$ is known and a slightly different technique is needed to estimate confidence limits for a sample mean with $\bar{x}$ and the standard deviation $s$. Theoretical statistics require Student's $t$ distribution which converges to the normal as $n$ grows large. It is one of a whole class of sampling distributions that are beyond the scope of this text. For more information in the field of inferential statistics—in particular, hypothesis testing and statistical decision theory—the reader must turn to basic sources.

Approximate error limits or control curves can be placed on frequency curves. A method proposed by Beard[9] involves placing lines above and below the fitted curve to form a reliability band. Table 5-9

**Table 5-9** *Error Limits for Flood Frequency Curves*

| Years of Record ($n$) | Exceedence Frequency (%, at 5% level) | | | | | | |
|---|---|---|---|---|---|---|---|
| | 99.9 | 99 | 90 | 50 | 10 | 1 | 0.1 |
| 5 | 1.22 | 1.00 | 0.76 | 0.95 | 2.12 | 3.41 | 4.41 |
| 10 | 0.94 | 0.76 | 0.57 | 0.58 | 1.07 | 1.65 | 2.11 |
| 15 | 0.80 | 0.65 | 0.48 | 0.46 | 0.79 | 1.19 | 1.52 |
| 20 | 0.71 | 0.58 | 0.42 | 0.39 | 0.64 | 0.97 | 1.23 |
| 30 | 0.60 | 0.49 | 0.35 | 0.31 | 0.50 | 0.74 | 0.93 |
| 40 | 0.53 | 0.43 | 0.31 | 0.27 | 0.42 | 0.61 | 0.77 |
| 50 | 0.49 | 0.39 | 0.28 | 0.24 | 0.36 | 0.54 | 0.67 |
| 70 | 0.42 | 0.34 | 0.24 | 0.20 | 0.30 | 0.44 | 0.55 |
| 100 | 0.37 | 0.29 | 0.21 | 0.17 | 0.25 | 0.36 | 0.45 |
| | 0.1 | 1 | 10 | 50 | 90 | 99 | 99.9 |
| | Exceedence frequency (%, at 95% level) | | | | | | |

*Note:* Tabular values are multiples of the standard deviation of the variate. Five percent error limits are added to the flood value from the fitted curve at the same exceedence frequency and the sum plotted. Ninety-five % limits are substracted from the flood value at the same exceedence frequency. Log values are added or subtracted before antilogging and plotting.

shows the factors by which the standard deviations of the variate must be multiplied to mark off a 90% reliability band above and below the frequency curve. The 5% level, for example, means that only 5% of future values should fall higher than the limit, and, similarly, only 5% should fall under the 95% limit. Nine of ten should fall within the band.

***Example 5-6*** The maximum annual instantaneous flows from the Maury River near Lexington, Virginia, for a 26-yr period are listed in Table 5-10.

***Table 5-10***

| Water (yr) | Discharge (cfs) | Water (yr) | Discharge (cfs) | Water (yr) | Discharge (cfs) |
|---|---|---|---|---|---|
| 1926 | 6,730 | 1935 | 13,800 | 1944 | 6,680 |
| 1927 | 9,150 | 1936 | 40,000 | 1945 | 6,540 |
| 1928 | 6,310 | 1937 | 10,200 | 1946 | 5,560 |
| 1929 | 10,000 | 1938 | 13,400 | 1947 | 7,700 |
| 1930 | 15,000 | 1939 | 8,950 | 1948 | 8,630 |
| 1931 | 2,950 | 1940 | 11,900 | 1949 | 14,500 |
| 1932 | 8,650 | 1941 | 5,840 | 1950 | 23,700 |
| 1933 | 11,100 | 1942 | 20,700 | 1951 | 15,100 |
| 1934 | 6,360 | 1943 | 12,300 | | |

Plot the Log Pearson III curve of best fit and determine the magnitude of the flood to be equaled or exceeded once in 5, 10, 50, and 100 yr. Using Table 5-9, also plot the upper and lower confidence limits.

*Solution*

1. The basic statistical calculations are summarized as follows:

| | Arithmetic | Log |
|---|---|---|
| Mean: $\bar{x}$ | 11,606 | 4.001 |
| Variance: $s^2$ | $53.87 \times 10^6$ | .0516 |
| Skew coefficient: $C_s$ | 2.4 | .38 |

2. After forming an array and computing plotting positions, the data are plotted on Fig. 5-10.

3. Plotting data for Log Pearson III (Table 5-11) are developed from Table II, Appendix C. Confidence limits are plotted in Fig. 5-10 using Table 5-9.

**Table 5-11**

| Chance (%) | ($C_s = .38$) $K$ | ($\bar{y} = 4.001$) ($s_y = .227$) $\bar{y} + Ks_y = \log Q$ | $Q$ |
|---|---|---|---|
| 99 | −2.050 | 3.535 | 3,424 |
| 95 | −1.532 | 3.653 | 4,496 |
| 90 | −1.241 | 3.719 | 5,235 |
| 80 | −0.855 | 3.760 | 5,752 |
| 50 | −0.062 | 3.987 | 9,700 |
| 20 | 0.818 | 4.187 | 15,370 |
| 10 | 1.315 | 4.300 | 19,930 |
| 4 | 1.872 | 4.426 | 26,670 |
| 2 | 2.248 | 4.511 | 32,470 |
| 1 | 2.597 | 4.590 | 38,950 |
| .5 | 2.924 | 4.665 | 46,290 |

## 5-11 Additional Frequency Analyses

In earlier examples of frequency analysis, only the series of annual maximum or minimum occurrences in the hydrologic record have been described. These extremes constitute an *annual series* that is consistent with frequency analysis and the manipulation of annual probabilities of occurrence. All of the observed data—say all floods or all the daily streamflows—would constitute a *complete series.* Any subset of the complete series is a *partial series.* The partial series examined most commonly is a subset other than the annual series. In selecting the maximum annual events from a record, it often happens that the second greatest event in one year exceeds the annual maximum in some other year. Analysis of the annual series neglects such events. Although they generally contain the same number of events, the extreme values analyzed without regard for the period (that is, year) of occurrence, is usually termed the *partial duration series.*

In Table 5-12 the maximum rainfall depths that occurred for any 30-min period during excessive rainfalls at Baltimore, Maryland, 1945–1954, are shown in the order of occurrence. The 65 observations represent a complete series. The 11 maximum annual events are underlined and represent the annual series. The greatest 11 events throughout the record are identified by an asterisk and represent the partial duration series.

The larger numbers occur in both series, and hence recurrence intervals for the less frequent events are the same. The theoretical differences in recurrence intervals based on annual- and partial duration series of the same length are shown in Table 5-13. The difference for intervals greater than 10 yr is negligible. The following example is illustrative.

***Example 5-7*** Perform a frequency analysis of the 30-min Baltimore rainfall data in Table 5-12 as an annual and a partial duration series and plot the results.

*Solution*

See Table 5-14. The data are plotted in Fig. 5-11.

The preceding example leads to consideration of the frequency analysis of rainfall depth or intensity for various durations of rainfall. Design problems often require the estimation of expected intensities for a critical time period. Frequency analysis of the rainfall record for periods other than the 30-min duration, for example, the maximum 5-, 10-, 20-, and 60-min occurrences, would yield a family of curves

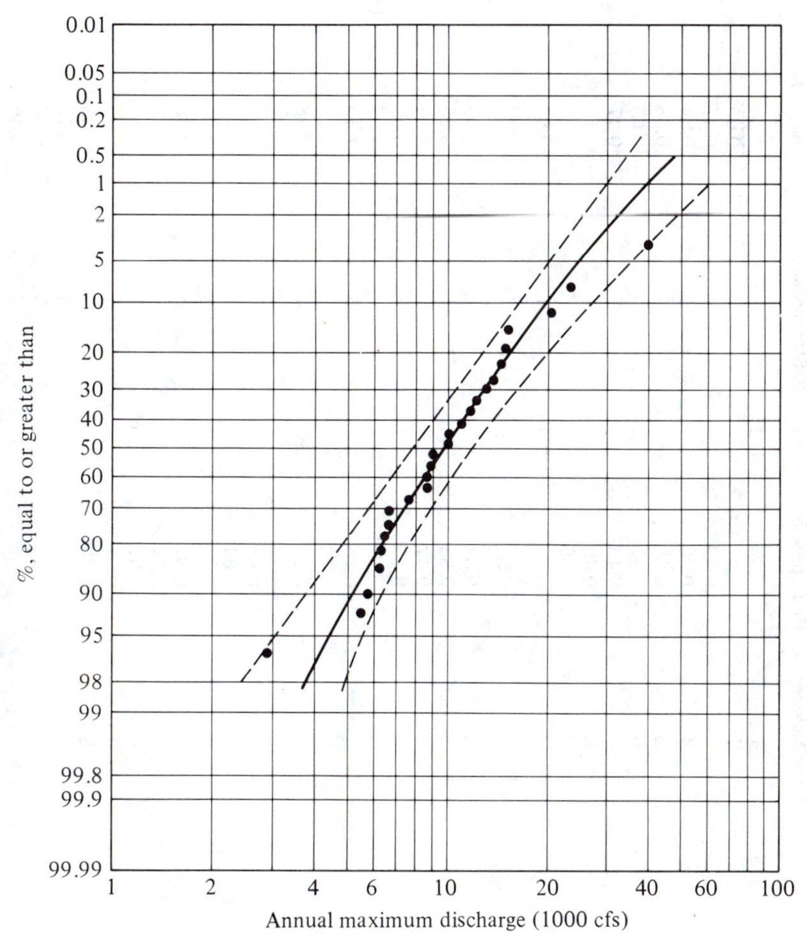


**Fig. 5-10.** *Maximum instantaneous annual flows, Maury River, Lexington, Virginia.*

**Table 5-12** *Maximum 30-Min Rainfall Depths, Baltimore, Md, 1945–1954*

| Year | Storm Number | RF Depth (in.) | Year | Storm Number | RF Depth (in.) | Year | Storm Number | RF Depth (in.) |
|---|---|---|---|---|---|---|---|---|
| 1945 | 1 | 0.38 | 1948 | 1 | 1.33* | 1953 | 1 | 0.40 |
| | 2 | 0.47 | | 2 | 0.65 | | 2 | 0.45 |
| | 3 | 0.39 | | 3 | 0.47 | | 3 | 0.53 |
| | 4 | 0.76 | | 4 | 0.84 | | 4 | 2.50* |
| | 5 | 0.56 | | 5 | 0.68 | | 5 | 1.03 |
| | 6 | 0.35 | | 6 | 0.63 | | 6 | 0.75 |
| | 7 | 0.43 | | 7 | 0.47 | | 7 | 0.70 |
| | 8 | 0.40 | | | | | 8 | 1.00* |
| | 9 | 0.36 | | | | | | |
| | | | 1949 | 1 | 0.52 | | | |
| | | | | 2 | 0.49 | | | |
| 1946 | 1 | 0.62 | | | | 1954 | 1 | 0.42 |
| | 2 | 0.55 | | | | | 2 | 0.70 |
| | 3 | 0.88 | 1950 | 1 | 0.55 | | 3 | 0.85 |
| | 4 | 0.47 | | 2 | 0.63 | | 4 | 0.60 |
| | 5 | 0.36 | | 3 | 0.69 | | | |
| | 6 | 1.15* | | 4 | 1.27* | 1955 | 1 | 0.70 |
| | 7 | 0.75 | | 5 | 1.10* | | 2 | 0.95 |
| | 8 | 1.53* | | | | | 3 | 1.02 |
| | 9 | 0.51 | 1951 | 1 | 0.88 | | 4 | 0.50 |
| | | | | 2 | 0.97 | | 5 | 0.65 |
| 1947 | 1 | 0.88 | | 3 | 0.59 | | 6 | 0.55 |
| | 2 | 2.04* | | 4 | 0.46 | | 7 | 0.52 |
| | 3 | 0.76 | | 5 | 0.50 | | 8 | 0.45 |
| | 4 | 0.97 | | 6 | 0.55 | | 9 | 0.54 |
| | 5 | 0.71 | | | | | 10 | 0.60 |
| | 6 | 1.07* | 1952 | 1 | 0.47 | | 11 | 0.80 |
| | 7 | 0.94 | | 2 | 1.20* | | 12 | 0.95 |
| | 8 | 1.20* | | 3 | 0.93 | | | |
| | | | | 4 | 0.70 | | | |
| | | | | 5 | 0.57 | | | |
| | | | | 6 | 0.46 | | | |
| | | | | 7 | 0.48 | | | |
| | | | | 8 | 1.30* | | | |

*Note:* Underlined items are the annual series. Asterisks identify the partial duration series.

***Table 5-13*** *Relationship Between the Partial Duration Series and the Annual Series*

| Recurrence Interval (yr) | |
|---|---|
| *Partial Duration Series* | *Annual Series* |
| 0.5 | 1.2 |
| 1.0 | 1.6 |
| 1.5 | 2.0 |
| 2.0 | 2.5 |
| 5.0 | 5.5 |
| 10.0 | 10.5 |

similar to those of Fig. 5-11. The usual method of presenting these data is to convert depth in inches to an intensity in in./hr and to summarize the data in intensity-duration-frequency curves as shown in Fig. 5-12. These curves are typical of the point analysis of rainfall data. It should be emphasized that the frequency curves join occurrences that are not necessarily from the same storm; that is, they do not represent a sequence of intensities during a single storm but only the average intensity expected for a specific duration. Fitting the curves and consideration of regional effects is covered in Chapter 6.

Figures 5-13 through 5-16 illustrate further applications of frequency analysis. Figures 5-13 and 5-14 represent standard point frequency analyses of the annual series of high and low flows for different durations. Figure 5-15 is a low-flow-duration-frequency curve based on the same data as Fig. 5-14. Figure 5-16 is based on an analysis of the *complete* series of daily flows although all observed values

***Table 5-14***

| | Depth (in.) | | Recurrence Interval |
|---|---|---|---|
| Order | *Annual Series* | *Partial Series* | $(n + 1)/m$ |
| 1 | 2.50 | 2.50 | 12 |
| 2 | 2.04 | 2.04 | 6 |
| 3 | 1.53 | 1.53 | 4 |
| 4 | 1.33 | 1.33 | 3 |
| 5 | 1.30 | 1.30 | 2.4 |
| 6 | 1.27 | 1.27 | 2 |
| 7 | 1.02 | 1.20 | 1.7 |
| 8 | 0.97 | 1.20 | 1.5 |
| 9 | 0.85 | 1.15 | 1.3 |
| 10 | 0.76 | 1.10 | 1.2 |
| 11 | 0.52 | 1.07 | 1.1 |

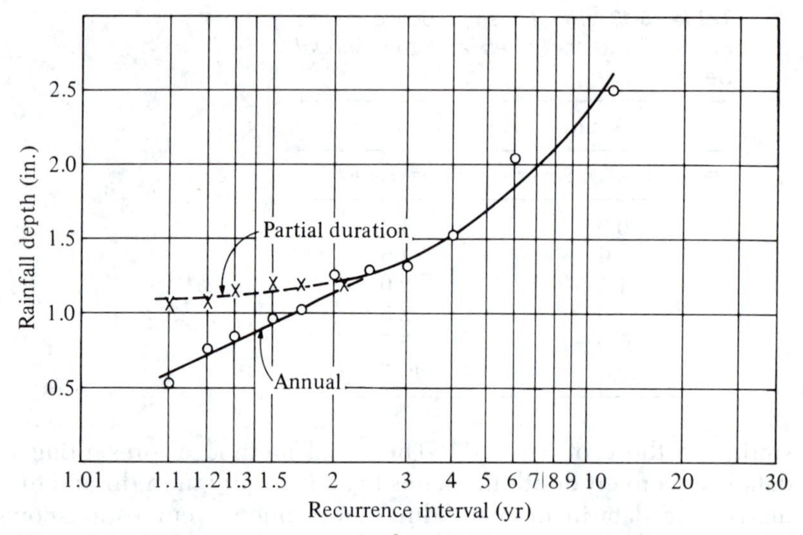


***Fig. 5-11.*** *Difference in annual and partial duration series 11-yr record of maximum 30-min durations, Baltimore, Maryland.*

are not plotted. Presented in this form such a curve usually is called a *duration* curve. Note that the probability scale must be labeled "percent of time," since the annual series was not used. Duration curves are useful in predicting the availability and variability of sustained flows, but, again, they do not represent the actual sequence of flows.

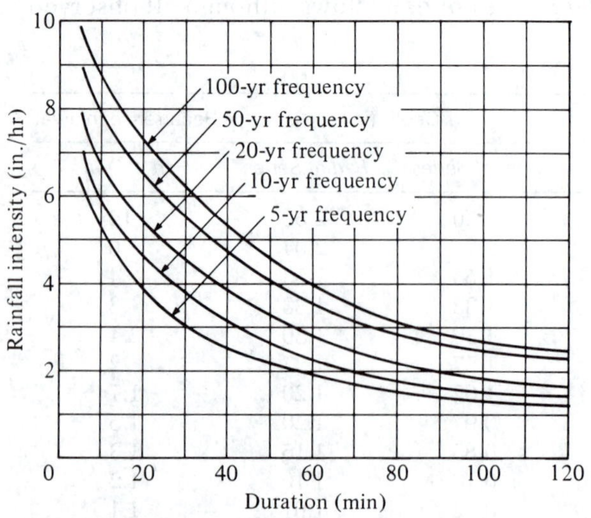


***Fig. 5-12.*** *Typical intensity-duration-frequency curves.*

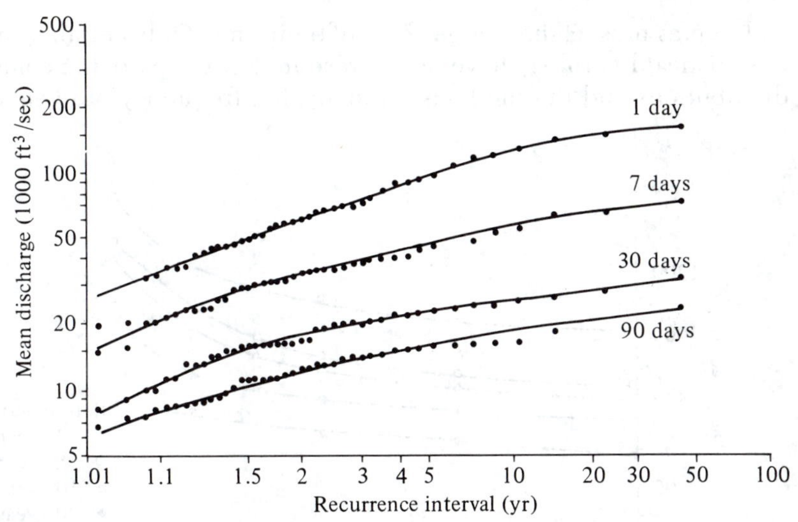


***Fig. 5-13.*** *High flow frequency curves, in the James River at Cartersville, Virginia. (Virginia Division of Water Resources.)*

## Concluding Comment

Statistics is not an end in itself, and the treatment in this chapter has been nothing more than an introduction to statistics. Serious students and practitioners must return again and again to the theory in standard works. They will find that evaluating new developments and

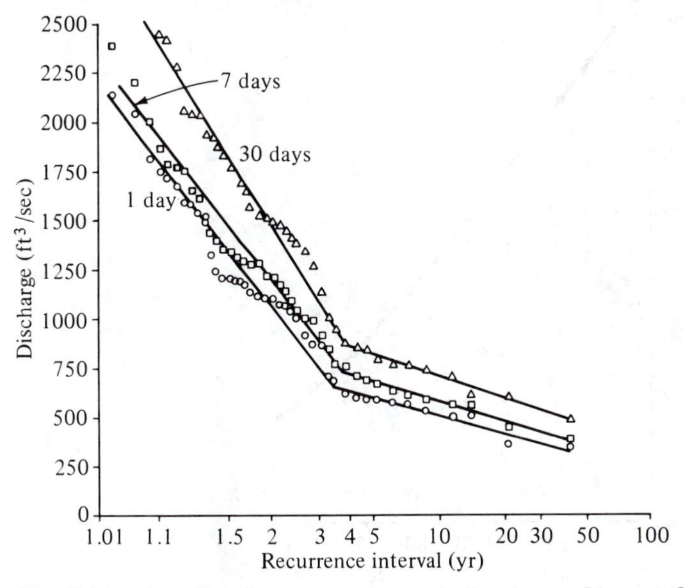


***Fig. 5-14.*** *Low flow frequency curves in the James River at Cartersville, Virginia. (Virginia Division of Water Resources.)*

techniques must claim a large share of their time. Only certain aspects of statistical hydrology have been presented, principally the common distributions and the methods for analyzing frequency of events ob-

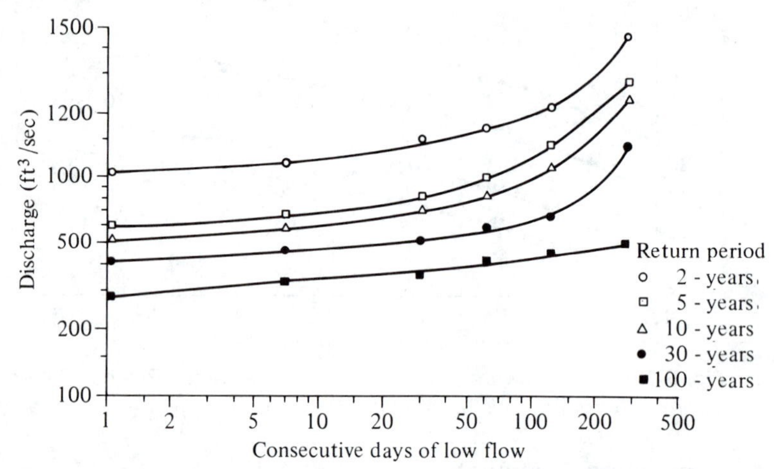


**Fig. 5-15.** *Low flow duration frequency curves in the James River Basin. Discharge to consecutive days of low flow James River at Cartersville, Virginia. Drainage area: 6242 mi². (Virginia Division of Water Resources.)*

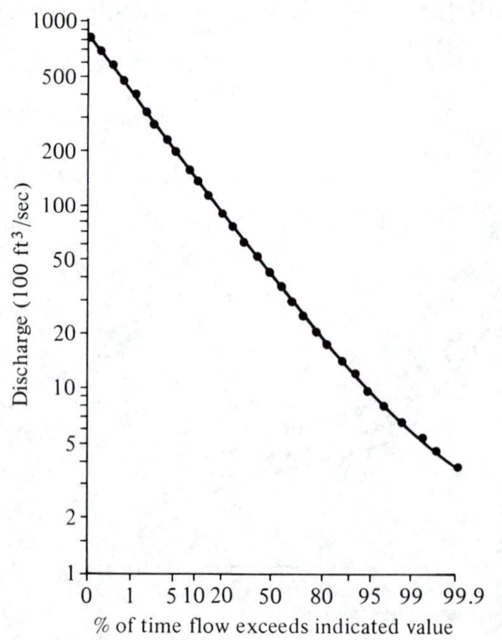


**Fig. 5-16.** *Flow duration curve in the James River at Cartersville, Virginia. (Virginia Division of Water Resources.)*

served at a single point. In the next chapter this information will be extended to joint variables and time series.

## Problems

5-1. The probabilities of events $E_1$ and $E_2$ are each .3. What is the probability that $E_1$ or $E_2$ will occur when (a) the events are independent but not mutually exclusive, and (b) when the probability of $E_1$, given $E_2$ is .1?

5-2. Events $A$ and $B$ are independent events having marginal probabilities of .4 and .5, respectively. Determine for a single trial: (a) the probability that both $A$ and $B$ will occur simultaneously, and (b) the probability that neither occurs.

5-3. The conditional probability, $P(E_1 \mid E_2)$, of a power failure (given that a flood occurs) is .9, and the conditional probability, $P(E_2 \mid E_1)$, of a flood (given that a power failure occurs) is .2. If the joint probability, $P(E_1 \text{ and } E_2)$, of a power failure and a flood is .1, determine the marginal probabilities, $P(E_1)$ and $P(E_2)$.

5-4. A temporary cofferdam is to be built to protect the 5-yr construction activity for a major cross valley dam. If the cofferdam is designed to withstand the 20-yr flood, what is the risk that the structure will be overtopped (a) in the first year, (b) in the third year exactly, (c) at least once in the 5-yr construction period, and (d) not at all during the 5-yr period?

5-5. What return period must an engineer use in his design of a bridge opening if he is willing to accept only a 50% risk that flooding will occur at least once in two successive years? Repeat for a risk of 100%.

5-6. A temporary flood wall has been constructed to protect several homes in the flood plain. The wall was built to withstand any discharge up to the 20-yr flood magnitude. The wall will be removed at the end of the 3-yr period after all the homes have been relocated. Determine the probability that

(a) The wall will be overtopped in any year.
(b) The wall will not be overtopped during the relocation operation.
(c) The wall will be overtopped at least once before all of the homes are relocated.
(d) The wall will be overtopped exactly once before all of the homes are relocated.
(e) The wall will be adequate for the first 2 yr and then overtopped in the third year.

5-7. Assume that the channel capacity of 12,000 cfs in Oak Creek near a private home on Capitol Beach Lake was equaled or exceeded in 3 of the past 60 yr. Find:

(a) The frequency of the 12,000-cfs value.
(b) The probability that the home will be flooded next year.
(c) The return period of the 12,000-cfs value.
(d) The probability that the home will not be flooded next year.
(e) The probability of two consecutive, safe years.
(f) The probability of a flood at least once in the next 20 yr.
(g) The probability of a flood in the second, but not the first, of 2 consecutive years.

(h) The probability of having floods in exactly 3 of the next 60 yr.
(i) The 20-yr flood risk.

5-8. The distribution of mean annual rainfall at 35 stations in the James River Basin, Va., is given in the following summary:

| Interval (2 in. groupings) | 36 or 37 in. | 38 or 39 in. | 40 or 41 in. | 42 or 43 in. |
|---|---|---|---|---|
| Number of observations | 2 | 4 | 7 | 9 |
| Interval (2 in. groupings) | 44 or 45 in. | 46 or 47 in. | 48 or 49 in. | 50 or 51 in. |
| Number of observations | 5 | 4 | 2 | 2 |

Compute the relative frequencies and plot the frequency distribution and the cumulative distribution. Estimate the probability that the mean annual rainfall will (a) exceed 40 in., (b) will exceed 50 in., and (c) will be between these values.

5-9. Show that Eq. 5-30 can be expressed in the following form for ease in computation:

$$s^2 = \frac{1}{n-1}\left\{ \sum_i x_i^2 - \frac{\left(\sum_i x_i\right)^2}{n} \right\}$$

Develop a similar form for Eq. 5-32.

5-10. Visit your computer center and inquire about the library programs available for statistical analysis. List the programs, the computations each will perform, and the input information required. Write a simple program to READ in $N$ data points and compute the mean, standard deviation, and skewness coefficient.

5-11. Complete the following mathematical statements about the properties of a PDF by inserting in the boxes on the left the correct item number from the right. Assume that $X$ is a series of annual occurrences from a normal distribution.

a. $\int_{-\infty}^{\infty} f(x)\,dx = \square$ — 1. Zero

b. $\int_{-\infty}^{m_1} f(x)\,dx = \square$ — 2. Unity

c. $\int_{\mu}^{\mu+\square} f(x)\,dx = .34$ — 3. Value with 5% chance of exceedance each year

d. $\int_{m_1}^{m_2} f(x)\,dx = \square$ — 4. 0.68

e. $\int_{-\infty}^{\square} f(x)\,dx = .5$ — 5. Value expected every 50 yr on the average

f. $\int_{\mu-\sigma}^{\mu+\sigma} f(x)\,dx = \square$ — 6. $P(X \leq m_1)$

g. $\int_{\square}^{\infty} f(x)\,dx = .02$ 7. $P(m_1 \leqslant X \leqslant m_2)$

h. $\int_{\mu}^{\mu} f(x)\,dx = \square$ 8. $P(m_1 \geqslant X \geqslant m_2)$

i. $1 - \int_{m_1}^{m_2} f(x)\,dx = \square$ 9. Median

j. $\int_{-\infty}^{\square} f(x)\,dx = 0.95$ 10. Standard deviation

5-12. The mean monthly temperature for September at the Tech first order weather station is found to be normally distributed. The mean is 65.5°F, the variance is 39.3°F$^2$, and the record is complete for 63 yr. With the aid of Table I, Appendix C, find (a) The midrange within which two-thirds of all future mean monthly values is expected to fall, (b) the midrange within which 95% of all future values is expected, (c) the limit below which 80% of all future values is expected, and (d) the highest and lowest values that are expected to be exceeded with a frequency of once in 10 yr and once in 100 yr. Verify the results by plotting the cumulative distribution on normal probability paper.

5-13. The total annual pumpage (in acre-ft) over the last 30 yr from a fully developed irrigation well field was observed as follows:

| | | | | |
|---|---|---|---|---|
| 2450 | 3300 | 3400 | 3650 | 3800 |
| 2650 | 3150 | 3100 | 3500 | 2850 |
| 3050 | 4300 | 3300 | 3300 | 3150 |
| 2100 | 3300 | 3650 | 3150 | 3550 |
| 2900 | 3250 | 3000 | 3400 | 3750 |
| 3900 | 3600 | 3150 | 3600 | 3000 |

Form an array of the data and plot on normal probability paper. Compute and plot the mean of the data and draw the line of best fit through the mean and as close as possible to the other points by eye. Estimate from the line the standard deviation and the highest value expected to be equaled or exceeded once in 50 yr.

5-14. Annual floods for a small river are reported to follow a normal probability distribution. The 2-yr flood for the basin has been estimated as 40,000 cfs and the 10-yr flood as 52,820 cfs. Using normal frequency factors, determine the magnitude of the 25-yr flood.

5-15. The total annual runoff from a small drainage basin is determined to be approximately normal with a mean of 14.0 in. and a variance of 9.0 in.$^2$. Determine the probability that the annual runoff from the basin will be less than 11.0 in. in all 3 of the next 3 consecutive years.

5-16. The 20-yr record of annual flood peaks on Furr's Run are

| Year | $Q$ (cfs) | Year | $Q$ (cfs) |
|---|---|---|---|
| 1951 | 1060 | 1961 | 1350 |
| 1952 | 2820 | 1962 | 1140 |
| 1953 | 1970 | 1963 | 2100 |
| 1954 | 1760 | 1964 | 1090 |
| 1955 | 1650 | 1965 | 2890 |
| 1956 | 1140 | 1966 | 1100 |
| 1957 | 1020 | 1967 | 1840 |
| 1958 | 2260 | 1968 | 1710 |
| 1959 | 1650 | 1969 | 1630 |
| 1960 | 870 | 1970 | 1260 |

Log the values of peak discharge and compute the mean, standard deviation, and skewness coefficient. Plot the data on lognormal probability paper. With the aid of Table 2, Appendix C, fit both the lognormal and Log Pearson III distributions. Compare estimates of the 50- and 100-yr events.

5-17. Compare the results of Prob. 5-16 by plotting on extreme-value paper. Compute the mean and variance of the discharges, and, using frequency factors from Table 5-6, find the Gumbel estimates of the 50- and 100-yr events.

5-18. The following parameters were computed for a stream near Lincoln, Neb.

Period of record = 1940 through 1959, inclusive
Mean annual flood = 7000 cfs
Standard deviation of annual floods = 1000 cfs
Skew coefficient of annual floods = 1.0
Mean of logarithms (base 10) of annual floods = 3.52
Standard deviation of logarithms = .50
Coefficient of skew of logarithms = 2.0

Determine the magnitude of the 25-yr flood by assuming that the peaks follow (a) the Log-Pearson Type III distribution, (b) the Gumbel distribution, and (c) the normal distribution.

5-19. Expand the computer program of Prob. 5-10 to include the computation of the mean, standard deviation, and skewness coefficient of the logarithms of the input data. Also, include a routine to sort the data by placing them in descending order and compute the corresponding plotting positions. Verify, using the data in Prob. 5-16.

5-20. Perform a complete frequency analysis on one of the three 33-yr records given in the table. Fit a Pearson Type III or Log-Pearson III and compare with the normal or lognormal of best fit. Plot the data and place control curves around the theoretical curve of best fit using Table 5-9.

| Year | Trempeuleau River Dodge, Wis. (DA = 643 mi²) $Q_{peak}$ (cfs) | Bow River Banff, Alberta, Canada (DA = 858 mi²) $Q_{peak}$ (cfs) | James River Scottsville, Va. (DA = 4570 mi²) $Q_{peak}$ (cfs) |
|---|---|---|---|
| 1928 | 3,700 | 10,200 | 75,600 |
| 1929 | 1,700 | 7,590 | 44,700 |
| 1930 | 3,360 | 9,280 | 45,800 |
| 1931 | 1,650 | 6,610 | 21,100 |
| 1932 | 3,600 | 9,850 | 31,400 |
| 1933 | 11,000 | 11,000 | 59,500 |
| 1934 | 2,570 | 9,490 | 38,800 |
| 1935 | 4,490 | 6,940 | 93,400 |
| 1936 | 7,180 | 7,720 | 126,000 |
| 1937 | 1,780 | 5,210 | 62,200 |
| 1938 | 3,170 | 7,770 | 87,400 |
| 1939 | 6,400 | 6,270 | 68,400 |
| 1940 | 3,120 | 7,220 | 130,000 |
| 1941 | 2,890 | 4,450 | 27,100 |
| 1942 | 5,680 | 5,850 | 80,600 |
| 1943 | 5,060 | 7,380 | 95,200 |
| 1944 | 2,040 | 5,590 | 133,000 |
| 1945 | 8,120 | 4,450 | 57,000 |
| 1946 | 4,570 | 7,210 | 41,200 |
| 1947 | 5,410 | 5,880 | 33,200 |
| 1948 | 4,840 | 10,320 | 59,600 |
| 1949 | 1,920 | 4,290 | 94,200 |
| 1950 | 3,600 | 10,080 | 73,300 |
| 1951 | 4,840 | 8,570 | 64,900 |
| 1952 | 6,950 | 5,460 | 54,500 |
| 1953 | 4,040 | 9,180 | 67,000 |
| 1954 | 5,710 | 10,120 | 62,900 |
| 1955 | 10,400 | 8,680 | 70,000 |
| 1956 | 17,400 | 9,060 | 20,400 |
| 1957 | 713 | 5,360 | 64,200 |
| 1958 | 1,140 | 6,730 | 44,500 |
| 1959 | 8,000 | 7,480 | 29,300 |
| 1960 | 1,480 | 6,440 | 64,200 |

5-21. Compare results of Prob. 5-20 with estimates by Gumbel's extreme-value distribution for the 50- and 100-yr events.

5-22. The pan-evaporation data (in.) for the month of July at a site in Missouri are

| | | | | |
|---|---|---|---|---|
| 9.7 | 11.7 | 11.2 | 11.3 | 11.5 |
| 11.2 | 8.8 | 11.4 | 11.8 | 8.9 |
| 9.3 | 9.2 | 9.3 | 9.3 | 10.4 |
| 9.8 | 8.7 | 11.5 | 10.9 | 10.2 |

Determine the mean, standard deviation, and coefficient of variation. What are the standard errors of these statistics? Establish the approximate 95% confidence limits of the mean.

5-23. A 40-yr record of rainfall indicates that the mean monthly precipitation during April is 3.85 in. with a standard deviation of 0.92. The distribution is normal. With 95% confidence, estimate the limits within which (a) next April's precipitation is expected to fall, and (b) the mean April precipitation for the next 40 yr.

5-24. Given the following values of peak flow rates for a small stream, determine the return period (yr) for a flood of 100 cfs both for an annual series and a partial duration series.

| Year | Date | Peak (cfs) | Year | Date | Peak (cfs) |
|---|---|---|---|---|---|
| 1963 | June 1 | 90 | 1968 | May 11 | 800 |
| | Aug. 3 | 300 | | June 8 | 700 |
| 1964 | June 7 | 60 | | Sept. 4 | 90 |
| 1965 | July 2 | 80 | 1969 | Aug. 8 | 400 |
| 1966 | May 18 | 100 | 1970 | May 9 | 30 |
| | June 3 | 90 | 1971 | Sept. 8 | 700 |
| 1967 | July 4 | 40 | 1972 | May 4 | 80 |

5-25. From the data given in the accompanying table of low flows, prepare a set of low flow frequency curves for the daily, weekly, and monthly durations.

*Lowest Mean Discharge (cfs) for the Following Number of Consecutive Days, Maury River near Buena Vista, Va.*

| Year | 1-Day | 7-Day | 30-Day |
|---|---|---|---|
| 1939 | 100.0 | 103.0 | 125.0 |
| 1940 | 167.0 | 171.0 | 194.0 |
| 1941 | 22.0 | 59.4 | 69.1 |
| 1942 | 101.0 | 127.0 | 173.0 |
| 1943 | 86.0 | 93.9 | 103.0 |
| 1944 | 62.0 | 65.9 | 77.4 |
| 1945 | 78.0 | 80.7 | 90.3 |
| 1946 | 76.0 | 78.6 | 87.1 |
| 1947 | 97.0 | 102.0 | 123.0 |
| 1948 | 154.0 | 176.0 | 215.0 |
| 1949 | 136.0 | 138.0 | 163.0 |
| 1950 | 113.0 | 125.0 | 139.0 |
| 1951 | 95.0 | 95.3 | 101.0 |
| 1952 | 115.0 | 116.0 | 120.0 |

| Year | 1-Day | 7-Day | 30-Day |
|---|---|---|---|
| 1953 | 85.0 | 86.1 | 90.8 |
| 1954 | 68.0 | 70.0 | 81.7 |
| 1955 | 83.0 | 96.1 | 99.9 |
| 1956 | 64.0 | 66.3 | 71.7 |
| 1957 | 62.0 | 64.1 | 75.8 |
| 1958 | 88.0 | 92.6 | 107.0 |
| 1959 | 76.0 | 80.9 | 117.0 |
| 1960 | 83.0 | 91.7 | 103.0 |
| 1961 | 99.0 | 103.0 | 152.0 |
| 1962 | 90.0 | 95.0 | 105.0 |
| 1963 | 60.0 | 60.6 | 70.8 |
| 1964 | 51.0 | 54.1 | 62.0 |
| 1965 | 64.0 | 68.7 | 76.2 |

5-26. For the 7-day low flows at Buena Vista given in Prob. 5-25, attempt to fit a straight-line frequency curve on lognormal or extreme-value probability paper, proceeding as follows: From the original plot of the data, estimate the lowest flow (say, $q$) at the high recurrence intervals; subtract this flow from all observed flows ($Q - q = Q^1$); and replot $Q^1$ versus the original recurrence intervals. Repeat if necessary. The best fitting curve will be a three-parameter frequency distribution.

5-27. The following table summarizes the number of occurrences of intensities of various durations for a 34-yr record of rainfall. Maximum intensities for the given durations were determined for all excessive storms and a count made of the exceedences. Interpolate for the average number of exceedences expected on a 5-yr frequency and plot the 5-yr intensity-duration-frequency curve.

| Duration (min) | Number of Times Stated Intensities Were Equaled or Exceeded | | | | | | |
|---|---|---|---|---|---|---|---|
| | Intensity (in./hr) | | | | | | |
| | *1.0* | *2.0* | *3.0* | *4.0* | *5.0* | *6.0* | *7.0* |
| 5 | | | 73 | 48 | 21 | 9 | 2 |
| 10 | | 68 | 51 | 26 | 11 | 3 | 1 |
| 15 | 72 | 35 | 23 | 11 | 5 | 1 | |
| 30 | 29 | 17 | 7 | 3 | 1 | | |
| 60 | 15 | 6 | 2 | 1 | | | |
| 120 | 8 | 1 | | | | | |

## References

1. Chow, Ven T., "Statistical and Probability Analysis of Hydrologic Data," Sec. 8-I, in V. T. Chow (ed.), *Handbook of Applied Hydrology* (New York: McGraw-Hill Book Company, 1964).

2. Fiering, M. B., "Information Analysis," in G. M. Fair, J. C. Geyer, and D. A. Okun, *Water Supply and Waste Water Removal* (New York: John Wiley & Sons, Inc., 1966), Chap. 4.
3. Benjamin, J. R., and C. Cornell, *Probability, Statistics and Decision for Civil Engineers* (New York: McGraw-Hill Book Company, 1969).
4. Mood, A. M., and F. A. Graybill, *Introduction to the Theory of Statistics*, 2nd ed. (New York: McGraw-Hill Book Company, 1963).
5. Duncan, A. J., *Quality Control and Statistics* (Homewood, Ill.: Richard D. Irwin, Inc., 1959).
6. Hoel, P. G., *Introduction to Mathematical Statistics*, 3rd ed. (New York: John Wiley & Sons, Inc., 1962).
7. Aitchison, J., and J. A. C. Brown, *The Log-Normal Distribution* (New York: Cambridge University Press, 1957).
8. Foster, H. A., "Theoretical Frequency Curves," *Trans. ASCE*, 87 (1924): 142–203.
9. Beard, L. R., *Statistical Methods in Hydrology*, Civil Works Investigations, U.S. Army Corps of Engineers, Sacramento Dictrict, 1962.
10. "A Uniform Technique for Determining Flood Flow Frequencies," Bull. No. 15, Water Resources Council, 1967.
11. "New Tables of Percentage Points of the Pearson Type III Distribution," Tech. Release No. 38, Central Technical Unit, U.S. Department of Agriculture, 1968.
12. Fiering, M. B., *Streamflow Synthesis* (Cambridge, Mass.: Harvard University Press, 1967).
13. Perkins, F. E., Simulation Lecture Notes, Summer Institute, "Applied Mathematical Programming in Water Resources," University of Nebraska, 1970.
14. Gumbel, E. J., "The Return Period of Flood Flows," *Ann. Math. Statist.*, 12, No. 2 (June 1941): 163–190.
15. Chow, Ven T., "The Log-Probability and Its Engineering Application," *Proc. ASCE*, 80, Paper No. 536 (Nov. 1954), 1–25.
16. "Probability Tables and Other Analysis of Extreme Value Data," Series 22, National Bureau of Standards Applied Mathematics, 1953.
17. Gumbel, E. J., *Statistics of Extremes* (New York: Columbia University Press, 1958).
18. Gumbel, E. J., "Statistical Theory of Extreme Values for Some Practical Application," National Bureau of Standards, Applied Mathematical Series 33, 1954.
19. Benson, M. A., "Plotting Positions and Economics of Engineering Planning," *ASCE, J. Hyd. Div.*, 88 (Nov. 1962): 57–71.
20. Gringorten, I. I., "A Plotting Rule for Extreme Probability Paper," *J. Geophys. Res.*, 68, No. 3 (Feb. 1963): 813–814.
21. Chow, Ven T., "A General Formula for Hydrologic Frequency Analysis," *Trans. Amer. Geophys. Union*, 32 (1951): 231–237.
22. Hazen, A., *Flood Flows* (New York: John Wiley & Sons, Inc., 1930).
23. Fair, G. M., J. C. Geyer, and D. A. Okun, *Water and Waste Water Engineering* (New York: John Wiley & Sons, Inc., 1966).
24. Kendall, G. R., "Statistical Analysis of Extreme Values," First Canadian Hydrology Symposium, National Research Council of Canada, Nov. 4 and 5, 1959.

# 6

# Regional Analysis: Joint Distributions and Time Series

In Chapter 5 the probability and statistical properties of random variables were introduced, but applications were restricted to a single variable and the analysis of point-frequency distributions. Complex hydrologic processes often require knowledge of the joint distribution of several random variables and consideration of the correlation between them. Dependence of sequential events in a time series is also an important concept in hydrology. Methods of analysis can be extended to the study of regional relationships, over both space and time, and techniques can be developed to produce synthetic sequences of variables which are artificial traces of the possible occurrence of hydrologic events that preserve the characteristics of the historical record. With the development of electronic computers, these methods and techniques have become a valuable element in planning and design.

In the present chapter an extension of probability theory leads to consideration of regression and correlation analysis, Markov and other time-dependent series, and the basis for generation of synthetic hydrologic variables.

## JOINT DISTRIBUTION AND CORRELATION

### 6-1 Additional Probability Concepts

Many hydrologic events must be studied as the joint occurrence of two or more random variables, and the probability or frequency analysis concerns their joint probability distribution. Joint probabilities and conditional probabilities were introduced in Chapter 5 but were not treated as complete distribution functions.

### Joint and Marginal Distributions

If an outcome results from the simultaneous occurrence of two discrete random variables $x$ and $y$, their joint probability function is $P(X = x_i \cap Y = y_j)$; or in shorter form, $P(x_i, y_j)$. For $x_i (i = 1, 2, \ldots, n)$ and $y_j (j = 1, 2, \ldots, m)$, the familiar rule for summing to unity holds.

$$\sum_{i=1}^{n} \sum_{j=1}^{m} P(x_i, y_j) = 1 \tag{6-1}$$

The summation can be carried to limiting values of $X$ and $Y$, say $x_k$ and $y_l$, to find the joint cumulative probability:

$$F(x_k, y_l) = \sum_{i=1}^{k} \sum_{j=1}^{l} P(x_i, y_j) \tag{6-2}$$

The concept of *marginal distributions* is defined by summing $P(x_i, y_j)$ over all values of one variable. The resulting marginal distribution is the probability function of the other variable without regard for the first. In other words, for two jointly distributed random variables, $X$ and $Y$, the distribution of $X$ is defined by finding the marginal distribution of the joint probability function. Thus

$$P(x_i) = \sum_{j=1}^{m} P(x_i, y_j) \tag{6-3}$$

and similarly,

$$P(y_j) = \sum_{i=1}^{n} P(x_i, y_j) \tag{6-4}$$

For the marginal CDF of $X$,

$$F(x_k) = \sum_{i=1}^{k} \sum_{j=1}^{m} P(x_i, y_j) \tag{6-5}$$

For continuous variables the analogous relationships follow directly. For the joint PDF $f(x, y)$

$$\int_{-\infty}^{\infty} \int_{-\infty}^{\infty} f(x, y)\, dy\, dx = 1 \tag{6-6}$$

and the cumulative function $F(x, y)$

$$F(x, y) = \int_{-\infty}^{x} \int_{-\infty}^{y} f(u, v)\, dv\, du \tag{6-7}$$

while the marginal distribution of, say, $X$ is

$$f(x) = \int_{-\infty}^{\infty} f(x, y)\, dy \qquad \textbf{(6-8)}$$

and the marginal CDF of $X$ is

$$F(x) = \int_{-\infty}^{x} f(u)\, du = \int_{-\infty}^{x}\int_{-\infty}^{\infty} f(u, y)\, dy\, du \qquad \textbf{(6-9)}$$

The marginal density functions are particularly helpful in finding conditional probabilities, $P(x \mid y)$ or $f(x \mid y)$. Recall

$$P(x \mid y) = \frac{P(X = x \cap Y = y)}{P(y)} \qquad \textbf{(6-10)}$$

and also

$$f(x \mid y) = \frac{f(x, y)}{f(y)} \qquad \textbf{(6-11)}$$

In both cases the joint probability is the product of the conditional and the marginal probability functions.

### Moments

Moments again provide the best description of the distribution. The joint moment of order $r$ and $s$ is defined as

$$\mu'_{r,s} = \int_{-\infty}^{\infty}\int_{-\infty}^{\infty} x^r y^s f(x, y)\, dy\, dx \qquad \textbf{(6-12)}$$

When $r = 1$ and $s = 0$,

$$\mu'_{1,0} = \int_{-\infty}^{\infty} x\left[\int_{-\infty}^{\infty} f(x, y)\, dy\right] dx \qquad \textbf{(6-13)}$$

and the expression in brackets is simply $f(x)$, the marginal PDF of $x$, as shown. Hence the moment is the mean of $x$:

$$\mu'_{1,0} = \mu_x = \int_{-\infty}^{\infty} x f(x)\, dx \qquad \textbf{(6-14)}$$

Similar results hold for $y$.

When the moments are written about the means,

$$\mu_{r,s} = \int_{-\infty}^{\infty}\int_{-\infty}^{\infty} (x - \mu_x)^r (y - \mu_y)^s f(x, y)\, dy\, dx \qquad \textbf{(6-15)}$$

With $r = 2$, $s = 0$, the result is $\sigma_x^2$, and for $r = 0$, $s = 2$, it is $\sigma_y^2$. Both are familiar quantities. But a new concept evolves from the third type of second order moment, that is, $r = 1$, $s = 1$. This moment is called the *covariance*, cov $(x, y)$, or $\sigma_{x,y}$.

$$\sigma_{x,y} = \int_{-\infty}^{\infty}\int_{-\infty}^{\infty} (x - \mu_x)(y - \mu_y) f(x, y)\, dy\, dx \qquad \textbf{(6-16)}$$

The correlation coefficient, usually identified as $\rho$, is a standardized and dimensionless form of the covariance as follows:

$$\rho_{x,y} = \frac{\text{cov}(x, y)}{\sigma_x \sigma_y} = \frac{\sigma_{x,y}}{\sigma_x \sigma_y} \tag{6-17}$$

The sample correlation coefficient, $r = s_{x,y}/s_x s_y$, is used to estimate $\rho$. Because the correlation coefficient varies between $\pm 1$, it is widely referred to as a measure of dependence or association between two variables. It actually indicates the degree of linear dependence. Thus a high value would indicate, for positive correlation, that large values of $x$ are associated with large values of $y$, and similarly for small values. A low value of the coefficient may indicate lack of association but not conclusively.[1] If $x$ and $y$ are independent, then $f(x, y) = f(x) \times f(y)$, and Eq. 6-16 gives a covariance of zero.

***Example 6-1*** For the joint distribution of quarterly flows (treated as discrete quantities) on two rivers as shown in Fig. 6-1, find the marginal distribution of $X$ and $Y$, and compute the means, variances, standard deviations, covariance, and correlation coefficient.

*Solution*

1. Summing over $Y$ gives the marginal distribution of $X$. Thus

| $x$ | 0.5 | 1.0 | 1.5 | 2.0 |
|---|---|---|---|---|
| $f(x)$ | 0.1 | 0.35 | 0.40 | 0.15 |

2. Summing over $X$ gives the marginal distribution of $Y$. Thus

| $y$ | 0.5 | 1.0 | 1.5 | 2.0 |
|---|---|---|---|---|
| $f(y)$ | 0.2 | 0.35 | 0.25 | 0.2 |

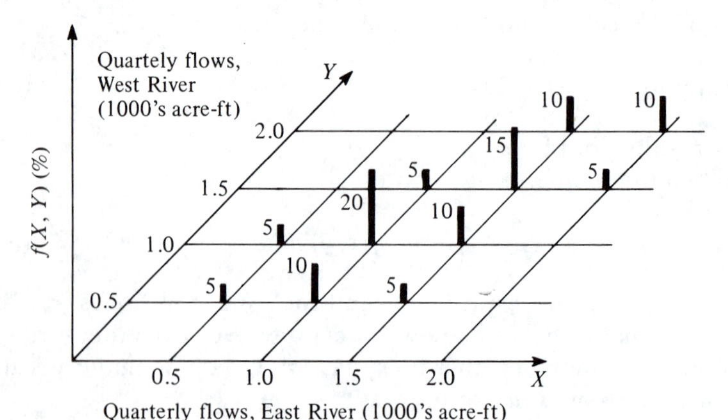


**Fig. 6-1.** *Joint discrete distribution of quarterly flows on East and West Rivers.*

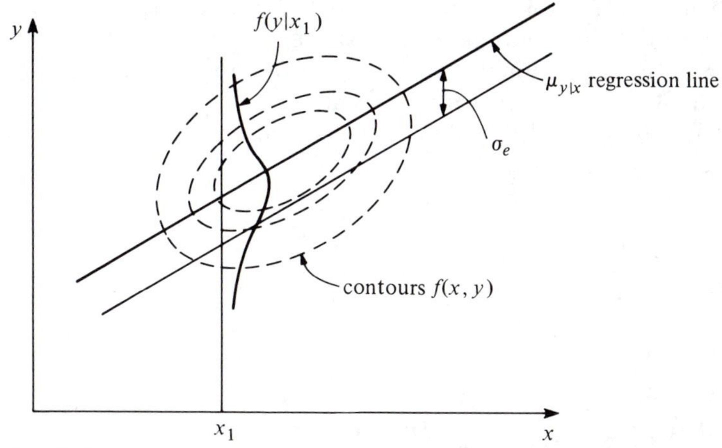


*Fig. 6-2.* *Bivariate density with conditional probability function.*

3. $\bar{x} = \sum xf(x) = 0.5(0.1) + 1.0(0.35) + 1.5(0.40) + 2.0(0.15) = 1.3$ KAF (thousands of acre-ft).
4. $s_x^2 = \sum (x - \bar{x})^2 f(x) = (0.5 - 1.3)^2(0.1) + (1.0 - 1.3)^2(0.35) + (1.5 - 1.3)^2 (0.40) + (2.0 - 1.3)^2(0.15) = 0.459$ KAF².
5. $s_x = 0.677$ KAF.
6. Similarly for $Y$, $\bar{y} = 1.23$ KAF; $s_y^2 = 0.262$ KAF²; $s_y = 0.512$ KAF.
7. $\text{cov}(x, y) = \sum (x - \bar{x})(y - \bar{y})f(x, y) = (0.5 - 1.3)(0.5 - 1.23)(0.05) + (0.5 - 1.3)(1.0 - 1.23)(0.05) + \cdots + (2.0 - 1.3)(2.0 - 1.23)(0.10) = 0.145$.
8. $r_{x,y} = \text{cov}(x, y)/(s_x)(s_y) = (0.145)/(0.677)(0.512) = 0.418$.

## 6-2 Multivariate Distributions

Joint density functions can be either discrete or continuous. A multivariate discrete distribution that is analogous to the binomial (Table 5-3) is termed the *multinomial.* Like the binomial, it also is essentially a counting technique and is treated in most standard texts on probability and statistics. Of more direct interest are the continuous multivariate distributions. The bivariate normal is used as an example. Figure 6-2 shows the density of $f(x, y)$. The PDF has the form

$$f(x, y) = K \exp\{M\} \quad \textbf{(6-18)}$$

where

$$K = \frac{1}{2\pi\sigma_x \sigma_y (1 - \rho^2)^{1/2}} \quad \textbf{(6-19)}$$

$$M = \left\{ -\left[\frac{1}{2(1 - \rho^2)}\right]\left[\left(\frac{x - \mu_x}{\sigma_x}\right)^2 - 2\rho\left(\frac{x - \mu_x}{\sigma_x}\right)\left(\frac{y - \mu_y}{\sigma_y}\right) + \left(\frac{y - \mu_y}{\sigma_y}\right)^2\right]\right\} \quad \textbf{(6-20)}$$

It is a five-parameter distribution, as follows: $\mu_x$, $\mu_y$, $\sigma_x$, $\sigma_y$, and $\rho$. The marginal distributions of $f(x, y)$ are univariate normal distributions and the conditional distribution is defined as $f(y \mid x) = f(x, y) \div f(x)$, which is also a univariate normal.

The conditional distribution has the form

$$f(y \mid x) = K' \exp \{M'\} \tag{6-21}$$

where

$$K' = \frac{1}{[\sqrt{2\pi}\,\sigma_y \sqrt{1 - \rho^2}\,]} \tag{6-22}$$

and

$$M' = \left\{ -\left[\frac{1}{2\sigma_y^2 (1 - \rho^2)}\right] \left[(y - \mu_y) - \rho \frac{\sigma_y}{\sigma_x} (x - \mu_x)\right]^2 \right\} \tag{6-23}$$

The distribution is normal with mean

$$\mu_{y \mid x} = \mu_y + \rho\frac{\sigma_y}{\sigma_x} (x - \mu_x) \tag{6-24}$$

and variance

$$\sigma_e^2 = \sigma_y^2(1 - \rho^2) \tag{6-25}$$

Both of these last two equations have great usefulness in regression and correlation studies. Equation 6-24 is linear and expresses the linear dependence between $y$ and $x$ as shown in Fig. 6-2. The mean value of $y$ can be computed for fixed values of $x$. Also, if the correlation between them is significant, we can predict the values of $y$ with less error than from the marginal distribution of $y$ alone. In fact, from Eq. 6-25, the fraction of the original variance explained or removed by the regression is

$$\rho^2 = 1 - \frac{\sigma_e^2}{\sigma_y^2} \tag{6-26}$$

It can be seen also from Eq. 6-24 that the slope of the regression line is

$$\rho \frac{\sigma_y}{\sigma_x} = \frac{\mu_{y \mid x} - \mu_y}{x - \mu_x} \tag{6-27}$$

or, if $x$ and $y$ are standardized, then $\rho$ itself is the slope, where

$$\rho = \frac{(\mu_{y \mid x} - \mu_y)/\sigma_y}{(x - \mu_x)/\sigma_x} \tag{6-28}$$

The bivariate case can be expanded generally to cover higher order, multivariate distributions.

## 6-3 Fitting Regression Equations

Regression lines as expressed by Eq. 6-24 and shown in Fig. 6-2 are useful in explaining linear dependence and, where significant correlation exists, in making predictions. For the bivariate case, in general, the procedure is to fit a linear model to a sample of random variables observed in pairs (see Fig. 6-3). The fitting technique is the method of

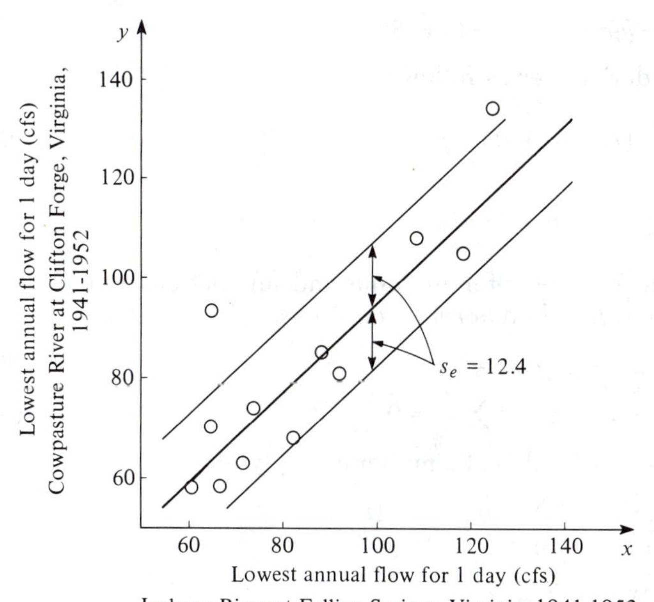


***Fig. 6-3.*** *Cross-correlation of low flows. Regression line:* $Y = 4.94 + 0.923X$; $r = 0.86$.

| *Year* | *Jackson River* | *Cowpasture River* |
|---|---|---|
| 41 | 61 | 58 |
| 42 | 92 | 81 |
| 43 | 65 | 70 |
| 44 | 72 | 63 |
| 45 | 82 | 68 |
| 46 | 67 | 58 |
| 47 | 74 | 74 |
| 48 | 118 | 105 |
| 49 | 124 | 134 |
| 50 | 108 | 108 |
| 51 | 65 | 93 |
| 52 | 88 | 85 |
| Mean = | 84.7 | 83.1 |
| Standard deviation = | 21.7 | 23.2 |

least squares which minimizes the sum of the residuals squared. Residuals as shown in the figure are the difference, vertically in this instance, from the mean value of $y$ predicted by the line and the $y$ value observed for the same corresponding value of $x$. The line to be fitted is

$$\bar{y}_i = \alpha + \beta x_i \tag{6-29}$$

The best estimates of $\alpha$ and $\beta$ are sought. Thus to minimize

$$\sum (y_i - \bar{y}_i)^2 = \sum [y_i - (\alpha + \beta x_i)]^2 \tag{6-30}$$

take partial derivatives as follows

$$\frac{\partial}{\partial \alpha} \{\sum [y_i - (\alpha + \beta x_i)]^2\} \tag{6-31}$$

$$\frac{\partial}{\partial \beta} \{\sum [y_i - (\alpha + \beta x_i)]^2\} \tag{6-32}$$

After carrying out the differentiations and summations, two equations result in $\alpha$ and $\beta$, called *normal equations.*

$$\sum y_i - n\alpha - \beta \sum x_i = 0 \tag{6-33}$$

$$\sum x_i y_i - \alpha \sum x_i - \beta \sum x_i^2 = 0 \tag{6-34}$$

Solving Eqs. 6-33 and 6-34 simultaneously yields

$$\alpha = \frac{\sum y_i}{n} - \frac{\beta \sum x_i}{n} \tag{6-35}$$

$$\beta = \frac{\sum x_i y_i - \sum x_i \sum y_i / n}{\sum x_i^2 - (\sum x_i)^2 / n} \tag{6-36}$$

Recall the slope is $\rho(\sigma_y/\sigma_x)$, or as estimated from sample data

$$\beta = r \frac{s_y}{s_x} \tag{6-37}$$

Also

$$\sigma_e^2 = \sigma_y^2(1 - \rho^2) \tag{6-38}$$

the square root of which is called the *standard error of estimate* and can be estimated from

$$s_e^2 = \frac{n-1}{n-2} s_y^2 (1 - r^2) \tag{6-39}$$

or

$$s_e^2 = \frac{1}{n-2} \sum (y_i - \bar{y}_i)^2 \tag{6-40}$$

where $y_i$ are the observed values and $\bar{y}_i$ are predicted by Eq. 6-29.

The bivariate example can be extended to multiple linear regressions. For example, the linear model in three variables, with $y$ the dependent variable and $x_1$ and $x_2$ the independent variables, has the form

$$y = \alpha + \beta_1 x_1 + \beta_2 x_2 \tag{6-41}$$

The normal equations are

$$\sum y = \alpha n + \beta_1 \sum x_1 + \beta_2 \sum x_2 \tag{6-42}$$

$$\sum yx_1 = \alpha \sum x_1 + \beta_2 \sum x_1^2 + \beta_2 \sum x_1 x_2 \tag{6-43}$$

$$\sum yx_2 = \alpha \sum x_2 + \beta_1 \sum x_1 x_2 + \beta_2 \sum x_2^2 \tag{6-44}$$

The square of the standard error of estimate is

$$s_e^2 = \frac{1}{n-3} \sum (y_i - \bar{y}_i)^2 \tag{6-45}$$

where $y_i$ are the observed values and $\bar{y}_i$ are predicted by Eq. 6-41. The *multiple correlation coefficient* is

$$R = \left(1 - \frac{s_e^2}{s_y^2}\right)^{1/2} \tag{6-46}$$

It should be noted again that a small correlation coefficient, or one of zero, signifies a lack of linear dependence, but it is possible that there is nonlinear dependence. Furthermore, transformations of an assumed model can be made functionally linear in many ways, the most familiar of which is reduction of a multiplicative relationship by logarithms, for example,

$$y = \alpha x_1^{\beta_1} x_2^{\beta_2} \tag{6-47}$$

which becomes linear when

$$\log y = \log \alpha + \beta_1 \log x_1 + \beta_2 \log x_2 \tag{6-48}$$

***Example 6-2*** The lowest annual flows for a 12-yr period on the Jackson and Cowpasture Rivers are tabulated in Fig. 6-3. The stations are upstream of the confluence of the two rivers that form the James River. Find the correlation between low flows and determine the regression equation.

*Solution*

1. The basic statistics are $\sum x = 1016$; $\sum y = 997$; $\sum x^2 = 91{,}216$; $\sum y^2 = 88{,}777$; and $\sum xy = 89{,}209$.
2. For the two-variable regression, $\alpha$ and $\beta$ are found from Eqs. 6-35 and 6-36.

$$\beta = \frac{[(89{,}209) - (1016)(997)/(12)]}{(91{,}216) - (1016)^2/(12)} = .923$$

$$\alpha = (997)/(12) - (0.923)(1016)/(12) = 4.91$$

The regression is $y = 4.91 + .923x$.

3. The correlation coefficient from Eq. 6-37 is

$$r = (.923)(21.7)/(23.2) = .86$$

4. From Eq. 6-39 the standard error of estimate, $s_e$, is 11.7 which is plotted as limits around the regression line in Fig. 6-3.

## 6-4 Stochastic Time Series

### Markov Processes

When dependence exists between sequential events in a random process, and the dependence is such that the next step depends only on the present state and not on other past states, the process is described as Markovian. In general, a Markov process describes only step-by-step dependence, called a *first order process,* or exhibiting lag-one serial correlation. The theory can be extended conceptually, however, to multiple-lag correlations which than are referred to as *higher order Markov processes.* In all cases, the process is defined in terms of discrete probabilities.

The probability relationships for a Markov process must provide for the conditional probabilities of the process moving from any state at period $t$ to any subsequent state at period $t + 1$. Thus

$$P(X_{t+1} = j \mid X_t = i) \tag{6-49}$$

expresses the conditional probability of "transitioning" from state $i$ at period $t$ to state $j$ at $(t + 1)$. Two conditions must be defined to describe completely the properties of the process.[2] One is the initial state, that is, the value of the variable at the beginning. The other is the complete matrix of transition probabilities. A third and resultant property is a statement, or matrix, of steady state probabilities which occur in all Markov processes after sufficient transitioning from state to state. The steady state probabilities are independent of the beginning state and the transition probabilities. The following example developed by Loucks[3] illustrates the principles and application.

***Example 6-3*** Figure 6-4 summarizes in discrete form the probability distribution of annual streamflows that vary from 60 to 100 KAF. Figure 6-4a shows the complete, unconditional probabilities of the historical record after determining the percent of past flows that fall into four discrete intervals. On analyzing the annual flow record further, it is found that serial correlation exists and can be summarized in the four conditional probability distributions shown in Fig. 6-4b through 6-4e. Develop the matrix of transition probabilities, and beginning in state 2 (flow interval, 70 to 80 KAF) show that the steady state probabilities are as given in Fig. 6-4a.

*Solution*

1. The steady state probabilities from Fig. 6-4a are $P_i = (.15, .31, .32, .22)$, a probability vector. Note that $\sum p_i = 1$.

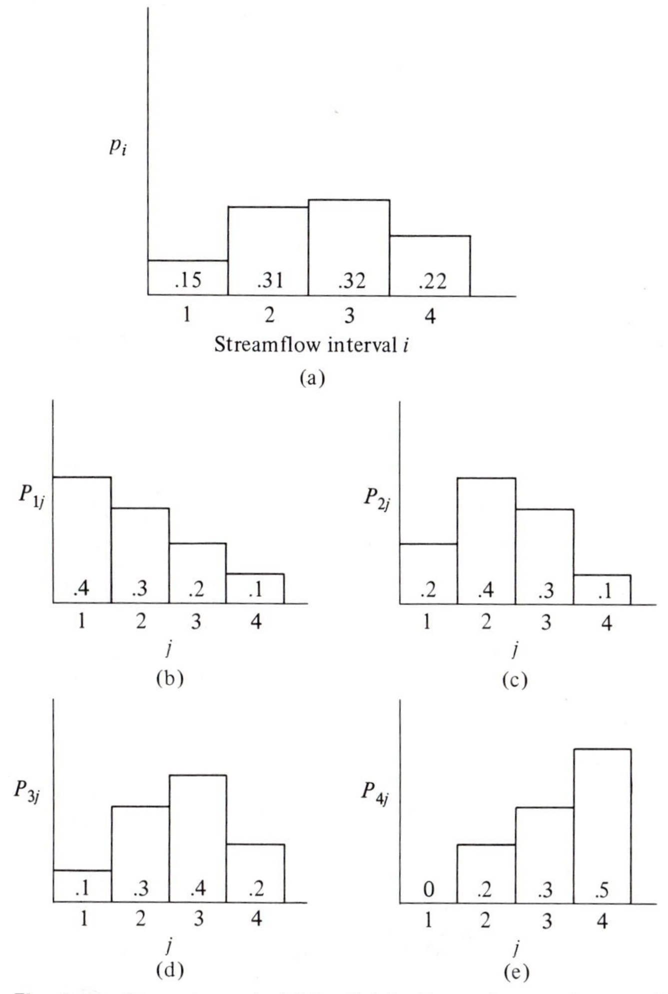


***Fig. 6-4.*** *Discrete probability distributions of annual streamflows: (a) unconditional distribution. Conditional distributions: (b) state 1 to state j; (c) state 2 to state j; (d) state 3 to state j; and (e) state 4 to state j.*

2. The transition probabilities, $p_{ij}$, expressing the probability of being in state $i$ and transitioning to state $j$ in the next year, are taken from Figs. 6-4b through 6-4e. Table 6-1 is the transition probability matrix, which can be expressed simply as $P_{ij}$. Note that the rows must sum to one.

***Table 6-1*** *Transition Probability Matrix*

| | | Flow State $j$ in Year $y+1$ | | | |
|---|---|---|---|---|---|
| | | 1 | 2 | 3 | 4 |
| Flow state $i$ in year $y$ | 1 | .4 | .3 | .2 | .1 |
| | 2 | .2 | .4 | .3 | .1 |
| | 3 | .1 | .3 | .4 | .2 |
| | 4 | .0 | .2 | .3 | .5 |

3. Starting in state 2 for year $y$, the probability state vector is $P_i^{(y)} =$ (0, 1, 0, 0). For transition to year $y + 1$, solve $P_i^{(y+1)} = P_i^{(y)} P_{ij}$. For example,

$$P_i^{(y+1)} = (0 \quad 1 \quad 0 \quad 0) \begin{pmatrix} .4 & .3 & .2 & .1 \\ .2 & .4 & .3 & .1 \\ .1 & .3 & .4 & .2 \\ 0 & .2 & .3 & .5 \end{pmatrix} = (.2 \quad .4 \quad .3 \quad .1)$$

Similarly, for $P_i^{(y+2)}$, the next transition

$$P_i^{(y+2)} = (.2 \quad .4 \quad .3 \quad .1)P_{ij} = (.19 \quad .33 \quad .31 \quad .17)$$

4. After nine transitions the steady state vector is reproduced as shown in Table 6-2.

***Table 6-2*** *Successive Streamflow Probability Vectors*

| | Streamflow State Probabilities | | | |
|---|---|---|---|---|
| Year $y$ | $P_1^{(y)}$ | $P_2^{(y)}$ | $P_3^{(y)}$ | $P_4^{(y)}$ |
| $y$ | .000 | 1.000 | .000 | .000 |
| $y + 1$ | .200 | .400 | .300 | .100 |
| $y + 2$ | .190 | .330 | .310 | .170 |
| $y + 3$ | .173 | .316 | .312 | .199 |
| $y + 4$ | .164 | .311 | .314 | .211 |
| $y + 5$ | .159 | .310 | .315 | .216 |
| $y + 6$ | .157 | .309 | .316 | .218 |
| $y + 7$ | .156 | .309 | .316 | .219 |
| $y + 8$ | .154 | .309 | .317 | .220 |
| $y + 9$ | .15 | .31 | .32 | .22 |
| $y + 10$ | .15 | .31 | .32 | .22 |

*Source:* After D. P. Loucks, "Stochastic Methods for Analyzing River Basin Systems," Tech. Rept. No. 16, Cornell University Resources and Marine Sciences Center, Ithaca, N.Y., 1969.

## Serial Correlation (Continuous Variables)

Sequential events in a time series represent a special case of joint distributions. They are not a true joint density because $x_t$ and $x_{t-1}$ are actually from the same distribution, $X$. But the density $f(x_t, x_{t-1})$ has meaning and usefulness in hydrology because of serial correlation. Some natural events exhibit a time-dependent trend, or persistence between successive occurrences, and their historical traces preserve the tendency. The analogy between joint and marginal distributions holds, however, because one variable can be "summed out."

$$f(x_t) = \int_{-\infty}^{+\infty} f(x_t, x_{t-1})\, dx_{t-1} \qquad \textbf{(6-50)}$$

Moments of the synthetic joint distribution are, by the principle of stationarity, independent of time no matter what the averaging period. The mean of $x_t$ or $x_{t-1}$ is $\mu_x$. In defining second moments, a covariance function appears that is a function of the time interval $t$. Thus

$$\phi(t) = \int_{-\infty}^{\infty} (x_t - \mu)(x_{t-1} - \mu) f(x_t)\, dx_t \qquad \textbf{(6-51)}$$

which for a stochastic time series is called the *autocovariance function.* When it is normalized by forming the ratio $\phi(t)/\phi(0) = \rho(t)$, where $\phi(0)$ is actually the variance of $X$, we have the definition for a parameter called the *autocorrelation function.* Autocorrelation is evident when $\rho(t)$ is plotted for all increments of $t$. Periodicity or lags at various time intervals are immediately apparent.

The typical model for an autoregressive process is an annual steamflow model of the first order, that is, considering only lag-one serial correlation. If autocorrelation is significant, a first order model is usually sufficient. Consider

$$(x_{i+1} - \mu) = c_1(x_i - \mu) + c_2 t \qquad \textbf{(6-52)}$$

a transformation from year $i$ to year $i + 1$. The mean is constant, and $t$ is a standardized random variate. If lag-one autocorrelation did not exist, $c_1$ obviously would be zero, and $c_2$ would have to be $\sigma$ of the $X_{i+1}$ and, by stationarity, the whole process. The model would be simply $x = \bar{x} + Ks$, with $K = t$. But, if correlation exists, it must follow the joint considerations outlined. It can be shown that

$$c_1 = \rho(1) \qquad \textbf{(6-53)}$$

and

$$c_2 = \sigma[1 - \rho^2(1)]^{1/2} \qquad \textbf{(6-54)}$$

Thus the model is

$$(x_{i+1} - \mu) = \rho(1)[(x_i - \mu)] + \sigma t[1 - \rho^2(1)]^{1/2} \qquad \textbf{(6-55)}$$

Such a model is like a Markov process because there is a transition from one state to the succeeding state without concern for lags greater than one. Variations of this model appear in later chapters for generating synthetic sequences.

## REGIONAL FREQUENCY ANALYSIS

In dealing with single historical sequences of hydrologic variables, the predictive value of fitted distributions is limited because the records are generally short and the sampling errors correspondingly large.[4–6] Additional and more reliable information often can be obtained within a homogeneous region by correlating dependent hydrologic variables with other causative or physically related factors. In such ways, hydrologic characteristics within the region can be summarized, and estimates of statistical parameters can be derived from general regional relationships. Typical examples are (1) the prediction of rainfall depths and frequencies in ungauged or incompletely gauged areas from characteristics at well-gauged sites in the same area; (2) the prediction of peak flood flows from the correlation of observed flows on measurable quantities such as drainage area size, precipitation, and stream slope; and (3) the prediction of the standard deviation of annual streamflows throughout a large basin based on a host of physical and climatic factors. Recourse to multiple regression techniques is apparent in such cases.

Definition of the regional boundaries depends on the parameters or variables to be estimated. In some cases, significant generalizations can be made over large physiographic regions (mean annual precipitation in the United States, for example); in other situations the region might have to be limited to drainage basins of certain sizes within smaller physiographic zones (for example, the peak rates of runoff for areas up to 1 $mi^2$ in the Piedmont Plateau). Homogeneity tests have been proposed for testing the significance of regional delineations.[7,8]

## 6-5 Generalized Hydrologic Characteristics

The rainfall atlases published by the U.S. Weather Bureau represent a convenient method for summarizing mean annual precipitation and storm rainfall depths expected with various frequencies. The controlling features of physiographic regions and climatic patterns are immediately apparent in such summaries, and estimates of mean values are easily interpolated between the isohyetals. Similar summaries can be made for other hydrologic variables—for example, mean annual evapotranspiration or runoff. The longer the records, the more reliable are the data.

Less widely tabulated are measures of variability: the variance or coefficient of variation, for example, of rainfall and runoff. But regional distinctions are still clearly established. Fair et al.[9] report that $C_v$ for the mean annual rainfall varies from 0.1 in well-watered regions to 0.5 in arid regions, whereas $C_v$ for the mean annual runoff ranges from 0.15 to 0.75. Since the runoff is usually considerably less than total rainfall in any region, the greater variability in runoff indicates the effect of other factors and presents difficulty in determining rainfall-runoff relationships. Maps showing isolines of the standard deviation of regional flood flows have been prepared by the Corps of Engineers.[4]

***Table 6-3***

| Duration | Skew Coefficient |
|---|---|
| Instantaneous | .00 |
| 1 day | −.04 |
| 3 day | −.12 |
| 10 day | −.23 |
| 30 day | −.32 |
| 90 day | −.37 |
| 1 year | −.40 |

As mentioned in Chapter 5, the large sampling errors in computing skewness from short records of hydrologic sequences suggest the use of regional values. Beard[4] proposed the figures in Table 6-3 for the logarithms of flood flows in the absence of regional studies. Methods of "smoothing" and averaging regional values of skewness have been proposed also.[5,10]

Many techniques used in the past for generalizing regional characteristics did not rely on statistical considerations. The so-called "station-year" method of extending rainfall records has proved helpful but has questionable statistical validity, especially if applied to dependent series or to stations in nonhomogeneous areas. The method has been used to combine, say, two 25-yr records to obtain a single 50-yr sequence. In practice, the analyst may have to use imagination and ingenuity to summarize regional characteristics, while remaining aware of actual and theoretical considerations.

### Regional Flood Characteristics

The *index-flood* method proposed by the U.S. Geological Survey is an example of summarizing regional characteristics successfully.[8,11] The method uses statistical data but combines them in graphical summaries. It can be supplemented and generally improved by using statistical methods, employing, for example, the regression techniques explained in the next article. The index method, as illustrated in Fig. 6-5, can be outlined as follows.

1. Prepare single-station, flood-frequency curves for each station within the homogeneous region (Fig. 6-5a).
2. Compute the ratio of flood discharges taken from the curves at various frequencies to the mean annual flood for the same station.
3. Compile ratios for all stations and find the median ratio for each frequency (Fig. 6-5b).
4. Plot the median ratios against recurrence interval to produce a regional frequency curve (Fig. 6-5c).

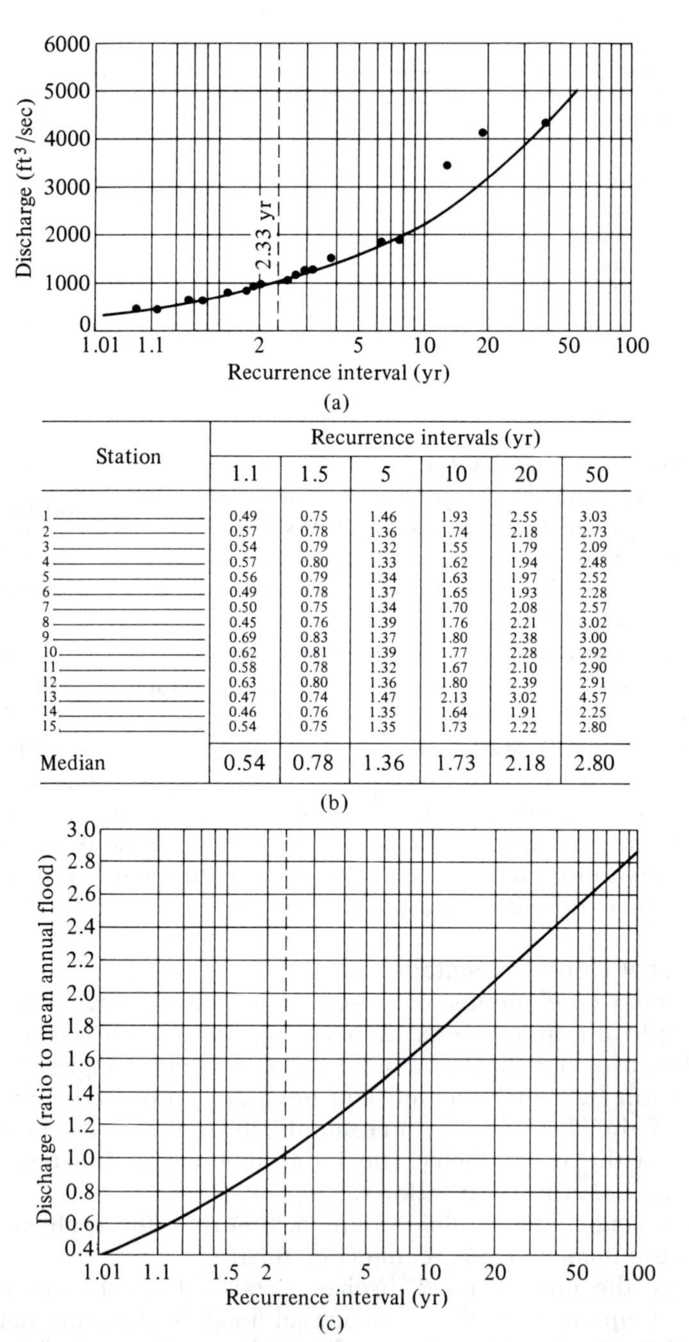


| Station | Recurrence intervals (yr) | | | | | |
|---|---|---|---|---|---|---|
| | 1.1 | 1.5 | 5 | 10 | 20 | 50 |
| 1 | 0.49 | 0.75 | 1.46 | 1.93 | 2.55 | 3.03 |
| 2 | 0.57 | 0.78 | 1.36 | 1.74 | 2.18 | 2.73 |
| 3 | 0.54 | 0.79 | 1.32 | 1.55 | 1.79 | 2.09 |
| 4 | 0.57 | 0.80 | 1.33 | 1.62 | 1.94 | 2.48 |
| 5 | 0.56 | 0.79 | 1.34 | 1.63 | 1.97 | 2.52 |
| 6 | 0.49 | 0.78 | 1.37 | 1.65 | 1.93 | 2.28 |
| 7 | 0.50 | 0.75 | 1.34 | 1.70 | 2.08 | 2.57 |
| 8 | 0.45 | 0.76 | 1.39 | 1.76 | 2.21 | 3.02 |
| 9 | 0.69 | 0.83 | 1.37 | 1.80 | 2.38 | 3.00 |
| 10 | 0.62 | 0.81 | 1.39 | 1.77 | 2.28 | 2.92 |
| 11 | 0.58 | 0.78 | 1.32 | 1.67 | 2.10 | 2.90 |
| 12 | 0.63 | 0.80 | 1.36 | 1.80 | 2.39 | 2.91 |
| 13 | 0.47 | 0.74 | 1.47 | 2.13 | 3.02 | 4.57 |
| 14 | 0.46 | 0.76 | 1.35 | 1.64 | 1.91 | 2.25 |
| 15 | 0.54 | 0.75 | 1.35 | 1.73 | 2.22 | 2.80 |
| Median | 0.54 | 0.78 | 1.36 | 1.73 | 2.18 | 2.80 |

(b)

***Fig. 6-5.*** *Index flood method of regional flood frequency analysis: (a) single-station flood frequency curve, (b) ratios $Q_t$ to $Q_{2.33}$ for 15 stations, and (c) regional flood frequency curve.*

Two statistical considerations involved are (1) a homogeneity test to justify definition of a region, and (2) a method for extending short records to place all stations on the same base period. A somewhat similar technique has been developed by Potter for the Bureau of Public Roads.[12] It relies on the graphical correlation of floods with physical and climatic variables, and is thus a technique that refers in part to the discussion in the next article. (see also Chapter 11).

### Regional Rainfall Characteristics

The variation of rainfall frequencies with duration was introduced in Chapter 5. Regression analysis can be used to fit intensity-duration-frequency curves similar to those shown in Fig. 5-12, and the constants interpreted as regional characteristics. Many formulas have been used in the past to fit these curves, but most of them are in a form with intensity ($i$) inversely proportional to duration ($t$). Steel[13] has used a model of the form $i = A/(t + B)$ to fit rainfall data throughout the United States. The constants $A$ and $B$ therefore serve as characteristic features of both the rainfall region and the frequency of occurrence in each area.

***Example 6-4*** Fit the following rainfall data to determine the 10-yr intensity-duration-frequency curve.

| | | | | | | |
|---|---|---|---|---|---|---|
| $t$ = duration (min) | 5 | 10 | 15 | 30 | 60 | 120 |
| $i$ = intensity (in./hr) | 7.1 | 5.9 | 5.1 | 3.8 | 2.3 | 1.4 |
| $1/i$ | .14 | .17 | .20 | .26 | .43 | .71 |

*Solution*

1. A model of the form $i = A/(t + B)$ can be expressed in linear form as $(1/i) = (t/A) + (B/A)$.
2. The regression of $(1/i)$ versus $t$ yields $(1/i) = .005t + 0.12$, from which $A = 200$ and $B = 24$.
3. Thus the rainfall formula is $i = 200/(t + 24)$. The correlation coefficient is $-0.997$ and the fit when plotted on the 10-yr curve of Fig. 5-12 is seen to be very good.

Maximum average rainfall depths have been published by the U.S. Weather Bureau[14] for durations between 30 min and 24 hr and for recurrence intervals between 1 yr and 100 yr. Depth-duration-frequency curves can be constructed for any location by plotting successive values from the various rainfall maps, preferably on logarithmic paper to facilitate fitting flatter curves. Correction factors are given to permit estimates of depths for duration less than 30 min. Special attention must be given to the extrapolation of point rainfall data to account for spatial variations. Duration frequency analyses

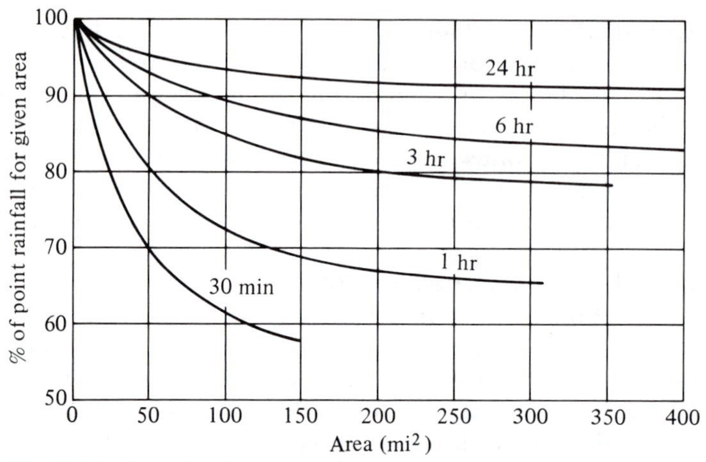


***Fig. 6-6.*** *Area-depth curves for use with duration frequency values (U.S. Weather Bureau.)*

based on maximum point gauge data exhibit the average area-depth relationship shown in Fig. 6-6.

The accuracy of area rainfall data depends heavily upon the density and location of gauges throughout the area considered. The simple averaging of the accumulation in all gauges gives no consideration to the effective area around each gauge or to the storm pattern. Two methods are available in calculating the weighted average of gauge records, the Thiessen polygon method and the isohyetal method. The Thiessen method assumes a linear variation of rainfall

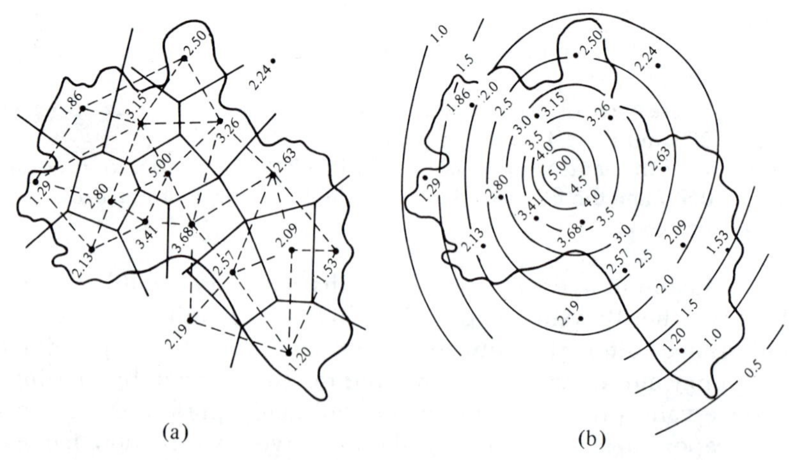


***Fig. 6-7.*** *Methods of determining rainfall averages: (a) Thiessen network (24-hr total; average basin precipitation = 2.54 in.); in (b) isohyetal map (24-hr total; average basin precipitation = 2.50 in.). The arithmetic average over the basin = (39.10/15) = 2.61 in.*

between each pair of gauges. Perpendicular bisectors of the connecting lines form polygons around each gauge (or partial polygons within the area boundary). If a sufficient number of gauges are available to construct contours of rainfall depth (isohyets), the weighting process can be carried out by using the average depth between isohyets and the area included between the isohyets and the area boundaries. Figure 6-7 shows both schemes.

An example of the effect of gauge location and density is shown in Figs. 6-8a and b. Figure 6-8a shows the increase in variability

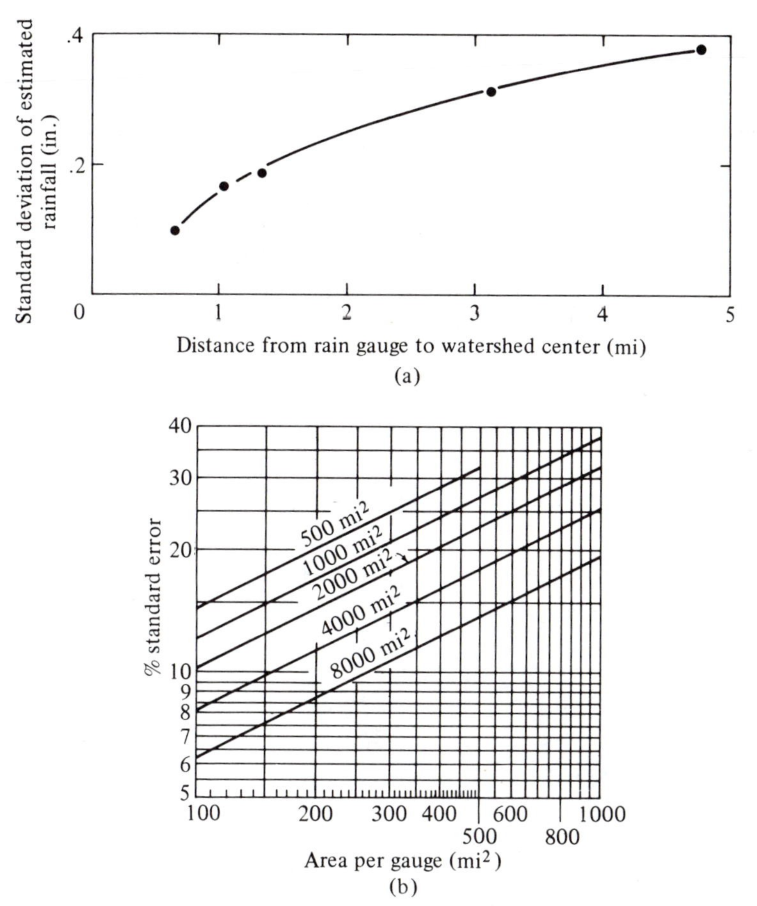


***Fig. 6-8.*** *6-8. (a) Relation between the standard deviation of the watershed and the point rainfall and distance of that point from the center of the watershed. (After W. G. Knisel, Jr., and R. W. Baird, in* ARS Precipitation Facilities and Related Studies. *Washington, D.C.: U.S. Department of Agriculture, Agricultural Research Service, 1971, Chap. 14.) (b) Standard error of average precipitation as a function of the network density an and drainage are for the Muskingum Basin. (U.S. Weather Bureau.)*

between Thiessen-weighted storm rainfalls and rainfall at a single gauge as the distance of single gauges from the watershed center increases. Figure 6-8b shows the effect of gauge density and total area on the standard error of the mean. Complete studies of precipitation patterns over large areas require detailed analysis of depth-area-duration data that depend upon the mass curves of accumulation from a network of gauges. The method is described in detail in other references.[15,17] Figure 6-9 depicts the depth-area relationship for the 24-hr storm shown in Fig. 6-7. It also required observations taken at various durations and the successive determination of average depths by the isohyetal method.

## 6-6 Multiple Regression Techniques

The regression techniques developed in Sec. 6-3 provide a powerful means for performing regional frequency analyses. These techniques are used to identify the mathematical dependence between observed values of physically related variables and thus can account for the additional information contained in correlated sequences of events. Sampling errors are reduced and the reliability of estimates is improved. In addition to predicting the mean or expected value of a hydrologic variable such as rainfall, runoff, or peak flows, the technique can be used to predict the expected value of other statistical parameters, for example, standard deviation, skewness, or autocorrela-

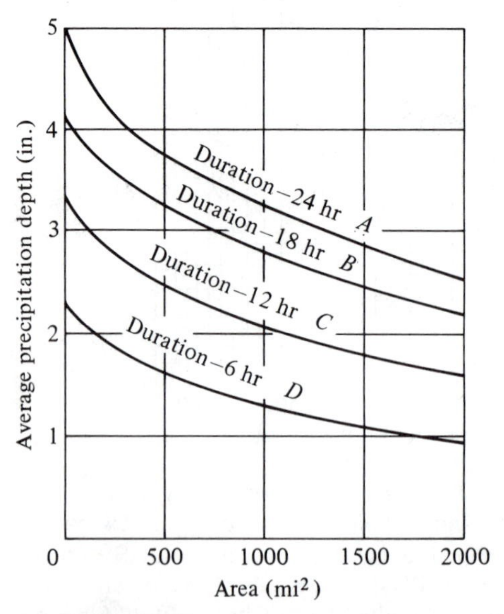


**Fig. 6-9.** *Depth-area-duration curves for 24-hr storm of Fig. 6-7. (From* Hydrology Handbook, *ASCE Manual of Practice, No. 28, 1949.)*

tion. The correlation is determined between the desired statistical parameter as dependent variable, and the appropriate physical and climatic variables within the basin or region as the independent variables. The procedures are significantly better than using relatively short historical sequences and point-frequency analysis.[4,5,11] Not only does the method reduce the inherently large sampling errors but it furnishes a means to estimate parameters at ungauged locations.

There are limitations to the techniques of Sec. 6-3. First, the analyst assumes the form of the model that can express only linear, or logarithmically linear, dependence. Second, he selects the independent variables to be included in the regression analysis. And, third, the theory assumes that the independent variables are indeed independent and are observed or determined without error. Advanced statistical methods that are beyond the scope of this text offer means to overcome some of these limitations but in practice it may be impossible to satisfy them. Therefore, care must be exercised in selecting the model and in interpreting results.

Accidental or casual correlation may exist between variables that are not functionally correlated. For this reason, correlation should be determined between hydrologic variables only when a physical relationship can be presumed. Because of the natural dependence between many factors treated as independent variables in hydrologic studies, the correlation between the dependent variable and each of the independent variables is different from the relative effect of the same independent variables when analyzed together in a multivariate model. One way to guard against this effect is by screening the variables initially by graphical methods. Another is to examine the results of the final regression equation to determine physical relevance.

Alternatively, regression techniques themselves aid in screening significant variables. When electronic computation is available, a procedure can be followed in which successive independent variables are added to the multiple regression model, and the relative effect of each is judged by the increase in the multiple correlation coefficient. Although statistical tests can be employed to judge significance, it is useful otherwise to specify that any variable remain in the regression equation if it contributes or explains, say, 1% or 5% of the total variance, or of $R^2$. A frequently used method is to compute the *partial correlation coefficients* for each variable. This statistic represents the relative decrease in the variance remaining $(1 - R^2)$ by the addition of the variable in question. If the variance remaining with the variable included in the regression is $(1 - R^2) = D^2$ and the variance remaining after removal is $(1 - R'^2) = D'^2$, then the partial regression correlation coefficient is $(D'^2 - D^2)/D'^2$.

At any modern electronic computer facility, a multiple regression program is normally included in the standard library routines. Most are extremely flexible, requiring minimal instructions and input data other than raw data. Special manipulations can effect an interchange of dependent and independent variables, bring one variable at a time into the regression equation, rearrange the independent variables in order of significance, and perform various statistical tests.

### Extending Hydrologic Records

Regression techniques frequently can be used to extend short records if significant correlation exists between the station of short record and a nearby station with a longer record. In Example 6-2, if the Jackson River records were complete from 1941 to date but the Cowpasture records were incomplete after 1952, the cross-correlation could be used to estimate the missing years by solving the regression equation for $Y$ from 1953 on using the $X$ flows as observed. The reliability of such methods depends on the correlation coefficient and the length of the concurrent records. If the concurrent record is too short or the correlation weak, the standard error of the parameter to be estimated can be increased and nothing is gained. The limiting value of cross-correlation for estimating means is approximately $\rho = 1/\sqrt{n}$, where $n$ is the length of the concurrent record.[18] Thus any correlation above .3 would improve the Cowpasture records. Estimates of other parameters with larger standard errors require higher cross-correlation for significant improvement. Extending or filling in deficient records often is necessary for regional studies in which every record used should be adjusted to the same length.

### Predicting Regionalized Hydrologic Variables

Cruff and Rantz[11] studied various methods of regional flood analysis and found the multiple regression technique a better predicter than either the index-flood method or the fitting of theoretical frequency distributions to individual historical records. They first used regression techniques to extend all records to a common base length. Next they extrapolated by various methods to estimate the 50- and 100-yr flood events and with multiple correlation examined several dependent variables including the drainage area $A$, the basin-shape factor (the ratio of the diameter of a circle of size $A$ to the length of the basin measured parallel to the main channel) $S_h$, channel shape $S$, the annual precipitation $P$, and others. They found only $A$ and $S_h$ to be significant, which resulted in prediction equations of the form $Q_t = cA^a S_h^b$. These equations were superior to those of the other techniques. The multiple correlation coefficient was as high as .954. It is interesting that regression techniques were employed in still a third way, that is, to estimate regional values of the mean and standard deviation after adjusting the record length. Example 6-5 illustrates the application of regression analysis to regionalize the standard deviation of annual maximum flow logarithms as a function of the drainage area size and the number of rainy days each year.

***Example 6-5*** In the following exhibit (Table 6-4) prepared by Beard,[4] the regional correlation is sought of the standard deviation of flow logarithms with the logarithms of the drainage area size and the number of rainy days

***Table 6-4*** *Logarithmic Data for 50 Gauging Stations*

$X_1 = 1 + \log s$ $X_2 = \log DA$ $X_3 = \log$ number of rainy days per year

| *Station Number (1)* | $X_2$ *(2)* | $X_3$ *(3)* | $X_1$ *(4)* | *Station Number (5)* | $X_2$ *(6)* | $X_3$ *(7)* | $X_1$ *(8)* |
|---|---|---|---|---|---|---|---|
| 1 | 1.61 | 2.11 | 0.29 | 33 | 1.94 | 1.87 | 0.20 |
| 2 | 2.89 | 2.12 | 0.18 | 34 | 2.73 | 1.36 | 0.58 |
| 3 | 4.38 | 2.11 | 0.17 | 35 | 3.63 | 1.81 | 0.64 |
| 4 | 3.20 | 2.04 | 0.44 | 36 | 1.91 | 1.58 | 0.37 |
| 5 | 3.92 | 2.07 | 0.38 | 37 | 2.26 | 1.48 | 0.27 |
| 6 | 1.61 | 2.04 | 0.37 | 38 | 2.97 | 1.89 | 0.54 |
| 7 | 3.21 | 2.09 | 0.30 | 39 | 0.70 | 1.32 | 0.63 |
| 8 | 3.65 | 1.99 | 0.35 | 40 | 0.30 | 1.54 | 0.78 |
| 9 | 3.23 | 2.15 | 0.16 | 41 | 3.38 | 1.62 | 0.46 |
| 10 | 4.33 | 2.08 | 0.11 | 42 | 2.87 | 2.03 | 0.44 |
| 11 | 1.60 | 2.09 | 0.32 | 43 | 2.42 | 2.26 | 0.24 |
| 12 | 2.82 | 2.00 | 0.34 | 44 | 4.53 | 1.93 | −0.03 |
| 13 | 2.40 | 2.00 | 0.25 | 45 | 3.04 | 1.78 | 0.30 |
| 14 | 3.69 | 2.09 | 0.43 | 46 | 4.13 | 2.00 | 0.17 |
| 15 | 2.18 | 2.19 | 0.27 | 47 | 1.49 | 2.01 | 0.14 |
| 16 | 2.09 | 2.17 | 0.25 | 48 | 5.37 | 1.95 | 0.10 |
| 17 | 4.48 | 1.91 | 0.52 | 49 | 1.36 | 2.11 | 0.27 |
| 18 | 4.95 | 1.95 | 0.18 | 50 | 2.31 | 2.23 | 0.18 |
| 19 | 2.21 | 1.97 | 0.39 | | | | |
| 20 | 3.41 | 2.08 | 0.40 | $\sum X$ | 147.55 | 96.24 | 17.89 |
| 21 | 4.82 | 1.88 | 0.25 | $\bar{X}$ | 2.951 | 1.925 | 0.358 |
| 22 | 1.78 | 1.93 | 0.23 | $\sum XX_2$ | 503.7779 | 285.5627 | 51.1527 |
| 23 | 4.39 | 1.74 | 0.54 | $\sum X \sum X_2/n$ | 435.4200 | 284.0042 | 52.7934 |
| 24 | 3.23 | 2.01 | 0.51 | $\sum xx_2$ | 68.3579 | 1.5585 | −1.6407 |
| 25 | 3.58 | 2.04 | 0.45 | | | | |
| 26 | 1.64 | 1.78 | 0.63 | $\sum XX_3$ | | 187.5912 | 33.2598 |
| 27 | 4.58 | 1.76 | 0.45 | $\sum X \sum X_3/n$ | | 185.2428 | 34.4347 |
| 28 | 3.26 | 1.93 | 0.59 | $\sum xx_3$ | 1.5585 | 2.3484 | −1.1749 |
| 29 | 4.29 | 1.81 | 0.46 | | | | |
| 30 | 1.23 | 1.89 | 0.32 | $\sum XX_1$ | | | 8.1635 |
| 31 | 3.44 | 1.48 | 0.96 | $\sum X \sum X_1/n$ | | | 6.4010 |
| 32 | 2.11 | 1.97 | 0.12 | $\sum xx_1$ | −1.6407 | −1.1749 | 1.7625 |

*Note:* $x = (X - \bar{X})$.

per year; $X_1$ is set equal to $(1 + \log s)$ to avoid negative values. Find the regression equation and the multiple correlation coefficient.

*Solution*

1. From Eqs. 6-42, 6-43, and 6-44, the parameters are

$$\alpha = 1.34; \quad \beta_1 = -0.013; \quad \beta_2 = -0.49$$

and the regression equation is

$$X_1 = 1.34 - 0.013X_2 - 0.49X_3$$

or

$$\log s = 0.34 - 0.013 \log (\text{DA}) - 0.49 \log (\text{days})$$

2. The multiple correlation coefficient from Eqs. 6-45 and 6-46 is $R = .56$.

## Problems

6-1. Observations for the past 10 years of withdrawals and estimated recharge of an artesian aquifer are given in the following table. Find the means, variances, standard deviations, covariance, and correlation coefficient.

| Measured Discharge (1000 acre-ft) | Estimated Recharge (1000 acre-ft) |
|---|---|
| 12.2 | 12.0 |
| 10.4 | 9.8 |
| 10.6 | 11.0 |
| 12.6 | 13.2 |
| 14.2 | 14.6 |
| 13.0 | 14.0 |
| 14.0 | 14.0 |
| 12.0 | 12.4 |
| 10.4 | 10.4 |
| 11.4 | 11.6 |

6-2. Fit a regression equation to the data in Prob. 6-1, treating discharge as the dependent variable. Compute the standard error of estimate. Estimate the expected discharge when recharge is 13 KAF. What would be the estimate of discharge if no information were available on recharge? What is the relative improvement provided by the regression estimate?

6-3. Prepare a computer program for simple, two-variable, linear regression. The program should read in $N$ pairs of observations, $Y$ and $X$. (a) Compute the means, variances, and standard deviations of both $Y$ and $X$, and (b) find the regression constants, the standard error of estimate, and the correlation coefficient. Verify with the data in Prob. 6-1.

6-4. From the following observations of variation of the mean annual rainfall with the altitude of the gauge, determine a linear prediction equation for the catchment. What is the correlation between rainfall and altitude?

| Gauge Number | Mean Annual Rainfall (in.) | Altitude of Gauge (1000 ft) |
|---|---|---|
| 1 | 22 | 4.2 |
| 2 | 28 | 4.4 |
| 3 | 25 | 4.5 |
| 4 | 31 | 5.4 |
| 5 | 32 | 5.6 |
| 6 | 37 | 5.6 |
| 7 | 36 | 5.8 |
| 8 | 35 | 6.0 |
| 9 | 36 | 6.6 |
| 10 | 46 | 6.6 |
| 11 | 41 | 6.8 |
| 12 | 41 | 7.0 |

6-5. Estimate the expected rainfall in Prob. 6-4 for a gauge to be installed at an altitude of 5500 ft.

6-6. The least-squares estimates of $A$ and $B$ in the bivariate regression equation $Y = A + BX$ are $A = 2.0$ and $B = 3.0$, where $Y$ is a transformation defined as $\log_{10} y$ and $X$ is defined as $\log_{10} x$. If $y$ and $x$ are related by $y = ax^b$, determine the values of $a$ and $b$.

6-7. The time of rise of flood hydrographs ($T_r$), defined as the time for a stream to rise from low water to maximum depth following a storm, is related to the stream length ($L$) and the average slope ($S$). From the information given below for 11 watersheds in Texas, New Mexico, and Oklahoma, derive a functional relation of the form $T_r = aL^bS^c$.

| Watershed Number | $T_r$ (min) | $L$ (1000 ft) | $S$ (ft/1000 ft) |
|---|---|---|---|
| 1 | 150 | 18.5 | 7.93 |
| 2 | 90 | 14.2 | 19.0 |
| 3 | 60 | 25.3 | 12.0 |
| 4 | 60 | 11.7 | 13.3 |
| 5 | 100 | 9.7 | 11.0 |
| 6 | 75 | 8.1 | 15.0 |
| 7 | 90 | 21.7 | 16.7 |
| 8 | 30 | 3.9 | 146.0 |
| 9 | 30 | 1.2 | 20.0 |
| 10 | 45 | 3.3 | 64.0 |
| 11 | 50 | 3.5 | 33.0 |

6-8. Repeat the exercise in Prob. 6-7 by fitting the relationship $T_r = dF^e$, where $F = L/\sqrt{S}$ with $L$ in mi and $S$ in ft/mi. Plot the results on log-log paper.

6-9. In Example 6-3, verify the eventual generation of the steady state probabilities by beginning in state 3 and transitioning to succeeding states according to the given transition probability matrix.

6-10. Using the information in Example 6-3 and Prob. 6-9, compile a probability matrix that shows the conditional probability after five transitions, that is, the probabilities of being in state $i$ at period $y$ and state $j$ at period $y + 5$.

6-11. Given the following transition probability matrix, find the steady state probability vector.

| | | State $j$ in period $y + 1$ | | |
|---|---|---|---|---|
| | | 1 | 2 | 3 |
| State $i$ in in period $y$ | 1 | 0.5 | 0.3 | 0.2 |
| | 2 | 0.3 | 0.6 | 0.1 |
| | 3 | 0.1 | 0.4 | 0.5 |

6-12. From the partial record of streamflows in Plum Creek near Meadville, Neb., find the monthly serial correlation coefficient between October and November. Repeat for April and May. Comment on the hydrologic reasons for any difference in the correlation between the two pairs of data.

*Streamflow Data, Plum Creek at Meadville, Neb. (100 acre-ft)*

| Year | Oct. | Nov. | Apr. | May | Annual Total |
|---|---|---|---|---|---|
| 1949 | 49.9 | 47.5 | 97.5 | 63.7 | 710.5 |
| 1950 | 48.5 | 47.9 | 76.8 | 106.5 | 702.8 |
| 1951 | 55.8 | 52.2 | 74.1 | 88.1 | 897.9 |
| 1952 | 72.2 | 73.2 | 116.4 | 163.7 | 1012.0 |
| 1953 | 55.0 | 52.0 | 71.0 | 115.2 | 810.6 |
| 1954 | 71.1 | 55.4 | 67.9 | 92.7 | 795.7 |
| 1955 | 56.3 | 58.4 | 62.8 | 61.9 | 730.0 |
| 1956 | 56.1 | 52.4 | 59.6 | 64.0 | 664.8 |
| 1957 | 55.7 | 58.6 | 72.2 | 102.3 | 761.6 |
| 1958 | 57.8 | 59.6 | 161.7 | 77.0 | 849.3 |
| 1959 | 60.2 | 57.8 | 69.9 | 63.6 | 707.5 |
| 1960 | 54.7 | 56.8 | 89.0 | 101.5 | 781.9 |
| 1961 | 57.9 | 56.1 | 55.7 | 67.6 | 689.2 |
| 1962 | 55.5 | 54.1 | 59.3 | 101.0 | 1010.0 |
| 1963 | 62.3 | 61.3 | 74.8 | 76.3 | 882.7 |
| 1964 | 60.5 | 57.9 | 89.5 | 90.0 | 792.1 |
| 1965 | 52.3 | 55.2 | 62.1 | 63.9 | 705.6 |
| 1966 | 58.8 | 54.4 | 71.1 | 63.8 | 731.9 |
| 1967 | 58.0 | 55.3 | 54.7 | 64.0 | 788.3 |

6-13. For the total annual flow given in Prob. 6-12, determine the serial correlation between successive years. Explain how you would use this information (a) to predict the expected flow in any future year when the preceding year's flow is 75,000 acre-ft, and (b) to add a random component to the estimate in (a) for a simulation study.

6-14. The results of a multiple regression analysis of over 200 flood records in Virginia led to the following regional flood frequency equations:

$$\begin{aligned}
Q_{1.2\text{-yr}} &= 9.13 \quad A^{.909} \quad S^{.293} \\
Q_{2.33\text{-yr}} &= 20.8 \quad A^{.861} \quad S^{.309} \\
Q_{5\text{-yr}} &= 38.1 \quad A^{.830} \quad S^{.300} \\
Q_{10\text{-yr}} &= 63.0 \quad A^{.802} \quad S^{.283} \\
Q_{25\text{-yr}} &= 104 \quad A^{.779} \quad S^{.266} \\
Q_{50\text{-yr}} &= 118 \quad A^{.795} \quad S^{.279}
\end{aligned}$$

where the flood discharge for the given frequency is in cfs, $A$ is the drainage area in mi$^2$, and $S$ is the channel slope in ft/mi (measured between the points that are 10 and 85% of the total river miles upstream of the gauging station to the drainage divide). Devise a staisfactory method for graphically portraying these regional flood frequency relations. (Note that there are four factors, $Q$, $T$, $A$, and $S$.)

6-15. Using the regression equations in Prob. 6-14, find the predicted floods for the North Fork, Shenandoah River, at Cootes Store. Drainage area = 215 mi$^2$ and channel slope = 44.3 ft/mi.

6-16. Compare the predictions from the regression equations in Prob. 6-14 with the values estimated by the frequency analysis in Example 5-6. Drainage area = 487 mi$^2$ and channel slope = 21.1 ft/mi.

6-17. Referring to Fig. 2-17(b), compare the average storm rainfall over the city of Baltimore on Sept. 10, 1957, computed by the isohyetal method with the simple average of total accumulation at the rain gauges within the city. Neglect the area to the south of the 1.0-in. isohyet.

6-18. Fit the formula $i = A/(t + B)$ to the data derived in Prob. 5-27 for the 5-yr intensity-duration-frequency curve.

6-19. Develop a regional flood frequency curve for the Rappahannock River Basin from the flood frequency data given in the following table.

*Peak Flood Frequency Discharges (ft³/sec) for Stations in the Rappahannock River Basin*

| Station | Drainage Area (mi) | Type of Series | 2.33 (mean) | Return Period in Years 5 | 10 | 25 | 50 |
|---|---|---|---|---|---|---|---|
| Rappahannock River near Warrenton, Va. | 192 | Annual | 4,150 | 8,350 | 9,000 | 14,000 | 19,250 |
| | | Partial | 4,600 | 8,650 | 9,200 | 14,000 | 19,250 |
| Rush River at Washington, Va. | 15.2 | Annual | 530 | 860 | 1,290 | 2,100 | 3,000 |
| | | Partial | 610 | 900 | 1,310 | 2,100 | 3,000 |
| Thornton River near Laurel Mills, Va. | 142 | Annual | 5,900 | 11,500 | 19,900 | 34,000 | 48,000 |
| | | Partial | 7,200 | 12,500 | 20,500 | 34,000 | 48,000 |
| Hazel River at Rixeyville, Va. | 286 | Annual | 7,300 | 11,800 | 17,200 | 25,000 | 41,000 |
| | | Partial | 8,300 | 12,400 | 18,000 | 25,500 | 41,000 |
| Rappahonnock River at Remington, Va. | 616 | Annual | 11,000 | 14,500 | 18,100 | 24,500 | 31,000 |
| | | Partial | 12,000 | 15,200 | 18,900 | 25,000 | 31,000 |

| Station | Drainage Area (mi) | Type of Series | 2.33 (mean) | Return Period in Years 5 | 10 | 25 | 50 |
|---|---|---|---|---|---|---|---|
| Rappahannock River at Kellys Ford, Va. | 641 | Annual | 12,300 | 19,000 | 26,800 | 42,000 | 57,500 |
| | | Partial | 14,000 | 20,000 | 27,500 | 42,000 | 57,500 |
| Mountain Run near Culpeper, Va. | 14.7 | Annual | 750 | 1,750 | 3,350 | 6,000 | 10,000 |
| | | Partial | 950 | 1,900 | 3,550 | 6,000 | 10,000 |
| Rapidan River near Ruckersville, Va. | 111 | Annual | 3,950 | 7,100 | 11,600 | 21,000 | 34,000 |
| | | Partial | 4,700 | 7,700 | 12,000 | 21,000 | 34,000 |
| Robinson River near Locust Dale, Va. | 180 | Annual | 4,600 | 7,000 | 9,800 | 15,400 | 21,500 |
| | | Partial | 5,150 | 7,300 | 10,100 | 15,800 | 21,500 |
| Rapidan River near Culpeper, Va. | 456 | Annual | 9,100 | 16,400 | 26,900 | 50,000 | 78,000 |
| | | Partial | 10,800 | 17,600 | 27,600 | 50,000 | 78,000 |
| Rappahannock River near Fredericksburg, Va. | 1,599 | Annual | 26,000 | 39,900 | 55,000 | 85,000 | 117,000 |
| | | Partial | 29,300 | 42,000 | 57,500 | 85,000 | 117,000 |

6-20. From the information given in Prob. 6-19, find the relationship between the mean annual flow and the drainage area for the Rappahannock River Basin. (Note that the functional expression should be of the form $Q_{2.33} = rA^s$.)

6-21. Using the results of Probs. 6-19 and 6-20, estimate the 30-yr flood for an ungauged watershed with a drainage area of 540 mi².

## References

1. Benjamin, J. R., and C. Cornell, *Probability, Statistics and Decision for Civil Engineers* (New York: McGraw-Hill Book Company, 1959).
2. Feller, W., *An Introduction to Probability Theory* (New York: John Wiley & Sons, Vol. 1, 1957, and Vol. 2, 1966).
3. Loucks, D. P., "Stochastic Methods for Analyzing River Basin Systems," Tech Rept. No. 16, Cornell University Resources and Marine Sciences Center, Ithaca, N.Y., 1969.
4. Beard, L. R., *Statistical Methods in Hydrology,* Civil Works Investigations; Sacramento District, U.S. Army Corps of Engineers, 1962.
5. Benson, M. A., and N. C. Matalas, "Synthetic Hydrology Based on Regional Statistical Parameters," *Water Resources Research,* 3, No. 4 (1967).
6. Matalas, N. C., "Mathematical Assessment of Synthetic Hydrology," *Water Resources Research,* 3, No. 4 (1967).

7. Chow, Ven T., "Statistical and Probability Analysis of Hydrologic Data," Sec. 8-I, V. T. Chow (ed.), *Handbook of Applied Hydrology* (New York: McGraw-Hill Book Company, 1964).
8. Dalrymple, T., "Flood-Frequency Analysis," *Manual of Hydrology,* Part 3, U.S. Geological Survey Water-Supply Paper 1543-A. Washington, D.C.: Government Printing Office, 1960.
9. Fair, G. M., J. C. Geyer, and D. A. Okun, *Water and Waste Water Engineering* (New York: John Wiley & Sons, Inc., 1966).
10. "Monthly Stream Simulation," Hydrologic Engineering Center, Computer Program 23-C-L267, Sacramento District, U.S. Army Corps of Engineers, July 1967.
11. Cruff, R. W., and S. E. Rantz, "A Comparison of Methods Used in Flood-Frequency Studies for Coastal Basins in California," *Flood Hydrology,* U.S. Geological Survey Water-Supply Paper 1580. Washington, D.C.: Government Printing Office, 1965.
12. Potter, W. D., "Peak Rates of Runoff from Small Watersheds," Hydraulic Design Series No. 2, Bureau of Public Roads. Washington, D.C.: Government Printing Office, Apr. 1961.
13. Steel, E. W., *Water Supply and Sewerage,* 4th ed. (New York: McGraw-Hill Book Company, 1960).
14. Hershfield, D. M., "Rainfall Frequency Atlas of the United States," Tech. Paper No. 40, U.S. Weather Bureau, 1961.
15. *Hydrology Handbook,* ASCE Manual of Practice, No. 28, 1949.
16. Knisel, W. G., Jr., and R. W. Baird, in *ARS Precipitation Facilities and Related Studies.* Washington, D.C.: U.S. Department of Agriculture, Agricultural Research Service, 1971, Chap. 14.
17. Linsley, R. K., Jr., M. A. Kohler, and J. L. H. Paulhus, *Applied Hydrology* (New York: McGraw-Hill Book Company, 1949).
18. Fiering, M. B., "Information Analysis," in G. M. Fair, J. C. Geyer, and D. A. Okun, *Water Supply and Waste Water Removal* (New York: John Wiley & Sons, Inc., 1966), Chap. 14.

# 7

# Hydrologic and Hydraulic Routing

Flood forecasting, reservoir design, watershed simulation, and comprehensive water resources projects generally utilize some form of routing technique. Routing is used to predict the temporal and spatial variations of a flood wave as it traverses a river reach or reservoir, or can be employed to predict the outflow hydrograph from a watershed subjected to a known amount of precipitation. Routing techniques may be classified into two categories—*hydrologic routing* and *hydraulic routing*.

Hydrologic routing employs the equation of continuity with either an analytic or an assumed relationship between storage and discharge within the system. Hydraulic routing, on the other hand, uses both the equation of continuity and the equation of motion, customarily the momentum equation. This particular form utilizes the partial differential equations for unsteady flow in open channels. It more adequately describes the dynamics of flow than does the hydrologic-routing technique and is treated in the latter portion of this chapter.

Applications of hydrologic routing techniques to problems of flood prediction, evaluations of flood control measures, and the effects of urbanization are numerous. Most flood warning systems instituted by NOAA and the Corps of Engineers incorporate some form of this technique to predict flood stages in advance of a severe storm. It is the method most frequently used to size spillways for small, intermediate, and large dams. Additionally, the synthesis of runoff hydrographs from gauged and ungauged watersheds is possible by the use of basic assumptions inherent in this approach. River, reservoir, and watershed routing techniques are presented in separate sections of this chapter.

## 7-1 Hydrologic River Routing

The first reference to real application of routing a flood hydrograph from one river station to another was by Graeff in 1833.[1] The technique was based on the use of wave velocity and a rating curve. Hydrologic river routing techniques are all founded upon the equation of continuity

$$I - O = \frac{dS}{dt} \tag{7-1}$$

where

$I$ = the inflow rate to the reach (cfs)
$O$ = the outflow rate from the reach (cfs)
$dS/dt$ = the rate of change of storage within the reach

Two hydrologic river routing techniques are described in subsequent paragraphs. Other techniques are introduced in Chapter 10, Sec. 10-5.

### Muskingum Method

Storage in a stable river reach can be expected to depend primarily on the discharge into and out of a reach, and upon hydraulic characteristics of the channel section. The storage within the reach at a given time can be expressed as[3]

$$S = \frac{b}{a}\left[xI^{m/n} + (1 - x)O^{m/n}\right] \tag{7-2}$$

Constants $a$ and $n$ reflect the stage discharge characteristics of control sections at each end of the reach, and $b$ and $m$ mirror the stage volume characteristics of the section. The factor $x$ defines the relative weights given to inflow and outflow within the reach.

The Muskingum method assumes that $m/n = 1$ and lets $b/a = K$, resulting in

$$S = K[xI + (1 - x)O] \tag{7-3}$$

where

$K$ = the storage time constant for the reach
$x$ = the weighting factor which varies between 0 and 0.5 for a given river section

Application of this equation has shown that $K$ is usually reasonably close to the travel time within the reach and $x$ will average about 0.2.

Behavior of the flood wave due to changes in the value of the weighting factor $x$ is readily apparent from examination of Fig. 7-1. The resulting downstream floodwave is commonly described by the amount of translation—that is, the time lag—and by the amount of attenuation, or reduction in peak discharge. As can be noted from Fig. 7-1, the value $x = 0.5$ results in a pure translation of the flood wave.

Application of Eqs. 7-1 and 7-3 to a river reach is a straightforward procedure if $K$ and $x$ are known. The routing procedure begins by dividing time into a number of equal increments, $\Delta t$, and expressing Eq. 7-1 in finite difference form

$$\frac{I_1 + I_2}{2} - \frac{O_1 + O_2}{2} = \frac{S_2 - S_1}{\Delta t} \tag{7-4}$$

The routing time interval $\Delta t$ is normally assigned any convenient value between the limits of $K/3$ to $K$.

Using subscripts 1 and 2 to denote the beginning and ending times for the interval $\Delta t$, the storage change in the river reach during the routing interval from Eq. 7-3 is

$$S_2 - S_1 = K\{x(I_2 - I_1) + (1 - x)(O_2 - O_1)\} \tag{7-5}$$

and substituting this into Eq. 7-4 results in the Muskingum routing equation

$$O_2 = C_0 I_2 + C_1 I_1 + C_2 O_1 \tag{7-6}$$

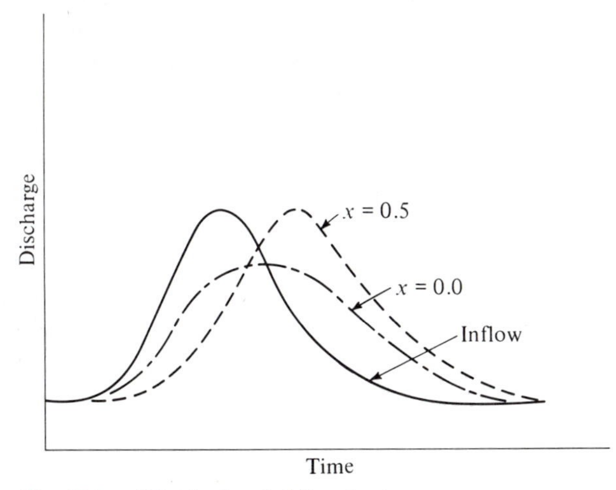


***Fig. 7-1.*** *Effect of weighting factor.*

in which

$$C_0 = \frac{-Kx + 0.5\Delta t}{K - Kx + 0.5\Delta t}$$

$$C_1 = \frac{Kx + 0.5\Delta t}{K - Kx + 0.5\Delta t}$$

$$C_2 = \frac{K - Kx - 0.5\Delta t}{K - Kx + 0.5\Delta t}$$

Note that $K$ and $\Delta t$ must have the same time units and also that the coefficients sum to 1.0.

Since $I_1$ and $I_2$ are known for every time increment, routing is accomplished by solving Eq. 7-6 for successive time increments using each $O_2$ as $O_1$ for the next time increment. Example 7-1 illustrates this row-by-row computation.

***Example 7-1*** Perform the flood routing for a reach of river given $x = 0.2$ and $K = 2$ days. The inflow hydrograph using $\Delta t = 1$ day is shown in column 1. Assume equal inflow and outflow rates on the 16th.

*Solution*
If $\Delta t = 1$ day and $x = 0.2$ and $K = 2$ days, then Eq. 7-6 gives $C_0 = 0.0477$, $C_1 = 0.428$, and $C_2 = 0.524$. Row-by-row computation is given in Table 7-1.

### Determination of $K$ and $x$

Values of $K$ and $x$ for Muskingum routing are commonly estimated using $K$ = travel time in the reach and the average value of $x = 0.2$. If both the inflow and outflow hydrograph records are available for one or more floods, the routing process is easily reversed to provide better values of $K$ and $x$ for the reach, or else the approximate method described in the Sec. 7-3 discussion on hydrologic watershed routing can be used. To illustrate, instantaneous values of $S$ versus $xI + (1 - x)O$ are first graphed for several selected values of $x$ as shown in Example 7-2. Because $S$ and $xI + (1 - x)O$ are assumed to be linearly related via Eq. 7-3, the accepted value of $x$ is that which gives the best linear plot (the narrowest loop). After plotting, the value for $K$ is determined as the reciprocal of the slope through the narrowest loop, since from Eq. 7-3

$$K = \frac{S}{xI + (1 - x)O} \tag{7-7}$$

Instantaneous values of $S$ for the graphs in Example 7-2 were determined by solving for $S_2$ in Eq. 7-4 for successive time increments. A value of $S_1 = 0$ was used for the initial increment, but the value is arbitrary since only the slope and not the intercept of Eq. 7-3 is desired.

**Table 7-1**

| Date | Inflow (1) | $C_0I_2$ (2) | $C_1I_1$ (3) | $C_2O_1$ (4) | Computed Outflow (5) |
|---|---|---|---|---|---|
| 16 | 4,260 | — | — | — | 4,260 |
| 17 | 7,646 | 364 | 1,823 | 2,232 | 4,419 |
| 18 | 11,167 | 532 | 3,272 | 2,315 | 6,119 |
| 19 | 16,730 | 798 | 4,779 | 3,206 | 8,783 |
| 20 | 21,590 | 1,029 | 7,160 | 4,602 | 12,791 |
| 21 | 20,950 | 999 | 9,240 | 6,702 | 16,941 |
| 22 | 26,570 | 1,267 | 8,966 | 8,877 | 19,110 |
| 23 | 46,000 | 2,194 | 11,371 | 10,013 | 23,578 |
| 24 | 59,960 | 2,860 | 19,688 | 12,355 | 34,903 |
| 25 | 57,740 | 2,754 | 25,662 | 18,289 | 46,705 |
| 26 | 47,890 | 2,284 | 24,712 | 24,473 | 51,469 |
| 27 | 34,460 | 1,643 | 20,496 | 26,970 | 49,109 |
| 28 | 21,660 | 1,033 | 14,748 | 25,733 | 41,514 |
| 29 | 34,680 | 1,654 | 9,270 | 21,753 | 32,677 |
| 30 | 45,180 | 2,155 | 14,843 | 17,122 | 34,120 |
| 31 | 49,140 | 2,343 | 19,337 | 17,879 | 39,559 |
| 1 | 41,290 | 1,969 | 21,031 | 20,729 | 43,729 |
| 2 | 33,830 | 1,613 | 17,672 | 22,914 | 42,199 |
| 3 | 20,510 | 978 | 14,479 | 22,112 | 37,569 |
| 4 | 14,720 | 702 | 8,778 | 19,686 | 29,166 |
| 5 | 11,436 | 545 | 6,300 | 15,283 | 22,128 |
| 6 | 9,294 | 443 | 4,894 | 11,595 | 16,932 |
| 7 | 7,831 | 373 | 3,977 | 8,872 | 13,222 |
| 8 | 6,228 | 297 | 3,351 | 6,928 | 10,576 |
| 9 | 6,083 | 290 | 2,665 | 5,542 | 8,497 |

***Example 7-2*** Given inflow and outflow hydrographs on the Muckwamp River, determine $K$ and $x$ for the river reach. (See Table 7-2.)

*Solution*

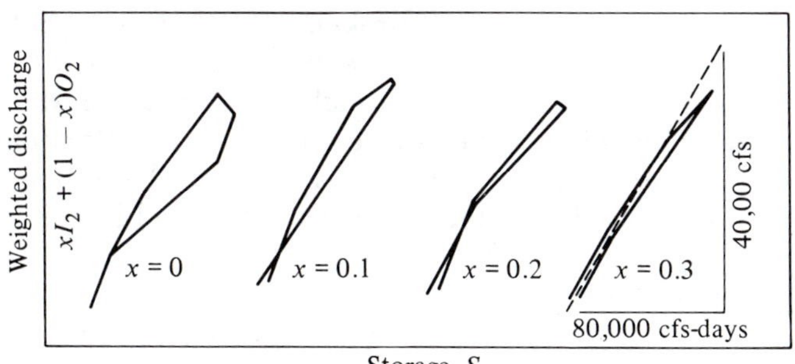

**Table 7-2**

| Date | $\bar{I} = \frac{I_1 + I_2}{2}$ (cfs) | $\bar{O} = \frac{O_1 + O_2}{2}$ (cfs) | $S_2$ (1000 cfs-days) | Weighted Discharge (cfs) $xI_2 + (1 - x)O_2$ $x = 0.1$ | $x = 0.2$ | $x = 0.3$ |
|---|---|---|---|---|---|---|
| 3-16 | 5,870 | 4,180 | 1.7 | 4,350 | 4,520 | 4,690 |
| 17 | 9,310 | 6,970 | 4.0 | 7,200 | 7,440 | 7,670 |
| 18 | 12,900 | 7,560 | 9.4 | 8,090 | 8,630 | 9,160 |
| 19 | 20,500 | 14,200 | 15.7 | 14,800 | 15,500 | 16,100 |
| 20 | 21,000 | 18,300 | 18.4 | 18,600 | 18,800 | 19,100 |
| 21 | 23,400 | 18,500 | 23.3 | 19,000 | 19,500 | 20,000 |
| 22 | 32,500 | 21,300 | 34.5 | 22,400 | 23,500 | 24,700 |
| 23 | 55,400 | 29,300 | 60.6 | 31,900 | 34,500 | 37,100 |
| 24 | 62,700 | 39,700 | 83.6 | 42,000 | 44,300 | 46,600 |
| 25 | 52,600 | 48,700 | 97.5 | 49,100 | 49,500 | 50,000 |
| 3-26 | 43,200 | 53,300 | 87.4 | 52,300 | 51,300 | 50,300 |
| 27 | 25,200 | 48,700 | 73.9 | 46,400 | 44,000 | 41,700 |
| 28 | 22,800 | 37,100 | 59.6 | 35,700 | 34,200 | 32,800 |
| 29 | 41,200 | 35,800 | 65.0 | 36,300 | 36,900 | 37,400 |
| 30 | 50,400 | 35,800 | 79.6 | 37,300 | 38,700 | 40,200 |
| 31 | 45,300 | 35,800 | 89.1 | 36,800 | 37,700 | 38,600 |
| 4- 1 | 38,800 | 42,700 | 85.2 | 42,300 | 41,900 | 41,500 |
| 2 | 27,000 | 44,100 | 68.0 | 42,400 | 40,800 | 39,000 |
| 3 | 16,200 | 35,400 | 48.9 | 33,500 | 31,600 | 29,600 |
| 4 | 12,400 | 25,200 | 36.1 | 23,900 | 22,600 | 21,400 |
| 5 | 10,200 | 16,400 | 29.9 | 15,800 | 15,200 | 14,500 |
| 6 | 8,080 | 11,500 | 26.5 | 11,200 | 10,800 | 10,500 |
| 7 | 6,010 | 9,380 | 23.1 | 9,040 | 8,710 | 8,370 |
| 8 | 5,050 | 7,860 | 20.3 | 7,300 | 7,300 | 7,020 |

Selecting the narrowest loop gives $x = 0.3$; $K = 80{,}000$ cfs-days/40,000 cfs $= 2.0$ days. These values could now be used to route other floods through the reach as in Example 7-1.

Inherent in this procedure is the postulation that the water surface in the reach is a uniform unbroken surface profile between upstream and downstream ends of the section. Additionally, it is presupposed that $K$ and $x$ are constant throughout the range in the stage of investigation. If significant departures from these restrictions are present, it may be necessary to work with shorter reaches of the river, or to employ a more sophisticated approach to the problem.

### Muskingum Crest Segment Routing

Sometimes it is desirable to route only a portion of an inflow hydrograph by the Muskingum method (for example, the crest segment when only the peak outflow is desired). This is easily accomplished by successively numbering the inflow rates as $I_1, I_2, I_3, \ldots, I_n, I_{n+1}, \ldots$ and rewriting Eq. 7-6 as

$$O_n = C_0 I_n + C_1 I_{n-1} + C_2 O_{n-1} \tag{7-8}$$

where $O_n$ is the outflow rate at any time $n$. The outflow $O_{n-1}$ is next eliminated from Eq. 7-8 by making the substitution

$$O_{n-1} = C_0 I_{n-1} + C_1 I_{n-2} + C_2 O_{n-2} \tag{7-9}$$

By repeated substitutions for the right-side outflow term, $O_{n-2}, O_{n-3}, \ldots$, can each be eliminated and $O_n$ can be expressed as a function only of the first $n$ inflow rates or, finally,

$$O_n = K_1 I_n + K_2 I_{n-1} + K_3 I_{n-2} + \cdots + K_n I_1 \tag{7-10}$$

where

$$K_1 = C_0$$
$$K_2 = C_0 C_2 + C_1$$
$$K_3 = K_2 C_2$$
$$K_i = K_{i-1} C_2 \text{ for } i > 2$$

To illustrate using data from Example 7-1, to find the outflow rate on the twenty-sixth

$$K_1 = C_0 = 0.0477$$
$$K_2 = C_0 C_2 + C_1 = 0.0477(0.524) + 0.428 = 0.4530$$
$$K_3 = K_2 C_2 = 0.4530(0.524) = 0.2374$$
$$\vdots$$
$$K_{11} = K_{10} C_2 = 0.0013$$

Thus the outflow on the twenty-sixth is calculated as $O_{11} = 0.0477 \times (47{,}890) + 0.4530(57{,}740) + \cdots + 0.0013(4260) = 51{,}469$ cfs.

### Graphical Integration Method

Many different graphical methods are used to solve the equation of continuity. A method shown here is rapid and produces a satisfactory outflow discharge hydrograph.[4] It deals with two factors $K$ and $L$; $L$ represents the lag time and $K$ is a coefficient which when multiplied by the change in outflow with respect to time gives the difference in storage with respect to time (or the storage time constant of Eq. 7-3 when $x = 0$).

$$K\frac{dO}{dt} = \frac{dS}{dt} \tag{7-11}$$

or, substituting Eq. 7-1 results in

$$\frac{I - O}{K} = \frac{dO}{dt} \tag{7-12}$$

where the difference between the inflow and the outflow divided by $K$ represents the change in outflow with respect to time. The graphical determination of $K$ and $L$ values for a particular river reach permits subsequent construction of outflow hydrographs from given or synthesized inflow hydrographs.

The time difference between points $A$ and $B$, $L$, is obtained from a previous flood wave of record as shown in Fig. 7-2a. The procedure to get $K$ for a known flood wave is as follows.

1. Lag the inflow hydrograph $L$ hours as in Fig. 7-2b.

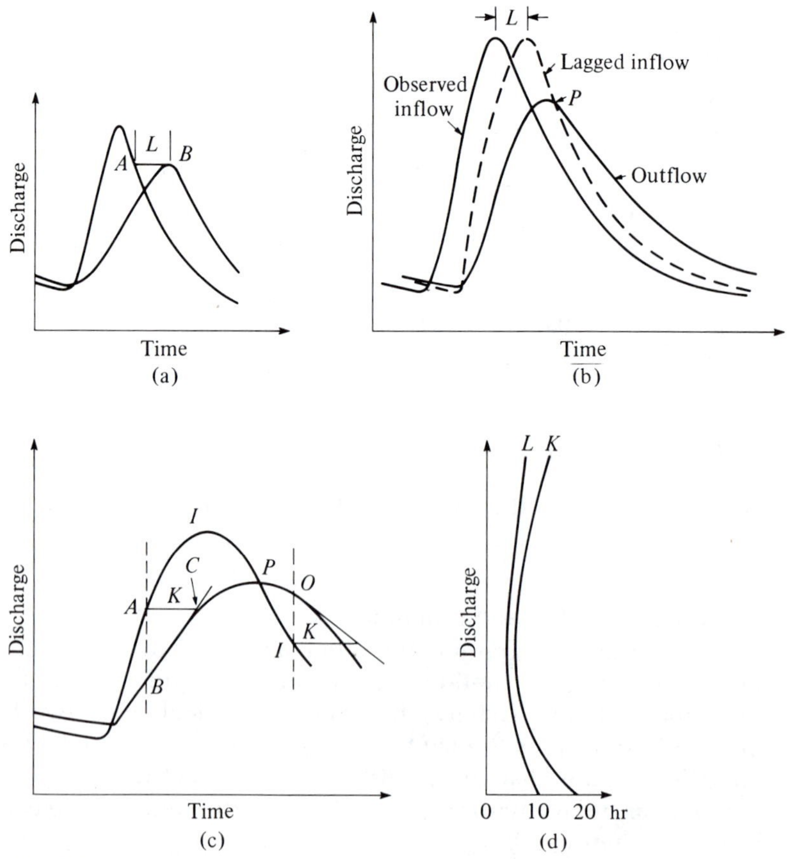


**Fig. 7-2.** *Determining K and L. (U.S. Weather Bureau, NOAA.)*

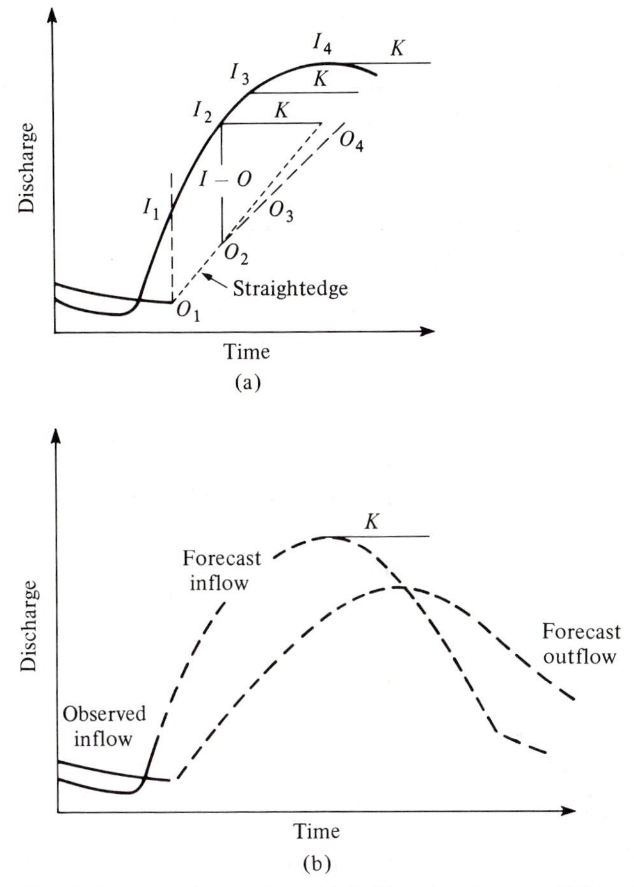


***Fig. 7-3.*** *K and L routing. (U.S. Weather Bureau, NOAA.)*

2. Select point $A$ on the rising side of the inflow hydrograph Fig. 7-2c.
3. Drop a vertical line downward through point $A$ to a point $B$ on the outflow hydrograph.
4. Draw a horizontal line through point $A$ to the right until it intersects the outflow hydrograph.
5. From point $B$ construct a line tangent to the outflow hydrograph upward until it intersects the horizontal line at a point $C$.
6. The distance in hours from $A$ to $C$ is $K$ for the value of outflow at point $B$.
7. Compute $K$ for several points along the outflow hydrograph and plot as shown in Fig. 7-2d.

The application of known $K$ and $L$ values to an inflow hydrograph can now be used to forecast an outflow hydrograph. Steps in the procedure are

1. Lag the inflow $L$ hours, Fig. 7-3a.

2. Plot a point $K$ hours from $I_2$ with the same discharge value.
3. A straightedge from this point to $O_1$ gives the slope at $O_2$.
4. Draw a short segment of the outflow hydrograph at time $O_2$.
5. The lagged inflow and graphically derived outflow for an entire storm are shown in Fig. 7-3b.

## 7-2 Hydrologic Reservoir Routing by the Storage Indication Method

The *storage indication* method of routing a hydrograph through a reservoir is also called the *modified Puls* method. A flood wave passing through a storage reservoir is both delayed and attenuated as it enters and spreads over the pool surface. Water stored in the reservoir is gradually released as pipe flow through turbines or outlet works, called *principal* spillways, or in extreme floods, over an *emergency* spillway. Figure 2 in Chapter 12 illustrates these reservoir components.

Flow over an ungated emergency spillway weir section can be described from energy, momentum, and continuity considerations by the form

$$O = CYH^x \tag{7-13}$$

where

$O$ = the outflow rate (cfs)
$Y$ = the length of the spillway crest (ft)
$H$ = reservoir depth above the spillway crest (ft)
$C$ = the discharge coefficient for the weir or section, theoretically 3.0
$x$ = exponent, theoretically $\frac{3}{2}$

Flow through a free outlet discharge pipe is similarly described by Eq. 7-13 where

$Y$ = the cross-sectional area of the discharge pipe ($ft^2$)
$H$ = head above the free outlet elevation (ft)
$C$ = the pipe discharge coefficient, theoretically 8.0
$x$ = exponent, theoretically $\frac{1}{2}$

Flow equations for other outlet conditions are available in most hydraulics textbooks. Storage values for various pool elevations in a reservoir are readily determined from computations of volumes confined between various pool areas measured from topographic maps. Since storage and outflow both depend only on pool elevation, the resulting storage-elevation curve and the outflow-elevation relationship (Eq. 7-13) can be easily combined to form a storage-outflow graph. Thus storage in a reservoir depends only on the outflow, contrasted to the dependence on the inflow and outflow in river routing (Eq. 7-3).

For convenience, $S$ is often defined as the "surcharge storage" or the storage above the emergency spillway crest, which simply means that the overflow rate is zero when $S$ is zero. If the graphed storage-outflow relationship is found to be linear, and if the slope of the line is defined as $K$, then

$$S = KO \tag{7-14}$$

and the reservoir is called a *linear reservoir.* Note that routing through a linear reservoir is a special case of Muskingum river routing shown in Fig. 7-1 using $x = 0.0$ in Eq. 7-3. Note also that the outflow rate in Fig. 7-1 is increasing only while the inflow exceeds the outflow. This observation is consistent with the assumptions that the inflow immediately goes into storage over the entire pool surface and that the outflow depends only on this storage.

Routing through a linear reservoir is easily accomplished by first dividing time into a number of equal increments and then substituting $S_2 = KO_2$ into Eq. 7-4 and solving for $O_2$, which is the only remaining unknown for each time increment.

To route an emergency flood through a nonlinear reservoir, the storage-outflow relationship and the continuity equation, Eq. 7-4, are combined to determine the outflow and storage at the end of each time increment $\Delta t$. Equation 7-4 can be rewritten as

$$I_n + I_{n+1} + \left(\frac{2S_n}{\Delta t} - O_n\right) = \frac{2S_{n+1}}{\Delta t} + O_{n+1} \tag{7-15}$$

in which the only unknown for any time increment is the term on the right side. Pairs of trial values of $S_{n+1}$ and $O_{n+1}$ could be generated that satisfy Eq. 7-15 and checked in the storage-outflow curve for confirmation. Rather than resort to this trial procedure, a value of $\Delta t$ is selected and points on the storage outflow curve are replotted as the "storage indication" curve shown in Fig. 7-4. This graph allows a *direct* determination of the outflow $O_{n+1}$ once a value of the ordinate $2S_{n+1}/\Delta t + O_{n+1}$ has been calculated from Eq. 7-15. The second unknown, $S_{n+1}$, can be read from the $S$-$O$ curve or found from Eq. 7-15. This row-by-row numerical integration of Eq. 7-15 with Fig. 7-4 is illustrated using $\Delta t = 1$ hr in Example 7-3.

***Example 7-3*** Given the triangular-shaped inflow hydrograph and the $2S/\Delta t + O$ curve of Fig. 7-4, find the outflow hydrograph for the reservoir assuming it to be completely full at the beginning of the storm. (See Table 7-3.)

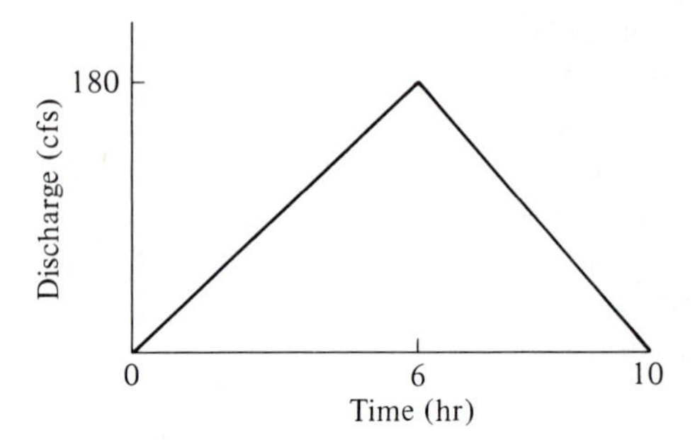


In selecting a routing period $\Delta t$, generally at least five points on the rising limb of the inflow hydrograph are employed in the calculations. An increased number of points on the rising limb, that is, a small $\Delta t$, improves the accuracy, since as $\Delta t \rightarrow 0$ the numerical integration approaches the true limit of the function being integrated, in this case $dS/dt$.

As can be seen from Example 7-3, the tabulation of column 3 comes from the given inflow hydrograph, column 4 is simply the addition of $I_n + I_{n+1}$, and columns 5 and 7 are initially zero, since in this problem no turbine or pipe flow occurs and the reservoir is assumed full at the commencement of inflow. Therefore, there is no available storage.

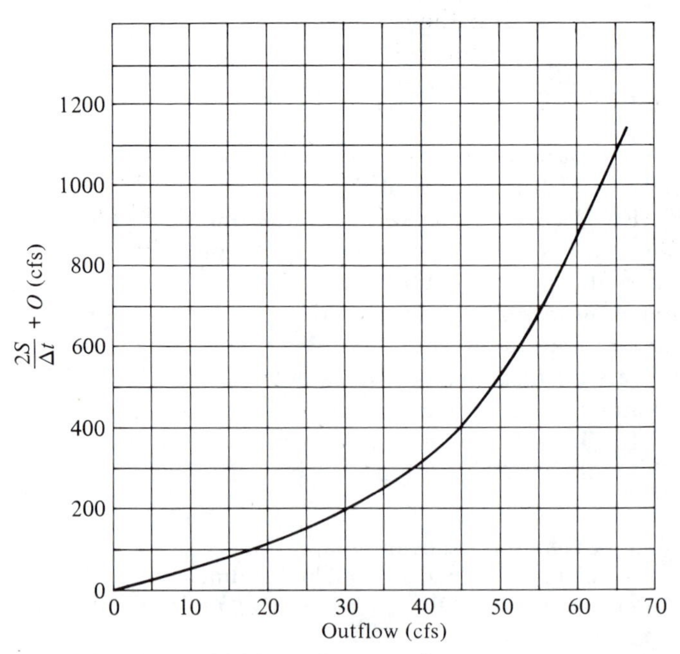


***Fig. 7-4.*** *Curve of $2S/\Delta t + O$ versus $O$.*

**Table 7-3** *Routing Table*

| (1) Time (hr) | (2) $n$ | (3) $I_n$ (cfs) | (4) $I_n + I_{n+1}$ (cfs) | (5) $\frac{2S_n}{\Delta t} - O_n$ (cfs) | (6) $\frac{2S_{n+1}}{\Delta t} + O_{n+1}$ (cfs) | (7) $O_{n+1}$ (cfs) |
|---|---|---|---|---|---|---|
| 0 | 1 | 0 | 30 | 0 | | 0 |
| 1 | 2 | 30 | 90 | 20 | 30 | 5 |
| 2 | 3 | 60 | 150 | 74 | 110 | 18 |
| 3 | 4 | 90 | 210 | 160 | 224 | 32 |
| 4 | 5 | 120 | 270 | 284 | 370 | 43 |
| 5 | 6 | 150 | 330 | 450 | 554 | 52 |
| 6 | 7 | 180 | 315 | 664 | 780 | 58 |
| 7 | 8 | 135 | 225 | 853 | 979 | 63 |
| 8 | 9 | 90 | 135 | 948 | 1078 | 65 |
| 9 | 10 | 45 | 45 | 953 | 1085 | 65 |
| 10 | 11 | 0 | 0 | 870 | 998 | 64 |
| 11 | 12 | 0 | 0 | 746 | 870 | 62 |
| 12 | 13 | 0 | 0 | 630 | 746 | 58 |

The starting value for column 6 may be readily computed as 30 cfs from Eq. 7-15, since

$$I_1 + I_2 + \frac{2S_1}{\Delta t} - O_1 = \frac{2S_2}{\Delta t} + O_2$$

$$30 + 0 = \frac{2S_2}{\Delta t} + O_2$$

The answer $O_2 = 5$ cfs corresponding to $2S_2/\Delta t + O_2 = 30$ cfs is taken from Fig. 7-4. A value in column 5 for $n = 2$ can be obtained, since $O_2$ is now known: subtract two times $O_2$ from $2S_2/\Delta t + O_2$, resulting in 20 cfs for the solution of $2S_2/\Delta t - O_2$ shown in column 5.

The stepwise procedure used to get outflow figures for all $n$ can be summarized as

1. Entries in columns 1 and 3 are known from the given inflow hydrograph.
2. Entries in column 4 are the additions of $I_n + I_{n+1}$ in column 3.
3. The $2S/\Delta t + O$ versus $O$ plot is entered with known values of $2S/\Delta t + O$ to find values of $O$.
4. From the tabulations in column 6 subtract twice the number in column 7 to enter a $2S/\Delta t - O$ in column 5.
5. Add the value in column 5 to the advanced sum in column 4 and enter the result in column 6 for the new period under consideration.
6. The new outflow is again found from the relation of $2S/\Delta t + O$ as in Fig. 7-4.

7. Steps 3 through 6 are repeated until the entire outflow hydrograph is generated.

## 7-3 Hydrologic Watershed Routing

Watershed runoff can be thought of as being modified by two types of storage, channel and reservoir. This concept allows the construction of conceptual models to describe the resulting runoff hydrographs. The simulation of watershed characteristics by a series of linear reservoirs was first proposed by Nash[5] although Zoch[6] utilized the concept with a single routing in 1934.

### Linear Reservoirs

Following the conceptual development of Nash, the watershed can be considered as $n$ serially arranged reservoirs each having a relationship between storage and outflow satisfying Eq. 7-14. If we assume that the first reservoir is instantaneously full and discharges into the second, the second into the third, and so on, combination of Eqs. 7-1 and 7-14 results in

$$I - O = K\frac{dO}{dt} \tag{7-16}$$

Since for the first reservoir $I$ = zero for $t > 0$,

$$-O_1 = K\frac{dO_1}{dt} \tag{7-17}$$

The solution of this differential equation is

$$O_1 = \frac{1}{K}\, e^{-t/K} \tag{7-18}$$

which represents the outflow from the first reservoir.

By definition of the mathematical model, the outflow from the first reservoir is the inflow to the second, and Eq. 7-16 gives

$$O_1 - O_2 = K\frac{dO_2}{dt} \tag{7-19}$$

and solving,

$$O_2 = \frac{1}{K}\left(\frac{t}{K}\right) e^{-t/K} \tag{7-20}$$

For the third reservoir the outflow is

$$O_3 = \frac{1}{2K}\left(\frac{t}{K}\right)^2 e^{-t/K} \tag{7-21}$$

and from the $n$th it is

$$O_n = \frac{1}{K\Gamma(n)} \left(\frac{t}{K}\right)^{n-1} e^{-t/K} \tag{7-22}$$

where $\Gamma(n)$ is the gamma function of $n$. Equation 7-22 describes the outflow from routing through $n$ linear reservoirs as an instantaneous unit hydrograph (IUH). For the application of this technique and for a discussion of IUH the reader is referred to Chapter 4.

Computations with Eq. 7-22 require estimates of $K$ and $n$. The product $nK$ can be approximated by the watershed lag time, computed from any of the methods described in Chapter 4. The best estimates of $K$ (time units) and $n$ (integer) for a watershed are found by selecting trial combinations and comparing hydrographs synthesized from the IUH with recorded hydrographs. Once the watershed is calibrated for one or more storms of record, other storm hydrographs can be synthesized.

### Muskingum Method

The Muskingum method can also be used to derive an outflow hydrograph from a watershed using Eq. 7-6 and watershed $K$ and $x$ values. A procedure was utilized by Overton[7] to analyze runoff from small ARS watersheds, assuming an isosceles triangular inflow such as shown in Fig. 7-5. Combining Eqs. 7-1 and 7-3 results in

$$I - O = Kx\frac{dI}{dt} + K(1-x)\frac{dO}{dt} \tag{7-23}$$

or

$$\frac{dO}{dt} + \frac{O}{K(1-x)} = \frac{I}{K(1-x)} - \left(\frac{x}{1-x}\right)\frac{dI}{dt} \tag{7-24}$$

The solution of this equation for the assumed inflow of Fig. 7-5 is

$$O = I'\left\{t - K\left[1 - \exp\frac{-t}{K(1-x)}\right]\right\} \qquad \text{for } 0 < t < t_p \tag{7-25}$$

where

$$I' = \frac{I_p}{t_p}$$

and

$$O = I'\left\{2t_p - t + K - K\left[2 - \exp\frac{-t_p}{K(1-x)}\right] \cdot \exp\left[-\frac{t-t_p}{K(1-x)}\right]\right\} \qquad \text{for } t_p < t < 2t_p \tag{7-26}$$

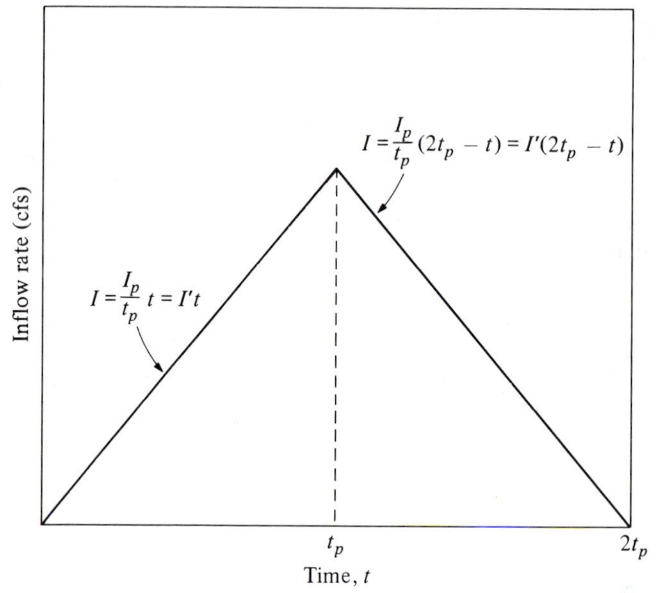


**Fig. 7-5.** *Triangular inflow for Muskingum method.*

The peak outflow $O_p$ and the time to its peak $T_p$ can be computed by setting the first derivative of Eq. 7-26 equal to zero, solving for $T_p$ and combining with Eq. 7-26 for the value of $O_p$. This results in

$$T_p = t_p - K(1-x)\ \ln_e\left\{\frac{1-x}{2 - \exp\left[-t_p/K(1-x)\right]}\right\} \qquad \textbf{(7-27)}$$

and

$$O_p = I'\{t_p - (T_p - t_p) + Kx\} \qquad \textbf{(7-28)}$$

The simplification of Eq. 7-27 can be made for $K < t_p$ by using an average value of the term $(1-x)\ln\{(1-K)/2\}$ equal to $-0.71$ for $x$ in the range $0 \leqslant x \leqslant 0.5$. Equation 7-27 reduces to

$$T_p = t_p + 0.71K \qquad \textbf{(7-29)}$$

and $K$ is described by

$$K = \frac{T_p - t_p}{0.71} = 1.41(T_p - t_p) \qquad \textbf{(7-30)}$$

The result for $x$ is obtained from Eq. 7-28 as

$$x = 0.71 - \frac{t_p}{K}\left\{\frac{I_p - O_p}{I_p}\right\} \qquad \textbf{(7-31)}$$

Values for $K$ and $x$ can now be determined by comparing the peak, and the time to peak, of the inflow and outflow to a watershed. Application of this method is similar to the Muskingum method for river reaches, since it is necessary only to calculate $C_0$, $C_1$, and $C_2$ and apply the results to a known inflow to generate a given outflow. The procedure described can be used to fill in missing peak records from a gauged watershed or to generate additional runoff records from prior rainfall records.

### Time-Area Curves

The outflow hydrograph from a watershed can also be described by the concept of routing a *time-area histogram*. This method assumes that the outflow hydrograph for any storm is characterized by separable watershed translation and storage effects. Pure translation of the direct runoff to the outlet via the drainage network is described using the channel travel time, resulting in an outflow hydrograph that ignores watershed storage effects. The storage effect is then incorporated by routing the translated hydrograph as an inflow hydrograph through a hypothetical reservoir, usually linear, located at the watershed outlet.

To apply the method, the basin is first divided into a number of time zones, Fig. 7-6a, separated by *isochrones*, or lines of equal travel time from the watershed outlet. The areas between isochrones are then determined and plotted versus the travel time as shown in Fig. 7-6b.

The translated reservoir inflow hydrograph ordinates $I_i$ for any selected design hyetograph, Fig. 7-7, can now be determined. Each block of rain in Fig. 7-7 is applied to the entire watershed; the runoff from each subarea reaches the outflow at lagged intervals defined by the time-area histogram. The simultaneous arrival of the runoff from

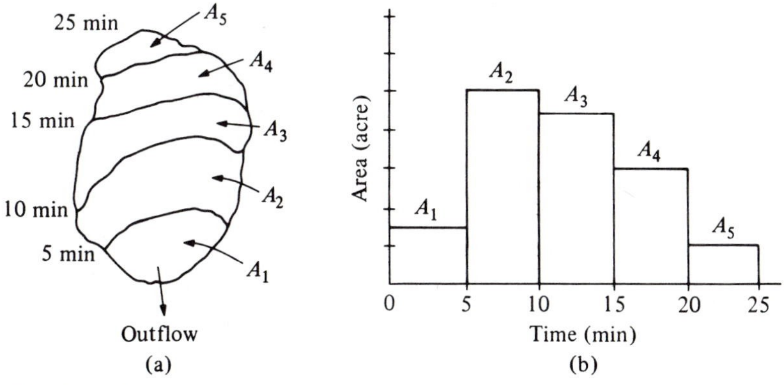


**Fig. 7-6.** *Time-area construction: (a) basin isochrones; and (b) time-area histogram.*

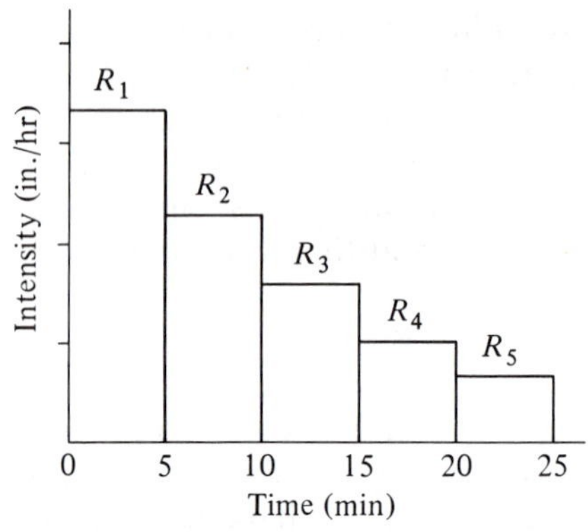


**Fig. 7-7.** *Design storm hyetograph.*

areas $A_1, A_2, \ldots$ for various storms $R_1, R_2, \ldots$ is determined by properly lagging and adding contributions, or

$$I_i = R_i A_1 + R_{i-1} A_2 + \cdots + R_1 A_i \tag{7-32}$$

where

$I_i$ = the reservoir inflow hydrograph ordinates (cfs)
$R_i$ = excess rainfall hyetograph ordinates (in./hr)
$A_i$ = time-area histogram ordinates (acres)

For example, the runoff from storms $R_1$ on $A_3$, $R_2$ on $A_2$, and $R_3$ on $A_1$ arrive at the outlet simultaneously, and $I_3$ is the total flow. Using Eq. 7-32 for $i = 1, 2, \ldots, 10$ results in the inflow hydrograph shown in Fig. 7-8.

The necessary routing through linear or nonlinear reservoir-type storage can now be performed using the methods described in Sec.

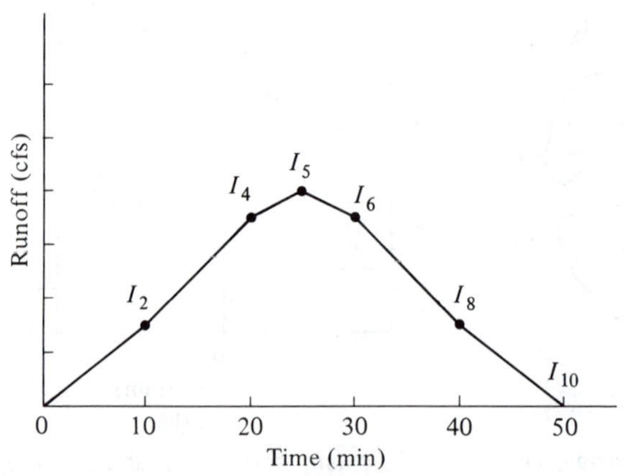


**Fig. 7-8.** *Inflow hydrograph.*

7-2. If Eq. 7-14 is used, the *watershed* storage coefficient $K$ must be estimated. If hydrograph data are not available, $K$ can be equated with the basin lag time using any applicable empirical equation. If an outflow hydrograph is available, the outflow rate and slope of the recession limb *at the inflection point* are determined, and $K$ is calculated from

$$K = \frac{-O}{dO/dt} \tag{7-33}$$

This equation comes from substituting $S = KO$ and $I = 0.0$ into Eq. 7-1. The recession limb inflection point is commonly considered to occur when the total inflow to the channel network, including the base flow, is zero.

An application of time-area curves in urban drainage design is called the Road Research Laboratory Method.[8] It employs the time-area histogram, but only contributions to the runoff from the impervious tracts in the urban location are considered. Storage routing is accomplished by the storage-indication method using the total storage available in the urban storm drain system. The depth of flow throughout the system can be assumed proportional to flow at the outlet. The discharge and depth of flow at the outlet can be presumed related by the Manning equation

$$O = AV = \frac{1.49}{n} AR^{2/3}S^{1/2} \tag{7-34}$$

where

$O$ = outflow discharge (cfs)
$A$ = area of the storm drain ($ft^2$)
$V$ = the average velocity of flow (ft/sec)
$R$ = the hydraulic radius (ft)
$S$ = the slope of the energy gradient (ft/ft)

Construction of the discharge-storage relation involves selecting various flow depths, calculating the discharge, the storage, and $2S/\Delta t + O$ values needed to draw a curve similar to that of Fig. 7-4. The inflow hydrograph of Fig. 7-8 is then routed by the use of Eq. 7-15.

## 7-4 Hydraulic River Routing

As noted previously, hydraulic routing employs both the equation of continuity and the equation of motion. Closed-form solutions to the complete hydraulic routing equations do not exist. Thus the application of these techniques demands computer operations. Numerous approaches are available for numerical integration of the equations.

Two basic systems will be used in the following discussion. The first is termed the *explicit method* because values of velocity and depth or discharge are explicitly calculated based on previously known information on the river, watershed, or reservoir. The second system is the *characteristic method*. Its name is derived from the fact that computations are performed along the so-called characteristic curves well known in the study of water waves and gas dynamics.

Hydraulic routing techniques are helpful in solving river routing problems, overland flow or sheet flow, and in systems simulation of a composite watershed.[9] The main disadvantage of the hydraulic routing approach is the need for a high speed digital computer. Generally, this drawback is not severe, since the majority of sophisticated routing needs are undertaken in conjunction with other studies requiring the use of a computer.

Hydraulic routing proceeds from the simultaneous solution of expressions of continuity and momentum. The general form of the statements is called the *spatially varied unsteady flow equation*.

These equations also apply to sheet flow or overland flow and include terms for laterally incoming rainfall. They can be simplified and used to resolve river routing problems. For completeness of presentation, a general form of the spatially varied unsteady flow equations will be presented first.

### Equation of Continuity

The equation of continuity states simply that inflow minus outflow equals the rate of change of storage. To relate this concept to a river section under a condition of rainfall or lateral inflow, consider an element of length $\Delta x$ and unit width as shown in Fig. 7-9, where

$\rho$ = the density of water
$V$ = the average velocity
$y$ = the depth
$A$ = the cross-sectional area
$i$ = the lateral inflow per elemental $\Delta x$
$S$ = the slope of the river bottom

The total inflow is

$$\rho\left(V - \frac{\partial V}{\partial x}\frac{\Delta x}{2}\right)\left(A - \frac{\partial A}{\partial x}\frac{\Delta x}{2}\right)\Delta t + \rho\int_{x}^{x+\Delta x}\int_{t}^{t+\Delta t} i(x,t)\,dt\,dx \tag{7-35}$$

The total outflow is

$$\rho\left(V + \frac{\partial V}{\partial x}\frac{\Delta x}{2}\right)\left(A + \frac{\partial A}{\partial x}\frac{\Delta x}{2}\right)\Delta t \tag{7-36}$$

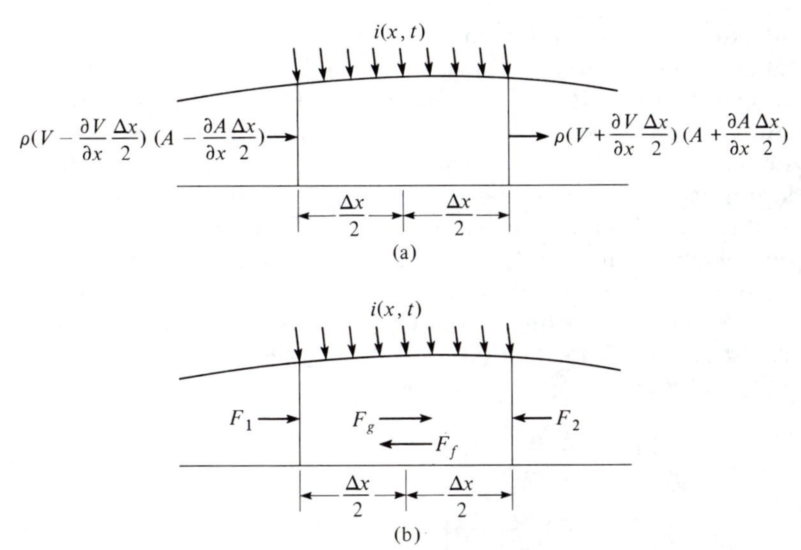


**Fig. 7-9.** *Continuity and momentum elements.*

The rate of change in storage is

$$\rho \frac{\partial A}{\partial t} \Delta x \, \Delta t \tag{7-37}$$

Consequently, continuity gives

$$-\rho\left(A \frac{\partial V}{\partial x} \Delta x + V \frac{\partial A}{\partial x} \Delta x\right) \Delta t + \rho \bar{i} \, \Delta x \, \Delta t - \rho \frac{\partial A}{\partial t} \Delta x \, \Delta t = 0 \tag{7-38}$$

where $\bar{i}$ = the average lateral inflow resulting from rainfall over $\Delta x$ and $\Delta t$.

The continuity equation of unsteady flow with lateral inflow is obtained by simplifying Eq. 7-38.

$$A \frac{\partial V}{\partial x} + V \frac{\partial A}{\partial x} + \frac{\partial A}{\partial t} = \bar{i} \tag{7-39}$$

For a unit width, $A = y$ and Eq. 7-39 takes the form

$$H_2 = y \frac{\partial V}{\partial x} + V \frac{\partial y}{\partial x} + \frac{\partial y}{\partial t} - \bar{i} = 0 \tag{7-40}$$

where $H_2$ is an abbreviation introduced here for use in subsequent paragraphs.

### Momentum Equation

In accordance with Newton's second law of motion, the change of momentum per unit of time on a body is equal to the resultant of all external forces acting on the body. The following derivation of the momentum equation of spatially varied unsteady flow is presented subject to the following assumptions: (1) the flow is unidirectional and velocity uniform across the flow section; (2) the pressure is hydrostatic; (3) the slope of the river bottom is relatively small; (4) the Manning formula may be used to evaluate the friction loss due to shear at the channel wall; (5) lateral inflow enters the stream with no velocity component in the direction of flow; and (6) the value of $\bar{i}$ represents the spatial and time variations of lateral inflow.

Forces acting on an element of length $\Delta x$ and unit width are shown in Fig. 7-9b. The forces $F_1$ and $F_2$ represent hydrostatic forces on the element and are expressed as

$$F_1 = \gamma\left\{\bar{y}A - \frac{\partial(\bar{y}A)}{\partial x}\frac{\Delta x}{2}\right\} \tag{7-41}$$

$$F_2 = \gamma\left\{\bar{y}A + \frac{\partial(\bar{y}A)}{\partial x}\frac{\Delta x}{2}\right\} \tag{7-42}$$

where $\bar{y}$ is the distance from the water surface to the centroid of the area. The resultant hydrostatic force is $F_1 - F_2$, or

$$F_p = -\gamma\frac{\partial(\bar{y}A)}{\partial x}\,\Delta x \tag{7-43}$$

By assuming a small slope for the river bottom, the gravitational force is given by

$$F_g = \gamma AS\,\Delta x \tag{7-44}$$

The frictional force along the bottom is equal to the friction slope $S_f$ multiplied by the weight of water in an element $\Delta x$.

$$F_f = \gamma AS_f\,\Delta x \tag{7-45}$$

The rate of change of momentum in the length $\Delta x$ may be expressed as

$$\frac{d(mV)}{dt} = m\frac{dV}{dt} + V\frac{dm}{dt} \tag{7-46}$$

in which $m$ is the mass of fluid.

If it is assumed that the incoming lateral inflow enters the moving fluid with no velocity component in the direction of flow and $\bar{i}$ represents the spatial and time variations of the lateral inflow, the rate of change of momentum for the element can be expressed as

$$\frac{d(mV)}{dt} = \rho A\,\Delta x\,\frac{dV}{dt} + \rho V\bar{i}\,\Delta x \tag{7-47}$$

where $dV/dt$ represents

$$\frac{dV}{dt} = \frac{\partial V}{\partial t} + V\frac{\partial V}{\partial x} \tag{7-48}$$

The rate of change of momentum is therefore

$$\rho A \, \Delta x \left\{ \frac{\partial V}{\partial t} + V \frac{\partial V}{\partial x} \right\} + \rho V \bar{i} \, \Delta x \tag{7-49}$$

Equating the rate of change of momentum to all external forces acting on the element results in

$$\frac{\partial V}{\partial t} + V \frac{\partial V}{\partial x} + \frac{g}{A} \frac{\partial(\bar{y}A)}{\partial x} + \frac{V\bar{i}}{A} = g(S - S_f) \tag{7-50}$$

Now for a unit width element, the relationship simplifies to

$$H_1 = \frac{\partial V}{\partial t} + V \frac{\partial V}{\partial x} + g \frac{\partial y}{\partial x} + \frac{V}{y} \bar{i} - g(S - S_f) = 0 \tag{7-51}$$

where $H_1$ is an abbreviation.

Equations 7-40 and 7-51 form a set of simultaneous expressions that can be solved for $V$ and $y$ subject to the appropriate boundary conditions. Unraveling these equations is extremely difficult without the aid of a high speed digital computer.

### Explicit Method

Using an explicit method of fixed time intervals requires rewriting Eqs. 7-40 and 7-51 in finite difference form so that values of velocity and depth for each succeeding time interval can be calculated, based upon previously computed figures for velocity and depth. A scheme employed for this purpose is indicated in Fig. 7-10.

Finite difference approximations for the terms in Eqs. 7-40 and 7-51 appear, using subscripts from Fig. 7-10, as

$$\left(\frac{\partial V}{\partial x}\right)_M = \frac{V_R - V_L}{2\Delta x} \tag{7-52}$$

$$\left(\frac{\partial V}{\partial t}\right)_P = \frac{V_P - V_M}{\Delta t} \tag{7-53}$$

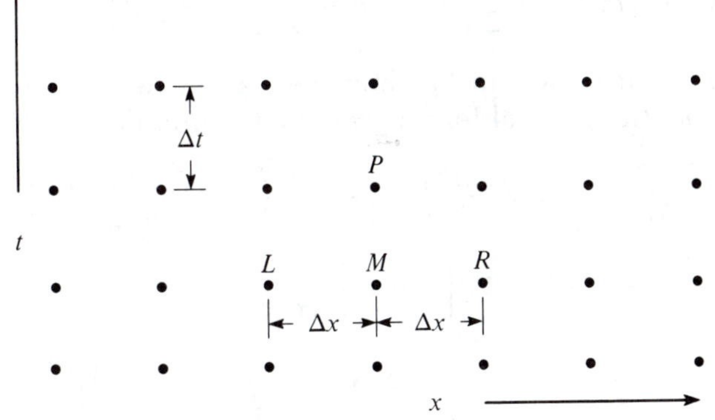


**Fig. 7-10.** *Finite difference network.*

$$\left(\frac{\partial y}{\partial x}\right)_M = \frac{Y_R - Y_L}{2\Delta x} \tag{7-54}$$

$$\left(\frac{\partial y}{\partial t}\right)_P = \frac{Y_P - Y_M}{\Delta t} \tag{7-55}$$

The substitution of these values in the equation of continuity with lateral inflow, Eq. 7-40 results in

$$Y_P = Y_M + \frac{\Delta t}{2\Delta x}\,[V_M(Y_L - Y_R) + Y_M(V_L - V_R)] + \bar{i}\,\Delta t \tag{7-56}$$

and similarly, the spatially varied unsteady flow equation, Eq. 7-51, appears as

$$\frac{V_P - V_M}{\Delta t} + V_M\frac{(V_R - V_L)}{2\Delta x} + g\frac{(Y_R - Y_L)}{2\Delta x} + \frac{V_P}{Y_P}\bar{i} = g(S - S_f) \tag{7-57}$$

The only existing unknown quantity at this phase in development of the equations is the term $S_f$, or the friction slope. If it is assumed that unsteady effects can be ignored in this term, the Chezy-Manning relationship

$$S_f = \frac{V\,|\,V\,|\,n^2}{2.2082Y^{4/3}} \tag{7-58}$$

can be solved for the frictional slope; thus, in finite difference form, evaluated at point $P$, the friction slope is

$$S_f = \frac{V_P\,|\,V_P\,|\,n^2}{2.2082Y_P^{4/3}} \tag{7-59}$$

Substituting this relationship in Eq. 7-57, and simplifying results,

$$V_P^2 + \frac{2.2082}{n^2g\,\Delta t}\,Y_P^{4/3}\left(1 + \bar{i}\,\frac{\Delta t}{Y_P}\right)V_P - \frac{2.2082}{n^2g\,\Delta t}\,Y_P^{4/3} \times\left\{V_M + \Delta t\left[V_M\frac{(V_L - V_R)}{2\Delta x} + g\frac{(Y_L - Y_R)}{2\Delta x} + gS\right]\right\} = 0 \tag{7-60}$$

This equation is quadratic in $V_P$; however, an answer for $V_P$ may be obtained from the general form for quadratic formula, resulting in

$$V_P = \frac{-2.2082}{2n^2g\,\Delta t}\,Y_P^{4/3}\left(1 + \bar{i}\,\frac{\Delta t}{Y_P}\right) + \left\{\frac{2.2082}{n^2g\,\Delta t}\,Y_P^{4/3}\left(\frac{2.2082}{4n^2g\,\Delta t}\,Y_P^{4/3} \times\left(1 + \bar{i}\,\frac{\Delta t}{Y_P}\right)^2 + V_M\left[1.0 + \frac{\Delta t}{2\Delta x}(V_L - V_R)\right] + g\,\Delta tS + g\frac{\Delta t}{2\Delta x}(Y_L - Y_R)\right)\right\}^{1/2} \tag{7-61}$$

The solution of the finite difference Eqs. 7-56 and 7-61 proceeds by finding a value for each $Y_P$ in the forward time line by Eq. 7-56, substituting this figure in Eq. 7-61, and computing each value of $V_P$ in the time line. The discharge for a 1-ft width at the end of the overland flow plane is simply the product $V_P Y_P$ for each time line. A discharge hydrograph is prepared by storing the product $Q_P$ for selected times and plotting the resulting curve. In prismatic channels the lateral inflow term may be omitted if it is not significant. The discharge would simply be a multiple of the width for a rectangular channel. Routing the flood wave entering the upper end of the channel is contingent upon the proper description of the boundary conditions, as will be described shortly.

### Characteristic Method

Before dealing with boundary conditions or the inherent problems of stability encountered in the explicit method, it is well to look at the characteristic solution to spatially varied unsteady flow equations. Equations 7-40 and 7-51 for $H_1$ and $H_2$ can be employed to form a linear combination with an unknown multiplier $\lambda$ as follows.

$$H = H_1 + \lambda H_2 \tag{7-62}$$

It is obvious that any two real distinct values of $\lambda$ will produce two equations in $V$ and $y$ retaining all the attributes of the original two equations $H_1$ and $H_2$. Often the use of this technique simplifies the solution for $V$ and $y$.

Combining Eqs. 7-40 and 7-51 according to Eq. 7-62 and grouping terms results in

$$H = \left\{ \frac{\partial V}{\partial x} (V + \lambda y) + \frac{\partial V}{\partial t} \right\} + \lambda \left\{ \frac{\partial y}{\partial x} \left( V + \frac{g}{\lambda} \right) + \frac{\partial y}{\partial t} \right\} - \bar{i}\lambda - g(S - S_f) + \frac{V}{y}\bar{i} \tag{7-63}$$

The second and first terms of Eq. 7-63 correspond to the total derivatives of $dy/dt$ and $dV/dt$ which are

$$\frac{dy}{dt} = \frac{\partial y}{\partial x}\frac{dx}{dt} + \frac{\partial y}{\partial t} \tag{7-64}$$

and

$$\frac{dV}{dt} = \frac{\partial V}{\partial x}\frac{dx}{dt} + \frac{\partial V}{\partial t} \tag{7-65}$$

Thus $H$ of Eq. 7-63 can now be rewritten as

$$H = \frac{dV}{dt} + \lambda \frac{dy}{dt} - \lambda \bar{i} - g(S - S_f) + \frac{V}{y}\bar{i} \tag{7-66}$$

Equating the results of $dx/dt$ of Eqs. 7-64 and 7-65 gives

$$\frac{dx}{dt} = V + \lambda y \tag{7-67}$$

and

$$\frac{dx}{dt} = V + \frac{g}{\lambda} \tag{7-68}$$

Solving these two equations for $\lambda$ produces

$$\lambda = \pm\sqrt{\frac{g}{y}} \tag{7-69}$$

The two real distinct roots for $\lambda$ obtained will convert the two partial differential equations into a pair of ordinary differential equations in $y$ and $V$ subject to the restraint of Eqs. 7-67 and 7-68. Substituting Eq. 7-69 into Eq. 7-66 produces

$$dV + \sqrt{\frac{g}{y}}\, dy + dt\left\{ g(S_f - S) + \frac{V}{y}\,\bar{i} - \sqrt{\frac{g}{y}}\,\bar{i}\right\} \tag{7-70}$$

$$dx = (V + \sqrt{gy})\, dt \tag{7-71}$$

$$dV - \sqrt{\frac{g}{y}}\, dy + dt\left\{ g(S_f - S) + \frac{V}{y}\,\bar{i} + \sqrt{\frac{g}{y}}\,\bar{i}\right\} = 0 \tag{7-72}$$

$$dx = (V - \sqrt{gy})\, dt \tag{7-73}$$

The solution of Eqs. 7-70 through 7-73 can be carried out numerically to completely describe the movement of a flood wave from one point in the river to another.

The curve described by Eq. 7-71 is the positive characteristic emanating from point $L$ in Fig. 7-11; that for Eq. 7-73 is the negative

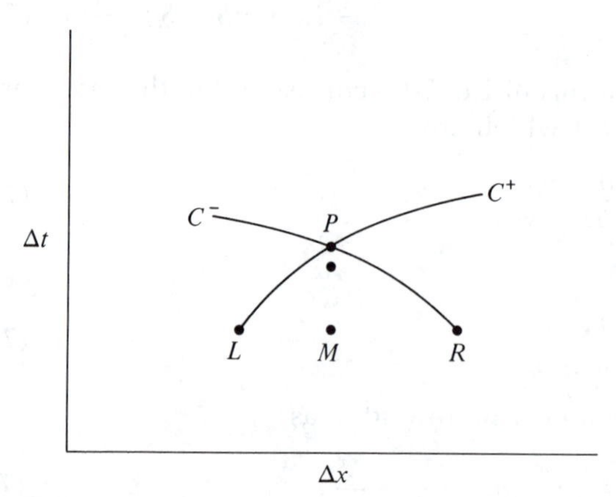


**Fig. 7-11.** *Characteristic curves.*

characteristic issuing from point $R$. Thus solving the equations for values of $V$ and $y$ at various points $P$ throughout the $x$-$t$ plane gives a complete picture of the flood wave movement.

Substitution of first order finite difference forms for $V$, $y$, and $t$ into Eqs. 7-70 through 7-73 results in

$$V_P - V_L + \sqrt{\frac{g}{Y_L}}(Y_P - Y_L) + (t_P - t_L)\left\{g(S_{fL} - S) + \frac{V_L}{Y_L}\bar{i} - \sqrt{\frac{g}{Y_L}}\bar{i}\right\} = 0 \quad C^+ \tag{7-74}$$

$$x_P - x_L = (V_L + \sqrt{gY_L})(t_P - t_L) \tag{7-75}$$

$$V_P - V_R - \sqrt{\frac{g}{Y_R}}(Y_P - Y_R) + (t_P - t_R)\left\{g(S_{fR} - S) + \frac{V_R}{Y_R}\bar{i} + \sqrt{\frac{g}{Y_R}}\bar{i}\right\} = 0 \quad C^- \tag{7-76}$$

$$x_P - x_R = (V_R - \sqrt{gY_R})(t_P - t_R) \tag{7-77}$$

These equations allow computation of the time step along the characteristic by subtracting Eq. 7-77 from Eq. 7-75, yielding

$$t_P = \{x_L - x_R + t_R(V_R - \sqrt{gY_R}) - t_L(V_L + \sqrt{gY_L})\} \div (V_R - V_L - \sqrt{gY_L} - \sqrt{gY_R}) \tag{7-78}$$

The distance along the $x$ plane is obtained from Eq. 7-75.

$$x_P = x_L + (V_L + \sqrt{gY_L})(t_P - t_L) \tag{7-79}$$

Equation 7-76 can now be subtracted from Eq. 7-74 to get the following expression

$$Y_P = V_L - V_R + (\sqrt{gY_L})(Y_L) + \left(\sqrt{\frac{g}{Y_R}}\right)(Y_R) - (t_P - t_L)\left\{g(S_{fL} - S) + \frac{V_L}{Y_L}\bar{i} - \left(\sqrt{\frac{g}{Y_L}}\right)(\bar{i})\right\} + (t_P - t_R)\left\{g(S_{fR} - S) + \frac{V_R}{Y_R}\bar{i} + \left(\sqrt{\frac{g}{Y_R}}\right)(\bar{i})\right\} / (\sqrt{g/Y_L} + \sqrt{g/Y_R}) \tag{7-80}$$

The value of $V_P$ found by Eq. 7-74 is

$$V_P = V_L - \sqrt{\frac{g}{Y_L}}(Y_P - Y_L) - (t_P - t_L)\left\{g(S_{fL} - S + \frac{V_L}{Y_L}\bar{i} - \left(\sqrt{\frac{g}{Y_L}}\right)(\bar{i})\right\} \tag{7-81}$$

### Initial and Boundary Conditions

The initial state of a river reach must be known or assumed before a hydraulic routing technique is appropriate. Steady uniform flow, backwater conditions, control by weirs or dams, or a nonuniform lateral inflow all are initial and/or boundary effects that must be incorporated into the numerical solution. State conditions, that is, subcritical or supercritical flow, dictate the type of finite difference network to be employed in the explicit method and determine the direction of the curves in the method of characteristics. Simple subcritical steady uniform flow is treated here; more complex initial and boundary conditions are covered in texts and articles on the subject of hydraulic routing.

The distinction between initial and boundary conditions is essentially one of position on the $x$ plane at the commencement of the solution procedure. The term *initial condition* describes the flow depth, velocity, or discharge at all points in the channel at time $t = 0$. *Boundary condition* refers to the depth, velocity, or discharge at the upper and lower ends of the river reach at all times $t > 0$. For subcritical flow, boundary conditions are necessary at both ends of the river reach whereas only two upstream boundary conditions are necessary in supercritical flow, since downstream effects cannot be propagated backward; that is, $V \pm c$ is always positive.

The general case of river routing has an initial condition of uniform steady flow. Steady uniform flow in an open channel can be calculated by the Chézy equation

$$Q = CA\sqrt{RS} \tag{7-82}$$

where

$Q$ = discharge (cfs)
$C$ = the factor of flow resistance, Chézy's $C$
$A$ = the cross-sectional area ($ft^2$)
$R$ = hydraulic radius (ft)
$S$ = slope of the energy line

The Manning equation provides another method of calculating steady uniform flow

$$Q = \frac{1.486}{n} AR^{2/3}S^{1/2} \tag{7-83}$$

where

$Q$ = discharge (cfs)
$n$ = the factor of flow resistance, Manning's $n$
$A$ = the cross-sectional area ($ft^2$)
$R$ = the hydraulic radius (ft)
$S$ = slope of the energy line

Thus given the geometric properties of a channel, estimating the factor of flow resistance allows the computation of the normal flow for a given depth of water. For initial conditions of flow in the channel it is necessary only to obtain the boundary conditions. Examples of boundary conditions for both the explicit and characteristics methods are shown in Examples 7-4 and 7-5.

***Example 7-4*** A 20-ft-wide rectangular channel 2 mi long having uniform flow of 6-ft depth is subjected to an upstream increase in flow to 2000 cfs in a period of 20 min. This flow then decreases uniformly to the initial flow depth in an additional period of 40 min. The channel has a bottom slope of 0.0015 ft/ft and an estimated Manning's $n$ of 0.02. Calculate the explicit solution of hydraulic routing for this situation.

*Solution*

Since all parameters for calculating initial uniform flow are given, computation readily made by Eq. 7-83 gives values for depth and velocity at every point in the channel. The upstream boundary must then be ascertained by modifying Eq. 7-56 to read

$$Y_P = Y_M + \frac{\Delta t}{\Delta x}\,[V_M(Y_M - Y_R) + Y_M(V_M - V_R)]$$

and

$$V_P = \frac{Q}{BY_P}$$

since values of $Y_L$ and $V_L$ are not available to the left of the upstream boundary. The downstream boundary computation is similarly

$$Y_P = Y_M + \frac{\Delta t}{\Delta x}\,[V_M(Y_L - Y_M) + Y_M(V_L - V_M)]$$

and

$$V_P = C_M\left(\frac{BY_P}{B + 2Y_P}\right)^{2/3}$$

where $C_M = 1.486S^{1/2}/n$.

Interior points are found by Eqs. 7-56 and 7-61 modified to eliminate the lateral flow terms. A Fortran IV computer program is presented in Table 7-4 to demonstrate this example.

Note that a primary disadvantage of the explicit method is its stability problem. Unstable solutions ensue if selection of $\Delta t$ is too large relative to $\Delta x$. The Courant condition $\Delta t \leq \Delta x/(V + c)$ provides a guide in selecting values for $\Delta t$. However, it is generally found for this type of explicit formulation that values of $\Delta t$ should be approximately 20% of those given by the Courant condition. Another way to achieve more stable solutions is by a diffusing difference approxima-

**Table 7-4**

```
FORTRAN IV G LEVEL  20                MAIN                DATE = 71314          18/39/21                PAGE 0001
                C       EXPLICIT METHOD
                C       RIVER ROUTING PORBLEM : RECTANGULAR CHANNEL WITH UNIFORM NORMAL
                C          FLOW IS SUBJECTED TO A FLOOD WAVE AT THE UPSTREAM REACH OF
                C          CHANNEL BEING CONSIDERED.      INITIAL UNIFORM DEPTH IS 6.0 FT.
                C          THE FLOOD WAVE IS SUCH THAT THE DISCHARGE AT THE UPSTREAM POINT
                C          INCREASES TO A MAXIMUN OF 2000 CFS IN A 20 MINUTE PERIOD AND
                C          THEN DECREASES TO THE IINITIAL UNIFORM FLOW IN A PERIOD OF 40
                C          MINUTES.
                C       DIMENSIONS ARE AS FOLLOWS : TIME (SECONDS),LENGTH (FT.),
                C          DEPTH (FT.),VELOCITY (FT. PER SEC),DISCHARGE (CFS)
                C       B IS THE WIDTH OF THE CHANNEL
                C       SO IS THE CHANNEL SLOPE
                C       RN IS THE MANNING ROUGHNESS COEFICIENT
                C       TTL IS THE LENGTH OF CHANNEL REACH BEING INVESTIGATED
                C       G IS THE ACCELERATION OF GRAVITY,32.2 FT. PER SEC PER SEC
                C       YO IS THE INITIAL UNIFORM NORMAL DEPTH
                C       NM IS THE NUMBER OF INCREMENTAL REACHES ALONG THE LENGTH OF RIVER
                C       FR IS THE ERROR TOLERANCE FOR ITERATIVE SOLUTIONS OF EQUATIONS
                C       TTP IS THE TIME PERIOD BETWEEN QO AND QMAX
                C       QMAX IS THE PEAK FLOOD DISCHARGE OF INFLOW HYDROGRAPH
                C       TAP IS THE TIME PERIOD AFTER QMAX UNTIL QO
                C       Q IS THE DISCHARGE AT THE UPSTREAM BOUNDARY POINT
                C       QOUT IS THE DISCHARGE AT THE DOWNSTREAM BOUNDARY POINT
                C       Y IS THE DEPTH OF FLOW
                C       V IS THE VELOCITY OF FLOW
                C       DT IS THE INCREMENTAL TIME STEP
                C       I IS THE TIME ASSOCIATED WITH ANY VALUES OF DISCHARGE ,DEPTH,OR
                C          VELOCITY CALCULATED
 0001                  DIMENSION Y(2,41),V(2,41),QQ(2,41)
 0002                  READ(1,950) RN,SO,TTL,B,ER,YO,G
 0003                  READ(1,951) TPRINT,DTP,TMAX,DT,NM
 0004                  READ(1,952) QMAX,TTP,TAP
 0005                  WRITE(3,955) RN,SO,TTL,B,ER,YO
 0006                  WRITE(3,956) TPRINT,DTP,TMAX,DT,NM
 0007                  WRITE(3,957) QMAX,TTP,TAP
 0008                  DX=TTL/NM
 0009                  N=NM
 0010                  M=N+1
 0011                  EFFF=1./3.
 0012                  EF=4./3.
 0013                  EFF=2./3.
 0014                  CN=RN*RN/2.2082
 0015                  CM=1.486/RN*SQRT(SO)
                C       CALCULATE THE INITIAL UNIFORM NORMAL DISCHARGE AND VELOCITY
 0016                  QO=CM*B*YO*(B*YO/(B+2.*YO))**EFF
 0017                  VO=QO/(B*YO)
 0018                  WRITE(3,960) YO,VO,QO
 0019                  WRITE(3,963)
 0020                  WRITE(3,965)
                C       SPECIFY THE INITIAL CONDITIONS
 0021                  DO 5 J=1,M
 0022                  Y(1,J)=YO
 0023                  V(1,J)=VO
 0024                5 CONTINUE
 0025                  DTX=DT/DX
 0026                  DTX2=DTX/2.
 0027                  GDT=G*DT
 0028                  GTS=GDT*SO
```

```
0029          GTN=GDT*CN
0030          T=0.0
0031        9 T=T+DT
0032          TD=T-TTP
0033          IF(TD) 16,16,17
0034       16 Q=QO+(QMAX-QO)/TTP*T
0035          GO TO 20
0036       17 IF(TD-TAP) 18,18,19
0037       18 Q=QMAX-(QMAX-QO)/TAP*TD
0038          GO TO 20
0039       19 Q=QO
0040       20 CONTINUE
      C         CALCULATE THE UPSTREAM BOUNDARY POINT
0041          Y(2,1)=Y(1,1)+DTX*(V(1,1)*(Y(1,1)-Y(1,2))+Y(1,1)*(V(1,1)-V(1,2)))
0042          V(2,1)=Q/(B*Y(2,1))
      C         CALCULATE THE INTERIOR POINTS
0043          DO 100 J=2,N
0044          J1=J-1
0045          J2=J+1
0046          YD=Y(1,J1)-Y(1,J2)
0047          VD=V(1,J1)-V(1,J2)
0048          FI=-V(1,J)+DTX2*(-V(1,J)*VD-G*YD)-GTS
0049          Y(2,J)=Y(1,J)+DTX2*(V(1,J)*YD+Y(1,J)*VD)
0050          BYFP=(B*Y(2,J)/(B+2.*Y(2,J)))**EF
0051          V(2,J)=-BYFP/(2.*GTN)+0.5*SQRT(BYFP/GTN*BYFP/GTN-4.*FI*BYFP/GTN)
0052          QQ(2,J)=V(2,J)*Y(2,J)*B
0053      100 CONTINUE
      C         CALCULATE THE DOWNSTREAM BOUNDARY POINT
0054          Y(2,M)=Y(1,M)+DTX*(V(1,M)*(Y(1,N)-Y(1,M))+Y(1,M)*(V(1,N)-V(1,M)))
0055          V(2,M)=CM*(B*Y(2,M)/(B+2.*Y(2,M)))**EFF
0056          QOUT=Y(2,M)*V(2,M)*B
0057          DO 130  J=1,M
0058          Y(1,J)=Y(2,J)
0059      130 V(1,J)=V(2,J)
0060          IF(T-TPRINT+0.002) 140,135,135
0061      135 TPRINT=TPRINT+DTP
0062          TT=T/60.0
0063          WRITE(3,970)TT,Q,Y(2,1),V(2,1),QQ(2,11),Y(2,11),V(2,11),QOUT,
             1Y(2,M),V(2,M)
0064      140 IF(T-TMAX) 9,299,299
0065      299 STOP
0066      950 FORMAT(7F10.6)
0067      951 FORMAT(4F10.4,I10)
0068      952 FORMAT(3F10.4)
0069      955 FORMAT(2X,'RN=',F10.4,5X,'SO=',F10.6,5X,'TTL=',F10.2,5X,'B=',
             1F10.2,5X,'ER=',F10.6,5X,'YO=',F10.2)
0070      956 FORMAT(2X,'TPRINT=',F10.2,5X,'DTP=',F10.2,5X,'TMAX=',F10.2,5X,
             1'DT=',F10.2,5X,'NM=',I5)
0071      957 FORMAT(2X,'QMAX=',F10.2,5X,'TTP=',F10.2,5X,'TAP=',F10.2)
0072      960 FORMAT(2X,'INITIAL UNIFORM DEPTH=',F10.4,5X,'INITIAL VELOCITY=',
             1F10.4,5X,'INITIAL DISCHARGE=',F10.4)
0073      963 FORMAT(25X,'UPSTREAM',24X,'MIDPOINT',24X,'DOWNSTREAM')
0074      965 FORMAT(1X,'TIME(MIN)',7X,'DISCHARGE',4X,'DEPTH',3X,'VELOCITY',6X,
             1'DISCHARGE',4X,'DEPTH',3X,'VELOCITY',6X,'DISCHARGE',4X,'DEPTH',
             23X,'VELOCITY')
0075      970 FORMAT(2X,F10.2,5X,3F10.4,5X,3F10.4,5X,3F10.4)
0076          END
```

**Table 7-4** *(cont.)*

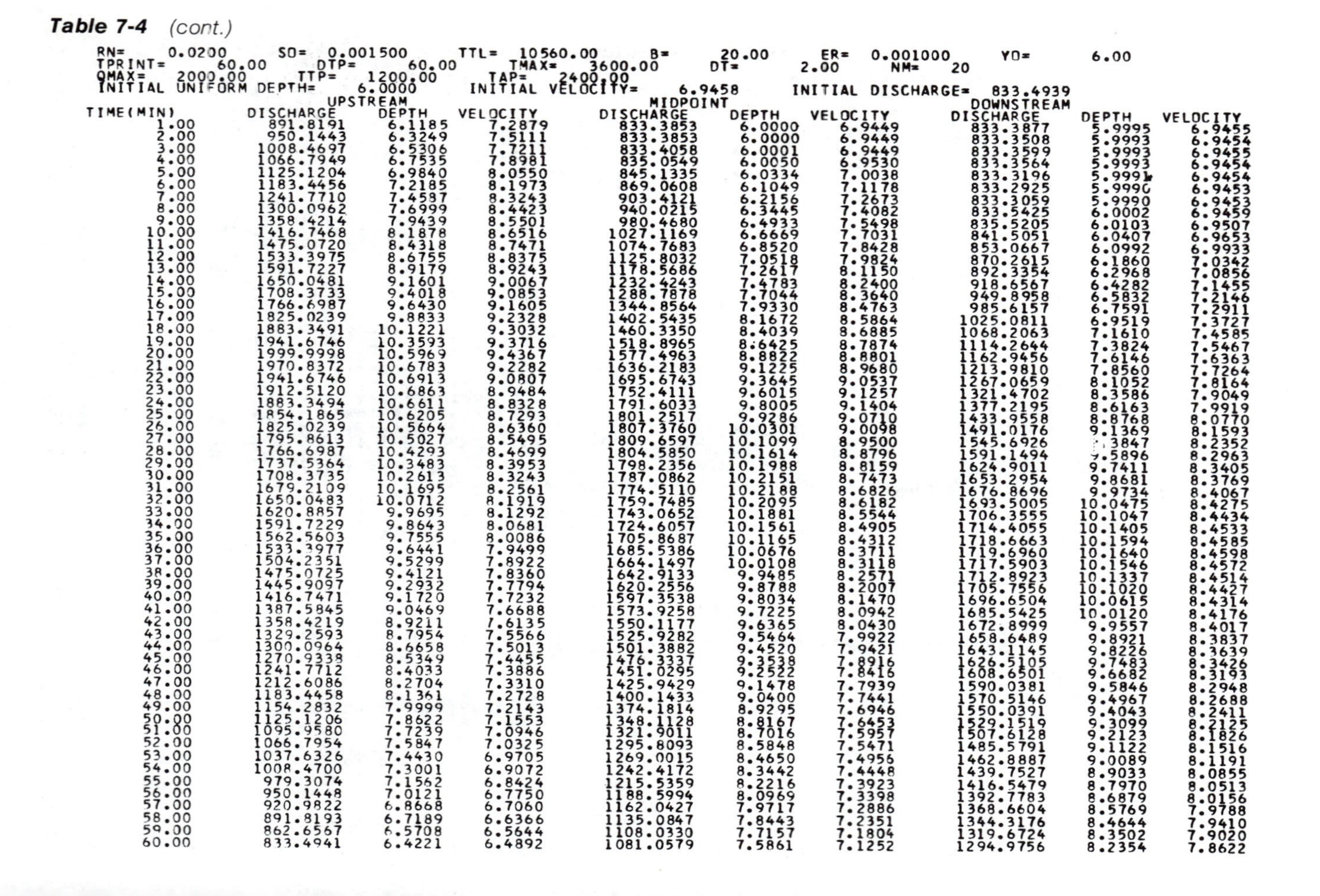

RN= 0.0200 SO= 0.001500 TTL= 10560.00 B= 20.00 ER= 0.001000 YO= 6.00
TPRINT= 60.00 DTP= 60.00 TMAX= 3600.00 DT= 2.00 NM= 20
QMAX= 2000.00 TTP= 1200.00 TAP= 2400.00
INITIAL UNIFORM DEPTH= 6.0000 INITIAL VELOCITY= 6.9458 INITIAL DISCHARGE= 833.4939

| | UPSTREAM | | | MIDPOINT | | | DOWNSTREAM | | |
|---|---|---|---|---|---|---|---|---|---|
| TIME(MIN) | DISCHARGE | DEPTH | VELOCITY | DISCHARGE | DEPTH | VELOCITY | DISCHARGE | DEPTH | VELOCITY |
| 1.00 | 891.8191 | 6.1185 | 7.2879 | 833.3853 | 6.0000 | 6.9449 | 833.3877 | 5.9995 | 6.9455 |
| 2.00 | 950.1443 | 6.3249 | 7.5111 | 833.3853 | 6.0000 | 6.9449 | 833.3508 | 5.9993 | 6.9454 |
| 3.00 | 1008.4697 | 6.5306 | 7.7211 | 833.4058 | 6.0001 | 6.9449 | 833.3599 | 5.9993 | 6.9455 |
| 4.00 | 1066.7949 | 6.7535 | 7.8981 | 835.0549 | 6.0050 | 6.9530 | 833.3564 | 5.9993 | 6.9454 |
| 5.00 | 1125.1204 | 6.9840 | 8.0550 | 845.1335 | 6.0334 | 7.0038 | 833.3196 | 5.9991 | 6.9454 |
| 6.00 | 1183.4456 | 7.2185 | 8.1973 | 869.0608 | 6.1049 | 7.1178 | 833.2925 | 5.9990 | 6.9453 |
| 7.00 | 1241.7710 | 7.4587 | 8.3243 | 903.4121 | 6.2156 | 7.2673 | 833.3059 | 5.9990 | 6.9453 |
| 8.00 | 1300.0962 | 7.6999 | 8.4423 | 940.0215 | 6.3445 | 7.4082 | 833.5425 | 6.0002 | 6.9459 |
| 9.00 | 1358.4214 | 7.9439 | 8.5501 | 980.4680 | 6.4933 | 7.5498 | 835.5205 | 6.0103 | 6.9507 |
| 10.00 | 1416.7468 | 8.1878 | 8.6516 | 1027.1169 | 6.6669 | 7.7031 | 841.5051 | 6.0407 | 6.9653 |
| 11.00 | 1475.0720 | 8.4318 | 8.7471 | 1074.7683 | 6.8520 | 7.8428 | 853.0667 | 6.0992 | 6.9933 |
| 12.00 | 1533.3975 | 8.6755 | 8.8375 | 1125.8032 | 7.0518 | 7.9824 | 870.2615 | 6.1860 | 7.0342 |
| 13.00 | 1591.7227 | 8.9179 | 8.9243 | 1178.5686 | 7.2617 | 8.1150 | 892.3354 | 6.2968 | 7.0856 |
| 14.00 | 1650.0481 | 9.1601 | 9.0067 | 1232.4243 | 7.4783 | 8.2400 | 918.6567 | 6.4282 | 7.1455 |
| 15.00 | 1708.3733 | 9.4018 | 9.0853 | 1288.7878 | 7.7044 | 8.3640 | 949.8958 | 6.5832 | 7.2146 |
| 16.00 | 1766.6987 | 9.6430 | 9.1605 | 1344.8564 | 7.9330 | 8.4763 | 985.6157 | 6.7591 | 7.2911 |
| 17.00 | 1825.0239 | 9.8833 | 9.2328 | 1402.5435 | 8.1672 | 8.5864 | 1025.0811 | 6.9519 | 7.3727 |
| 18.00 | 1883.3491 | 10.1221 | 9.3032 | 1460.3350 | 8.4039 | 8.6885 | 1068.2063 | 7.1610 | 7.4585 |
| 19.00 | 1941.6746 | 10.3593 | 9.3716 | 1518.8965 | 8.6425 | 8.7874 | 1114.2644 | 7.3824 | 7.5467 |
| 20.00 | 1999.9998 | 10.5969 | 9.4367 | 1577.4963 | 8.8822 | 8.8801 | 1162.9456 | 7.6146 | 7.6363 |
| 21.00 | 1970.8372 | 10.6783 | 9.2282 | 1636.2183 | 9.1225 | 8.9680 | 1213.9810 | 7.8560 | 7.7264 |
| 22.00 | 1941.6746 | 10.6913 | 9.0807 | 1695.6743 | 9.3645 | 9.0537 | 1267.0659 | 8.1052 | 7.8164 |
| 23.00 | 1912.5120 | 10.6863 | 8.9484 | 1752.4111 | 9.6015 | 9.1257 | 1321.4702 | 8.3586 | 7.9049 |
| 24.00 | 1883.3494 | 10.6611 | 8.8328 | 1791.6033 | 9.8005 | 9.1404 | 1377.2195 | 8.6163 | 7.9919 |
| 25.00 | 1854.1865 | 10.6205 | 8.7293 | 1801.2517 | 9.9286 | 9.0710 | 1433.9558 | 8.8768 | 8.0770 |
| 26.00 | 1825.0239 | 10.5664 | 8.6360 | 1807.3760 | 10.0301 | 9.0098 | 1491.0176 | 9.1369 | 8.1593 |
| 27.00 | 1795.8613 | 10.5027 | 8.5495 | 1809.6597 | 10.1099 | 8.9500 | 1545.6926 | 9.3847 | 8.2352 |
| 28.00 | 1766.6987 | 10.4293 | 8.4699 | 1804.5850 | 10.1614 | 8.8796 | 1591.1494 | 9.5896 | 8.2963 |
| 29.00 | 1737.5364 | 10.3483 | 8.3953 | 1798.2356 | 10.1988 | 8.8159 | 1624.9011 | 9.7411 | 8.3405 |
| 30.00 | 1708.3735 | 10.2613 | 8.3243 | 1787.0862 | 10.2151 | 8.7473 | 1653.2954 | 9.8681 | 8.3769 |
| 31.00 | 1679.2109 | 10.1695 | 8.2561 | 1774.5110 | 10.2188 | 8.6826 | 1676.8696 | 9.9734 | 8.4067 |
| 32.00 | 1650.0483 | 10.0712 | 8.1919 | 1759.7485 | 10.2095 | 8.6182 | 1693.5005 | 10.0475 | 8.4275 |
| 33.00 | 1620.8857 | 9.9695 | 8.1292 | 1743.0652 | 10.1881 | 8.5544 | 1706.3555 | 10.1047 | 8.4434 |
| 34.00 | 1591.7229 | 9.8643 | 8.0681 | 1724.6057 | 10.1561 | 8.4905 | 1714.4055 | 10.1405 | 8.4533 |
| 35.00 | 1562.5603 | 9.7555 | 8.0086 | 1705.8687 | 10.1165 | 8.4312 | 1718.6663 | 10.1594 | 8.4585 |
| 36.00 | 1533.3977 | 9.6441 | 7.9499 | 1685.5386 | 10.0676 | 8.3711 | 1719.6960 | 10.1640 | 8.4598 |
| 37.00 | 1504.2351 | 9.5299 | 7.8922 | 1664.1497 | 10.0108 | 8.3118 | 1717.5903 | 10.1546 | 8.4572 |
| 38.00 | 1475.0725 | 9.4121 | 7.8360 | 1642.9133 | 9.9485 | 8.2571 | 1712.8923 | 10.1337 | 8.4514 |
| 39.00 | 1445.9097 | 9.2932 | 7.7794 | 1620.2556 | 9.8788 | 8.2007 | 1705.7556 | 10.1020 | 8.4427 |
| 40.00 | 1416.7471 | 9.1720 | 7.7232 | 1597.3538 | 9.8034 | 8.1470 | 1696.6504 | 10.0615 | 8.4314 |
| 41.00 | 1387.5845 | 9.0469 | 7.6688 | 1573.9258 | 9.7225 | 8.0942 | 1685.5425 | 10.0120 | 8.4176 |
| 42.00 | 1358.4219 | 8.9211 | 7.6135 | 1550.1177 | 9.6365 | 8.0430 | 1672.8999 | 9.9557 | 8.4017 |
| 43.00 | 1329.2593 | 8.7954 | 7.5566 | 1525.9282 | 9.5464 | 7.9922 | 1658.6489 | 9.8921 | 8.3837 |
| 44.00 | 1300.0964 | 8.6658 | 7.5013 | 1501.3882 | 9.4520 | 7.9421 | 1643.1145 | 9.8226 | 8.3639 |
| 45.00 | 1270.9338 | 8.5349 | 7.4455 | 1476.3337 | 9.3538 | 7.8916 | 1626.5105 | 9.7483 | 8.3426 |
| 46.00 | 1241.7712 | 8.4033 | 7.3886 | 1451.0295 | 9.2522 | 7.8416 | 1608.6501 | 9.6682 | 8.3193 |
| 47.00 | 1212.6086 | 8.2704 | 7.3310 | 1425.9429 | 9.1478 | 7.7939 | 1590.0381 | 9.5846 | 8.2948 |
| 48.00 | 1183.4458 | 8.1361 | 7.2728 | 1400.1433 | 9.0400 | 7.7441 | 1570.5146 | 9.4967 | 8.2688 |
| 49.00 | 1154.2832 | 7.9999 | 7.2143 | 1374.1814 | 8.9295 | 7.6946 | 1550.0391 | 9.4043 | 8.2411 |
| 50.00 | 1125.1206 | 7.8622 | 7.1553 | 1348.1128 | 8.8167 | 7.6453 | 1529.1519 | 9.3099 | 8.2125 |
| 51.00 | 1095.9580 | 7.7239 | 7.0946 | 1321.9011 | 8.7016 | 7.5957 | 1507.6128 | 9.2123 | 8.1826 |
| 52.00 | 1066.7954 | 7.5847 | 7.0325 | 1295.8093 | 8.5848 | 7.5471 | 1485.5791 | 9.1122 | 8.1516 |
| 53.00 | 1037.6326 | 7.4430 | 6.9705 | 1269.0015 | 8.4650 | 7.4956 | 1462.8887 | 9.0089 | 8.1191 |
| 54.00 | 1008.4700 | 7.3001 | 6.9072 | 1242.4172 | 8.3442 | 7.4448 | 1439.7527 | 8.9033 | 8.0855 |
| 55.00 | 979.3074 | 7.1562 | 6.8424 | 1215.5359 | 8.2216 | 7.3923 | 1416.5479 | 8.7970 | 8.0513 |
| 56.00 | 950.1448 | 7.0121 | 6.7750 | 1188.5994 | 8.0969 | 7.3398 | 1392.7783 | 8.6879 | 8.0156 |
| 57.00 | 920.9822 | 6.8668 | 6.7060 | 1162.0427 | 7.9717 | 7.2886 | 1368.6604 | 8.5769 | 7.9788 |
| 58.00 | 891.8193 | 6.7189 | 6.6366 | 1135.0847 | 7.8443 | 7.2351 | 1344.3176 | 8.4644 | 7.9410 |
| 59.00 | 862.6567 | 6.5708 | 6.5644 | 1108.0330 | 7.7157 | 7.1804 | 1319.6724 | 8.3502 | 7.9020 |
| 60.00 | 833.4941 | 6.4221 | 6.4892 | 1081.0579 | 7.5861 | 7.1252 | 1294.9756 | 8.2354 | 7.8622 |

```
           C     CHARACTERISTICS METHOD
           C        RIVER ROUTING PROBLEM : RECTANGULAR CHANNEL WITH UNIFORM NORMAL
           C           FLOW IS SUBJECTED TO A FLOOD WAVE AT THE UPSTREAM REACH OF
           C           CHANNEL BEING CONSIDERED.      INITIAL UNIFORM DEPTH IS 6.0 FT.
           C           THE FLOOD WAVE IS SUCH THAT THE DISCHARGE AT THE UPSTREAM POINT
           C           INCREASES TO A MAXIMUN OF 2000 CFS IN A 20 MINUTE PERIOD AND
           C           THEN DECREASES TO THE INITIAL UNIFORM FLOW IN A PERIOD OF 40
           C           MINUTES.
           C        DIMENSIONS ARE AS FOLLOWS : TIME (SECONDS),LENGTH (FT.),
           C           DEPTH (FT.),VELOCITY (FT. PER SEC),DISCHARGE (CFS)
           C        B IS THE WIDTH OF THE CHANNEL
           C        SO IS THE CHANNEL SLOPE
           C        RN IS THE MANNING ROUGHNESS COEFICIENT
           C        TTL IS THE LENGTH OF CHANNEL REACH BEING INVESTIGATED
           C        G IS THE ACCELERATION OF GRAVITY,32.2 FT. PER SEC PER SEC
           C        YO IS THE INITIAL UNIFORM NORMAL DEPTH
           C        NM IS THE NUMBER OF INCREMENTAL REACHES ALONG THE LENGTH OF RIVER
           C        M IS EQUAL TO NM+1
           C        ER IS THE ERROR TOLERANCE FOR ITERATIVE SOLUTIONS OF EQUATIONS
           C        TTP IS THE TIME PERIOD BETWEEN QO AND QMAX
           C        QMAX IS THE PEAK FLOOD DISCHARGE OF INFLOW HYDROGRAPH
           C        TAP IS THE TIME PERIOD AFTER QMAX UNTIL QO
           C        Q IS THE DISCHARGE AT THE UPSTREAM BOUNDARY POINT
           C        QOUT IS THE DISCHARGE AT THE DOWNSTREAM BOUNDARY POINT
           C        Y IS THE DEPTH OF FLOW
           C        V IS THE VELOCITY OF FLOW
           C        X IS THE DISTANCE FROM THE UPSTREAM BOUNDARY TO ANY POINT ALONG
           C           THE RIVER
           C       T IS THE TIME ASSOCIATED WITH ANY VALUES OF DISCHARGES,DEPTH, OR
           C           VELOCITY CALCULATED
           C        SFL AND SFR ARE THE FRICTION SLOPE OF POINTS ADJACENT TO THE
           C           POINT BEING CALCULATED.  SFL AND SFR ARE EVALUATED AT THE TIME
           C           INCREMENT BEFORE THAT FOR THE CURRENT CALCULATIONS
0001             DIMENSION X(2,21),T(2,21),Y(2,21),V(2,21)
0002             READ(1,950) RN,SO,TTL,G,TMAX
0003             READ(1,951) QMAX,TTP,TAP,ER,YO,B
0004             READ(1,953) NM,M
0005             WRITE(3,955) RN,SO,TTL,TMAX
0006             WRITE(3,956) QMAX,TTP,TAP,ER,YO,B
0007             WRITE(3,958) NM,M
0008             DX=TTL/NM
0009             GSO=G*SO
0010             CN=RN*RN/2.2082
0011             EF1=1./3.
0012             EF2=2./3.
0013             EF4=4./3.
0014             CM=1.486/RN*SQRT(SO)
           C        CALCULATE THE INITIAL UNIFORM NORMAL DISCHARGE AND VELOCITY
0015             QO=CM*B*YO*(B*YO/(B+2.*YO))**EF2
0016             VO=QO/(B*YO)
0017             WRITE(3,960) YO,VO,QO
0018             WRITE(3,965)
0019             WRITE(3,967)
           C        SPECIFY THE INITIAL CONDITIONS
0020             Y(1,1)=YO
0021             V(1,1)=VO
0022             T(1,1)=0.0
0023             X(1,1)=0.0
```

*Table 7-4* (cont.)

```
FORTRAN IV G LEVEL  20                MAIN              DATE = 71314          18/32/50              PAGE 0002

0024              DO 6 J=2,M
0025              Y(1,J)=YO
0026              V(1,J)=VO
0027              T(1,J)=0.0
0028            6 X(1,J)=X(1,J-1)+DX
0029              L=0
0030            9 IF(L) 10,25,10
0031           10 NI=3
0032              N=NM-1
0033              NCI=1
0034              NCE=M
          C         CALCULATE THE UPSTREAM BOUNDARY POINT
0035              F2=V(1,2)-SQRT(G*Y(1,2))
0036              X(2,1)=0.0
0037              T(2,1)=T(1,2)-X(1,2)/F2
0038              TH1=T(2,1)/60.0
0039              TD=T(2,1)-TTP
0040              IF(TD) 16,16,17
0041           16 Q=QO+(QMAX-QO)/TTP*T(2,1)
0042              GO TO 20
0043           17 IF(TD-TAP) 18,18,19
0044           18 Q=QMAX-(QMAX-QO)/TAP*TD
0045              GO TO 20
0046           19 Q=QO
0047           20 CONTINUE
0048              F6=SQRT(G/Y(1,2))
0049              SFR=CN*V(1,2)*V(1,2)/(B*Y(1,2)/(B+2.*Y(1,2)))**EF4
0050              FU=-V(1,2)+F6*Y(1,2)+(T(2,1)-T(1,2))*(G*SFR-GSO)
0051              Y(2,1)=FU/(2.*F6)+0.5*SQRT(FU*FU/(F6*F6)+4.*Q/(B*F6))
0052              V(2,1)=Q/(B*Y(2,1))
0053              GO TO 30
0054           25 NI=2
0055              N=NM
0056              NCI=2
0057              NCE=NM
          C         CALCULATE THE INTERIOR POINTS
0058           30 DO 100 J=NI,N,2
0059              J1=J-1
0060              J2=J+1
0061              F1=V(1,J1)+SQRT(G*Y(1,J1))
0062              F2=V(1,J2)-SQRT(G*Y(1,J2))
0063              F3=X(1,J1)-F1*T(1,J1)
0064              F4=X(1,J2)-F2*T(1,J2)
0065              T(2,J)=(F3-F4)/(F2-F1)
0066              X(2,J)=F4+F2*T(2,J)
0067              F5=SQRT(G/Y(1,J1))
0068              F6=SQRT(G/Y(1,J2))
0069              SFL=CN*V(1,J1)*V(1,J1)/(B*Y(1,J1)/(B+2.*Y(1,J1)))**EF4
0070              SFR=CN*V(1,J2)*V(1,J2)/(B*Y(1,J2)/(B+2.*Y(1,J2)))**EF4
0071              F7=V(1,J1)+F5*Y(1,J1)+(T(1,J1)-T(2,J))*(G*SFL-GSO)
0072              F8=V(1,J2)-F6*Y(1,J2)+(T(1,J2)-T(2,J))*(G*SFR-GSO)
0073              Y(2,J)=(F7-F8)/(F6+F5)
0074              V(2,J)=F8+F6*Y(2,J)
0075          100 CONTINUE
0076          108 IF(L) 110,120,110
0077          110 J=NM
          C         CALCULATE THE DOWNSTREAM BOUNDARY POINT
0078              F1=V(1,J)+SQRT(G*Y(1,J))
```

```
0079          SFL=CN*V(1,J)*V(1,J)/(B*Y(1,J)/(B+2.*Y(1,J)))**EF4
0080          X(2,M)=TTL
0081          T(2,M)=T(1,J)+(TTL-X(1,J))/F1
0082          TH2=T(2,M)/60.0
0083          F5=SQRT(G/Y(1,J))
0084          FD=-V(1,J)-F5*Y(1,J)+(T(2,M)-T(1,J))*(G*SFL-GSO)
      C        SOLVE NON-LINEAR EQUATION BY NEWTON'S ITERATION METHOD
0085          YPO=Y(2,N)
0086          DO 115 K1=1,25
0087          FBY=B*YPO/(B+2.*YPO)
0088          FB=FBY/YPO
0089          YP1=YPO-(CM*FBY**EF2+F5*YPO+FD)/(EF2*CM/FBY**EF1*FB*FB+F5)
0090          IF(ABS(YP1-YPO)-ER) 117,117,115
0091      115 YPO=YP1
0092      117 Y(2,M)=YP1
0093          V(2,M)=CM*(B*Y(2,M)/(B+2.*Y(2,M)))**EF2
0094          QOUT=V(2,M)*Y(2,M)*B
0095          WRITE(3,970) TH1,Q,Y(2,1),V(2,1),TH2,QOUT,Y(2,M),V(2,M)
0096          L=0
0097          GO TO 121
0098      120 L=1
      C        CHANGE RECENTLY CALCULATED ( T+DT ) VALUES TO ( T ) VALUES
0099      121 DO 122 J=NCI,NCE,2
0100          X(1,J)=X(2,J)
0101          T(1,J)=T(2,J)
0102          V(1,J)=V(2,J)
0103      122 Y(1,J)=Y(2,J)
0104          IF(T(2,NCE)-TMAX) 9,123,123
0105      123 STOP
0106      950 FORMAT(5F10.6)
0107      951 FORMAT(6F10.6)
0108      953 FORMAT(2I10)
0109      955 FORMAT(2X,'RN=',F10.4,5X,'SO=',F10.6,5X,'TTL=',F10.2,5X,'TMAX=',
         1F10.2)
0110      956 FORMAT(2X,'QMAX=',F10.2,5X,'TTP=',F10.2,5X,'TAP=',F10.2,5X,'ER=',
         1F10.6,5X,'YO=',F10.4,5X,'B=',F10.2)
0111      958 FORMAT(2X,'NM=',I5,5X,'M=',I5)
0112      960 FORMAT(2X,'INITIAL UNIFORM DEPTH=',F10.4,5X,'INITIAL VELOCITY=',
         1F10.4,5X,'INITIAL DISCHARGE=',F10.4)
0113      965 FORMAT(19X,'UPSTREAM',55X,'DOWNSTREAM')
0114      967 FORMAT(6X,'TIME(MIN',1X,'DISCHARGE',2X,'DEPTH',4X,'VELOCITY',22X,
         1'TIME(MIN)',1X,'DISCHARGE',2X,'DEPTH',4X,'VELOCITY')
0115      970 FORMAT(5X,F10.2,3F10.4,20X,F10.2,3F10.4)
0116          END
```

**Table 7-4** *(cont.)*

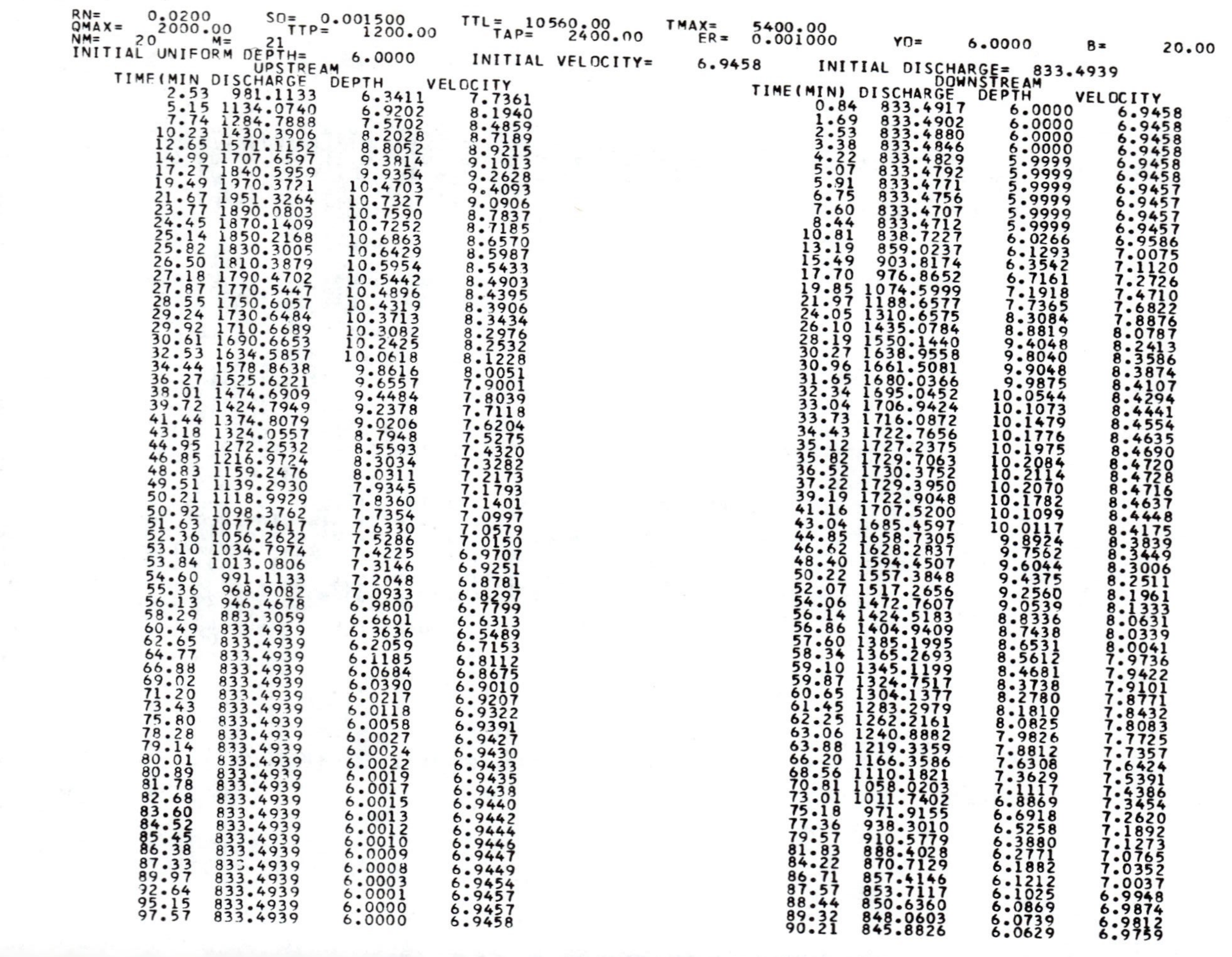

```
RN=     0.0200      SO=  0.001500     TTL=  10560.00     TMAX=   5400.00
QMAX=   2000.00     TTP=  1200.00     TAP=   2400.00     ER=   0.001000     YO=   6.0000     B=   20.00
NM=   20     M=   21
INITIAL UNIFORM DEPTH=   6.0000     INITIAL VELOCITY=   6.9458     INITIAL DISCHARGE=  833.4939
```

UPSTREAM

| TIME(MIN | DISCHARGE | DEPTH | VELOCITY |
|---|---|---|---|
| 2.53 | 981.1133 | 6.3411 | 7.7361 |
| 5.15 | 1134.0740 | 6.9202 | 8.1940 |
| 7.74 | 1284.7888 | 7.5702 | 8.4859 |
| 10.23 | 1430.3906 | 8.2028 | 8.7189 |
| 12.65 | 1571.1152 | 8.8052 | 8.9215 |
| 14.99 | 1707.6597 | 9.3814 | 9.1013 |
| 17.27 | 1840.5959 | 9.9354 | 9.2628 |
| 19.49 | 1970.3721 | 10.4703 | 9.4093 |
| 21.67 | 1951.3264 | 10.7327 | 9.0906 |
| 23.77 | 1890.0803 | 10.7590 | 8.7837 |
| 24.45 | 1870.1409 | 10.7252 | 8.7185 |
| 25.14 | 1850.2168 | 10.6863 | 8.6570 |
| 25.82 | 1830.3005 | 10.6429 | 8.5987 |
| 26.50 | 1810.3879 | 10.5954 | 8.5433 |
| 27.18 | 1790.4702 | 10.5442 | 8.4903 |
| 27.87 | 1770.5447 | 10.4896 | 8.4395 |
| 28.55 | 1750.6057 | 10.4319 | 8.3906 |
| 29.24 | 1730.6484 | 10.3713 | 8.3434 |
| 29.92 | 1710.6689 | 10.3082 | 8.2976 |
| 30.61 | 1690.6653 | 10.2425 | 8.2532 |
| 32.53 | 1634.5857 | 10.0618 | 8.1228 |
| 34.44 | 1578.8638 | 9.8616 | 8.0051 |
| 36.27 | 1525.6221 | 9.6557 | 7.9001 |
| 38.01 | 1474.6909 | 9.4484 | 7.8039 |
| 39.72 | 1424.7949 | 9.2378 | 7.7118 |
| 41.44 | 1374.8079 | 9.0206 | 7.6204 |
| 43.18 | 1324.0557 | 8.7948 | 7.5275 |
| 44.95 | 1272.2532 | 8.5593 | 7.4320 |
| 46.85 | 1216.9724 | 8.3034 | 7.3282 |
| 48.83 | 1159.2476 | 8.0311 | 7.2173 |
| 49.51 | 1139.2930 | 7.9345 | 7.1793 |
| 50.21 | 1118.9929 | 7.8360 | 7.1401 |
| 50.92 | 1098.3762 | 7.7354 | 7.0997 |
| 51.63 | 1077.4617 | 7.6330 | 7.0579 |
| 52.36 | 1056.2622 | 7.5286 | 7.0150 |
| 53.10 | 1034.7974 | 7.4225 | 6.9707 |
| 53.84 | 1013.0806 | 7.3146 | 6.9251 |
| 54.60 | 991.1133 | 7.2048 | 6.8781 |
| 55.36 | 968.9082 | 7.0933 | 6.8297 |
| 56.13 | 946.4678 | 6.9800 | 6.7799 |
| 58.29 | 883.3059 | 6.6601 | 6.6313 |
| 60.49 | 833.4939 | 6.3636 | 6.5489 |
| 62.65 | 833.4939 | 6.2059 | 6.7153 |
| 64.77 | 833.4939 | 6.1185 | 6.8112 |
| 66.88 | 833.4939 | 6.0684 | 6.8675 |
| 69.02 | 833.4939 | 6.0390 | 6.9010 |
| 71.20 | 833.4939 | 6.0217 | 6.9207 |
| 73.43 | 833.4939 | 6.0118 | 6.9322 |
| 75.80 | 833.4939 | 6.0058 | 6.9391 |
| 78.28 | 833.4939 | 6.0027 | 6.9427 |
| 79.14 | 833.4939 | 6.0024 | 6.9430 |
| 80.01 | 833.4939 | 6.0022 | 6.9433 |
| 80.89 | 833.4939 | 6.0019 | 6.9435 |
| 81.78 | 833.4939 | 6.0017 | 6.9438 |
| 82.68 | 833.4939 | 6.0015 | 6.9440 |
| 83.60 | 833.4939 | 6.0013 | 6.9442 |
| 84.52 | 833.4939 | 6.0012 | 6.9444 |
| 85.45 | 833.4939 | 6.0010 | 6.9446 |
| 86.38 | 833.4939 | 6.0009 | 6.9447 |
| 87.33 | 833.4939 | 6.0008 | 6.9449 |
| 89.97 | 833.4939 | 6.0003 | 6.9454 |
| 92.64 | 833.4939 | 6.0001 | 6.9457 |
| 95.15 | 833.4939 | 6.0000 | 6.9457 |
| 97.57 | 833.4939 | 6.0000 | 6.9458 |

DOWNSTREAM

| TIME(MIN) | DISCHARGE | DEPTH | VELOCITY |
|---|---|---|---|
| 0.84 | 833.4917 | 6.0000 | 6.9458 |
| 1.69 | 833.4902 | 6.0000 | 6.9458 |
| 2.53 | 833.4880 | 6.0000 | 6.9458 |
| 3.38 | 833.4846 | 6.0000 | 6.9458 |
| 4.22 | 833.4829 | 5.9999 | 6.9458 |
| 5.07 | 833.4792 | 5.9999 | 6.9458 |
| 5.91 | 833.4771 | 5.9999 | 6.9457 |
| 6.75 | 833.4756 | 5.9999 | 6.9457 |
| 7.60 | 833.4707 | 5.9999 | 6.9457 |
| 8.44 | 833.4712 | 5.9999 | 6.9457 |
| 10.81 | 838.7227 | 6.0266 | 6.9586 |
| 13.19 | 859.0237 | 6.1293 | 7.0075 |
| 15.49 | 903.8174 | 6.3542 | 7.1120 |
| 17.70 | 976.8652 | 6.7161 | 7.2726 |
| 19.85 | 1074.5999 | 7.1918 | 7.4710 |
| 21.97 | 1188.6577 | 7.7365 | 7.6822 |
| 24.05 | 1310.6575 | 8.3084 | 7.8876 |
| 26.10 | 1435.0784 | 8.8819 | 8.0787 |
| 28.19 | 1550.1440 | 9.4048 | 8.2413 |
| 30.27 | 1638.9558 | 9.8040 | 8.3586 |
| 30.96 | 1661.5081 | 9.9048 | 8.3874 |
| 31.65 | 1680.0366 | 9.9875 | 8.4107 |
| 32.34 | 1695.0452 | 10.0544 | 8.4294 |
| 33.04 | 1706.9424 | 10.1073 | 8.4441 |
| 33.73 | 1716.0872 | 10.1479 | 8.4554 |
| 34.43 | 1722.7656 | 10.1776 | 8.4635 |
| 35.12 | 1727.2375 | 10.1975 | 8.4690 |
| 35.82 | 1729.7063 | 10.2084 | 8.4720 |
| 36.52 | 1730.3752 | 10.2114 | 8.4728 |
| 37.22 | 1729.3950 | 10.2070 | 8.4716 |
| 39.19 | 1722.9048 | 10.1782 | 8.4637 |
| 41.16 | 1707.5200 | 10.1099 | 8.4448 |
| 43.04 | 1685.4597 | 10.0117 | 8.4175 |
| 44.85 | 1658.7305 | 9.8924 | 8.3839 |
| 46.62 | 1628.2837 | 9.7562 | 8.3449 |
| 48.40 | 1594.4507 | 9.6044 | 8.3006 |
| 50.22 | 1557.3848 | 9.4375 | 8.2511 |
| 52.07 | 1517.2656 | 9.2560 | 8.1961 |
| 54.06 | 1472.7607 | 9.0539 | 8.1333 |
| 56.14 | 1424.5183 | 8.8336 | 8.0631 |
| 56.86 | 1404.9409 | 8.7438 | 8.0339 |
| 57.60 | 1385.1995 | 8.6531 | 8.0041 |
| 58.34 | 1365.2693 | 8.5612 | 7.9736 |
| 59.10 | 1345.1199 | 8.4681 | 7.9422 |
| 59.87 | 1324.7517 | 8.3738 | 7.9101 |
| 60.65 | 1304.1377 | 8.2780 | 7.8771 |
| 61.45 | 1283.2979 | 8.1810 | 7.8432 |
| 62.25 | 1262.2161 | 8.0825 | 7.8083 |
| 63.06 | 1240.8882 | 7.9826 | 7.7725 |
| 63.88 | 1219.3359 | 7.8812 | 7.7357 |
| 66.20 | 1166.3586 | 7.6308 | 7.6424 |
| 68.56 | 1110.1821 | 7.3629 | 7.5391 |
| 70.81 | 1058.0203 | 7.1117 | 7.4386 |
| 73.01 | 1011.7402 | 6.8869 | 7.3454 |
| 75.18 | 971.9155 | 6.6918 | 7.2620 |
| 77.36 | 938.3010 | 6.5258 | 7.1892 |
| 79.57 | 910.5779 | 6.3880 | 7.1273 |
| 81.83 | 888.4028 | 6.2771 | 7.0765 |
| 84.22 | 870.7129 | 6.1882 | 7.0352 |
| 86.71 | 857.4146 | 6.1212 | 7.0037 |
| 87.57 | 853.7117 | 6.1025 | 6.9948 |
| 88.44 | 850.6360 | 6.0869 | 6.9874 |
| 89.32 | 848.0603 | 6.0739 | 6.9812 |
| 90.21 | 845.8826 | 6.0629 | 6.9759 |

tion—simply substituting the following expressions into Eqs. 7-40 and 7-51

$$\left(\frac{\partial V}{\partial t}\right)_M = \frac{V_P - \frac{1}{2}(V_L + V_R)}{\Delta t} \qquad \left(\frac{\partial V}{\partial x}\right)_M = \frac{V_L - V_R}{2\Delta x}$$

$$\left(\frac{\partial y}{\partial t}\right)_M = \frac{Y_P - \frac{1}{2}(Y_L + Y_R)}{\Delta t} \qquad \left(\frac{\partial y}{\partial x}\right)_M = \frac{Y_L - Y_R}{2\Delta x}$$

$$S_f = \frac{S_{f_L} + S_{f_R}}{2}$$

This exchange allows significant increases in the $\Delta t$ interval to be used; however, $\Delta t$ is still bounded by the Courant condition as well as an additional limitation arising from the influence of the term $S_f$ on the stability of an explicit numerical solution of the unsteady flow equations. The latter condition is known as the "friction criteria" and is given by

$$\Delta t \leq \frac{2.21 R^{4/3}}{g n^2 \, |V|}$$

The lesser value of $\Delta t$, as computed from the Courant condition and the friction criteria, determines the maximum allowable $\Delta t$ interval.

***Example 7-5*** Investigate the same channel geometry and imposed flow defined in Example 7-4 by the characteristic method of hydraulic routing.

*Solution*

Initial conditions of the flow depth and velocity are calculated using Eq. 7-83. The upstream boundary point is then found from Eq. 7-77:

$$x_P - x_R = F_2(t_P - t_R)$$

where

$$F_2 = V_R - \sqrt{gY_R}$$

Since $x_P = 0$ at the upstream boundary,

$$t_P = t_R - \frac{x_R}{F_2}$$

Now, applying Eq. 7-76,

$$V_P - V_R - F_6(Y_P - Y_R) + (t_P - t_R)g(S_{f_R} - S) = 0$$

in which $F_6 = \sqrt{g/Y_R}$.

Next, substituting $V_P = Q/BY_P$ and simplifying,

$$\frac{Q}{BY_P} - F_6 Y_P + FU = 0$$

where

$$FU = -V_R + F_6 y_R + (t_P - t_R)g(S_{f_R} - S)$$

Rearranging to make the equation quadratic in $Y_P$ results in

$$Y_P = \frac{FU}{2F_6} + \frac{1}{2}\sqrt{\frac{(FU)^2}{(F_6)^2} + \frac{4Q}{F_6 B}}$$

and computing the upstream velocity is

$$V_P = \frac{Q}{BY_P}$$

The downstream boundary values are worked out in a similar manner. From Eq. 7-75,

$$t_P = t_L + \frac{(L - x_L)}{F_1}$$

with $L$ = length of reach;

$$F_1 = V_L + \sqrt{gY_L}$$

Then from Eq. 7-74,

$$V_P - V_L + F_5(Y_P - Y_L) + (t_P - t_L)g(S_{f_L} - S) = 0$$

or the reduced downstream boundary figures for velocity and depth follow the relation

$$V_P + F_5 Y_P + FD = 0$$

where

$$F_5 = \sqrt{g/Y_L}$$

$$FD = -V_L - F_5 Y_L + (t_P - t_L)g(S_{f_L} - S)$$

However, since Manning's equations can be utilized to find the velocity

$$V_P = \frac{1.486}{n} S^{1/2}\left(\frac{BY_P}{B + 2Y_P}\right)^{2/3} = C_M\left(\frac{BY_P}{B + 2Y_P}\right)^{2/3}$$

Rearranging and substituting in the downstream boundary equation yields the following expression for the depth

$$CM\left(\frac{BY_P}{B + 2Y_P}\right)^{2/3} + F_5 Y_P + FD = 0$$

This equation is nonlinear in $Y_P$ and can quickly be evaluated by Newton's iteration technique.[10] The general form of this procedure is

$$Y_{P_{k+1}} = Y_{P_k} - \frac{f(Y_{P_k})}{f'(Y_{P_k})}$$

where the prime denotes the first derivative of the function of $Y_P$ and the nonlinear equation is expressed by $f(Y_P) = 0$. The first approximation for $Y_P$, that is, $Y_{P1}$ is taken as $Y_M$. Calculation of the interior points continues as outlined in Eqs. 7-78 through 7-81.

## 7-5 Hydraulic Reservoir Routing

The application of hydraulic routing techniques in the design of reservoirs is relatively new. The TVA[11] used an explicit method to (1) investigate flow conditions at the cooling water facilities of a nuclear power plant; (2) determine velocity and stage variations in a narrow winding river below an existing hydroelectric station; (3) compute time-space variations in velocity, water surface elevation, and discharge in a reservoir subject to a proposed pumping schedule; (4) study the interrelationship of two connecting reservoirs; and (5) simulate the passage of a flood wave through a proposed reservoir. One advantage of the explicit scheme is its latitude in preselecting time and space intervals for which the values of depth, velocity, and discharge are calculated. However, a distinct drawback occurs in restricting the time intervals for a stable solution.

The method of characteristics can overcome the limitation imposed on the small time intervals. However, values of depth, velocity, and discharge must be interpolated for selected locations, a manageable procedure easily incorporated if desired.

Initial conditions within the reservoir may include a steady flow profile, a transient profile, or a flat pool. Boundary conditions vary with discharge relationships, or outlet controls, that is, overflow spillways, morning-glory spillways, gates, and the like. Additionally, local inflows, resistance coefficients, and geometric properties of the channel and reservoir must be prescribed, generally from field-measured data.

An example of a general reservoir problem is presented as Example 7-6. It employs the method of characteristics and demonstrates routing a flood hydrograph through a reservoir of known physical dimensions.

***Example 7-6*** A trapezoid-shaped reservoir of dimensions shown in Fig. 7-12 has an initial spillway flow of 100 cfs. A sudden flood results in a wave that increases the upstream flow to 2500 cfs in a period of 1 hr after which it decreases to the original 100 cfs in the next hour. The reservoir is assumed to have a roughness coefficient $n = 0.02$, a length of reservoir of $TTL = 5280$,

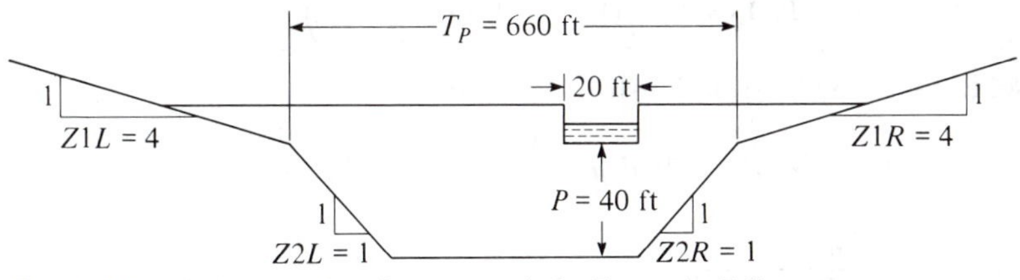


**Fig. 7-12.** *Valley section for reservoir in Example 7-6.*

and a bottom slope $S = 0.002$. The 20-ft-wide free overflow spillway is assumed to have a coefficient of discharge of 3.5; thus $Q_{\text{spillway}} = CYH^{3/2}$.

*Solution*

The initial backwater profile can be calculated using Newton's iteration technique as described in Example 7-5 to prescribe the initial conditions of flow prior to the flood. Computations can then be performed to obtain the upstream boundary point by Newton's iteration procedure, remembering that cross-sectional area changes as the depth of the water in the reservoir exceeds the crest of the overflow spillway. The upstream boundary point, by the use of Eq. 7-77, is

$$x_P - x_R = F_2(t_P - t_R)$$

where

$$F_2 = V_R - \sqrt{\frac{gT}{A_R}}$$

Since $x_P = 0$ at the upstream boundary,

$$t_P = t_R - \frac{x_R}{F_2}$$

Now, from Eq. 7-76,

$$V_P - V_R - \sqrt{\frac{gT}{A_R}}\,(Y_P - Y_R) + (t_P - t_R)(S_{f_R} - S)g = 0$$

and substituting $V_P = Q/A_P$ results in

$$\frac{Q}{A_P} - \sqrt{\frac{gT_R}{A_R}}\,Y_P + FU = 0$$

in which

$$\text{FU} = -V_R + F_6 Y_R + (t_P - t_R)(gS_{f_R} - gS)$$

and

$$F_6 = \sqrt{\frac{gT}{A_R}}$$

Next, if $Y_R > P$, that is, the crest of the spillway

$$A_P = A_c + T_P(Y_P - P) + (Y_P - P)^2\left(\frac{Z1R + Z1L}{2}\right)$$

with the following relationships defined as

$$A_c = [T_P - (Z2R + Z2L)P]P + \frac{P^2}{2}(Z2R + Z2L)$$

$$S_{f_R} = \frac{n^2}{2.208}\,\frac{V_R\,|V_R|}{R_R^{4/3}}$$

$$R_R = \frac{A_R}{P_c + (Y_R - P)(\sqrt{1 + Z1R^2} + \sqrt{1 + Z1L^2})}$$

$$A_R = A_c + T_P(Y_R - P) + (Y_R - P)^2\left(\frac{Z1R + Z1L}{2}\right)$$

$$P_c = T_P - P(Z2R + Z2L) + P(\sqrt{1 + Z1R^2} + \sqrt{1 + Z2L^2})$$

The iterative solution for the depth $Y_P$ is

$$Y_{P_{k+1}} = Y_{P_k} - \frac{Q + (FU - F_6 Y_{P_k})A_1}{(FU - F_6 Y_{P_k})A_2 - F_6 A_1}$$

where

$$A_1 = A_c + T_P(Y_{P_k} - P) + (Y_{P_k} - P^2)\left(\frac{Z1R + Z1L}{2}\right)$$

$$A_2 = T_P + (Z1R + Z1L)(Y_P - P)$$

Then for the case $Y_R < P$,

$$Y_{P_{k+1}} = Y_{P_k} - \frac{Q + (FU - F_6)A_1}{(FU - F_6 y_{P_k})A_2 - F_6 A_1}$$

with

$$A_1 = T_P Y_{P_k} - \left(\frac{Z2R + Z2L}{2}\right) Y_{P_k}^2$$

$$A_2 = T_P - (Z2R + Z2L)Y_{P_k}$$

Adopting the continuity equation $Y_P = Q/A_1$ provides a means of determining the velocity. Interior points are calculated from Eqs. 7-78 through 7-81 as in Example 7-5. The downstream boundary, however, must again be computed through application of Newton's iteration technique. Newton's technique is discussed in most elementary calculus textbooks.

$$t_P = t_L + \frac{TTL - x_L}{V_L - \sqrt{\frac{gA_L}{T}}}$$

and

$$V_P + \sqrt{\frac{gT}{A_L}}\, Y_P + FD = 0$$

where

$$FD = -V_L - \sqrt{\frac{gT}{A_L}}\, Y_L + (t_P - t_L)(S_{f_L} - S)g$$

and

$$S_{f_L} = \frac{n^2 V_L\,|V_L|}{2.21 R_L^{4/3}}$$

$$R_L = \frac{A_L}{T_P - (Z2R + Z2L)P + Y_L(\sqrt{1 + Z2R^2} + \sqrt{1 + Z2L^2})}$$

$$A_L = T_P Y_L - \tfrac{1}{2}(Z2R + Z2L)Y_L^2$$

but the free overflow discharge is

$$Q = 70.0(Y_{P_k} - P)^{3/2}$$

Substituting in the preceding equation produces

$$\frac{70(Y - P)^{3/2} + F_5 Y_{P_k}}{A_c + T_p(Y_{P_k} - P)^2[(Z1R + Z1L)/2]} + FD = 0$$

where

$$F_5 = \sqrt{\frac{gT}{A_L}}$$

Thus the iterative solution for depth $Y_P$ is

$$Y_{P_{k+1}} = Y_{P_k} - [70(Y_{P_k} - P)^{3/2} + (F_5 Y_{P_k} + FD)A_1] \div [110(Y_{P_k} - P)^{1/2} + (F_5 Y_{P_R} + FD)A_2 + F_5 A_1]$$

in which

$$A_1 = T_P Y_{P_k} - \tfrac{1}{2}(Z2R + Y2L)Y_{P_k}^2$$

The velocity $V_P$ is computed from continuity $A_2 = [T_P - (Z2R + Z2L)Y_{P_k}]Y_{P_k}$. A Fortran IV program is presented as Table 7-5 to show the use of these equations in hydraulic reservoir routing by the method of characteristics. Numerous other boundary conditions could be employed by manipulating the desired boundary equations.

## 7-6 Hydraulic Watershed Routing

A watershed is composed of two components of routed flow, the overland and channel flow portions. A watershed can thus be "built up" or described mathematically by portraying the various phases of flow of the effective rainfall throughout its boundaries. This delineation can lead to a complex computer routine so that most applications of this technique have used simplifications in routing the overland flow phase of the watershed.

An investigation by Machmeier and Larson[12] employed this concept to develop hydrographs for a 21.35-mi² watershed. The smallest size unit used in the buildup of the total watershed measured 800 ft by 1750 ft or 0.05 mi². A diagram of the conceptual watershed is shown in Fig. 7-13. Each branch in Fig. 7-13b is identical to the 2.75-mi² subwatershed detailed in Fig. 7-13a.

Typical cross-sectional dimensions of main channels in southeastern Minnesota[12] are given by

$$A = 6.79yM^{0.3} + 1.92y^2M^{0.06} \qquad \textbf{(7-84)}$$

**Table 7-5**

```
FORTRAN IV G LEVEL  20                MAIN                DATE = 71314        18/28/08                PAGE 0001

              C        CHARACTERISTICS METHOD
              C        RESERVOIR ROUTING PROBLEM :  TRAPEZOIDAL RESERVOIR WITH INITIAL
              C           STEADY FLOW OF 100 CFS IS SUBJECTED TO A FLOOD WAVE AT THE
              C           UPSTREAM END OF THE RESERVOIR.  THE FLOOD INCREASES THE FLOW
              C           AT THE UPSTREAM END FROM THE INITIAL STEADY FLOW TO A MAXIMUN
              C           OF 2500 CFS IN 1 HR. AND THEN DECREASES TO THE INITIAL FLOW
              C           DURING ANOTHER 1 HR. PERIOD.
              C        DIMENSIONS ARE AS FLLLOWS : TIME(SECONDS),LENGTH(FEET),
              C           DEPTH(FEET),VELOCITY(FEET PER SECOND),DISCHARGE(CFS)
              C        QO IS THE INITIAL STEADY RESERVOIR DISCHARGE
              C        QMAX IS THE PEAK FLOOD DISCHARGE OF INFLOW HYDROGRAPH
              C        TTP IS THE TIME PERIOD BETWEEN QO AND QMAX
              C        TAP IS THE TIME PERIOD AFTER QMAX UNTIL QO
              C        SPL IS THE LENGTH OF THE RESERVOIR SPILLWAY CREST
              C        DISCC IS THE SPILLWAY DISCHARGE COEFFICIENT
              C        Z1R IS THE SIDE SLOPE OF THE RESERVOIR ABOVE THE ELEVATION OF THE
              C            TOP OF THE DAM (RIGHT SIDE LOOKING UPSTREAM FROM THE DAM).
              C        Z1L IS THE SAME AS Z1R EXCEPT IT IS FOR THE LEFT SIDE.
              C        Z2R IS THE SIDE SLOPE OF THE RESERVOIR BELOW THE DAM ( LOOKING
              C             UPSTREAM).
              C        Z2L IS THE SAME AS Z2R EXCEPT IT IS FOR THE LEFT SIDE.
              C        P IS THE HEIGHT OF THE DAM AND ALSO THE ELEVATION AT WHICH THE
              C           SIDE SLOPES OF THE RESERVOIR CHANGE FROM Z1 TO Z2
              C        TP IS THE WIDTH OF THE RESERVOIR AT THE TOP OF THE DAM.
              C        TTL IS THE LENGTH OF THE RESERVOIR.
              C        YA AND VA ARE THE DEPTH AND VELOCITY OF THE RESERVOIR AT A
              C             CROSS-SECTION LOCATED UPSTREAM FROM THE DAM BEYOND ANY
              C             SIGNIFICANT SURFACE DRAWDOWN.
              C        T(Y) CALCULATES THE TOP WIDTH FOR A DEPTH ( Y )
              C        A(Y) CALCULATES THE CROSS-SECTIONAL AREA FOR A DEPTH ( Y )
              C        R(Y) CALCULATES THE HYDRAULIC RADIUS FOR A DEPTH ( Y )
              C        B IS THE WIDTH OF THE RESERVOIR BOTTOM
              C        SO IS THE RESERVOIR BOTTOM SLOPE
              C        RN IS THE MANNING ROUGHNESS COEFICIENT
              C        ALPHA IS THE VELOCITY DISTRIBUTION ENERGY COEFFICIENT
              C        G IS THE ACCELERATION OF GRAVITY,32.2 FT. PER SEC PER SEC
              C        NM IS THE NUMBER OF INCREMENTAL REACHES ALONG THE LENGTH OF RIVER
              C        ER IS THE ERROR TOLERANCE FOR ITERATIVE SOLUTIONS OF EQUATIONS
              C        Q IS THE DISCHARGE AT THE UPSTREAM BOUNDARY POINT
              C        QOUT IS THE DISCHARGE AT THE DOWNSTREAM BOUNDARY POINT
              C        Y IS THE DEPTH OF FLOW
              C        V IS THE VELOCITY OF FLOW
              C        X IS THE DISTANCE FROM THE UPSTREAM BOUNDARY TO ANY POINT ALONG
              C           THE RESERVOIR
              C        T IS THE TIME ASSOCIATED WITH ANY VALUES OF DISCHARGE ,DEPTH,OR
              C           VELOCITY CALCULATED
              C        SFL AND SFR ARE THE FRICTION SLOPE OF POINTS ADJACENT TO THE
              C           POINT BEING CALCULATED.  SFL AND SFR ARE EVALUATED AT THE TIME
              C           INCREMENT BEFORE THAT FOR THE CURRENT CALCULATIONS.
              C        COMPUTED VALUES FOR THE INLET AND OUTLET ARE PRINTED EACH 20 TH
              C           TIME THEY ARE COMPUTED.
0001                  COMMON/AAA/X(2,21),T(2,21),V(2,21),Y(2,21),TTL,QO,P,EF1,EF2,EF4,
                     1EF10,EF13,AP,PP,TP,Z1R,Z1L,ZH1R,ZH1L,Z2R,Z2L,ZH2R,ZH2L,YA,VA,SO,
                     2CN,ALPHA,B,TMAX,DX,G,M
0002                  A1(Y)=AP+TP*(Y-P)+(Y-P)*(Y-P)/2.*(Z1R+Z1L)
0003                  P1(Y)=PP+(Y-P)*(ZH1R+ZH1L)
0004                  T1(Y)=TP+(Y-P)*(Z1R+Z1L)
0005                  A2(Y)=B*Y+Y*Y/2.*(Z2R+Z2L)
```

**Table 7-5** *(cont.)*

```
FORTRAN IV G LEVEL  20                MAIN              DATE = 71314        18/28/08             PAGE 0002
 0006                P2(Y)=B+Y*(ZH2R+ZH2L)
 0007                T2(Y)=B+Y*(Z2R+Z2L)
 0008                READ(1,950) RN,SO,TTL,TMAX,G,NM
 0009                READ(1,951) QO,QMAX,TTP,TAP,ER
 0010                READ(1,952) TP,ALPHA,SPL,DISCC,P
 0011                READ(1,953) Z1R,Z1L,Z2R,Z2L
 0012                WRITE(3,955) RN,SO,TTL,TMAX,NM
 0013                WRITE(3,956) QO,QMAX,TTP,TAP,ER
 0014                WRITE(3,957) TP,ALPHA,SPL,DISCC,P
 0015                WRITE(3,958) Z1R,Z1L,Z2R,Z2L
 0016                ZH1R=SQRT(Z1R*Z1R+1.)
 0017                ZH1L=SQRT(Z1L*Z1L+1.)
 0018                ZH2R=SQRT(Z2R*Z2R+1.)
 0019                ZH2L=SQRT(Z2L*Z2L+1.)
 0020                B=TP-(Z2R+Z2L)*P
 0021                AP=B*P+P*P/2.*(Z2R+Z2L)
 0022                PP=B+P*(ZH2R+ZH2L)
 0023                DISCF=DISCC*SPL
 0024                DX=TTL/NM
 0025                GSO=G*SO
 0026                CN=RN*RN/2.2082
 0027                M=NM+1
 0028                EF1=1./3.
 0029                EF2=2./3.
 0030                EF4=4./3.
 0031                EF10=10./3.
 0032                EF13=13./3.
 0033                CM=1.486/RN*SQRT(SO)
            C         CALCULATE THE GRADUALLY VARIED FLOW PROFILE ALONG THE RESERVOIR
            C              AND SPECIFY INITIAL CONDITIONS
 0034                YOD=P+(QO/DISCF)**EF2
            C         SOLVE NON-LINEAR EQUATION BY NEWTON'S ITERATION METHOD
 0035                YAO=YOD
 0036                DO 5 KO=1,15
 0037                YP=YAO-P
 0038                YPS=SQRT(YP)
 0039                YA1=YAO-(QO-DISCF*YP*YPS)/(-1.5*DISCF*YPS)
 0040                IF(ABS(YA1-YAO)-ER) 6,6,5
 0041              5 YAO=YA1
 0042              6 YA=YA1
 0043                VA=QO/A1(YA)
 0044                WRITE(3,960) YA,VA,QO
 0045                CALL GVFP
 0046                WRITE(3,961)
 0047                DO 7 J=1,M
 0048                WRITE(3,962) X(1,J),Y(1,J),V(1,J)
 0049              7 T(1,J)=0.0
 0050                WRITE(3,965)
 0051                WRITE(3,967)
 0052                KT=20
 0053                L=0
 0054              9 IF(L) 10,27,10
 0055             10 NI=3
 0056                NCI=1
 0057                NCE=M
 0058                N=NM-1
            C         CALCULATE THE UPSTREAM BOUNDARY POINT
 0059                YR=Y(1,2)
```

```
0060            VR=V(1,2)
0061            IF(YR-P) 11,11,12
0062         11 F2=VR-SQRT(G*A2(YR)/T2(YR))
0063            GO TO 13
0064         12 F2=VR-SQRT(G*A1(YR)/T1(YR))
0065         13 X(2,1)=0.0
0066            T(2,1)=T(1,2)-X(1,2)/F2
0067            TH1=T(2,1)/60.0
0068            TD=T(2,1)-TTP
0069            IF(TD) 16,16,17
0070         16 Q=QO+(QMAX-QO)/TTP*T(2,1)
0071            GO TO 20
0072         17 IF(TD-TAP) 18,18,19
0073         18 Q=QMAX-(QMAX-QO)/TAP*TD
0074            GO TO 20
0075         19 Q=QO
0076         20 CONTINUE
0077            IF(YR-P) 21,21,24
0078         21 F6=SQRT(G*T2(YR)/A2(YR))
0079            SFR=CN*VR*VR*(P2(YR)/A2(YR))**EF4
0080            FU=-VR+F6*YR+(T(2,1)-T(1,2))*(G*SFR-GSO)
      C           SOLVE NON-LINEAR EQUATION BY NEWTON'S ITERATION METHOD
0081            YPO=YR
0082            DO 22 KO=1,25
0083            FAD1=B+YPO*(Z2R+Z2L)
0084            YP1=YPO-(Q/A2(YPO)-F6*YPO+FU)/(-Q*FAD1/(A2(YPO)*A2(YPO))-F6)
0085            IF(ABS(YP1-YPO)-ER) 23,23,22
0086         22 YPO=YP1
0087         23 Y(2,1)=YP1
0088            V(2,1)=Q/A2(YP1)
0089            GO TO 30
0090         24 F6=SQRT(G*T1(YR)/A1(YR))
0091            SFR=CN*VR*VR*(P1(YR)/A1(YR))**EF4
0092            FU=-VR+F6*YR+(T(2,1)-T(1,2))*(G*SFR-GSO)
      C           SOLVE NON-LINEAR EQUATION BY NEWTON'S ITERATION METHOD
0093            YPO=YR
0094            DO 25 KO1=1,25
0095            YP=YPO-P
0096            FAD1=TP+(YPO-P)*(Z1R+Z1L)
0097            YP1=YPO-(Q/A1(YPO)-F6*YPO+FU)/(-Q*FAD1/(A1(YPO)*A1(YPO))-F6)
0098            IF(ABS(YP1-YPO)-ER) 26,26,25
0099         25 YPO=YP1
0100         26 Y(2,1)=YP1
0101            V(2,1)=Q/A1(YP1)
0102            GO TO 30
0103         27 NI=2
0104            NCI=2
0105            NCE=NM
0106            N=NM
      C           CALCULATE THE INTERIOR POINTS
0107         30 DO 100 J=NI,N,2
0108            J1=J-1
0109            J2=J+1
0110            YR=Y(1,J2)
0111            VR=V(1,J2)
0112            YL=Y(1,J1)
0113            VL=V(1,J1)
0114            IF(YR-P) 32,32,33
```

*Table 7-5* (cont.)

```
FORTRAN IV G LEVEL  20                  MAIN              DATE = 71314        18/28/08              PAGE 0004

0115          32 F2=VR-SQRT(G*A2(YR)/T2(YR))
0116             F6=SQRT(G*T2(YR)/A2(YR))
0117             SFR=CN*VR*VR*(P2(YR)/A2(YR))**EF4
0118             GO TO 34
0119          33 F2=VR-SQRT(G*A1(YR)/T1(YR))
0120             F6=SQRT(G*T1(YR)/A1(YR))
0121             SFR=CN*VR*VR*(P1(YR)/A1(YR))**EF4
0122          34 IF(YL-P) 35,35,36
0123          35 F1=VL+SQRT(G*A2(YL)/T2(YL))
0124             F5=SQRT(G*T2(YL)/A2(YL))
0125             SFL=CN*VL*VL*(P2(YL)/A2(YL))**EF4
0126             GO TO 37
0127          36 F1=VL+SQRT(G*A1(YL)/T1(YL))
0128             F5=SQRT(G*T1(YL)/A1(YL))
0129             SFL=CN*VL*VL*(P1(YL)/A1(YL))**EF4
0130          37 F3=X(1,J1)-F1*T(1,J1)
0131             F4=X(1,J2)-F2*T(1,J2)
0132             T(2,J)=(F3-F4)/(F2-F1)
0133             X(2,J)=F4+F2*T(2,J)
0134             F7=V(1,J1)+F5*Y(1,J1)+(T(1,J1)-T(2,J))*(G*SFL-GSO)
0135             F8=V(1,J2)-F6*Y(1,J2)+(T(1,J2)-T(2,J))*(G*SFR-GSO)
0136             Y(2,J)=(F7-F8)/(F6+F5)
0137             V(2,J)=F8+F6*Y(2,J)
0138         100 CONTINUE
0139         108 IF(L) 110,120,110
0140         110 J=NM
           C      CALCULATE THE DOWNSTREAM BOUNDARY POINT
0141             YL=Y(1,J)
0142             VL=V(1,J)
0143             F1=VL+SQRT(G*A1(YL)/T1(YL))
0144             F5=SQRT(G*T1(YL)/A1(YL))
0145             SFL=CN*VL*VL*(P1(YL)/A1(YL))**EF4
0146             X(2,M)=TTL
0147             T(2,M)=T(1,J)+(TTL-X(1,J))/F1
0148             TH2=T(2,M)/60.0
0149             FD=-VL-F5*YL+(T(2,M)-T(1,J))*(G*SFL-GSO)
           C      SOLVE NON-LINEAR EQUATION BY NEWTON'S ITERATION METHOD
0150             YPO=Y(2,N)
0151             DO 115 K1=1,25
0152             YP=YPO-P
0153             YPS=SQRT(YP)
0154             FBD1=TP+YP*(Z1R+Z1L)
0155             YP1=YPO-(DISCF*YPS*YP/A1(YPO)+F5*YPO+FD)/(DISCF*YPS/A1(YPO)*
               1(1.5-YP*FBD1/A1(YPO))+F5)
0156             IF(ABS(YP1-YPO)-ER) 117,117,115
0157         115 YPO=YP1
0158         117 Y(2,M)=YP1
0159             V(2,M)=DISCF*(YP1-P)**1.5/A1(YP1)
0160             IF(KT-20) 119,118,119
0161         118 QOUT=V(2,M)*A1(YP1)
0162             WRITE(3,970) TH1,Q,Y(2,1),V(2,1),TH2,QOUT,Y(2,M),V(2,M)
0163             KT=1
0164             L=0
0165             GO TO 121
0166         119 KT=KT+1
0167             L=0
0168             GO TO 121
0169         120 L=1
```

```
                  C        CHANGE RECENTLY CALCULATED ( T+DT ) VALUES TO ( T ) VALUES
 0170               121 DO 122 J=NCI,NCE,2
 0171                   X(1,J)=X(2,J)
 0172                   T(1,J)=T(2,J)
 0173                   Y(1,J)=Y(2,J)
 0174               122 V(1,J)=V(2,J)
 0175                   IF(T(2,NCI)-TMAX) 9,299,299
 0176               299 STOP
 0177               950 FORMAT(5F10.6,I10)
 0178               951 FORMAT(5F10.6)
 0179               952 FORMAT(7F10.6)
 0180               953 FORMAT(4F10.6)
 0181               955 FORMAT(2X,'RN=',F10.4,5X,'SO=',F10.6,5X,'TTL=',F10.2,5X,'TMAX=',
                       1F10.2,5X,'NM=',I5)
 0182               956 FORMAT(2X,'QO=',F10.2,5X,'QMAX=',F10.2,5X,'TTP=',F10.2,5X,'TAP=',
                       1F10.2,5X,'ER=',F10.6)
 0183               957 FORMAT(2X,'TP=',F10.2,5X,'ALPHA=',F10.2,5X,'SPL=',F10.2,5X,'DISCC=
                       1',F10.3,5X,'P=',F10.2)
 0184               958 FORMAT(5X,'Z1R=',F10.2,5X,'Z1L=',F10.2,5X,'Z2R=',F10.2,5X,'Z2L=',
                       1F10.2)
 0185               960 FORMAT(2X,'INITIAL OUTLET DEPTH=',F10.4,5X,'INITIAL OUTLET VELOCIT
                       1Y=',F10.4,5X,'INITIAL OUTLET DISCHARGE=',F10.4)
 0186               961 FORMAT(2X,'STEADY STATE CONDITIONS:',10X,'X-DISTANCE',9X,'DEPTH',
                       18X,'VELOCITY')
 0187               962 FORMAT(30X,3F15.4)
 0188               965 FORMAT(21X,'INLET',60X,'OUTLET')
 0189               967 FORMAT(7X,'TIME(MIN)','DISCHARGE',2X,'DEPTH',4X,'VELOCITY',23X,
                       1'TIME(MIN)','DISCHARGE',2X,'DEPTH',4X,'VELOCITY')
 0190               970 FORMAT(5X,F10.2,3F10.4,20X,F10.2,3F10.4)
 0191                   END
```

**Table 7-5** *(cont.)*

```
FORTRAN IV G LEVEL  20                    GVFP                DATE = 71314         18/28/08                  PAGE 0001

 0001                  SUBROUTINE GVFP
                C       GVFP ( GRADUALLY VARIED FLOW PROFILE ) IS A SUBROUTINE TO COMPUTE
                C         DEPTHS AND VELOCITIES OF FLOW AT SPECIFIED LOCATIONS ALONG
                C         A STEADY STATE GRADUALLY VARIED FLOW.
 0002                  COMMON/AAA/X(2,21),T(2,21),V(2,21),Y(2,21),TTL,QO,P,EF1,EF2,EF4,
                      1EF10,EF13,AP,PP,TP,Z1R,Z1L,ZH1R,ZH1L,Z2R,Z2L,ZH2R,ZH2L,YA,VA,SO,
                      2CN,ALPHA,B,TMAX,DX,G,M
 0003                  GA1(W)=AP+TP*(W-P)+(W-P)*(W-P)/2.*(Z1R+Z1L)
 0004                  GP1(W)=PP+(W-P)*(ZH1R+ZH1L)
 0005                  GDA1(W)=TP+(W-P)*(Z1R+Z1L)
 0006                  GA2(W)=B*W+W*W/2.*(Z2R+Z2L)
 0007                  GP2(W)=B+W*(ZH2R+ZH2L)
 0008                  GDA2(W)=B+W*(Z2R+Z2L)
 0009                  ER=0.00001
 0010                  ALF=ALPHA/(2.*G)
 0011                  Y(1,M)=YA
 0012                  V(1,M)=VA
 0013                  X(1,M)=TTL
 0014                  F1=ALF*QO*QO
 0015                  F2=CN*QO*QO*DX/2.
 0016                  DSX=TTL
 0017                  MM=M
 0018                  DDY=0.0
 0019                5 YAO=YA-DDY
 0020                  IF(YA-P) 7,6,6
 0021                6 F3=SO*DX-YA-F1/GA1(YA)**2.-F2*GP1(YA)**EF4/GA1(YA)**EF10
 0022                  GO TO 8
 0023                7 F3=SO*DX-YA-F1/GA2(YA)**2.-F2*GP2(YA)**EF4/GA2(YA)**EF10
 0024                8 DO 14 K=1,15
 0025                  IF(YAO-P) 12,10,10
 0026               10 YA1=YAO-(YAO+F1/GA1(YAO)**2.-F2*GP1(YAO)**EF4/GA1(YAO)**EF10+F3)/
                      1(1.-2.*F1*GDA1(YAO)/GA1(YAO)**3.-F2*EF2*GP1(YAO)**EF1/GA1(YAO)**
                      2EF13*(2.*(ZH1R+ZH1L)*GA1(YAO)-5.*GDA1(YAO)*GP1(YAO)))
 0027                  GO TO 13
 0028               12 YA1=YAO-(YAO+F1/GA2(YAO)**2.-F2*GP2(YAO)**EF4/GA2(YAO)**EF10+F3)/
                      1(1.-2.*F1*GDA2(YAO)/GA2(YAO)**3.-F2*EF2*GP2(YAO)**EF1/GA2(YAO)**
                      2EF13*(2.*(ZH2R+ZH2L)*GA2(YAO)-5.*GDA2(YAO)*GP2(YAO)))
 0029               13 IF(ABS(YA1-YAO)-ER) 15,15,14
 0030               14 YAO=YA1
 0031               15 MM=MM-1
 0032                  DDY=YA-YA1
 0033                  DSX=DSX-DX
 0034                  X(1,MM)=DSX
 0035                  Y(1,MM)=YA1
 0036                  IF(YA1-P) 17,16,16
 0037               16 V(1,MM)=QO/GA1(YA1)
 0038                  GO TO 18
 0039               17 V(1,MM)=QO/GA2(YA1)
 0040               18 YA=Y(1,MM)
 0041                  VA=V(1,MM)
 0042                  IF(MM-1) 20,20,5
 0043               20 RETURN
 0044                  END
```

```
RN=     0.0200      SO=   0.002000       TTL=    5280.00      TMAX=  10800.00       NM=    20
QO=     100.00      QMAX=    2500.00       TTP=    3600.00       TAP=   3600.00       ER=  0.001000
TP=     660.00      ALPHA=       1.00       SPL=      20.00      DISCC=       3.500      P=      40.00
   Z1R=        4.00       Z1L=        4.00       Z2R=       1.00       Z2L=       1.00
INITIAL OUTLET DEPTH=   41.2684        INITIAL OUTLET VELOCITY=     0.0039     INITIAL OUTLET DISCHARGE=  100.0000
STEADY STATE CONDITIONS:
```

| X-DISTANCE | DEPTH | VELOCITY |
|---|---|---|
| 0.0 | 30.7084 | 0.0053 |
| 264.0000 | 31.2364 | 0.0052 |
| 528.0000 | 31.7644 | 0.0051 |
| 792.0000 | 32.2924 | 0.0051 |
| 1056.0000 | 32.8204 | 0.0050 |
| 1320.0000 | 33.3484 | 0.0049 |
| 1584.0000 | 33.8764 | 0.0048 |
| 1848.0000 | 34.4044 | 0.0047 |
| 2112.0000 | 34.9324 | 0.0047 |
| 2376.0000 | 35.4604 | 0.0046 |
| 2640.0000 | 35.9884 | 0.0045 |
| 2904.0000 | 36.5164 | 0.0044 |
| 3168.0000 | 37.0444 | 0.0044 |
| 3432.0000 | 37.5724 | 0.0043 |
| 3696.0000 | 38.1004 | 0.0042 |
| 3960.0000 | 38.6284 | 0.0042 |
| 4224.0000 | 39.1564 | 0.0041 |
| 4488.0000 | 39.6844 | 0.0041 |
| 4752.0000 | 40.2124 | 0.0040 |
| 5016.0000 | 40.7404 | 0.0040 |
| 5280.0000 | 41.2684 | 0.0039 |

| INLET | | | | OUTLET | | | |
|---|---|---|---|---|---|---|---|
| TIME(MIN) | DISCHARGE | DEPTH | VELOCITY | TIME(MIN) | DISCHARGE | DEPTH | VELOCITY |
| 0.28 | 111.3357 | 30.7090 | 0.0059 | 0.25 | 99.9998 | 41.2684 | 0.0039 |
| 5.62 | 324.7297 | 30.7194 | 0.0173 | 5.59 | 101.1768 | 41.2784 | 0.0039 |
| 10.95 | 538.0654 | 30.7488 | 0.0287 | 10.92 | 104.6819 | 41.3077 | 0.0041 |
| 16.28 | 751.2932 | 30.7976 | 0.0399 | 16.25 | 110.5658 | 41.3563 | 0.0043 |
| 21.61 | 964.3567 | 30.8649 | 0.0511 | 21.57 | 118.7943 | 41.4228 | 0.0046 |
| 26.93 | 1177.2097 | 30.9507 | 0.0623 | 26.90 | 129.6592 | 41.5082 | 0.0050 |
| 32.24 | 1389.7981 | 31.0551 | 0.0732 | 32.21 | 143.2280 | 41.6117 | 0.0055 |
| 37.55 | 1602.0845 | 31.1761 | 0.0841 | 37.52 | 159.6167 | 41.7324 | 0.0061 |
| 42.85 | 1814.0161 | 31.3146 | 0.0948 | 42.81 | 179.2083 | 41.8714 | 0.0069 |
| 48.14 | 2025.5728 | 31.4701 | 0.1053 | 48.10 | 201.9385 | 42.0265 | 0.0077 |
| 53.42 | 2236.7019 | 31.6419 | 0.1156 | 53.38 | 228.2163 | 42.1987 | 0.0087 |
| 58.68 | 2447.3657 | 31.8303 | 0.1257 | 58.65 | 258.0596 | 42.3864 | 0.0098 |
| 63.94 | 2342.4756 | 32.0196 | 0.1195 | 63.90 | 290.3877 | 42.5818 | 0.0109 |
| 69.18 | 2132.7659 | 32.1911 | 0.1082 | 69.14 | 319.6682 | 42.7525 | 0.0120 |
| 74.41 | 1923.4792 | 32.3375 | 0.0971 | 74.37 | 345.3154 | 42.8979 | 0.0129 |
| 79.64 | 1714.5237 | 32.4609 | 0.0862 | 79.60 | 367.5969 | 43.0212 | 0.0137 |
| 84.85 | 1505.8542 | 32.5623 | 0.0755 | 84.81 | 386.4294 | 43.1236 | 0.0144 |
| 90.07 | 1297.3831 | 32.6451 | 0.0649 | 90.03 | 401.9836 | 43.2068 | 0.0149 |
| 95.27 | 1089.0889 | 32.7080 | 0.0543 | 95.23 | 413.7549 | 43.2691 | 0.0153 |
| 100.48 | 880.9092 | 32.7513 | 0.0439 | 100.44 | 422.1350 | 43.3131 | 0.0156 |
| 105.68 | 672.8206 | 32.7753 | 0.0335 | 105.64 | 426.5693 | 43.3363 | 0.0158 |
| 110.88 | 464.7686 | 32.7803 | 0.0231 | 110.84 | 427.4624 | 43.3409 | 0.0158 |
| 116.08 | 256.7085 | 32.7662 | 0.0128 | 116.04 | 424.8496 | 43.3273 | 0.0157 |
| 121.29 | 100.0000 | 32.7363 | 0.0050 | 121.25 | 418.8760 | 43.2960 | 0.0155 |
| 126.49 | 100.0000 | 32.7002 | 0.0050 | 126.45 | 412.0562 | 43.2602 | 0.0153 |
| 131.70 | 100.0000 | 32.6647 | 0.0050 | 131.66 | 405.4060 | 43.2250 | 0.0150 |
| 136.91 | 100.0000 | 32.6295 | 0.0050 | 136.87 | 398.9153 | 43.1905 | 0.0148 |
| 142.12 | 100.0000 | 32.5956 | 0.0050 | 142.08 | 392.7324 | 43.1574 | 0.0146 |
| 147.33 | 100.0000 | 32.5628 | 0.0050 | 147.29 | 386.3982 | 43.1234 | 0.0144 |
| 152.55 | 100.0000 | 32.5298 | 0.0050 | 152.51 | 380.1658 | 43.0897 | 0.0141 |
| 157.77 | 100.0000 | 32.4965 | 0.0050 | 157.72 | 374.1973 | 43.0573 | 0.0139 |
| 162.99 | 100.0000 | 32.4646 | 0.0050 | 162.94 | 368.1040 | 43.0240 | 0.0137 |
| 168.21 | 100.0000 | 32.4337 | 0.0050 | 168.16 | 362.2380 | 42.9918 | 0.0135 |
| 173.43 | 100.0000 | 32.4028 | 0.0050 | 173.39 | 356.5552 | 42.9604 | 0.0133 |
| 178.65 | 100.0000 | 32.3721 | 0.0050 | 178.61 | 350.9080 | 42.9291 | 0.0131 |

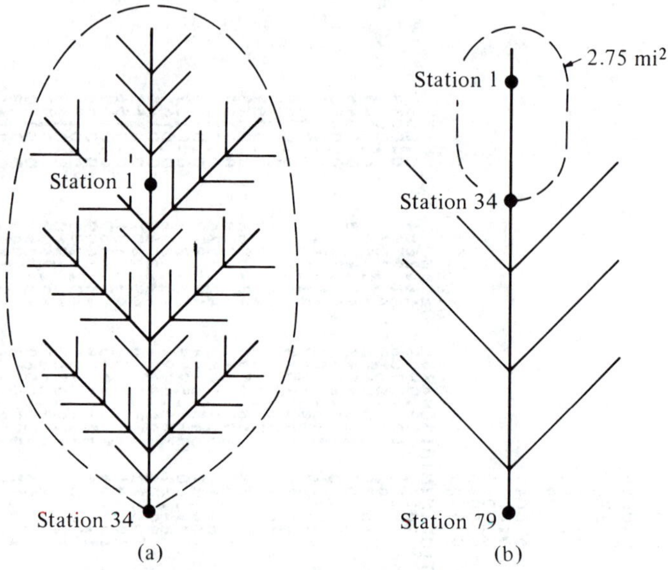


*Fig. 7-13.* *Diagram of model watershed showing (a) complete channel system for single 2.75-mi² subwatershed; and (b) arrangement of 2.75 mi² subwatersheds. (After R. E. Machmeier and C. L. Larson, "Runoff Hydrographs for Mathematical Watershed Model,"* Proc. ASCE, J. Hyd. Div. *(Nov. 1968): 1453–1474.)*

and

$$P = 6.79M^{0.3} + 2.0y\,(3.69M^{0.12} + 1)^{1/2} \tag{7-85}$$

where

$A$ = the cross-sectional area (ft²)
$y$ = the depth of the cross section (ft)
$M$ = the drainage area above the point in the watershed (mi²)
$P$ = wetted perimeter (ft)

Equations 7-84 and 7-85 assist in identifying cross-sectional characteristics based only upon a known location within the watershed. Additional channel information is necessary in the form of the definition of channel slope at any point. This relationship is given by

$$S = 0.003155M^{-0.2}$$

where

$S$ = the slope (ft/ft)
$M$ = the drainage area above the point in the watershed

The smallest input in the total watershed is from a 0.35-mi² subwatershed. This unit is composed of seven 0.05-mi² units and its outflow was obtained by lagging the outflow hydrographs according to their respective travel times to the outlet. The outflow for this 0.35-mi$^2$ unit is shown in Fig. 7-14.

Basic equations for hydraulic routing, Eqs. 7-39 and 7-50, are used to formulate the solution to this watershed routing problem. They can be reduced to

$$(AV)_x + A_t = 0 \tag{7-86}$$

and

$$V_t + VV_x = g(S - S_f) \tag{7-87}$$

in which the subscript indicates partial derivatives with respect to $x$ or $t$. Since routing is to be performed only past the 0.35 mi$^2$ watershed, the lateral inflow term $i$ is zero and the expression, $g\ \partial y/\partial x$ in Eq. 7-51 is assumed to be negligible. Simultaneous solution of $Q = AV$, and the upstream boundary condition, insures that the effect of the depth and velocity at the next station downstream enters the calculations for the depth and velocity at the upstream boundary.

An explicit routing is performed for this watershed having a main stream length of 7.433 mi. The reaches and contributing areas are listed in Table 7-6. Results of the routings for various durations and supply rates are shown in Figs 7-15 and 7-16.

Needless to say, numerous applications of hydraulic routing techniques appear in the literature; each is generally structured for a specific situation. The need to perform hydraulic watershed routing is frequently undertaken in conjunction with a simulation study as will

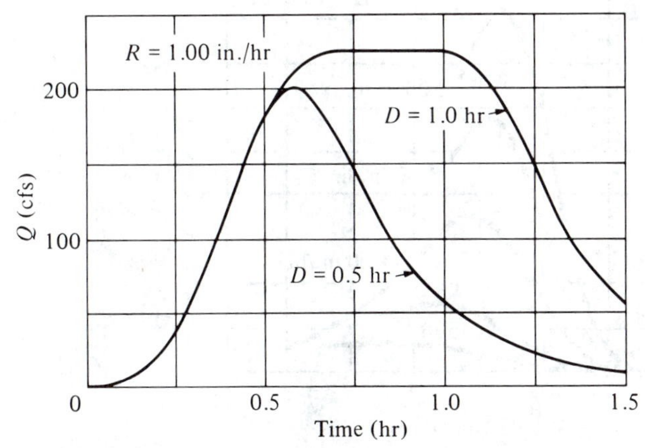


**Fig. 7-14.** *Hydrographs for a 0.35-mi$^2$ watershed with supply rate of 1.00 in./hr and two different durations. (After R. E. Machmeier and C. L. Larson, "Runoff Hydrographs for Mathematical Watershed Model*, Proc. ASCE, J. Hyd. Div. *(Nov. 1968): 1453–1474.)*

**Table 7-6** *Station Numbers and Reach Characteristics for Model Watershed*

| Reach Number (1) | Station Number (2) | Watershed Area for Reach, ($mi^2$) (3) | Length of Reach, (mi) (4) |
|---|---|---|---|
| 1 | 1–6 | 0.35 | 0.332 |
| 2 | 7–17 | 1.15 | 0.665 |
| 3 | 18–28 | 1.95 | 0.665 |
| 4 | 29–47 | 2.75 | 1.197 |
| 5 | 48–51 | 8.25 | 0.552 |
| 6 | 52–59 | 8.95 | 1.218 |
| 7 | 60–63 | 14.45 | 0.522 |
| 8 | 64–71 | 15.15 | 1.218 |
| 9 | 72–75 | 20.65 | 0.522 |
| 10 | 76–79 | 21.35 | 0.522 |

*Source:* After R. E. Machmeier and C. L. Larsen, "Runoff Hydrographs for Mathematical Watershed Model," *Proc. ASCE, J. Hyd. Div.*, Nov. 1968.

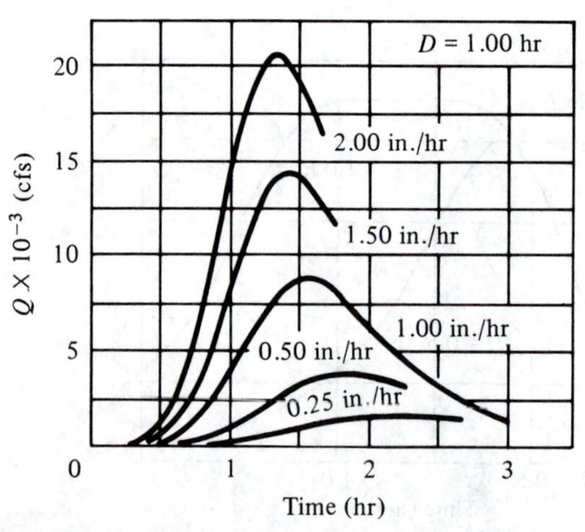


**Fig. 7-15.** *Model watershed hydrographs for various supply rates and a duration of 1.00 hr. (After R. E. Machmeier and C. L. Larson, "Runoff Hydrographs for Mathematical Watershed Model,"* Proc. ASCE, J. Hyd. Div. *(Nov. 1968): 1453–1474.)*

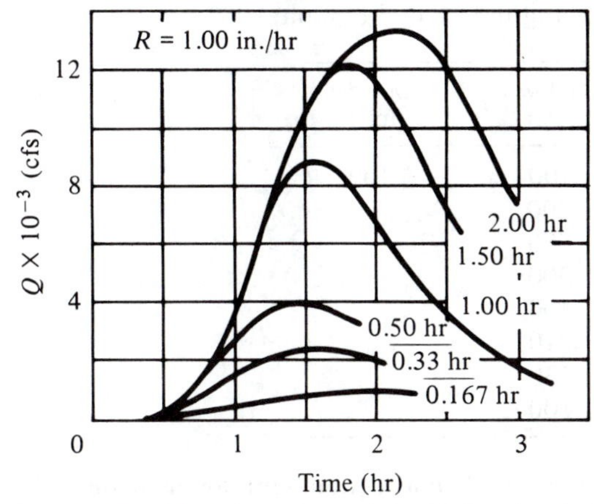


***Fig. 7-16.*** *Model watershed hydrographs for various durations and a supply rate of 1.00 in./hr. (After R. E. Machmeier and C. L. Larson, "Runoff Hydrographs for Mathematical Watershed Model,"* Proc. ASCE, J. Hyd. Div. *(Nov. 1968): 1453–1474.)*

be further discussed in Chapter 10. Material presented here and in Chapter 10 is by no means exhaustive but rather is presented so that an interested student can understand the structuring processes of hydrologic modeling.

## Problems

7-1. Discuss the main differences between hydrologic and hydraulic routing techniques.

7-2. The Muskingum river routing equation, $O_2 = C_0 I_2 + C_1 I_1 + C_2 O_1$, was derived by substituting the storage equation $S_i = K[xI_i + (1-x)O_i]$ where $S_1 = K[xI_1 + (1-x)O_1]$ and $S_2 = K[xI_2 + (1-x)O_2]$ into the continuity equation $\bar{I} = \bar{O} + \Delta S/\Delta t$ and combining like terms. In these equations, $I_1$, $O_1$, and $S_1$ are the inflow, outflow, and storage at the beginning of the time period; and $I_2$, $O_2$, and $S_2$ are the corresponding values at the end of the time period. The terms $\bar{I}$ and $\bar{O}$ are the average inflow and outflow during the time period, and $\Delta S$ is the change in storage. Perform the described derivation and verify the equations for $C_0$, $C_1$, and $C_2$.

7-3. If the Muskingum $K$ value is 12 hr for a reach of a river, and if the $x$ value is 0.2, what would be a reasonable value of $\Delta t$ for routing purposes?

7-4. A river reach has a storage relationship given by $S_i = aI_i + bO_i$. Derive a routing equation for $O_2$ analogous to the Muskingum equation 7-6. Give equations for the coefficients of $I_1$, $O_1$, and $I_2$.

7-5. List the steps (starting with a measured inflow and outflow hydrograph for a river reach) necessary to determine the Muskingum $K$ and $x$ values. If the inflow and outflow are recorded in cfs, state the units that would result for $K$ and $x$ if your list of steps is followed.

7-6. Given the following inflow hydrograph:

| Hour | Inflow (cfs) | Outflow (cfs) |
|---|---|---|
| 6 A.M. | 100 | 100 |
| Noon | 300 | |
| 6 P.M. | 680 | |
| Midnight | 500 | |
| 6 A.M. | 400 | |
| Noon | 310 | |
| 6 P.M. | 230 | |
| Midnight | 100 | |

Assume that the outflow hydrograph at a section 3 mi downstream is desired.

a. Compute the outflow hydrograph by the Muskingum method using values of $K = 11$ hr and $x = 0.13$.
b. Plot the inflow and outflow hydrographs on a single graph.
c. Repeat steps (a) and (b) using $x = 0.00$.

7-7. Given the following values of measured discharges at both ends of a 30-mi river reach:

| Time | Inflow (cfs) | Outflow (cfs) |
|---|---|---|
| 6 A.M. | 10 | 10 |
| Noon | 30 | 12.9 |
| 6 P.M. | 68 | 26.5 |
| Midnight | 50 | 43.1 |
| 6 A.M. | 40 | 44.9 |
| Noon | 31 | 41.3 |
| 6 P.M. | 23 | 35.3 |
| Midnight | 10 | 27.7 |
| 6 A.M. | 10 | 19.4 |
| Noon | 10 | 15.1 |
| 6 P.M. | 10 | 12.7 |
| Midnight | 10 | 11.5 |
| 6 A.M. | 10 | 10.8 |

a. Determine the Muskingum $K$ and $x$ values for this reach.
b. Holding $K$ constant (at your determined value), use the given inflow hydrograph to determine and plot three outflow hydrographs for values of $x$ equal to the computed value, 0.5, and 0.0. Plot the actual outflow and numerically compare the root-mean-square of residuals when each of the three calculated hydrographs are compared with the measured outflow.

7-8. Select a rural watershed in your geographic region that has rainfall runoff records. Use the Muskingum method of watershed routing to find $K$ and $x$.

7-9. Precipitation began at noon on June 14 and caused a flood hydrograph in a stream. As the hydrograph passed, the following measured streamflow data at cross sections $A$ and $B$ were obtained.

| Time June 14–17 | Inflow, Section $A$ (cfs) | Outflow, Section $B$ (cfs) |
|---|---|---|
| 6 A.M. | 10 | 10 |
| Noon | 10 | 10 |
| 6 P.M. | 30 | 13 |
| Midnight | 70 | 26 |
| 6 A.M. | 50 | 43 |
| Noon | 40 | 45 |
| 6 P.M. | 30 | 41 |
| Midnight | 20 | 35 |
| 6 A.M. | 10 | 28 |
| Noon | 10 | 19 |
| 6 P.M. | 10 | 15 |
| Midnight | 10 | 13 |
| 6 A.M. | 10 | 11 |
| Noon | 10 | 10 |

a. Determine the Muskingum $K$ and $x$ values for the river reach.
b. Determine the hydrograph at section $B$ if a different storm produced the following hydrograph at section $A$:

| Time | Inflow (cfs) | Time (cont.) | Inflow (cfs) |
|---|---|---|---|
| 6 A.M. | 100 | Noon | 400 |
| Noon | 100 | 6 P.M. | 300 |
| 6 P.M. | 200 | Midnight | 200 |
| Midnight | 500 | 6 A.M. | 100 |
| 6 A.M. | 600 | Noon | 100 |

7-10. The outflow rate (cfs) and storage (cfs-hr) for an emergency spillway of a certain reservoir are linearly related by $O = S/3$, where the number 3 has units of hr. Use this and the continuity equations $S_2 + O_2\Delta t/2 = \bar{I}\,\Delta t + S_1 - O_1\,\Delta t/2$ to determine the peak outflow rate from the reservoir for the following inflow event:

| Time (hr) | $I$ (cfs) | $S$ (cfs-hr) | $O$ (cfs) |
|---|---|---|---|
| 0 | 0 | 0 | 0 |
| 2 | 400 | | |
| 4 | 600 | | |
| 6 | 200 | | |
| 8 | 0 | | |

7-11. A simple reservoir has a linear storage indication curve defined by the equation

$$O = \frac{O}{2} + \frac{S}{\Delta t}$$

where $\Delta t$ is equal to 1.0 hr. If $S$ at 8 A.M. is zero cfs-hr, use the continuity equation to route the following hydrograph through the reservoir:

| Time | 8 A.M. | 9 A.M. | 10 A.M. | 11 A.M. | Noon | 1 P.M. |
|---|---|---|---|---|---|---|
| $I$ (cfs) | 0 | 200 | 400 | 200 | 0 | 0 |

7-12. For a vertical-walled reservoir with a surface area, $A$, show how the two routing equations 7-13 and 7-15 could be written to contain only $O_2$, $S_2$, and known values (computed from $O_1$, $S_1$, and so on). Eliminate $H$ from all the equations. How could these two equations be solved for the two unknowns?

7-13. Given: Vertical-walled reservoir, surface area = 1000 acres; emergency spillway width = 97.1 ft (ideal spillway); $H$ = water surface elevation (ft) above the spillway crest; and initial inflow and outflow are both 100 cfs.

a. In acre-ft and cfs-days, determine the values for reservoir storage $S$ corresponding to the following values of $H$: 0, 0.5, 1, 1.5, 2, 3, 4 ft.
b. Determine the values of the emergency spillway $Q$ corresponding to the depths named in part (a).
c. Carefully plot and label the discharge-storage curve (cfs versus cfs-days) and the storage-indication curve (cfs versus cfs, Fig. 7-4) on rectangular coordinate graph paper.
d. Determine the outflow rates over the spillway at the ends of successive days corresponding to the following inflow rates (instantaneous rates at the ends of successive days): 100, 400, 1200, 1500, 1100, 700, 400, 300, 200, 100, 100, 100. Use a routing table similar to the one used in Example 7-3 and continue the rotating procedure until the outflow drops below 10 cfs.
e. Plot the inflow and outflow hydrographs on a single graph. Where should these curves cross?

7-14. Route the given inflow hydrograph through the reservoir by assuming the initial water level is at the emergency spillway level (1160 ft) and that the principal spillway is plugged with debris. The reservoir has a 500-ft-wide ideal emergency spillway ($C = 3.0$) located at the 1160-ft elevation. Storage-area-elevation data are

| Elevation (ft) | Area of Pool ($ft^2$) | Storage ($ft^3$) |
|---|---|---|
| 1110 | 0 | 0 |
| 1120 | $0.85 \times 10^6$ | $4.25 \times 10^6$ |
| 1140 | 3.75 | 50.25 |
| 1158 | 9.8 | 172.15 |
| 1160 | 10.8 | 192.75 |

| Elevation (ft) | Area of Pool (ft²) | Storage (ft³) |
|---|---|---|
| 1162 | 11.8 | 215.35 |
| 1164 | 12.8 | 239.95 |
| 1166 | 13.8 | 266.55 |
| 1168 | 14.85 | 295.20 |
| 1180 | 25.0 | 528.55 |

The inflow hydrograph data are

| Time (hr) | $I$ (cfs) |
|---|---|
| 0.0 | 0 |
| 0.5 | 3630 |
| 1.0 | 10920 |
| 1.5 | 10720 |
| 2.0 | 5030 |
| 2.5 | 1600 |
| 3.0 | 460 |
| 3.5 | 100 |
| 4.0 | 10 |
| 4.5 | 0 |

7-15. Given the following data, route a storm hydrograph through a full reservoir and plot on the same graph the inflow and resulting outflow hydrograph for the Green-Acre Branch watershed. The bottom of the rectangular spillway is placed at elevation 980.0. Given: Area = 0.64 mi²; length = 1.10 mi; $L_{ca}$ = 0.53 mi; $C_t$ = 2.00; $C_p$ = 0.62; outflow = $CYH^{3/2}$: $C$ = 3.5, $L$ = 10 ft; and the following storage elevation curve table.

| Elevation (ft) | Incremental Storage $10^{-4}$ (ft³) | Total storage $10^{-4}$ (ft³) |
|---|---|---|
| 960 | | 0 |
| | 40 | |
| 970 | | 40 |
| | 210 | |
| 980 | | 250 |
| | 590 | |
| 990 | | 840 |
| | 1240 | |
| 1000 | | 2080 |

Find

a. The 15-min unit hydrograph by Snyder's method.
b. The 30-min unit hydrograph using the lag method.
c. The storm hydrograph that results from 1.8 in. rain for the first 30 min and 0.63 in. for the next 15 min.

d. Develop a $2S/\Delta t + O$ versus $O$ curve using a routing period of 15 min and the outflow and storage curves provided.
e. Route the storm hydrograph through the reservoir assuming it is full to the bottom of the spillway elevation 980.
f. Indicate maximum height of water in the reservoir.
g. At what elevation should the top of the dam be placed to obtain 5 ft of freeboard?

7-16. Repeat Prob. 7-15 with the reservoir initially empty.

7-17. A flood hydrograph is to be routed by the Muskingum method through a 10-mi reach with $K = 2$ hr. Into how many subreaches must the 10-mi river reach be divided in order to use $\Delta t = 0.5$ hr and still satisfy the stability criteria $K/3 \leq \Delta t \leq K$.

7-18. Repeat Prob. 7-6a by dividing the 3-mi reach into two subreaches with equal $K$ values of 5.5 hr. Compare the results.

7-19. Discuss the problems associated with the use of a reservoir routing technique such as the storage-indication method in routing a flood through a river reach.

7-20. Verify Eq. 7-51.

7.21. Verify Eq. 7-61.

7-22. Rewrite Eqs. 7-51 and 7-61 for the case of no lateral inflow.

7-23. Verify Eqs. 7-70 through 7-73.

7-24. Rework Example 7-4 with values of Mannings $n = 0.01$.

7-25. Rework Examples 7-4 and 7-5 for a trapezoidal channel with 2:1 side slopes.

7-26. Assume that Example 7-6 has a further restriction imposed by the downstream river channel. Rework this example to provide a maximum outflow of 325 cfs.

## References

1. Graeff, "Traité d'hydraulique." Paris: (1883), pp. 438–443.
2. G. T. McCarthy, "The Unit Hydrographs and Flood Routing," unpublished manuscript, North Atlantic Division, U.S. Army Corps of Engineers, June 1938.
3. Chow, V. T., *Open Channel Hydraulics* (New York: McGraw-Hill Book Company, 1959).
4. U.S. Weather Bureau, "Elements of River Forecasting," Tech. Mem., WBTMHYDRO-4, U.S. Department of Commerce, Oct. 1967.
5. Nash, J. E., "The Form of the Instantaneous Unit Hydrograph," *Int. Assoc. Sci. Hydrol.*, Pub. 45, 3 (1957): 14–121.
6. Zoch, R. T., "On the Relation Between Rainfall and Streamflow," *Monthly Weather Review*, 62 (1934): 315–322.
7. Overton, D. E., "Muskingum Flood Routing of Upland Streamflow," *J. Hydrol.*, 4, No. 3, (1966): 185–200.
8. Terstriep, M. L., and J. B. Stall, "Urban Runoff by Road Research Laboratory Method," *Proc. ASCE, J. Hyd. Div.*, Nov. 1969.
9. Harbaugh, T. E., "Numerical Techniques for Spatially Varied Unsteady Flow," University of Missouri Water Resources Center, Rept. No. 3, 1967.

10. Fread, D. L., and T. E. Harbaugh, "Gradually Varied Flow Profiles by Newton's Iteration Technique," *J. Hydrol.*, 2 (1971): 129–139.
11. Garrison, J. M., J. Granju, and J. T. Price, "Unsteady Flow Simulation in Rivers and Reservoirs," *Proc. ASCE, J. Hyd. Div.*, Sept. 1969; pp. 1559–1576.
12. Machmeier, R. E., and C. L. Larson, "Runoff Hydrographs for Mathematical Watershed Model," *Proc. ASCE, J. Hyd. Div.*, Nov. 1968, pp. 1453–1474.

# 8

# Groundwater Hydrology

## 8-1 Introduction

The amount of water stored below ground in the United States exceeds by a significant amount all above ground storage in streams, rivers, reservoirs, and lakes including the Great Lakes.[6] This enormous reservoir sustains streamflow during precipitation-free periods and constitutes the major source of fresh water for many arid localities. Figure 8-1 indicates the distribution and nature of primary groundwater areas of the United States.

The quantification of the volume and rate of flow of groundwater in various regions is an exceedingly difficult task because volumes and flow rates are determined to a considerable extent by the geology of the region. The character and arrangement of rocks and soils are important factors, and these are often highly variable within a groundwater reservoir. An additional difficulty is the inability to measure directly many critical geologic and hydraulic reservoir characteristics.

In spite of these predicaments, hydrologists are continually developing new techniques for measurement and analysis that are contributing to an extensive body of knowledge in the field of groundwater hydrology. Many practical problems can be adequately solved by employing these techniques. This chapter presents the fundamentals of flow in a porous medium and shows how they are applied to the solution of various hydrologic problems.

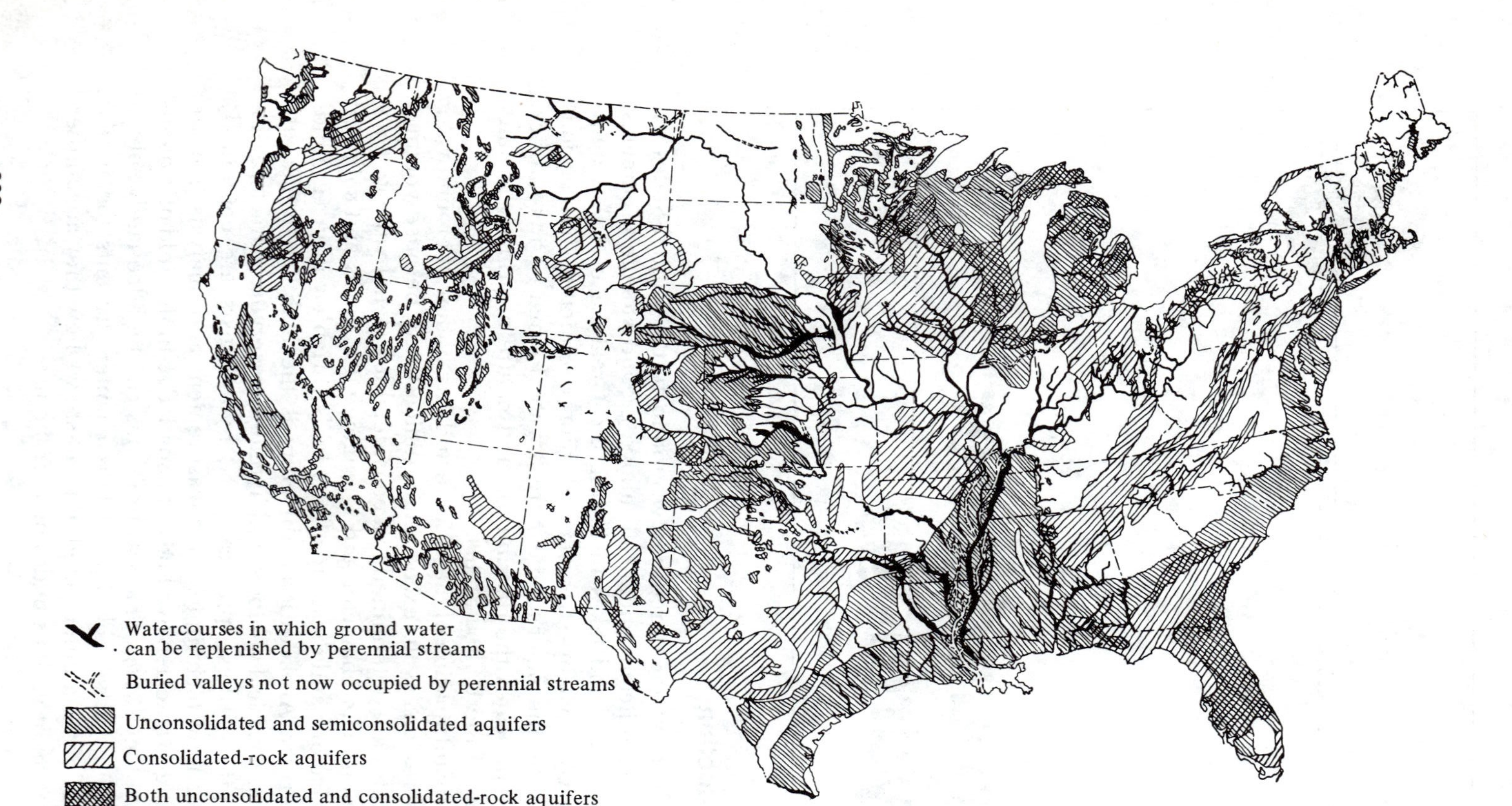


***Fig. 8-1.*** *Groundwater areas in the United States. Patterns show that areas underlaid by aquifers are generally capable of yielding to individual wells 50 gpm or more of water containing not more than 2000 ppm of dissolved solids (includes some areas where more highly mineralized water is actually used). (From H. E. Thomas, "Underground Sources of Water"* Water, The Yearbook of Agriculture, *Washington, D.C.: U.S. Department of Agriculture 1955.)*

## 8-2 Groundwater Flow—General Properties

Understanding the movement of groundwater requires a knowledge of the time and space dependency of the flow, the nature of the porous medium and fluid, and the boundaries of the flow system.

Groundwater flows are usually three-dimensional. Unfortunately, the solution of such problems by analytic methods is intensely complex unless the system is symmetric.[2] In other cases, space dependency in one of the coordinate directions may be so slight that assumption of two-dimensional flow is satisfactory. Many problems of practical importance fall into this class. Sometimes one-dimensional flow can be assumed, thus further simplifying the solution.

Fluid properties such as velocity, pressure, temperature, density, and viscosity often vary in time and space. When time dependency occurs, the issue is termed an *unsteady flow problem* and solutions are usually difficult. On the other hand, situations where space dependency alone exists are *steady flow problems.* Only homogeneous (single-phase) fluids will be considered here. For a discussion of multiple phase flow, Ref. 2 is recommended.

Boundaries to groundwater flow systems may be fixed geologic structures or free water surfaces that are dependent for their position on the state of the flow. A hydrologist must be able to define these boundaries mathematically if he is to solve groundwater flow problems.

Porous media through which groundwaters flow may be classified as isotropic, anisotropic, heterogeneous, homogeneous, or several possible combinations of these. An *isotropic* medium has uniform properties in all directions from a given point. *Anisotropic* media have one or more properties that depend on a given direction. For example, permeability of the medium might be greater along a horizontal plane than along a vertical one. *Heterogeneous* media have nonuniform properties of anisotropy or isotropy, while *homogeneous* media are uniform in their characteristics.

## 8-3 Subsurface Distribution of Water

Groundwater distribution may be generally categorized into zones of aeration and saturation. The saturated zone is one in which all voids are filled with water under hydrostatic pressure. In the zone of aeration, the interstices are filled partly with air, partly with water. The saturated zone is commonly called the *groundwater zone.* The zone of aeration may ideally be subdivided into several subzones. Todd classifies these as follows.[4]

1. *Soil water zone.* A soil water zone begins at the ground sur-

face and extends downward through the major root band. Its total depth is variable and dependent upon soil type and vegetation. The zone is unsaturated except during periods of heavy infiltration. Three categories of water classification may be encountered in this region: hygroscopic water, which is adsorbed from the air; capillary water, held by surface tension; and gravitational water, which is excess soil water draining through the soil.

2. *Intermediate zone.* This belt extends from the bottom of the soil-water zone to the top of the capillary fringe and may change from nonexistence to several hundred feet in thickness. The zone is essentially a connecting link between a near-ground surface region and the near-water-table region through which infiltrating fluids must pass.

3. *Capillary zone.* A capillary zone extends from the water table (Figure 8-2) to a height determined by the capillary rise that can be generated in the soil. The capillary band thickness is a function of soil texture and may fluctuate not only from region to region but also within a local area.

4. *Saturated zone.* In the saturated zone, groundwater fills the pore spaces completely and porosity is therefore a direct measure of storage volume. Part of this water (specific retention) cannot be removed by pumping or drainage because of molecular and surface tension forces. Specific retention is the ratio of volume of water retained against gravity drainage to gross volume of the soil.

Water that can be drained from a soil by gravity is known as the *specific yield.* It is expressed as the ratio of the volume of water that can be drained by gravity to the gross volume of the soil. Values of

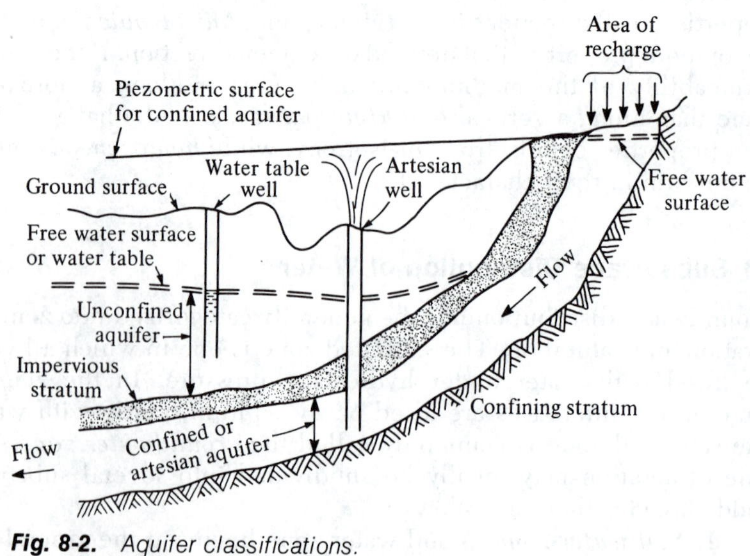


**Fig. 8-2.** *Aquifer classifications.*

specific yield depend upon the soil particle size, shape and distribution of pores, and degree of compaction of the soil. Average values for alluvial aquifers range from 10 to 20 %. Meinzer and others have developed procedures for determining the specific yield.[46]

## 8-4 Geologic Considerations

The determination of groundwater volumes and flow rates requires a thorough knowledge of the geology of a groundwater basin. In bedrock areas, hydrologic characteristics of the rocks, that is, their location, size, orientation, and ability to store or transmit water, must be known. In unconsolidated rock areas, basins often contain hundreds to thousands of feet of semiconsolidated to unconsolidated fill deposits that originated from the erosion of headwater areas. Such fills often contain extensive quantities of stored water. The characteristics of these basin fills must be evaluated.

A knowledge of the distribution and nature of geohydrologic units such as *aquifers*, *aquifuges*, and *aquicludes* is essential to proper planning for development or management of groundwater supplies. In addition, bedrock basin boundaries must be located and an evaluation made of their leakage characteristics.

An aquifer is a water-bearing stratum or formation that is capable of transmitting water in quantities sufficient to permit development. Aquifers may be considered as falling into two categories, confined and unconfined, depending upon whether or not a water table or free surface exists under atmospheric pressure. Storage volume within an aquifer is changed whenever water is recharged to, or discharged from, an aquifer. In the case of an unconfined aquifer this may be easily determined as

$$\Delta S = S_y \, \Delta V \tag{8-1}$$

where

$\Delta S$ = the change in storage volume
$S_y$ = the average specific yield of the aquifer
$\Delta V$ = the volume of the aquifer lying between the original water table and the water table at some later specified time

For saturated, confined aquifers, pressure changes produce only slight modifications in the storage volume. In this case, the weight of the overburden is supported partly by hydrostatic pressure and somewhat by solid material in the aquifer. When hydrostatic pressure in a confined aquifer is reduced by pumping or other means, the load on the aquifer increases, causing its compression, with the result that some water is forced out. Decreasing the hydrostatic pressure also causes a small expansion which in turn produces an additional release of water. For confined aquifers, water yield is expressed in terms of a *storage coefficient* $S_c$, defined as the volume of water an aquifer takes in or releases per unit surface area of aquifer per unit change in head normal to the surface. Figure 8-2 illustrates the classifications of aquifers.

In addition to water-bearing strata exhibiting satisfactory rates of yield, there are also nonwater-bearing and impermeable strata that may contain large quantities of water but whose transmission rates are not high enough to permit effective development. An aquifuge is a formation impermeable and devoid of water; an aquiclude is an impervious stratum.

## 8-5 Fluctuations in Groundwater Level

Any circumstance that alters the pressure imposed on underground water will also cause a variation in the groundwater level. Seasonal factors, changes in stream and river stages, evapotranspiration, atmospheric pressure changes, winds, tides, external loads, various forms of withdrawal and recharge, and earthquakes all may produce fluctuations in the water table level or piezometric surface, depending upon whether the aquifer is free or confined.[4] It is important that an engineer concerned with the development and utilization of groundwater supplies be aware of these factors. He should also be able to evaluate their importance relative to operation of a specific groundwater basin.

## 8-6 Groundwater-Surfacewater Relationships

Notwithstanding that water resource development has often been based on the predominant use of either surface or groundwaters, it must be emphasized that these two components of the total water resource are interdependent. Changes in one component can have far-reaching effects on the other. Coordinated development and management of the combined resource is critical. Linkage between surface and groundwaters should be investigated in all regional studies so that adverse effects can be noted if they exist and opportunities for joint management understood.

In Chapter 4 it was shown how surface stream flows are sustained by the groundwater resource, and it was also pointed out that groundwaters are replenished by infiltration derived from precipitation on the earth's surface.

Underground reservoirs often are extensive and can serve to store water for a multitude of uses. If withdrawals from these reservoirs consistently exceed recharge, *mining* occurs and ultimate depletion of the resource results. By properly coordinating the use of surface and groundwater supplies, optimum regional water resource development seems most likely to be assured. Several studies directed toward this coordinated use have been initiated.[1,5,18]

## 8-7 Hydrostatics

Water located in pore spaces of a saturated medium is under pressure (called *pore pressure*) which can be determined by inserting a piezometer in the medium at a point of interest. If location $A$ (Fig. 8-3) is considered, it can be seen that pore pressure is given by

$$p = h_a \gamma \tag{8-2}$$

where

$p$ = the pore pressure (gauge pressure)
$h_a$ = the head measured from the point to the water table
$\gamma$ = the specific weight of water

Pore pressure is considered positive or negative, depending upon whether the pressure head is measured above (positive) or below (negative) the point under consideration. If an arbitrary datum is established, the total head or piezometric head above the datum is

$$P_p = z + h \tag{8-3}$$

where $P_p$ is known as the piezometric potential. In Fig. 8-3 this is equal to $h_a + z_a$ for point $A$ in the saturated zone and $z_b - h_b$ for point $B$ in the unsaturated zone. The term $h_a$ is the pore pressure of $A$ while $-h_b$ denotes tension or vacuum (negative pore pressure) at $B$.

## 8-8 Groundwater Flow

Analogies can be drawn between flow in pipes under pressure and in fully saturated confined aquifers. The flow of groundwater with a free

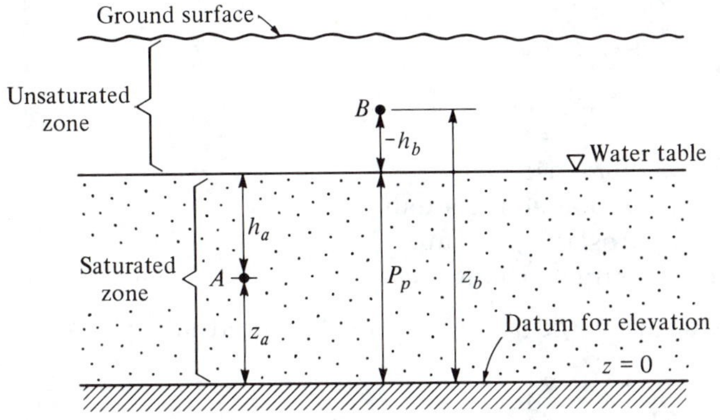


**Fig. 8-3.** *Definition sketch showing hydrostatic pressures in a porous medium.*

surface is also similar to that in an open channel. A major difference is the geometery of a groundwater system flow channel as compared with common hydraulic pipe flow or channel systems. The problem can easily be recognized by envisioning a discharging cross section composed of a number of small openings, each with its own geometry, orientation, and size so that the flow velocity issuing from each pore varies in both magnitude and direction. Difficulties in analyzing such systems are apparent. Computations are usually based on macroscopic averages of fluid and medium properties over a given cross-sectional area.

Unknown quantities to be determined in groundwater flow problems are density, pressure, and velocity if constant temperature conditions are assumed to exist. In general, water is considered incompressible, so the number of working variables is reduced. An exception to this is discussed later relative to the storage coefficient for a confined aquifer. Primary emphasis here will be placed on the flow of water in a saturated porous medium.

## 8-9 Darcy's Law

Darcy's law for fluid flow through a permeable bed is stated as[19]

$$Q = -KA\frac{dh}{dx} \tag{8-4}$$

where

$A$ = the total cross-sectional area including the space occupied by the porous material
$K$ = the hydraulic conductivity of the material
$Q$ = the flow across the control area $A$

In Eq. 8-4,

$$h = z + \frac{p}{\gamma} + C \tag{8-5}$$

where

$h$ = the piezometric head
$z$ = the elevation above a datum
$p$ = the hydrostatic pressure
$C$ = an arbitrary constant

If the specific discharge $q = (Q/A)$ is substituted in Eq. 8-4,

$$q = -K\frac{d}{dx}\left(z + \frac{p}{\gamma}\right) \tag{8-6}$$

Note that $q$ also equals the porosity $n$ multiplied by the pore velocity

$V_p$. Darcy's law is widely used in groundwater flow problems. Several applications will be illustrated in later sections.

Darcy's law is limited in applicability to cases where the Reynolds number is on the order of 1. For Reynolds numbers less than 1, Darcy's law may be considered valid. Deviations from Darcy's law have been shown to occur at Reynolds numbers as low as 2, depending upon such factors as grain size and shape. The Reynolds number $N_R$ is defined herein as

$$N_R = \frac{\rho q d}{\mu} \tag{8-7}$$

where

$q$ = the specific discharge
$d$ = the mean grain diameter
$\rho$ = fluid density
$\mu$ = dynamic viscosity

For many conditions of practical importance (zones lying adjacent to collecting devices are an exception), Darcy's law has been found to apply.

Of special interest is the fact that the Darcy equation is analogous to Ohm's law

$$i = \left(\frac{1}{R}\right) E \tag{8-8}$$

where

$i$ = the current
$R$ = the resistance
$E$ = the voltage

Current and velocity are analogous as are $K$ and $1/R$, and $E$, and $dh/dx$. The similarity of the two equations is the basis for electric analog models of groundwater flow systems.[4,38]

***Example 8-1*** Water temperature in an aquifer is 60°F and the rate of water movement = 1.2 ft/day. The average particle diameter in the porous medium is 0.08 in. Find the Reynolds number and indicate whether Darcy's law is applicable.

Equation 8-7 gives the Reynolds number as

$$N_R = \frac{\rho q d}{\mu}$$

This may also be written as

$$N_R = \frac{q d}{\nu}$$

From Table 1 in Appendix B, $\nu$ is found to be $1.21 \times 10^{-5}$ ft²/sec. Converting the velocity $q$ into units of ft/sec gives $q = 1.2/86{,}400 = 1.39 \times 10^{-5}$. The mean grain diameter in ft = 0.08/12 = 0.0067. Substituting these values in the equation, we obtain

$$N_R = \frac{1.39 \times 10^{-5} \times 0.0067}{1.21 \times 10^{-5}}$$

$$= 0.0077$$

Since $N_R < 1.0$, Darcy's law does apply.

## 8-10 Permeability

The hydraulic conductivity $K$ is an important parameter that is often separated into two components, one related to the medium, the other to the fluid. The product

$$k = Cd^2 \tag{8-9}$$

called the *specific* or *intrinsic permeability*, is a function of the medium only. In Eq. 8-9, $d$ represents the mean grain diameter of the particles; and $C$ is a constant shape factor associated with packing, size distribution, and other factors.[2,4] Using this definition, hydraulic conductivity, also known as the *coefficient of permeability*, can be written

$$K = \frac{k\gamma}{\mu} \tag{8-10}$$

Dimensions of intrinsic permeability are $L^2$. Since values of $k$ given as ft² or cm² are extremely small, a unit of measure known as the *darcy* has been widely adopted.

$$1 \text{ darcy} = 0.987 \times 10^{-8} \text{ cm}^2 \qquad \text{or} \qquad 1.062 \times 10^{-11} \text{ ft}^2$$

Several ways of expressing hydraulic conductivity are reported in the literature. The U.S. Geological Survey has defined the standard coefficient of permeability $K_s$ as the number of gallons per day of water passing through 1 ft² of medium under a unit hydraulic gradient at a temperature of 60°F. Another measure, called the *field coefficient of permeability* $K_f$, is defined as

$$K_f = K_s \left( \frac{\mu_{60}}{\mu_f} \right) \tag{8-11}$$

where

$\mu_{60}$ = the dynamic viscosity of water at 60°F
$\mu_f$ = the dynamic viscosity at the prevailing field temperature

The temperature effect on density is neglected, since it is usually quite small over the range of groundwater temperatures encountered in practice.

It is often convenient to use the coefficient of transmissivity

$$T = K_f b \tag{8-12}$$

where

$K_f$ = the field hydraulic conductivity
$b$ = the saturated depth of the aquifer

Table 8-1 gives the values of the intrinsic permeability and the standard coefficient of permeability for several classes of materials. Considerable variation within divisions can occur; hence a careful geologic survey should accompany all groundwater studies.

***Example 8-2*** Laboratory tests of an aquifer material give a standard coefficient of permeability $K_s = 3.78 \times 10^2$ gpd/ft². If the prevailing field temperature is 50°F, find the field coefficient of the permeability $K_f$.

Using Eq. 8-11, we obtain

$$K_f = K_s \left( \frac{\mu_{60}}{\mu_f} \right)$$

From Table 1, Appendix B, the kinematic viscosity at 60°F = $1.21 \times 10^{-5}$ ft²/sec and at 50°F it is $1.41 \times 10^{-5}$ ft²/sec. For constant density,

$$K_f = \frac{3.78 \times 10^2 \times 1.21 \times 10^{-5}}{1.41 \times 10^{-5}}$$

and

$$K_f = 3.24 \times 10^2 \text{ gpd/ft}^2$$

***Table 8-1*** *Some Values of the Standard Coefficient of Permeability and Intrinsic Permeability for Several Classes of Materials*

| Material | Approximate Range $K_s$ (gal/day/ft²) | Approximate Range $k$ (darcys) |
|---|---|---|
| Clean gravel | $10^6$–$10^4$ | $10^5$–$10^3$ |
| Clean sands; mixtures of clean gravels and sands | $10^4$–10 | $10^3$–1 |
| Very fine sands; silts; mixtures of sands, silts, clays; stratified clays | 10–$10^{-3}$ | 1–$10^{-4}$ |
| Unweathered clays | $10^{-3}$–$10^{-4}$ | $10^{-4}$–$10^{-5}$ |

## 8-11 Velocity Potential

Potential theory is directly applicable to groundwater flow computations. The *velocity potential* $\phi$ is a scalar function of time and space. The potential is defined by

$$\phi(x, y, z) = -K\left(z + \frac{p}{\gamma}\right) + C \tag{8-13}$$

where $C$ is an arbitrary constant. By definition, its derivative with respect to any given direction is the velocity of flow in that direction. Thus it may be written that

$$u = \frac{\partial \phi}{\partial x} \qquad v = \frac{\partial \phi}{\partial y} \qquad w = \frac{\partial \phi}{\partial z} \tag{8-14}$$

where

$u$, $v$, and $w$ = velocities in the $x$, $y$, and $z$ directions, respectively

and $K$ is assumed constant. In vector notation this becomes

$$V = \text{grad } \phi = \nabla\phi \tag{8-15}$$

with $V$ the combined velocity vector and

$$\text{grad } \phi = \frac{\partial \phi}{\partial x}\,\mathbf{i} + \frac{\partial \phi}{\partial y}\,\mathbf{j} + \frac{\partial \phi}{\partial z}\,\mathbf{k} = \nabla\phi \tag{8-16}$$

## 8-12 Hydrodynamic Equations

The determination of values for the variables $u$, $v$, $w$, and $h$ is the target of most groundwater flow problems. The first three variables are the specific discharge components in the $x$, $y$, and $z$ directions, respectively, while $h$ is the total head at a specified point in the flow domain. To effect a solution, four equations involving these variables are needed. These are the equations of motion in each direction plus the continuity equation.

The equations of motion are based on Newton's second law,

$$F = ma \tag{8-17}$$

where

$F$ = the force
$m$ = the mass
$a$ = the acceleration

Considering forces acting on a fluid element, accelerations in the three coordinate directions may be determined according to Eq. 8-17. If frictionless flow is assumed (reasonable for many cases of flow in porous media), the body forces plus the surface force (pressure) must be equivalent to the total force in each direction. In the manner of Harr,[3] the following equations (Euler's equations) in the three coordinate directions are obtained:

$$\frac{\partial u}{\partial t} + u\frac{\partial u}{\partial x} + v\frac{\partial u}{\partial y} + w\frac{\partial u}{\partial z} = X - \frac{1}{\rho}\frac{\partial p}{\partial x} \tag{8-18}$$

$$\frac{\partial v}{\partial t} + u\frac{\partial v}{\partial x} + v\frac{\partial v}{\partial y} + w\frac{\partial v}{\partial z} = Y - \frac{1}{\rho}\frac{\partial p}{\partial y} \tag{8-19}$$

$$\frac{\partial w}{\partial t} + u\frac{\partial w}{\partial x} + v\frac{\partial w}{\partial y} + w\frac{\partial w}{\partial z} = Z - \frac{1}{\rho}\frac{\partial p}{\partial z} - g \tag{8-20}$$

where $X$, $Y$, and $Z$ are body forces per unit mass in each coordinate direction. For steady flow [$u$, $v$, $w$, and $h, \neq f(t)$] the first terms in the left-hand side of each equation vanish. With laminar groundwater flow in the range of validity of Darcy's law, velocities are small (often on the order of 5 ft/yr to 5 ft/day).[4] Thus for steady laminar flow, Eqs. 8-18 through 8-20 reduce to

$$X = \frac{1}{\rho}\frac{\partial p}{\partial x} \qquad Y = \frac{1}{\rho}\frac{\partial p}{\partial y} \qquad Z = \frac{1}{\rho}\frac{\partial p}{\partial z} + g \tag{8-21}$$

In most groundwater flow problems the velocity head is negligible; thus $p$ may be given as $\rho g(h - z)$. Then Eq. 8-21 becomes

$$X = g\frac{\partial h}{\partial x} \qquad Y = g\frac{\partial h}{\partial y} \qquad Z = g\frac{\partial h}{\partial z} \tag{8-22}$$

Remembering that Darcy's law defines $\partial h/\partial x = -u/K$, and so on, it follows that

$$X = -\frac{gu}{K} \qquad Y = -\frac{gv}{K} \qquad Z = -\frac{gw}{K} \tag{8-23}$$

For steady laminar flow, the body forces are linear functions of velocity and Eqs. 8-18 to 8-20 may be written as

$$g\frac{\partial h}{\partial x} = -g\frac{u}{K} \tag{8-24}$$

$$g\frac{\partial h}{\partial y} = -g\frac{v}{K} \tag{8-25}$$

$$g\frac{\partial h}{\partial z} = -g\frac{w}{K} \tag{8-26}$$

where

$$u = -\frac{K\partial h}{\partial x}$$

$$v = -\frac{K\partial h}{\partial y} \tag{8-27}$$

$$w = -\frac{K\partial h}{\partial z}$$

This demonstrates that the equations of motion fit Darcy's law for steady laminar flow.

The continuity equation may be stated as[20]

$$\frac{\partial \rho}{\partial t} + \partial \frac{(\rho u)}{\partial x} + \partial \frac{(\rho v)}{\partial y} + \partial \frac{(\rho w)}{\partial z} = 0 \tag{8-28}$$

This equation is valid for a compressible fluid with time-dependent properties. In steady compressible flow the first term becomes zero, and for steady incompressible flow the equation becomes

$$\frac{\partial u}{\partial x} + \frac{\partial v}{\partial y} + \frac{\partial w}{\partial z} = 0 \tag{8-29}$$

Now since $u = \partial \phi / \partial x$, and so on, Eq. 8-29 becomes

$$\nabla^2 \phi = \frac{\partial^2 \phi}{\partial x^2} + \frac{\partial^2 \phi}{\partial y^2} + \frac{\partial^2 \phi}{\partial z^2} = 0 \tag{8-30}$$

which is known as the *Laplace equation*. With steady state laminar flow, groundwater motion is completely described by the continuity equation subject to appropriate boundary conditions.

If the hydraulic conductivity $K$ is constant, Eq. 8-30 can be written as

$$\nabla^2 h = 0 \tag{8-31}$$

the expression of steady incompressible flow in a homogeneous isotropic porous medium.

For unsteady flow, the compressibility of both aquifer and water are pertinent. Consider a small element of porous medium that has a volume $\Delta x\ \Delta y\ \Delta z$. Then the term in a continuity equation representing a change in storage is defined by

$$\frac{\partial(\rho n\ \Delta x\ \Delta y\ \Delta z)}{\partial t} \tag{8-32}$$

Presupposing that compressive forces are predominant in the vertical ($z$) direction, lateral changes can be neglected. Thus in terms of the element described, only $\Delta z$ is considered variable. A storage expression written as the sum of three terms involving partial derivatives of the variables $\Delta z$, $\rho$, and porosity $n$ is[2]

$$\frac{\partial(\rho n\ \Delta x\ \Delta y\ \Delta z)}{\partial t} = \left( n\rho \frac{\partial(\Delta z)}{\partial t} + \rho\, \Delta z \frac{\partial n}{\partial t} + n\, \Delta z \frac{\partial \rho}{\partial t} \right) \Delta x\ \Delta y \tag{8-33}$$

The three elements on the right can be expressed in terms of pore pressure $p$, the aquifer compressibility $\alpha$, and the fluid compressibility $\beta$.[2,4]

Fluid compressibility is defined as the reciprocal of its bulk modulus of elasticity. It is given by[4]

$$\beta = -\frac{\partial V/V}{\partial p} \tag{8-34}$$

where

$V$ = the volume
$p$ = the pore pressure

If the piezometric surface of a confined aquifer is lowered a distance of one unit, the amount of water released from a column of aquifer of unit horizontal cross-sectional area is defined as the storage coefficient $S$. This is analogous to the specific yield $S_y$ of an unconfined aquifer. Obviously, in Eq. 8-34 $S$ is equivalent to $\partial V$. Further, if the aquifer column is of height $b$, $V = b$. The change in pressure $\partial p$ is equivalent to the negative product of the change in head (one unit) and specific weight of water. Making these substitutions in Eq. 8-34, we find that

$$\beta = \frac{S}{\gamma b} \tag{8-35}$$

Now if the aquifer material is considered elastic, that is, if $\Delta z$ and $n$ can be modified, the volume change can be expressed in terms of alteration in the density of the material due to the difference in packing. Thus

$$\left(\frac{\partial V}{V}\right) = -\left(\frac{\partial \rho}{\rho}\right) \tag{8-36}$$

Introducing Eqs. 8-35 and 8-36 into Eq. 8-34 gives

$$\partial \rho = \frac{\rho S}{b\gamma}\,\partial p \tag{8-37}$$

Next, substituting this expression for $\partial \rho$ in Eq. 8-28, we obtain

$$\frac{\partial(\rho u)}{\partial x} + \frac{\partial(\rho v)}{\partial y} + \frac{\partial(\rho w)}{\partial z} = -\frac{\rho S}{b\gamma}\,\frac{\partial p}{\partial t} \tag{8-38}$$

The left-hand side of this equation can be expanded to

$$\rho\left(\frac{\partial u}{\partial x} + \frac{\partial v}{\partial y} + \frac{\partial w}{\partial z}\right) + \left(u\,\frac{\partial \rho}{\partial x} + v\,\frac{\partial \rho}{\partial y} + w\,\frac{\partial \rho}{\partial z}\right) \tag{8-39}$$

The second term is normally very small compared with the first and can be neglected. The validity of this assumption improves as the flow angle decreases. Using Eq. 8-39 and the foregoing assumption, Eq. 8-38 becomes

$$\frac{\partial u}{\partial x}+\frac{\partial v}{\partial y}+\frac{\partial w}{\partial z}=-\frac{S}{b\gamma}\frac{\partial p}{\partial t} \tag{8-40}$$

or if isotropic conditions prevail,

$$K\nabla^2 h=\frac{S}{b\gamma}\frac{\partial p}{\partial t} \tag{8-41}$$

since from Eq. 8-27 $u=-K\,\partial h/\partial x$, and so on. Inserting $\gamma h$ for $p$ and the transmissivity $T$ for $Kb$ produces

$$\nabla^2 h=\frac{S}{T}\frac{\partial h}{\partial t} \tag{8-42}$$

which is the general equation for unsteady flow in a confined aquifer of constant thickness $b$.

The storage coefficient $S$ and the transmissivity are commonly called the *formation constants* of a confined aquifer. For an unconfined aquifer Eq. 8-42 reverts to

$$\nabla^2 h=\frac{S}{Kb}\frac{\partial h}{\partial t} \tag{8-43}$$

since $b$ is a function of the change in head. The unsteady flow equation for an unconfined aquifer is nonlinear in form. The solution of such an equation is discussed by Jacob.[22] Where variations in saturated thickness of unconfined aquifers are minor, Eq. 8-42 may be used as an approximation.[4]

For unconfined aquifers, the right-hand side of Eq. 8-43 is often negligible so that the equation

$$\nabla^2 h=0 \tag{8-31}$$

is frequently valid for both steady and unsteady flow.

## 8-13 Flowlines and Equipotential Lines

Many problems of practical interest in groundwater hydrology can be considered two-dimensional flow problems. The equation of continuity for steady incompressible flow in an isotropic medium then becomes

$$\frac{\partial u}{\partial x}+\frac{\partial v}{\partial y}=0 \tag{8-44}$$

and

$$\nabla^2 h=\frac{\partial^2 h}{\partial x^2}+\frac{\partial^2 h}{\partial y^2}=0 \tag{8-45}$$

$$\nabla^2\phi = \frac{\partial^2\phi}{\partial x^2} + \frac{\partial^2\phi}{\partial y^2} = 0 \qquad (8\text{-}46)$$

The Laplace equation is satisfied by two conjugate harmonic functions $\phi$ and $\psi$.[3,4] Curves $\phi(x, y)$ = constant are orthogonal to the curves $\psi(x, y)$ = constant. The function $\phi(x, y)$ is the velocity potential, the function $\psi(x, y)$ is known as the *stream function* and is defined by

$$u = \frac{\partial\psi}{\partial y} \qquad v = -\frac{\partial\psi}{\partial x} \qquad (8\text{-}47)$$

Substituting Eq. 8-47 into Eq. 8-44 yields

$$\frac{\partial^2\psi}{\partial x\,\partial y} - \frac{\partial^2\psi}{\partial y\,\partial x} = 0 \qquad (8\text{-}48)$$

It has already been shown that

$$u = \frac{\partial\phi}{\partial x} \qquad v = \frac{\partial\phi}{\partial y}$$

so we can write

$$\frac{\partial\phi}{\partial x} = \frac{\partial\psi}{\partial y} \qquad \frac{\partial\phi}{\partial y} = -\frac{\partial\psi}{\partial x} \qquad (8\text{-}49)$$

These are known as the *Cauchy-Riemann equations*. The stream function satisfies both the equation of continuity and the equations of Cauchy-Riemann. It can also be shown that the Laplace equation is satisfied and therefore[2,3]

$$\nabla^2\psi = \frac{\partial^2\psi}{\partial x^2} + \frac{\partial^2\psi}{\partial y^2} = 0 \qquad (8\text{-}50)$$

Refer now to Fig. 8-4. If V is a velocity vector tangent to a particle

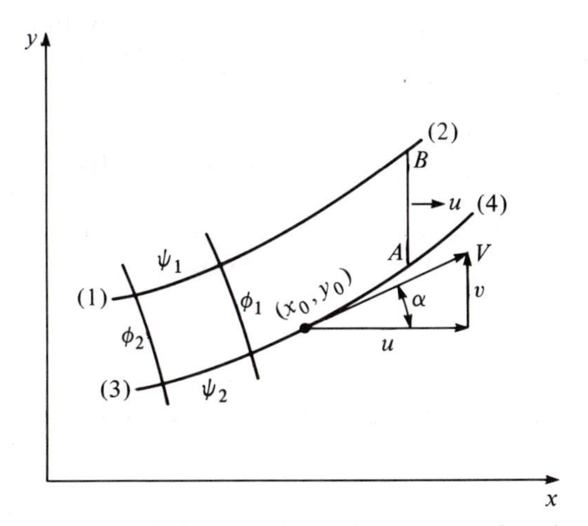


**Fig. 8-4.** *Definition sketch for a stream function.*

flow path 3–4, then it can be decomposed into two components $u$ and $v$.[20] By geometry of the figure

$$\frac{v}{u} = \frac{dy}{dx} = \tan \alpha \tag{8-51}$$

and thus

$$v\,dx - u\,dy = 0 \tag{8-52}$$

If Eqs. 8-47 are substituted into Eq. 8-51, then

$$\frac{\partial \psi}{\partial x}\,dx + \frac{\partial \psi}{\partial y}\,dy = 0 \tag{8-53}$$

The total differential $d\psi$ is equal to zero, and $\psi$ must be a constant. A series of curves $\psi(x, y)$ equal to a succession of constants can be drawn and will be tangent at all points to the velocity vectors. These curves trace the flow path of a fluid particle and are known as *streamlines* or *flowlines*. An important property of the stream function is demonstrated with the aid of Fig. 8-4. Consider the flow crossing a vertical section $AB$ between streamlines defined as $\psi_1$ and $\psi_2$. If the discharge across the section is designated as $Q$, it is apparent that

$$Q = \int_{\psi_2}^{\Psi_1} u\,dy \tag{8-54}$$

or

$$Q = \int_{\Psi_2}^{\Psi_1} d\psi \tag{8-55}$$

and

$$Q = \psi_1 - \psi_2 \tag{8-56}$$

Equation 8-56 illustrates the important property that flow between two streamlines is constant. Streamline spacing reveals the relative magnitudes of flow velocities between them. Higher values are associated with narrower spacings, and vice versa.

The curves in Fig. 8-4 designated as $\phi_1$ and $\phi_2$, called *equipotential lines*, are determined by velocity potentials $\phi(x, y) =$ constant. These curves interesect the flowlines at right angles, illustrated in the following way. The total differential $d\phi$ is given by

$$d\phi = \frac{\partial \phi}{\partial x}\,dx + \frac{\partial \phi}{\partial y}\,dy \tag{8-57}$$

Substituting for terms $\partial\phi/\partial x$ and $\partial\phi/\partial y$ their equivalents $u$ and $v$ makes

$$u\,dx + v\,dy = 0 \tag{8-58}$$

and

$$\frac{dy}{dx} = -\frac{u}{v} \tag{8-59}$$

Thus equipotential lines are normal to flowlines. The system of flowlines and equipotential lines forms a flow net.

One significant point of difference between $\phi$ and $\psi$ functions is that equipotential lines exist only when the flow is irrotational. For two-dimensional flow the condition of irrotationality is said to exist when the $z$ component of vorticity $\zeta_z$ is zero, or

$$\zeta_z = \left(\frac{\partial v}{\partial x} - \frac{\partial u}{\partial y}\right) = 0 \tag{8-60}$$

Proof of this is given by Eskinazi.[20] Substituting for $u$ and $v$ in Eq. 8-60 in terms of $\phi$, we obtain

$$\frac{\partial^2 \phi}{\partial x\, \partial y} - \frac{\partial^2 \phi}{\partial y\, \partial x} = 0 \tag{8-61}$$

This indicates that when the velocity potential exists, the criterion for irrotationality is satisfied.

Once either streamlines or equipotential lines in a flow domain are determined, the other is automatically known because of the relationships in Eq. 8-49. Thus

$$\psi = \int \left(\frac{\partial \phi}{\partial x}\, dy - \frac{\partial \phi}{\partial y}\, dx\right) \tag{8-62}$$

and

$$\phi = \int \left(\frac{\partial \psi}{\partial y}\, dx - \frac{\partial \psi}{\partial x}\, dy\right)$$

It is enough then to determine only one of the functions, since the other can be obtained using relations Eq. 8-62. The complex potential given by

$$w = \phi + i\psi \tag{8-63}$$

where $i$ is the square root of $-1$ is widely used in analytic flow net analyses.[2,3] Of special importance is the fact that

$$\nabla^2 w = \nabla^2 \phi + i\nabla^2 \psi = 0 \tag{8-64}$$

satisfies the conditions of continuity and irrotationality simultaneously.

Equations presented in this section have been limited to the case of two-dimensional flow. Extension to three dimensions would be obtained in a similar fashion.

## 8-14 Boundary Conditions

To solve groundwater flow problems it is necessary that appropriate boundary conditions be specified. Some of the more commonly encountered ones are described in this section; more comprehensive discussions will be found elsewhere.[13,23]

Boundary conditions discussed can be categorized as follows: impervious boundaries, surfaces of seepage, constant head boundaries, and lines of seepage (free surfaces).

Impervious boundaries may be man-made objects such as concrete dams, rock strata, or soil strata that are highly impervious. In Fig. 8-5 the impervious boundary $AB$ represents such a limit. Since flow cannot cross an impervious boundary, velocity components normal to it vanish and the impervious boundary is a streamline. In other words, at the boundary, $\psi$ = constant.

Next look at the upstream face of the earth dam $BC$. At any point of elevation $y$ along $BC$ the pressure can be assumed hydrostatic, or

$$p = \gamma(h - y) \tag{8-65}$$

The definition of a velocity potential states that

$$\phi = -K\left(\frac{p}{\gamma} + y\right) + C \tag{8-66}$$

Substituting for pressure in Eq. 8-66 yields

$$\phi = -K\left(\frac{\gamma(h - y)}{\gamma} + y\right) + C \tag{8-67}$$

and

$$\phi = -Kh + C \tag{8-68}$$

Thus for a constant reservoir level $h$ and an isotropic medium,

$$\phi = \text{constant}$$

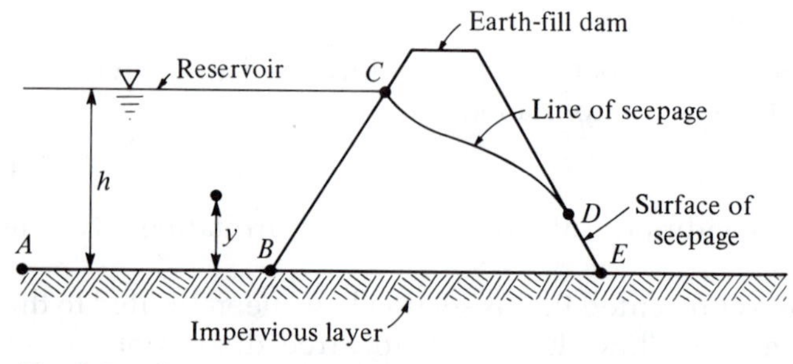


*Fig. 8-5.* Some common boundary conditions.

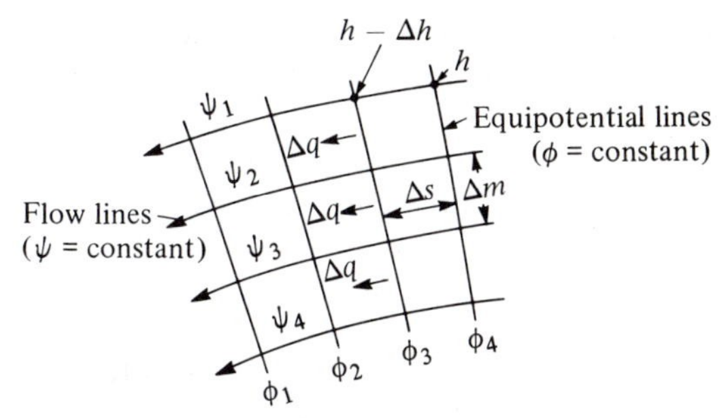


***Fig. 8-6.*** *Segment of an orthogonal flow net.*

and surface $BC$, often termed a *reservoir boundary,* is an equipotential line.

The free surface or line of seepage $CD$ in Fig. 8-5 is seen to be a boundary between the saturated and unsaturated zones. Since flow does not occur across this boundary, it is obviously also a streamline. Pressure along this free surface must be constant, and therefore along $CD$

$$\phi + Ky = \text{constant} \tag{8-69}$$

This is a linear relationship in $\phi$, and therefore equal vertical falls along $CD$ must be associated with successive equipotential drops. One important groundwater flow problem is to determine the location of the line of seepage.

The surface of seepage $DE$ of Fig. 8-5 represents the location at which water seeps through the downstream face of the dam and trickles toward point $E$. The pressure along $DE$ is atmospheric. The surface of seepage is neither a flowline nor an equipotential line.

## 8-15 Flow Nets

*Flow nets,* or graphical representations of families of streamlines and equipotential lines, are widely used in groundwater studies to determine quantities, rates, and directions of flow. The use of flow nets is limited to steady incompressible flow at constant viscosity and density for homogeneous media or for regions that can be compartmentalized into homogeneous segments. Darcy's law must be applicable to the flow conditions.

The manner in which a flow net can be used in problem solving is best explained with the aid of Fig. 8-6. This diagram shows a portion of a flow net constructed so that each unit bounded by a pair of streamlines and equipotential lines is approximately square. The reason for this will be clear later.

A flow net can be determined exactly if functions $\phi$ and $\psi$ are known beforehand. This is often not the case, and as a result, graphically constructed flow nets have been much used. The preparation of a flow net requires application of the concept of square elements and adherence to boundary conditions. Graphical flow nets are usually difficult for a beginner to create, but with reasonable practice an acceptable net can be drawn. Various mechanical methods for graphical flow net construction are presented in the literature and will not be discussed here.[3,23]

After a flow net has been constructed, it can be analyzed using geometry of the net and by applying Darcy's law.

Remembering that $h = (p/\gamma + z)$, we find that Fig. 8-6 shows that the hydraulic gradient $G_h$ between two equipotential lines is given by

$$G_h = \frac{\Delta h}{\Delta s} \tag{8-70}$$

Then applying Darcy's law, in the manner of Todd,[4] the flow increment between adjacent streamlines is

$$\Delta q = K\,\Delta m\left(\frac{\Delta h}{\Delta s}\right) \tag{8-71}$$

where $\Delta m$ represents the cross-sectional area for a net of unit width normal to the plane of the diagram. If the flow net is constructed in an orthogonal manner and composed of approximately square elements,

$$\Delta m \approx \Delta s$$

and

$$\Delta q = K\,\Delta h \tag{8-72}$$

Now if there are $n$ equipotential drops between the equipotential lines, it is evident that

$$\Delta h = \frac{h}{n}$$

where $h$ is the total head loss over the $n$ spaces. If the flow is divided into $m$ sections by the flowlines, then the discharge per unit width of the medium will be

$$Q = \sum_{i=1}^{m} \Delta q = \frac{Kmh}{n} \tag{8-73}$$

When the medium's hydraulic conductivity is known, the discharge can be computed using Eq. 8-73 and a knowledge of flow net geometry.

Where the flow net has a free surface or line of seepage, the entrance and exit conditions given in Fig. 8-5 will be useful. A more comprehensive discussion of these conditions is given in Ref. 24.

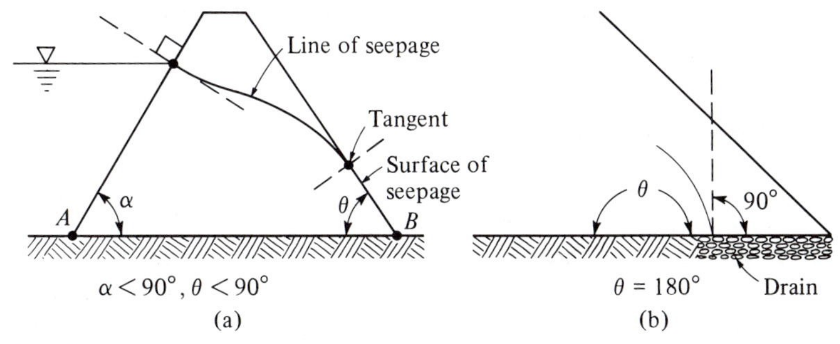


**Fig. 8-7.** *Some entrance and exit conditions for the line of seepage. (After A. Casagrande, "Seepage Through Dams," in* Contributions to Soil Mechanics, 1925–1940, *Boston: Boston Society of Civil Engineers, 1940.)*

Some trouble arises in flow net construction at locations where the velocity becomes infinite or vanishes. Such points are known as *singular points* and according to DeWiest may be placed in three separate categories.[2] In the first classification flowlines and equipotential lines do not intersect at right angles. Such a situation often occurs when a boundary coincides with a flowline; point *A* in Fig. 8-7 is an example.

The second classification has a discontinuity along the boundary that abruptly changes the slope of the streamline. In Fig. 8-8, points *A*, *B*, and *C* represent such discontinuities. At points *A* and *C* the velocity is infinite, while at point *B* it is zero. If the angle of discontinuity measured in a counterclockwise direction inside the flow field is less than 180°, the velocity is zero; if larger than 180°, it is infinite. The angle at *A* is 270°, for example.

The third category includes the case where a source or sink exists in the flow net. Under these circumstances the velocity is infinite, since squares of the flow net approach zero size as the source or sink

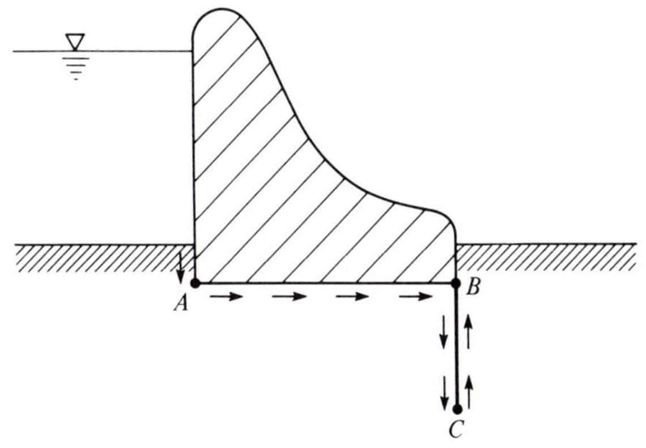


**Fig. 8-8.** *Flowline slope discontinuities.*

is approached. Wells and recharge wells represent sinks and sources in a practical sense and will be discussed later.

## 8-16 Variable Hydraulic Conductivity

It is common for flow within a porous medium of one hydraulic conductivity to enter another region with a different hydraulic conductivity. When such a boundary is crossed, flowlines are refracted. The change in direction that occurs can be determined as a function of the two permeabilities involved in the manner of Todd and DeWiest.[2,4] Figure 8-9 illustrates this.

Consider two soils of permeabilities $K_1$ and $K_2$ which are separated by the boundary $LR$ shown in Fig. 8-9. The direction of the flowlines before and after crossing the boundary is defined by angles $\theta_1$ and $\theta_2$.

For continuity to be preserved, the velocity components in media $K_1$ and $K_2$, which are normal to the boundary, must be equal, since the cross-sectional area at the boundary is $AB$ for a unit depth. Using Darcy's law and noting the equipotential drops $h_a$ and $h_b$,

$$K_1 \frac{\Delta h_a}{AC} \cos \theta_1 = K_2 \frac{\Delta h_b}{BD} \cos \theta_2 \tag{8-74}$$

From the geometry of the figure it is apparent that

$$AC = AB \sin \theta_1$$

$$BD = AB \sin \theta_2$$

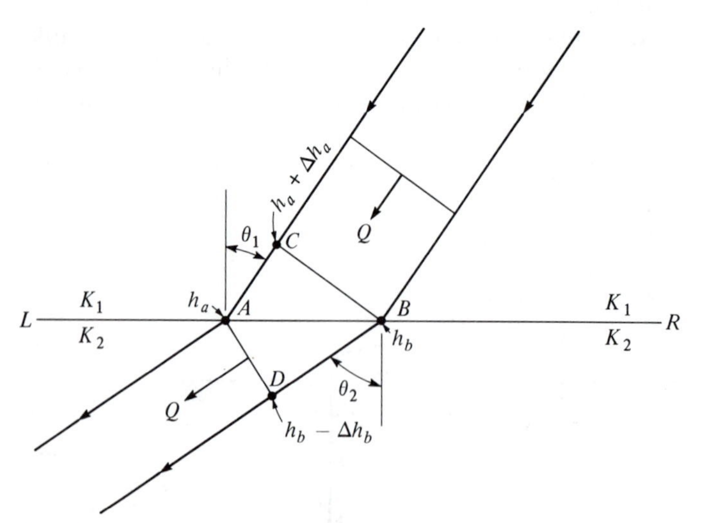


**Fig. 8-9.** *Flowline refraction.*

The head loss between $A$ and $B$ is shown on the figure to be equal to both $\Delta h_a$ and $\Delta h_b$, and since there can be only a single value,

$$\Delta h_a = \Delta h_b$$

Introducing these expressions in Eq. 8-74 produces

$$\frac{K_1}{\tan \theta_1} = \frac{K_2}{\tan \theta_2} \tag{8-75}$$

For refracted flow in a saturated porous medium, the ratio of the tangents of angles formed by the intersection of flowlines with normals to the boundary is given by the ratio of hydraulic conductivities. As a result of refraction, the flow net on the $K_2$ side of the boundary will no longer be squares if the equipotential line spacing $DB$ is maintained. To adjust the net on the $K_2$ side, the relation

$$\frac{\Delta h_b}{\Delta h_a} = \frac{K_1}{K_2} \tag{8-76}$$

can be used where $\Delta h_b \neq \Delta h_a$.

Equipotential lines are also refracted in crossing permeability boundaries. The relationship for this is

$$\frac{K_1}{K_2} = \frac{\tan \alpha_2}{\tan \alpha_1} \tag{8-77}$$

where $\alpha$ is the angle between the equipotential line and a normal to the boundary of permeability.[2]

## 8-17 Anisotropy

In many cases hydraulic conductivity is dependent upon the direction of flow within a given layer of soil. This condition is said to be anisotropic. Sedimentary deposits often fit this aspect, with flow occurring more readily along the plane of deposition than across it. Where the permeability within a plane is uniform but very small across it as compared to that along the plane, a flow net can still be used after proper adjustments are made. A discussion of this is given elsewhere.[2,3,33] Nonhomogeneous aquifers require special consideration but may sometimes be analyzed by using representative or average parameters. A detailed study is outside the scope of this book.[2,3,38]

## 8-18 Dupuit's Theory

Groundwater flow problems in which one boundary is a free surface can be analyzed on the basis of Dupuit's theory of unconfined flow. This theory is founded on two assumptions made by Dupuit in 1863.[14] First, if the line of seepage is only slightly inclined, streamlines may be considered horizontal and, correspondingly, equipotential lines will be essentially vertical. Second, slopes of the line of seepage and the hydraulic gradient are equal. When field conditions are known to be satisfactorily represented by these assumptions, the results obtained according to Dupuit's theory compare very favorably with those arrived at by more rigorous techniques.

Figure 8-10 is useful in translating the foregoing assumptions into a mathematical statement. Consider an element given in the figure which has a base area $dx\,dy$ and a vertical height $h$. Writing the continuity equation in the $x$ direction and considering steady flow to be the case,

$$\text{inflow}_{x_0} = \text{velocity}_{x_0} \times \text{area}_{x_0} \tag{8-78}$$

The velocity at $x = 0$ is given by Darcy's law as

$$u_{x_0} = -K\frac{\partial h}{\partial x} \tag{8-79}$$

Thus the discharge across the element at $x = 0$ is

$$Q_0 = -K\frac{\partial h}{\partial x}\,h\,dy \tag{8-80}$$

The outflow at $x = dx$ is obtained by a Taylor's series expansion as

$$Q_{dx} = -K\frac{\partial h}{\partial x}\,h\,dy + dx\frac{\partial}{\partial x}\left(-K\frac{\partial h}{\partial x}\,h\,dy\right) + \cdots \tag{8-81}$$

Subtracting the outflow from the inflow if $K$ is considered constant, we obtain

$$I_x - O_x = K\,dx\,dy\frac{\partial}{\partial x}\left(h\frac{\partial h}{\partial x}\right) \tag{8-82}$$

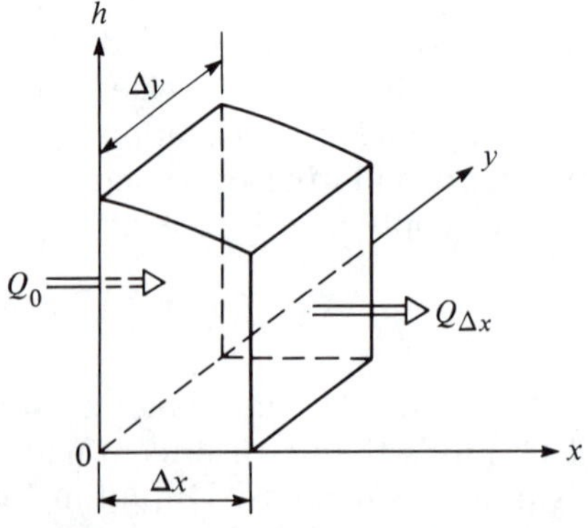


***Fig. 8-10.*** *Definition sketch for development of Dupuit's equation.*

or

$$I_x - O_x = \frac{K\,dx\,dy}{2}\frac{\partial}{\partial x}\left(\frac{\partial h^2}{\partial x}\right) \qquad \textbf{(8-83)}$$

where $dx$ and $dy$ are considered fixed lengths. A similar consideration in the $y$ direction yields

$$I_y - O_y = \frac{K\,dx\,dy}{2}\frac{\partial}{\partial y}\left(\frac{\partial h^2}{\partial y}\right) \qquad \textbf{(8-84)}$$

Assuming that there is no movement in the vertical direction, these are the only components of the inflow and outflow. Further, still dealing with steady flow, the change in storage must be zero. As a result,

$$\frac{K\,dx\,dy}{2}\frac{\partial}{\partial x}\left(\frac{\partial h^2}{\partial x}\right) + \frac{K\,dx\,dy}{2}\frac{\partial}{\partial y}\left(\frac{\partial h^2}{\partial y}\right) = 0 \qquad \textbf{(8-85)}$$

and since $(K\,dx\,dy)/2$ is constant, this reduces to

$$\frac{\partial^2 h^2}{\partial x^2} + \frac{\partial^2 h^2}{\partial y^2} = 0 \qquad \textbf{(8-86)}$$

or

$$\nabla^2 h^2 = 0 \qquad \textbf{(8-87)}$$

Consequently, according to Dupuit's assumptions, Laplace's equation for the function $h^2$ must be satisfied.[25]

In the particular case where recharge is occurring as a result of infiltrated water reaching the water table, a simple adjustment may be made to Eq. 8-86. If the recharge intensity (dimensionally $LT^{-1}$) is specified as $R$, then the total recharge to the element of Fig. 8-10 will be $R\,dx\,dy$ and the continuity equation for steady flow becomes

$$K\frac{dx\,dy}{2}\left(\frac{\partial^2 h^2}{\partial x^2} + \frac{\partial^2 h^2}{\partial y^2}\right) + R\,dx\,dy = 0 \qquad \textbf{(8-88)}$$

or more simply,

$$\nabla^2 h^2 + \frac{2}{K}R = 0 \qquad \textbf{(8-89)}$$

Now, applying Dupuit's theory to the flow problem illustrated on Fig. 8-11b, and assuming one-dimensional flow in the $x$ direction only, the discharge per unit width of the aquifer given by Darcy's law is

$$Q = -Kh\frac{dh}{dx} \qquad \textbf{(8-90)}$$

In this instance $h$ is the height of the line of seepage at any position $x$

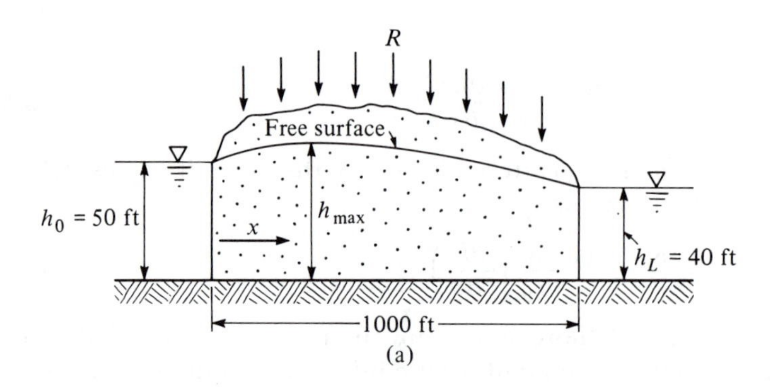


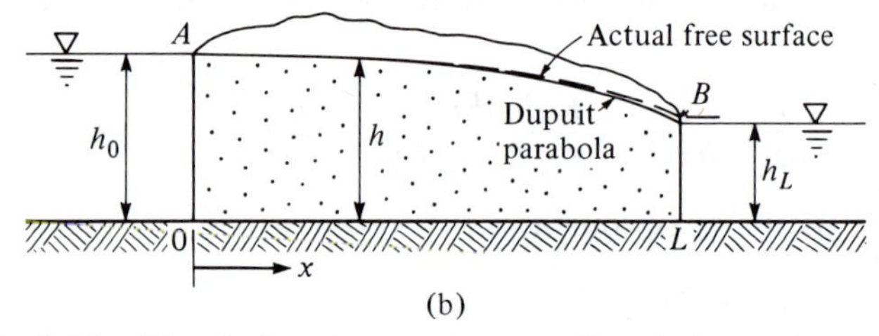


***Fig. 8-11.*** *Steady flow in a porous medium between two water bodies: (a) free surface with infiltration; and (b) free surface without infiltration.*

along the impervious boundary. For the one-dimensional example considered here, Eq. 8-86 becomes

$$\frac{d^2h^2}{dx^2} = 0 \tag{8-91}$$

Upon integration,

$$h^2 = ax + b \tag{8-92}$$

where $a$ and $b$ are constants.

Then for boundary conditions at $x = 0$, $h = h_0$,

$$b = h_0^2 \tag{8-93}$$

Differentiation of Eq. 8-92 yields

$$2h\frac{dh}{dx} = a \tag{8-94}$$

Also from Darcy's equation, $h\,dh/dx = -Q/K$. Making this substitution,

$$a = \frac{-2Q}{K} \tag{8-95}$$

and inserting the values of the constants in Eq. 8-92, we obtain

$$h^2 = -2\frac{Q}{K}x + h_0^2 \tag{8-96}$$

This is the equation of a free surface. It is a parabola (often called *Dupuit's parabola*). If the existence of a surface of seepage at $B$ is ignored, and noting that at $x = L$, $h = h_L$, we find that Eq. 8-96 becomes

$$h_L^2 = -\frac{2QL}{K} + h_0^2 \qquad \textbf{(8-97)}$$

or

$$Q = \frac{K}{2L}(h_0^2 - h_L^2) \qquad \textbf{(8-98)}$$

which is known as the *Dupuit equation.*

***Example 8-3*** Refer to Fig. 8-11a. Given the dimensions shown and a recharge intensity $R$ of 0.01 ft/day, find the discharge at $x = 1000$ ft using Dupuit's equation. Assume that $K = 8$.

*Solution*
Note that

$$\frac{dQ}{dx} = R$$

or

$$Q = Rx + C$$

At $x = 0$,

$$Q = Q_0$$

therefore,

$$Q = R_x + Q_0$$

Also,

$$Q = -Kh\frac{dh}{dx}$$

$$-Kh\frac{dh}{dx} = Rx + Q_0$$

Integrating yields

$$\frac{-Kh^2}{2}\bigg|_{h_0}^{h_L} = \frac{Rx^2}{2}\bigg|_0^L + Q_0x\bigg|_0^L$$

and inserting the limits,

$$\frac{-K(h_L^2 - h_0^2)}{2} = \frac{RL^2}{2} + Q_0L$$

$$Q_0 = \frac{K(h_0^2 - h_L^2)}{2L} - \frac{RL}{2}$$

Then since $Q = Rx + Q_0$,

$$Q = R\left(x - \frac{L}{2}\right) + \frac{K(h_0^2 - h_L^2)}{2L}$$

$$R = 0.01 \times 7.5 = 0.075 \text{ gpd/ft}^2$$

$$Q = 0.075(1000 - 500) + \frac{8(50^2 - 40^2)}{2000}$$

$$= 0.075 \times 500 + \frac{8 \times 900}{2000}$$

$$= 37.5 + 3.6$$

$$= 41.1 \text{ gpd/ft}^2$$

## 8-19 Methods for Developing Groundwater Supplies

Development of groundwater supplies is accomplished mainly through wells or infiltration galleries. Many factors are involved in the performance of these collection works, and a thorough knowledge of groundwater flow mechanics and regional geology is essential. Some groundwater flow problems can be solved by applying relatively simple mathematical tools. Other problems require more rigorous analyses. Graphical studies and model analyses are also widely employed.

### Flow to Wells

A well system can be considered as composed of three elements —the well structure, pump, and discharge piping.[15] The well itself contains an open section through which water enters and a casing to transport the flow to the ground surface. The open section is usually a perforated casing or slotted metal screen permitting water to enter and at the same time preventing collapse of the hole. Occasionally, gravel is placed at the bottom of the well casing around the screen.

When a well is pumped, water is removed from the aquifer immediately adjacent to the screen. Flow then becomes established at locations some distance from the well in order to replenish this withdrawal. Because of flow resistance offered by the soil, a head loss results and the piezometric surface adjacent to the well is depressed, producing a cone of depression (Fig. 8-12), which spreads until equilibrium is reached and steady state conditions are established.

The hydraulic characteristics of an aquifer (which are described by the storage coefficient and aquifer permeability) can be determined by laboratory or field tests. The three most commonly used field methods are the application of tracers, the use of field permeameters, and aquifer performance tests.[4] A discussion of aquifer performance tests will be given here along with the development of flow equations for wells.[11,15,16]

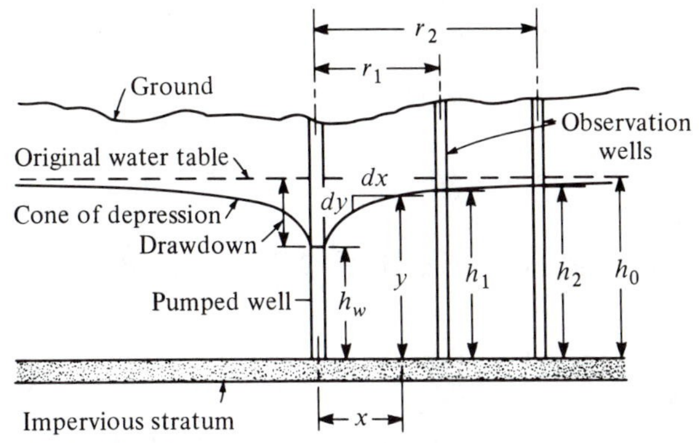


***Fig. 8-12.*** *Well in an unconfined aquifer.*

Aquifer performance tests may be either (1) equilibrium or (2) nonequilibrium tests. In (1) the cone of depression must be stabilized for a flow equation to be derived. For (2) the derivation includes a condition that steady state conditions have not been reached. Adolph Thiem published the first performance tests based on equilibrium conditions in 1906.[8]

## 8-20 Steady Unconfined Radial Flow Toward a Well

The basic equilibrium equation for an unconfined aquifer can be derived using the notation of Fig. 8-12. Here flow is assumed to be radial; the original water table is considered to be horizontal; the well is presumed to fully penetrate the aquifer of infinite aereal extent; and steady state conditions must prevail. Then flow toward the well at any distance $x$ away must equal the product of the cylindrical element of area at that section and the flow velocity. With Darcy's law this becomes

$$Q = 2\pi x y K_f \frac{dy}{dx} \tag{8-99}$$

where

$2\pi xy$ = the area through any cylindrical shell, in ft² with the well as its axis

$K_f$ = the hydraulic conductivity (ft/sec)

$dy/dx$ = the water table gradient at any distance $x$

$Q$ = the well discharge (ft³/sec)

Integrating over the limits specified, we find that

$$\int_{r_1}^{r_2} Q \frac{dx}{x} = 2\pi K_f \int_{h_1}^{h_2} y\, dy \tag{8-100}$$

$$Q \log_e \frac{r_2}{r_1} = \frac{2\pi K_f(h_2^2 - h_1^2)}{2} \tag{8-101}$$

and

$$Q = \frac{\pi K_f(h_2^2 - h_1^2)}{\log_e (r_2/r_1)} \tag{8-102}$$

Converting $K_f$ to the field units of gpd/ft², $Q$ to gpm, and $\log_e$ to $\log_{10}$, we can rewrite Eq. 7-102 as

$$K_f = \frac{1055Q \log_{10} (r_2/r_1)}{h_2^2 - h_1^2} \tag{8-103}$$

If the drawdown in the well does not exceed one-half of the original aquifer thickness $h_0$, reasonable estimates of $Q$ or $K_f$ can be obtained by using Eq. 8-102 or 8-103, even if the height $h_1$ is measured at the well periphery where $r_1 = r_w$, the radius of the well boring.

***Example 8-4*** An 18-in. well fully penetrates an unconfined aquifer of 100-ft depth. Two observation wells located 100 and 235 ft from the pumped well are known to have drawdowns of 22.2 and 21 ft, respectively. If the flow is steady and $K_f$ = 1320 gpd/ft², what would be the discharge?

*Solution*
Equation 8-102 is applicable, and for the given units this is

$$Q = \frac{K(h_2^2 - h_1^2)}{1055 \log_{10} (r_2/r_1)}$$

$$\log_{10} (r_2/r_1) = \log_{10} (235/100) = 0.37107$$

$$h_2 = 100 - 21 = 79 \text{ ft}$$

$$h_1 = 100 - 22.2 = 77.8 \text{ ft}$$

$$Q = \frac{1320(79^2 - 77.8^2)}{1055 \times 0.37107}$$

$$= 634.44 \text{ gpm}$$

## 8-21 Steady Confined Radial Flow Toward a Well

The basic equilibrium equation for a confined aquifer can be obtained in a similar manner, using the notation of Fig. 8-13. The same assumptions apply. Mathematically, the flow in ft³/sec is found from

$$Q = 2\pi x m K_f \frac{dy}{dx} \tag{8-104}$$

Integrating, we obtain

$$Q = 2\pi K_f m \frac{h_2 - h_1}{\log_e (r_2/r_1)} \tag{8-105}$$

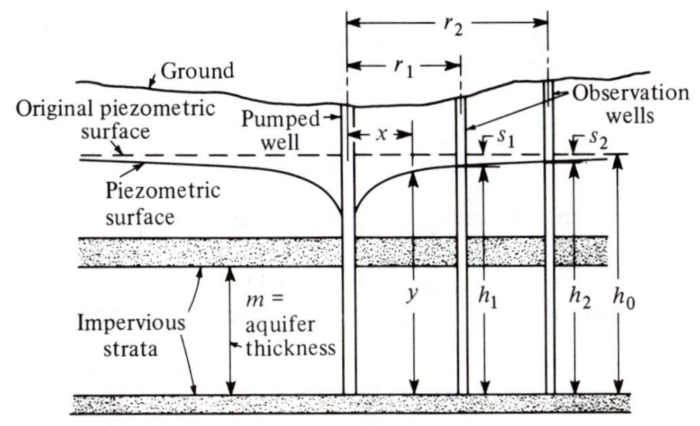


**Fig. 8-13.** *Radial flow to a well in a confined aquifer.*

The coefficient of permeability may be determined by rearranging Eq. 8-105 to the form

$$K_f = \frac{528Q \log_{10}(r_2/r_1)}{m(h_2 - h_1)} \tag{8-106}$$

where

$Q$ = gpm
$K_f$ = the permeability (gpd/ft²)
$r$ and $h$ = ft

***Example 8-5*** Determine the permeability of an artesian aquifer being pumped by a fully penetrating well. The aquifer is 90 ft thick and composed of medium sand. The steady state pumping rate is 850 gpm. The drawdown of an observation well 50 ft away is 10 ft; in a second observation well 500 ft away it is 1 ft.

*Solution*

$$K_f = \frac{528Q \log_{10}(r_2/r_1)}{m(h_2 - h_1)}$$

$$= \frac{528 \times 850 \times \log_{10}(10)}{90 \times (10 - 1)}$$

$$= 554 \text{ gpd/ft}^2$$

## 8-22 Well in a Uniform Flow Field

For a steady state well in a uniform flow field where the original piezometric surface is not horizontal, a somewhat different situation from that previously assumed prevails. Consider the artesian aquifer shown in Fig. 8-14. The heretofore assumed circular area of influence becomes distorted in this case. A solution is possible by applying potential theory; by using graphical means; or, if the slope of the

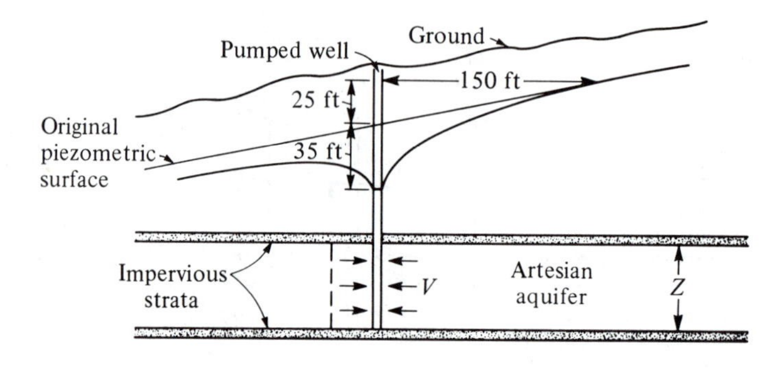


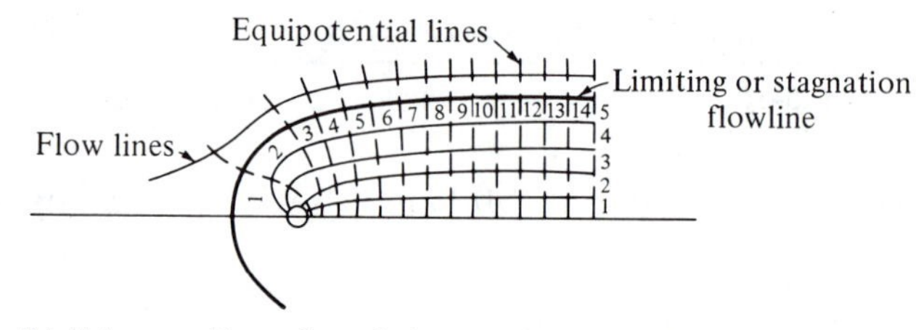


*Fig. 8-14. Well in a uniform flow field and flow net definition.*

piezometric surface is very slight, Eq. 8-105 may be employed without serious error.

Figure 8-14 provides a graphical solution to a uniform flow field problem. First, an orthogonal flow net consisting of flowlines and equipotential lines must be constructed. This should be done so that the completed flow net will be composed of a number of elements that approach little squares in shape. Once the net is complete, it can be analyzed by considering the net geometry and using Darcy's law in the manner of Todd.[4]

***Example 8-6*** Find the discharge to the well of Fig. 8-14 by using an applicable flow net. Consider the aquifer to be 35 ft thick, $K_f = 3.65 \times 10^{-4}$ fps, and other dimensions as shown.

*Solution*

Using Eq. 8-73, we find that

$$q = \frac{Kmh}{n}$$

where

$$h = (35 + 25) = 60 \text{ ft}$$

$$m = 2 \times 5 = 10$$

$$n = 14$$

$$q = \frac{3.65 \times 10^{-4} \times 60 \times 10}{14}$$

$$= 0.0156 \text{ cfs per unit thickness of the aquifer}$$

The total discharge $Q$ is thus

$$Q = 0.0156 \times 35 = 0.55 \text{ cfs or } 245 \text{ gpm}$$

## 8-23 Well Fields

When more than one unit in a well field is pumped, there is a composite effect on the free water surface. This consequence is illustrated by Fig. 8-15 in which the cones of depression are seen to overlap. The drawdown at a given location is equal to the sum of the individual drawdowns.

If within a particular well field, pumping rates of the pumped wells are known, the composite drawdown at a point can be determined. In like manner, if the drawdown at one point is known, the well flows can be calculated.

If the drawndown at a given point is designated as $m$, and subscripts $1, 2, \ldots, n$ are used to relate this drawdown to a particular well (for example, $m_1$ refers to the drawdown for $W_1$) for the total drawdown $m_T$ at some location,[4]

$$m_T = \sum_{i=1}^{n} m_i \tag{8-107}$$

The number of wells, their rate of pumping, and well-field geometry and characteristics determine the total drawdown at a specified location.

Again considering Eq. 8-102, we obtain

$$h_0^2 - h^2 = \frac{Q}{\pi K} \log_e \left(\frac{r_0}{r}\right) \tag{8-108}$$

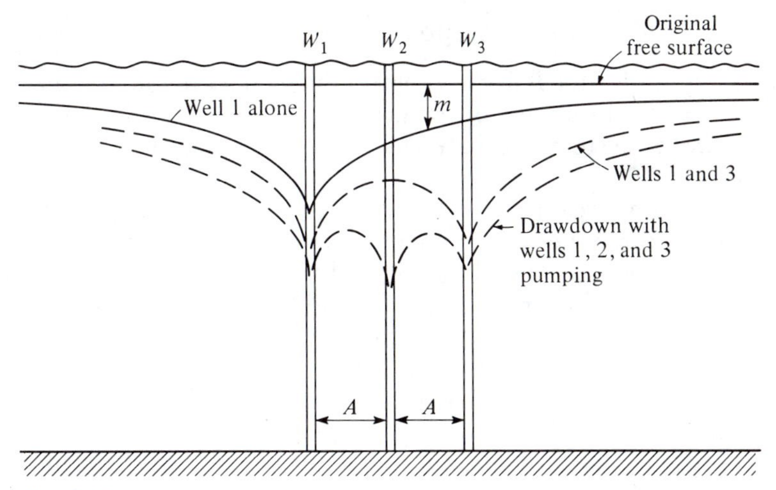


***Fig. 8-15.*** *Combined effect of pumping several wells at equal rates.*

It can be seen that the drawdown for a well pumped at rate $Q$ can be computed if $h_0$, $r_0$, and $r$ are known. It follows then from Eq. 8-107 that for $n$ pumped wells in an unconfined aquifer,

$$h_0^2 - h^2 = \sum_{i=1}^{n} \frac{Q_i}{\pi K} \log_e \left( \frac{r_{0i}}{r_i} \right) \tag{8-109}$$

where

$h_0$ = the original height of the water table
$h$ = the combined effect height of the water table after pumping $n$ wells
$Q_i$ = the flow rate of the $i$th well
$r_{0i}$ = distance of the $i$th well to a location at which the drawdown is considered negligible
$r_i$ = the distance from well $i$ to the point at which the drawdown is being investigated

Todd indicates that values of $r_0$ used in practice often range from 500 to 1000 ft.[4] The impact of this assumption is softened because $Q$ in Eq. 8-108 is not very sensitive to $r_0$. Equation 8-109 should be used only where drawdowns are relatively small.

For flow in a confined aquifer the expression for combined drawdown becomes

$$h_0 - h = \sum_{i=1}^{n} \frac{Q_i}{2\pi K m} \log_e \left( \frac{r_{0i}}{r_i} \right) \tag{8-110}$$

Equations for well flow covering a variety of particular well-field patterns are reported in the literature.[4] Those given here are applicable for steady flow in a homogeneous isotropic medium.

## 8-24 The Method of Images

Some groundwater flow problems subjected to boundary conditions negating the direct use of radial flow equations can be transformed into infinite systems fitting these equations by applying the method of images.[15,22,49]

When a stream is located near a pumped well and the stream and aquifer are interconnected, the drawdown curve of a pumped well may be affected as shown in Fig. 8-16. Another boundary condition often affecting the drawdown of a well is an impervious formation that limits the extent of the aquifer. The cone of depression of a pumped well is not affected until the boundary is intersected. After that, the shape of the drawdown curve will be changed by the boundary. Boundary effects can frequently be evaluated by means of so-called "image wells." The boundary condition is replaced by either a recharging or a discharging well which is pumped or re-

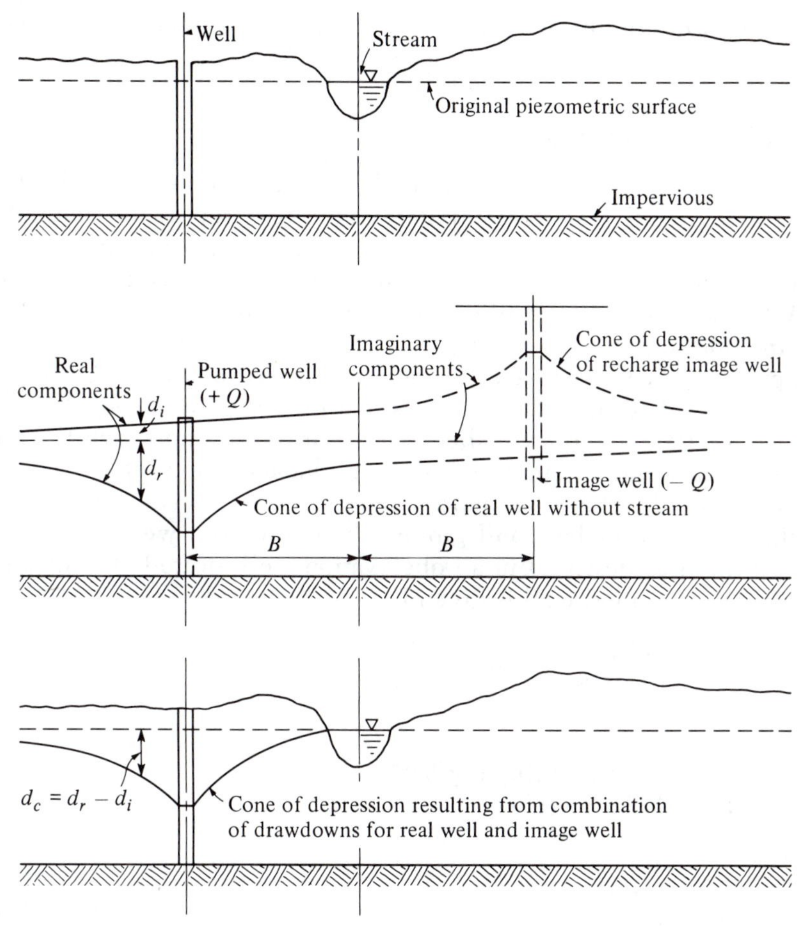


***Fig. 8-16.*** *Drawdown in a pumping well whose aquifer is connected to a stream.*

charged at a rate equivalent to that of the pumped well. That is, in an infinite aquifer, drawdowns of the real and image wells would be identical. The image well is located at a distance from the boundary equal to that of the real well but on the opposite side (Fig. 8-16). Streams are replaced by recharge wells while impermeable boundaries are supplanted by pumped image wells. Computations for the case of a well and impervious boundary directly follow the procedures outlined under the section on well fields. For the well and stream system, the recharge image well is considered to have a negative discharge. The heads are then added according to this sign convention.

The procedure for combining drawdown curves of real and image wells to obtain an actual drawdown curve is illustrated graphically for the example shown in Fig. 8-16. More detailed information on other cases can be found elsewhere.[2,49]

## 8-25 Unsteady Flow

When a new well is first pumped, a large portion of the discharge comes directly from the storage volume released as the cone of depression develops. Under these circumstances the equilibrium equations overestimate permeability and therefore the yield of the well. When steady state conditions are not encountered—as is usually the situation in practice—a nonequilibrium equation must be used. Two approaches can be taken, the rather rigorous method of C. V. Theis or a simplified procedure such as that proposed by Jacob.[9,10]

In 1935 Theis published a nonequilibrium approach that takes into consideration time and storage characteristics of the aquifer.[9] His method utilizes an analogy between heat transfer described by the Biot-Fourier law, and groundwater flow to a well. Theis states that the drawdown (s) in an observation well located at a distance $r$ from the pumped well is given by

$$s = \frac{114.6Q}{T} \int_u^\infty \frac{e^{-u}}{u}\, du \qquad \text{(8-111)}$$

where

$T$ = transmissibility (gpd/ft)
$Q$ = discharge (gpm)

and

$$u = \frac{1.87r^2S_c}{Tt} \qquad \text{(8-112)}$$

where

$S_c$ = the storage coefficient
$t$ = time in days since the start of pumping

The integral in Eq. 8-111 is usually known as the *well function* of $u$ and commonly written as $W(u)$. It may be evaluated from the infinite series

$$W(u) = -0.577216 - \log_e u + u - \frac{u^2}{2 \times 2!} + \frac{u^3}{3 \times 3!} \cdots \qquad \text{(8-113)}$$

The basic assumptions employed in the Theis equation are essentially the same as those in equation 8-102 except for the nonsteady state condition. Some values of this function are given in Table 8-2.

Equations 8-111 and 8-112 can be solved by comparing a log-log

**Table 8-2** *Values of $W(u)$ for Various Values of $u$*

| $u$ | 1.0 | 2.0 | 3.0 | 4.0 | 5.0 | 6.0 | 7.0 | 8.0 | 9.0 |
|---|---|---|---|---|---|---|---|---|---|
| $\times 1$ | 0.219 | 0.049 | 0.013 | 0.0038 | 0.0011 | 0.00036 | 0.00012 | 0.000038 | 0.000012 |
| $\times 10^{-1}$ | 1.82 | 1.22 | 0.91 | 0.70 | 0.56 | 0.45 | 0.37 | 0.31 | 0.26 |
| $\times 10^{-2}$ | 4.04 | 3.35 | 2.96 | 2.68 | 2.47 | 2.30 | 2.15 | 2.03 | 1.92 |
| $\times 10^{-3}$ | 6.33 | 5.64 | 5.23 | 4.95 | 4.73 | 4.54 | 4.39 | 4.26 | 4.14 |
| $\times 10^{-4}$ | 8.63 | 7.94 | 7.53 | 7.25 | 7.02 | 6.84 | 6.69 | 6.55 | 6.44 |
| $\times 10^{-5}$ | 10.94 | 10.24 | 9.84 | 9.55 | 9.33 | 9.14 | 8.99 | 8.86 | 8.74 |
| $\times 10^{-6}$ | 13.24 | 12.55 | 12.14 | 11.85 | 11.63 | 11.45 | 11.29 | 11.16 | 11.04 |
| $\times 10^{-7}$ | 15.54 | 14.85 | 14.44 | 14.15 | 13.93 | 13.75 | 13.60 | 13.46 | 13.34 |
| $\times 10^{-8}$ | 17.84 | 17.15 | 16.74 | 16.46 | 16.23 | 16.05 | 15.90 | 15.76 | 15.65 |
| $\times 10^{-9}$ | 20.15 | 19.45 | 19.05 | 18.76 | 18.54 | 18.35 | 18.20 | 18.07 | 17.95 |
| $\times 10^{-10}$ | 22.45 | 21.76 | 21.35 | 21.06 | 20.84 | 20.66 | 20.50 | 20.37 | 20.25 |
| $\times 10^{-11}$ | 24.75 | 24.06 | 23.65 | 23.36 | 23.14 | 22.96 | 22.81 | 22.67 | 22.55 |
| $\times 10^{-12}$ | 27.05 | 26.36 | 25.96 | 25.67 | 25.44 | 25.26 | 25.11 | 24.97 | 24.86 |
| $\times 10^{-13}$ | 29.36 | 28.66 | 28.26 | 27.97 | 27.75 | 27.56 | 27.41 | 27.28 | 27.16 |
| $\times 10^{-14}$ | 31.66 | 30.97 | 30.56 | 30.27 | 30.05 | 29.87 | 29.71 | 29.58 | 29.46 |
| $\times 10^{-15}$ | 33.96 | 33.27 | 32.86 | 32.58 | 32.35 | 32.17 | 32.02 | 31.88 | 31.76 |

*Source:* After L. K. Wenzel, "Methods for Determining Permeability of Water Bearing Materials with Special Reference to Discharging Well Methods," U.S. Geological Survey, Water-Supply Paper 887, Washington, D.C., 1942.

plot of $u$ versus $W(u)$ known as a *type curve*, with a log-log plot of the observed data $r^2/t$ versus $s$. In plotting type curves, $W(u)$ and $s$ are ordinates, $u$ and $r^2/t$ are abscissas. The two curves are superimposed and moved about until segments coincide. In this operation the axes must remain parallel. A coincident point is then selected on the matched curves and both plots marked. The type curve then yields values of $u$ and $W(u)$ for the desired point. Corresponding values of $s$ and $r^2/t$ are determined from a plot of the observed data. Inserting these values in Eqs. 8-111 and 8-112 and rearranging, values for transmissibility $T$ and storage coefficient $S_c$ can be found.

Often this procedure can be shortened and simplified. When $r$ is small and $t$ large, Jacob found that values of $u$ are generally small.[10]

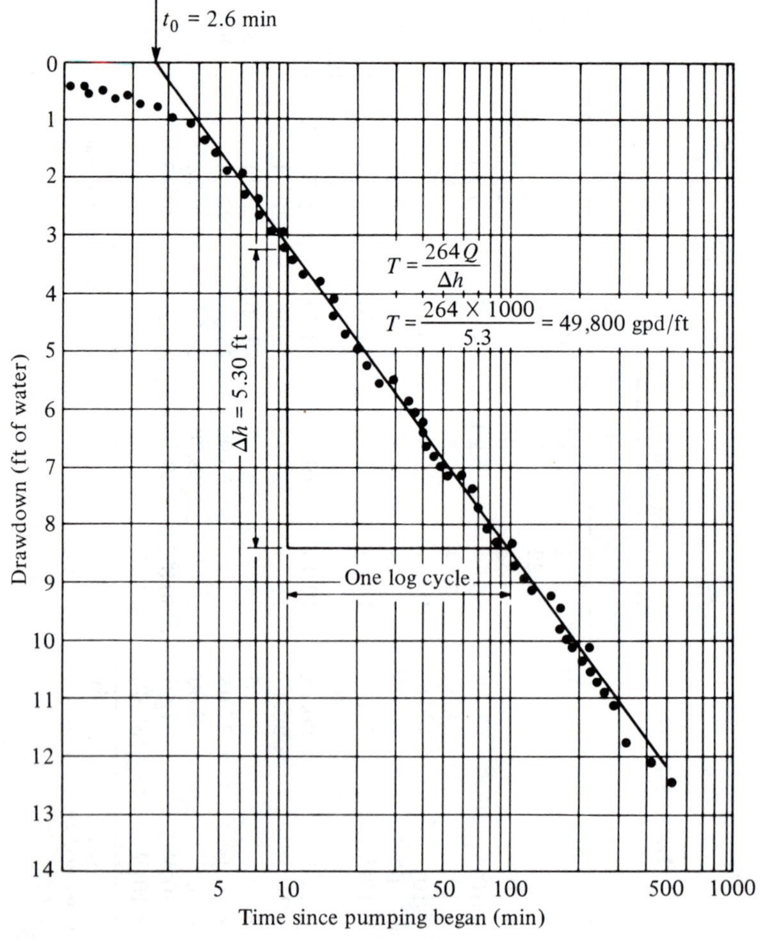


**Fig. 8-17.** *Pumping test data, Jacob method.*

Thus terms in the series of Eq. 8-113 beyond the second one become negligible and the expression for $T$ becomes

$$T = \frac{264Q(\log_{10} t_2 - \log_{10} t_1)}{h_0 - h} \qquad \textbf{(8-114)}$$

which can be further reduced to

$$T = \frac{264Q}{\Delta h} \qquad \textbf{(8-115)}$$

where

$\Delta h$ = the drawdown per log cycle of time $[(h_0 - h)/(\log_{10} t_2 - \log_{10} t_1)]$
$Q$ = well discharge (gpm)
$h_0$ and $h$ = as defined in Fig. 8-13
$T$ = the transmissibility (gpd/ft)

Field data on drawdown $(h_0 - h)$ versus $t$ are drafted on semilogarithmic paper. The drawdown is plotted on an arithmetic scale, Fig. 8-17. This plot forms a straight line whose slope permits computing formation constants using Eq. 8-115 and

$$S_c = \frac{0.3Tt_0}{r^2} \qquad \textbf{(8-116)}$$

with $t_0$ the time corresponding to zero drawdown.

***Example 8-7*** Using the following data, find the formation constants for an aquifer using a graphical solution to the Theis equation. Discharge equals 540 gpm.

| Distance from Pumped Well, $r$ (ft) | $r^2/t$ | Average Drawdown, $s$ (ft) |
|---|---|---|
| 50 | 1,250 | 3.04 |
| 100 | 5,000 | 2.16 |
| 150 | 11,250 | 1.63 |
| 200 | 20,000 | 1.28 |
| 300 | 45,000 | 0.80 |
| 400 | 80,000 | 0.51 |
| 500 | 125,000 | 0.33 |
| 600 | 180,000 | 0.22 |
| 700 | 245,000 | 0.15 |
| 800 | 320,000 | 0.10 |

*Solution*
Plot $s$ versus $r^2/t$ and $W(u)$ versus $u$ as shown in Fig. 8-18. Determine the match point as noted and compute $S_c$ and $T$ using Eqs. 8-111 and 8-112,

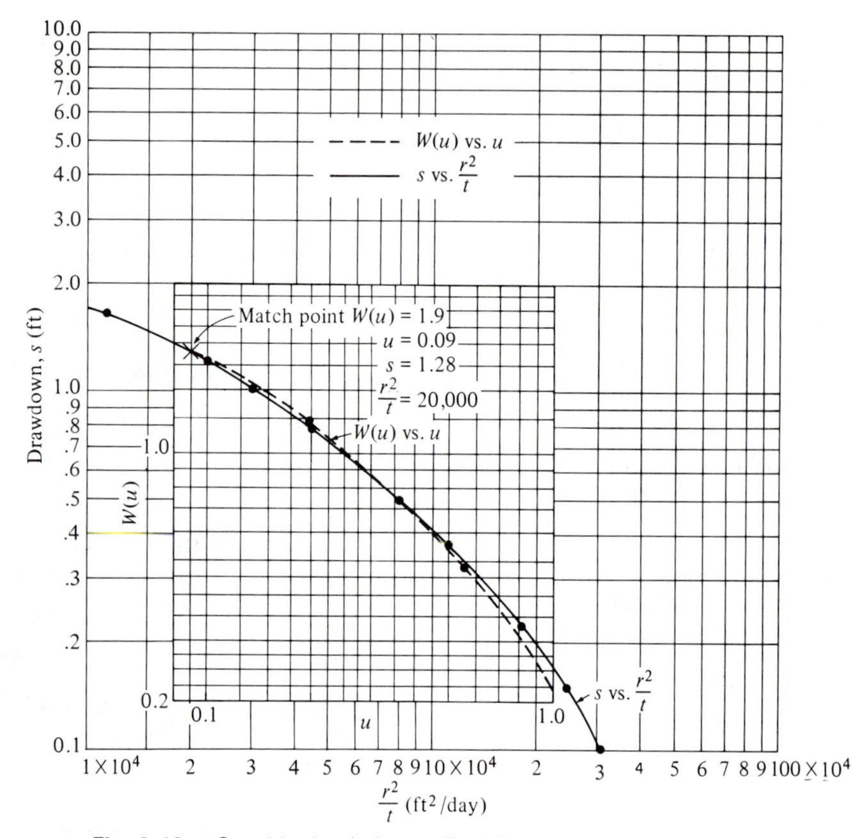


**Fig. 8-18.** *Graphical solution to Theis' equation.*

$$T = \frac{114.6Q}{s} W(u)$$

$$= \frac{114.6 \times 540}{1.28} \times 1.9 = 99{,}500 \text{ gpd/ft}$$

$$S_c = \frac{uT}{1.87r^2/t}$$

$$= \frac{0.09 \times 99{,}500}{1.87 \times 20{,}000} = 0.240$$

***Example 8-8*** Using the data given in Fig. 8-17, find the coefficient of transmissibility $T$ and storage coefficient $S_c$ for an aquifer. Given $Q = 1000$ gpm and $r = 300$ ft.

*Solution*

Find the value of $\Delta h$ from the graph, 5.3 ft. Then by Eq. 8-115

$$T = \frac{264Q}{\Delta h} = \frac{264 \times 1000}{5.3}$$

$$= 49{,}800 \text{ gpd/ft}$$

Using Eq. 8-116, we find that

$$S_c = \frac{0.3Tt_0}{r^2}$$

Note from Fig. 8-16 that $t_0 = 2.6$ min. Converting to days, we find that this becomes

$$t_0 = 1.81 \times 10^{-3} \text{ days}$$

and

$$S_c = \frac{0.3 \times 49{,}800 \times 1.81 \times 10^{-3}}{(300)^2}$$

$$= 0.0003$$

## 8-26 Leaky Aquifers

The foregoing analyses have dealt with free aquifers or those confined between impervious strata. In reality, many cases exist wherein the confining strata are not completely impervious and water is actually transferred from them to the productive aquifer. The flow regime is altered and computations must include leakage. Since about 1930, leaky aquifers have been the subject of research by investigators such as De Glee, Jacob, Hantush, DeWiest, Walton, and others.[27–36,38] A thorough treatment of their work is beyond the scope of this book; interested readers should consult the indicated references.

## 8-27 Partially Penetrating Wells

In many actual situations there is only partial penetration of the well. The question then arises as to the applicability of procedures developed previously for full penetration.

Numerous studies of this problem have been conducted.[13,37,39] In 1957 Hantush reported that steady flow to a well just penetrating an infinite leaky aquifer becomes very nearly radial at a distance from the well of about 1.5 times the aquifer thickness.[39] As depth of penetration increases, the approach to radial flow becomes increasingly apparent. Therefore, computation of drawdowns for partially penetrating wells are made using equations for total penetration with relative safety, provided that the distance from the pumped well is greater than 1.5 times the aquifer thickness. At points closer to the well, it is frequently possible to use a flow net or other relationships developed for this region.[2,4,38]

## 8-28 Salt-Water Intrusion

The contamination of fresh groundwater by the intrusion of salt water often presents a serious quality problem. Islands and coastal regions are particularly vulnerable. Aquifers located inland sometimes contain highly saline waters as well. Fresh water is lighter than salt water (specific gravity of the latter is about 1.025) and forms a fresh water layer above the underlying salt water. This equilibrium is disturbed when an aquifer is pumped, since salt water replaces the fresh water removed. Under equilibrium conditions, a drawdown of 1 ft in a fresh water table corresponds to a rise of about 40 ft by salt water. Wells subjected to salt water intrusion obviously have limited pumping rates.

Recharge wells have been drilled in coastal areas to maintain a head sufficient to preclude sea water intrusion, a practice employed effectively in Southern California.

## 8-29 Computers and Numerical Methods in Groundwater Hydrology

Many advances in the application of computers and numerical methods have taken place since the late 1940s.[41–45]

Electric analogs solve a wide variety of groundwater flow problems,[2,43,45] and consist essentially of a resistance-capacitance network. Table 8-3 indicates the manner in which components of the electric analog and the actual flowfield are related.

Digital computers have also proved to be versatile tools for use in groundwater studies.[2,44,47] The applicable mathematical model is usually written in finite difference or finite element form. Groundwater simulation models will be discussed further in Chapter 10.

***Table 8-3*** *Elements of a Groundwater Reservoir and an Electric Analog Compared*

| Groundwater Reservoir Component | Corresponding Electric Analog Component |
|---|---|
| Hydraulic conductivity | Resistivity |
| Aquifer storage | Capacitance |
| Head | Voltage |
| Volumetric flow rate | Amperage |

***Table 8-4*** *Some Important Forms of Recharge and Discharge*

| Recharge | Discharge |
|---|---|
| Seepage from streams, ponds, lakes | Seepage to lakes, streams, springs |
| Subsurface inflows | Subsurface outflows |
| Infiltrated precipitation | Evapotranspiration |
| Water recharged artificially | Pumping or other artificial means of collection |

## 8-30 Groundwater Basin Development

To utilize groundwater resources efficiently while simultaneously permitting the maximum development of the resource, equilibrium must be established between withdrawals and replenishments. Economic, legal, political, social, and water quality aspects require full consideration.

Lasting supplies of groundwater will be assured only when long-term withdrawals are balanced by recharge during the corresponding period. The potential of a groundwater basin can be assessed by employing the water budget equation,

$$\sum I - \sum O = \Delta S$$

where the inflow $\sum I$ includes all forms of recharge, the total outflow $\sum O$ includes every kind of discharge, and $\Delta S$ represents the change in storage during the accounting period. The most significant forms of recharge and discharge are those listed in Table 8-4.

A groundwater hydrologist must be able to estimate the quantity of water that can be economically and safely produced from a groundwater basin in a specified time period. He should also be competent to evaluate the consequences of imposing various rates of withdrawal on an underground supply.

Development of groundwater basins should be based on careful study, since groundwater resources are finite and exhaustible. If the various types of recharge balance the withdrawals from a basin over a period of time, no difficulty will be encountered. Excessive drafts, however, can deplete underground water supplies to a point where economic development is not feasible. The mining of water will ultimately deplete the entire supply.

## Problems

8-1. What is the Reynolds number for flow in a soil when the water temperature is 50°F, the velocity 0.6 ft/day, and the mean grain diameter 0.08 in.?

8-2. A 12-in. well fully penetrates a confined aquifer 100 ft thick. The coefficient of permeability is 600 gpd/ft². Two test wells located 40 and 120 ft away show a difference in drawdown between them of 9 ft. Find the rate of flow delivered by the well.

8-3. Determine the permeability of an artesian aquifer being pumped by a fully penetrating well. The aquifer composed of medium sand is 130 ft thick. The steady state pumping rate is 1300 gpm. The drawdown in an observation well 65 ft away is 12 ft, and in a second observation well 500 ft away 1.2 ft. Find $K_f$ in gpd/ft².

8-4. Consider a confined aquifer with a coefficient of transmissibility $T = 700$ ft³/(day)(ft). At $t = 5$ min the drawdown = 5.1 ft; at 50 min, $s = 20.0$ ft; at 100 min, $s = 26.2$ ft. The observation well is 60 ft from the pumping well. Find the discharge of the well.

8-5. Assume that an aquifer being pumped at a rate of 300 gpm is confined and pumping test data are given as follows. Find the coefficient of transmissibility $T$ and the storage coefficient $S$. Assume $r = 55$ ft.

| Time since pumping started (min) | 1.3 | 2.5 | 4.2 | 8.0 | 11.0 | 100.0 |
|---|---|---|---|---|---|---|
| Drawdown $s$ (ft) | 4.6 | 8.1 | 9.3 | 12.0 | 15.1 | 29.0 |

8-6. Given the following data:

$Q = 60{,}000$ ft³/day  $t = 30$ days, $r = 1$ ft
$T = 650$ ft³/(day)(ft)  $S_c = 6.4 \times 10^{-4}$

Assume this to be a nonequilibrium problem. Find the drawdown $s$. Note for

$u = 8.0 \times 10^{-9}$  $W(u) = 18.06$
$u = 8.2 \times 10^{-9}$  $W(u) = 18.04$
$u = 8.6 \times 10^{-9}$  $W(u) = 17.99$

8-7. The water temperature in an aquifer is 58°F and the rate of water movement 1.2 ft/day. The average particle diameter in a porous medium is 0.06 in. Find the Reynolds number and indicate whether Darcy's law applies.

8-8. A laboratory test of a soil gives a standard coefficient of permeability $K_s = 3.78 \times 10^2$ gpd/ft². If the prevailing field temperature is 60°F, find the field coefficient of permeability $K_f$.

8-9. An 18-in. well fully penetrates an unconfined aquifer 100 ft deep. Two observation wells located 90 and 235 ft from the pumped well are known to have drawdowns of 22.5 and 20.6 ft, respectively. If the flow is steady and $K_f = 1300$ gpd/ft², what would be the discharge?

8-10. A confined aquifer 80 ft deep is being pumped under equilibrium conditions at a rate of 700 gpm. The well fully penetrates the aquifer. Water levels in observation wells 150 and 230 ft are 95 and 97 ft, respectively. Find the field coefficient of permeability.

8-11. Given the well and flow net data in the following figure, find the discharge using a flow net solution. The well is fully penetrating and the confined aquifer 50 ft thick; $K_f = 2.87 \times 10^{-4}$ ft/sec.

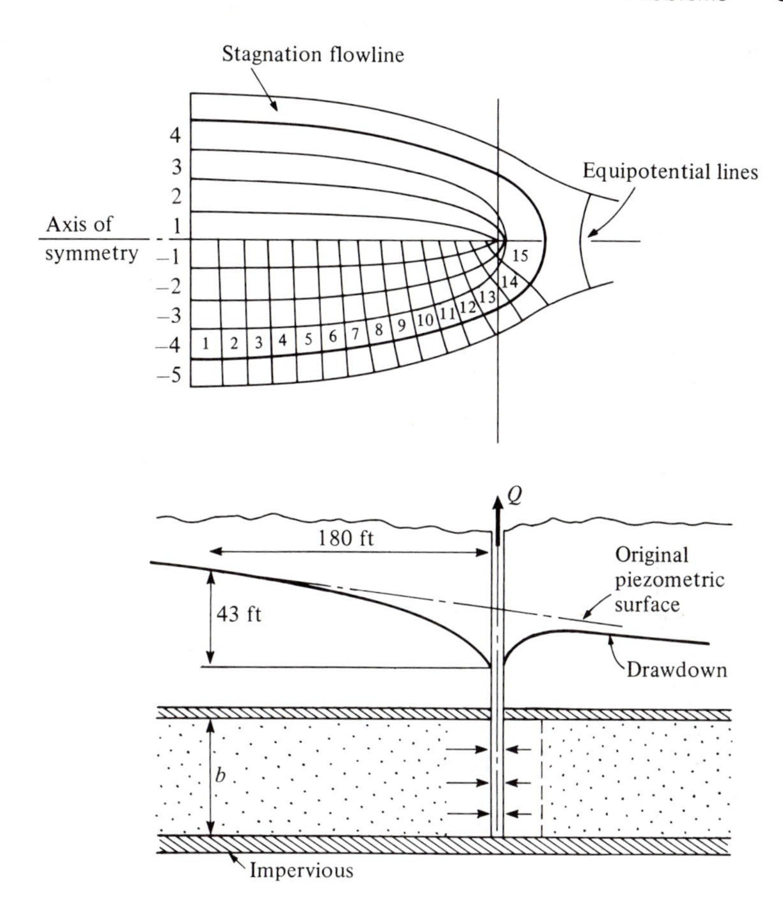


8-12. A well is pumped at the rate of 500 gpm under nonequilibrium conditions. For the data listed, find the formation constants $S$ and $T$. Use the Theis method.

| $r^2/t$ | Average Drawdown, $h$ (ft) |
|---|---|
| 1,250 | 3.24 |
| 5,000 | 2.18 |
| 11,250 | 1.93 |
| 20,000 | 1.28 |
| 45,000 | 0.80 |
| 80,000 | 0.56 |
| 125,000 | 0.38 |
| 180,000 | 0.22 |
| 245,000 | 0.15 |
| 320,000 | 0.10 |

8-13. Given a well pumping at a rate of 590 gpm. An observation well is located at $r = 180$ ft. Find $S$ and $T$ using the Jacob method for the following test data.

| Drawdown (ft) | Time (min) |
|---|---|
| 0.43 | 26 |
| 0.94 | 78 |
| 1.08 | 99 |
| 1.20 | 131 |
| 1.34 | 173 |
| 1.46 | 218 |
| 1.56 | 266 |
| 1.63 | 303 |
| 1.68 | 331 |
| 1.71 | 364 |
| 1.85 | 481 |
| 1.93 | 573 |
| 2.00 | 661 |
| 2.06 | 732 |
| 2.12 | 843 |
| 2.15 | 926 |
| 2.20 | 1034 |
| 2.23 | 1134 |
| 2.28 | 1272 |
| 2.30 | 1351 |
| 2.32 | 1419 |
| 2.36 | 1520 |
| 2.38 | 1611 |

8-14. By employing a finite difference method, find and plot the flowline in the following figure for $\psi = 0.5$.

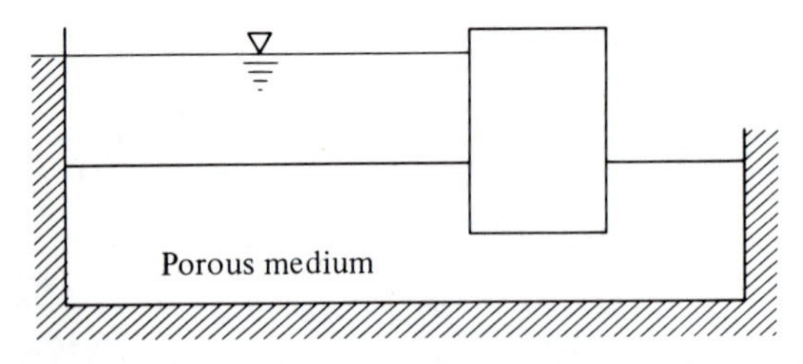


8-15. A 24-in. diameter well penetrates the full depth of an unconfined aquifer. The original water table and a bedrock aquifuge were located 50 ft and 150 ft, respectively, below the land surface. After pumping at a rate of 1700 gpm continuously for 1920 days, equilibrium drawdown conditions were established, and the original water levels in observation wells located 1000 and 100 ft from the center of the pumped well were lowered 10 and 20 ft, respectively. (a) Determine the field permeability (gpd/ft²) of the aquifer.

(b) For the same well, zero drawdown occurred outside a circle with a 10,000-ft radius measured from the center of the pumped well. Inside the circle, the average drawdown in the water table was observed to be 10 ft. Determine the coefficient of storage of the aquifer.

8-16. A well fully penetrates the 100-ft depth of a saturated unconfined aquifer. The drawdown at the well casing is 40 ft when equilibrium conditions are established using a constant discharge of 50 gpm. What is the drawdown when equilibrium is established using a constant discharge of 66 gpm?

8-17. After a long rainless period, the flow in Wahoo Creek decreases by 8 cfs from Memphis downstream 8 mi to Ashland. The stream penetrates an unconfined aquifer, where the water table contours near the creek parallel the west bank and slope to the stream by 0.00020, while on the east side the contours slope away from the stream toward the Lincoln wellfield at 0.00095. Compute the transmissivity of the aquifer knowing $Q = TIL$ where $I$ is the slope and $L$ is the length.

8-18. The time-drawdown data for an observation well located 300 ft from a pumped artesian well (500 gpm) are given in the following table. Find the coefficient of storage ($ft^3$ of water/$ft^3$ of aquifer) and the transmissivity (gpd/ft) of the aquifer by the Theis Method. Use $3 \times 3$ cycle log paper.

| Time (hr) | Drawdown (ft) | Time (hr) | Drawdown (ft) |
|---|---|---|---|
| 1.8 | 0.27 | 9.8 | 1.09 |
| 2.1 | 0.30 | 12.2 | 1.25 |
| 2.4 | 0.37 | 14.7 | 1.40 |
| 3.0 | 0.42 | 16.3 | 1.50 |
| 3.7 | 0.50 | 18.4 | 1.60 |
| 4.9 | 0.61 | 21.0 | 1.70 |
| 7.5 | 0.84 | 24.4 | 1.80 |

8-19. Over a 100-$mi^2$ surface area, the average level of the water table for an unconfined aquifer has dropped 10 ft because of the removal of 128,000 area-ft of water from the aquifer. Determine the storage coefficient for the aquifer. The specific yield is 0.2 and the porosity is 0.22.

8-20. Over a 100-$mi^2$ surface area, the average level of the piezometric surface for a confined aquifer in the Denver area has declined 400 ft as a result of long-term pumping. Determine the amount of water (acre-ft) pumped from the aquifer. The porosity is 0.3 and the coefficient of storage is 0.0002.

## References

1. Clark, J. W., W. Viessman, Jr., and M. J. Hammer, *Water Supply and Pollution Control*, 2nd ed. (New York: Thomas Y. Crowell Company, 1965).
2. DeWiest, R. J. M., *Geohydrology* (New York: John Wiley & Sons, Inc., 1965).
3. Harr, M. E., *Groundwater and Seepage* (New York: McGraw-Hill Book Company, 1962).

4. Todd, D. K., *Groundwater Hydrology* (New York: John Wiley & Sons, Inc., 1960).
5. Buras, Nathan, "Conjunctive Operation of Dams and Aquifers," *ASCE, J. Hydr. Div. Proc.*, 89, No. HY6 (Nov. 1963).
6. Ferris, J. G., "Ground Water," *Mech. Eng.*, Jan. 1960.
7. Thomas, H. E., "Underground Sources of Water," in *Water, The Yearbook of Agriculture.* Washington, D.C.: U.S. Department of Agriculture, 1955.
8. Thiem, G., *Hydrologische Methodern* (Leipzig: Gebhardt, 1906), p. 56.
9. Theis, C. V., "The Relation Between the Lowering of the Piezometric Surface and the Rate and Duration of Discharge of a Well Using Ground Water Storage," *Trans. Am. Geophys. Union,* 16 (1935): 519–524.
10. Cooper, H. H., Jr., and C. E. Jacob, "A Generalized Graphical Method for Evaluating Formation Constants and Summarizing Well-Field History," *Trans. Am. Geophys. Union,* 27 (1946): 526–534.
11. Hoffman, John F., "How Underground Reservoirs Provide Cool Water for Industrial Uses," *Heating, Piping, and Air Conditioning,* Oct. 1960.
12. Meinzer, O. E., "Outline of Methods for Estimating Groundwater Supplies," U.S. Geological Survey, Water-Supply Paper 638-C, Washington, D.C., 1932.
13. Muskat, M., *The Flow of Homogeneous Fluids Through Porous Media* (Ann Arbor, Mich.: J. W. Edwards, Inc., 1946).
14. Dupuit, Jules, *Etudes théoriques et pratiques sur le mouvement des eau dans les canaux de couverts et à travers les terrains perméables,* 2nd ed. (Paris: Dunod, 1863).
15. Hoffman, John F., "Field Tests Determine Potential Quantity, Quality of Ground Water Supply," *Heating, Piping, and Air Conditioning,* Aug. 1961.
16. Hoffman, John F., "Well Location and Design," *Heating, Piping, and Air Conditioning,* Aug. 1963.
17. Richter, R. C., and R. Y. D. Chun, "Artificial Recharge of Ground Water Reservoirs in California," *Proc. ASCE, J. Irrigation and Drainage Div.,* 85, No. IR4 (Dec. 1959).
18. Clendenen, F. B., "A Comprehensive Plan for the Conjunctive Utilization of a Surface Reservoir with Underground Storage for Basin-Wide Water Supply Development: Solano Project California," Doctor of Eng. thesis, University of California, Berkeley, 1959.
19. Darcy, Henri, *Les fontaines publiques de la ville de Dijon* (Paris: V. Dalmont, 1856).
20. Eskinazi, Salamon, *Principles of Fluid Mechanics* (Boston: Allyn and Bacon, Inc., 1962).
21. Kaplan, W., *Advanced Calculus* (Reading, Mass.: Addison-Wesley Publishing Company, Inc., 1952).
22. Jacob, C. E., "Flow of Groundwater," in Hunter Rouse (ed.), *Engineering Hydraulics* (New York: John Wiley & Sons, Inc., 1950).
23. Taylor, D. W., *Fundamentals of Soil Mechanics* (New York: John Wiley & Sons, Inc., 1948).
24. Casagrande, A., "Seepage Through Dams," in *Contributions to Soil Mechanics, 1925–1940* (Boston: Boston Society of Civil Engineers, 1940).

25. Polubarinova-Kochina, P. Ya. *Theory of Groundwater Movement* (Princeton, N.J.: Princeton University Press, 1962).
26. Wenzel, L. K., "Methods for Determining Permeability of Water Bearing Materials with Special Reference to Discharging Well Methods," U.S. Geological Survey, Water-Supply Paper 887, Washington, D.C., 1942.
27. De Glee, G. J., *Over Grondwaterstromingen by Wateronttrekking by middel van Plutten* (Delft: T. Waltman, Jr., 1930, p. 175).
28. Jacob, C. E., "Radial Flow in a Leaky Artesian Aquifer," *Trans. Am. Geophys. Union,* 27 (1946): 198–205.
29. Hantush, M. S., "Plain Potential Flow of Groundwater with Linear Leakage," Ph.D. dissertation, University of Utah, 1949.
30. Hantush, M. S., and C. E. Jacob, "Nonsteady Radial Flow in an Infinite Leaky Aquifer and Nonsteady Green's Functions for an Infinite Strip of Leaky Aquifer," *Trans. Am. Geophys. Union,* 36 (1955): 95–112.
31. Hantush, M. S., and C. E. Jacob, "Flow to an Eccentric Well in a Leaky Circular Aquifer," *J. Geophys. Res.,* 65 (1960): 3425–3431.
32. Hantush, M. S., "Analysis of Data from Pumping Tests in Leaky Aquifers," *Trans. Am. Geophys. Union,* 37 (1956): 702–714.
33. Hantush, M. S., "Modification of the Theory of Leaky Aquifers," *J. Geophys. Res.,* 65 (1960): 3713–3725.
34. DeWiest, R. J. M., "On the Theory of Leaky Aquifers," *J. Geophys. Res.,* 66 (1961): 4257–4262.
35. DeWiest, R. J. M., "Flow to an Eccentric Well in a Leaky Circular Aquifer with Varied Lateral Replenishment," *Geofis. Pura e Aplic.,* 54 (1963): 87–102.
36. Walton, W. C., "Leaky Artesian Aquifer Conditions in Illinois," Report of Investigation No. 39, Illinois State Water Survey, 1960.
37. Kirkham, D., "Exact Theory of Flow into a Partially Penetrating Well," *J. Geophys. Res.,* 64 (1959): 1317–1327.
38. Walton, William C., *Groundwater Resource Evaluation* (New York: McGraw-Hill Book Company, 1970).
39. Hantush, M. S., "Nonsteady Flow to a Well Partially Penetrating an Infinite Leaky Aquifer," *Proc. Isaqi Sci. Soc.,* 1 (1957): 10–19.
40. Todd, David K., "Ground Water Has To Be Replenished," *Chem. Eng. Prog.,* 59, No. 11 (November 1963).
41. Stallman, R. W., "From Geologic Data . . . To Aquifer Analog Models," *Geotimes,* 5, No. 5 (Apr. 1961).
42. Remson, Irwin, Charles A. Appel, and Raymond A. Webster, "Groundwater Models Solved by Digital Computer," *Proc. ASCE, J. Hydr. Div.,* 91, No. HY 3 (May 1965).
43. Stallman, Robert W., "Electric Analog of Three-Dimensional Flow to Wells and Its Application to Unconfined Aquifers," United States Department of the Interior Geological Survey, Open File Report, July 26, 1961.
44. Fayers, F. J., and J. W. Sheldon, "The Use of a High-Speed Digital Computer in the Study of the Hydrodynamics of Geologic Basins," *J. Geophys. Res.,* 67, No. 6 (June 1962): 2421–2431.
45. Brown, Russell H., "Progress in Ground Water Studies with the Electric-Analog Model," *JAWWA* (Aug. 1962): 943–958.

46. Meinzer, O. E., "The Occurrence of Groundwater in the United States," U. S. Geological Survey, Water-Supply Paper No. 489, 1923.
47. Robertson, J. M., *Hydrodynamics in Theory and Application* (Englewood Cliffs, N.J.: Prentice-Hall, Inc., 1965).
48. Shahbazi, M., and D. K. Todd, "Analytic Techniques for Determining Ground Water Flow Fields," Water Resources Center Contribution No. 117, Hydraulic Laboratory, University of California, Berkeley, Aug. 1967.
49. Wisler, C. O., and E. F. Brater, *Hydrology* (New York: John Wiley & Sons, Inc., 1959).

# 9

# Snow Hydrology

## 9-1 Introduction

In many areas snow is the dominant source of streamflow. Mountainous areas in the west are prime examples. Goodell has indicated that about 90% of the yearly water supply in the high elevations of the Colorado Rockies is derived from snowfall.[2] Equally high proportions are also likely in the Sierras of California and numerous regions in the Northwest. A significant but lesser share of the annual water yield in the Northeast and Lake States also originates in snow. It is important that the hydrologist understand the nature and distribution of snowfall and the mechanisms involved in the snowmelt process if he is to make adequate estimates of the streamflow derived from this source.

Snowmelt usually begins in the spring. The runoff derived is normally out of phase with the periods of greatest water need; therefore, various control schemes such as storage reservoirs have been developed to minimize this problem. An additional point of significance is that some of the greatest floods result from combined large-scale rainstorms and snowmelt. Streamflow forecasting is highly dependent upon adequate knowledge of the extent and characteristics of snow fields within the watershed. The water yield from snowfall can be increased by minimizing the vaporization of snow and melt water. Timing the yield can be managed within limits by controlling the rate of snowmelt. Early results can be obtained by speeding the melt process whereas the snowmelt period can be extended or delayed by retarding it. The annual snowfall distribution in the United States is shown in Fig. 9-1.

An adequate understanding of meteorological factors is as much a prerequisite in considering the snowmelt process as it is in dealing with evapotranspiration. The atmosphere supplies moisture for both snowfall and condensation of water vapor on the snowpack, regulates the exchange of energy within a watershed, and is a controlling factor in snowmelt rates.

As in the rainfall-runoff process, geographic, geologic, topographic, and vegetative factors also are operative in the snow accumulation-snowmelt runoff process.[7,8]

For rainfall-runoff relationships, point rainfall measures are used in estimating areal and time distribution over the basin. A similar approach is taken in snow hydrology although the point-areal relationships are usually more complex. Soundly based mathematical equations can be used to determine the various components of snowmelt at a given location. Adequate measures of depth and other snowpack properties can also be obtained at specific locations. The use of these measurements in estimating amount and distribution in area and time of snow over large watershed areas is a much less rigorous procedure. Usually, average conditions related to particular areal subdivisions over time are used as the foundation for basin-wide hydrologic estimates. Such procedures are often in the category of index methods (Sec. 9-6).

## 9-2 Snow Accumulation and Runoff

Under the usual conditions encountered in regions with heavy snowfall, the runoff from the snowpack is the last occurrence in a series of events beginning when the snowfall reaches the ground. The time interval from the start to the end of the process might vary from as little as a day or less to several months or more. Newly fallen snow has a density of about 10% (the percentage of snow volume its water equivalent would occupy) but as the snow depth enlarges, settling and compaction increase the density.[8]

The temperature in a deep layer of accumulated snow is often well below freezing after prolonged cold periods. When milder weather sets in, melting occurs first at the snowpack surface. This initial meltwater moves only slightly below the surface and again freezes through contact with colder underlying snow. During the refreezing process, the heat of fusion released from melt water raises the snowpack temperature. Heat is also transferred to the snowpack from overlying air and the ground. During persistent warm periods, the temperature of the entire snowpack continually rises and finally

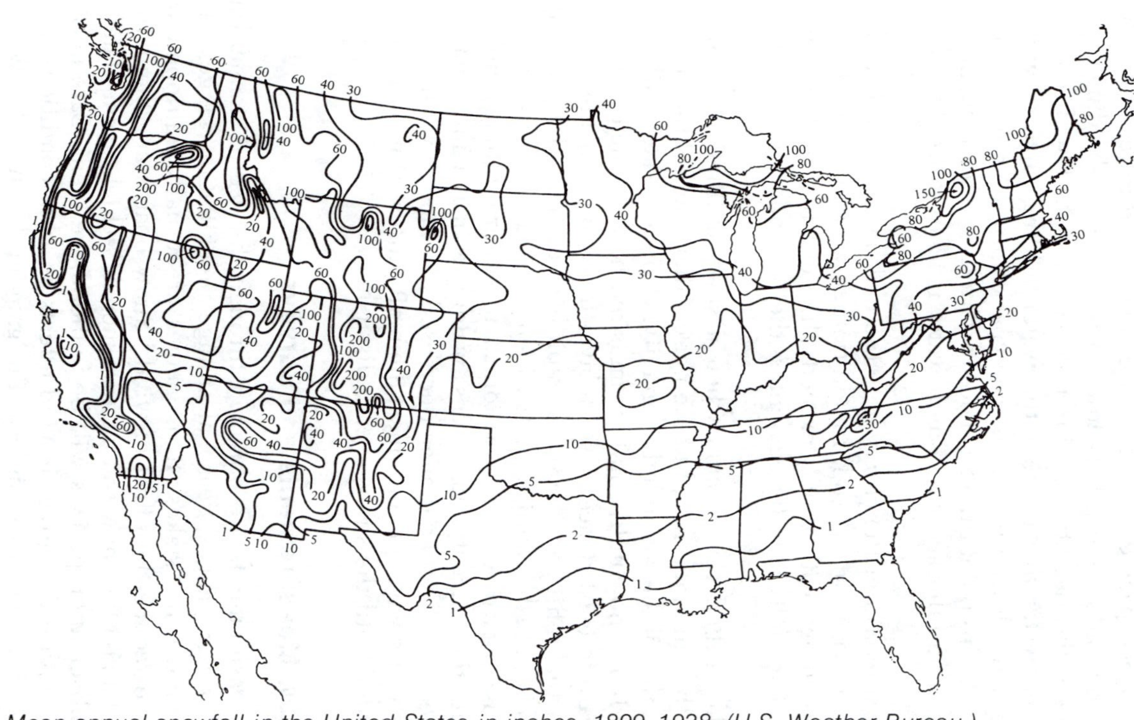

***Fig. 9-1.*** *Mean annual snowfall in the United States in inches, 1899–1938. (U.S. Weather Bureau.)*

reaches 32°F. With continued melting, water begins flowing down through the pack. The initial melt component is retained on snow crystals in capillary films. Once the liquid water-holding capacity of the snow is reached, the snow is said to be *ripe*. Throughout the foregoing process, pack density increases due to the refreezing of meltwater and buildup of capillary films. After the water-holding capacity is reached, the density remains relatively constant with continued melt. Melt water that exceeds the water-holding capacity will continue to move down through the snowpack until the ground is finally reached. At this point runoff can occur. Three situations that may exist at the ground interface when melt water reaches it are described by Horton.[9,8]

First, consider the case where the melt rate is less than the infiltration capacity of the soil. In addition, downward capillary pull of the soil coupled with gravity exceeds the same pull of the snow less gravity. The meltwater directly enters the soil and a slush layer is not formed.

The second case occurs when a soil's infiltration capacity is greater than the melt rate, but the net capillary pull of the snowpack exceeds that of the soil aided by gravity. Capillary water builds up in the overlying snow until equilibrium is reached at which upward and downward forces balance. A slush layer forms and provides a supply of water that infiltrates the soil as rapidly as it enters the slush layer.

The final situation is one in which the melt rate exceeds the infiltration capacity. A slush layer forms and water infiltrates the soil at the infiltration capacity rate. Excess water acts in a manner analogous to surface runoff but at a much decreased overland flow rate.

As warm weather continues, the melt process is maintained and accelerated until the snowcover is dissipated.

## 9-3 Snow Measurements and Surveys

Snow measurements are obtained through the use of standard and recording rain gauges, seasonal storage precipitation gauges, snow boards, and snow stakes. Rain gauges are usually equipped with shields to reduce the effect of wind.[3] Snow boards are about 16 in. square, laid on the snow so that new snowfall which accumulates between observation periods will be found above them. Care must be taken to assure that adverse wind effects or other conditions do not produce an erroneous sample at the gauging location. Snow stakes are calibrated wooden posts driven into the ground for periodic observation of the snow depth or inserted into the snowpack to determine its depth.

Direct measurements of snow depth at a single station are generally not very useful in making estimates of the distribution over large areas, since the measured depth may be highly unrepresentative because of drifting or blowing. To circumvent this problem, snow-surveying procedures have been developed. Such surveys provide information on the snow depth, water equivalent, density, and quality at various points along a snow course. All these measures are of direct use to a hydrologist.

The water equivalent is the depth of water that would weigh the same amount as that of the sample. In this way snow can be described in terms of inches of water. Density is the percentage of snow volume that would be occupied by its water equivalent. The quality of the snow relates to the ice content of the snowpack and is expressed as a decimal fraction. It is the ratio of the weight of the ice content to the total weight. Snow quality is usually about 0.95 except during periods of rapid melt, when it may drop to 0.70 to 0.80 or less. The thermal quality of snow, $Q_t$, is the ratio of heat required to produce a particular amount of water from the snow, to the quantity of heat needed to produce the same amount of melt from pure ice at 32°F. Values of $Q_t$ may exceed 100% at subfreezing temperatures. The density of dry snow is approximately 10% but there is considerable variability between samples. With aging, the density of snow increases to values on the order of 50% or greater.

A snow course includes a series of sampling locations, normally not less than 10 in number.[1] The various stations are spaced about 50 to 100 ft apart in a geometric pattern designed in advance. Points are permanently marked so that the same locations will be surveyed each year—very important if snow course memoranda are to be correlated with areal snow cover and depth, expected runoff potential, or other significant factors. Survey data are obtained directly by foresters and others, by aerial photographs and observations, and by automatic recording stations that telemeter information to a central processing location.

In the western United States the Soil Conservation Service coordinates many snow surveys. Various states, federal agencies, and private enterprises are also engaged in this type of activity. Sources of snow survey data are summarized in Ref. 1.

## 9-4 Point and Areal Snow Characteristics

The estimation of areal snow depth and water equivalent from point measurement data is highly important in hydrologic forecasting.

### Estimates of Areal Distribution of Snowfall

Normally, taking arithmetic averages or using Thiessen polygons does not provide reliable results for estimating areal snow distribution from point gaugings. This is because orographic and topographic effects are often pronounced, and gauging networks frequently are not dense enough to permit the straightforward use of normal averaging techniques. Significantly however, regional orographic effects are relatively constant from year to year and storm to storm for tracts small when compared with the areal extent of general storms occurring in the region.[16] This factor permits many useful approaches in estimating the areal snow distribution once the basic pattern has been found for a region.

One method used to estimate basin precipitation from point observations assumes that the ratio of station precipitation to basin precipitation is approximately constant for a storm or storms. This can be stated as[16]

$$\frac{P_b}{P_a} = \frac{N_b}{N_a} \tag{9-1}$$

or

$$P_b = \frac{P_a N_b}{N_a} \tag{9-2}$$

where

$P_b$ = the basin precipitation
$P_a$ = the observed precipitation at a point or group of stations
$N_b$ = the annual precipitation for the basin
$N_a$ = the normal annual precipitation for the control station or stations

The normal annual precipitation is determined from a map (carefully prepared if it is to be representative) displaying the mean annual isohyets for the region. The precipitation is determined by planimetering areas between the isohyets. If the number of stations used and their distribution adequately depict the basin, Eq. 9-2 can provide a good approximation. For stations not uniformly distributed, weighting coefficients based on the percentage of the basin area portrayed by a gauge are sometimes used in determining $N_a$ for the group.

Another system used in estimating areal snowfall is the isopercental method. In this approach, the storm or annual station precipitation is expressed as a percent of the normal annual total. Isopercental lines are drawn and can be superimposed on a normal annual precipitation map (NAP) to produce new isohyets representing the storm of interest. A NAP map indicates the general nature of the basin's topographic effects, while the isopercental map shows the deviations from this pattern. The advantage of this method over preparing an isohyetal map directly is that relatively consistent storm pattern features of the NAP can be taken into consideration as well as observed individual storm variations.

### Estimates of Basin-Wide Water Equivalent

A hydrologist must be concerned not only with the amount and areal distribution of snowfall, but also with estimating the water equivalent of this snowpack over the basin, since in the final analysis it is this factor which determines runoff. Basin water equivalent may be given as an index or reported in a quantitative manner such as inches depth for the watershed.

The customary procedure for determining the basin water equivalent is to take observed data from snow course stations and to provide an index of basin conditions. Various procedures employ averages, weighted averages, and other approaches to accomplish this.[16] The important point to remember is that the usefulness of any index will be based on how well it represents the overall basin conditions, not on how favorably it describes a particular point value. Indexes do not actually provide a quantitative evaluation of the property they cover. Instead they give relative changes in the factor. By introducing additional data however, an index can be utilized in a prediction equation. For example, if the basin water equivalent can be estimated by subtracting the runoff and loss components from the precipitation input, the index can be correlated with actual basin water equivalent in a quantitative manner.

### Areal Snow Cover

Estimates of the areal distribution of snowfall are very helpful in making hydrologic forecasts. A knowledge of actual areal extent of snow cover on the ground at any given time is also applied in hydrograph synthesis and in making seasonal volumetric forecasts of the runoff. Observations of snow cover are generally obtained by ground and air reconnaissance and photography. Between snow-cover surveys, approximations of the extent of the snow cover are based on available hydrometeorological data. Snowcover depletion patterns within a given basin are normally relatively uniform from year to year; thus snow cover indexes can often be developed from data gathered at a few representative stations.

## 9-5 The Snowmelt Process

The snowmelt process converts ice content into water within the snowpack. Rates differ widely due to variations in causative factors to be discussed later. These divergencies are not as strikingly apparent when considering drainage from the snowpack, however, since the pack itself tends to filter out these nonuniformities so that the drainage exhibits a more consistent rate.

### Energy Sources for Snowmelt

The heat necessary to induce snowmelt is derived from short- and long-wave radiation, condensation of vapor, convection, air and ground conduction, and rainfall. The most important of these sources are convection, vapor condensation, and radiation. Rainfall ranks about fourth in importance while conduction is usually a negligible source.

### Energy Budget Considerations

If snowmelt is considered as a heat transfer process, an energy budget equation can be written to determine the heat equivalent of the snowmelt. Such an equation is of the form[16]

$$H_m = H_{r1} + H_{rs} + H_c + H_e + H_g + H_p + H_q \tag{9-3}$$

where

$H_m$ = the heat equivalent of snowmelt
$H_{r1}$ = net long-wave radiation exchange between the snowpack and surroundings
$H_{rs}$ = the absorbed solar radiation
$H_c$ = the heat transferred from the air by convection
$H_e$ = the latent heat of vaporization derived from condensation
$H_g$ = the heat conduction from the ground
$H_p$ = the rainfall heat content
$H_q$ = the internal energy change in the snowpack

In this equation $H_{rs}$, $H_g$, and $H_p$ are all positive; $H_{r1}$ is usually negative in the open; $H_e$ and $H_q$ may take on positive or negative values; and $H_c$ is normally positive. The actual amount of melt from a snowpack for a given total heat energy is a function of the snowpack's thermal quality. The heat energy required to produce a centimeter of water from pure ice at 32°F is 80 langleys (g-cal/cm$^2$). Therefore, 203.2 langleys are needed to get an inch of runoff from a snowpack of 100% thermal quality. If the term $H_m$ represents the combined total heat input in langleys, an equation for snowmelt $M$ in inches is[16]

$$M = \frac{H_m}{203.2Q_t} \tag{9-4}$$

where $Q_t$ = the thermal quality of the snowpack. Figure 9-2 gives a graphical solution to this equation for several values of $Q_t$.

### Turbulent Exchange

The quantity of heat transferred to a snowpack by convection and condensation is commonly determined from turbulent exchange equations. Such an approach has been widely used, since measurements of temperature and vapor pressure must be made in the tur-

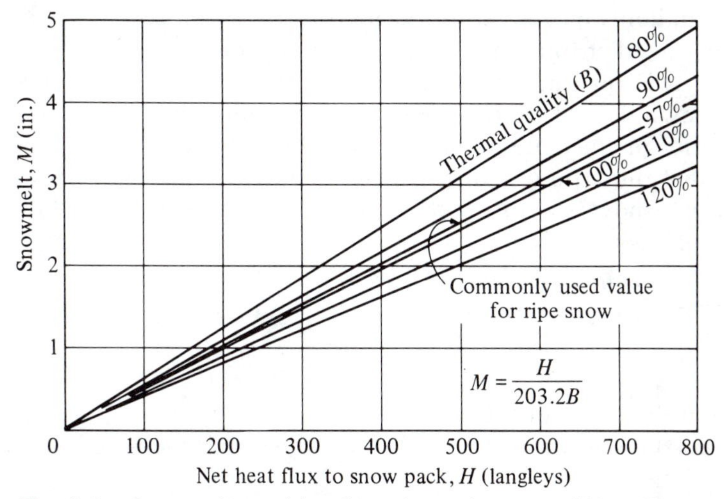


***Fig. 9-2.*** *Snowmelt resulting from thermal energy. (After "Snow Hydrology," North Pacific Division, U.S. Army Corps of Engineers, Portland, Ore., June 30, 1956.)*

bulent zone where vertical water vapor, temperature, and wind velocity gradients are controlled by the action of eddies. In the following two subsections several practical equations for estimating condensation and convection melt will be given. Here a combined theoretical equation is presented to better acquaint the reader with the theory of turbulent exchange.

The basic turbulent exchange equation can be written as[16]

$$Q = A\frac{dq}{dz} \tag{9-5}$$

where

$Q$ = the time rate of flow of a specified property of the air such as water vapor through a unit horizontal area
$dq/dz$ = the vertical gradient of the property
$q$ = the property
$z$ = the elevation
$A$ = an exchange coefficient

Property $q$ must be unaffected by the vertical transport. Properties pertinent to this discussion are the air temperature, water vapor, and wind velocity. Theoretically, the potential temperature should be utilized but air temperatures measured at normal distances above the snowpack will not cause serious errors. The potential temperature of dry air is that which the air would take if brought adiabatically from its actual pressure to a standard pressure.

Gradients of the various properties of importance here follow a power law distribution where conditions of atmospheric stability exist.[17] This qualification is characteristic of the atmosphere's state over snowfields.[16] Logarithmic profiles are more nearly representative of neutral or unstable atmospheric conditions. The power law provides that the ratio of values of a property determined at two levels above the snow is equivalent to the ratio of the levels raised to some power. Thus

$$\frac{q_2}{q_1} = \left(\frac{z_2}{z_1}\right)^{1/n} \tag{9-6}$$

where

$q$ = the value of the property
$z$ = the elevation (with subscripts denoting the level)
$n$ = the power law exponent

If $z_1$ is made equal to 1, $q_1$ assumed to be the property value at this height, and the subscript dropped for the second level, Eq. 9-6 becomes

$$q = q_1 z^{1/n} \tag{9-7}$$

The magnitude of $q$ is taken as the difference in values of $q$ measured at the level $z$ and the snow surface. For example, if $T = 38°F$ at height $z$, and temperature is the property of interest, then $q = (38 - 32) = 6°F$. The gradient of the property $dq/dz$ can be obtained by differentiating Eq. 9-7:

$$\frac{dq}{dz} = \left(\frac{q_1}{n}\right) z^{(1-n)/n} \tag{9-8}$$

If this expression is substituted in the basic turbulent exchange equation (9-5), the following relationship obtains:

$$Q = A\left(\frac{q_1}{n}\right) z^{(1-n)/n} \tag{9-9}$$

Thus eddy exchange of the property at a specified elevation $z$ is determined from observations of the property at unity level. The exchange coefficient is also related to elevation $z$. For equilibrium conditions up to the usual levels of measurement of moisture and temperature, gradients of these variables are such that the eddy transfer of moisture and heat is constant with height. Then the exchange coefficient A must be inversely related to the property gradient, or

$$\frac{A}{A_1} = \frac{\left(\frac{dq}{dz}\right)_1}{dq/dz} \tag{9-10}$$

Substitution in Eq. 9-8 for $dq/dz$ gives the following result:

$$A = A_1 z^{(n-1)/n} \tag{9-11}$$

since $(dq/dz)_1 = (q_1/n)$ for $z = 1$. Now, if the value of $A$ from Eq. 9-11 is inserted in Eq. 9-9,

$$Q = A_1 \left(\frac{q_1}{n}\right) \tag{9-12}$$

The exchange coefficient at an observation level has been shown to be directly proportional to the wind velocity measured at that elevation.[18] Therefore, it may be written that

$$A_1 = kv_1 \tag{9-13}$$

where

$v_1$ = the wind velocity at level one
$k$ = a constant of proportionality

Substituting for $A_1$ in Eq. 9-12 gives

$$Q = \left(\frac{k}{n}\right) q_1 v_1 \tag{9-14}$$

Using the power law equation 9-7 we find that

$$q_1 = qz^{-1/n} \tag{9-15}$$

and

$$v_1 = vz^{-1/n} \tag{9-16}$$

After making these substitutions in Eq. 9-14 and denoting the observation level of $v$ as $z_b$, and that of $q$ as $z_a$ (since these may be different), Eq. 9-14 becomes

$$Q = \left(\frac{k}{n}\right)(z_a z_b)^{-1/n} q_a v_b \tag{9-17}$$

This is a generalized turbulent exchange equation. Consideration will now be given to developing specific theoretical equations for condensation and convection melt.

First, consider the case of the condensation melt. The property to be transported in this case is water vapor, and since the exchange coefficient expresses the transfer of an air mass, it is necessary to determine the moisture content of the air mass. This can be accomplished by using the specific humidity which gives the weight of the water vapor contained in a unit weight of moist air. Equation 9-18 can be used to calculate specific humidity:

$$q = \frac{0.622e}{p_a} \tag{9-18}$$

where

$e$ = the vapor pressure
$p_a$ = the total pressure of the moist air

Inserting this expression in Eq. 9-17 for $q_a$ yields[16]

$$Q_e = \left(\frac{k}{n}\right) (z_a z_b)^{-1/n} \left(\frac{0.622}{p}\right) e_a v_b \tag{9-19}$$

where $Q_e$ = the moisture transfer to the snow surface per unit time.

The vapor pressure property $e_a$ is the difference in vapor pressures between the level $z_a$ and the snow surface. The value of $e_a$ may thus be either positive (condensation) or negative (evaporation). In addition to the condensate on the snow surface, release of the latent heat of vaporization (about 600 cal/g) from the condensate will melt 7.5 g of snow per gram of condensate if the thermal quality is 100%. Multiplying moisture transfer $Q_e$ by 8.5 (melt plus condensate per gram of condensate), the time rate of condensation melt ($M_e$) becomes

$$M_e = 8.5 \left(\frac{k}{n}\right) (z_a z_b)^{-1/n} \left(\frac{0.622}{p}\right) e_a v_b \tag{9-20}$$

The proportionality constant $k$ is a complex function related to the air density and other factors. Since the density of air is a function of elevation, $k$ also varies with height. The constant may be made independent of density, and therefore of elevation, by introducing a factor to compensate directly for the density-elevation relationship. Atmospheric pressure serves to accomplish this, and the equation can be adjusted by multiplying by the ratio $p/p_0$ where $p$ is the pressure at the snowfield elevation and $p_0$ the sea level pressure. Introducing this ratio in Eq. 9-20 and a new constant $K_1$, which is related to sea level pressure, gives[16]

$$M_e = 8.5 \left(\frac{k_1}{n}\right) (z_a z_b)^{-1/n} \left(\frac{0.622}{p_0}\right) e_a v_b \tag{9-21}$$

For convection melt, the property of importance in Eq. 9-17 is air temperature. To convert air temperature into thermal units, the specific heat of air $c_p$ must be introduced. Putting these values in Eq. 9-17, heat transfer by eddy exchange $H_c$ converts to[16]

$$H_c = \left(\frac{k}{n}\right) (z_a z_b)^{-1/n} c_p T_a v_b \tag{9-22}$$

Since the latent heat of fusion is 80 cal/gram, convective snowmelt $M_c$ in grams in the cgs system is given by $H_c/80$, or

$$M_c = \left(\frac{1}{80}\right) \left(\frac{k}{n}\right) (z_a z_b)^{-1/n} c_p T_a v_b \tag{9-23}$$

Introducing the elevation density correction $p/p_0$,

$$M_c = \left(\frac{1}{80}\right)\left(\frac{k_1}{n}\right)(z_a z_b)^{-1/n}\left(\frac{p}{p_0}\right) c_p T_a v_b \tag{9-24}$$

Equations 9-21 and 9-24 can be combined into a single convection-condensation melt $M_{ce}$ equation of the form[16]

$$M_{ce} = \frac{k_1}{n}(z_a z_b)^{-1/n}\left(\frac{1}{80}\frac{p}{p_0} c_p T_a + 8.5\,\frac{0.622}{p_0} e_a\right) v_b \tag{9-25}$$

This is a generalized theoretical equation for snowmelt which results from the turbulent transfer of water vapor and heat to the snowpack. It is assumed that the exchange coefficients for heat and water vapor are equal. Their evaluation is accomplished by experimentation.[16]

A combined physical equation of the general nature of Eq. 9-25 has been developed by Light.[10] Widely used, its individual convection and condensation melt components will be discussed in following sections. The combined form of the Light equation is

$$D = \frac{\rho k_0^2}{80 \ln (a/z_0) \ln (b/z_0)} U\left[c_p T + (e - 6.11)\frac{423}{p}\right] \tag{9-26}$$

where

- $D$ = the effective snowmelt (cm/sec)
- $\rho$ = air density
- $k_0$ = von Kármán's coefficient = 0.38
- $z_0$ = the roughness parameter = 0.25
- $a, b$ = the levels at which the wind velocity, temperature, and vapor pressure are measured, respectively
- $U$ = the wind velocity
- $c_p$ = the specific heat of air
- $T$ = the air temperature
- $e$ = the vapor pressure of the air
- $p$ = the atmospheric pressure

### Convection

Heat for snowmelt is transferred from the atmosphere to the snowpack by convection. The amount of snowmelt by this process is related to temperature and wind velocity. The following equation can be used to estimate the 6-hr depth of snowmelt in inches by convection:[11]

$$D = KV(T - 32) \tag{9-27}$$

where

- $V$ = the mean wind velocity (mph)
- $T$ = the air temperature (°F)

On the basis of the theory of air turbulence and heat transfer (turbulent exchange), a theoretical value for the exchange coefficient $K$ of $0.00184 \times 10^{-0.0000156h}$ has been given by Light.[10] In this relationship $h$, the elevation in feet, is used to reflect the change in barometric pressure due to difference in altitude. The expression is said to represent conditions for an open, level snowfield where measurements of wind and temperature are made at heights of 50 and 10 ft, respectively, above the snow. Values of the expression $10^{-0.0000156h}$ vary from 10.0 at sea level to 0.70 at 10,000 ft of elevation. The actual values of $K$ are normally less than the theoretical figure due to such factors as forest cover. Empirical 6-hr $K$ values have been reported in the literature.[11]

Anderson and Crawford give an expression for the hourly snowmelt due to convection as

$$M = \frac{cV(T_a - 32)}{Q_t} \tag{9-28}$$

where

$M$ = the hourly melt (in.)
$V$ = the wind velocity (mi/hr)
$T_a$ = the surface air temperature (°F)
$Q_t$ = the snow quality
$c$ = a turbulent exchange coefficient determined empirically[8]

Temperature measurements are at 4 ft, wind gauged at 15 ft. The corresponding value of $c$ is reported as 0.0002.

### Condensation

Heat given off by condensing water vapor in a snowpack is often the most important heat source, particularly when temperatures are in the higher ranges (50–60°F). To melt a pound of ice at 32°F, a thermal input of 144 Btu is required. A pound of moisture originating from the condensation process at 32°F, produces about 1073 Btu. On this basis, 1 in. of condensate will produce approximately 7.5 in. of water from the snow. A total yield of around 8.5 in. of snowmelt including the condensate will thus be derived.

A water vapor supply at the snow surface is formed by the turbulent exchange process; consequently, a mass transfer equation similar to those presented for evaporation studies fits the melt process. An equation for hourly snowmelt from condensation takes the form[8]

$$M = \frac{bV}{Q_t}(e_a - 6.11) \tag{9-29}$$

where

$b$ = an empirical constant

$e_a$ = the vapor pressure of the air (mb) the numerical value
6.11 = the saturation vapor pressure (mb) over ice at 32°F

Also, $M$, $Q_t$, and $V$ are as previously defined. The constant $b$ has a value of 0.001 for temperature and wind measurements at 4 and 15 ft, respectively.[8]

A similar expression but for 6-hr snowmelt ($D$) is given as

$$D = K_1 V(e_a - 6.11) \tag{9-30}$$

where the theoretical value of $K_1$ is said by Light to equal 0.00578 if wind and temperature data are obtained at the 50- and 10-ft levels, respectively, and the snowfield is level and open.[10] Actual figures based on observation are generally lower than this due mainly to forest influences. A value of 0.0032 has been reported by Wilson for three study basins in Wyoming. For condensation melt to occur, the dewpoint temperature must exceed 32°F. When it drops below that level, evaporation occurs at the snow surface. An equation for snow evaporation takes the form

$$E = \frac{kV(e_a - e_s)}{Q_t} \tag{9-31}$$

where

$E$ = the hourly evaporation in inches
$e_s$ = the saturation vapor presure over the snow
$k$ = an empirical constant

Also, $V$, $e_a$, and $Q_t$ are as defined before.[8] In the expression $k = 0.0001$, temperature and wind measurements are taken as for Eq. 9-30, and the temperature of the air is assumed equal to that of the snow surface for temperatures below 32°F.

### Radiation Melt

The net amount of short- and long-wave radiation received by a snowpack can be a very important source of heat energy for snowmelt. Under clear skies, the most significant variables in radiation melt are insolation, reflectivity or albedo of the snow, and air temperature. Humidity effects, while existent, are usually not important. When cloud cover exists, striking changes in the amount of radiation from an open snow field are in evidence. The general nature of these effects is illustrated in Fig. 9-3.[16] Combined short- and long-wave radiation exchange as a function of cloud height and cover is represented. Radiation melt is shown to be more significant in the spring than in the winter. It should also be noted that winter radiation melt tends to increase with cloud cover and decreasing cloud height as a result of the more dominant role played by long-wave radiation during that period.

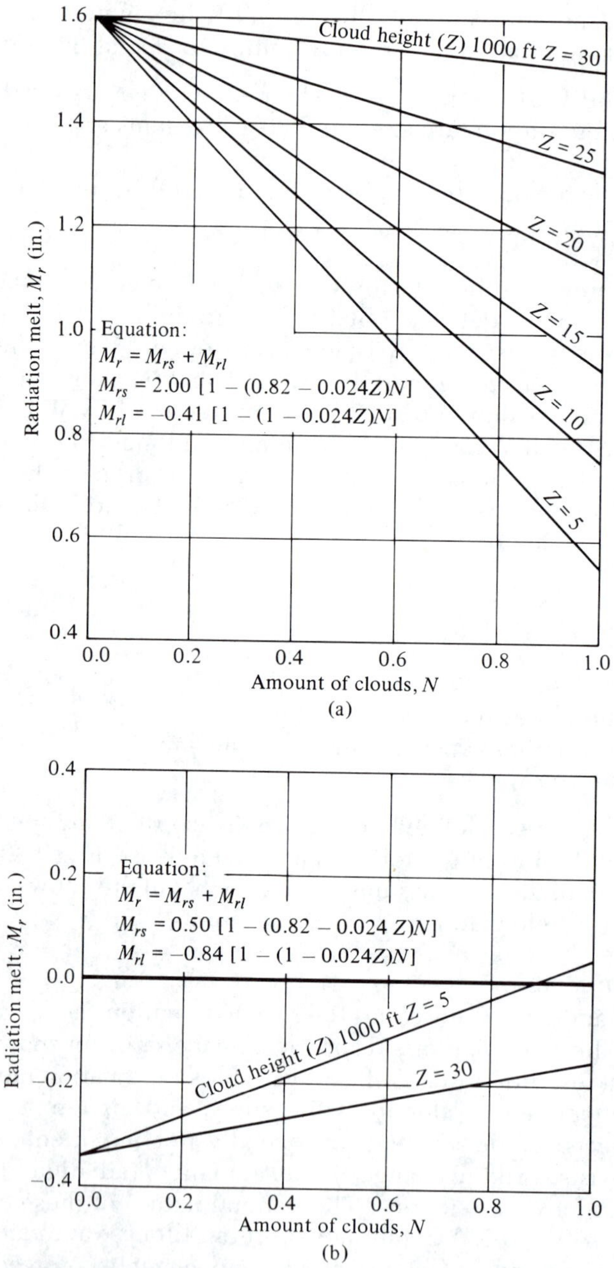


**Fig. 9-3.** *Daily radiation melt in the open with cloudy skies: (a) during spring, May 20; and (b) during winter, February 15. (After "Snow Hydrology," North Pacific Division, U.S. Army Corps of Engineers, Portland, Ore., June 30, 1956.)*

Forest canopies also exhibit important characteristics in regulating radiative heat exchange. These effects differ somewhat from those exhibited by the cloud cover, especially where short-wave radiation is concerned. Clouds and trees both limit insolation, but clouds are very reflective while a large amount of the intercepted insolation is absorbed by the forest. Consequently, the forest is warmed and part of the incident energy directly transferred to snow in the form of long-wave radiation; an additional fraction is transferred indirectly by air also heated by the forest.

Figure 9-4 illustrates some effects of forest canopy on radiation snowmelt. The figure typifies average conditions for a coniferous cover in the middle latitudes.[16] In winter, the maximum radiation melt is associated with complete forest cover, and in spring the greatest radiation melt occurs in the open. Generalizations should not be drawn from these curves which indicate relative seasonal effects of forest cover on radiation melt for the conditions described. Another factor affecting radiation melt is the land slope and its aspect (orientation). Radiation received by north-facing slopes is less than that for south exposure inclines in the northern hemisphere, for example.

Solar energy provides an important source of heat for snowmelt. Above the earth's atmosphere, the thermal equivalent of solar radiation normal to the radiation path is 1.94 langleys/min (1 langley is approximately $3.97 \times 10^{-3}$ Btu/cm$^2$). The actual amount of radiation reaching the snow pack is modified by many factors such as the degree of cloudiness, topography, and vegetal cover. The importance of vegetal cover in influencing snowmelt, long recognized, has prompted many forest management schemes to regulate snowmelt.[1,2,12]

Two basic laws are applicable to radiation. Planck's law states that the temperature of a black body is related to spectral distribution of energy which it radiates. Integration of Planck's law for all wavelengths produces Stefan's law,

$$R_a = \sigma T^4 \tag{9-32}$$

where

$R_a$ = the total radiation
$\sigma$ = Stefan's constant ($0.826 \times 10^{-10}$ langley/min/(°K)$^{-4}$
$T$ = the temperature (°K)

Because snow radiates as a black body, the amount of radiation is related to its temperature (Planck's law), and total energy radiated is according to Stefan's law. Long-wave radiation by a snow pack is determined in a complex fashion through the interactions of temperature, forest cover, and cloud conditions.

Direct solar short-wave radiation received at the snow surface is not all transferred to sensible heat. Part of the radiation is reflected

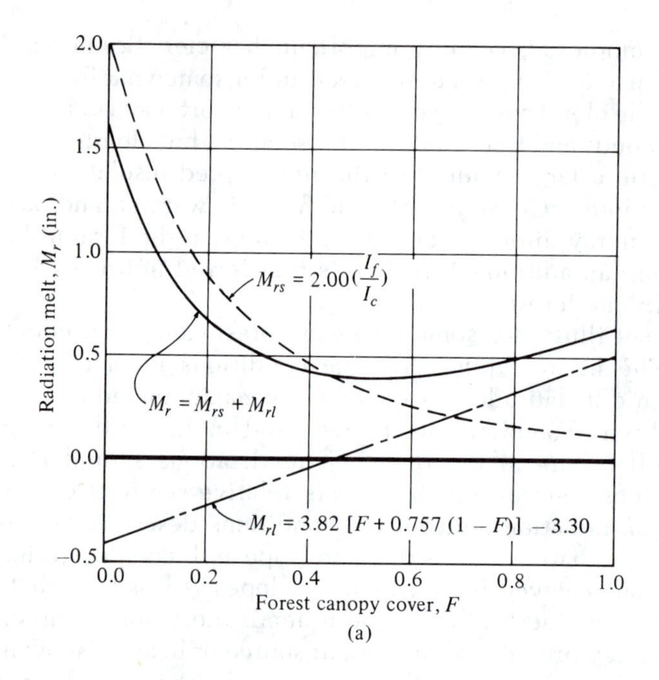


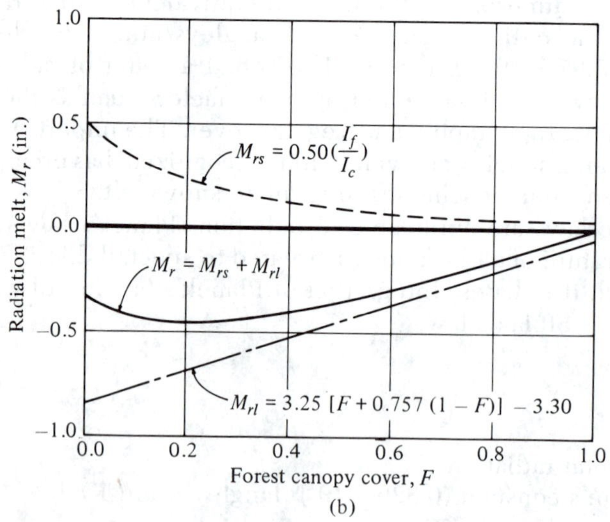


***Fig. 9-4.*** *Daily radiation melt in the forest with clear skies: (a) during spring, May 20; and (b) during winter, February 15. (After "Snow Hydrology," North Pacific Division, U.S. Army Corps of Engineers, Portland, Ore., June 30, 1956.)*

and thus lost for melt purposes. Short-wave reflection is known as *albedo* and ranges from about 40% for melting snow late in the season to approximately 80% for newly fallen snow. Values as high as 90% have also been reported in several cases.[13] This property of the snow pack to reflect large fractions of the insolation explains why the covers persist and air temperatures remain low during clear, sunny, winter periods.

That portion of short-wave radiation not reflected and available for snowmelt may become long-wave radiation, or be conducted within the snowpack. Some heat may also be absorbed by the ground with no resultant melt if the ground is frozen.

An expression for hourly short-wave radiation snowmelt is given as[16]

$$M = \frac{H_m}{203.2Q_t} \tag{9-33}$$

where

$H_m$ = the net absorbed radiation (langleys)
203.2 = a conversion factor for changing langleys to inches of water

When the snow quality is 1, long-wave radiation is exchanged between the snow cover and its surroundings. Snowmelt from net positive long-wave radiation follows Eq. 9-33. If the net long-wave radiation is negative (back radiation), there is an equivalent heat loss from the snowpack.

An approximate method of estimating 12-hr snowmelt $D_{12}$ (periods midnight to noon, noon to midnight) from direct solar radiation has been given by Wilson.[11] The relationship is of the form

$$D_{12} = D_0(1 - 0.75m) \tag{9-34}$$

where

$D_0$ = the snowmelt occurring in a half-day in clear weather
$m$ = the degree of cloudiness (0 for clear weather, 1.0 for complete overcast).

Suggested values for $D_0$ are 0.35 in. (March), 0.42 in. (April), 0.48 in. (May), 0.53 in. (June) within latitudes 40 to 48 degrees.[11]

#### Rainfall

Heat derived from rainfall is generally small, since during those periods when rainfall occurs on a snowpack, the temperature of the rain is probably quite low. Nevertheless, at higher temperatures, rainfall may constitute a significant heat source; it affects the aging process of the snow and frequently is very important in this respect. An equation for hourly snowmelt from rainfall is[8]

$$M = \frac{P(T_w - 32)}{144Q_t} \tag{9-35}$$

where

$P$ = the rainfall (in.)
$T_w$ = the wet bulb temperature assumed to be that of the rain

This equation is based upon the relationship between heat required to melt ice (144 Btu per pound of ice) and the amount of heat given up by a pound of water when its temperature is decreased by one degree.

Daily snowmelt by rainfall estimates are given by

$$M_d = 0.007\, P_d(T_a - 32) \tag{9-36}$$

where

$M_d$ = the daily snowmelt (in.)
$P_d$ = the daily rainfall (in.)
$T_a$ = the mean daily air temperature (°F) of saturated air taken at the 10 ft level[14]

### Conduction

Major sources of heat energy to the snowpack are radiation, convection, and condensation. Under usual conditions, the reliable determination of hourly or daily melt quantities can be founded on these heat sources plus rainfall if it occurs. An additional source of heat, negligible in daily melt computations but perhaps significant over an entire melt season, is ground conduction.

Ground conduction melt is the result of upward transfer of heat from ground to snowpack due to thermal energy which was stored in the ground during the preceding summer and early fall. This heat source can produce meltwater during winter and early spring periods when snowmelt at the surface does not normally occur. Heat transfer by ground conduction can be expressed by the relationship[16]

$$H_q = K\frac{dT}{dz} \tag{9-37}$$

where

$K$ = the thermal conductivity of soil
$dT/dz$ = the temperature gradient perpendicular to soil surface

The snowmelt from ground conduction is generally exceedingly small. Wilson notes that after about 30 days of continuous snow cover, heat transferred from the ground to the snow is insignificant.[11] The amount of snowmelt from ground conduction during a snowmelt season has been estimated at approximately 0.02 in./day.[14] Ground conduction does act to provide moisture to the soil; thus when other favorable conditions for snowmelt occur, a more rapid development of runoff can be expected.

This section has emphasized the physics of snowmelt. The manner in which heat can be provided to initiate the melt process was discussed. Equations 9-27 through 9-31 and 9-33 to 9-37 inclusive can be used to estimate the melt at a given point. The task of computing runoff from snowmelt in a basin cannot be approached in such a simple fashion, since there are many complex factors operative. The remainder of this chapter will be devoted to the general subject of runoff from snowmelt investigations. Figure 9-5 illustrates hourly variation in the principal heat fluxes to a snowpack for a cloudy day.

## 9-6 Snowmelt Runoff Determinations

Various approaches to runoff determination from snowmelt have been followed. They range from relatively simple correlation analyses that completely ignore the physical snowmelt process to relatively sophisticated methods using physical equations. Most techniques can be considered as based on degree-day correlations, analyses of recession curves, correlation analyses, physical equations, or various indexes. Each will be discussed in the following paragraphs.

### Purposes of Snowmelt Runoff Estimates

Snowmelt runoff estimates are extremely important for many regions of the United States and other countries in (1) forecasting seasonal water yields for a diversity of water supply purposes, (2) regulating rivers and storage works, (3) implementing flood control programs, and (4) selecting design floods for particular watersheds. Maximum floods in many areas are often due to a combination of rainfall and snowmelt runoff. In effect, the determination of snowmelt runoff has the same utility as the calculation of runoff from rainfall. In some areas it will, in fact, be the more important of the two.

### Snowpack Condition

The manner in which runoff from either rainfall or snowmelt is affected by conditions prevalent within the snowpack is of primary interest to a hydrologist. Various views on storage characteristics of a snowpack have been advanced.[16] These range from the concept that a snowpack can retain large amounts of liquid water to the hypothesis that snowpack storage is negligible. There is no universally applicable relationship, and it becomes important to base any runoff considerations on a knowledge of the character of a snowpack at the time

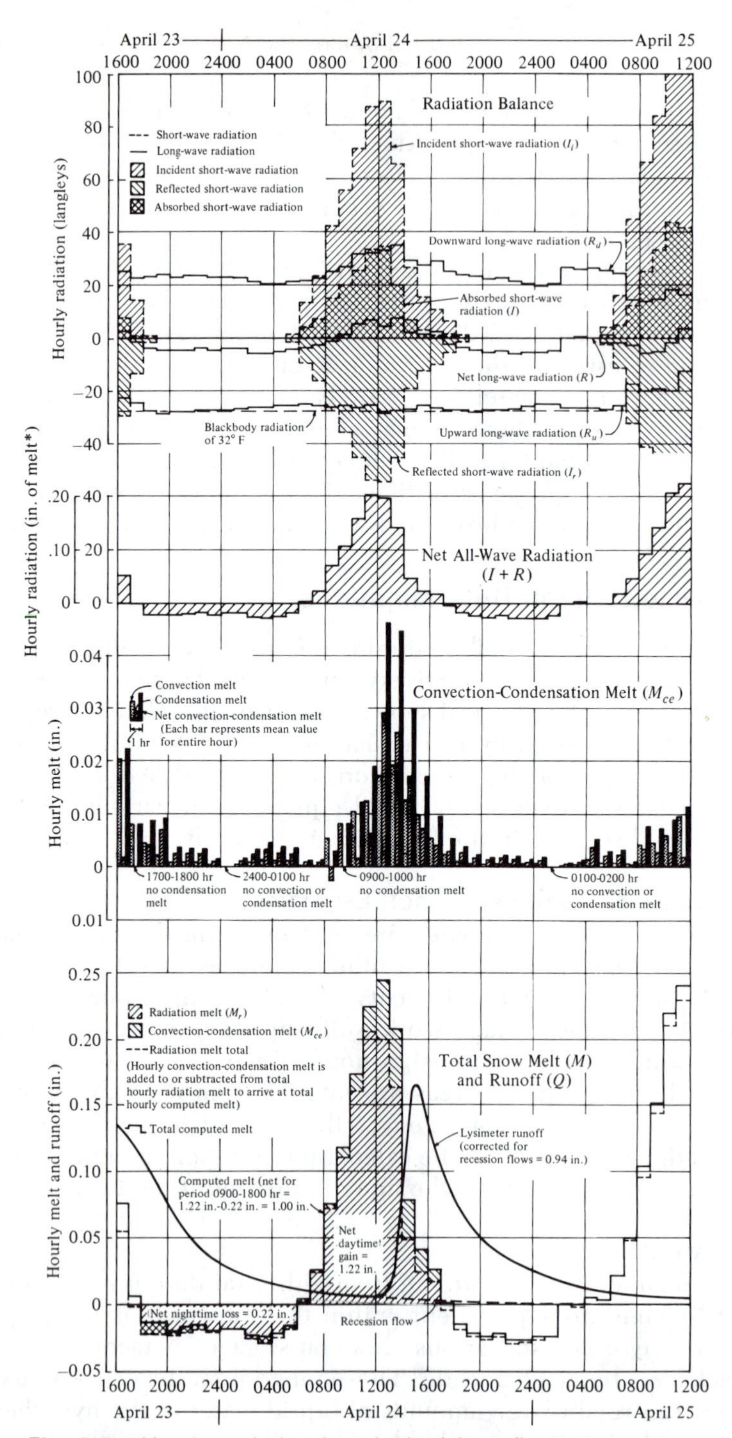


***Fig. 9-5.*** *Hourly variation in principal heat fluxes to a snowpack for a cloudy day. (After "Snow Hydrology," North Pacific Division, U.S. Army Corps of Engineers, Portland, Ore., June 30, 1956.)*

of study. Winter runoff is related to a snowpack's condition whereas in the spring, once active melt begins, little or no delay in the transport of melt or rainfall through the snowpack occurs.

For drainage basins in mountainous areas, snowpack storage effects may be approximated by subdividing the watershed into relatively uniform areas. Normally, this will be accomplished by using elevation zones. Snowpack at the lowest levels may be conditioned to readily transmit rain or meltwater, whereas in higher elevations a liquid water deficit may prevail. At uppermost elevations, the snowpack may be very dry and cold and thus in a condition for the optimum storage of water. The storage potential of the watershed zones must be based on representative measurements of the snow depth, density, temperature, water equivalent, moisture content, and snowpack character. The snowpack character relates to the physical structure of the pack. Unfortunately, adequate measures of all of these factors are not always available or easily obtained. Estimates of changes between sampling periods are usually indexed to readily observed meteorologic variables.

The formulation of snowpack storage and time delay characteristics can be fashioned by assuming a homogeneous pack. In this case storage is related directly to the liquid water deficit and cold content of the pack. Time delay is a function of the inflow rate. It is considered that the snowpack storage potential must be entirely satisfied before runoff begins. In reality this is not the case, but the assumption permits an analysis to be made. As melt proceeds, the storage potential of any snowpack diminishes.

Storage of a snowpack before runoff commences is considered to be the sum of the equivalent water requirement to raise the temperature of the snowpack to 0°C (cold content $W_c$), and the liquid water-holding capacity of the snowpack. If the cold content is given in inches of water needed to bring a snowpack temperature to 0°C, it may be represented by[16]

$$W_c = \frac{W_0 T_s}{160} \tag{9-38}$$

where

$T_s$ = the mean snowpack temperature below zero degrees
$W_0$ = the initial water equivalent of the snowpack in inches for an assumed specific heat of ice of 0.5

The time $t_c$ in hours needed to raise the snowpack temperature to zero is thus given by

$$t_c = \frac{W_0 T_s}{160(i + m)} \tag{9-39}$$

where $i$ is the rainfall intensity (in./hr) and $m$ the rate of melt (in./hr). Storage required to meet the liquid water deficit of the snowpack is given by

$$S_f = \frac{f_p}{100}(W_0 + W_c) \tag{9-40}$$

where

$S_f$ = the amount of water stored (in.)
$f_p$ = the percent deficiency in liquid water of the snowpack

The time in hours $t_f$ needed to fill the storage $S_f$ is given by

$$t_f = \frac{f_p(W_0 + W_c)}{100(i + m)} \tag{9-41}$$

It has been specified that the total storage potential $S_p$ to be met prior to the runoff is given as

$$S_p = W_c + S_f \tag{9-42}$$

This is also known as "permanent" storage, since it is not available to the runoff until the snowpack has finally melted. An additional storage component transitory storage $S_t$ is that water stored in the snowpack while moving through it to become runoff. Until initiation of runoff, the transitory storage in inches can be expressed as

$$S_t = \frac{D(i + m)}{V} \tag{9-43}$$

where

$D$ = the depth of the snowpack (ft)
$V$ = the rate of transmission through the snowpack (ft/hr)

The delay time of water in passing through the snowpack $t_t$ is thus

$$t_t = \frac{D}{V} \tag{9-44}$$

for $t_t$ in hours. Assuming that $W_c$ is very small compared with $W_0$, the depth of the snowpack is given by

$$D = \frac{W_0}{\rho_s} \tag{9-45}$$

with $\rho_s$ the density of the snowpack. Then

$$t_t = \frac{W_0}{\rho_s V} \tag{9-46}$$

Before the runoff commences, the total water $S$ stored in the snowpacks, in inches, is given by

$$S = W_c + S_f + S_t \tag{9-47}$$

which can also be written as

$$S = W_0 \left\{ \frac{T_s}{160} + \frac{f_p}{100} + \frac{(i + m)}{\rho_s V} \right\} \tag{9-48}$$

The total time in hours which passes before runoff is produced is thus[16]

$$t = t_c + t_f + t_t \tag{9-49}$$

or

$$t = W_0 \left\{ \frac{T_s}{160(i + m)} + \frac{f_p}{100(i + m)} + \frac{1}{\rho_s V} \right\} \tag{9-50}$$

After establishing the active runoff from the snowpack, the only significant term in Eq. 9-49 is $t_t$, and this is usually small compared with the overall basin lag and can be neglected. With increased snowmelt and runoff, additional increments of water previously withheld by snow blockage to drainage outlets and other factors are released. Adequate quantification of this cannot be accomplished at present.[16] A deep snowpack, say 15 ft, having a mean temperature of −5°C, could store about 4 in. of liquid water before the onset of runoff. Figure 9-6 illustrates the nature of the water balance in a snowpack during a rainstorm.

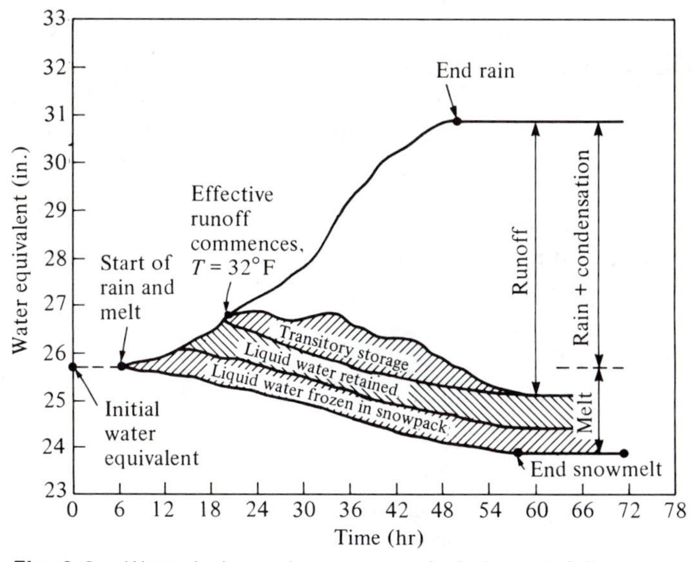


*Fig. 9-6. Water balance in a snowpack during rainfall.*

### Indexes

Hydrologic indexes are made up of hydrologic or meteorologic variables to describe their functioning. The index variable is more easily measured or handier than the element it represents. When mean fixed relationships are known to exist between point measurements and watershed values, indexes can be used to record both areal and temporal aspects of basin values. Indexes serve to permit (1) readily obtainable observations to depict hydrologic variables or processes which themselves cannot be easily measured, and (2) simplification of computational methods by allowing individual observations or groups of observations to replace watershed values in time and space. The adequacy of an index is based on (1) the ability of the index to adequately describe the physical process it represents, (2) the random variability of the observation, (3) the degree to which the point observation is typical of actual conditions, and (4) the nature of variability between the point measurement and basin means.[16] Indexes may be equations or simple coefficients, and variable or constant.

The types of data required to make comprehensive thermal budget studies are normally unavailable in whole or part for watersheds other than ones which themselves are experimental areas. As a result, a hydrologist must make the best use of information at hand. The most commonly available data are daily maximum and minimum temperature, humidity, and wind velocity. Less prevalent are continuous measurements of these data, and few stations record solar radiation or the duration of sunshine. Hourly cloudiness data can sometimes be obtained from local airport weather stations.

A completely general index for reliably describing snowmelt-runoff relationships for all basins has not been established. Most indexes include coefficients valid only for specific topographic, meteorologic, hydrologic, and seasonal conditions and are therefore limited in applicability to other watersheds. Table 9-1 shows some types of indexes that have been successfully used in snowmelt investigations.

The snowmelt runoff equation stated in terms of thermal budget indexes is

$$Y = a + \sum_{i=1}^{n} b_i X_i \tag{9-51}$$

where

$Y$ = the snowmelt runoff
$a$ = a regression constant
the $b_i$'s = the regression coefficients
the $X_i$'s = individual indexes

**Table 9-1** *Some Indexes Used to Describe Thermal Budget Variables*

| Thermal Budget Component | Index |
|---|---|
| Absorbed short-wave radiation | Duration of sunshine data<br>Diurnal temperature range |
| Long-wave radiation* | Air temperature for heavy forested areas<br>For open areas long-wave radiation should be estimated |
| Convective heat exchange | $(T_a - T_b)V$, where $T_a$ is air temperature, $T_b$ the snow surface temperature or base temperature, and $V$ the wind speed |
| Heat of condensation | $(e_a - e_s)V$, where $e_a$ and $e_s$ are vapor pressures of air and snow surface or a base value, and $V$ is wind speed |

* Figure 9-7 illustrates an approximate linear relationship between melt and long-wave radiation used by the U.S. Army Corps of Engineers for index purposes.

Various indexes usable to represent the terms of Eq. 9-51 are selected and a standard regression analysis performed to determine $a$ and $b_i$. It should be noted that every term in the heat budget equation is not always significant for a particular analysis, and thus the number of $X_i$'s will vary for different basins and conditions. A final melt equation developed by the Corps of Engineers for the partly forested Boise River basin above Twin Springs, Idaho, was

$$Q = 0.00238G + 0.0245(T_{max} - 77) \tag{9-52}$$

where

$Q$ = the daily snowmelt runoff (in.) over the snow covered area

$G$ = an estimated value of the daily all-wave radiation exchange in the open (langleys)

$T_{max}$ = the daily maximum temperature at Boise (°F)[16]

The equation is said to predict the daily snowmelt runoff values within 0.11 in. of observed values about 67% of the time. Figure 9-8 illustrates this relationship.

In attempting to develop suitable indexes for snowmelt, a hydrologist should seek the approach most closely resembling the thermal budget of the area, within the limitations of available data.

### Temperature Indexes

The atmospheric temperature is an extremely useful parameter in snowmelt determination. It reflects the extent of radiation and the

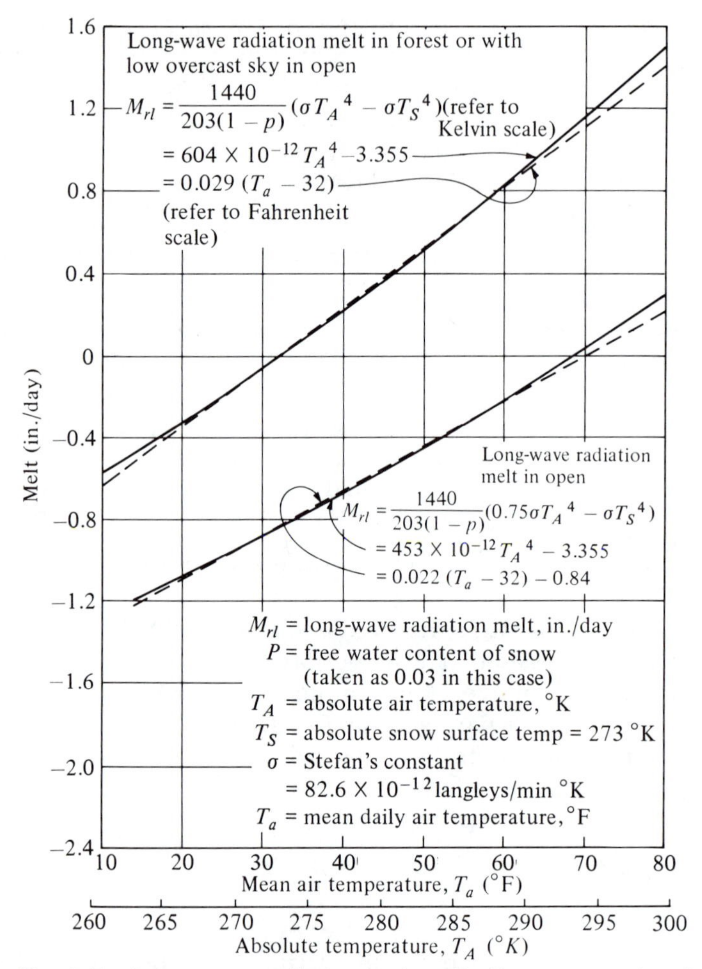


***Fig. 9-7.*** *Long-wave radiation melt, with linear approximation. (After "Snow Hydrology," North Pacific Division, U.S. Army Corps of Engineers, Portland, Ore., June 30, 1956.)*

vapor pressure of the air; it is also sensitive to air motion. Frequently, it is the only adequate meteorological variable regularly on hand so that widespread use has been made of degree-day relationships in snowmelt computations.

A *degree day* is defined as a deviation of 1° from a given datum temperature consistently over a 24-hr period. In snowmelt computations, the reference temperature is usually 32°F. If the mean daily temperature is 43°F, for example, this is equivalent to 11 degree days above 32°F. If the temperature does not drop below freezing during the 24-hr period, there will be 24 degree hr for each degree departure

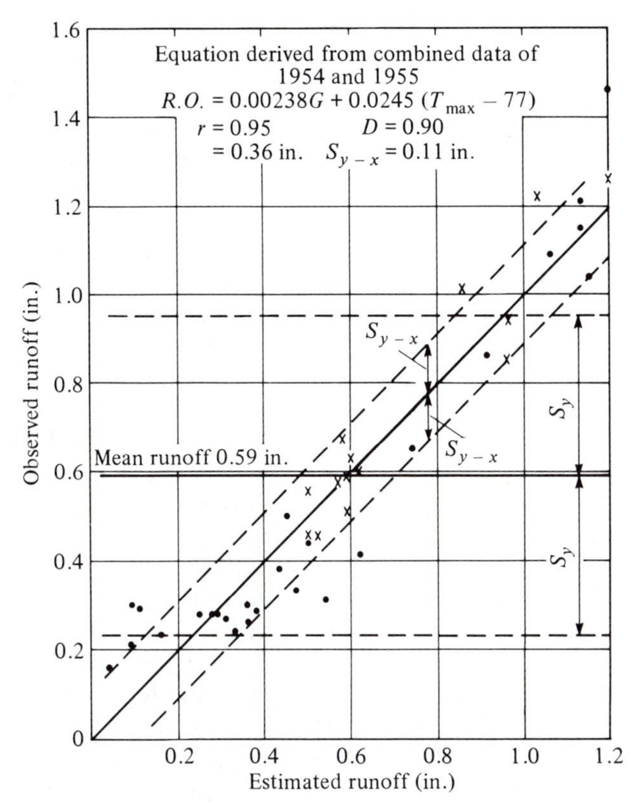


*Fig. 9-8.* *Observed versus estimated runoff.*
*Legend:*
x 1954
• 1955

R.O. = the daily generated snowmelt runoff (in.) depth over a snow-covered area
$G$ = the daily net all-wave radiation absorbed by snow in the open (langleys)
$T_{max}$ = the daily maximum temperature, Boise (°F)
$r$ = the coefficient of correlation
$D$ = the coefficient of determination
$s_y$ = the standard deviation of observed runoff (in.)
$S_{yx}$ = the standard error of the estimated runoff (in.)

*(After U.S. Army Corps of Engineers)*

above 32°F. In this example there would be 264 degree hr for the day of observation.

Various ways of estimating the mean temperature have enabled investigators to take several approaches. One method is to simply average the maximum and minimum daily temperatures. Bases other than 32°F are also used. Regardless of the particular attack employed, a degree hour or degree day is an index to the amount of heat present for snowmelt or other purposes and has proved useful in point-snowmelt and runoff from snowmelt determinations.

The standard practice in developing snowmelt relationships on the basis of temperature is to correlate degree days or degree hours with the snowmelt or basin runoff. In some cases, other factors are introduced to define forest cover effects and/or other influences. Another approach often used is to calculate a degree-day factor—the ratio of runoff or snowmelt to accumulated degree days which produced the runoff or melt. Unfortunately, the degree-day factor has been found to vary seasonally and between basins; therefore, single representative values should be used with caution. Point-degree-day factors for snow-covered basins range from 0.015 to 0.20 in. per degree per day when melting occurs. Gartska states than an average point value of 0.05 can be used to represent spring snowmelt, provided that caution is used.[20] Linsley and others state that basin mean degree-day factors are usually between 0.06 and 0.15 in./degree day under conditions of continuous snow cover and at melting temperatures. Figure 9-9 illustrates temperature index equations for springtime snowmelt or clear and forested areas.[16]

### Generalized Basin Snowmelt Equations

Extensive studies by the U.S. Army Corps of Engineers at various laboratories in the West have produced several general equations for snowmelt during (1) rain-free periods, and (2) periods of rain.[16] When rain is falling, heat transfer by convection and condensation is of prime importance. Solar radiation is slight, and long-wave radiation can be readily determined from theoretical considerations. When rain-free periods prevail, both solar and terrestrial radiation become significant and may require direct evaluation. Convection and condensation are usually less critical during rainless intervals. The equations are summarized as follows:[16]

1. Equations for periods with rainfall.
   a. For open (cover below 10%) or partly forested (cover from 10 to 60%) watersheds,

$$M = (0.029 + 0.0084kv + 0.007\,P_r)(T_a - 32) + 0.09 \qquad \textbf{(9-53)}$$

   b. For heavily forested areas (over 80% cover),

$$M = (0.074 + 0.007P_r)(T_a - 32) + 0.05 \qquad \textbf{(9-54)}$$

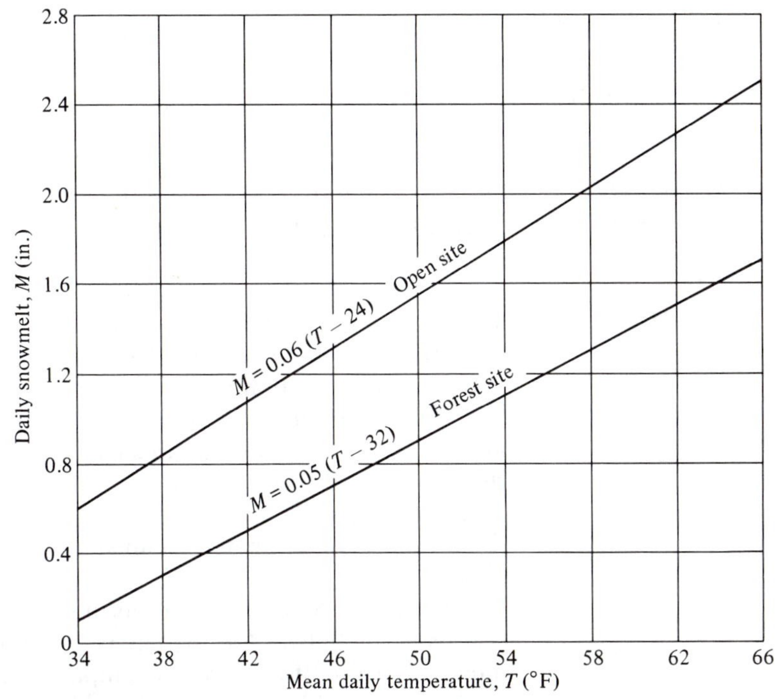


***Fig. 9-9.*** *Mean temperature index. Note: The equations are applicable only for the range of temperatures shown in the diagram. (After "Snow Hydrology," North Pacific Division, U.S. Army Corps of Engineers, Portland, Ore., June 30, 1956.)*

where

$M$ = the daily snowmelt (in./day)
$P_r$ = the rainfall intensity (in./day)
$T_a$ = the temperature of saturated air at 10-ft level (°F)
$v$ = the average wind velocity at 50-ft level (mph)
$k$ = the basin constant, which includes forest and topographic effects, and represents average exposure of the area to wind

Values decrease from 1.0 for clear plains areas to about 0.3 for dense forests.

2. Equations for rain-free periods.
   a. For heavy forested areas,

$$M = 0.074(0.53\,T_a' + 0.47\,T_d') \tag{9-55}$$

   b. For forested areas (cover of 60 to 80%),

$$M = k(0.0084v)(0.22T_a' + 0.78T_d') + 0.029T_a \tag{9-56}$$

c. For partly forested areas,

$$M = k'(1 - F)(0.0040I_i)(1 - a) + k(0.0084v)(0.22T'_a + 0.78T'_d) + F(0.029T'_a) \quad \textbf{(9-57)}$$

d. For open areas,

$$M = k'(0.00508I_i)(1 - a) + (1 - N)(0.0212T'_a - 0.84) + N(0.029T'_c) + k(0.0084v)(0.22T'_a + 0.78T'_d) \quad \textbf{(9-58)}$$

where

$M$, $v$, and $k$ = as previously described
$T'_a$ = the difference between the 10-ft air and the snow surface (°F) temperatures
$T'_d$ = the difference between the 10-ft dewpoint and snow-surface temperatures (°F)
$I_i$ = the observed or estimated insolation (langleys)
$a$ = the observed or estimated mean snow surface albedo
$k'$ = the basin short-wave radiation melt factor (varies from 0.9 to 1.1) which is related to mean exposure of open areas compared to an unshielded horizontal surface
$F$ = the mean basin forest-canopy cover (decimal fraction)
$T'_c$ = the difference between the cloud-base and snow-surface temperatures (°F)
$N$ = the estimated cloud cover (decimal fraction)

Note that the use of equations of the type given must be related to the areal extent of the snow cover if realistic values are to be obtained. Present methods of determining this are not totally adequate.

### The Water Budget

The water budget can be used to estimate the snowmelt runoff from a watershed.[16] Such an approach has particular merit for areas where hydrometeorological records are short. Difficulty with the method is the usual lack of satisfactory data to properly quantify the various components. A hydrologic budget equation for the earth's surface (Eq. 1-2) can be written as

$$R = P - L - \Delta S \quad \textbf{(9-59)}$$

where

$P$ = the gross precipitation
$R$ = the runoff

$L$ = the losses
$\Delta S$ = the change in storage

For snowmelt computations this equation is modified somewhat.

Gross precipitation for a given period $P$ will now be defined as the sum of precipitations in the form of snow $P_s$ and rain $P_r$, or

$$P = P_r + P_s \tag{9-60}$$

This may also be written as

$$P = P_r + P_s = P_n + L_i \tag{9-61}$$

where

$P_n$ = net precipitation
$L_i$ = interception loss

A further refinement yields

$$P = P_{rn} + L_{ri} + P_{sn} + L_{si} \tag{9-62}$$

where

$P_{rn}$ and $P_{sn}$ = net rainfall and snowfall
$L_{ri}$ and $L_{si}$ = the rain and snow interception

Figure 9-10 indicates the nature of snow interception by forested areas. Additional information on interception can be found in Chapter 3.

The total loss $L$ will be

$$L = L_{si} + L_{ri} + L_e + Q_{sm} \tag{9-63}$$

where

$L_e$ = the evapotranspiration loss
$Q_{sm}$ = the change in available soil moisture. The storage term $\Delta S$ is then given as

$$\Delta S = (W_2 - W_1) + Q_g \tag{9-64}$$

where

$W_2$ and $W_1$ = the final and initial water equivalents of the snowpack
$Q_g$ = the ground and channel storage.

Inserting values for $P$, $L$, *and* $\Delta S$ from Eqs. 9-62 through 9-64 in Eq. 9-59 gives

$$R = P_{rn} + L_{ri} + P_{sn} + L_{si} - L_{si} - L_{ri} - L_e - Q_{sm} - (W_2 - W_1) - Q_g \tag{9-65}$$

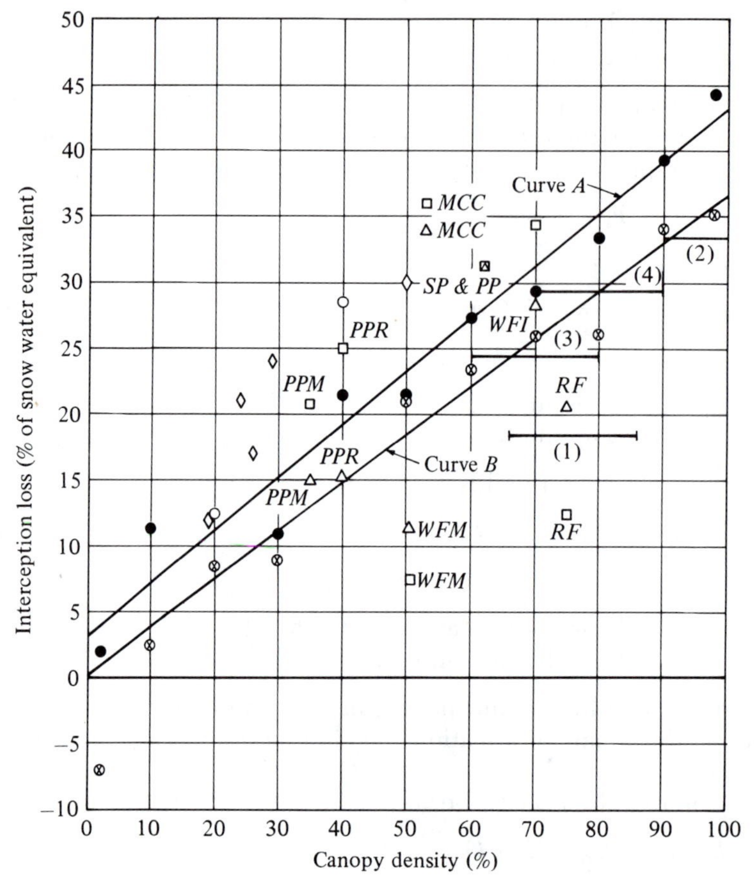


***Fig. 9-10.*** *Snowfall interception loss. (After "Snow Hydrology, North Pacific Division, U.S. Army Corps of Engineers, Portland, Ore., June 30, 1956.)*

and canceling positive and negative values of $L_{ri}$ and $L_{si}$ produces

$$R = P_{rn} + P_{sn} - (W_2 - W_1) - Q_{sm} - Q_g - L_e \qquad \textbf{(9-66)}$$

The expression $P_{sn} - (W_2 - W_1)$ represents the snowmelt $M$; therefore,

$$R = P_{rn} + M - Q_{sm} - Q_g - L_e \qquad \textbf{(9-67)}$$

If reliable estimates of the terms in Eq. 9-67 can be secured, the basin discharge $R$ is computable.

### Hydrograph Recessions

Recession curves have been discussed in Chapter 4 and take the general form

$$Q = Q_0 e^{-kt} \qquad \textbf{(9-68)}$$

where

$Q$ = the discharge at time $t$
$Q_0$ = the initial rate of flow
$k$ = a recession constant

Studies of daily streamflow by hydrographs permit evaluating the amount of runoff derived from snowmelt. The technique used is essentially one of separation of the daily hydrographs. Figure 9-11 (not to scale and oversimplified) illustrates the procedure. Assume that the first, second, and succeeding peaks, respectively, fit snowmelt days. If the ultimate recession curve is extended backward in time, at a point $A$ the recession curve from hydrograph 2 will intersect it. The area between recessions from hydrograph 1 and hydrograph 2 (shown cross-hatched) is the melt attributed to day 1. In like manner, a series of snowmelt hydrographs can be studied to determine their individual melt components. By observing such hydrograph features as the height to peak $X$, the height to trough $Y$, and the form of the recession, volumetric and rate forecasts of snowmelt runoff can be made. A more comprehensive treatment of this subject can be found in Ref. 20.

### Hydrograph Synthesis

Syntheses of runoff hydrographs associated with snow hydrology are of two types. The first is a short-term forecast. The second kind is the development of flow distribution for a complete melt season or a particular rain-on-snow occurrence. Short-term forecasting is very

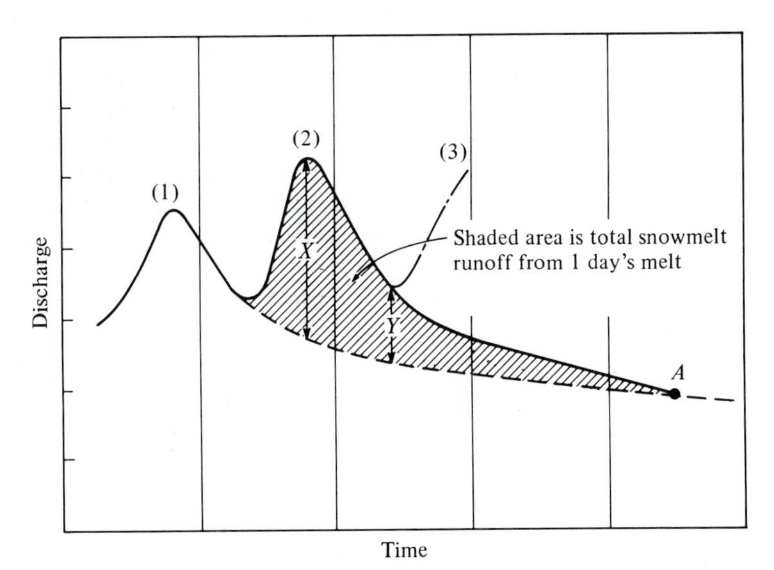


**Fig. 9-11.** *Separation of a snowmelt hydrograph.*

helpful in preparing plans to operate reservoirs or other flow controls, while the synthesis of particular storm period hydrographs is basic to calculating design floods. To forecast a few days in advance, only the present state of a snowfield and streamflow need be known to make an estimate. For long-term forecasting, it is necessary to have the reliable prediction of various meteorological parameters in addition to a knowledge of the initial conditions. Known historic parameters can be used for reconstructing historic flows whereas assumed or generated parameters satisfy design flood syntheses. Figure 9-12 displays some common hydrometeorological data.

In snowmelt hydrograph syntheses, several factors (not of great concern where only rainfall exists) must be carefully considered. First, a drainage basin with snow cover cannot be accepted as a homogeneous system, since the areal extent of the blanket is highly important. Where only snowmelt flows are developed, the contributing area need not be the entire drainage—only that portion with snow cover. If rainfall occurs during the snow-cover period, contributions can come from bare areas while other expanses may produce combined runoff. The nature of losses in such cases may differ greatly for nonsnow overlayed and covered locations.

The altitude is an exceedingly pertinent factor in the hydrology of tracts subjected to snowfall. Rates of snowmelt decrease with elevation due to a general reduction in temperature with height. Orographic effects and the temperature-elevation relationships tend to raise the amount of precipitation with altitude. Greater snow-cover depth occurs because of increased precipitation and reduced melt rates. As a result, the basin-wide melt and cover-area increase with height as the snowline is approached, then diminish with elevation over the higher places normally completely snow-covered until late in the season. A snowpack exhibits another important trait in relation to rainstorms. In the spring, relatively little runoff occurs from snow-free regions compared with that from a snowfield for moderate rainfalls. During very cold weather, the situation during heavy rains is often reversed, since a dry snowpack can retain significant amounts of water.

Two basic approaches introduce elevation effects into procedures for hydrograph synthesis.[16] The first divides the basin into a series of elevation zones where the snow depth, precipitation losses, and melt are assumed uniform. A second method considers the watershed as a unit, so adjustments are made to account for the areal extent of the snow cover, varying melt rates, precipitation, and other factors.

To synthesize a snowmelt hydrograph, information on the precipitation losses, snowmelt, and time distribution of the runoff are needed. Snowmelt is generally estimated by index methods for forecasting, but in design flood synthesis the heat budget approach is the

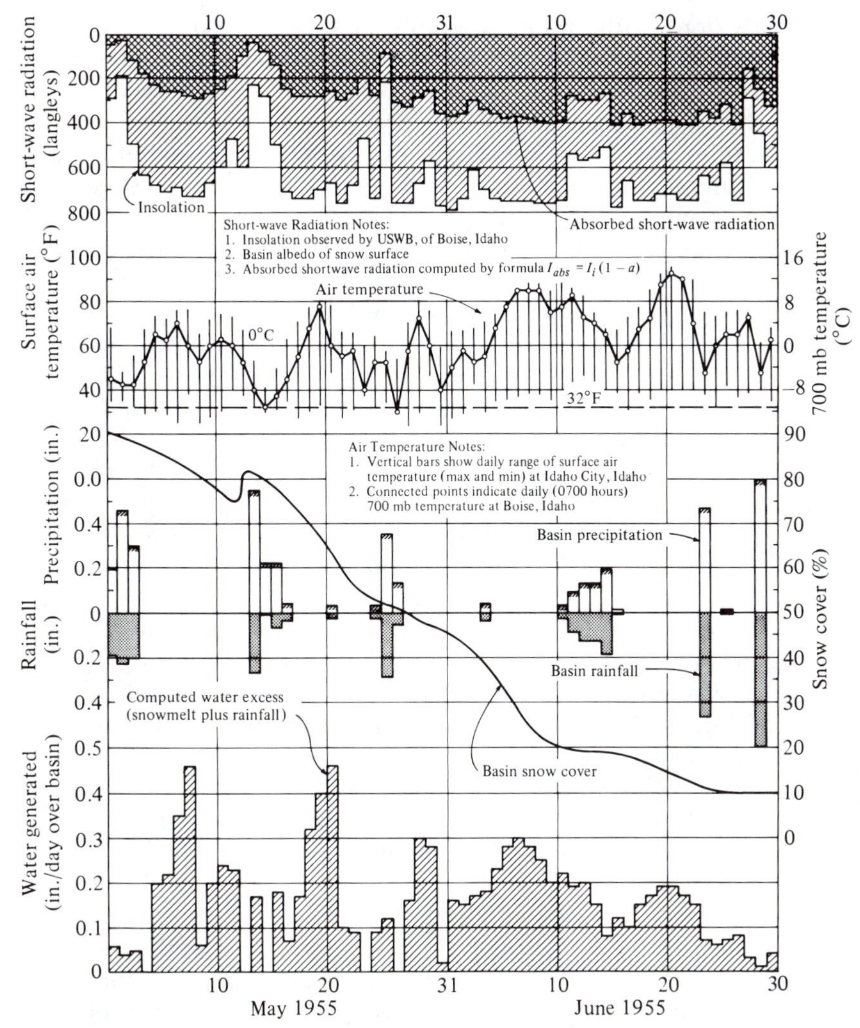


***Fig. 9-12.*** *Hydrometeorological data and computation of water generated. (After "Snow Hydrology," North Pacific Division, U.S. Army Corps of Engineers, Portland, Ore., June 30, 1956.)*

most used. Precipitation is determined from gaugings and historic or generated data. Losses are defined in two ways where snowmelt is involved. For rain-on-snow hydrographs all the water is considered a loss if delayed very long in reaching a stream. This is basically the concept of direct runoff employed in rainstorm hydrograph analyses. For hydrographs derived principally from snowmelt, only that part of the water which becomes evapotranspiration, or deep percolation, or permanently retained in the snowpack is considered to be lost. Assessing the time distribution of runoff from snow-covered areas is commonly done with unit hydrographs or storage routing techniques. For rain-on-snow events, normal rainfall-type unit hydrographs are applied; for the distribution of strictly snowmelt excess, special long-tailed unit graphs are employed. Storage routing techniques are widely exercised to synthesize spring snowmelt hydrographs, perhaps dividing them into several components and different representative storage times.

The time distribution of snowmelt runoff differs from that of rainstorms due mainly to large contrasts in the rates of runoff generation. For flood flows associated with rainfall only, direct runoff is the prime concern, and time distribution of base flow is only approximated. Big errors in estimates of base flow are not generally of any practical significance where major rainstorm floods occur. In rainstorm flows, infiltrated water is treated as part of the base flow component and little effort is directed toward determining its time distribution when it appears as runoff. In using the unit hydrograph approach to estimate snowmelt hydrographs, it is customary to separate the surface and subsurface components and route them independently.

Storage routing has been used extensively for routing floods through reservoirs or river reaches. It is also applicable in preparing runoff hydrographs. In snowmelt runoff estimates, the rainfall and meltwater are treated as inputs to be routed through the basin, using storage times selected from the hydrologic characteristics of the watershed. Two basic hydrologic routing approaches are related to the assumption of (1) reservoir-type storage, or (2) storage that is a function of inflow and outflow. These methods were treated in depth in Chapter 7.

Storage routing techniques that accommodate surface and groundwater components, assign different empirically derived storage times to each, and then route them separately have been employed.[21] An additional system uses a multiple storage, reservoir-type storage scheduling.[16] In this method inflow is routed through two or more stages of storage successively. Figure 9-13 illustrates such an approach. Any desired travel time can be obtained by properly selecting the storage time and the number of stages. Retention times between steps may also be varied to reflect various basin hydrologic characteristics. Clark has suggested that the use of single-stage routing after translating input in time permits computations to be simplified.[22]

The most practiced method for synthesizing snowmelt runoff hydrographs has been the unit hydrograph. The character of snowmelt unit graphs differs primarily in time base length from that of rainstorm unit plots. As discussed in Chapter 4 rainstorm unit hydrographs often are derived from single isolated storm events. In snowmelt runoff,

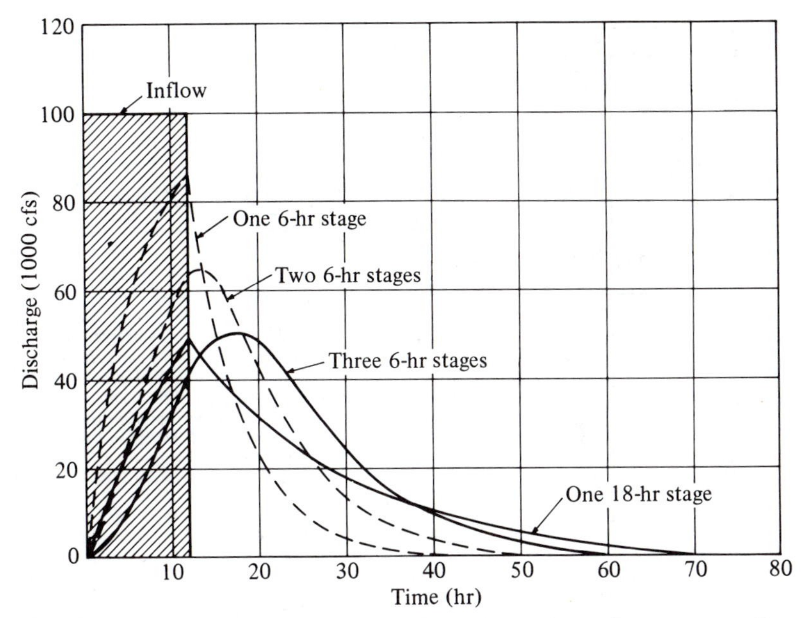


***Fig. 9-13.*** *Example of multiple stage reservoir-type storage routing. Note: This figure illustrates the use of multiple storage reservoir-type storage routing for evaluating the time distribution of runoff in a manner analogous to unit hydrographs. (After "Snow Hydrology," North Pacific Division, U.S. Army Corps of Engineers, Portland, Ore., June 30, 1956.)*

rates of water excess are small and approximately continuous. As a result, the use of S-hydrographs is indicated.[16]

The S-hydrograph method has considerable utility, since it allows (1) adjustments to the derived unit hydrograph for nonuniform generation rates, (2) adjusting the observed time period to a desired interval, (3) ready adjustments of the area under the unit hydrograph to unit volume, (4) averaging several hydrographs to get a unit hydrograph, and (5) a particular unit hydrograph to be separated into two unit graphs of unequal generation periods (a particularly useful technique in snow hydrology). Figure 9-14 shows an example of the S-hydrograph method in adjusting for nonuniform generation rates of water excess.

Once a percentage S-hydrograph is derived, a unit hydrograph of any desired period can be obtained as indicated in Figure 9-15. Ordinates of the S-hydrograph are relative to the equilibrium rate associated with a specified water excess for the selected period. The choice of a "best period" for a snowmelt-unit hydrograph creates some problems. The rising portion is best defined by a relatively short period, whereas the long tail is correlated with a long interval. A method of choosing two points has been devised, one to represent

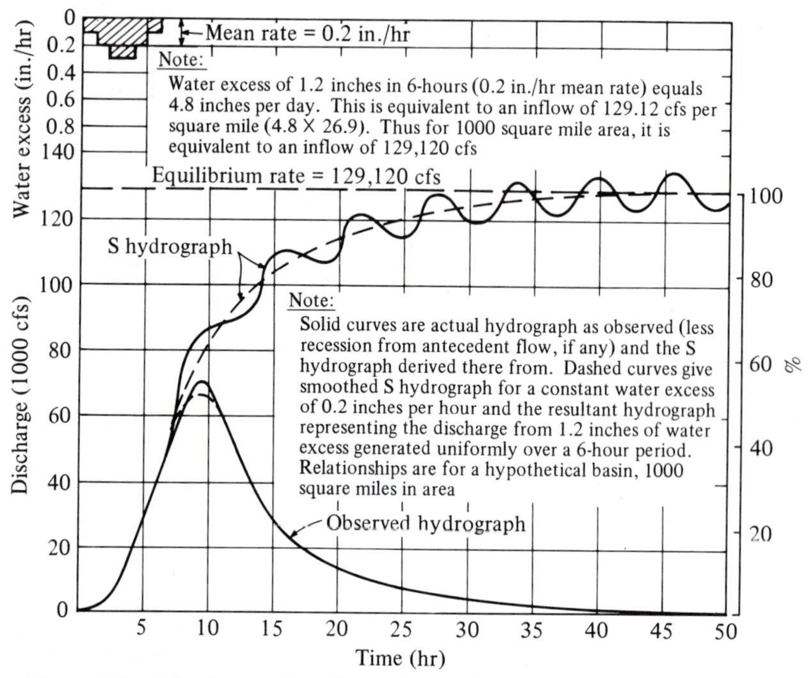


*Fig. 9-14.* *S-hydrograph–unit hydrograph relationships. (After "Snow Hydrology," North Pacific Division, U.S. Army Corps of Engineers, Portland, Ore., June 30, 1956.)*

the early periods of runoff and one to represent the later periods.[16] The procedure is illustrated on Fig. 9-16, although the recession time chosen is unrealistically shortened for purposes of explanation. Assume that it is desirable to separate the S-hydrograph at point *D*. A horizontal line drawn through *D* cuts the original S-curve into two S-hydrographs, *CDB* and *ADE*. By choosing any desired time periods, unit hydrographs for each S-curve can be obtained by taking incremental differences and multiplying them by the applicable conversion factors. The figure describes the procedure covering a 3-hr period for curve *CDB* and a 6-hr period for curve *ADE*.

Figures 9-17 and 9-18 illustrate the reconstitution of several hydrographs using methods discussed in this section.

## Runoff Forecasting

Dependable seasonal forecasts are essential to many aspects of water resources management and planning. In particular, they are indispensable to adequately planning multiple purpose reservoirs where the various uses generally are in competition and sometimes not compatible. The increased emphasis being placed on water man-

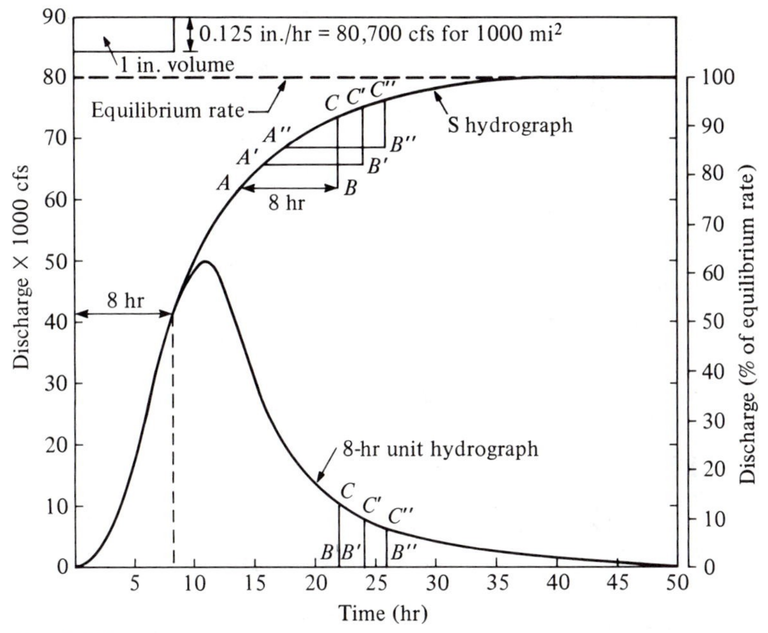


***Fig. 9-15.*** *Derivation of unit hydrograph from S-hydrograph. The ordinates of the 8-hr UHG are determined as shown by incremental differences taken 8 hr apart. (After "Snow Hydrology," North Pacific Division, U.S. Army Corps of Engineers, Portland, Ore., June 30, 1956.)*

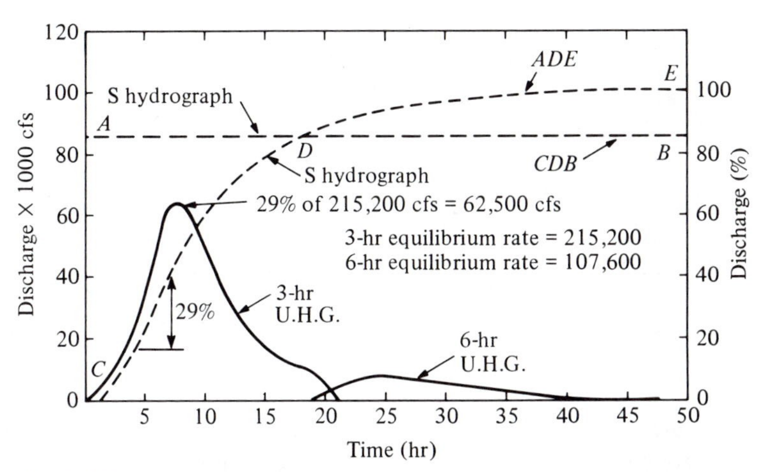


***Fig. 9-16.*** *Derivation of unit hydrographs having different periods from a divided S-hydrograph. Note: S-hydrograph CDB used to derive 3-hr UHG; S-hydrograph ADE used to derive 6-hr UHG. (After "Snow Hydrology," North Pacific Division, U.S. Army Corps of Engineers, Portland, Ore., June 30, 1956.)*

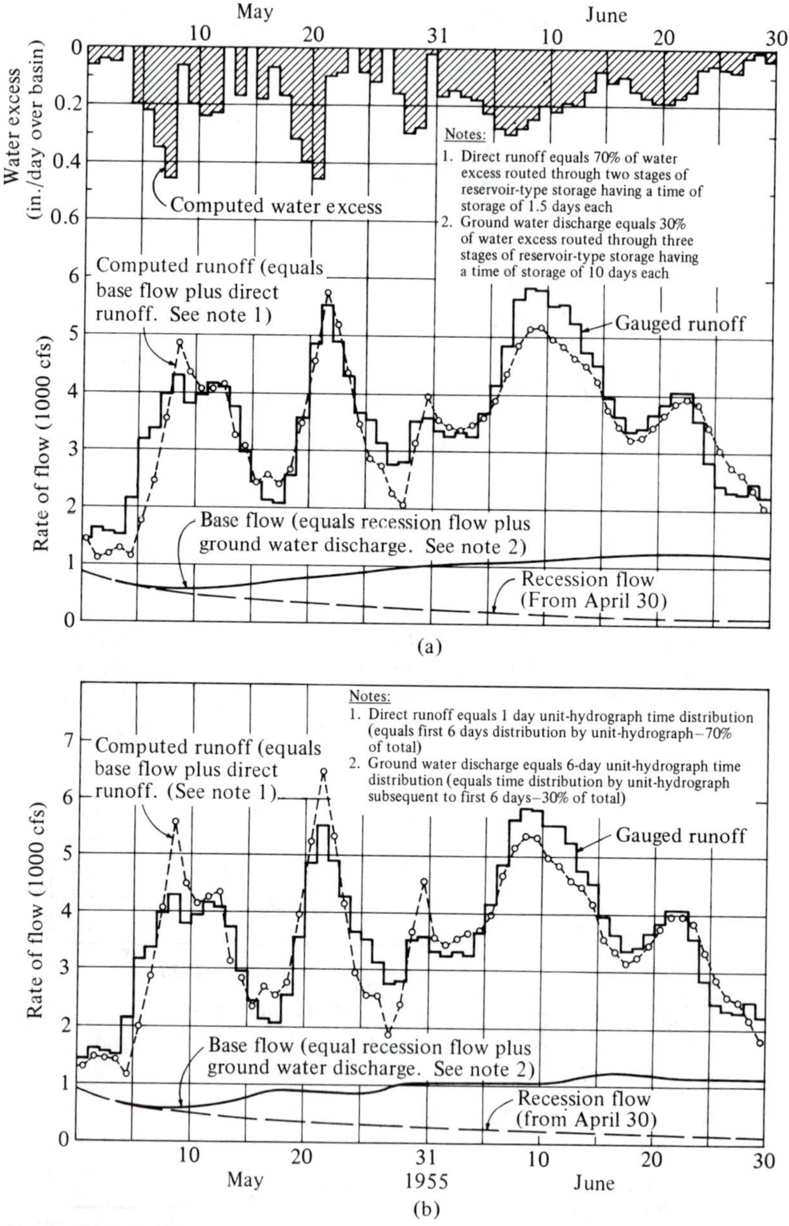


*Fig. 9-17.* *Reconstitution of seasonal hydrograph of mean daily flows: (a) storage routing method; and (b) unit hydrograph method. (After "Snow Hydrology," North Pacific Division, U.S. Army Corps of Engineers, Portland, Ore., June 30, 1956.)*

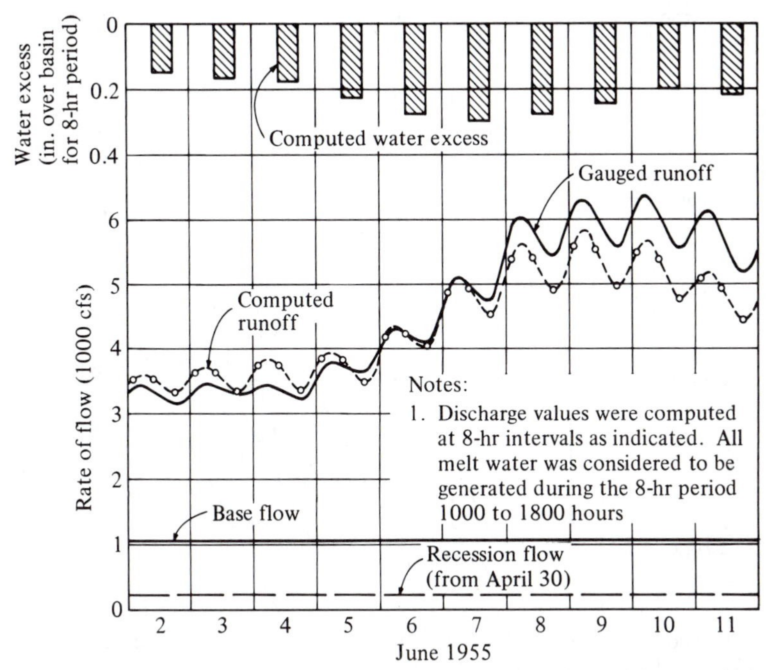


***Fig. 9-18.*** *Reconstitution of hydrograph showing diurnal flow variation. Note: Discharge values were computed at 8-hr intervals as indicated. All melt water was considered to be generated during the 8-hr period 1000 to 1800 hr. (After "Snow Hydrology," North Pacific Division, U.S. Army Corps of Engineers, Portland, Ore., June 30, 1956.)*

agement for quality control also supports the need for better techniques in predicting seasonal flows.

The sparsity of data and inability to properly evaluate controlling factors in runoff processes are contributing causes to limitations on forecasting methods in many cases. Nevertheless, satisfactory forecasts can be made if proper attention is given to the significant variables. When factors of importance are reliably determined at the time of the forecast, most errors result from conditions that set in after observations were made. Subsequent events may have little effect on forecasts in some basins where approximate divergences can be fairly well quantified. Note also that an acceptable magnitude of error will vary with the circumstances for which the prediction is made.

Principal forecasting tools are the water budget, index methods, and combinations of the two. The water budget has already been discussed in detail, as have a number of snowmelt methods that usually involve correlation of specific variables with historic runoff records. Indexes represent factors such as precipitation, water equivalent, soil moisture, groundwater, and evapotranspiration. Figure 9-19 plots an

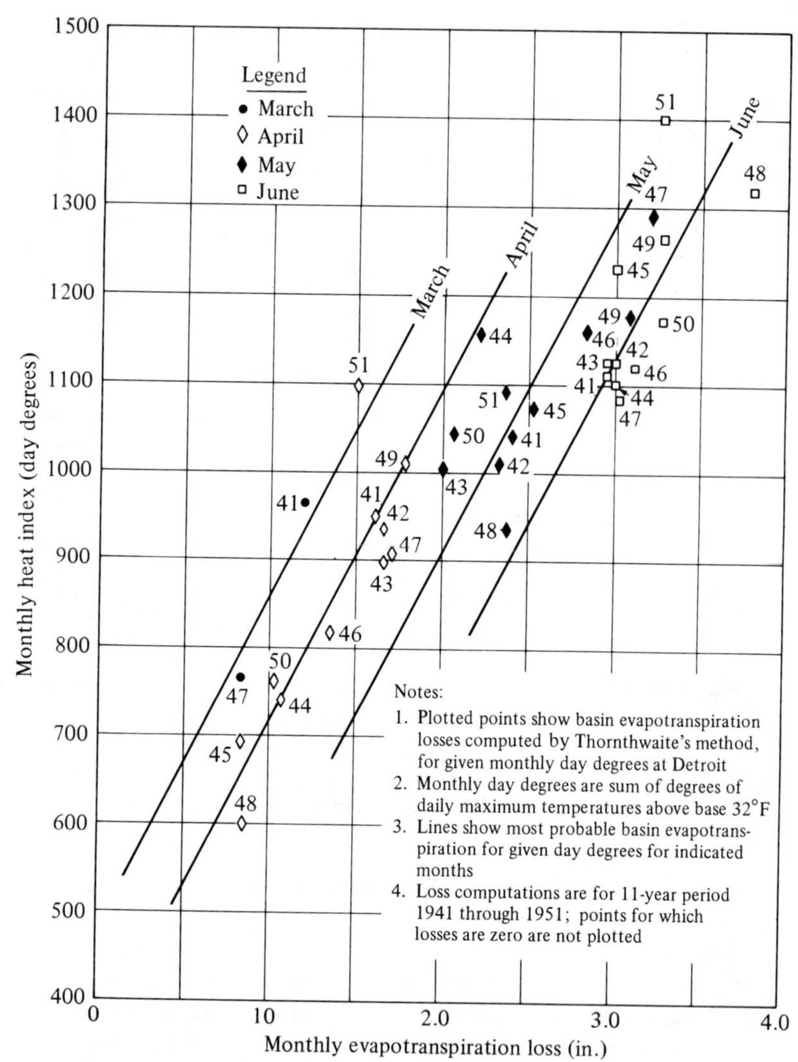


***Fig. 9-19.*** *Loss by evapotranspiration. Legend: ●, March; ◇, April; ◆, May; □, June. Notes: 1. Plotted points show the basin evapotranspiration losses computed by Thornthwaite's method, for given monthly day degrees at Detroit. 2. Monthly day degrees are the sum of degrees of the daily maximum temperatures above base 32°F. 3. Lines show the most probable basin evapotranspiration for given day degrees for indicated months. 4. Loss computation are for an 11-yr period 1941 through 1951; points for which losses are zero are not plotted. (After "Snow Hydrology," North Pacific Division, U.S. Army Corps of Engineers, Portland, Ore., June 30, 1956.)*

evapotranspiration index. The reliability of the water balance and index methods is fixed in large measure by the amount and adequacy of available hydrologic data.[19] Final proof of the method is its ability to accurately model the simulated system. Adequate forecasting by index methods normally requires records for approximately 25 yr.

## Problems

9-1. Are the snowmelt effects of condensation convection, radiation, warm rain, and conduction additive? Answer by analyzing the conditions that produce large amounts of snowmelt by each process and examine the conditions to determine if the effects are additive.

9-2. The following are obtained from a core-sample survey of a snowpack:

air temperature = 68°F
relative humidity = 20%
snowpack density = 0.2
snowpack depth = 10 ft
snowpack temperature = 22°F

(a) What is the vapor pressure of the air? (b) Will condensation on the snowpack occur, based on the vapor pressure? (c) What is the cold content of 1 $ft^2$ of surface area of the snowpack? (d) Is the snowpack ripe?

9-3. During a completely cloudy 12-hr period during April, the following averages existed for an average, ripe snowpack located 10,000 ft above sea level at a latitude of 44°N:

air temperature = 50°F
mean wind velocity = 10 mph
relative humidity = 50%
average rainfall intensity = 0.03 in./hr for 12 hr
wet bulb psychrometer reading = 48°F

Estimate the snowmelt in inches of water for each of the methods (convection, condensation, radiation, and warm rain) and tabulate the reading including percentage contributions of each source during the 12-hr period.

9-4. Given a snowpack with thermal quality of 0.90, determine the snowmelt in inches if the total input is 137 langleys.

9-5. Use Eq. 9-53 to estimate the snowmelt at an elevation of 3000 ft in a partly forested area if the rainfall intensity is 0.3 in./day, the wind velocity is 20 mph, and the temperature of the saturated air is 42°F.

9-6. Solve Prob. 9-5 for a dense forest cover.

## References

1. Chow, Ven Te (ed.), *Handbook of Applied Hydrology* (New York: McGraw-Hill Book Company, 1964).
2. Goodell, B. C., "Snowpack Management for Optimum Water Benefits," Conference Preprint 379, ASCE Water Resources Engineering Conference, Denver, Colo., May 1966.

3. Butler, S. S., *Engineering Hydrology* (Englewood Cliffs, N.J.: Prentice-Hall, Inc., 1957).
4. Yen, Y. C., "Heat Transfer by Vapor Transfer in Ventilated Snow," *J. Geophys. Res.*, 68, No. 4 (Feb. 15, 1963).
5. Hobbs, P. V., "The Effect of Time on the Physical Properties of Deposited Snow," *J. Geophys. Res.*, 70, No. 16 (Aug. 15, 1966).
6. Martinelli, M., Jr., "Some Hydrologic Aspects of Alpine Snowfields Under Summer Conditions," *J. Geophys. Res.*, 64, No. 4 (Apr. 1959).
7. Crawford, N. H., and R. K. Linsley, Jr., "Digital Simulation in Hydrology: Stanford Watershed Model IV," Department of Civil Engineering, Stanford University, Stanford, Calif., Tech. Rept. No. 39, July 1966.
8. Anderson, E. A., and N. H. Crawford, "The Synthesis of Continuous Snowmelt Runoff Hydrographs on a Digital Computer," Department of Civil Engineering, Stanford University, Stanford, Calif., Tech. Rept. No. 36, June 1964.
9. Horton, R. E., "Phenomena of the Contact Zone Between the Ground Surface and a Layer of Melting Snow," Transactions of Meetings of International Commission of Snow and Glaciers, Edinburgh, Sept. 1936, International Association of Hydrology Bulletin 23.
10. Light, P., "Analysis of High Rates of Snow Melting," *Trans. Am. Geophys. Union,* 22, Part 1 (1941).
11. Wilson, W. T., "An Outline of the Thermodynamics of Snowmelt," *Trans. Am. Geophys. Union,* 22, Part 1 (1941).
12. Kittredge, J., *Forest Influences* (New York: McGraw-Hill Book Company, 1948).
13. Mantis, H. T., et al., "Review of the Properties of Snow and Ice," U.S. Army Corps of Engineers, Snow, Ice, and Permafrost Research Establishment, SIPRE Rept. 4, July 1951.
14. "Runoff from Snowmelt," U.S. Army Corps of Engineers, Engineering and Design Manuals, EM 1110-2-1406, Jan. 1960.
15. Linsley, R. K., Jr., M. A. Kohler, and J. L. H. Paulhus, *Hydrology for Engineers* (New York: McGraw-Hill Book Company, 1958).
16. "Snow Hydrology," North Pacific Division, U.S. Army Corps of Enneers, Portland, Ore., June 30, 1956.
17. The John Hopkins University, Laboratory of Climatology, "Micrometeorology of the Surface Layer of the Atmosphere; the Flux of Momentum, Heat, and Water Vapor, Final Report," *Publications in Climatology,* 7, No. 2 (1954).
18. Sverdrup, H. U., "The Eddy Conductivity of the Air over a Smooth Snow Field," *Geofys. Publik.*, 11, No. 7 (1936).
19. Quick, M. C., "River Flood Flows: Forecasts and Probabilities," *Proc. ASCE, J. Hyd. Div.*, 91, No. HY3 (May 1965).
20. Gartska, W. U., L. D. Love, B. C. Goodell, and F. A. Bertle, "Factors Affecting Snowmelt and Streamflow," U.S. Bureau of Reclamation and U.S. Forest Service, 1958.
21. Zimmerman, A. L., "Reconstruction of the Snow-Melt Hydrograph in the Payette River Basin," Proceedings of the Western Snow Conference, Portland, Ore., Apr. 1955.
22. Clark, C. O., "Storage and the Unit Hydrograph," *Trans. ASCE,* 110 (1945).

# 10

# Hydrologic Synthesis and Simulation

Information regarding flow rates at any point of interest along a stream is necessary in the analysis and design of many types of water projects. Although many streams have been gauged to provide continuous records of streamflow, planners and engineers are sometimes faced with little or no available streamflow information and must rely on *synthesis* and *simulation* as tools for use in rationalizing decisions regarding structure sizes, the effects of land use, flood control measures, water supplies, and the effects of natural or induced watershed or climatic changes.

The problems of decision making in both the design and operation of large-scale systems of flood control reservoirs, canals, aqueducts, and water supply operations have resulted in a need for mathematical approaches such as simulation and synthesis to investigate the total project. Simulation is simply the mathematical description of the response of a hydrologic water resource system to a series of events during a selected time period. For example, simulation can mean calculating daily, monthly, or seasonal streamflow based on rainfall; or computing the discharge hydrograph resulting from a known or hypothetical storm; or simply filling in the missing values in a streamflow record.

Stochastic techniques used to extend records, either rainfall or streamflow, fit mainly under the method classified as *synthesis*. This procedure relies on the statistical properties of an existing record or regional estimates of these parameters.

## 10-1 Hydrologic Synthesis

Synthesis enables hydrologists to deal with data inadequacies, particularly if record lengths are not sufficiently extensive. Short historical records, such as streamflow, are extended to longer sequences using hydrologic synthesis and other techniques of the broad science known as *operational hydrology.* These new, synthetic sequences either preserve the statistical character of the historical records or follow a prescribed probability distribution, or both. When coupled with computer simulation techniques, the techniques provide hydrologists with improved design and analysis capabilities.

Hydrologic synthesis techniques are classified as (a) *mass curve analyses* which assume that historical records will repeat themselves exactly, (b) *random generation* techniques which assume that successive flows are independent and distributed according to a known probability distribution, and (c) *Markov generation* techniques which assume that flows in sequence are dependent and that the next flow in sequence is influenced by some subset of the previous flows. Mass curve or random generation techniques are normally applied only to annual or seasonal flows. Successive flows for shorter time intervals are usually correlated, necessitating analysis by the Markov generation method.

### Mass Curve Analysis

One of the earliest and simplest synthesis techniques was devised by Rippl[1] to investigate reservoir storage capacity requirements. His analysis assumes that the future inflows to a reservoir will be a duplicate of the historical record repeated in its entirety as many times end to end as is necessary to span the useful life of the reservoir. Sufficient storage is then selected to hold surplus waters for release during critical periods when inflows fall short of demands. Reservoir size selection is easily accomplished from an analysis of peaks and troughs in the mass curve of accumulated synthetic inflow versus time.[2,3,4] A concern for realistic magnitudes of future flows is a strong argument against mass curve analysis because future flows can be similar but are unlikely to be identical to past flows. Random generation and Markov modeling techniques produce sequences that are different from historical flows although still representative.

***Example 10-1.*** Streamflows past a proposed reservoir site during a 5-yr period of record were, respectively, in each year, 14,000, 10,000, 6000, 8000, and 12,000 acre-ft. Use Rippl's mass curve method to determine the size of reservoir needed to provide a yield of 9000 acre-ft in each of the next 10 yr.

A 10-yr sequence of synthetic flows, using Rippl's assumptions, is shown in Table 10-1. Inflows are set equal to the historial record repeated twice.

***Table 10-1***

| | Flows (thousands of acre-ft) | | | | | | | | | |
|---|---|---|---|---|---|---|---|---|---|---|
| Year | 1 | 2 | 3 | 4 | 5 | 6 | 7 | 8 | 9 | 10 |
| Inflow | 14 | 10 | 6 | 8 | 12 | 14 | 10 | 6 | 8 | 12 |
| Cum. inflow | 14 | 24 | 30 | 38 | 50 | 64 | 74 | 80 | 88 | 100 |

When cumulative inflow and cumulative draft are plotted, the maximum deficiency shown in the accompanying figure is 4000 acre-ft. Thus a reservoir with a 4000 acre-ft capacity should be placed in the stream. Starting with a full reservoir at the beginning of year 1, the reader should verify the adequacy of the reservoir by "simulating" a draft of 9000 for 10 yr.

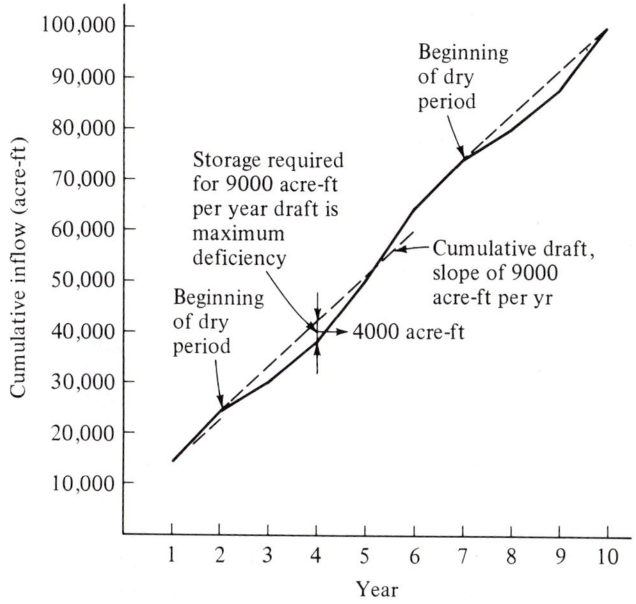


*Mass curve for example 10-1. ——— Cumulative inflow; --- cumulative draft.*

## Random Generation

This method of generating sequences of future flows is a simple random rearrangement of past records. If the stream is ungauged and records are not available, a probability distribution is selected and a sequence of future flows that follow the distribution is generated.

Whenever historical flows are available, a sequence of future flows can be synthesized by first consulting a table of random numbers, selecting a number, matching this with the rank-in-file number of a past flow, and listing the corresponding flow as the first value in the new sequence. The next random number would be used in a similar fashion to generate the next flow, and so on. Random numbers having no corresponding flows are neglected and the next random number is selected. Table C-3 in Appendix C is a table of uniformly distributed random numbers (each successive number has an equal probability of taking on any of the possible values). To illustrate its use in the random generation process, the first three years of a synthetic flow sequence would be generated by selecting the fifty-third, seventy-fourth, and twenty-third from the list of past flows. Alternatively, the flows in 19<u>53</u>, 19<u>74</u>, and 19<u>23</u> could also be selected as the new random sequence.

Most digital computer installations have random number generation capabilities in their system libraries. Rather than storing large tables of numbers such as Table C-3, successive random integers are usually generated by the *mixed congruential method*[5] using

$$p_{i+1} = \text{modulo}\ \frac{ap_i + b}{c} \tag{10-1}$$

where

| | |
|---|---|
| $p_i$ = | the $i$th random number in the sequence ($p_1$ is any initial random integer) |
| $a$, $b$, and $c$ = | positive integers |
| $p_{i+1}$ = | the $(i+1)$st random number, calculated as the arithmetic remainder when $ap_i + b$ is divided by $c$ |

Thus the modulo term is an operator signifying that $p_{i+1}$ is the remainder formed when $ap_i + b$ is divided by $c$. For example, if $p_1 = 10$, $a = 3$, $b = 5$, and $c = 10$, then $p_2 = \text{modulo}\ [(30 + 5)/10] = (3 \text{ plus a remainder of } 5) = (3\text{ R }5)$ giving $p_2 = 5$. The reader should verify that $p_3 = 0$.

A large integer constant $c$ is usually selected because $c$ is the greatest possible value of $p_i$. Also, the sequence repeats itself after the $c$th value is calculated, and a large $c$ is desirable. Constants $a$ and $b$ are normally selected as odd integers with values less than $c$. The sequence is initiated with $p_1$ known. Additional numbers are then sequentially determined from Eq. 10-1 as demonstrated in Example 10.2.

***Example 10-2.*** Annual flows in Crooked Creek were 19,000, 14,000, 21,000, 8000, 11,000, 23,000, 10,000, and 9000 acre-ft, respectively, for years 1, 2, 3, 4, 5, 6, 7, and 8. Generate a 5-yr sequence of annual flows, $Q_j$, by matching five random numbers with these year numbers, using the mixed congruential method to generate the random numbers.

*Solution:*

Random integers between 0 and $10^6$ are generated from Eq. 10-1 using $c = 10^6$, $a = 41$, and $b = 3$. The initial or *seed* random number is $p_1 = 123{,}456$. The $Q_j$ values in Table 10-2 are selected from the eight given flows by matching the respective year number with the random last digit of each random number $p_{i+1}$. In the first row, $[41(123, 456)+3]/10^6$ is $5{,}061{,}699/10^6$, or 5 with a remainder of 61,699. Thus $p_2$ is the remainder 61,699. The last digit 9 has no corresponding flow in the 8-yr sequence, so the next random number, 2, places the 14,000 cfs flow in year 2 in the first position of the synthetic 5-yr sequence.

**Table 10-2**

| $i$ | $p_i$ | $(41\,p_i + 3)/10^6$ | $p_{i+1}$ | Last Digit | Counter $j$ | $Q_j$ (acre-ft) |
|---|---|---|---|---|---|---|
| 1 | 123,456 | 5 R 061699 | 061,699 | 9 | — | Skip |
| 2 | 061,699 | 2 R 529662 | 529,662 | 2 | 1 | 14,000 |
| 3 | 529,662 | 21 R 716145 | 716,145 | 5 | 2 | 11,000 |
| 4 | 716,145 | 29 R 361948 | 361,948 | 8 | 3 | 9,000 |
| 5 | 361,948 | 14 R 839871 | 839,871 | 1 | 4 | 19,000 |
| 6 | 839,871 | 34 R 434714 | 434,714 | 4 | 5 | 8,000 |

Streamflow synthesis for ungauged streams requires either a regional frequency curve or an equation of the cumulative distribution function. The synthetic sequence can be generated using an assumed CDF shape with regional estimates of distribution parameters such as the mean and standard deviation.

The random generation process for gauged or ungauged basins begins with the selection of a sequence of random numbers $p_i$ from Table C-3 or from Eq. 10-1 if digital computers are used. Each $p_i$ is made fractional by placing a decimal point in front, and the resulting decimal fraction is treated either as $G(x)$ or $F(x)$ (see Chapter 5) in solving for streamflow $x$.

***Example 10-3.*** Generate a 5-yr sequence of annual flows in Crooked Creek having a Pearson Type III CDF with a mean of 14,000 acre-ft, a standard deviation of 5000 acre-ft, and a skew coefficient of 0.2.

*Solution:*

Successive random values of $p_i$ from Example 10-2 are treated as exceedance probabilities $G(Q)$, which in turn are entered into Table C-2 to find the interpolated Pearson Type III frequency factor $K$ (see Table 10-3). Corresponding flows are found from Eq. 5-46, or

$$Q_j = \bar{Q} + K_j\,(s) \quad \textbf{(5-46)}$$

or

$$Q_j = 14{,}000 + K_j\,(5000) \quad \textbf{(10-2)}$$

### Markov Generation

The synthesis of sequences of streamflows by random generation ignores the existence of *persistence* present to some extent in most hydrologic sequences. (Persistence is the tendency for high flows to

**Table 10-3**

| $j$ | $p_i$ (Example 10.2) | G(Q) (%) | $j$ | $K_j$ (Table C-2) | $Q_j$ (Eq. 10-2) (acre-ft) |
|---|---|---|---|---|---|
| 1 | 123,456 | 12.3 | 1 | 1.17 | 19,850 |
| 2 | 61,699 | 6.1 | 2 | 1.56 | 21,800 |
| 3 | 529,662 | 52.9 | 3 | −0.11 | 13,450 |
| 4 | 716,145 | 71.6 | 4 | −0.62 | 10,900 |
| 5 | 361,948 | 36.1 | 5 | 0.36 | 15,800 |

be followed by high flows, and low flows to be followed by low flows.) Markov Models assume that each flow is dependent only on one or more of the most recent flows.

*Single-period* and *multiperiod* Markov Models view streamflows as chains of serially dependent values, where each value has a deterministic part and a random error part. To illustrate, the *lag-one single-period* Markov chain assumes that the next flow in sequence, $Q_{i+1}$, is the sum of the mean $\bar{Q}_{j+1}$ plus a dependent fractional part of the deviation of the previous flow $Q_i$ from its mean, $\bar{Q}_j$, plus a random component $e$. Expressed as an equation, the lag-one Markov Model[6] is

$$Q_{i+1} = \bar{Q}_{j+1} + r_j(Q_i - \bar{Q}_j) + e \qquad \textbf{(10-3)}$$

where

$Q_{i+1}$ = the generated streamflow in period $i + 1$ (day, month, year)
$\bar{Q}_{j+1}$ = the mean of observed flows in period $j + 1$
$Q_i$ = the generated flow in period $i$
$\bar{Q}_j$ = the mean of observed flows in period $j$
$r_j$ = the correlation coefficient for the relation of flows for period $j + 1$ to flows for period $j$
$e$ = the simulation error due to unexplained variance

Now, since $e$ represents the error in the relationship, rewriting Eq. 10-3, we obtain

$$Q_{i+1} = \bar{Q}_{j+1} + r_j(Q_i - \bar{Q}_j) + t_i\sigma_{j+1}(1 - r_j^2)^{1/2} \qquad \textbf{(10-4)}$$

where

$t_i$ = a random number selected from a normal distribution having a zero mean and a unit variance
$\sigma_{j+1}$ = the standard deviation of observed flows for the period $j + 1$

To apply Eq. 10-4, it is necessary to calculate the required means, variances, and correlations and then to initiate the synthetic sequence of flows with some value $Q_i$ with $i = 1$. The mean of observed flows is often selected as the initial flow in the random synthetic runoff sequence, but the choice affects all future flows in the sequence. Fiering and Jackson[4] recommended that the first 50 flows in the sequence be discarded as a means of providing a warm-up period to eliminate starting condition bias. Standard normal random deviates, $t_i$, are generated using the random generation procedures described earlier.

Whenever the single-period Markov generator is used for annual flows, the period $j + 1$ flows and the period $j$ flows have equal means, or $\bar{Q}_{j+1} = \bar{Q}_j$. This assumption is not valid for smaller periods, such as monthly or daily subdivisions. For example, October and November flows seldom have equal means. To reflect different seasonal or monthly means, the *multiperiod* Markov Model is used. This requires a double indexing subscript in Eq. 10-4 or

$$Q_{i,j} = \bar{Q}_j + b_j(Q_{i-1,j-1} - \bar{Q}_{j-1}) + t_i\sigma_j(1 - r_j^2)^{1/2} \tag{10-5}$$

where $j$ is the number of seasonal periods in the year and the other terms have been defined except for $b_j = r_j(\sigma_j/\sigma_{j-1})$ which must be employed, since $\bar{Q}_{j+1} \neq \bar{Q}_j$. This model has been used extensively in streamflow analysis. The single- and multiperiod Markov generation procedures sometimes result in negative flows. These flows must be retained for generating the next flows in sequence and then they may be discarded.

***Example 10-4.*** The parameters describing quarterly flows in million gallons per day for the Patapsco River are listed in Table 10-4. Apply the information in conjunction with a table of random numbers and create a 2-yr record of quarterly flows. The solution is shown in Table 10-5. A flow chart for the process is sketched in Fig. 10-1.

### Normal Distributions

The model in Fig. 10-1 is applicable to simulating flows at a single site where the population of flows is normally distributed for each calendar month. The equation as stated preserves the mean, the variance, and the correlation between successive flows of the historic monthly streamflow sequence. The simulation procedure

***Table 10-4*** *Quarterly Streamflow Parameters, Patapsco River*

| $j$ | $b_j$ | $\sigma_j$ | $r_j$ | $(1 - r_j^2)^{1/2}$ | $\bar{Q}_j$ |
|---|---|---|---|---|---|
| 1 | .66 | 73 | .57 | .82 | 68 |
| 2 | .43 | 85 | .41 | .91 | 137 |
| 3 | .14 | 90 | .43 | .90 | 183 |
| 4 | .91 | 29 | .36 | .93 | 107 |

**Table 10-5** *Tabular Generation of Synthetic Quarterly Streamflows*

| $i$ (1) | $j$ (2) | $\bar{Q}_j$ (3) | $\bar{Q}_{j-1}$ (4) | $Q_{i-1,j-1}$ (5) | $k$ (6) | $t_i$ (7) | $b_j(Q_{i-1,j-1} - \bar{Q}_{j-1})$ (8) | $t_i\sigma_j(1 - r_j^2)^{1/2}$ (9) | $Q_{i,j}$ (10) [(3) + (8) + (9)] |
|---|---|---|---|---|---|---|---|---|---|
| 1 | 1 | 68 | 107 | 107 | 0.5374 | .094 | 0.0 | 5.6 | 73.6 |
| 2 | 2 | 137 | 68 | 73.6 | 0.6338 | .342 | 2.4 | 26.5 | 165.9 |
| 3 | 3 | 183 | 137 | 165.9 | 0.3530 | −.377 | 4.0 | −30.6 | 156.4 |
| 4 | 4 | 107 | 183 | 156.4 | 0.6343 | .343 | −24.2 | 9.3 | 92.1 |
| 5 | 1 | 68 | 107 | 92.1 | 0.0263 | −1.94 | −9.8 | −116.4 | 0.0 |
| 6 | 2 | 137 | 68 | 0.0 | 0.6455 | .373 | −29.2 | 28.9 | 136.7 |
| 7 | 3 | 183 | 137 | 136.7 | 0.8507 | 1.04 | −0.04 | 84.5 | 267.5 |
| 8 | 4 | 107 | 183 | 267.5 | 0.3485 | .39 | 76.8 | 10.6 | 194.5 |

Column 1. Period index, $i = 1$ to 8 quarters for a 2-yr run.
Column 2. Quarterly index, $j = 1$ to 4, two cycles.
Column 3. Mean quarterly flow.
Column 4. Preceding mean quarterly flow.
Column 5. The generated flow for the preceding quarter. Note that the first entry is the same as column 4, that is, generation begins using the mean flow from the preceding quarter.
Column 6. Uniform random digits from a random number table. Could be produced simply by drawing randomly from a card deck, with replacement.
Column 7. Random normal deviates from Appendix C, Table 1. If column 6 is less than 0.5, find $z$ for $(0.5 - k)$ and set $t = -z$; if $k > 0.5$, find $z$ for $(k - 0.5)$ and set $t = z$.
Column 8. Regression component of Eq. 10-5.
Column 9. Random component of Eq. 10-5.
Column 10. Generated flow.

begins by assuming an initial value of $Q_i$, computing $Q_{i+1}$, and repeating by replacing $Q_i$ by $Q_{i+1}$. In extending a record, the initial assumption of $Q_i$ is the last recorded monthly streamflow.

### Lognormal Distributions

Monthly flows frequently do not appear to be from a normally distributed population. Experience has shown, however, that annual flows are generally normal and monthly flows are lognormal or Pearson III. A recursive equation for lognormal simulation presented by Matalas[7] preserves the mean $\mu_x$, standard deviation $\sigma_x$, skewness $\gamma_x$, and lag-one correlation coefficient $r_x$ for the monthly flows $Q$. Using $Y = \log(Q - a)$, the recursive Eq. 10-4 appears as

$$(Y_{i+1} - \mu_y) = r_y(Y_i - \mu_y) + t\sigma_y(1 - r_y^2)^{1/2} \qquad \textbf{(10-6)}$$

where

$Y_{i+1}$ = the generated lognormal flow in month $i + 1$
$\mu_y$ = the mean of the lognormal flow in month $i + 1$
$r_y$ = the correlation coefficient for lognormal flow in month $i + 1$ with month $i$
$\sigma_y$ = the standard deviation of observed flows in month $i$
$t$ = the normally distributed random number with zero mean and unit variance

Values for $a$, $\mu_y$, $\sigma_y$, and $r_y$ must be found in the following manner to insure preserving a synthetic sequence that will resemble the historic sequence in terms of $\mu_x$, $\sigma_x$, $\gamma_x$, and $r_x$. Relationships for $\mu_x$, $\sigma_x^2$, $\gamma_x$, and $r_x$ are given as

$$\mu_x = a + \exp(1/2\sigma_y^2 + \mu_y) \tag{10-7}$$

$$\sigma_x^2 = \exp(2(\sigma_y^2 + \mu_y)) - \exp(\sigma_y^2 + 2\mu_y) \tag{10-8}$$

$$r_x = (\exp(\sigma_y^2 r_y) - 1)/(\exp(\sigma_y^2) - 1) \tag{10-9}$$

$$\gamma_x = \frac{\exp(3\sigma_y^2) - 3\exp(\sigma_y^2) + 2}{(\exp(\sigma_y^2) - 1)^{3/2}} \tag{10-10}$$

Since the values of $\mu_x$, $\sigma_x$, $\gamma_x$, $r_x$ are known, Eqs. 10-7 through 10-10 can be solved for $\mu_y$, $\sigma_y$, $r_y$, and $a$. Equation 10-6 may then be utilized to step forward in monthly increments finding values of $Y_{i+1}$. To get the synthetic sequence of streamflow, the value of $a$ is added to the antilog of each value of $Y_{i+1}$.

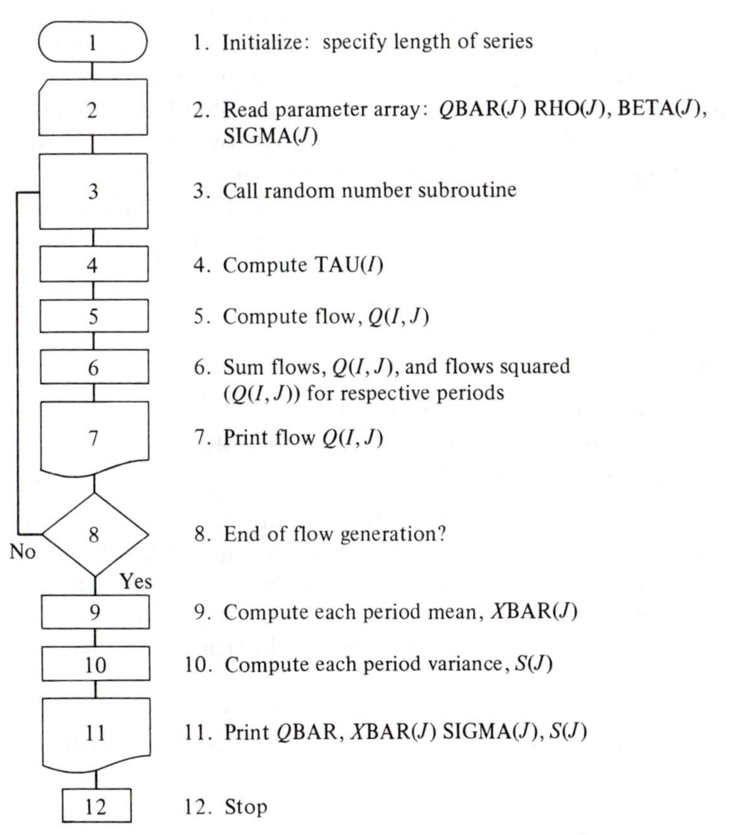


**Fig. 10-1.** *Flow chart for computing synthetic streamflow.*

### Pearson III Distributions

The Pearson III distribution for simulating daily streamflows in reservoir studies has been advocated by Beard[8] and is recommended when simulating daily flows for critical flood months. The reason for this is obvious—there would be an immense number of calculations to simulate daily flows for any lengthy period. The basic recursion formula is

$$Q_{i+2} = b_1 Q_{i+1} + b_2 Q_i + X_r(1 - r^2) \qquad \textbf{(10-11)}$$

where

$Q_{i+2}, Q_{i+1}, Q_i$ = the normal standardized variates for successive days
$X_r$ = the random standardized variate
$b_1, b_2$ = regression coefficients
$r$ = the correlation coefficient for the regression equation

This simulation model is a second-order Markov chain; hence each daily flow is assumed to be partially dependent on flows in each of the previous 2 days. Normal standardized variates for successive days are found by first classifying streamflows into months, and determining the logarithmic monthly mean, standard deviation, and skew coefficient.* Next, logarithmic daily flows are standardized by subtracting the mean monthly logarithm from each daily value and dividing by the monthly standard deviation. This standardized variate is transformed to normal by the Pearson type III approximation

$$t = \frac{6}{g}\left\{\left[\frac{g}{2}\,k + 1\right]^{1/3} - 1\right\} + \frac{g}{6} \qquad \textbf{(10-12)}$$

where

$t$ = the standard normal deviate
$k$ = the Pearson Type III standard deviate
$g$ = the skew coefficient

A typical simulation with this technique first requires calculation of the following statistics from the existing streamflow record.

1. The logarithmic mean, serial correlation coefficient, and skew coefficient of daily flows for each month of the record period.
2. The correlation coefficients between logarithmic flows of successive days.
3. A regression coefficient of the standard deviation for the daily flow logarithms within each month of record to the logarithm of the monthly total flow.

* In order to avoid zero flows, 0.01 times the monthly mean is added to each daily flow before taking the logarithm.

Given the calculated statistics, the simulation of daily flows could be structured in the following manner:

1. Generate standardized variates from Eq. 10-11.
2. Use the logarithm of the monthly mean flow as an initial estimate of the mean of the logarithmic daily flows for the month.
3. Calculate the standard deviation of flow logarithms by the previously determined regression equation.
4. Apply the inverse of Eq. 10-12

$$k = \frac{2}{g}\left[\frac{g}{6}\left(t - \frac{g}{6}\right) + 1\right]^3 - \frac{2}{g} \tag{10-13}$$

to transform the standardized variates to flows, multiplying by the appropriate standard deviation and adding the mean.

5. Add the difference between the total monthly flow generated and the given monthly flow to the given monthly flow, and repeat the simulation.
6. Multiply daily results of the second simulation by the ratio of the given monthly total to the generated monthly total.

Each simulation technique commonly requires some modifications when applied to individual problems. Methods outlined thus far can be utilized as guides in establishing a procedure to follow in constructing a sequential simulation program. Only the simulation of flows for a given station have been presented with serial correlation, skewness, means, and standard deviations maintained. When generating streamflow sequences for regional areas, the preservation of cross-correlation, that is, between stations, becomes a significant factor.

Synthetically generated runoff sequences are employed to determine the capacities of reservoirs to satisfy specified demands. Individually generated flow magnitudes are uncertain, as are the synthetically generated sequences of flows. Hydrologists can, however, estimate probabilities of flows by generating several equally likely sequences of flows. Herein lies one of the most useful applications of Markov generating techniques.

## 10-2 Hydrologic Simulation

In this chapter, *simulation* of all or parts of a surface, groundwater, or combined system implies the use of digital computational methods to imitate historical events or preview the future response of the physical system to a specific plan or action. Physical, analog, hybrid, or other models for simulating the behavior of hydraulic and hydrologic systems and system components have had, and will continue to have, application in imitating prototype behavior but are not discussed herein.

### Types of Simulation Models

In recent years the science of computer simulation of ground and surface water resource systems has passed from scattered academic interests to a practical engineering procedure. The varied nature of developed and applied simulation models has caused a proliferation of categorization attempts. A few of the most descriptive classifications are presented for comparison.

One of the earliest classifications separates simulation models into *physical* and *mathematical* categories. Physical models include analog technology and principles of similitude applied to small-scale models. In contrast, mathematical models rely on mathematical statements to represent the system. A laboratory flume may be a 1:10 physical model of a stream, while the unit hydrograph theory of Chapter 4 is a mathematical model of the response of a watershed to various effective storms.

A second classification is achieved by considering physical, analog, and some digital models as *continuous* because the processes occur and are observed continuously. Many digital simulation models rely on the necessity and advantages of slicing space and time into finite increments, and thus qualify as *discrete* models. A well-known example of the latter is the storage-indication method for routing a flood hydrograph through a reservoir by generating instantaneous reservoir discharge rates at the end points of equally spaced intervals in time.

Processes that involve changes over time and time-varying interactions can be simulated by *dynamic* models. In contrast, models that examine time-independent processes are frequently called *static*. Few hydrologic simulation models fall into the latter category.

Simulation models are also classified as *descriptive* and *conceptual*. The former have had the greatest application and are of particular interest to practicing hydrologists because they are designed to account for observed phenomena through empiricism and the use of basic fundamentals such as continuity or momentum conservation assumptions. Conceptual models, on the other hand, rely heavily on theory to interpret phenomena rather than to represent the physical process. Examples of the later include models based on probability theory.

Models that ignore spatial variations in parameters throughout an entire system are classified as *lumped parameter* models. An excellent example is the use of a unit hydrograph for predicting time distributions of surface runoff for different storms on a homogeneous drainage area. The "lumped parameter" is the time of concentration, which is held constant for all storms. *Distributed parameter* models account for behavior variations from point to point throughout the system. Most modern groundwater simulation models are distributed in that they allow variations in storage and transmissivity parameters over a grid or node system superimposed over the plan of an aquifer.

Another pair of terms commonly used to describe simulation models are *black-box* and *structure-imitating*. Both accept input and transform it into output. In the former case, the transformation is accomplished by techniques that have little or no known physical basis. The alchemist's ability to transform lead into gold or plants into medicine was accomplished in a black-box fashion, that is, lead input gave gold output. In hydrology, black-box models may also sometimes transform "plants" into "medicine" even though the reasons for success are not clearly understood. For example, a model that accepts a sequence of numbers, reduces each by 20%, and outputs the results might be entirely adequate for predicting the attenuation of a flood wave as it travels through a reach of a given stream. In contrast, a structure-imitating model would be designed to use meaningful principles of fluid mechanics and hydraulics to facilitate the transformation.

Other terms commonly used to describe simulation models compare *stochastic* models with *deterministic* models. Many stochastic processes are approximated by deterministic approaches if they exclude all consideration of random parameters or inputs. For example, the simulation of a reservoir system operating policy for water supply would properly include considerations of uncertainties in natural inflows, yet many water supply systems are designed on a deterministic basis by mass curve analyses which assume that sequences of historical inflows are repetitive.

Deterministic methods of modeling hydrologic behavior of a watershed have become popular. Basically *deterministic simulation* describes the behavior of the hydrologic cycle in terms of mathematical relationships outlining the interactions of various phases of the hydrologic cycle. Frequently, the models are structured to simulate a streamflow value, hourly or daily, from given rainfalls within the watershed boundaries. The model is "verified" or "calibrated" by comparing results of the simulation with existing records. Once the model is adjusted to fit the known period of data, additional periods of streamflow can be generated.

As a final comment on the classifications of models, hydrologic systems can be investigated in greater detail if the time frame of simulation is shortened. Many short-term hydrologic models could be classified as *event simulation* models as contrasted with *sequential* models. An example of the former is the Corps of Engineers single-event model, HEC-1,[9] and an example of the latter is the Stanford Watershed Model developed by Crawford and Linsley,[10] which is normally applied to systems that are operated for three, four, five, or more years. In order to make the Stanford model manageable, a lumped parameter approach is used along with the extended time frame. Event simulation models, in contrast, allow greater flexibility for the use of distributed parameters and shorter time increments. A typical event simulation model might generate input to, and output from, a reservoir for a single storm event, using a time increment of 1 hr or perhaps even 1 min.

### Limitations of Simulation

Because simulation entails a mathematical abstraction of real world systems, some degree of misrepresentation of system behavior can occur. The extent to which the model and system outputs vary depends on many factors. The test of a developed simulation model consists of verification by demonstrating that the behavior is consistent with the known behavior of the physical system.

Verified simulation models have limitations that should be considered in uses for water resources planning and analysis. A particularly troublesome trait of simulation models is the difficulty encountered in using the models to generate optimal plans for development and management. They will allow performance assessments of specific schemes but cannot be used efficiently to generate schemes, particularly optimal plans, for stated objectives. Once a near-optimal plan is formulated by some other technique, a limited number of simulation runs are normally effective for testing and improving the plan by modifying combinations of decision variables using random or systematic sampling techniques. Some of the techniques for generating optimal plans are described in Sec. 10-8.

Another limitation of many simulation models involves the inflexibility of changing the operating procedures for potential or existing components of the system being modeled. Programming a computer to handle reservoir storage and release processes, for example, requires large portions of the program to define the operating rules, and considerable reprogramming is required if other operating procedures are to be investigated.

A fourth limitation of simulation models is the potential overreliance on sophisticated output when hydrologic and economic inputs are inadequate. The techniques of operational hydrology can be used to obviate data inadequacies, but these also require input. The current unresolved controversy over the use of synthetic data centers on the question of whether or not operational hydrology provides better information than simpler approaches such as mass curve analyses.

### Advantages of Simulation

Computer simulation of hydrologic processes has several important advantages that should be recognized whenever considering the merits of a simulation approach to a problem that has other possible solutions. One alternative to digital simulation is to build and operate either the prototype system or a physically scaled version. Simulation by physical modeling has found application to many components of systems such as the design of hydraulic structures or the investigation of stream bank stability. However, for the analysis of complex water resource systems comprised of many interacting components, computer simulation often proves to be the only feasible tool.

Another alternative to digital simulation is a hand solution of the governing equations. Simulation models, once formulated, can accomplish identical results in less time. Also, solutions that would be impossible to achieve by hand are frequently achieved by simulation. In addition, the system can be nondestructively tested; proposed modifications of the designs of system elements can be tested for feasibility or compared with alternatives; and many proposals can be studied in a short time period.

An often overlooked advantage of simulation includes the insight gained by gathering, organizing, and processing the data, and by mentally and mathematically formulating the model algorithms that reproduce behavior patterns in the prototype.

### Steps in Digital Simulation

A simulation model is a set of equations and algorithms (for example, operating policies for reservoirs) that describe the real system and imitate the behavior of the system. A fundamental first step in organizing a simulation model involves a detailed analysis of all existing and proposed components of the system and the collection of pertinent data. This step is called the system identification or inventory phase. Included items of interest are site locations, reservoir characteristics, rainfall and streamflow histories, water and power demands, and so forth. Typical inventory items required for a simulation study and data needs that are specific to some of the models are detailed in subsequent paragraphs.

The second phase is model conceptualization, which often provides feedback to the first phase by defining actual data requirements for the planner and identifying system components that are important to the behavior of the system. This step involves (1) selecting a technique or techniques that are to be used to represent the system elements, (2) formulating the comprehensive mathematics of the techniques, and (3) translating the proposed formulation into a working computer program that interconnects all the subsystems and algorithms.

Following the system identification and conceptualization phases are several steps of the implementation phase. These include (1) validating the model, preferably by demonstrating that the model reproduces any available observed behavior for the actual or a similar system; (2) modifying the algorithms as necessary to improve the accuracy of the model; and (3) putting the model to work by carrying out the simulation experiments.

### General Components of Hydrologic Simulation Models

Numerous mathematical models have been developed for the purpose of simulating various hydrologic phenomena and systems. A general conceptual model including most of the important components is shown in Fig. 10-2; several others are described subsequently. Imported water in the lower left could be input to reservoir or groundwater storage or channel flow, or it might be guided directly to water allocations on the far right if either storage or distribution were deemed unnecessary. The routing of channel flow or overland flow could be accomplished by simple lumped parameter techniques, or solutions of the unsteady-state flow equations for discrete segments of the channel could be used. In other words, the selection of techniques and algorithms to represent each component depends on the degree of refinement desired as output and also on knowledge of the system. A distributed parameter approach is justified only when available information is adequate. All components are evident in the models described in Sec. 10-3 to the end of the chapter.

### Data Needs for Hydrologic Simulation

The simulation of all or parts of a water system requires a resource inventory as part of the initial planning process. Most model input data requirements (90% or more) are map or field-available, or can be empirically determined or obtained from engineering handbooks and equations. A reasonably general list of data inventory topics that encompasses most hydrologic-economic modeling needs follows.

A. Basin and Subbasin Characteristics
   1. Lag times, travel times in reaches, times of concentration.
   2. Contributing areas, mean overland flow distances and slopes.
   3. Design storm abstractions: evapotranspiration, infiltration, depression, and interception losses. Composite curve numbers, infiltration capacities and parameters, $\phi$-indices.
   4. Land use practices, soil types, surface and subsurface divides.
   5. Water use sites for recreation, irrigation, flood damage reduction, diversions, flow augmentation, and pumping.
   6. Numbering system for junctions, subareas, gauging and precipitation stations.
   7. Impervious areas, forested areas, areas between isochrones, irrigable acreages.

B. Channel Characteristics
   1. Channel bed and valley floor profiles and slopes.

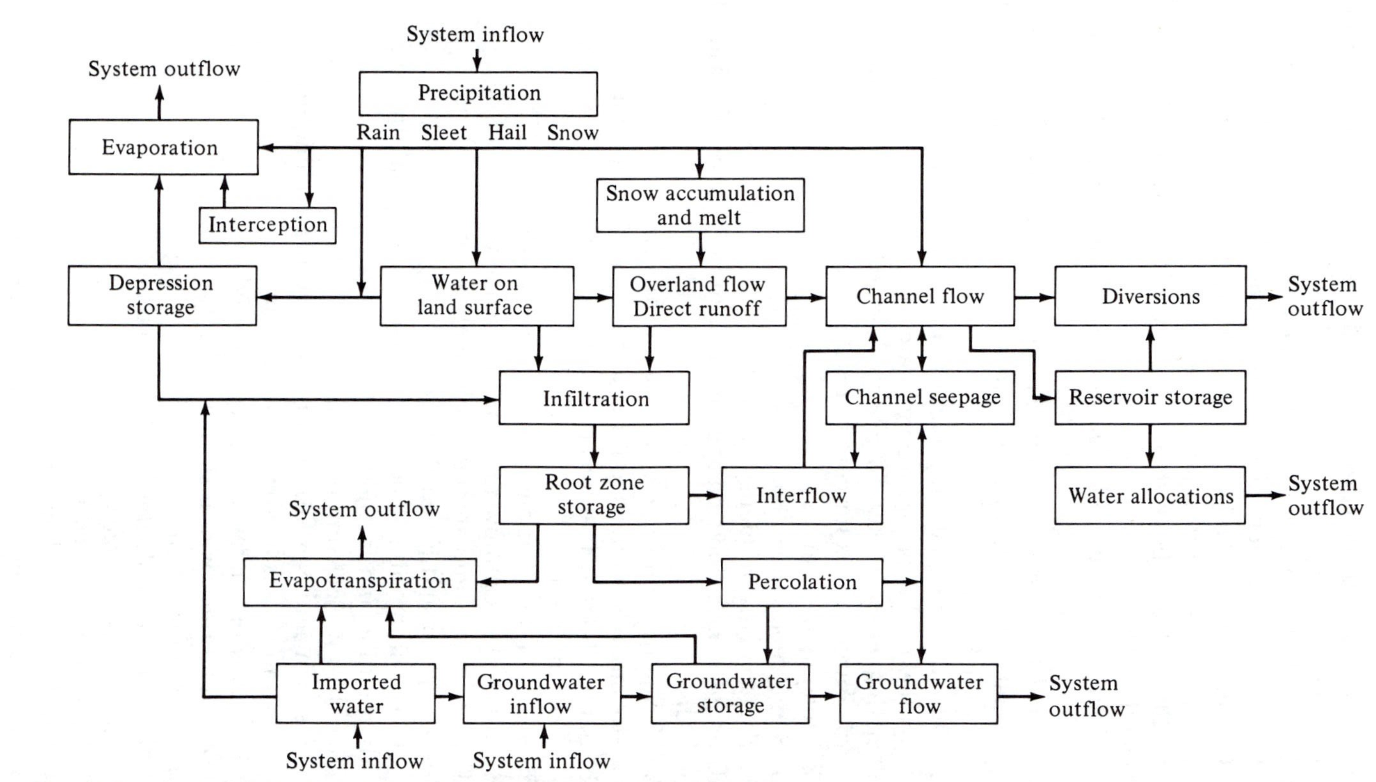


**Fig. 10-2.** *Components of a joint hydrologic water resource system.*

2. Manning's or Chézy coefficients for various reaches, or hydraulic or field data from which these coefficients could be estimated.
3. Channel and valley cross-sectional data for each river reach.
4. Seepage information; channel losses and base flows.
5. Channel and overbank storage characteristics, existing or proposed channelization and levee data.
6. Sediment loads, bank stability, and vegetative growth.

C. Streamflow Data
1. Hourly, daily, monthly, annual streamflow data at all gauging stations, including statistical analyses.
2. Flood frequency data and curves at gauging stations, or regional curves for ungauged sites, preferably on an annual and seasonal basis.
3. Flow duration data and curves at gauging stations (also any synthesized data for ungauged areas).
4. Rating curves; stage–discharge, velocity–discharge, depth–discharge curves for certain reaches.
5. Flooded area curves.
6. Stage versus area flooded.
7. Stage versus frequency curves.
8. Stage versus flood damage curves, preferably on a seasonal basis.
9. Hydraulic radius versus discharge curves.
10. Streamflows at ungauged sites as fractions of gauged values.
11. Return flows as fractions of water use allocations diverted for consumptive use.
12. Seasonal distributions of allocations to users.
13. Minimal streamflow to be maintained at each site.
14. Mass curves and storage–yield analyses at gauged sites.

D. Design Floods and Flood Routing
1. Design storm and flood determination; temporal and spatial distribution and intensity.
2. Standard project storms and floods.
3. Selection and verification of flood routing techniques to be used and necessary routing parameters.
4. Base flow estimates during design floods.
5. Available records of historic floods.

E. Reservoir Information
1. List of potential sites and location data.
2. Elevation—storage curves.
3. Elevation—area curves.
4. Normal, minimal, other pool levels.
5. Evaporation and seepage loss data or estimates.
6. Sediment, dead storage requirements.

7. Reservoir economic life.
8. Flood control operating policies and rule curves.
9. Outflow characteristics, weir and outlet equations, controls.
10. Reservoir-based recreation benefit functions.
11. Costs versus reservoir storage capacities.
12. Purposes of each reservoir and beneficiaries and benefits.

F. Preliminary Analyses
1. Identify water-user groups and all basin sites for hydropower production, reservoir-based recreation, irrigation, flood damage reduction from reservoir capacity, industrial and municipal water supply, diversions, and flow augmentation.
2. Compile annual and seasonal streamflows and flood records at each gauged site for the period of record at each site.
3. Determine the fraction of the allocation to each consumptive use that is assumed to return to the stream at each user site in the basin.
4. Perform frequency analyses of annual streamflow and flood values at each gauge site in the basin.
5. Determine for each reservoir site the evaporation and seepage losses.
6. Select mean probabilities to be used in the firm and secondary yield analyses.
7. Develop flood peak probability distributions at each potential flood damage center or reach in the basin.
8. Determine the fraction of water to be allocated during each period to each water use site in the basin.
9. Determine existing and proposed hydropower plant capacities and load factors.
10. Identify any minimal allowable streamflows to be maintained for flow augmentation at each flow augmentation reach in the basin.
11. Specify any maximal or minimal constraints on any of the annual or seasonal water allocations, storage capacities, or target yields.
12. Specify any constraints on maximal or minimal dead storage, active storage, flood control storage, or total storage capacities at any or all of the reservoir sites in the basin.
13. Determine annual capital, operation, maintenance, and replacement (OMR) costs at each reservoir site as functions of a range of total reservoir capacities or scales of development.
14. Determine benefits as functions of energy produced.
15. Determine annual capital and OMR costs at each hydropower production site as functions of various plant capacities.
16. Determine benefit-loss functions for a variety of allocations to domestic, commercial, industrial, and diversion uses.

17. Determine short-run losses as functions of deviations (both deficit and surplus) in planned or target allocations to user sites.
18. Develop benefit functions at each irrigation site in the basin. This analysis requires information on the area of land that can be irrigated per unit of water allocated, the quantities of each crop that can be produced per unit area of land, the total fixed and variable costs of producing each crop, and the unit prices that will clear the market of any quantity of each crop.
19. Develop flood-damage-reduction benefit functions at each potential flood damage site. This analysis requires records of historical and/or simulated floods, channel storage capacities, and flood control reservoir operating policies.
20. Develop reservoir-based recreation benefit functions at each recreation site in the basin.

## 10-3 Major Hydrologic Simulation Models

A few of the numerous event, continuous, and urban runoff digital models for simulating the hydrologic cycle are compared in Table 10-6. As shown in the table, most of the models were developed for, or by, universities or federal agencies. All have moderate-to-extensive input data requirements, and all have from 1 to 10% of inputs that are largely judgment parameters. These are normally validated by repeated trials with the models. The urban runoff models are primarily event simulation models but have been isolated in Table 10-6 because the descriptions of urban models are deferred to Chapter 11.

Several of the major event and continuous streamflow simulation models shown in Table 10-6 are described in varied detail here. The Stanford and HEC-1 models are emphasized as good examples, and applications of both are described in Secs. 10-4 and 10-5. For further reference, all the models listed in Table 10-6 are briefly described, along with about 100 other models, in the publication, "Models and Methods Applicable to Corps of Engineers Studies."[11] Fleming's text also presents complete descriptions of the SSARR, SWM, RRL, HSP, USDAHL, and other models.[12]

### API Model

This model was one of the earliest structured to give a deterministic simulation of a continuous streamflow hydrograph. It has been tested on watersheds of 68 and 817 mi$^2$ but must be calibrated to each watershed to obtain a reliable method of simulating the streamflow.[13] A flow diagram showing the structure is given in Fig. 10-3. Four basic components describe the interrelationships pertaining to this model of streamflow in a river: a unit hydrograph, an API rainfall-runoff

**Table 10-6** *Major Digital Simulation Models of Hydrologic Processes*

| Code Name | Model Name | Agency or Organization | % Inputs by Judgment[a] | Date of Original Development |
|---|---|---|---|---|
| | *Rainfall-Runoff Event Simulation Models* | | | |
| HEC-1 | HEC-1 Flood Hydrograph Package | Corps | 1 | 1973 |
| TR-20 | Computer Program for Project Hydrology | SCS | 5 | 1965 |
| USGS | USGS Rainfall-Runoff Model | USGS | 10 | 1972 |
| HYMO | Hydrologic Model Computer Language | ARS | 1 | 1972 |
| SWMM | Storm Water Management Model | EPA | 5 | 1971 |
| | *Continuous Streamflow Simulation Models* | | | |
| API | Antecedent Precipitation Index Model | Private | 1 | 1969 |
| USDAHL | 1970, 1973, 1974 Revised Watershed Hydrology | ARS | 1 | 1970 |
| SWM-IV | Stanford Watershed Model IV | Stanford Univ. | 10 | 1959 |
| KWM | Kentucky Watershed Model | Univ. Kentucky | 10 | 1969 |
| OPSET | Self-optimizing Hydrologic Simulation Model | Univ. Kentucky | 5 | 1970 |
| HSP | Hydrocomp Simulation Program | Private | 10 | 1967 |
| TWM | Texas Watershed Model | Texas Tech Univ. | 10 | 1970 |
| NWSRFS | National Weather Service Runoff Forecast System | | 10 | 1972 |
| SSARR | Streamflow Synthesis and Reservoir Regulation | Corps | 3 | 1958 |
| OSUSWM | Ohio State University Version of SWM-IV | OSU | 10 | 1972 |
| | *Urban Runoff Simulation Models* | | | |
| UCUR | University of Cincinnati Urban Runoff Model | Univ. Cincinnati | 2 | 1972 |
| NERO | Chicago Hydrograph Method | City of Chicago | 5 | 1970 |
| STORM | Quantity and Quality of Urban Runoff | Corps | 3 | 1974 |
| RRL | Road Research Laboratory Model | British | 1 | 1962 |
| MITCAT | MIT Catchment Model | MIT | 5 | 1970 |
| SWMM | Storm Water Management Model | EPA | 5 | 1971 |

[a] Judgment percentages are from U.S. Army Waterways Experiment Station, "Models and Methods Applicable to Corps of Engineers Urban Studies," Miscellaneous Paper H-74-8, National Technical Information Service, Aug. 1974.

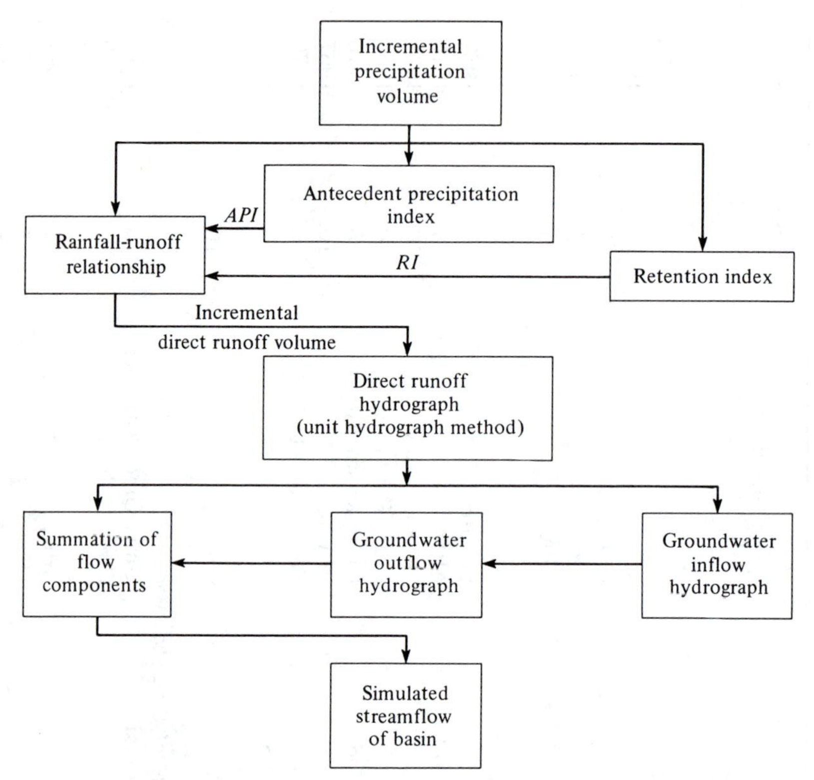


***Fig. 10-3.*** *Schematic diagram of API-type hydrologic model. (After W. T. Sittner, C. E. Schauss, and J. C. Monro, "Continuous Hydrograph Synthesis with an API-Type Hydrologic Model,"* Water Resources Research, *5, No. 5 (1969): 1007–1022.)*

connection, a relation for groundwater recession, and a relation for computing the groundwater flow hydrograph as a function of the direct runoff hydrograph. This model generates both groundwater flow and direct runoff discharge from precipitation values.

### Stanford Watershed Model IV (SWM-IV)

Crawford and Linsley designed this digital computer program to simulate portions (the land phase) of the hydrologic cycle for an entire watershed.[10] It has been widely accepted as a tool to synthesize a continuous hydrograph of hourly or daily streamflows at a watershed outlet. A lumped parameter approach is used and data requirements are much less than for distributed models. Hourly and daily precipitation data, daily evaporation data, and a variety of watershed parameters are input. Complete details of the model structure are included in the paragraphs that follow. A thorough description of the application of SWM-IV to two watersheds is included in Sec. 10-4.

The relationships and linkage of the various components of

SWM-IV are shown in Fig. 10-4. Hydrologic fundamentals are used at each point to transform the input data into a hydrograph of streamflow at the basin outlet. Rainfall and evaporation data are first entered into the program. Incoming rainfall is distributed, as shown in Fig. 10-4, among interception, impervious areas such as lakes and streams, and water destined to be infiltrated or to appear in the upper zone as surface runoff or interflow, both of which contribute to the channel inflow. The infiltration and upper zone storage eventually percolate to lower zone storage and to active and inactive groundwater storage. User-assigned parameters govern the rate of water movement between the storage zones shown in Fig. 10-4.

Three zones of moisture regulate soil moisture profiles and groundwater conditions. The rapid runoff response encountered in smaller watersheds is accounted for in the upper zone, while both upper and lower zones control such factors as overland flow, infiltration, and groundwater storage. The lower zone is responsible for longer-term infiltration and groundwater storage that is later released as base flow to the stream. The total stream flow is a combination of overland flow, groundwater flow, and interflow.

**Model Structure** The SWM-IV is made up of a sequence of computation routines for each process in the hydrologic cycle (interception, infiltration, routing, and so on). Separate discussions of each component are provided in the following paragraphs. Actual calculations proceed from circled process to circled process as illustrated by the arrows in Fig. 10-4. All the moisture that was originally stored in the watershed or was input as precipitation during any time period is balanced in the continuity equation

$$P = E + R + \Delta S \quad \textbf{(10-14)}$$

where

$P$ = precipitation
$E$ = evapotranspiration
$R$ = runoff
$\Delta S$ = the total change in storage in the upper, lower, and groundwater storage zones

The change in storage for each zone is calculated as the difference between the volumes of inflow and outflow. Further, all hydrologic activity in a time interval is simulated and balanced before the program proceeds to the next time interval to repeat the computations. The simulation terminates when no additional data are input.

**Interception** Interception is the first of several abstractions modeled by the SWM-IV. All incoming precipitation is intercepted unless the precipitation intensity exceeds the interception rate or if

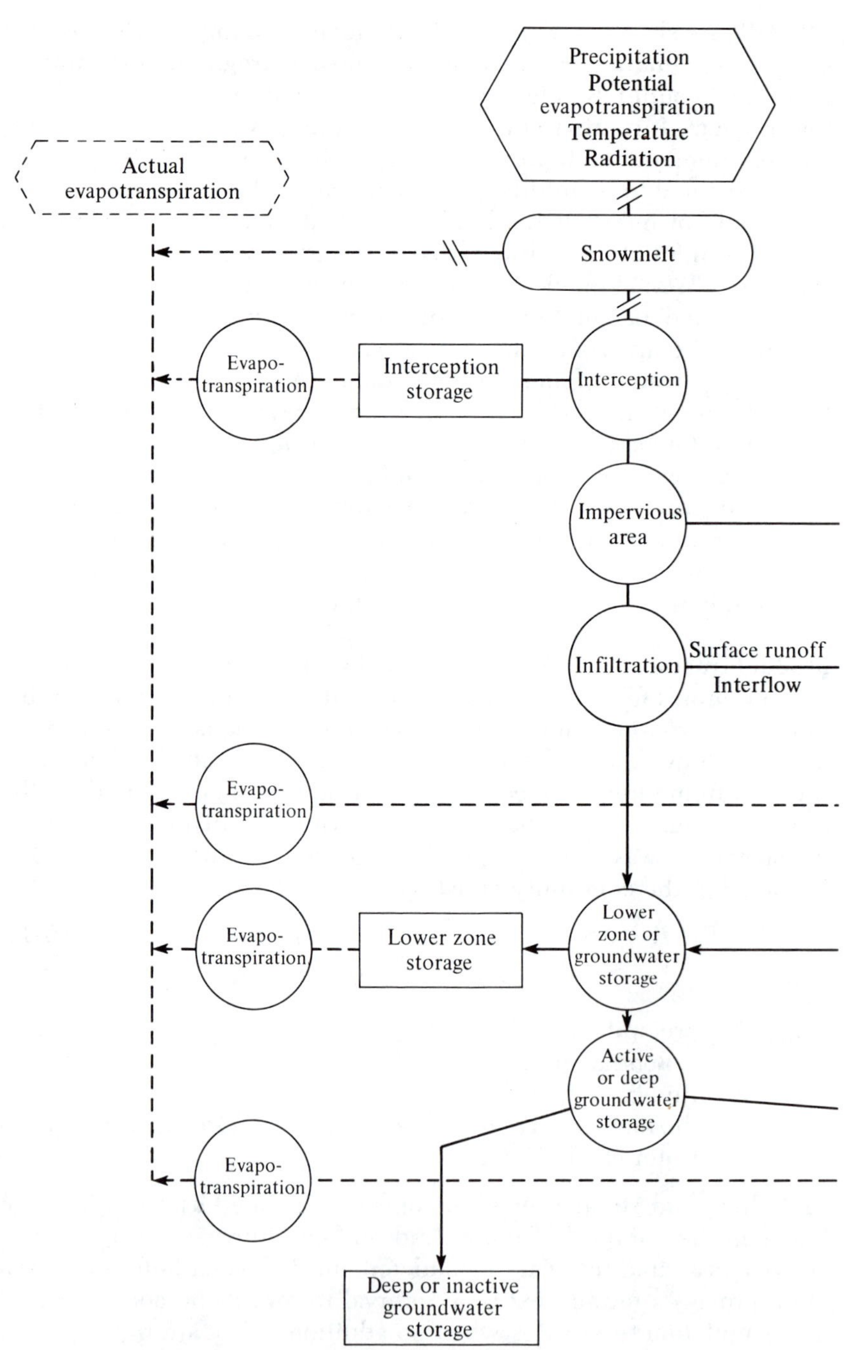


**Fig. 10-4.** *Stanford Watershed Model IV flow chart. (After N. H. Crawford and R. K. Linsley, Jr., "Digital Simulation in Hydrology: Stanford Watershed Model IV," Department of Civil Engineering, Stanford University, Stanford, Calif., Tech. Rept. No. 39, july 1966.)*

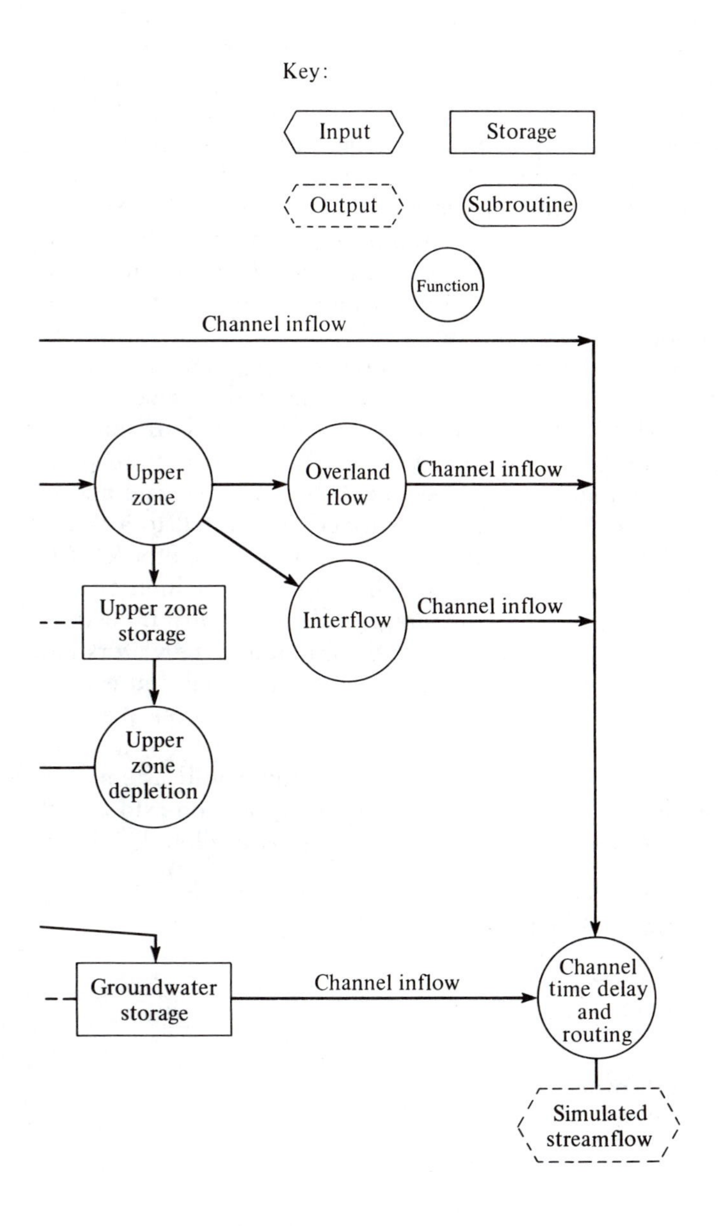
Key:
Input
Storage
Output
Subroutine
Function
Channel inflow
Upper zone
Overland flow
Channel inflow
Upper zone storage
Interflow
Channel inflow
Upper zone depletion
Groundwater storage
Channel inflow
Channel time delay and routing
Simulated streamflow

the interception storage fills. Interception rates depend on the precipitation rate and on the watershed cover. Typical values of interception maximums are provided in Table 10-7. The variable name EPXM is an identifier used in the ALGOL computer program.

**Evapotranspiration** The determination of evapotranspiration (ET) rates by SWM-IV was described in Sec. 3-4. Evapotranspiration occurs, as shown in Fig. 10-4, from interception storage, upper zone storage, lower zone storage, stream and lake surfaces, and from groundwater storage. Evapotranspiration from interception and upper zone storage is set equal to the potential rate, $E_p$, which is assumed to be the lake evaporation rate calculated as the product of a pan coefficient times the input values of the evaporation pan data. The evaporation of any intercepted water is assumed to occur at a rate equal to the potential evapotranspiration rate and ceases when the interception storage has been depleted.

Evaporation from stream and lake surfaces also occurs at the potential rate. The total volume is governed by the total surface area of streams and lakes (ETL) defined as the ratio of the total stream and lake area in the watershed to the total watershed area. Evapotranspiration from groundwater storage also occurs at the potential rate and is calculated in a similar fashion using a surface area equal to a factor $K24EL$ multiplied by the watershed area. Thus the parameter $K24EL$ represents the fraction of the total watershed area over which evapotranspiration from the groundwater storage will occur. Most investigators set this parameter at a value equal to the fraction of the watershed area covered by phreatophytes. Its value is normally small but can be large, for example, in an agricultural area that has many acres of alfalfa subirrigation.

If interception storage is depleted, the model will attempt to satisfy the potential for ET by drawing from the upper zone storage at the potential rate. Once the upper zone storage is depleted, ET oc-

***Table 10-7*** *Typical Maximum Interception Rates*

| Watershed Cover | EPXM (in./hr) |
|---|---|
| Grassland | 0.10 |
| Moderate forest cover | 0.15 |
| Heavy forest cover | 0.20 |

*Source:* After N. H. Crawford and R. K. Linsley, Jr., "Digital Simulation in Hydrology: Stanford Watershed Model IV," Department of Civil Engineering, Stanford University, Tech. Rept. No. 39, Stanford, Calif., July 1966.

curs from the lower zone but not at the potential rate; the ET rate from the lower zone is always less than $E_p$. When interception and the upper zone storage do not satisfy the potential, any excess enters as $E_p$ in Fig. 3-4, and the rate of evapotranspiration from the lower zone is determined from the shaded area, or

$$E = E_p - \frac{E_p^2}{2r} \tag{10-15}$$

The variable $r$ is the evapotranspiration opportunity, defined as the maximum water amount available for ET at a particular location during a prescribed time period. This factor varies from point to point over any watershed from zero to a maximum value of

$$r = K3 \frac{LZS}{LZSN} \tag{10-16}$$

where

$LZS$ = the current soil moisture storage in the lower zone (in.)
$LZSN$ = a nominal storage level, normally set equal to the median value of the lower zone storage (in.). See pages 447, 448.
$K3$ = an input parameter that is a function of watershed cover as shown in Table 10-8

The ratio $LZS/LZSN$ is known as the lower zone soil moisture ratio and is used to compare the actual lower zone storage with the nominal value at any time. Values of ET opportunity are assumed to vary over a watershed from zero to $r$ along the straight line shown in Fig. 3-4. This assumed linear cumulative distribution of the parameter over an area is also used in evaluating areal distributions of infiltration rates.

**Infiltration** Like the evapotranspiration opportunity, the infiltration capacity of a watershed is highly variable from point to point, and is

***Table 10-8*** *Typical Lower Zone Evapotranspiration Parameters*

| Watershed Cover | *K*3 |
|---|---|
| Open land | 0.20 |
| Grassland | 0.23 |
| Light forest | 0.28 |
| Heavy forest | 0.30 |

*Source:* After N. H. Crawford and R. K. Linsley, Jr., "Digital Simulation in Hydrology: Stanford Watershed Model IV." Department of Civil Engineering, Stanford University, Tech. Rept. No. 39, Stanford, Calif., July 1966.

assumed to be distributed according to a linear cumulative distribution function shown as a line from the origin to point $b$ in Fig. 10-5.

Infiltration into the lower and groundwater storage zones is determined as a function of the moisture supply, $\bar{x}$, available for infiltration. Steps to determine infiltration for a given moisture supply, $\bar{x}$, are

1. The net infiltration is determined from the area labeled "infiltration" in Fig. 10-5. This water is assumed to infiltrate into the lower and groundwater storage zones. The area enclosed by the trapezoid is given by the equations in the first row of Table 10-9. If the moisture supply, $\bar{x}$, exceeds the maximum infiltration capacity, $b$, the maximum allowed net infiltration is $b/2$, which is the median infiltration capacity.
2. Some of the moisture supply contributes to an increase in the interflow detention during any time increment and is calculated as the region indicated by an arrow in Fig. 10-5. Equations for this area using various ranges in $\bar{x}$ are provided in the second row of Table 10-9. The volume of water in a state of being transported as interflow at any instant is called the interflow detention or detained interflow.
3. Any remaining moisture supplied, $\Delta D$ in Fig. 10-5, contributes to increasing the surface detention during the time increment. Equations for this triangular-shaped area are included in Table 10-9 for various values of $\bar{x}$.

The quantity of net infiltration is controlled largely by the maximum infiltration capacity $b$, while the parameter $c$ significantly affects hydrograph shapes because the parameter controls the amount of water detained during the time increment. The values of $b$ and $c$ for any time interval depend on the soil moisture ratio, $LZS/LZSN$, and on the input parameters $CB$ and $CC$; $CB$ is an index that controls the

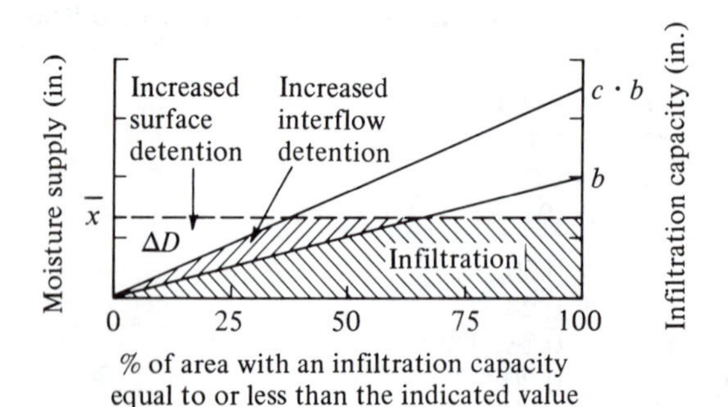


***Fig. 10-5.*** *Assumed linear areal variation of infiltration capacity over a watershed. (After N. H. Crawford and R. K. Linsley, Jr., "Digital Simulation in Hydrology: Stanford Watershed Model IV," Department of Civil Engineering, Stanford University, Stanford, Calif., Tech. Rept. No. 39, July 1966.)*

**Table 10-9** *Equations for the Shaded Areas in Fig. 10-5*

| Component | $\bar{x} < b$ | $b < \bar{x} < c \cdot b$ | $\bar{x} > c \cdot b$ |
|---|---|---|---|
| Net infiltration | $\bar{x} - \frac{\bar{x}^2}{2b}$ | $\frac{b}{2}$ | $\frac{b}{2}$ |
| Increase in interflow detention | $\frac{\bar{x}^2}{2b}\left(1 - \frac{1}{c}\right)$ | $\bar{x} - \frac{b}{2} - \frac{\bar{x}^2}{2c \cdot b}$ | $\frac{b}{2}(c - 1)$ |
| Increase in surface detention | $\frac{\bar{x}^2}{2c \cdot b}$ | $\frac{\bar{x}^2}{2c \cdot b}$ | $\bar{x} - \frac{c \cdot b}{2}$ |
| Percentage of increased detention assigned to interflow | $100\left(1 - \frac{1}{c}\right)$ | $100\left\{1 - \frac{\bar{x}^2}{2c \cdot b[x - (b/2)]}\right\}$ | $\left[100 \frac{c - 1}{(2\bar{x}/b) - 1}\right]$ |

*Source:* After N. H. Crawford and R. K. Linsley, Jr., "Digital Simulation in Hydrology: Stanford Watershed Model IV," Department of Civil Engineering, Stanford University, Tech. Rept. No. 39, Stanford, Calif., July 1966.

rate of infiltration and depends on the soil permeability and the volume of moisture that can be stored in the soil. Values in the range from 0.3 to 1.2 are common. The parameter $CC$ is an input value that fixes the level of interflow relative to the overland flow.

If the soil moisture ratio is less than 1.0, the variable $b$ is found from

$$b = \frac{CB}{2^{(4 \cdot LZS/LZSN)}} \tag{10-17}$$

and when $LZS/LZSN$ is greater than 1.0, the equation for $b$ is

$$b = \frac{CB}{2^{\{4.0+2 \cdot [(LZS/LZSN)-1.0]\}}} \tag{10-18}$$

These equations were developed by Crawford and Linsley from numerous trials using SWM-IV in many different watersheds. When the soil moisture ratio reaches a value of 2.0, the variable $b$ reaches its minimum value of $\frac{1}{64}$ of $CB$. The parameter $c$ is determined from

$$c = (CC)\ 2^{(LZS/LZSN)} \tag{10-19}$$

Variations in parameters $b$ and $c$ with changes in $LZS/LZSN$ are shown in Figs. 10-6 and 10-7. Mid-range values of $CB = 1.0$ and $CC = 1.0$ were used in developing these curves.

Figure 10-8 is a graph of the distribution of water among infiltration, interflow, and overland flow for various values of the moisture supply, $\bar{x}$. Different values of $b$ and $c$ would produce a different set of curves.

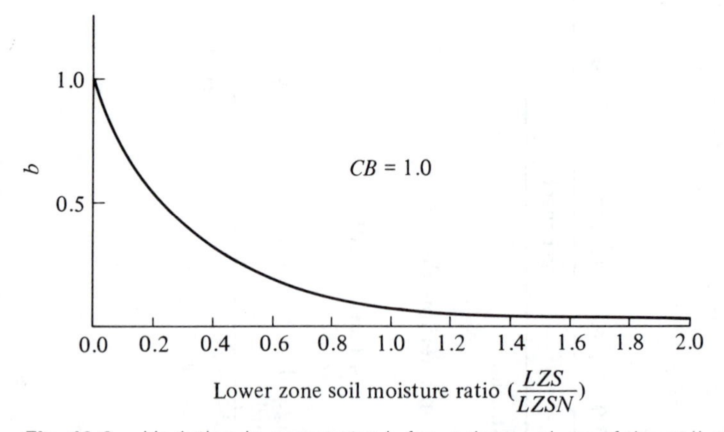


*Fig. 10-6.* *Variation in parameter b for various values of the soil moisture ratio. (After N.H. Crawford and R. K. Linsley, Jr., "Digital Simulation in Hydrology: Stanford Watershed Model IV," Department of Civil Engineering, Stanford University, Stanford, Calif., Tech. Rept. No. 39, July 1966.)*

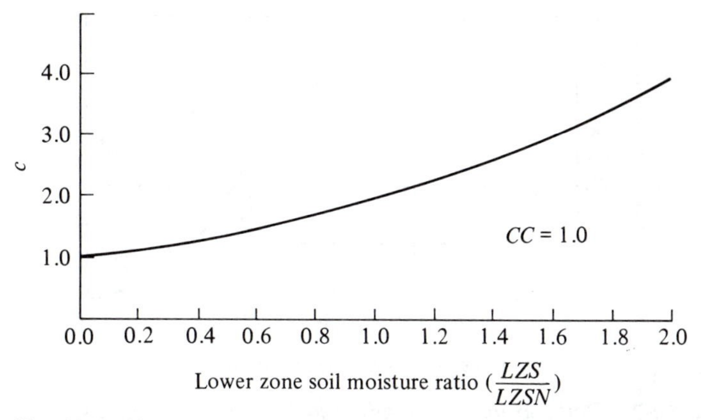


***Fig. 10-7*** *Variation in parameter c for various values of the soil moisture ratio. (After N. H. Crawford and R. K. Linsley, Jr., "Digital Simulation in Hydrology: Stanford Watershed Model IV," Department of Civil Engineering, Stanford University, Stanford, Calif., Tech. Rept. No. 39, July 1966.)*

Water stored as overland flow surface detention will either contribute to streamflow or enter the upper zone storage as depicted in Fig. 10-4. That portion which enters the upper zone storage is called delayed infiltration and is a function of the upper zone soil moisture ratio, $UZS/UZSN$, as shown in Fig. 10-9. The inflection point occurs at a soil moisture ratio of 2.0. If the ratio is less than 2.0, the percent retained by the upper zone is given by

$$P_r = 100\left[1.0 - \left(\frac{UZS}{2UZSN}\right)\left(\frac{1.0}{1.0 + UZI1}\right)^{UZI1}\right] \tag{10-20}$$

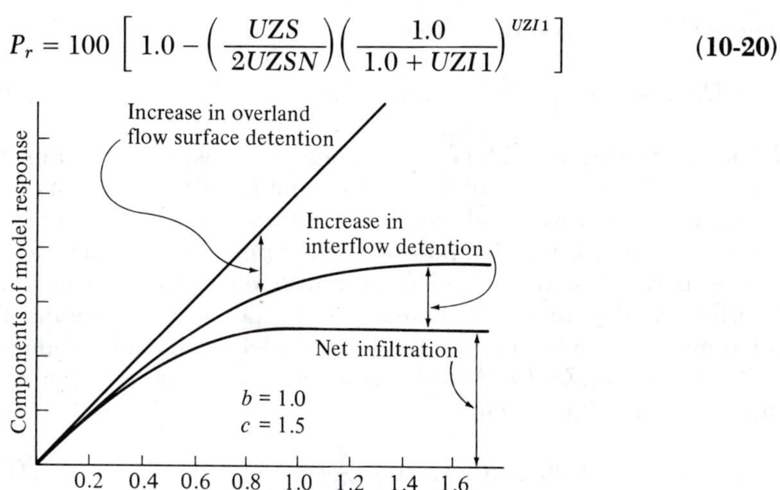


***Fig. 10-8.*** *Typical SWM-IV response to moisture supply variations. (After N. H. Crawford and R. K. Linsley, Jr., "Digital Simulation in Hydrology: Stanford Watershed Model IV," Department of Civil Engineering, Stanford University, Stanford, Calif., Tech. Rept. No. 39, July 1966.)*

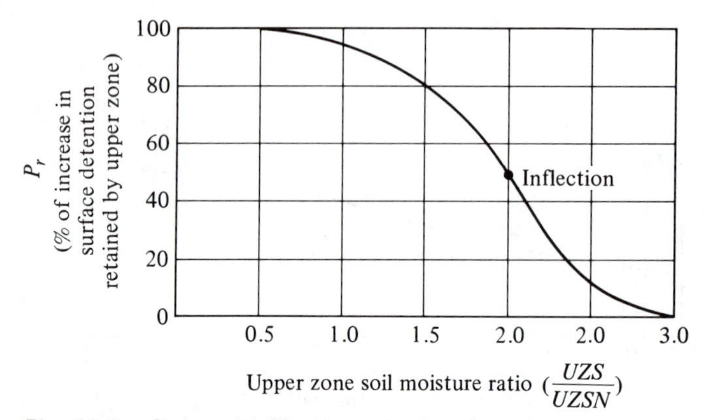


***Fig. 10-9.*** *Delayed infiltration as a function of upper zone soil moisture ratio. (After N. H. Crawford and R. K. Linsley, Jr., "Digital Simulation in Hydrology: Stanford Watershed Model IV," Department of Civil Engineering, Stanford University, Stanford, Calif., Tech. Rept. No. 39, July 1966.)*

where $UZI1$ is determined from

$$UZI1 = 2.0\left|\left(\frac{UZS}{2UZSN}\right) - 1.0\right| + 1.0 \qquad \textbf{(10-21)}$$

The curve is defined to the right of the inflection point by

$$P_r = 100\left[1.0 - \left(\frac{1.0}{1.0 + UZI2}\right)^{UZI2}\right] \qquad \textbf{(10-22)}$$

where $UZI2$ is determined from

$$UZI2 = 2.0\left|\left(\frac{UZS}{UZSN}\right) - 2.0\right| + 1.0 \qquad \textbf{(10-23)}$$

**Upper Zone Storage** The upper storage zone, as shown in Fig. 10-4, receives a large portion of the rain during the first few hours of the storm, while the lower and groundwater storage zones may or may not receive any moisture. That portion of the upper zone storage which is not evaporated or transpired is proportioned to the surface runoff, interflow, and percolation. Percolation (upper zone depletion) from the upper zone to the lower zone in Fig. 10-4 occurs only when $UZS/UZSN$ exceeds $LZS/LZSN$. When this occurs, the percolation rate in in./hr is determined from

$$PERC = 0.003(CB)(UZSN)\left(\frac{UZS}{UZSN} - \frac{LZS}{LZSN}\right)^3 \qquad \textbf{(10-24)}$$

where $CB$ is an index that controls the rate of infiltration. It ranges from 0.3 to 1.2 depending on the soil permeability and on the volume of moisture that can be stored in the soil. The variables $UZS$ and $UZSN$ are defined as the actual and nominal soil moisture storage amounts in the upper zone. The nominal value of *UZSN* is approximately a function of watershed topography and cover, and is always considered to be much smaller than the nominal *LZSN* value. The initial estimates of *UZSN* relative to *LZSN* are found from Table 10-10.

The parameters *LZSN* and *CB* must also be estimated at the beginning of a simulation study. The combination that will most satisfactorily reproduce both long- and short-term historical responses to hydrologic inputs can be determined by the following procedure:[10]

1. Assume an initial value for *LZSN* equal to one-quarter of the mean annual rainfall plus 4 in. (used in arid and semiarid regions), or one-eighth of the annual mean rainfall plus 4 in. (used in coastal, humid or subhumid climates).
2. Determine the initial value of *UZSN* from Table 10-10.
3. Assume a value for *CB* in the normal range from 0.3 to 1.2.
4. Simulate a period of record using the streamflow, rainfall, and evaporation data; and systematically adjust *LZSN, UZSN, CB,* and other parameters until agreement between synthesized and recorded streamflows is satisfactory. If the annual water budgets do not balance, *LZSN* is adjusted; *CB* is adjusted on the basis of comparisons between synthesized and recorded flow rates for individual storms.

**Lower Zone Storage and Groundwater** The lower groundwater storage zone in Fig. 10-4 receives water from the net infiltration and from percolation. The percolation rate is determined from Eq. 10-24. The percentage of net infiltration that reaches groundwater storage depends on the soil moisture ratio *LZS*/*LZSN* as shown in Fig. 10-10. If this ratio is less than 1.0, the percentage $P_g$ is found from

$$P_g = 100\left[\frac{LZS}{LZSN}\left(\frac{1.0}{1.0 + LZI}\right)^{LZI}\right] \qquad \textbf{(10-25)}$$

***Table 10-10*** *Values of* UZSN *as a function of* LZSN *for Initial Estimates in Simulation with SWM-IV*

| Watershed | *UZSN* |
|---|---|
| Steep slopes, limited vegetation, low depression storage. | 0.06*LZSN* |
| Moderate slopes, moderate vegetation, moderate depression storage. | 0.08*LZSN* |
| Heavy vegetal or forest cover, soils subject to cracking, high depression storage, very mild slopes. | 0.14*LZSN* |

*Source:* After N. H. Crawford and R. K. Linsley, Jr., "Digital Simulation in Hydrology: Stanford Watershed Model IV," Department of Civil Engineering, Stanford University, Tech. Rept. No. 39, Calif., July 1966.

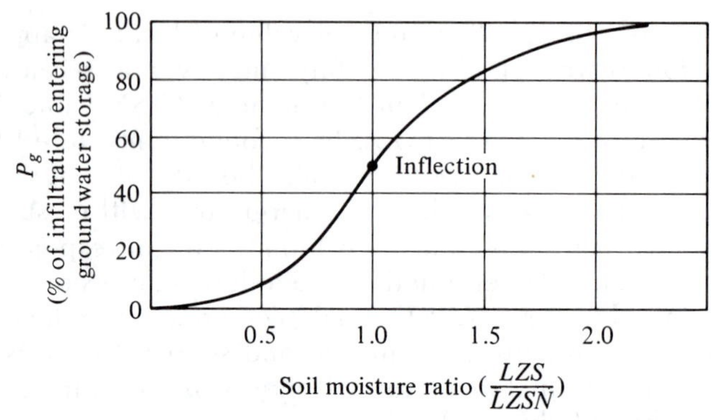


***Fig. 10-10.*** *Percentage of infiltrated water that reaches groundwater storage. (After N. H. Crawford and R. K. Linsley, Jr., "Digital Simulation in Hydrology: Stanford Watershed Model IV," Department of Civil Engineering, Stanford University, Stanford, Calif., Tech. Rept. No. 39, July 1966.)*

and if $LZS/LZSN$ is greater than 1.0, the percentage is

$$P_g = 100\left[1.0 - \left(\frac{1.0}{1.0 + LZI}\right)^{LZI}\right] \tag{10-26}$$

In both equations, the variable $LZI$ is defined as

$$LZI = 1.5\left|\frac{LZS}{LZSN} - 1.0\right| + 1.0 \tag{10-27}$$

Note from Fig. 10-10 that the nominal storage $LZSN$ equals the lower zone storage $LZS$ when half or 50% of all the incoming moisture enters groundwater storage.

The outflow from the groundwater storage, $GWF$, at any time is based on the commonly used linear semilogarithmic plot of base flow discharge versus time. This technique was described in Sec. 4-7 and illustrated in Fig. 4-15. In modified form the base flow equation is

$$GWF = (LKK4)[(1.0 + KV(GWS)](SGW) \tag{10-28}$$

where $LKK4$ is defined by

$$LKK4 = 1.0 - (KK24)^{1/96} \tag{10-29}$$

in which $KK24$ is the minimum of all the observed daily recession constants (see Sec. 4-7), where each constant is the ratio of the groundwater discharge rate to the groundwater discharge rate 24 hr earlier. Thus the recession constant $KK24$ ($K$ in Eq. 4-13) is determined using $t = 1$ day. The variable $GWS$ in Eq. 10-28 has values that depend on the long-term inflows to groundwater storage. Its value on any given day (for example, the $i$th day) is calculated as 97% of the previous day's value, adjusted for any inflow to groundwater storage, or $GWS_i = 0.97\ (GWS_{i-1} + \text{inflow to groundwater storage during day } i)$.

In Eq. 10-28, *SGW* is a groundwater storage parameter that reflects the fluctuations in the volume of water stored, and ranges from 0.10 to 3.90 in. The term *KV* in Eq. 10-28 allows for changes that are known to exist in the groundwater recession rates as time passes. When *KV* is zero, Eq. 10-28 reduces to Eq. 4-13 and the groundwater recession follows the linear semilog relationship. If the usual dry season recession rate *KK*24 is too large for wet periods (when groundwater storages are being recharged by seepage from the streams) the parameter *KV* is hand-adjusted so that the term $1.0 + KV(GWS)$ will reduce the effective rate to some desired value during recharge periods. Table 10-11 illustrates this computation by showing effective recession rates for various combinations of *KK*24 and *GWS* when *KV* is set equal to 1.0.

The fraction of active or deep groundwater storage that is either lost to deep or inactive groundwater storage (Fig. 10-4) or is diverted as flow across the drainage basin boundary is input as parameter *K24L*. This fraction is the total inflow to groundwater and represents all of the active groundwater storage that does not contribute to streamflow.

**Overland Flow** The overland flow process has been studied by many investigators. A wide range of methods for estimating the velocities and depths of sheet flow over a rough land surface has been applied and falls in the hydraulic-hydrologic categories of Chapter 7. Hydraulic overland flow methods involve finite difference and other numerical techniques to solve at various points the partial differential equations of continuity and momentum for unsteady overland flow. The hydrologic methods, including those adopted in SWM-IV, approximate the velocities and depths for unsteady overland flows by a lumped parameter approach that requires much less data than the hydraulic techniques.

***Table 10-11*** *Effective Recession Rates for Various Combinations of KK24 and GWS when KV = 1.0*

| | GWS | | | |
|---|---|---|---|---|
| KK24 | 0.0 | 0.5 | 1.0 | 2.0 |
| 0.99 | 0.99 | 0.985 | 0.98 | 0.97 |
| 0.98 | 0.98 | 0.970 | 0.96 | 0.94 |
| 0.97 | 0.97 | 0.955 | 0.94 | 0.91 |
| 0.96 | 0.96 | 0.940 | 0.92 | 0.88 |

*Source:* After N. H. Crawford and R. K. Linsley, Jr., "Digital Simulation in Hydrology: Stanford Watershed Model IV," Department of Civil Engineering, Stanford University, Rept. No. 39, Stanford, Calif., July 1966.

Average values of lengths, slopes, and roughnesses of overland flow in the Manning and continuity equations are used in SWM-IV to continuously calculate the surface detention storage $D_e$. The overland flow discharge rate $q$ is then related to $D_e$.

As the rain supply rate continues in time, the amount of water detained on the surface increases until an equilibrium depth is established. The amount of surface detention at equilibrium estimated by SWM-IV is

$$D_e = \frac{0.000818\, i^{0.6}\, n^{0.6}\, L^{1.6}}{S^{0.3}} \tag{10-30}$$

where

$D_e$ = the surface detention at equilibrium (ft³/ft) of overland flow width
$i$ = the rain rate (in./hr)
$S$ = the slope (ft/ft)
$L$ = the length of overland flow (ft)

The overland flow discharge rate is next determined as a function of detention storage from

$$q = \frac{1.486}{n} S^{1/2} \left(\frac{D}{L}\right)^{5/3} \left(1.0 + 0.6\left(\frac{D}{D_e}\right)^3\right)^{5/3} \tag{10-31}$$

where

$q$ = the overland flow discharge rate (cfs per ft of width)
$D$ = the average detention storage during the time interval

The equation also applies during the recession that occurs after rain ceases, but the ratio $D/D_e$ is assumed to be 1.0. Typical overland flow roughness coefficients are provided in Table 10-12.

***Table 10-12*** *Typical Manning Equation Overland Flow Roughness Parameters, NN*

| Watershed Cover | Manning's *n* for Overland Flow |
|---|---|
| Smooth asphalt | 0.012 |
| Asphalt or concrete paving | 0.014 |
| Packed clay | 0.03 |
| Light turf | 0.20 |
| Dense turf | 0.35 |
| Dense shrubbery and forest litter | 0.40 |

*Source:* After N. H. Crawford and R. K. Linsley, Jr., "Digital Simulation in Hydrology: Stanford Watershed Model IV," Department of Civil Engineering, Stanford University, Tech. Rept. No. 39, Stanford, Calif., July 1966.

The time at which detention storage reaches an equilibrium is determined from

$$t_e = \frac{0.94\, L^{3/5}\, n^{3/5}}{i^{2/5}\, S^{3/10}} \tag{10-32}$$

where $t_e$ is the time to equilibrium (min). Crawford and Linsley show that these equations very accurately reproduce measured overland flow hydrographs.[10]

For each time interval, $\Delta t$, an end-of-interval surface detention $D_2$ is calculated from the initial value $D_1$ plus any water added $\Delta D$ (Fig. 10-5) to surface detention storage during the time interval, less any overland flow discharge $\bar{q}$ that escapes from detention storage during the time interval. This is simply an expression of continuity, or

$$D_2 = D_1 + \Delta D - \bar{q}\Delta t \tag{10-33}$$

The discharge $\bar{q}$ is found from Eq. 10-31 using a value of $D = (D_1 + D_2)/2$. Equations 10-30 through 10-33 allow the complete determination of overland flow using easily found basin-wide values of the average length, slope, and roughness of overland flow.

**Interflow** The water temporarily detained as interflow storage is treated in the same fashion as overland flow detention storage. The inflow to interflow detention was defined in Fig. 10-5. The outflow is simulated using a daily recession constant similar to that defined for groundwater discharge. The interflow recession constant *IRC* is the average ratio of the interflow discharge at any time to the interflow discharge 24 hours earlier. For each 15-min time interval modeled, the outflow from detention storage is

$$INTF = LIFC4(SRGX) \tag{10-34}$$

where

$$LIRC4 = 1.0 - (IRC)^{1/96} \tag{10-35}$$

The variable *SRGX* is the water stored in the interflow detention at any time. Its value continuously changes when the continuity equation is applied to each time interval. The end-of-interval value of *SRGX* depends, according to continuity, on the value at the beginning of the interval and any inflow to or discharge from the interflow detention during the interval.

**Channel Translation and Routing** The Stanford Watershed Model utilizes a hydrologic watershed routing technique to translate the channel inflow to the watershed outlet. The time-area method described in Sec. 7-3 is adopted almost as presented in Chapter 7. The only difference is that the design storm hyetograph of Fig. 7-7 is not used in the Stanford Channel routing model. In place of the net rain hyetograph, the Stanford model views the sum of all channel inflow components as an "inflow" hyetograph. This inflow is then translated in time through the channel to the basin outlet, where it is next routed through an equivalent storage system to account for the attenuation caused by storage in the channel system. Routing through the linear reservoir (linear in the sense that storage is assumed to be directly proportional to the outflow, Eq. 7-14) is accomplished from

$$O_2 = \bar{I} - KS1\,(\bar{I} - O_1) \qquad \textbf{(10-36)}$$

where

$O_2$ = the outflow rate at the end of the time interval
$O_1$ = the outflow rate at the beginning
$\bar{I}$ = the average inflow rate during the time interval

Also,

$$KS1 = \frac{(1/K - \Delta t/2)}{(1/K + \Delta t/2)} \qquad \textbf{(10-37)}$$

Equation 7-33 is used to determine the storage coefficient $K$. Examples of the determination of this and other necessary parameters from watershed data are included in Sec. 10-4.

**Applications of the SWM-IV** Applications of the model typically begin with data for a 3- to 6-yr calibration period for which rainfall and runoff data are available. These data are used to allow successive adjustments of several parameters until the simulated and recorded hydrographs of the streamflow agree. If sufficient data are available, a second period of record may be reserved for use as a control to check the accuracy of the parameters derived from a calibration with the first half of the data.

The Stanford Watershed Model was originally developed in 1959 and has undergone several modifications since that time. James[14] translated the Crawford and Linsley model from ALGOL to FORTRAN. Several modifications of the FORTRAN version have evolved from a variety of investigations. Included among these are the Kentucky Watershed Model (KWM), the Kentucky self-calibrating version (OPSET), the Ohio State University version, the Texas version, the HYDROCOMP Simulation Program (HSP), and the National Weather Service Runoff Forecasting Model. Brief descriptions of several of these are included in subsequent paragraphs.

### Kentucky Watershed Model (KWM)

This revision of the Stanford model is described and documented by Liou[15] and is called the Kentucky Watershed Model

(KWM), or simply the Kentucky version. Differences between the Stanford and Kentucky versions are largely adaptations that make the Kentucky model applicable to the climate and geography of Kentucky and other parts of the humid eastern portion of the United States. Other modifications in computational efficiency and output format are included in the Kentucky version.

James describes a number of the many world-wide applications of the KWM and its adapted forms.[16] In its original form, the model use is limited to homogenous subwatersheds over 1 $mi^2$ in area but no larger than 2000 $mi^2$.

### Kentucky Self-calibrating Version (OPSET)

Both the Stanford and Kentucky Watershed models require a trial-and-adjustment technique to determine values of several parameters that are difficult to estimate from available data. Reasonable values of the parameters are generally estimated in initial calibration trials and are then adjusted in successive trials to produce a better match of simulated and recorded streamflows. Considerable judgment and experience in using this trial-and-adjustment technique are required for efficient calibration of the Stanford and Kentucky versions.

Recognizing the availability of optimization techniques for obtaining an optimal set (OPSET) of parameters, Liou developed and documented a self-calibrating watershed model.[15] The approach used is an analytical matching of synthesized and recorded streamflows which automatically estimates 13 parameters that previously were determined by the trial-and-adjustment technique. Included among these parameters are indices of the watershed storage capacities, infiltration rates, evapotranspiration rates, and other factors used in translating ground and surface waters into streamflow. These parameters are described further in Sec. 10-4.

### Ohio State University Version

Several significant contributions to the capabilities of the Stanford Watershed Model are included in the Ohio State University version developed by Ricca.[17] To facilitate the analysis of previously tabulated results, Ricca added several subroutines that graphically plot recorded and synthesized hydrographs along with rainfall hyetographs (graphs of intensity versus time) superimposed over selected storm hydrographs.

Additional modifications contained in the OSU version include the addition of input parameters that control amounts of water entering soil cracks and amounts of water used to recharge watershed swamps and marshes that dry up in middle or late summer. Provisions for introducing several groundwater recession constants (compared to a single value used by the Stanford and Kentucky versions) for geographic areas of stratified geology have also been included.

Another major revision included in the OSU version is the capability of using streamflow routing and other time increments smaller than the minimum 15-min increment of the original versions. This modification implies that the OSU version may be used to simulate streamflows for small watersheds with a time of concentration less than 15 min. The OSU version also includes a new snowmelt component for applications in regions with small amounts of snow and relatively little snow survey data.

#### Revised Model of Watershed Hydrology (USDAHL)

Growing interest in the effects of soil types, vegetation, pavements, and farming practices on infiltration and overland flow has resulted in the growth of the USDAHL continuous simulation model. The 1974 version[18] of this model was developed by investigators at the Agricultural Research Service Hydrograph Laboratory, primarily to serve the purposes of agricultural watershed engineers.

Input data to the model is relatively extensive. Continuous records of the precipitation; the weekly average temperatures; the weekly average pan-evaporation amounts; and detailed data on soils, vegetation, land use, and cultural practices are required.

The study watershed is initially divided into as many as four distinct land use or soil-type zones. Fourteen subroutines and a main calling routine compute for each zone the snowmelt, infiltration, overland flow, channel flow, evapotranspiration, groundwater evaporation and movement, groundwater recharge, and return flow.

Evapotranspiration potentials are estimated by applying assigned coefficients to pan-evaporation data. Infiltration for each soil or land use zone is computed using a modified Holtan equation. Water stored in cracks in dry soils is simulated as a function of soil moisture. Manning's equation and the continuity equation are used to route overland flow. The streamflow is routed by a simultaneous solution of the continuity equation and a storage function. Groundwater movements are calculated by Darcy's equation. The daily snowmelt on each zone is calculated as a function of the temperature at which snowmelt starts, the weighted average vegetative density for the zone, the weekly average air temperature, and the potential snowmelt per day in the zone snowpack. Precipitation falling during a snowmelt day also contributes to the snowmelt equation.

Among other uses, the model has been applied by the Soil Conservation Service in preparing environmental impact statements. Figure 10-11 shows the results of applying the 1974 version to annual runoff from four widely separated and widely diversified ARS experimental watersheds. In addition to the runoff, the model computes the evapotranspiration amounts, soil moisture changes, return flows, and groundwater recharge depths for each of the zones.

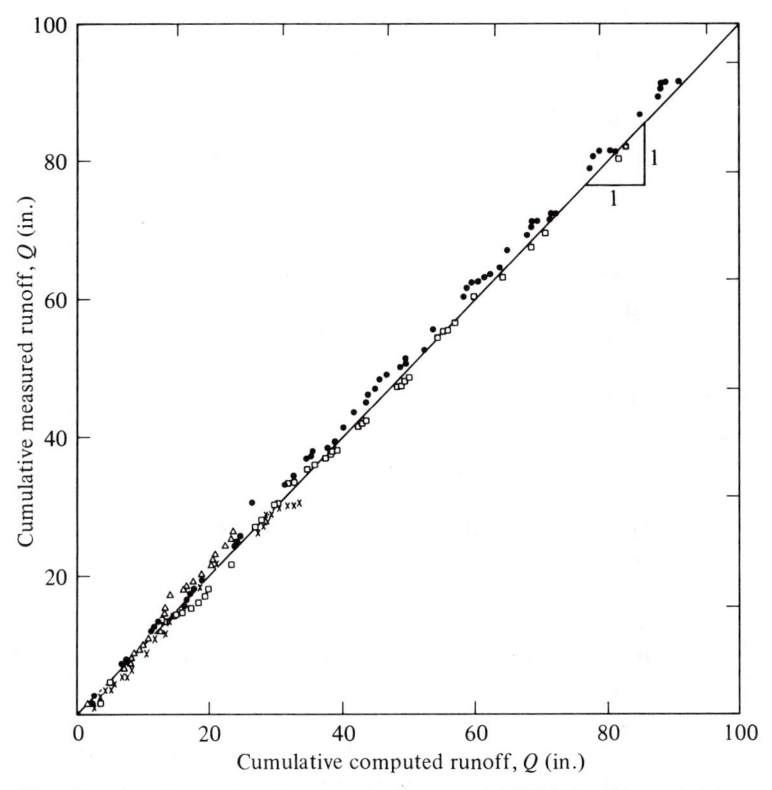


***Fig. 10-11.*** *Chart showing the accuracy of USDAHL–74 model for estimating the cumulative computed runoff as compared with the cumulative measured runoff at four watersheds. ● W-97, Coshocton, Ohio; △ W-11, Hastings, Nebr.; □ W-3, Ft. Lauderdale, Fla.; × W-G, Riesel, Tex. (After H. N. Holtan and N. C. Lopez, "USDAHL–73 Revised Model of Watershed Hydrology," U.S. Department of Agriculture, Plant Physiology Institute, Rept. No. 1, 1973.)*

Although other modifications are possible, the USDAHL model is specifically designed for relatively small rural watersheds.

### Texas Watershed Model (TWM)

Claborn and Moore developed the University of Texas version of the Stanford model, incorporating significant changes in the model time steps and input data.[19] Time increments may be considerably less than 1 hr (that is, 5 min in some cases), resulting in added detail of output.

In addition to allowing shorter time increments, the TWM incorporates new parameters to describe evaporation, infiltration, and soil water movement, all of which are related where possible to watershed physical properties. Claborn and Moore state that with the described changes, the TWM models small watersheds more accurately than the SWM-IV. Also, when measured flow is supplied, the TWM allows several types of statistical analyses.

Input data required by the TWM is very similar to that of the SWM-IV. The model requires the input rainfall for time increments that may be from 1 min to 1 day, and monthly evapotranspiration and pan-evaporation values are required. Other input includes the time-area histogram (see Chapter 7), initial watershed soil moisture and watershed parameters, and recorded streamflows for comparison with the simulated flows.

The interception loss for the TWM is based on seasonal variations, and depression storage is also treated in more detail than by the SWM-IV. Infiltration is calculated by Holtan's equation using the procedures of SWM-IV. However, two soil moisture zones rather than one lower zone are simulated by the TWM. The flow between the zones is determined by the conductivities. This change from one lower zone to two soil moisture zones necessitated the requirement-ment for monthly consumptive use in TWM.

In contrast to the KWM, the Texas version does not simulate snowmelt and should not be applied to areas where significant snowmelt runoff is prevalent. Other output from the TWM is similar to that of the SWM-IV.

#### National Weather Service River Forecast System (NWSRFS)

Yet another version of the Stanford Watershed Model was developed by the Hydrologic Research Laboratory staff at the National Weather Service Office of Hydrology.[20] The NWSRFS model was developed for use in forecasting river flows and stages by the National Weather Service. The model has been successfully applied to several river basins ranging in size from 70 mi$^2$ in North Carolina to 1000 mi$^2$ in Oklahoma. River forecasting in large river basins does not require the detail incorporated in SWM for smaller watersheds. For this reason, the NWS model includes two major changes involving the use of a longer time increment, simplified programming, fewer process computations, and a rapid procedure for determining optimal watershed parameters that allow the model to reproduce historical flows accurately.

A 6-hr time increment is used by the model, allowing fewer rainfall inputs and more important, fewer detailed calculations of processes such as overland flow that occur in shorter time periods. Iterations are thus completed more rapidly than with the SWM. As with the OPSET model, the National Weather Service optimization procedure for determining parameter values gives the model a strength not available with the SWM-IV.

Other modifications include hourly computations of infiltration, detention, upper and lower zone retention; and daily computations of percolation of water from the upper soil zone to groundwater storage. Evaporation from stream surfaces and groundwater evapotranspiration are handled separately by the SWM-IV, and are jointly computed in the NWS version. As mentioned, overland flow routing is eliminated and is replaced by three types of runoff: surface runoff, interflow, and groundwater flow; representing fast, medium, and slow response.

Input data for model calibration consist of mean daily discharges and instantaneous hydrographs for a few selected runoff events. Rainfall is input as a continuous record of 6-hr basin-wide means determined from areal averaging techniques. Because of the changes in the routing increment and in the detail of process simulation, the output from the NWS version is similar in makeup to the SWM-IV output.

**Streamflow Synthesis and Reservoir Regulation Model (SSARR)**
Another widely used continuous streamflow simulation model designed for large basins was developed by the Corps of Engineers.[21] The SSARR model was developed primarily for streamflow and flood forecasting and for reservoir design and operation studies. Prior to the development of NWSRFS, hydrologists at the National Weather Service used the SSARR model. The model has been applied to both rain and snowmelt events; recent versions were used in detailed hydrologic analyses of the Mekong River Basin in Southeast Asia.

Applications of the model begin with a subdivision of the drainage basin into homogeneous hydrologic units of a size and character consistent with subdivides, channel confluences, reservoir sites, diversion points, soil types, and other distinguishing features. The streamflows are computed for all significant points throughout the river system.

Rainfall data can be input at any number of stations in the basin. The part that will run off is divided into the base flow, subsurface or interflow, and surface runoff. The division is based on indices and on the intensity of the direct runoff. Each component is simply delayed according to different processes, and all are then combined to produce the final subbasin outflow hydrograph. This subarea runoff is then routed through stream channels and reservoirs to be combined with other subarea hydrographs, all of which become part of the output.

Routings through channels and reservoirs are accomplished by the same technique. This requires an assumption of short stream reaches, and occasional allowances for backwater effects are necessary in the channel routing process. Streamflows are synthesized on the basis of rainfall and snowmelt runoff. Snowmelt can be determined on the basis of the precipitation depth, elevation, air and dew point temperatures, albedo, radiation, and wind speed. Snowmelt options include the temperature index method or the energy budget method, and are discussed in detail in Sec. 10-7.

Input includes the precipitation depths; the watershed-runoff indexes for subdividing flow among the three processes; initial reservoir elevations and outflows; drainage areas; bounds on usable storage and allowable discharge from reservoirs; total computation periods; routing intervals; and other special instructions to control plots, prints, and other input-output alternatives.

This model was one of the earliest continuous streamflow simulation models using a lumped parameter representation, and has its primary strength in its verified accuracy indicated by tests conducted in several large drainage basins including the Columbia River Basin and the Mekong River Basin.

### Hydrocomp Simulation Program (HSP)

The most sophisticated and significant outgrowth of the Stanford Watershed Model was developed at Hydrocomp, Inc., and has been named the Hydrocomp Simulation Program.[22] Among several advantages incorporated in HSP are hydraulic reservoir routing techniques and kinematic-wave channel routing techniques. Other major changes include the addition of water quality simulation capabilities. Due to these additions, the model is often referred to as the Hydrocomp Water Quality Model.

The HSP model has been used routinely for several types of hydrologic studies including flood plain mapping, water quality studies, storm water and urban flooding studies, urban drainage facility design, and water quality aspects of urban runoff.

The model consists of three basic computer routines:

1. Library—allows the use of direct access disk storage to handle input data with efficient data management routines.
2. Lands—handles the usual SWM lands phase along with added processes in calculating soil moisture budgets, groundwater recharge and discharge, inflow to stream channels, and eutrophication.
3. Channel—is responsible for assembling and routing all channel inflow through channel networks, lakes, and reservoirs.

The HSP model incorporates a continuous water balance by tracking precipitation through all possible avenues of the hydrologic and water resource system. The groundwater inflow, interflow, and surface runoff are individually simulated, lagged, and combined at appropriate times as the channel inflow. The routing of computed inflow through the channel network utilizes a modified kinematic wave model. Water quality constituents are related to variable discharge rates and other water parameters so that the coupling of quantity and quality of the runoff is accomplished.

Inputs for simulating water quality include the temperature, radiation, wind, and humidity, and observed values of the factors under study which form the basis for calibration. At least 2 yr of data are preferable for calibration; however, calibration has been achieved with less than 1 yr of data. Other required input includes hourly precipitation in 5, 15, or 30-min, or greater, time increments and potential evapotranspiration; if snow or water quality simulation is desired, the temperature, radiation, wind, and humidity factors are needed.

Outputs from HSP can be obtained for any desired point within the watershed. Included in the output options are values of quality data at outfalls or other points; river stages; reservoir levels; hourly and mean daily discharge rates; stream and lake temperature; dissolved oxygen and total dissolved solids; algae counts; phosphorus; nitrate; nitrite; ammonia; total nitrogen; phosphate; pH; carbonaceous BOD; coliforms; conservative metals; and the usual daily, monthly and annual water budgets, snow depths, and end-of-period moisture equivalents.

Typical simulation periods in HSP applications range from 20 to 50 yr. Hour-by-hour data are not viewed as an exact sequence of future flows. Rather, the data are used for analysis of the probability of occurrences of ranges in the factors of interest. When used in this manner, the model is functioning with a purpose similar to that of some of the operational hydrology techniques described in Sec. 10-1.

## 10-4 Continuous Streamflow Simulation Model Studies

This section describes in detail two independent applications of the Kentucky version of the Stanford Watershed Model to small basins in Kentucky and Nebraska. Results obtained by Clarke[23] in modeling the Cave Creek (CC) watershed in Kentucky and by the authors using KWM for the Big Bordeaux Creek (BBC) watershed in Nebraska are compared. Both are small, homogeneous watersheds having relatively good records of precipitation and runoff.

### Selection of a Study Watershed and Study Period

Several guidelines exist for selecting a watershed and time period to be modeled in a simulation study. The use of relatively small, homogeneous watersheds or subdivisions of larger watersheds is recommended to minimize any difficulty caused by ignoring spatial variations in precipitation over larger areas. This also minimizes the effects of lumping watershed characteristics such as soil types, soil profiles, impervious areas, and land uses into single parameters representing the entire catchment. Ross suggests an upper limit of 25 mi² for study-watershed drainage areas.[24]

One difficulty in restricting watershed size arises from the fact that few streams for small drainage areas are continuously gauged, and reliable streamflow data are difficult to obtain. For example, the BBC watershed was selected because it was one of few small watersheds in Nebraska with sufficient records. The Cave Creek watershed was selected on the basis of the following criteria:[23]

1. A minimum of 10 yr of continuous runoff records in order to firmly establish the existing rainfall-runoff relationship.
2. A drainage area of less than 5 mi² so as to be representative of small drainage basins for which better runoff coefficients are needed.
3. A location in close proximity to a rain gauge for which hourly precipitation data are available.
4. The availability of soil surveys for the watershed under study.

### Data Sources

Hourly precipitation data were available at sites 1.20 mi from the CC watershed and 8.0 miles from the BBC gauging station. Soil survey records and runoff data were available for both watersheds. Daily pan evaporation data were available approximately 30 mi south of the BBC watershed and 25 miles south of the CC watershed. Drainage areas of 2.53 mi² for the CC watershed and 9.22 mi² for the BBC watershed were found from U.S. Geological Survey Quadrangle maps. Input parameters for the Stanford and Kentucky versions are compared and defined in Table 10-13. Numerical values, using the Stanford version parameter names, are tabulated for both watersheds in Table 10-14. Each parameter is described in detail in the following sections.

### Time-Area Histogram Data

The time-area histograms for the BBC and CC watersheds are developed in Figs. 10-12 and 10-13, respectively. Travel times and times of concentration for the Stanford Watershed Model are found from the Kirpich equation (for watersheds larger than 15 acres)[25], or

$$T_c = 0.0078\left[\frac{L}{(S)^{0.5}}\right]^{0.77} \qquad \textbf{(10-38)}$$

where

$T_c$ = the time of concentration (min)
$L$ = the horizontal projection of the channel length from the most distant point to the basin outlet (ft)
$S$ = the slope between the two points

In developing the time-area histogram for Big Bordeaux Creek,

***Table 10-13*** *Parameter Name Comparisons for Kentucky and Stanford Watershed Models*

| Parameter Name | | |
|---|---|---|
| *Kentucky Version* | *Stanford Version* | Parameter Description |
| *NCTRI* | *Z* | Integer number of elements in the time-area histogram |
| *CTRI* | *C* | Time-area histogram ordinate |
| *RMPF* | *MINH* | Discharge value below which no synthesized data is to be printed (cfs) |
| *RGPMB* | *K1* | Multiplication factor for precipitation data for a distant station |
| *AREA* | *AREA* | Watershed drainage area (mi²) |
| *FIMP* | *A* | Fraction of watershed area that is impervious |
| *FWTR* | *ETL* | Fraction of watershed area in lakes or swamps |
| *VINTMR* | *EPXM* | Maximum rate of vegetative interception for a dry watershed (in./hr) |
| *BUZC* | *CX* | Index for estimating surface storage capacity |
| *SUZC* | *EDF* | Index for estimating soil surface storage capacity during summers |
| *LZC* | *LZSN* | Index of moisture storage in soil profile above water table (in.) |
| *ETLF* | *K3* | Evapotranspiration parameter for lower zone soil moisture |
| *SUBWF* | *K24L* | Subsurface flow from the basin |
| *GWETF* | *K24EL* | Groundwater evapotranspiration by phreatophytes |
| *SIAC* | *EF* | Factor varying infiltration by season |
| *BMIR* | *CB* | Index of infiltration rate |
| *BIVF* | *CY* | Index of rate and quantity of water entering interflow |
| *OFSS* | *SS* | Average basin ground slope (ft/ft) |
| *OFSL* | *L* | Average overland flow distance (ft) |
| *OFMN* | *NN* | Manning roughness coefficient for overland flow |
| *OFMNIS* | *NNU* | Manning roughness coefficient for flow over impervious areas |
| *IFRC* | *IRC* | Daily interflow recession constant |
| *CSRX* | *KSC* | Streamflow routing parameter for low flows |
| *FSRX* | *KSF* | Streamflow routing parameter for flood flows |
| *CHCAP* | *CHCAP* | Index capacity of existing channel, bank-full (cfs) |
| *EXQPV* | *RFC* | Exponent of flow for nonlinear routing |
| *BFNLR* | *KV24* | Daily baseflow nonlinear recession adjustment factor |
| *BFRC* | *KK24* | Daily baseflow recession constant |
| *GWS* | *GWS* | Index of groundwater storage (in.) |
| *UZS* | *UZS* | Depth of interception and depression storage at beginning of year (in.) |
| *LZS* | *LZS* | Current equivalent depth of moisture in the soil profile (in.) |
| *BFNX* | — | Current value of *BFNLR* |
| *IFS* | — | Interflow storage |
| *CONOPT* | *DKN* | Control options for input, output, and internal branching |
| *QQQ* | *QQQ* | Title of computer simulation output, alphanumeric input |
| *DIV* | *DIV* | Mean daily diversion into or out of the basin (cfs) |

**Table 10-14** *Summary of Input Parameters for Big Bordeaux Creek and Cave Creek Watersheds*

| Parameter Name | Description | BBC Value(s) | Units | CC Value(s) |
|---|---|---|---|---|
| *TCONC* | Time of concentration | 105 | min | 60 |
| *TINC* | Routing interval | 15 | min | 15 |
| *Z* | Number of elements in the time-area histogram | 7 | — | 4 |
| *C* | Time-area histogram ordinates | 0.129 | — | 0.18 |
| | | 0.158 | — | 0.29 |
| | | 0.221 | — | 0.31 |
| | | 0.151 | — | 0.22 |
| | | 0.126 | — | — |
| | | 0.145 | — | — |
| | | 0.070 | — | — |
| *AREA* | Watershed drainage area | 9.22 | $mi^2$ | 2.53 |
| *A* | Impervious fraction of the watershed surface | 0.0 | — | 0.0 |
| *ETL* | Watershed stream and lake surface area as fraction of watershed area | 0.005 | — | 0.0 |
| *SS* | Average overland flow ground slope | 0.088 | ft/ft | 0.075 |
| *L* | Average length of overland flow | 183.2 | ft | 300.0 |
| *CHCAP* | Bank-full flow in channel at gauging station | 39 | cfs | 40 |
| *IRC* | Daily interflow recession constant | 0.485 | — | 0.75 |
| *KK24* | Daily base flow recession constant | 0.977 | — | 0.94 |
| *KSC* | Streamflow routing parameter for low flows | 0.989 | — | 0.90 |
| *KSF* | Streamflow routing parameter for flood flows | 0.989 | — | 0.90 |
| *MINH* | Minimum hourly flow rate to be printed | 1.0 | cfs | 0.2 |
| *K1* | Precipitation adjustment factor for distant gauge | 1.0 | — | 1.0 |
| *NN* | Manning roughness coefficient for overland flow | 0.37 | — | 0.10 |
| *NNU* | Manning roughness coefficient for impervious areas | 0.013 | — | 0.015 |

| | | | | | |
|---|---|---|---|---|---|
| *EMIN* | Factor for varying infiltration by seasons | | 0.5 | — | — |
| *EPXM* | Maximum interception rate for dry watershed | | 0.15 | in./hr | 0.10 |
| *CX* | Surface storage capacity index | | 0.80 | — | 0.90 |
| *EDF* | Soil surface moisture storage capacity | | 1.10 | — | 1.25 |
| *LZSN* | Soil profile moisture storage index | | 11.78 | in. | 4.85 |
| *K3* | Soil evaporation parameter | | 0.28 | — | 0.25 |
| *K24L* | Index of inflow to deep inactive groundwater | | 0.0 | — | 0.0 |
| *K24EL* | Fraction of watershed area in phreatophytes | | 0.0 | — | 0.0 |
| *EF* | Factor allowing for seasonal infiltration rates | | 1.0 | — | 0.15 |
| *CB* | Factor controlling infiltration rates | | 0.75 | — | 0.65 |
| *CY* | Index controlling water entering interflow | | 3.0 | — | 3.50 |
| *KV24* | Parameter for allowing nonlinear recession | | 1.0 | — | 0.99 |
| *SGW* | Groundwater storage volume parameter | | 0.1 | in. | — |
| *UZS* | Equivalent depth of upper zone storage | | 0.0 | in. | — |
| *LZS* | Equivalent depth of lower zone storage | | 7.0 | in. | — |
| *GWS* | Index of antecedent moisture conditions | | 0.2 | in. | — |
| *VOLUME* | Volume of water in swamp storage | | 0.0 | acre-ft | — |
| *EVCR* | Monthly evaporation pan coefficients | Jan. | 0.911 | — | — |
| | | Feb. | 0.911 | — | — |
| | | Mar. | 0.911 | — | — |
| | | Apr. | 0.911 | — | — |
| | | May | 0.552 | — | — |
| | | June | 0.677 | — | — |
| | | July | 0.654 | — | — |
| | | Aug. | 0.651 | — | — |
| | | Sep. | 0.642 | — | — |
| | | Oct. | 0.911 | — | — |
| | | Nov. | 0.911 | — | — |
| | | Dec. | 0.911 | — | — |

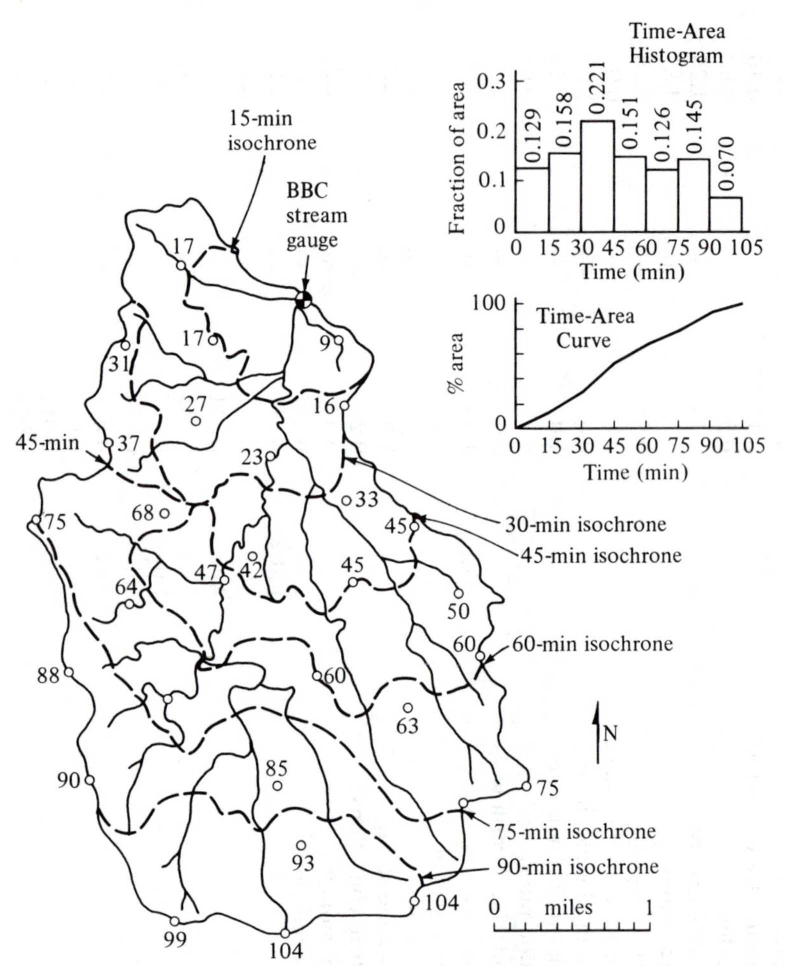


*Fig. 10-12.* *Time-area histogram development for Big Bordeaux Creek. (The values represent travel times (min.) to the outlet.)*

30 points and corresponding travel times (min) were plotted as shown in Figure 10-12. The dashed isochrones (lines of equal travel time to the basin outlet) were constructed by linear interpolation between the plotted points. Areas between pairs of isochrones were determined from planimetering. To contrast, the time-area histogram for Cave Creek was constructed, as shown in Fig. 10-13, by assuming that the flow velocity was constant everywhere, equal to the average streamflow velocity obtained by dividing $T_c$ into the channel length $L$. This procedure simply places all points along each isochrone at equal distances to the basin outlet.

### Watershed Parameters

The watershed drainage area (*AREA*), the impervious fraction of the watershed surface draining directly into the stream (*A*), the fractional stream and lake surface area (*ETL*), the average ground slope of overland flow perpendicular to the contours (*SS*), and the mean length of overland flow (*L*) for the Big Bordeaux Creek watershed were determined from areas, elevations and lengths measured from $7\frac{1}{2}$-min series USGS topographic maps (see Table 10-14 for BBC values). Other BBC parameter values are $ETL = 0.005$ (determined from

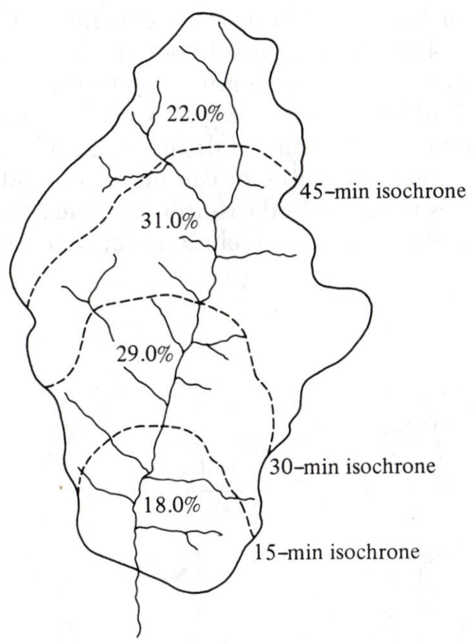


***Fig. 10-13.*** *Derivation of the time-area histogram for the Cave Creek Watershed. Legend: ----, isochrones.*

*Time-Area Histogram*

| Time (min) | Contributing Area (%) |
|---|---|
| 0–15 | 18.0 |
| 15–30 | 29.0 |
| 30–45 | 31.0 |
| 45–60 | 22.0 |

*(After D. K. Clarke, "Applications of Stanford Watershed Model Concepts to Predict Flood Peaks for Small Drainage Areas," Division of Research, Kentucky Department of Highways, 1968.)*

*ETL* = half the product of the stream length and channel width at the outlet), *SS* = 0.088 ft/ft (determined from Fig. 10-14 as the mean of 140 measured values between 20-ft contours at each gridline intersection), and *L* = 183.2 ft (determined from Fig. 10-14 as the average of 140 lengths measured perpendicular to contour lines from gridline intersection points to the nearest channel). The average overland flow distances can also be estimated as the reciprocal of twice the drainage density (see Chapter 4).

The bank-full channel capacity for Big Bordeaux Creek (*CHCAP*) was assumed to be equal to the discharge corresponding to the annual flood with a return period of 1.5 years. No criteria exists for easily obtaining *CHCAP* from maps, but Wolman and Leopold[26] have shown that the recurrence interval of a bank-full annual flood for many uncontrolled small and medium streams is nearly 1.5 yr. This finding allows a simple determination of the bank-full discharge if a flood frequency curve for the stream is available or can be developed. Regional flood frequency curves were applied resulting in an estimate of the bank-full BBC flood of 39 cfs. Another technique for determining

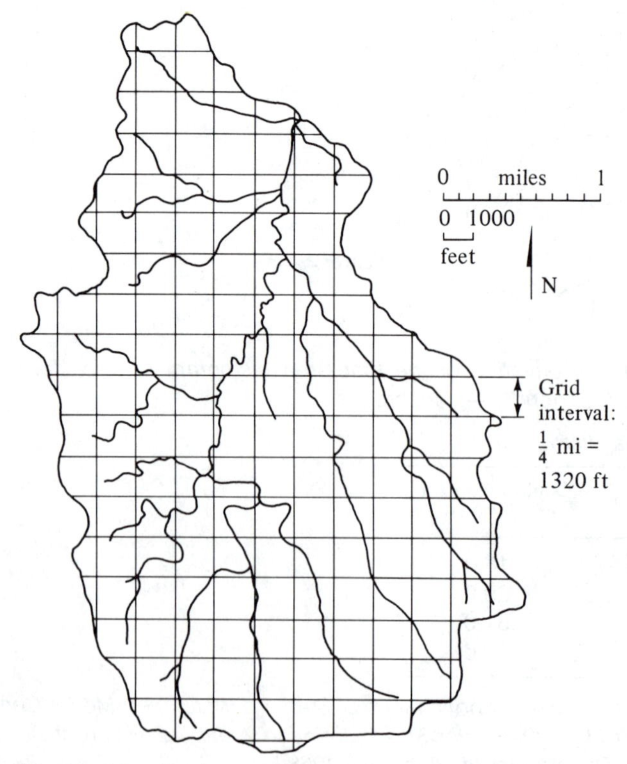


**Fig. 10-14.** *Grid overlay for determination of the mean overland slope and distance for Big Bordeaux Creek.*

*CHCAP* involves the use of the stream-gauging-station rating curve if the gauge height for bank-full flow is known or can be estimated from topographic maps. The Cave Creek value of 40 cfs was determined from a hydraulic analysis of the cross section and profile of the main channel. As a rule of thumb, *CHCAP* is selected to produce well-defined overbank and flood plain flow if $Q$ exceeds twice the value of *CHCAP*. Similarly, low, shallow flows are assumed whenever $Q$ is less than half of *CHCAP*.

### Streamflow Recession and Routing Parameters

The rate at which the model allows water to pass through the upper soil zones to the channels is controlled by the daily interflow recession constant (*IRC*). A graphical semilogarithmic technique of hydrograph analysis developed by Barnes is used to estimate this parameter.[27] The determination of the interflow recession constant for Big Bordeaux Creek is illustrated in Fig. 10-15, which is a semilogarithmic plot of average daily discharge (after deducting base flow) versus time for the flood event of July 20, 1969. The daily interflow recession constant (*IRC*) was determined as 0.485. Base flow was separated using the technique described in Chapter 4.

The daily base flow recession constant (*KK*24) controls the rate of discharge to the channels from the groundwater. A graphical technique similar to that used in determining the interflow recession constant is illustrated in Fig. 10-16 using the same flood event. The method reveals that the direct runoff and the interflow ceased on August 4 and that the base flow (by virtue of the linear plot on semilogarithmic paper) has a recession defined adequately by $Q_t = Q_0 (0.977)^t$, where $Q_t$ and $Q_0$ are base flows on successive days after August 4, and $t$ is 1 day for this analysis. The recession constant is simply determined as the slope of the base flow line in Fig. 10-16. An alternative method for determining *KK*24 using data collected several days after several flood peaks is illustrated for Cave Creek in Fig. 10-17. The line represents an envelope drawn to the right of the data points.

Two streamflow routing parameters are utilized by the simulation model to route inflow hydrographs to the point of interest for the basin (in this case to the stream gauging station for comparison with measured flows). The first parameter (*KSC*) is utilized in routing only if channel flows are less than half the channel capacity (*CHCAP*), and the second (*KSF*) accounts for channel and flood plain storage during flood flows (flows greater than twice the channel capacity). Respective data for Big Bordeaux Creek were determined using the two runoff hydrographs shown in Figs. 10-18 and 10-19, representing low and flood flows, respectively. Because the inflection point on the recession portion of each hydrograph represents the time at which a reversal of

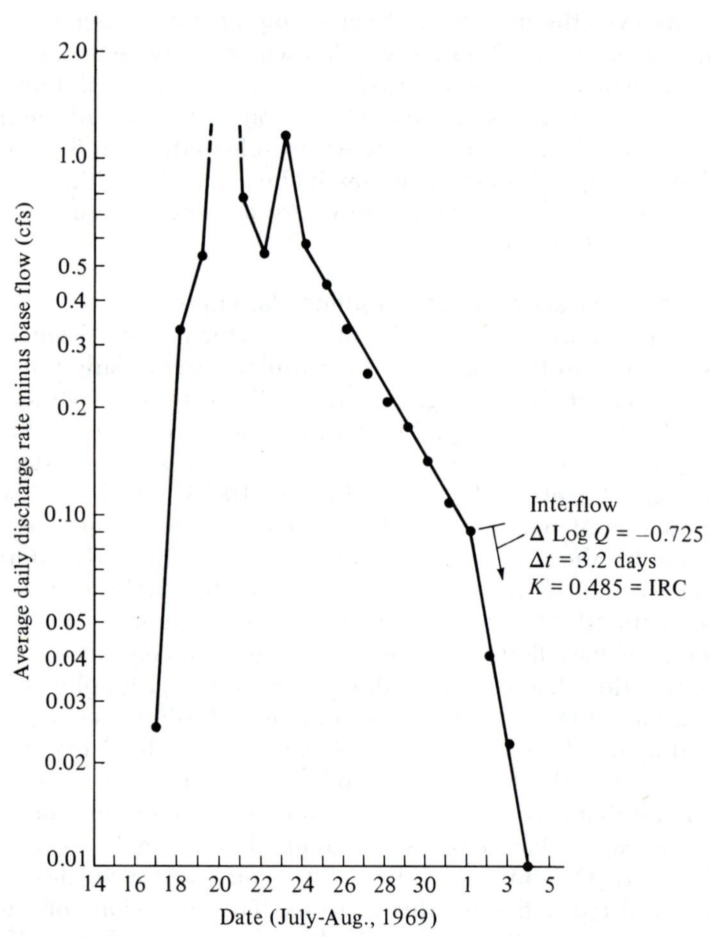


***Fig. 10-15.*** *Determination of the interflow recession constant (IRC) for Big Bordeaux Creek, flood event of July 20, 1969. The determination of* $K = IRC$.

$$Q_t = Q_0 K^t$$
$$t = 1 \text{ day}$$

flow direction through the channel banks occurs, then the parameters *KSC* and *KSF* represent daily recession constants for the water in storage in the channel. The equation used in determining both parameters is derived from a hydrologic routing technique equating $1/K$ times the difference in the average inflow and outflow during the routing period with the time rate of change of the outflow in the channel reach. As shown in the figures, the low flow and flood flow routing parameters for Big Bordeaux Creek are identical. Table 10-14 shows that a similar result was observed for Cave Creek.

## Hydrologic Parameters and Data

In addition to the described watershed characteristics, several hydrologic parameters and an impressive amount of hydrologic data are required as input to the simulation program. Due to the excessive bulk of hydrologic data for daily evaporation, hourly precipitation, and daily streamflow for a 4-yr period in BBC and a 10-yr period in CC, the actual hydrologic data values have been compiled but are not included in this text.

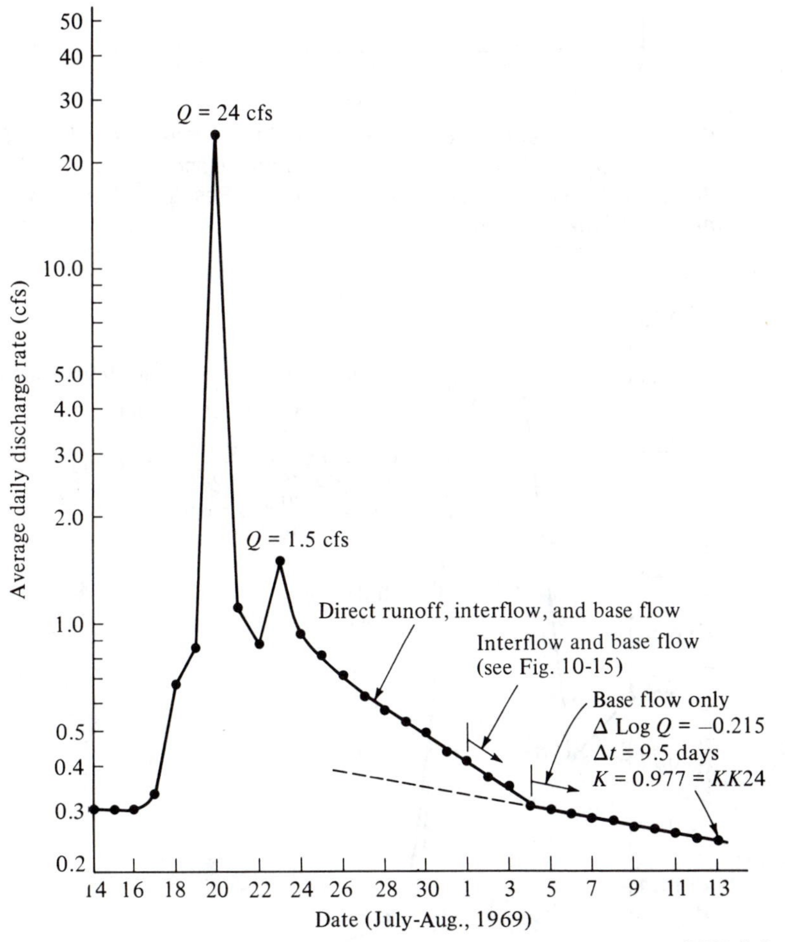


***Fig. 10-16.*** *Determination of the baseflow recession constant (KK24) for Big Bordeaux Creek, flood event of July 20, 1969. The determination of K = KK24.*

$$Q_t = Q_0 K^t$$
$$\text{Log } Q_t = \text{Log } Q_0 + t \text{ Log } K$$
$$\text{Log } K = \Delta \text{Log } Q / \Delta t$$

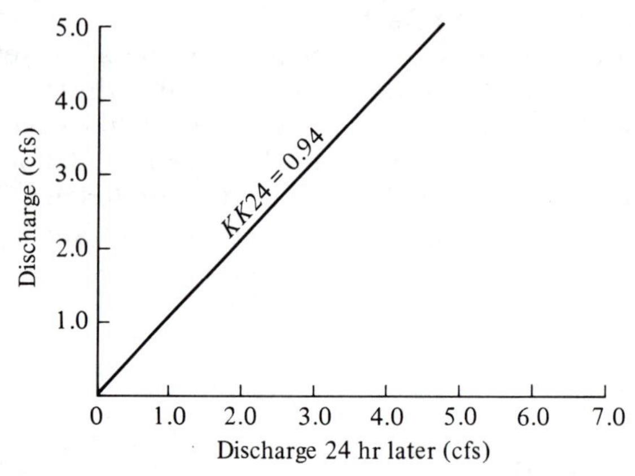


***Fig. 10-17.*** *Determination of KK24 for Cave Creek. (After D. K. Clarke, "Applications of Stanford Watershed Model Concepts to Predict Flood Peaks for Small Drainage Areas," Division of Research, Kentucky Department of Highways, 1968.)*

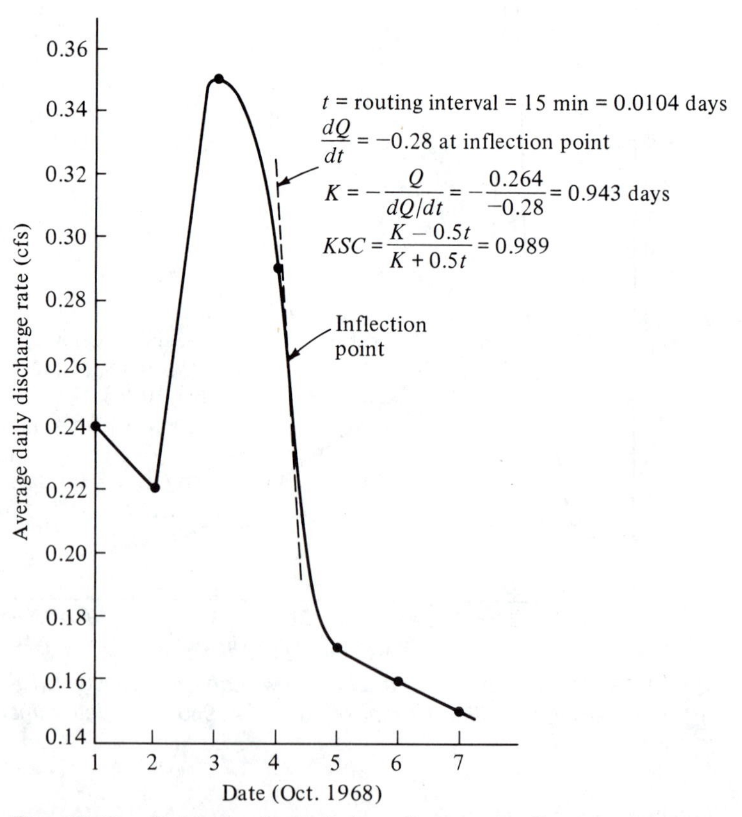


***Fig. 10-18.*** *Determination of low flow streamflow routing parameter (KSC) for Big Bordeaux Creek, low flow event of October 3, 1968.*

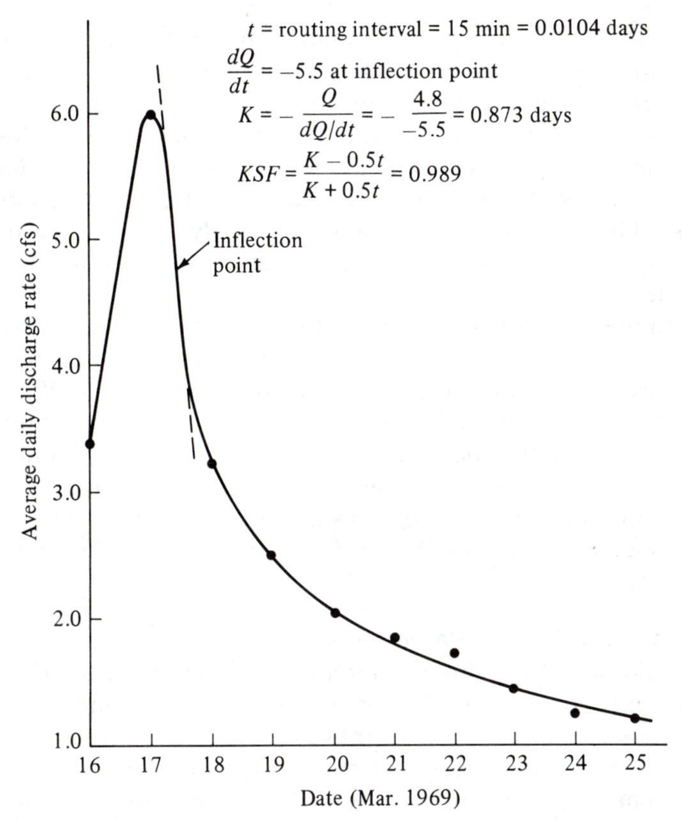


***Fig. 10-19.*** *Determination of flood-flow streamflow routing parameter (KSF) for Big Bordeaux Creek, flood-flow event of March 17, 1969.*

One useful hydrologic parameter used as input only in the BBC application is the hourly flow at the gauging station below which no printing of simulated discharges is desired. For the initial simulation of runoff from Big Bordeaux Creek, a *MINH* of 1.0 cfs was selected to minimize the printed output. This can easily be decreased in successive simulation trials, after the uncertainties in some of the other parameters have been reduced.

A precipitation-weighting adjustment parameter ($K1$), representing the long-term ratio of average precipitation over the basin to average precipitation at the precipitation gauge, might be greater or less than 1.0 if the gauge were located at any distance from the study basin. The recording gauges for both watersheds were within close proximity, and both $K1$ parameters were set to 1.0.

Monthly evaporation pan coefficients are used by the simulation program as multipliers in converting input pan-evaporation data to lake evaporation data (and also for determining potential evapotranspiration rates). The nearest evaporation pan for Big Bordeaux Creek is approximately 20 mi south at the Box Butte Reservoir, providing records of daily and monthly pan-evaporation depths for the months between and including May and September. Estimates of corresponding monthly lake evaporation amounts at Box Butte Reservoir were obtained from maps and charts developed by Shaffer.[28] Values listed in Table 10-14 are assumed to apply to each year of the simulation even though the simulation program allows changes from year to year.

Manning's roughness parameters for flow over soil and impervious surfaces are both required as input to the program. For the Big Bordeaux Creek area, the initial estimates for overland flow (*NN*) and impervious surface flow (*NNU*) were 0.37 and 0.013, representing coefficients for dense shrubbery and forest litter for overland flow and smooth concrete for impervious surface flow. The later $n$ is significant only if the fraction of impervious area ($A$) is nonzero. The Cave Creek analysis incorporated $NN = 0.10$ for light turf and $NNU = 0.015$ for concrete pavement.

### Trial and Adjustment Parameters

Several of the following parameters are determined by trial and adjustment until the comparison between the simulated and recorded streamflows is satisfactory. Guidelines for establishing initial values exist for only a few of the parameters, whereas most are initially determined from suggested ranges. The BBC data in Table 10-14 represent initial, unmodified estimates; those shown for CC are the optimal result of many repetitive runs.

One factor for varying infiltration by seasons (*EMIN*) ranges between 0.1 and 1.0 and has been shown to be significant in matching measured and simulated winter peak flow rates, Briggs.[29] Because no guidelines for estimating this parameter are presently available, the suggested midvalue of 0.5 was selected for Big Bordeaux Creek.

Several soil moisture and routing parameters require initial input values that are difficult to estimate from available data. Calibration procedures successively improve the initial estimates by trial and adjustment. For Big Bordeaux Creek, the parameters and initial estimates are as follows:

*EPXM*—the maximum interception rate (in./hr) for a dry watershed. Crawford and Linsley[10] suggest trial values of 0.10, 0.15, and 0.20 for grasslands, moderate forest covers, and heavy forest covers, respectively. The 0.15 in./hr value was selected for the moderate forest cover along Big Bordeaux Creek. Clarke used 0.10 in the Cave Creek study.

*CX*—an index of the surface capacity to store water as interception and depression storage. This parameter normally ranges from

0.10 to 1.65, and the midvalue of 0.80 was selected for Big Bordeaux Creek although a greater number might be indicative of the forest cover. Clarke independently arrived at a final, similar value (0.90) for Cave Creek.

*EDF*—an index of soil-surface moisture storage capacity, representing the additional moisture storage capacity available during warmer months due to vegetation. Depending on the soil type, the index ranges from 0.45 to 2.00. Sandy soils similar to those in the BBC area readily give up moisture to vegetation, resulting in increased storage capacity. Initial values of 1.10 and 1.25 were independently selected for the Big Bordeaux and Cave Creek areas.

*LZSN*—a soil-profile moisture storage index (in.) approximately equal to the volume of water stored above the water table and below the ground surface. This parameter is a major runoff-volume parameter, inversely related to the basin yields, interflow, and groundwater flow. The *LZSN* index, depending on porosity and the specific yield of the soil, ranges from 2.0 to 20.0, and 4 in. plus half the mean annual rainfall can be used as an initial estimate in areas experiencing seasonal rainfall. By the use of a 1931 to 1952 average of 15.55 in. of annual precipitation, the Chadron, Nebraska, precipitation station gives an initial *LZSN* = 11.78 in. for Big Bordeaux Creek. A similar analysis was used for Cave Creek.

*K*3—a soil evaporation parameter that controls the rate of evapotranspiration losses from the lower soil zone. The parameter ranges from 0.2 to 0.9 depending on the type and extent of the vegetative cover. As an initial estimate for the light forest cover in BBC, 0.28 was selected which agreed with the estimates suggested by Crawford and Linsley.[10] Also *K*3 is approximately equal to the fraction of the basin covered by forest and deep-rooted vegetation. Recommendations[10] for barren ground, grasslands, and heavy forests are, respectively, 0.20, 0.23, and 0.30; 0.25 was optimal for the Cave Creek study.

*K*24*L*—a parameter controlling the fraction of moisture lost or diverted from active groundwater storage through transverse flow across the drainage basin boundary. It also represents that portion of the inflow to the groundwater that percolates to the deep or inactive groundwater. The *K*24*L* parameter can be estimated from observed changes in deep groundwater levels, or it is often assumed to be zero because these losses are small compared to the magnitudes of rainfall and runoff.

*K*24*EL*—the fraction of the total watershed over which evapotranspiration from groundwater storage is assumed to occur at the potential rate. This parameter is assumed zero unless a significant quantity of vegetation draws water directly from the water table. Plants that seek phreatic water, such as cedar or cottonwood or alfalfa, are called phreatophytes.

*EF*—a factor, ranging from 0.1 to 4.0 that relates infiltration rates to evaporation rates. This parameter simply allows a more rapid infiltration rate recovery during warmer seasons. A normal starting value of 1.0 was selected for Big Bordeaux Creek, whereas the Cave Creek value was set much lower.

*CB*—an index that controls the rate of infiltration, depending on the soil permeability and the volume of moisture that can be stored in the soil. A midvalue of 0.75 (0.3 to 1.2) was selected for the sandy soils around Big Bordeaux Creek; a smaller value of 0.65 was optimal for the Cave Creek study.

*CY*—an index controlling the time distribution and quantities of moisture entering interflow. This index ranges from 0.55 to 4.5, and a moderately high value of 3.0 was selected for Big Bordeaux Creek because the watershed contains many pine needle mats. Clarke also selected a moderately high *CY* for Cave Creek.

*KV24*—a daily base flow recession adjustment factor used to produce a simulated curvilinear base flow recession. An initial value of 1.0 for Big Bordeaux Creek was selected because the base flow recession for the hydrograph in Fig. 10-16 is linear. Later adjustments might be required in matching simulated and recorded base flow recessions.

*SGW*—a groundwater storage increment (in.), reflecting the fluctuations in storage volume. Usually, an initial estimate (0.10 was selected for Big Bordeaux Creek) is made and adjusted after several simulation trials; *SGW* ranges from 0.10 to 3.90 in. This and the following four parameters were used only in the BBC study.

*UZS*—the current volume (in.) of soil surface moisture as interception and depression storage. Because the simulation begins on October 1 of the first calibration year, the parameter may initially be designated as zero unless precipitation occurs during the last few days of September.

*LZS*—the current volume (in.) of soil moisture storage between the land surface and the water table. Sixty percent of *LZSN*, or 7.0 in., was selected to initiate the Big Bordeaux Creek simulation.

*GWS*—the current groundwater slope index (in.). This index provides an indication of antecedent moisture conditions. Suggested initial values fall between 0.15 and 0.25, or the value assigned to *SGW* can be used. A midrange 0.20 in. was selected for Big Bordeaux Creek.

*VOLUME*—the volume (acre-ft) assigned to swamp storage and dry ground recharge, accounting for the runoff required to recharge swamps in late summer. Since no swamps were visible on the U.S. Geological Survey 7.5 min topographic maps, an initial value of 0.0 acre-ft was selected for Big Bordeaux Creek.

### Hydrologic Data

In addition to the parameter values of Table 10-14, the Kentucky version requires a large volume of hydrologic data for each water year of the simulation. The following description of required input data for each water year illustrates the data that were compiled and reduced to necessary input form for 4 water years beginning on October 1, 1968, and ending on September 30, 1972. Only the input for BBC is described. The input includes the following data for each water year.

1. The new year identification card containing the water year and the recorded annual streamflow. The values of the annual streamflow for Big Bordeaux Creek are listed in Table 10-15.

2. The stream gauge identification card containing a description of the stream gauging site.

3. The title to be applied to the ordinate for graphical plots of the simulated and recorded runoff hydrograph; namely, the daily average flow rate (cfs).

4. Pan-evaporation data, read as 365 or 366 single cards containing daily evaporation amounts (in.).

5. The monthly evaporation pan coefficients, read as 12 cards containing the data listed in Table 10-14, beginning with October.

6. Recorded daily streamflows (average flow for the day, cfs), read as 365 or 366 cards containing data for October 1 through September 30.

7. Hourly rainfall data, read as any number of cards for each water year. Two cards per day, each containing an identification of the gauge and date, are used to provide hourly depths in inches before noon on the first card and after noon on the second. Cards are required only for half-days experiencing precipitation. Because of the variable number of possible rainfall cards, a sentinel card with the year set equal to 1988 is placed at the end of the data, indicating that all the precipitation data for the water year has been read.

### Output from the Kentucky Version

Depending on which optional input, output and branching parameters are selected, a variety of output data is available from the

***Table 10-15***

| Water Years | Recorded Annual Streamflow (Acre-ft) |
|---|---|
| 1968–1969 | 434.0 |
| 1969–1970 | 333.2 |
| 1970–1971 | 465.4 |
| 1971–1972 | 296.4 |

Kentucky version, including plotted graphs of measured and synthesized daily streamflow rates. Options include

1. A table of synthesized average daily streamflow rates (cfs).
2. A table of monthly and annual totals of synthesized daily flow rates.
3. Synthesized monthly and annual totals of runoff in equivalent inches over the watershed.
4. Synthesized monthly and annual interflow amounts (in.) over the watershed.
5. Synthesized monthly and annual baseflow amounts (in.) over the watershed.
6. The volume of synthesized streamflow runoff from the watershed for the entire water year (acre-ft).
7. A summation of all the recorded daily streamflow rates (cfs) for each month and year.
8. The recorded annual total of runoff (in.) over the watershed.
9. The recorded volume of runoff from November through March (in.) over the watershed.
10. The amount of synthesized snow from November through March (in.) over the watershed.
11. The volume of the recorded annual streamflow (acre-ft).
12. The sum of the recorded precipitation for each month and for the year.
13. The synthesized monthly and annual totals of evapotranspiration (in.).
14. The monthly and annual recorded lake evaporation amounts (in.)
15. End-of-the-month levels of *UZS*, the current surface moisture storage (in.).
16. End-of the-month levels of *LZS*, the current soil moisture storage (in.).
17. End-of-the-month values of *SGW*, the current groundwater storage fluctuation (in.).
18. An annual moisture balance (in.), which represents the moisture not accounted for within the program. This is illustrated in the Cave Creek output.

### Output from the Kentucky Version for Cave Creek

Typical synthetic and actual flow rates at the Cave Creek gauging station are shown for a single day in Fig. 10-20. Other typical output for portions of one water year of the simulation is presented in Tables 10-16 and 10-17. The former provides an hour-by-hour listing of all flow rates in excess of the specified value of *MINH*, Table 10-14. Note that the streamflow was zero from October through December and exceeded *MINH* only during two days in January.

Table 10-17 contains most of the information described in the 18

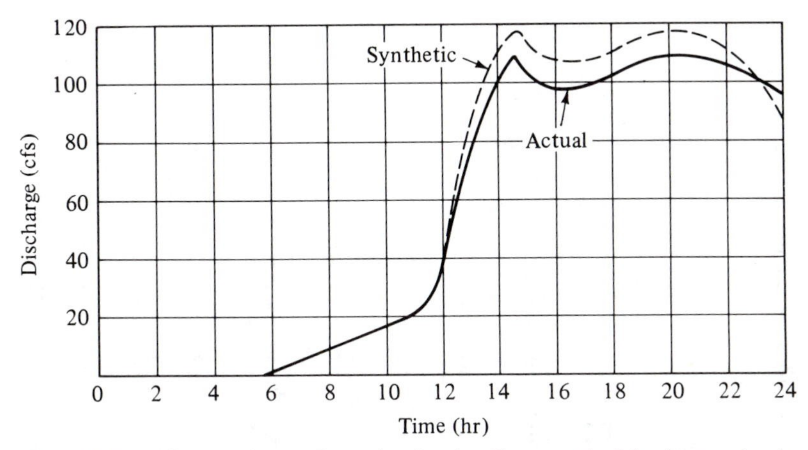


***Fig. 10-20.*** *Comparison of synthesized with recorded hydrographs for Cave Creek. (After D. K. Clarke, "Applications of Stanford Watershed Model Concepts to Predict Flood Peaks for Small Drainage Areas," Division of Research, Kentucky Department of Highways, 1968.)*

items of output for the Kentucky version. The daily flows are followed by the synthetic and recorded monthly totals; monthly interflow and baseflow amounts; monthly precipitation totals; monthly actual and potential ET amounts; and end-of-month storages in the soil profile and groundwater zones in inches. Of particular interest is the annual summary in the lower right. Of the 37.5 in. of precipitation, 23.8 in. went to ET, 11.6 in. ran off or was discharged from storage, and the remaining 2.1 in. recharged the soil profile. Moisture not accounted for during the year was 0.0844 in.

### Sensitivity of Model Response to Parameter Changes

One interesting and useful aspect of simulation is the ease with which changes in watershed parameters can be evaluated. Clarke tested the sensitivity of *KWM* by varying several of the parameters in Table 10-14 over reasonable ranges while holding all other parameters constant. The results of his analysis for Cave Creek on a typical day in March are summarized in Table 10-18. These observations were taken from graphs such as Figs. 10-21 and 10-22, which illustrate the sensitivity of flood magnitude and timing to changes in $L$ and $KSC$. These results and the summary in Table 10-18 are applicable only to the CC watershed and should not be viewed as generally applicable.

## 10-5 Rainfall-Runoff Event Simulation Models

Sequential simulation described in Secs. 10-3 and 10-4 consists of step-by-step computation through the entire time period. Simulation

## Table 10-16 *Typical Storm Hydrograph Output from Stanford Watershed Model for Cave Creek*

CAVE CREEK, PRM $ RUN 1, K. CLARKE ,OCT. 18,1966

CAVE CREEK NEAR LEXINGTON , KENTUCKY — WATER YEAR 1963-64 — KY. VERSION STANFORD WATERSHED MODEL

| Day | | | | | | | | | | | | | | |
|---|---|---|---|---|---|---|---|---|---|---|---|---|---|---|
| OCTOBER | | | | | | | | | | | | | | |
| 1 | | RECORDED FLOW= | | 0.0 | | | | | | | | | | |
| NOVEMBER | | | | | | | | | | | | | | |
| 1 | | RECORDED FLOW= | | 0.0 | | | | | | | | | | |
| DECEMBER | | | | | | | | | | | | | | |
| 1 | | RECORDED FLOW= | | 0.1 | | | | | | | | | | |
| JANUARY | | | | | | | | | | | | | | |
| 1 | | RECORDED FLOW= | | 0.0 | | | | | | | | | | |
| 20 | AM | 0.2 | 0.3 | 1.8 | 5.5 | 6.6 | 6.7 | 6.7 | 6.7 | 6.7 | 6.7 | 6.7 | 6.7 | |
| | PM | 6.7 | 6.7 | 6.7 | 6.7 | 6.6 | 6.6 | 6.6 | 6.5 | 6.5 | 6.4 | 6.4 | 6.3 | 5.8 |
| | | MAXIMUM= | 6.7 | C.F.S. | TIME | 10.00 | A.M. | | | | | | | |
| 21 | AM | 6.3 | 6.2 | 6.1 | 6.1 | 6.0 | 6.0 | 5.9 | 5.8 | 5.8 | 5.7 | 5.7 | 5.6 | |
| | PM | 5.5 | 5.5 | 5.4 | 5.3 | 5.3 | 5.2 | 5.1 | 5.1 | 5.0 | 5.0 | 4.9 | 4.8 | 5.6 |
| | | MAXIMUM= | 6.3 | C.F.S. | TIME | 0.15 | A.M. | | | | | | | |
| FEBRUARY | | | | | | | | | | | | | | |
| 1 | | RECORDED FLOW= | | 0.8 | | | | | | | | | | |
| 15 | AM | 4.3 | 4.2 | 4.2 | 4.2 | 4.1 | 4.1 | 4.0 | 4.0 | 4.0 | 3.9 | 3.9 | 3.8 | |
| | PM | 3.8 | 3.7 | 3.7 | 4.0 | 4.1 | 4.3 | 4.5 | 4.6 | 4.8 | 4.9 | 5.0 | 5.1 | 4.2 |
| | | MAXIMUM= | 5.1 | C.F.S. | TIME | 12.00 | P.M. | | | | | | | |
| 16 | AM | 5.2 | 5.2 | 5.3 | 5.4 | 5.4 | 5.4 | 5.5 | 5.5 | 5.5 | 5.5 | 5.5 | 5.5 | |
| | PM | 5.5 | 5.5 | 5.5 | 5.4 | 5.4 | 5.4 | 5.4 | 5.3 | 5.3 | 5.3 | 5.2 | 5.2 | 5.4 |
| | | MAXIMUM= | 5.5 | C.F.S. | TIME | 10.30 | A.M. | | | | | | | |
| 17 | AM | 5.1 | 5.1 | 5.0 | 5.0 | 5.0 | 4.9 | 4.9 | 4.8 | 4.8 | 4.7 | 4.7 | 4.6 | |
| | PM | 4.6 | 4.5 | 4.5 | 4.4 | 4.4 | 4.3 | 4.3 | 4.2 | 4.2 | 4.1 | 4.1 | 4.0 | 4.6 |
| | | MAXIMUM= | 5.2 | C.F.S. | TIME | 0.15 | A.M. | | | | | | | |
| MARCH | | | | | | | | | | | | | | |
| 1 | | RECORDED FLOW= | | 2.0 | | | | | | | | | | |
| 4 | AM | 3.1 | 3.1 | 3.0 | 3.0 | 3.0 | 3.4 | 6.8 | 10.2 | 13.2 | 16.4 | 23.1 | 42.3 | |
| | PM | 87.2 | 112.9 | 112.8 | 107.4 | 107.5 | 112.2 | 117.0 | 118.4 | 115.5 | 112.0 | 108.5 | 104.1 | 60.3 |
| | | MAXIMUM= | 118.7 | C.F.S. | TIME | 7.30 | P.M. | | | | | | | |
| 5 | AM | 99.8 | 95.6 | 91.7 | 88.1 | 84.6 | 81.4 | 78.3 | 75.4 | 72.6 | 70.0 | 67.6 | 65.2 | |
| | PM | 63.0 | 60.9 | 59.0 | 57.1 | 55.3 | 53.6 | 52.0 | 50.5 | 49.0 | 47.6 | 46.3 | 45.0 | 67.1 |
| | | MAXIMUM= | 101.4 | C.F.S. | TIME | 0.15 | A.M. | | | | | | | |
| 6 | AM | 43.8 | 42.7 | 41.6 | 40.5 | 39.5 | 38.5 | 37.6 | 36.7 | 35.9 | 35.1 | 34.3 | 33.5 | |
| | PM | 32.8 | 32.1 | 31.4 | 30.8 | 30.1 | 29.5 | 29.0 | 28.4 | 27.8 | 27.3 | 26.8 | 26.3 | 33.8 |
| | | MAXIMUM= | 44.3 | C.F.S. | TIME | 0.15 | A.M. | | | | | | | |
| 7 | AM | 25.8 | 25.3 | 24.9 | 24.4 | 24.0 | 23.6 | 23.2 | 22.8 | 22.4 | 22.0 | 21.6 | 21.3 | |
| | PM | 20.9 | 20.6 | 20.2 | 19.9 | 19.6 | 19.3 | 19.0 | 18.7 | 18.4 | 18.1 | 17.8 | 17.5 | 21.3 |
| | | MAXIMUM= | 26.0 | C.F.S. | TIME | 0.15 | A.M. | | | | | | | |
| 8 | AM | 17.2 | 17.0 | 16.7 | 16.5 | 16.2 | 16.0 | 16.0 | 16.7 | 24.6 | 41.3 | 47.0 | 46.0 | |
| | PM | 44.5 | 43.2 | 42.1 | 41.1 | 41.5 | 45.7 | 47.3 | 46.4 | 45.5 | 44.5 | 43.6 | 42.8 | 34.2 |
| | | MAXIMUM= | 47.4 | C.F.S. | TIME | 6.30 | P.M. | | | | | | | |
| 9 | AM | 41.9 | 41.1 | 40.3 | 39.6 | 39.1 | 40.9 | 41.4 | 40.9 | 40.3 | 39.8 | 39.2 | 38.7 | |
| | PM | 38.1 | 37.7 | 37.2 | 36.9 | 36.5 | 36.3 | 36.9 | 45.4 | 66.5 | 97.2 | 118.6 | 130.9 | 50.1 |
| | | MAXIMUM= | 134.9 | C.F.S. | TIME | 12.00 | P.M. | | | | | | | |

*Source:* After D. K. Clarke, "Applications of Stanford Watershed Model Concepts to Predict Flood Peaks for Small Drainage Areas," Division of Research, Kentucky Department of Highways, 1968.

**Table 10-17** *Typical Daily and Monthly Moisture Summary Output from Stanford Watershed Model*

CAVE CREEK, PRM G RUN 1, K. CLARKE ,OCT. 18,1966

| CAVE CREEK NEAR LEXINGTON, KENTUCKY DAY | OCT | NOV | DEC | JAN | FEB | MAR | APR | WATER YEAR 1963-64 MAY | JUN | JUL | AUG | KENTUCKY WATERSHED MODEL SEPT | ANNUAL | |
|---|---|---|---|---|---|---|---|---|---|---|---|---|---|---|
| 1 | C.C | 0.C | 0.0 | 0.0 | 0.3 | C.6 | 0.9 | 1.7 | 0.1 | 0.1 | 0.0 | 0.0 | | |
| 2 | C.C | 0.C | 0.0 | 0.0 | 0.2 | C.6 | 0.8 | 1.3 | 0.1 | 0.1 | 0.0 | 0.0 | | |
| 3 | C.0 | 0.C | 0.0 | 0.0 | 0.2 | 2.6 | 0.7 | 1.0 | 0.1 | 0.1 | 0.0 | 0.0 | | |
| 4 | C.C | 0.0 | 0.0 | 0.0 | 0.1 | 60.3 | 0.7 | 0.8 | C.1 | 0.1 | 0.0 | 0.0 | | |
| 5 | C.C | 0.0 | 0.0 | 0.C | 0.1 | 67.1 | 0.6 | 0.7 | 0.1 | 0.1 | 0.0 | 0.0 | | |
| 6 | C.C | 0.0 | 0.0 | 0.7 | 1.9 | 33.8 | 1.4 | 0.6 | C.1 | 0.1 | 0.0 | 0.0 | | |
| 7 | C.0 | 0.0 | 0.0 | 1.9 | 3.3 | 21.3 | 1.7 | 0.5 | 0.1 | 0.1 | 0.0 | 0.0 | | |
| 8 | C.0 | 0.C | 0.0 | 1.7 | 2.8 | 34.2 | 1.5 | 0.4 | 0.1 | 0.1 | 0.0 | 0.0 | | |
| 9 | C.0 | 0.0 | 0.0 | 1.5 | 2.1 | 50.1 | 1.2 | 0.4 | 0.1 | 0.1 | 0.0 | 0.0 | | |
| 1C | C.0 | 0.0 | 0.0 | 1.4 | 1.8 | 93.3 | 0.9 | 0.3 | 0.1 | 0.1 | 0.0 | 0.0 | | |
| 11 | C.C | 0.C | 0.0 | 1.0 | 2.8 | 44.9 | 0.8 | 0.3 | C.1 | 0.1 | 0.0 | 0.0 | | |
| 12 | C.C | 0.0 | C.0 | 0.8 | 2.5 | 27.7 | 0.6 | 0.3 | 0.1 | 0.1 | 0.0 | 0.0 | | |
| 13 | C.C | 0.0 | 0.0 | 1.1 | 3.2 | 18.9 | 0.6 | 0.2 | 0.1 | 0.1 | 0.0 | 0.0 | | |
| 14 | C.C | 0.0 | 0.0 | 0.9 | 4.6 | 14.0 | 0.5 | C.2 | 0.2 | 0.1 | C.0 | 0.0 | | |
| 15 | C.C | 0.0 | C.0 | 0.7 | 4.2 | 16.4 | 0.4 | 0.2 | 0.4 | 0.1 | 0.0 | 0.0 | | |
| 16 | C.C | 0.0 | 0.0 | 0.5 | 5.4 | 14.5 | 0.4 | 0.2 | 0.4 | 0.1 | 0.0 | 0.0 | | |
| 17 | C.C | 0.0 | 0.0 | 0.4 | 4.6 | 11.0 | 0.4 | 0.2 | 0.3 | 0.1 | 0.0 | 0.0 | | |
| 18 | C.C | 0.0 | 0.0 | 0.3 | 3.6 | 8.1 | 0.3 | 0.2 | 0.4 | 0.1 | 0.0 | 0.0 | | |
| 19 | C.0 | 0.0 | 0.0 | 0.3 | 3.1 | 6.0 | 0.3 | 0.2 | 1.1 | 0.1 | 0.0 | 0.0 | | |
| 20 | C.C | 0.0 | 0.0 | 5.8 | 2.5 | 5.7 | 0.5 | 0.2 | 1.0 | 0.1 | 0.0 | 0.0 | | |
| 21 | C.0 | 0.0 | 0.0 | 5.6 | 1.9 | 10.5 | 0.5 | 0.1 | C.8 | 0.1 | 0.0 | 0.0 | | |
| 22 | C.0 | 0.0 | 0.0 | 4.1 | 1.5 | 8.9 | C.7 | 0.1 | 0.6 | 0.1 | 0.0 | 0.0 | | |
| 23 | C.0 | 0.0 | C.0 | 2.9 | 1.1 | 6.8 | 0.8 | 0.1 | 0.5 | 0.1 | 0.0 | 0.0 | | |
| 24 | C.0 | 0.0 | 0.0 | 2.1 | 0.9 | 5.0 | 0.7 | 0.1 | 0.4 | 0.0 | 0.0 | 0.0 | | |
| 25 | C.0 | 0.0 | 0.0 | 1.7 | 0.7 | 3.8 | 0.6 | 0.1 | 0.3 | 0.0 | 0.0 | 0.0 | | |
| 26 | C.0 | 0.0 | 0.0 | 1.3 | 0.7 | 3.1 | 0.7 | 0.1 | 0.2 | 0.0 | 0.0 | 0.0 | | |
| 27 | C.0 | 0.0 | 0.0 | 1.0 | C.9 | 2.5 | 1.5 | 0.1 | 0.2 | 0.0 | 0.0 | 0.0 | | |
| 28 | C.0 | 0.0 | 0.0 | 0.7 | 0.8 | 2.0 | 2.0 | 0.1 | 0.1 | 0.0 | 0.0 | 0.5 | | |
| 29 | C.0 | 0.0 | 0.0 | 0.5 | 0.8 | 1.6 | 2.4 | 0.2 | C.1 | 0.0 | 0.0 | 1.4 | | |
| 30 | C.0 | 0.0 | 0.0 | 0.4 | | 1.3 | 2.1 | 0.2 | 0.1 | 0.0 | 0.0 | 1.4 | | |
| 31 | 0.0 | | 0.0 | 0.3 | | 1.1 | | 0.1 | | 0.0 | 0.0 | | | |
| SYNTHETIC | C. | C. | 0. | 40. | 59. | 578. | 27. | 11. | 9. | 2. | 1. | 3. | 730. | CFSD |
| TOTAL | 0.C04 | 0.001 | 0.004 | 0.589 | 0.863 | 8.492 | 0.401 | 0.164 | 0.127 | 0.029 | 0.008 | 0.051 | 10.73 | INCHES/DA |
| INTERFLOW | C.CCC | 0.000 | C.002 | 0.503 | 0.747 | 4.550 | 0.215 | 0.054 | 0.083 | 0.009 | 0.000 | 0.052 | 6.614 | INCHES/DA |
| BASE | 0.C04 | C.001 | C.C04 | C.047 | 0.116 | 0.434 | 0.195 | 0.106 | 0.051 | 0.029 | 0.008 | 0.009 | 1.003 | INCHES/DA |
| | | | | | | | | | | | | | 1448. | ACFT |
| RECORDED | C. | 3. | 2. | 59. | 117. | 536. | 30. | 14. | 8. | 5. | 0. | 10. | 786. | CFSD |
| | | | | | | | | | | | | | 11.55 | INCHES/DA |
| | | | | | | | | | | ( | 1543. | ) | 1558. | ACFT |
| PRECIP | 0.33 | 1.81 | 0.81 | 2.79 | 2.55 | 10.28 | 2.86 | 1.68 | 3.55 | 3.99 | 1.96 | 4.90 | 37.51 | INCHES/DA |
| EVP/TRAN-NET | 1.236 | 1.265 | C.875 | 0.702 | 0.576 | 0.693 | 1.851 | 3.149 | 4.777 | 4.396 | 2.925 | 1.344 | 23.790 | INCHES/DA |
| -POTENTIAL | 4.687 | 2.137 | C.875 | 0.702 | 0.576 | 0.693 | 1.871 | 4.346 | 5.865 | 5.814 | 7.841 | 5.704 | 41.111 | INCHES/DA |
| STORAGES-UZS | 0.050 | 0.761 | 0.371 | 0.440 | 0.454 | 0.348 | 0.854 | 0.947 | 0.C00 | 0.000 | 0.000 | 2.485 | | INCHES/DA |
| LZS | 0.951 | 0.784 | 1.102 | 2.488 | 3.524 | 4.640 | 4.743 | 3.150 | 2.753 | 2.322 | 1.355 | 2.281 | | INCHES/DA |
| SGW | C.0C1 | 0.001 | C.C02 | 0.024 | 0.077 | 0.164 | 0.120 | 0.032 | 0.019 | 0.008 | 0.002 | 0.050 | | INCHES/DA |
| INDICES-UZSN | 1.969 | 1.065 | C.805 | 0.490 | 0.361 | 0.328 | 0.722 | 1.579 | 2.404 | 2.461 | 3.113 | 1.616 | | |
| GWS | 0.014 | 0.006 | 0.C05 | 0.042 | 0.126 | 0.329 | 0.249 | 0.113 | C.067 | 0.036 | 0.015 | 0.058 | | |
| INF | 1.605 | 0.859 | 0.474 | 0.330 | 0.330 | 0.330 | 0.537 | 1.191 | 1.832 | 1.965 | 2.421 | 1.958 | | |
| BALANCE | -C.0844 | INCHES | | | | | | | | | | | | |

*Source:* After D. K. Clarke. "Applications of Stanford Watershed Model Concepts to Predict Flood Peaks for Small Drainage Areas," Division of Research, Kentucky Department of Highways, 1968.

**Table 10-18** *Effects of Increasing Various Parameters for a March Storm over Cave Creek*

| KWM Parameter Name | Representing | Range of Increase | Effect on Flood Peak | Effect on Runoff Volume | Effect on Flood Timing |
|---|---|---|---|---|---|
| *LZSN* | Soil moisture | 3–24 in. | Decreased | Reduced | Slight |
| *CY* | Moisture entering interflow | 0.3–3.0 | Decreased | Reduced | None |
| *CB* | Infiltration | 0.2–25 | Decreased | Reduced | None |
| *IRC* | Moisture leaving interflow | 0.2–0.6 | Decreased | Slight | None |
| *NN* | Overland flow resistance | Not specified | Decreased | Reduced | Delayed |
| *A* | Impervious fraction | Not specified | Increased | Increased | Delayed |
| *SS* | Overland flow slope | Not specified | Increased | Increased | Hastened |
| *CX* | Depression storage | Not specified | None | Reduced | None |
| *EDF* | Unfilled depression storage | 1–2 | Decreased | Slight | None |
| *TCONC* | Time of concentration | Not specified | Decreased | None | Delayed |
| *Z* | Number of histogram elements | 16–62 | Decreased | Reduced | Delayed |
| *KSF* | Channel routing parameter | 0.85–0.99 | Decreased | Reduced | Delayed |

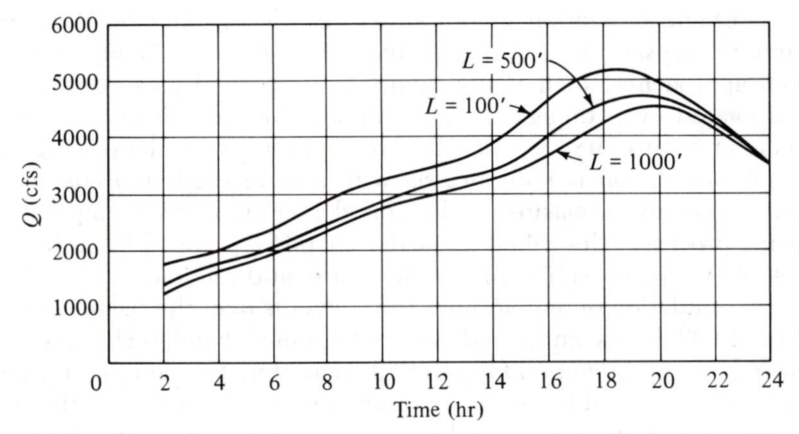


**Fig. 10-21.** *Sensitivity of model response to the length-of-overland-flow parameter. (After D. K. Clarke, "Applications of Stanford Watershed Model Concepts to Predict Flood Peaks for Small Drainage Areas," Division of Research, Kentucky Department of Highways, 1968.)*

of monthly or yearly streamflow or rainfall events is satisfactorily accomplished through this type of operation. However, for many hydrologic events that occur in shorter time intervals, the sheer amount of computation prohibits the sequential generation of these data for large-scale simulations. For example, the calculation of simulated rainfall depths hour by hour requires 8760 hourly items for each year of record. To achieve efficiency in computation, including only short events significantly reduces the length of simulation time. Such a technique is called ***event simulation.***

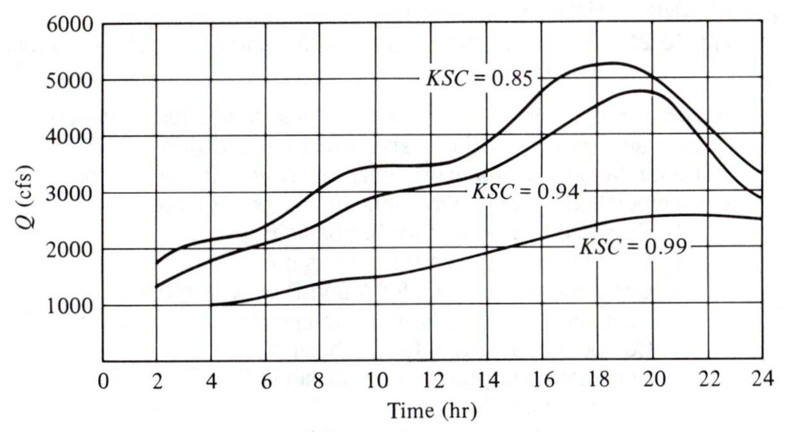


**Fig. 10-22.** *Sensitivity of model response to the channel routing parameter. (After D. K. Clarke, "Applications of Stanford Watershed Model Concepts to Predict Flood Peaks for Small Drainage Areas," Division of Research, Kentucky Department of Highways, 1968.)*

Event simulation model structures closely imitate the rainfall runoff processes developed in Chapters 3, 4, and 7. Lumped parameter approaches, such as unit hydrograph methods, are generally incorporated even though a greater opportunity to use distributed parameter approaches is present. Preparation for implementing most event simulation models begins with a watershed subdivision into homogeneous subbasins as illustrated in Fig. 10-23. Computations, using processes described in earlier chapters, proceed from the most remote upstream subbasin in a downstream direction.

Simulation of one or more storm events over the basin shown in Fig. 10-23 is accomplished by conventional hand techniques that have been programmed for a digital computer. The computer accepts parameters for subbasin $B$; computes the hydrograph resulting from the storm event; repeats the hydrograph computation for subbasin $A$; combines the two computed hydrographs into a single hydrograph; routes the hydrograph by conventional techniques through reach $C$ to

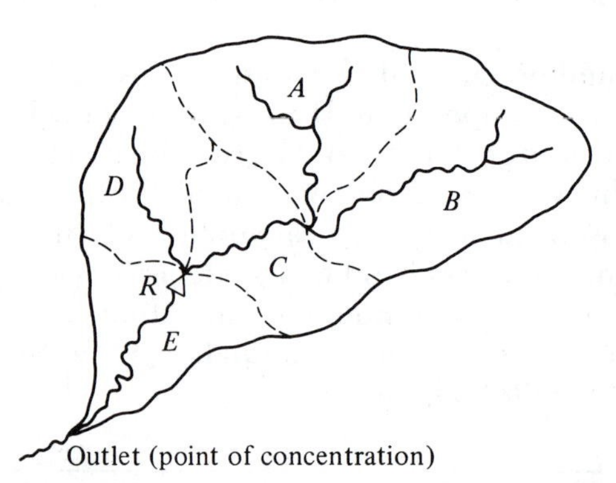


***Fig. 10-23.*** *Typical watershed subdivision and computation sequence for event simulation models.*

a. Subdivide basin to accommodate reservoir sites, damage centers, diversion points, surface and subsurface divides, gaging stations, precipitation stations, land uses, soil types, geomorphologic features.
b. Computation sequence in event simulation models:
   1. Compute hydrograph for Subbasin B.
   2. Compute hydrograph for Subbasin A.
   3. Combine hydrographs for A and B by superposition.
   4. Route combined hydrograph to upstream end of reservoir R.
   5. Compute hydrograph for Subbasin C.
   6. Compute hydrograph for Subbasin D.
   7. Combine three hydrographs at R.
   8. Route combined hydrograph through reservoir R.
   9. Route Reservoir outflow hydrograph to outlet.
   10. Compute hydrograph for Subbasin E.
   11. Combine two hydrographs at outlet.

the upstream end of reservoir $R$, where it is combined with the computed hydrograph for subbasin $C$; and so on through the procedure detailed in Fig. 10-23.

Hydrograph computations for subbasins are determined using unit hydrograph procedures as illustrated in Fig. 10-24. The precipitation hyetograph is input uniformly over the subbasin area, and precipitation losses are abstracted, leaving an excess precipitation hyetograph that is convoluted (see Chapter 4) with the prescribed unit hydrograph to produce a surface runoff hydrograph for the subbasin. The abstracted losses are divided among the loss components on the basis of prescribed parameters. Subsurface flows and waters derived from groundwater storage are transformed into a subsurface runoff hydrograph which, when combined with the surface runoff hydrograph, form the total streamflow hydrograph at the subbasin outlet. This hydrograph can then be routed downstream, combined with another contributing hydrograph, or simply output if this subbasin is the only, or the final, subbasin being considered.

The rainfall-runoff processes depicted in Figs. 10-23 and 10-24 are recognized by most of the event simulation models named in Table 10-6. Specific computation techniques for losses, unit hydrographs, river routing, reservoir routing, and base flow are compared in Table 10-19 for five of the major federal agency rainfall-runoff event simulation models. For example, HEC-1 will compute unit hydrographs by Clark's method[30] or a unit hydrograph may be input. All the models allow for some degree of selection among available techniques. Brief descriptions of each of these models are followed by an illustrative example of an application of the HEC-1 model to a single storm occurring over a 250 mi$^2$ watershed near Lincoln, Nebraska.

### U.S. Geological Survey Rainfall-Runoff Model

Carrigan calibrated the USGS Model to be used in evaluating short streamflow records and calculating peak flow rates for natural drainage basins.[31] The program actually monitors the daily moisture content of the subbasin soil and can be used as a continuous streamflow simulation model. The model is classified as an event simulation model because its calibration is based on short-term records of rainfall, evaporation, and discharges during a few documented floods.

Input to the model consists of initial estimates of 10 parameters which are modified by the model through an objective optimization fitting procedure that matches simulated and recorded flow rates. Other input includes daily rainfall and evaporation, close-interval (5 to 60 min) rainfall and discharge data, drainage areas, impervious areas, and base flow rates for each flood.

Phillip's[32] infiltration equation (Eq. 10-32) is used to determine a rainfall excess hyetograph which is translated to the subbasin outlet

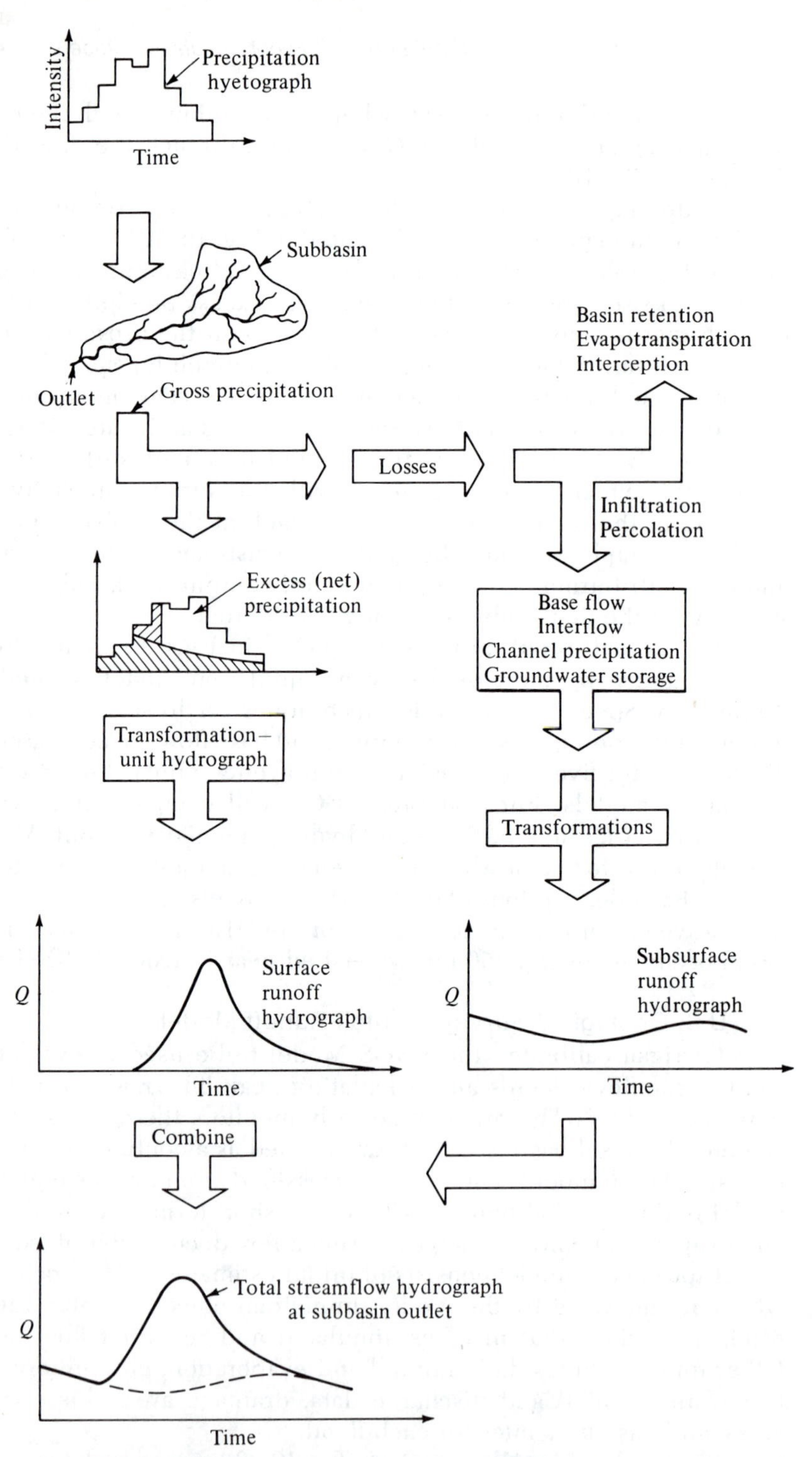


**Fig. 10-24.** *Typical lumped parameter event simulation model of the rainfall-runoff process.*

***Table 10-19*** *Hydrologic Processes and Options Used by Several Agency Rainfall-Runoff Event Simulation Models*

| Modeled Components | Model Code Names (see Table 10-6) | | | | |
|---|---|---|---|---|---|
| | *HEC-1 (Corps)* | *TR-20 (SCS)* | *USGS (USGS)* | *HYMO (ARS)* | *SWMM (EPA)* |
| Infiltration and Losses | | | | | |
| Holtan's equation | | | | | |
| Horton's equation | | | | | X |
| Phillip's equation | | | X | | |
| SCS curve number method | | X | | X | |
| Variable loss rate | X | | | | |
| Standard capacity curves | | | | | X |
| Unit Hydrograph | | | | | |
| Input | X | X | | | |
| Clark's | X | | X | | |
| Snyder's | X | | | | |
| 2-parameter gamma response | | | | X | |
| Dimensionless unit hydrograph | | X | | | |
| River Routing | | | | | |
| Hydraulic | | | | | X |
| Muskingum | X | | | | |
| Tatum | X | | | | |
| Straddle-stagger | X | | | | |
| Modified Puls | X | | | | |
| Working R&D | X | | | | |
| Variable storage coefficient | X | | | X | |
| Convex method | | X | | | |
| Translation only | | | X | | |
| Reservoir Routing | | | | | |
| Storage-indication | X | X | | X | |
| Base Flow | | | | | |
| Input | | | X | | |
| Constant value | | X | X | | |
| Recession equation | X | | | | |
| Snowmelt Routine | Yes | No | No | No | No |

and then routed through a linear reservoir, using the time-area watershed routing technique described in Chapter 7.

The USGS rainfall-runoff model can be used to simulate streamflows for relatively short-periods for small basins with approximately linear storage-outflow characteristics in regions where snowmelt or frozen ground is not significant. Output from the model includes a table showing peak discharges, storm runoff volumes, storm rainfall amounts, and an iteration-by-iteration printout of magnitudes of parameters and residuals in fitting volumes and peak flow rates.

### Computer Program for Project Formulation Hydrology (TR-20)

A particularly powerful hydrologic process and water surface profile computer program was developed by C-E-I-R, Inc.,[33] and is known by the code name TR-20 which is an acronym for the U.S. Soil Conservation Service Technical Release Number 20. The model is essentially a computer program of methods used by the Soil Conservation Service as presented in the *National Engineering Handbook*.[34]

The program is recognized as an engineer-oriented rather than computer-oriented package, having been developed with ease of use as a purpose. Input data sheets and output data are designed for ease in use and interpretation by field engineers, and the program contains a liberal number of operations that are user-accommodating, even at the expense of machine time.

The TR-20 was designed to use soil and land-use information to determine runoff hydrographs for known storms and to perform reservoir and channel routing of the generated hydrographs. The program has been used in all 50 states by SCS engineers for flood insurance and flood hazard studies, for the design of reservoir and channel projects, and for urban and rural watershed planning.

Surface runoff is computed from any historical or synthetic storm using the SCS curve number approach described in Chapter 12 to abstract losses. The standard dimensionless hydrograph shown in Fig. 4-26 is used to develop unit hydrographs for each subarea in the watershed. The excess rainfall hyetograph is constructed using the effective rain and a given rainfall distribution, and is then applied incrementally to the unit hydrograph to obtain the subarea runoff hydrograph for the storm.

As shown in Table 10-19, TR-20 uses the storage-indication method to route hydrographs through reservoirs (see Sec. 7-2). The base flow is added to the direct runoff hydrographs at any time to produce the total flow rates. The program uses the logic depicted in Fig. 10-23 by computing the total flow hydrographs, routing the flows through stream channels and reservoirs, combining the routed hydrographs with those from other tributaries, and routing the combined hydrographs to the watershed outlet. As many as 200 channel reaches and 99 reservoirs or flood-water-retarding structures can be accommodated in any single application of the model. To add to this capability, the program allows the concurrent input of up to 9 different storms over the watershed area.

Subdivision of the watershed is facilitated by determining the locations of control points. Control points are defined as stream locations corresponding to cross-sectional data, reservoir sites, damage centers, diversion points, gauging stations, or tributary confluences where hydrograph data may be desired. Subarea data requirements include the drainage area, the time of concentration, the reach lengths, structure data as described in Sec. 10-2, and either routing coefficients for each reach or cross-sectional data along the channels. Whenever cross-sectional data are provided, the model calculates the water surface elevations in addition to the peak flow rates and time of occurrence at each section. Subarea sizes are dictated by the locations of control points. To provide routing and flood hazard information, it is necessary to define enough control points so that the hydraulic characteristics of the stream are defined between control sections. Applications with TR-20 normally incorporate control points spaced between a few hundred feet to 2 mi or more apart. The resulting subareas that contribute runoff to a control point are usually less than 5 $mi^2$. Common subarea sizes for structures are less than 25 $mi^2$ even though there is no limitation on reach length or subarea size within the program.

Minimal input data requirements to TR-20 include the watershed characteristics, at least one actual or synthetic storm including the depth, duration, and distribution; the discharge, capacity, and elevation data for each structure; and the routing coefficients or cross-sectional data for each reach. Input can be described according to the following outline:

1. Watershed characteristics.
   a. The area in $mi^2$ contributing runoff to each reservoir and cross section.
   b. Runoff curve number *CN* for each subarea. (See Chapter 12.)
   c. The antecedent moisture condition associated with each subarea, coded as dry, normal, or wet.
   d. The time of concentration for each subarea (hr).
   e. The length of each channel routing reach and subarea mainstream.
2. Velocity-routing coefficient table.
   a. A table containing routing coefficients for a range of velocities (ft/sec). This table is contained within the program and need only be entered if the user desires different velocities.
3. Dimensionless hydrograph.
   a. This table is contained within the program and need only be entered if the user desires a different hydrograph.
4. Actual hydrograph.
   a. Actual hydrographs can be introduced at any point in the watershed. Hydrograph ordinates are read as discharge rates (cfs) spaced at equal time increments apart, up to a maximum of 300 entries.
   b. Base flow rates (cfs) can also be specified.
5. Base flow.
   a. In addition to the option of specifying the base flow rates asso-

ciated with a hydrograph that was input, the base flow can be specified or modified at any other control point.

6. Storm data.
   a. Storms are numbered from 1 to 9 and are input as cumulative depths at equally spaced time increments.
   b. As an alternative to specifying cumulative depths at various time increments, a dimensionless storm can be input, and up to 9 storms can be synthesized by specifying each storm depth and duration.
7. Stream cross-sectional data.
   a. Up to 200 cross sections may be input for a single run. Cross-sectional data consists of up to 20 pairs of values of the discharge versus flow area.
   b. If cross-sectional data are provided, the routing coefficients are determined from them. In the absence of such data, the user must specify a routing coefficient for each reach.
8. Structure data.
   a. The reservoir data consists of up to 20 pairs of outflow discharge rates (cfs) versus storage (acre-ft).
   b. A maximum of 99 structures are allowed in a run.

The desired output from TR-20 must be specified on a set of control cards. Hydrographs at each control point for each storm can be printed by specifying the control point identification in the control cards. Any combination of the following items can be produced at each control point:

1. The peak discharge rate, time of peak, and peak water surface elevations.
2. The discharge rates in tabular form for the entire hydrograph.
3. Water surface elevations for the entire duration of runoff.
4. The volume of direct runoff, determined from the area under the hydrograph.
5. Punched hydrograph ordinates in any specified format.
6. Summary tables containing water balance information.

Figure 10-25 summarizes the data required and the processes simulated by TR-20.[33] The maps, graphs, and tables are simplified representations of the various basic data needed by the computer program and are determined from field surveys. Rainfall frequency data are input from data in the U.S. Weather Bureau TP-40 report.[35] Peak-discharge and area-flooded information for present and future conditions for several return periods are punched by TR-20 in a form suitable for direct use in an economic evaluation model.

### Problem-Oriented Computer Language for Hydrologic Modeling (HYMO)

A unique computer language designed for use by hydrologists who have no conventional computer programming experience was de-

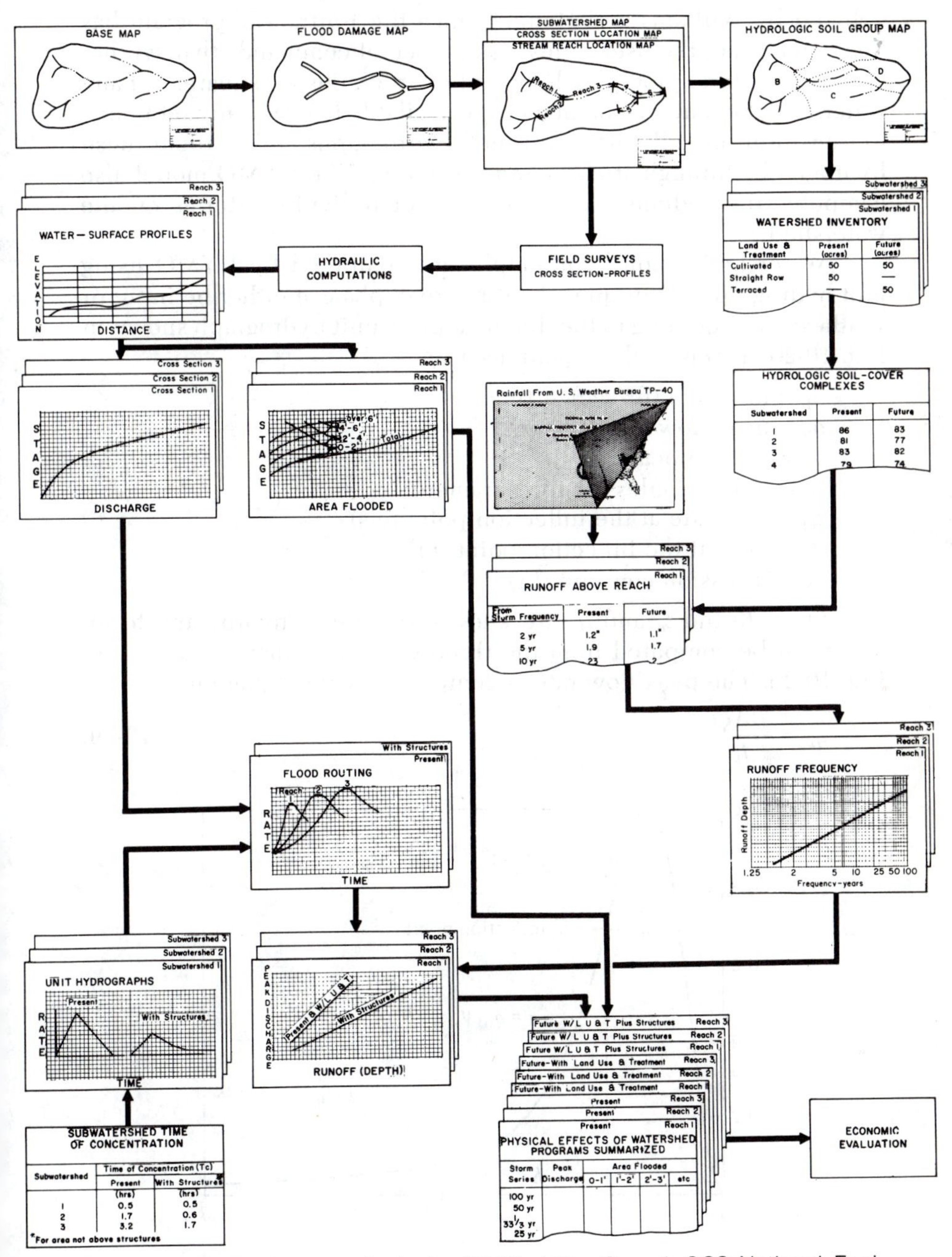


***Fig. 10-25.*** *Program logic for TR-20. (After Sec. 4,* SCS National Engineering Handbook.*)*

veloped by Williams and Hann.[36] Once the Fortran-IV program has been compiled, the user forms a sequence of commands that synthesize, route, store, plot, punch, or add hydrographs for subareas of any watershed. Seventeen commands are available to use in any sequence to transform rainfall data into runoff hydrographs and to route these hydrographs through streams and reservoirs. The HYMO model also computes the sediment yield in tons for individual storms on the watershed.

Watershed runoff hydrographs are computed by HYMO using unit hydrograph techniques. Unit hydrographs can either be input or synthesized according to the dimensionless unit hydrograph shown in Fig. 10-26. Terms in the equations are

$q$ = flow rate (ft³/sec) at time $t$
$q_p$ = peak flow rate (ft³/sec)
$t_p$ = time to peak (hr)
$n$ = dimensionless shape parameter
$q_0$ = flow rate at the inflection point (cfs)
$t_0$ = time at the inflection point (hr)
$K$ = recession constant (hr)

Once $K$ and $t_p$ and $q_p$ are known, the entire hydrograph to infinity can be computed from the three segment equations shown in Fig. 10-26. The peak flow rate is computed by the equation

$$q_p = \frac{BAQ}{t_p} \tag{10-39}$$

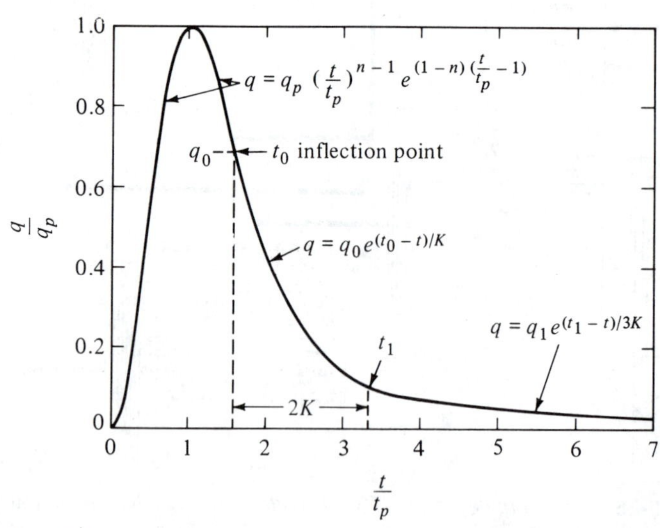


*Fig. 10-26. Dimensionless unit hydrograph used in HYMO. (After J. R. Williams and R. W. Hann, "HYMO: Problem-Oriented Computer Language for Hydrologic Modeling," U.S. Department of Agriculture, Agriculture Research Service, May 1973.)*

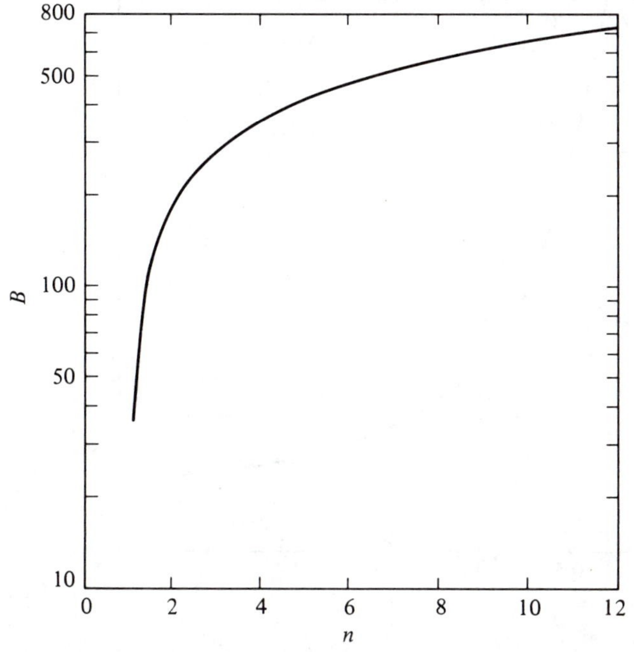


***Fig. 10-27.*** *Relationship between dimensionless shape parameter n and watershed parameter B. (After J. R. Williams and R. W. Hann, "HYMO: Problem-Oriented Computer Language for Hydrologic Modeling," U.S. Department of Agriculture, Agriculture Research Service, May 1973.)*

where

$B$ = a watershed parameter, related to $n$ as shown in Fig. 10-27
$A$ = watershed area (mi²)
$Q$ = volume of runoff (in.), determined by HYMO from the SCS rainfall-runoff equation described in Chapter 12

The runoff for a unit hydrograph would of course be 1.0 in. The parameter $n$ in Fig. 10-27 is obtained from Fig. 10-28. Parameters $K$ and $t_p$ for ungauged watersheds are determined from regional regression equations based on 34 watersheds located in Texas, Oklahoma, Arkansas, Louisiana, Mississippi, and Tennessee, ranging in size from 0.5 to 25 mi², or

$$K = 27.0A^{0.231}SLP^{-0.777}\left(\frac{L}{W}\right)^{0.124} \tag{10-40}$$

and

$$t_p = 4.63A^{0.422}SLP^{-0.46}\left(\frac{L}{W}\right)^{0.133} \tag{10-41}$$

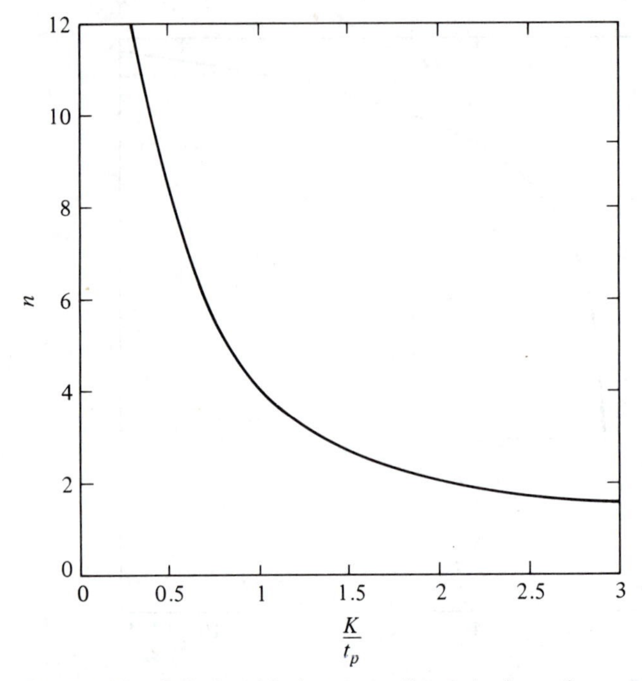


***Fig. 10-28.*** *Relationship between dimensionless shape parameter n and recession constant/time to peak. (After J. R. Williams and R. W. Hann, "HYMO: Problem-Oriented Computer Language for Hydrologic Modeling," U.S. Department of Agriculture, Agriculture Research Service, May 1973.)*

where

$SLP$ = the difference in elevation (ft), divided by flood plain distance (mi), between the basin outlet and the most distant point on the divide

$L/W$ = the basin length-width ratio

River routing is accomplished in HYMO by a revised *variable storage coefficient* (VSC) method.[37] The continuity equation, $I - O = dS/dt$, and the storage equation, $S = KO$, are combined and discretized according to the methods outlined in Chapter 7. The VSC method recognizes the variability in $K$ as the flow leaves the confines of the stream channel and inundates the flood plain and valley area. Relationships between $K$ and $O$ are determined by HYMO from the input cross-sectional data, or HYMO will calculate the relationship using Manning's equation if the flood plain and channel roughness coefficients are specified. The bed slope and reach length are also part of the required input.

The widely adopted storage-indication method (see Chapter 7) is used to route inflow hydrographs through reservoirs. The storage-outflow curve must be determined externally by the user and is input to the program as a table containing paired storages and outflows, using storage defined as zero whenever the outflow is zero.

The user-oriented commands and the data requirements for each command are as follows:

1. Start: the time rainfall begins on the watershed.
2. Store hydrograph: the time increment to be used, the lowest flow rate to be stored, the watershed area, and the successive flow rates spaced at the specified time increment.
3. Develop hydrograph: the desired time increment, the watershed area, the SCS runoff curve number $CN$, the watershed channel length and maximum difference in elevation, and the cumulative rainfall beginning with zero and accumulated at the end of each time increment until the end of the storm.
4. Compute the rating curve: cross-sectional identification number, number of points in the cross section, the maximum elevation of the cross section, the main channel and left and right flood plain slopes, Manning's $n$ for each segment, and finally pairs of horizontal and vertical coordinates of the points describing the cross section.
5. Reach computations: the number of cross sections in the routing reach, the time increment to be used in routing, the reach length, and the discharge rates for which the variable storage coefficient is to be computed.
6. Print hydrograph: the identification number of the cross section at which hydrographs are to be printed.
7. Punch hydrograph: the identification number of the cross section at which the hydrographs are to be punched.
8. Plot hydrograph: the identification number of the cross sections at which the hydrographs are to be plotted.
9. Add hydrographs: the identification numbers of the hydrographs to be added.
10. Route reservoir: the identification numbers of the locations of the outflow and inflow hydrographs, and the discharge-storage relationship for the reservoir.
11. Compute travel time: the reach identification number, reach length, and reach slope.
12. Sediment yield: several factors describing soil erodibility, cropping management, erosion control practices, slope length, and slope gradient.

Output from HYMO includes the synthesized or user-provided unit hydrographs, the storm runoff hydrographs, the river- and reservoir-routed hydrographs, and the sediment yield for individual storms on each subwatershed. Hydrographs computed by HYMO compared closely with measured hydrographs from the 34 test watersheds.

### Storm Water Management Model (SWMM)

The Environmental Protection Agency Model, SWMM,[38] is listed in Table 10-6 in two locations corresponding to rainfall-runoff event simulation and urban runoff simulation. The model is primarily an urban runoff simulation model, and is described in detail in Chapter 11.

In brief, SWMM's hydrograph and routing routines are hydraulic rather than hydrologic. A lumped parameter approach is used for subcatchments consisting of single parking lots, city lots, and so forth. Accumulated rainfall on these plots is first routed as overland flow to gutter or storm drain inlets, where it is then routed as open or closed channel flow to the receiving waters or to some type of treatment facility. Of the five event-simulation models compared in Table 10-19, the SWMM gives the greatest detail in simulation but cannot be used in large rural watershed simulations.

Overland flow depths and flow rates are computed for each time step using Manning's equation along with the continuity equation. The water depth over a subcatchment will increase without inducing an outflow until the depth reaches a specified detention requirement. If and whenever the resulting depth over the subcatchment, $D_r$, is larger than the specified detention requirement, $D_d$, an outflow rate is computed using Manning's equation

$$V = \frac{1.49}{n}(D_r - D_d)^{2/3}S^{1/2} \qquad \textbf{(10-42)}$$

and

$$Q_w = VW(D_r - D_d) \qquad \textbf{(10-43)}$$

where

$V$ = the velocity
$n$ = Manning's coefficient
$S$ = the ground slope
$W$ = the width of the overland flow
$Q_w$ = the outflow discharge rate

After flow depths and rates from all subcatchments have been computed, they are combined along with the flow from the immediate upstream gutter to form the total flow in each successive gutter.

The gutter and pipe flows are routed by the Manning and continuity equations to any points of interest in the network where they are added to produce hydrograph ordinates for each time step in the routing process. The time step is advanced in increments until the runoff from the storm is no longer being produced. The parameters of the gutter shape, slope, and length must be supplied by the user. Manning's roughness coefficients for the pipes or channels must also be supplied and are available in most hydraulics textbooks.

Other input required for a typical simulation with the SWMM model include

1. Watershed characteristics such as the infiltration parameters, percent impervious area, slope, area, detention storage depth, and Manning's coefficients for overland flow.
2. The rainfall hyetograph for the storm to be simulated.
3. The land use data, average market values of dwellings in subareas, and populations of subareas.
4. Characteristics of gutters such as the gutter geometry, slope, roughness coefficients, maximum allowable depths, and linkages with other connecting inlets or gutters.
5. Street cleaning frequency.
6. Treatment devices selected and their sizes.
7. Indexes for costs of facilities.
8. Boundary conditions in the receiving waters.
9. Storage facilities, location and volume.
10. Inlet characteristics such as surface elevations and invert elevations.
11. Characteristics of pipes such as type, geometry, slope, Manning's *n*, and downstream and upstream junction data.

### HEC-1 Flood Hydrograph Package (HEC-1)

The U.S. Army Corps of Engineers Hydrologic Engineering Center developed a series of comprehensive computer programs as computational aids for consultants, universities, federal, state, and local agencies. Programs for flood hydrograph computations, water surface profile computations, reservoir system analyses, monthly streamflow synthesis, and reservoir system operation for flood control comprise the series. Only the flood hydrograph package, HEC-1, is described herein.[9]

The HEC-1 model is essentially a general program consisting of a calling program and six subroutines. Two of these subroutines determine the optimal unit hydrograph, loss rate, or streamflow routing parameters by matching recorded and simulated hydrograph values. The other subroutines perform snowmelt computations, unit hydrograph computations, hydrograph routing and combining computations, and hydrograph balancing computations. In addition to being capable of simulating the usual rainfall-runoff event processes, HEC-1 will also simulate multiple floods for multiple basin development plans and perform the economic analysis of flood damages by numercially integrating areas under damage-frequency curves for existing and postdevelopment conditions.

In comparison to other event-simulation models, HEC-1 is relatively compact and still able to execute a variety of computational procedures in a single computer run. The model is applicable only to single-storm analysis because there is no provision for precipitation loss rate recovery during periods of little or no precipitation.

After dividing the watershed into subareas and routing reaches as shown in Fig. 10-23, the precipitation for a subarea can be determined by one of three methods: (1) nonrecording and/or recording precipitation station data, (2) basin mean precipitation, or (3) standard project or probable maximal hypothetical precipitation distributions. Either actual depths or net rain amounts may be input, depending on the user's choice of techniques for abstracting losses. The HEC-1 loss rate function is described in the following discussion and is easily by-passed if the net rain is available for direct input.

Figure 10-29 illustrates how HEC-1 separates jobs into either optimization jobs, types 1 and 2, or nonoptimization jobs, types 3, 4, and 5. The five available operations are named in the figure caption, and their logic can be traced by following the corresponding numbers through the flow chart. The nonoptimization jobs begin with precipitation in one of the forms described and then proceed to the branch labeled with a Roman numeral II. Hydrologic processes such as the subarea runoff computation, routing computation, combining, balancing, comparing, or summarizing are specified in the input in the sequence illustrated in Fig. 10-23. The completion of all except the summary component sends the program back to numeral II to determine if any additional steps in the sequence are desired.

The loss rate for the HEC-1 model is an exponential decay function that depends on the rainfall intensity and the antecedent losses. The instantaneous loss rate, in in./hr, is

$$L_t = K'P_t^E \qquad \textbf{(10-44)}$$

where

$L_t$ = the instantaneous loss rate (in./hr)
$P_t$ = intensity of the rain (in./hr)
$E$ = the exponent of recession (range of 0.5 to 0.9)
$K'$ = a coefficient, decreasing with time as losses accumulate

$$K' = K_0C^{-CUML/10} + \Delta K \qquad \textbf{(10-44a)}$$

where

$K_0$ = the loss coefficient at the beginning of the storm (when $CUML = 0$), an average value of 0.6
$CUML$ = the accumulated loss (in.) from the beginning of the storm to time $t$
$C$ = a coefficient, an average of 3.0

If $\Delta K$ is zero, the loss rate coefficient $K'$ becomes a parabolic function of the accumulated loss, *CUML*, and would thus plot as a straight line function of *CUML* on semilogarithmic graph paper if $K$ were plotted on the logarithmic scale. The straight-line relationship is depicted in Fig. 10-30, showing the decrease in the loss rate coefficient as the losses accumulate during any storm. Because loss rates typically decrease much more rapidly during the initial minutes of a storm, the loss rate $K'$ is increased above the straight-line amount by an amount equal to $\Delta K$, which in turn is made a function of the amount of losses that will accumulate before the $K'$ value is again equal to the straight-line value, $K$. This initial accumulated loss, $CUML_1$, is user-specified and is related to $\Delta K$ in such a fashion that the initial loss rate $K'$ is 20% times $CUML_1$ greater than $K_0$ (see Fig. 10-30). Initial loss coefficients $K_0$ are difficult to estimate, and standard curves in Chapters 3 and 11 are available to determine initial infiltration rates, $L_0$. For gauged basins, HEC-1 allows the user to input rainfall and runoff data from which the loss rate parameters are optimized to give a best fit to the information provided. Estimates of parameters for ungauged basins fall in the judgment realm noted in Table 10-6. An alternative to the described loss rate function is available in HEC-1, which is an initial abstraction followed by a constant loss rate, similar to a $\phi$-index.

The HEC-1 model provides for separate computations of snowmelt in up to 10 elevation zones. The precipitation in any zone is considered to be snow if the zone temperature is less than a base temperature, usually 32°F, plus 2°. The snowmelt is computed by the degree-day or energy budget methods whenever the temperature is equal to or greater than the base temperature. The elevation zones are usually considered in increments of 1000 feet although any equal increments can be used.

Unit hydrographs for each subarea can be provided by the user, or Clark's method[30] of synthesizing an instantaneous unit hydrograph (IUH) can be used. Clark's method is more commonly recognized as the time-area-curve method of hydrologic watershed routing described in Sec. 7-3. The time-area histogram, determined from an isochronal map of the watershed, is convoluted with a unit design-storm hyetograph using Eq. 7-32, as illustrated in Figs. 7-6 and 7-7. The methods described in Sec. 7-2 are then used to route the resulting hydrograph through linear reservoir storage using Eq. 7-14 with a watershed storage coefficient $K$. Input data for Clark's method consists of the time-area-curve ordinates, the time of concentration for the Clark unit graph, and the watershed storage coefficient $K$. If the time-area curve for the watershed under consideration is not available, the model provides a synthetic time-area curve at the user's request.

Because the Corps of Engineers commonly uses Snyder's method (Chapter 4, Sec. 4-16) in unit hydrograph synthesis for large

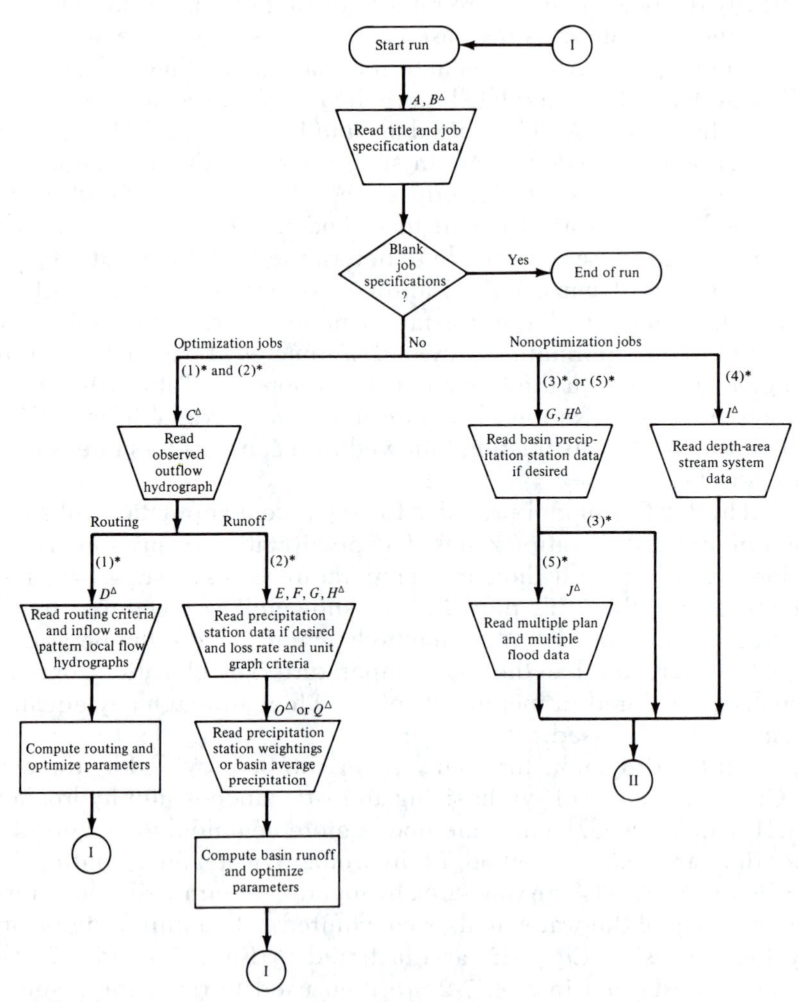


***Fig. 10-29.*** *HEC-1 flow chart and program logic. (After U.S. Army Corps of Engineers, "HEC-1 Flood Hydrograph Package," Users and Programmers Manuals, HEC Program 723-X6-L2010, Jan. 1973.)*

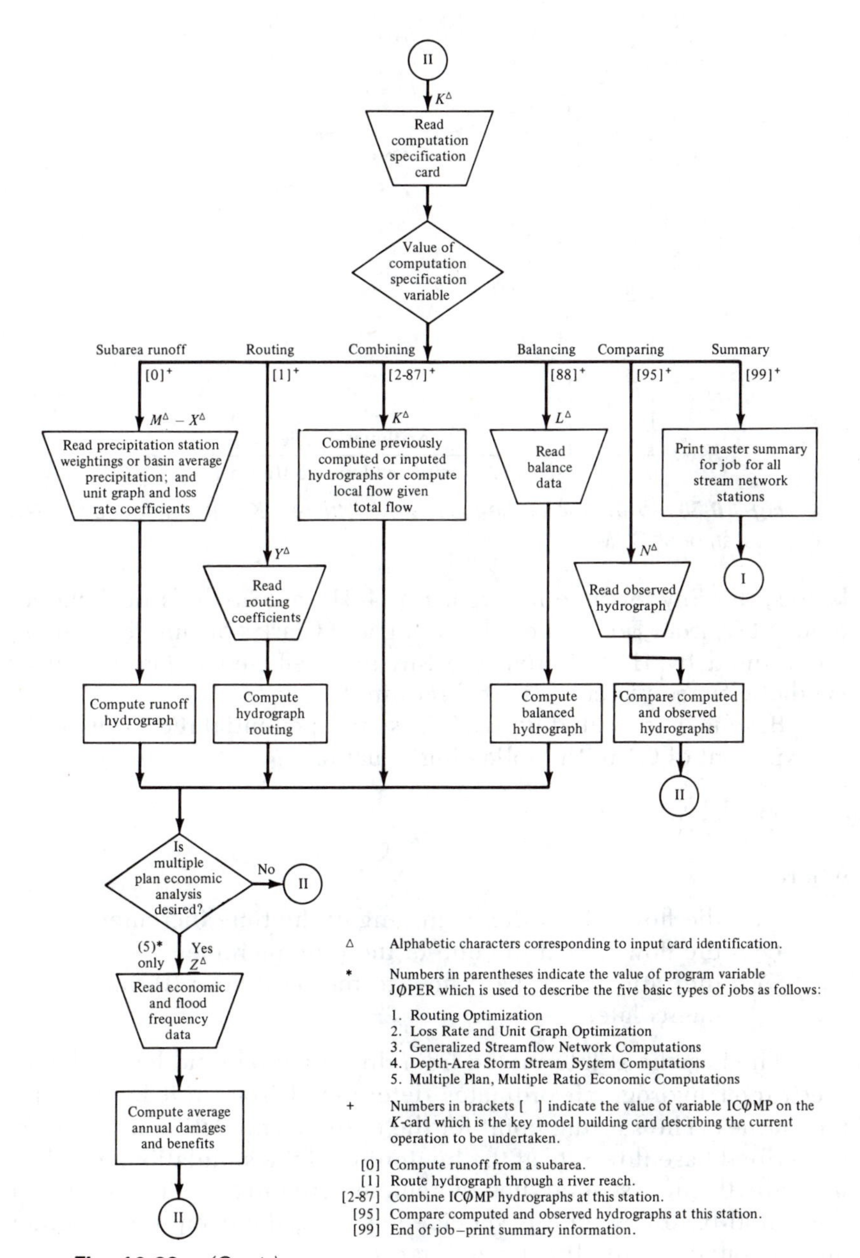


**Fig. 10-29.** *(Cont.)*

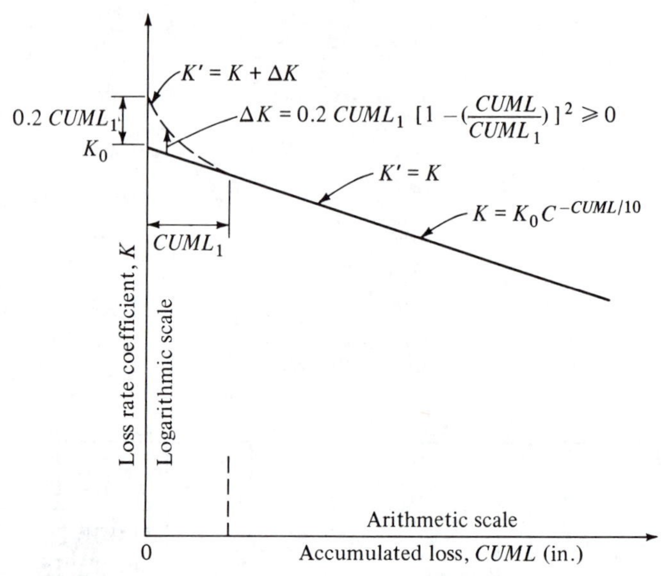


**Fig. 10-30.** *Variation of the loss rate coefficient K' with the accumulated loss amount CUML.*

basins, the Snyder time lag from Eq. 4-31 and Snyder's peaking coefficient $C_p$ from Eq. 4-34 can be input, and Clark's parameters will be determined by HEC-1 from the Snyder coefficients. The actual or synthetic time-area curve is still required.

Base flow is treated by HEC-1 as an exponential recession using an exponent of 0.1 in the following equation:

$$Q_2 = \frac{Q_1}{R^{0.1}} \tag{10-45}$$

where

$Q_1$ = the flow rate at the beginning of the time increment
$Q_2$ = the flow rate at the end of the time increment
$R$ = the ratio of the base flow to the base flow 10 time increments later

The base flow determined from this equation is added to the direct runoff hydrograph ordinates determined from unit hydrograph techniques. The starting point for the entire computation is the user-prescribed base flow rate at the beginning of the simulation, which is normally the flow several time increments prior to any direct runoff. If the initial base flow rate is specified as zero, the computer program output contains only direct runoff rates.

The HEC-1 package allows the user a choice of several hydrologic or "storage-routing" techniques for routing floods through river reaches and reservoirs. All use the continuity equation and some form of the storage-outflow relation; all are described in more detail in Chapter 7 and by Chow. The six routing procedures included in the program are

1. Modified Puls—this method is also called the Storage-Indication Method, and is a level-pool-routing technique normally reserved for use with reservoirs or flat streams. The technique was described in detail in Sec. 7-2.
2. Muskingum—described in detail in Sec. 7-1.
3. Working R&D—normally used if stream junctions are encountered in routing. This method allows for storage coefficient changes as stages increase. The technique requires estimates of Muskingum's $x$ and the storage-indication curve for the reach.
4. Straddle-Stagger—also known as the Progressive Average Lag method. The technique simply averages a subset of consecutive inflow rates, and the averaged inflow value is lagged a specified number of time increments to form the outflow rate.
5. Tatum—also known as the Successive Average Lag method. This method is similar to the Straddle-Stagger method, but only pairs of flows are averaged rather than larger subsets. Also, the reach is divided into many subreaches spaced so that the averaged pair of inflows becomes the outflow from each subreach and the inflow to the next subreach.
6. Multiple Storage—another averaging and lagging technique used successfully by the Corps of Engineers for routing hydrographs on the Columbia River in Washington and Oregon. Until exponents in the routing equation for other rivers become available, the technique probably has limited application.

The hydrograph balancing routine shown as a branch in Fig. 10-29 is included to convert any given hydrograph to one having user-specified runoff volumes within given durations. As shown, the hydrograph comparing routine reads an observed hydrograph at a particular basin location and compares the observed with the computed hydrographs in a tabulation of (1) instantaneous flow rates and (2) the numerical difference for each time.

Input to HEC-1 is facilitated by stacking data cards in a sequence compatible with the desired computation sequence. Card codes are used in the first column of each data card to identify the type of information contained on the card. The card codes and card contents are summarized in Table 10-20.

Output from HEC-1 is both complete and descriptive. Options are available for graphical and symbol map displays of intermediate or summary hydrographs or precipitation hyetographs. The extent of output from HEC-1 is further illustrated in Example 10-5.

**Table 10-20** *Input Data Requirements and Card Code for HEC-1*

| Card Code | Card Content |
|---|---|
| *A* | Job title |
| *B* | Job specifications |
| *C* | Observed hydrograph to be reconstituted |
| *D* | Routing optimization criteria and observed inflow hydrograph |
| *E, F* | Unit graph and loss rate optimization criteria |
| *G, H* | Station precipitation data for all subbasins of the watershed |
| *I* | Precipitation depth-drainage-area data |
| *J* | Multiflood, multiplan data |
| *K* | Computations specification for model building |
| *L* | Hydrograph balancing criteria |
| *M-X* | Subarea runoff computation data including precipitation, losses, and unit hydrograph information |
| *Y* | Individual reach routing criteria |
| *Z* | Economic and flood frequency data |

***Example 10-5.*** In June 1963 the Oak Creek Watershed shown in Fig. 10-31 experienced a severe flood-producing storm in a 6-hr period. Average excess rain depths over each of the nine subareas *A, B, . . ., H,* and *I* ranged from 1.0 in. in the southwest to 7.8 in. in the north. Crest-stage flood records show that the storm produced peak flows of 27,500 cfs at Agnew (point 3) and 21,600 cfs at the watershed outlet (point 8). The map in Fig. 10-31 gives the subbasin divides, reach lengths, and stream bed elevations (underlined).

A reservoir located at point 9 will store virtually any probable storm over subarea *I*. Using the remaining elongated watershed area *A* through *H* within the boldface border, use the June storm to simulate hydrographs at each of points 1 through 8 using a single run of HEC-1 and compare peak flows with recorded values at points 3 and 8.

The net storm depths shown in Fig. 10-31 are applied uniformly over each subarea. The records provide the time distribution for the 12 successive half-hour periods of the thunderstorm in percent: 3, 4, 5, 6, 9, 37, 10, 8, 6, 4, 5, 3, and 4% during the 12 successive half-hour periods of the thunderstorm. The net rain was determined from the measured depths using the SCS runoff equation[34] and a basin-wide composite curve number of 73 (see Chapter 12).

The computation logic to simulate runoff for this storm consists of the following 22 steps:

1. Compute the hydrograph for area *A* at point 1.
2. Route the *A* hydrograph from point 1 to point 2.
3. Compute the hydrograph for area *B* at point 2.
4. Combine the two hydrographs at point 2.
5. Route the combined hydrograph to point 3.
6. Compute the hydrograph for area *C* at point 3.
7. Combine the two hydrographs at point 3.
8. Route the combined hydrograph to point 4.

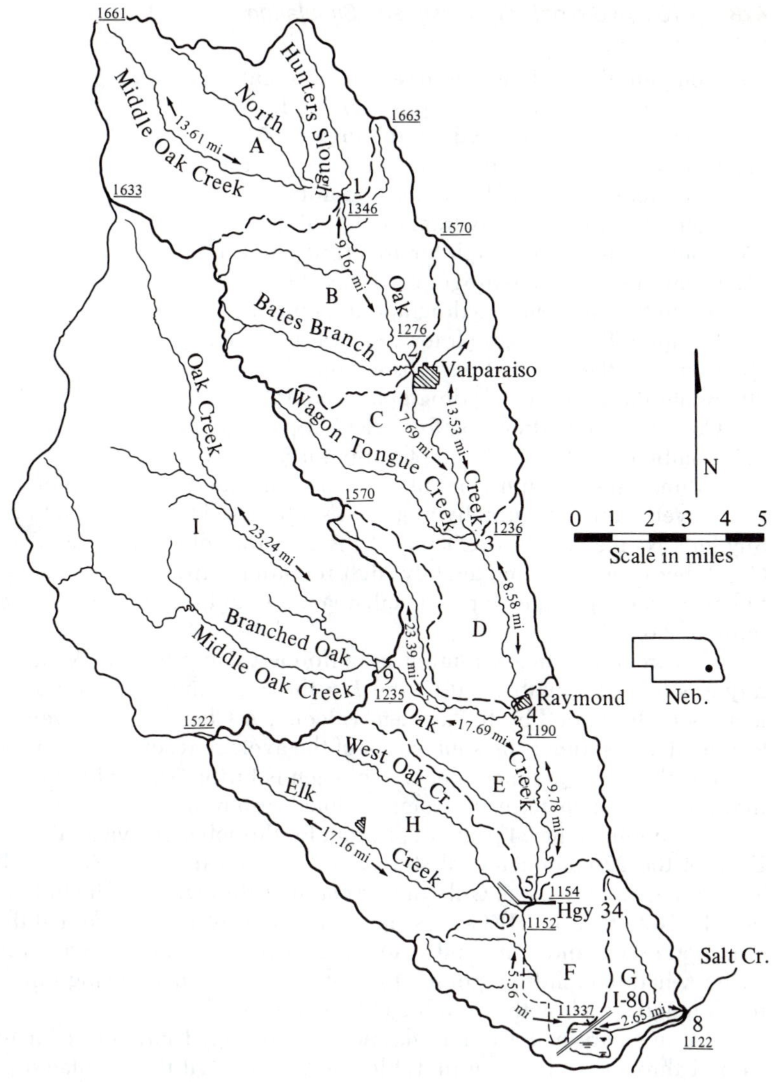


***Fig. 10-31.*** *Oak Creek watershed subarea map and data sheet, June 1963.*

| Area | Mi² | Net Rain (in.) |
|---|---|---|
| A | 33.4 | 7.8 |
| B | 26.9 | 4.3 |
| C | 27.3 | 4.1 |
| D | 9.2 | 2.8 |
| E | 28.3 | 2.4 |
| F | 17.0 | 1.7 |
| G | 5.0 | 1.7 |
| H | 28.0 | 1.0 |
| I | 82.9 | — |
| Total | 258.0 | |

9. Compute the hydrograph for area *D* at point 4.
10. Combine the two hydrographs at point 4.
11. Route the combined hydrograph to point 5.
12. Compute the hydrograph for area *E* at point 5.
13. Combine the two hydrographs at point 5.
14. Route the combined hydrograph to point 6.
15. Compute the hydrograph for area *H* at point 6.
16. Combine the two hydrographs at point 6.
17. Route the combined hydrograph to point 7.
18. Compute the hydrograph for area *F* at point 7.
19. Combine the two hydrographs at point 7.
20. Route the combined hydrograph to point 8.
21. Compute the hydrograph for area *G* at point 8.
22. Combine the two hydrographs at point 8.

Runoff hydrographs for subareas are simulated by convolving the net storm hyetograph with unit hydrographs synthesized by Clark's method using Snyder's coefficients (see Chapter 4). A $C_p$ value of 0.8 is applied for Oak Creek because of the moderately high retention capacity of the watershed. Subarea time lag values for each subarea are found from Eq. 4-31 using a $C_t$ value of 2.0.

Hydrograph stream routing is performed using the Muskingum technique (Chapter 7) with $x = 0.15$ and $K =$ the approximate reach travel time, using length divided by the average velocity. A Chézy average velocity determined as 100 times the square root of the average reach slope is used. If $K$ exceeds three routing increments, the reach is further subdivided by HEC-1 into shorter lengths to insure computational resolution.

A sample of the 24 pages of printout for this job is shown as Table 10-21. Each of the 22 computational steps are separated in sequence by a line of asterisks across the page. Only the first steps 1 through 5 are included in the sample. Note that the HEC-1 loss rate function was not used so that the end-of-period excess and rain depths are equal. Note also that hydrograph routing of the *A* hydrograph from point 1 to point 2 was facilitated using three equal reach lengths each with a *K*-value (*AMSKK*) of 1.2 hr.

A summary of HEC-1 peak and time-averaged flow rates for each of steps 1 through 22 is given in Table 10-22. Note that the simulated peak at point 3 is 27,533 cfs, which agrees very well with the recorded value of 27,500. The corresponding simulated and observed peak flows at point 8 are 22,453 and 21,600 cfs, respectively.

## 10-6 Groundwater Simulation

Digital simulation models are used in a different manner to study the storage and movement of water in a porous medium. Distributed rather than lumped parameter models are used to imitate observed events and to evaluate future trends in the development and management of groundwater systems. The equations describing the flow of water in a porous medium were derived in Chapter 8 and are not repeated here. This section deals primarily with solution techniques used in solving the hydrodynamic equations of motion and continuity, followed by brief discussions of (1) typical input requirements, (2) techniques of calibrating and verifying the models, and (3) the sensitivity of groundwater models to parameter changes.

With few exceptions, the hydrodynamic equations for groundwater flow have no analytical solutions, and groundwater modeling relies on *finite difference* and *finite element* methods to provide approximate solutions to a wide variety of groundwater problems. These solutions, as with streamflow simulation models, are facilitated by first subdividing the region to be modeled into subareas. Groundwater system subdivision depends more on geometric criteria and less on topographic criteria in the sense that the region is overlaid by a regular or semiregular pattern of node points at which (or between which) specific measures of aquifer and water system parameters are input and other parameters are calculated. Approximate solutions of simultaneous linear and nonlinear equations are found by making initial estimates of the solution values, testing the estimates in the equations of motion and continuity, adjusting the values, and finally accepting minor violations in the basic principles or making further adjustments of the parameters in an orderly and converging fashion.

The orderly solution of finite difference analogs of the steady or unsteady-state partial differential equation of motion for flow of groundwater in a confined aquifer (Eq. 8-42) or an unconfined aquifer (Eq. 8-43) is obtained by *relaxation* methods. An early relaxation solution of the equation is discussed by Jacob.[39] For two-dimensional problems, the iterative alternating-direction-implicit (ADI) method developed by Peaceman and Rachford[40] is adopted.

Prickett and Lonnquist[41] used the ADI technique to calculate fluctuations in water table elevations at all nodes in an aquifer model by proceeding through time in small increments from a known initial state. Their model is computationally efficient and readily applied, and is particularly attractive for use with problems involving time variables and numerous nodes. The primary aquifer parameters are the permeability and storage coefficient which, if assumed constant over the aquifer plan, result in a homogeneous and isotropic condition. Where data are available, spatial variations of $K$ and $S$ are easily incorporated in the Prickett and Lonnquist program. For those familiar with relaxation methods, the Gauss-Seidel and the successive-over-relaxation (SOR) methods have had application in solving difference equations.

Input to groundwater system models may be classified as spatial and temporal. Spatial input includes initial or projected water table maps, saturated thickness data over the region, land surface contour maps, transmissivity maps, regional variations in storage coefficients, locations and types of wells and canals, locations and types of aquifer

**Table 10-21** *Sample Output from HEC-1 for the Oak Creek Watershed*

```
******************************
HEC-1 VERSION DATED JAN 1973
UPDATED AUG 74
CHANGE NO. 01
******************************
                              OAK CREEK STUDY
              RUN 1 - TEST HEC-1 USING RECORD STORM OF JUNE 23-24, 1963
       USED QUARTILE 2, SNYDERS CP=.8, VEL=100 ROOT S, AVG LAG CHOW $ SNYD, X=.15

                                JOB SPECIFICATION
                 NQ    NHR  NMIN  IDAY    IHR  IMIN METRC  IPLT  IPRT NSTAN
                 75      0    30     1      0     0     0     0     0     0
                                  JOPER     NWT
                                      3       0


 **********         **********         **********         **********         **********

                             SUB-AREA RUNOFF COMPUTATION

                            COMPUTE HYDROGRAPH FOR AREA A
                  ISTAQ   ICOMP   IECON   ITAPE    JPLT    JPRT   INAME
                      1       0       0       0       0       0       1

                                  HYDROGRAPH DATA
           IHYDG    IUHG   TAREA    SNAP   TRSDA   TRSPC   RATIO   ISNOW   ISAME   LOCAL
               0       1   33.40     0.0     0.0     0.0     0.0       0       0       0

                                   PRECIP DATA
                             NP   STORM     DAJ     DAK
                             12    7.80     0.0     0.0
                                  PRECIP PATTERN
  0.03     0.05     0.06     0.09     0.37     0.10     0.08     0.06     0.04     0.05
  0.03     0.04

                                     LOSS DATA
          STRKR   DLTKR   RTIOL   ERAIN   STRKS   RTIOK   STRTL   CNSTL   ALSMX   RTIMP
            0.0     0.0    1.00     0.0     0.0    1.00     0.0     0.0     0.0     0.0

                               UNIT HYDROGRAPH DATA
                            TP#  2.93    CP#0.80    NTA#  0
```

```
                              RECESSION DATA
              STRTQ#     0.0      QRCSN#     0.0      RTIOR# 1.50
APPROXIMATE CLARK COEFFICIENTS FROM GIVEN SNYDER CP AND TP ARE TC# 7.47 AND R# 2.91 INTERVALS

        UNIT HYDROGRAPH 20 END-OF-PERIOD ORDINATES, LAG#   2.92 HOURS, CP# 0.79  VOL# 1.00
     437.      1545.      2926.      4317.      5394.      5861.      5701.      4812.      3544.      2505.
    1771.      1252.       885.       626.       442.       313.       221.       156.       110.        78.

                             END-OF-PERIOD FLOW
                        TIME    RAIN    EXCS     COMP Q
                      1  0 30   0.23    0.23       102.
                      1  0 60   0.39    0.39       532.
                      1  1 30   0.47    0.47      1492.
                      1  1 60   0.70    0.70      3181.
                      1  2 30   2.89    2.89      6661.
                      1  2 60   0.78    0.78     12350.
                      1  3 30   0.62    0.62     19097.
                      1  3 60   0.47    0.47     25789.
                      1  4 30   0.31    0.31     31109.
                      1  4 60   0.39    0.39     34061.
                      1  5 30   0.23    0.23     34457.
                      1  5 60   0.31    0.31     32147.
                      1  6 30   0.0     0.0      28099.
                      1  6 60   0.0     0.0      23842.
                      1  7 30   0.0     0.0      19844.
                      1  7 60   0.0     0.0      16156.
                      1  8 30   0.0     0.0      12773.
                      1  8 60   0.0     0.0       9754.
                      1  9 30   0.0     0.0       7173.
                      1  9 60   0.0     0.0       5115.
                      1 10 30   0.0     0.0       3603.
                      1 10 60   0.0     0.0       2525.
                      1 11 30   0.0     0.0       1759.
                      1 11 60   0.0     0.0       1205.
                      1 12 30   0.0     0.0        692.
                      1 12 60   0.0     0.0        446.
                      1 13 30   0.0     0.0        281.
                      1 13 60   0.0     0.0        173.
                      1 14 30   0.0     0.0        105.
                      1 14 60   0.0     0.0         53.
                      1 15 30   0.0     0.0         24.
                      1 15 60   0.0     0.0          0.
                      1 16 30   0.0     0.0          0.
                      1 16 60   0.0     0.0          0.
                      1 17 30   0.0     0.0          0.
```

***Table 10-21*** *(Continued)*

| | | | | | |
|---|---|---|---|---|---|
| 1 | 17 | 60 | 0.0 | 0.0 | 0. |
| 1 | 18 | 30 | 0.0 | 0.0 | 0. |
| 1 | 18 | 60 | 0.0 | 0.0 | 0. |
| 1 | 19 | 30 | 0.0 | 0.0 | 0. |
| 1 | 19 | 60 | 0.0 | 0.0 | 0. |
| 1 | 20 | 30 | 0.0 | 0.0 | 0. |
| 1 | 20 | 60 | 0.0 | 0.0 | 0. |
| 1 | 21 | 30 | 0.0 | 0.0 | 0. |
| 1 | 21 | 60 | 0.0 | 0.0 | 0. |
| 1 | 22 | 30 | 0.0 | 0.0 | 0. |
| 1 | 22 | 60 | 0.0 | 0.0 | 0. |
| 1 | 23 | 30 | 0.0 | 0.0 | 0. |
| 1 | 23 | 60 | 0.0 | 0.0 | 0. |
| 2 | 0 | 30 | 0.0 | 0.0 | 0. |
| 2 | 0 | 60 | 0.0 | 0.0 | 0. |
| 2 | 1 | 30 | 0.0 | 0.0 | 0. |
| 2 | 1 | 60 | 0.0 | 0.0 | 0. |
| 2 | 2 | 30 | 0.0 | 0.0 | 0. |
| 2 | 2 | 60 | 0.0 | 0.0 | 0. |
| 2 | 3 | 30 | 0.0 | 0.0 | 0. |
| 2 | 3 | 60 | 0.0 | 0.0 | 0. |
| 2 | 4 | 30 | 0.0 | 0.0 | 0. |
| 2 | 4 | 60 | 0.0 | 0.0 | 0. |
| 2 | 5 | 30 | 0.0 | 0.0 | 0. |
| 2 | 5 | 60 | 0.0 | 0.0 | 0. |
| 2 | 6 | 30 | 0.0 | 0.0 | 0. |
| 2 | 6 | 60 | 0.0 | 0.0 | 0. |
| 2 | 7 | 30 | 0.0 | 0.0 | 0. |
| 2 | 7 | 60 | 0.0 | 0.0 | 0. |
| 2 | 8 | 30 | 0.0 | 0.0 | 0. |
| 2 | 8 | 60 | 0.0 | 0.0 | 0. |
| 2 | 9 | 30 | 0.0 | 0.0 | 0. |
| 2 | 9 | 60 | 0.0 | 0.0 | 0. |
| 2 | 10 | 30 | 0.0 | 0.0 | 0. |
| 2 | 10 | 60 | 0.0 | 0.0 | 0. |
| 2 | 11 | 30 | 0.0 | 0.0 | 0. |
| 2 | 11 | 60 | 0.0 | 0.0 | 0. |
| 2 | 12 | 30 | 0.0 | 0.0 | 0. |
| 2 | 12 | 60 | 0.0 | 0.0 | 0. |
| 2 | 13 | 30 | 0.0 | 0.0 | 0. |
| | | SUM | 7.79 | 7.79 | 334600. |

```
                 PEAK    6-HOUR   24-HOUR   72-HOUR   TOTAL VOLUME
        CFS     34457.   24144.     6971.     4461.        334601.
     INCHES                6.72      7.77      7.77           7.77
      AC-FT              11978.    13834.    13834.         13834.

**********        **********        **********        **********        **********

                         HYDROGRAPH ROUTING

                   ROUTE 'A' HYD TO POINT 2
        ISTAQ   ICOMP   IECON   ITAPE    JPLT    JPRT   INAME
            2       1       0       0       0       0       1
                             ROUTING DATA
                QLOSS   CLOSS     AVG    IRES   ISAME
                  0.0     0.0     0.0       0       0

        NSTPS   NSTDL     LAG   AMSKK       X     TSK   STORA
            3       0       0   1.200   0.150     0.0      0.

                        ROUTED FLOWS AT       2
     102.      102.      104.      118.      183.      367.      782.     1615.     3083.     5333.
    8361.    11979.    15836.    19495.    22522.    24602.    25610.    25589.    24684.    23090.
   21011.    18641.    16157.    13707.    11403.     9320.     7492.     5928.     4620.     3550.
    2693.     2018.     1494.     1093.      791.      566.      401.      281.      195.      135.
      92.       63.       43.       29.       19.       13.        8.        6.        4.        2.
       2.        1.        1.        0.        0.        0.        0.        0.        0.        0.
       0.        0.        0.        0.        0.        0.        0.        0.        0.        0.
       0.        0.        0.        0.        0.

                 PEAK    6-HOUR   24-HOUR   72-HOUR   TOTAL VOLUME
        CFS     25610.   20912.     6986.     4471.        335335.
     INCHES                5.82      7.78      7.78           7.78
      AC-FT              10375.    13864.    13864.         13864.

**********        **********        **********        **********        **********

                      SUB-AREA RUNOFF COMPUTATION

                   COMPUTE HYD FOR B AT PT 2
        ISTAQ   ICOMP   IECON   ITAPE    JPLT    JPRT   INAME
            2       0       0       0       0       0       1

                            HYDROGRAPH DATA
   IHYDG    IUHG   TAREA    SNAP   TRSDA   TRSPC   RATIO   ISNOW   ISAME   LOCAL
       0       1   26.90     0.0     0.0     0.0     0.0       0       0       0
```

**Table 10-21** *(Continued)*

```
                                        PRECIP DATA
                                  NP   STORM     DAJ     DAK
                                  12   4.30      0.0     0.0
                                        PRECIP PATTERN

             0.03     0.05     0.06     0.09     0.37     0.10     0.08     0.06     0.04     0.05
             0.03     0.04

                                          LOSS DATA
               STRKR   DLTKR   RTIOL   ERAIN   STRKS   RTIOK   STRTL   CNSTL   ALSMX   RTIMP
                0.0     0.0     1.00    0.0     0.0     1.00    0.0     0.0     0.0     0.0

                                      UNIT HYDROGRAPH DATA
                                 TP#  2.51     CP#0.80     NTA#  0

                                        RECESSION DATA
                           STRTQ#     0.0     QRCSN#     0.0     RTIOR#  1.50
APPROXIMATE CLARK COEFFICIENTS FROM GIVEN SNYDER CP AND TP ARE TC# 6.79 AND R# 2.13 INTERVALS

               UNIT HYDROGRAPH 16 END-OF-PERIOD ORDINATES, LAG#   2.49 HOURS, CP# 0.79  VOL# 1.00
             528.     1819.     3341.     4720.     5520.     5509.     4679.     3270.     2025.     1254.
             776.      481.      298.      184.      114.       71.

                                      END-OF-PERIOD FLOW
                               TIME    RAIN    EXCS     COMP Q
                           1   0 30    0.13    0.13        68.
                           1   0 60    0.21    0.21       348.
                           1   1 30    0.26    0.26       958.
                           1   1 60    0.39    0.39      2001.
                           1   2 30    1.59    1.59      4133.
                           1   2 60    0.43    0.43      7530.
                           1   3 30    0.34    0.34     11318.
                           1   3 60    0.26    0.26     14693.
                           1   4 30    0.17    0.17     16824.
                           1   4 60    0.21    0.21     17301.
                           1   5 30    0.13    0.13     16120.
                           1   5 60    0.17    0.17     13725.
                           1   6 30    0.0     0.0      11193.
                           1   6 60    0.0     0.0       9027.
```

| | | | | | |
|---|---|---|---|---|---|
| 1 | 7 | 30 | 0.0 | 0.0 | 7144. |
| 1 | 7 | 60 | 0.0 | 0.0 | 5476. |
| 1 | 8 | 30 | 0.0 | 0.0 | 3989. |
| 1 | 8 | 60 | 0.0 | 0.0 | 2726. |
| 1 | 9 | 30 | 0.0 | 0.0 | 1741. |
| 1 | 9 | 60 | 0.0 | 0.0 | 1061. |
| 1 | 10 | 30 | 0.0 | 0.0 | 587. |
| 1 | 10 | 60 | 0.0 | 0.0 | 345. |
| 1 | 11 | 30 | 0.0 | 0.0 | 198. |
| 1 | 11 | 60 | 0.0 | 0.0 | 112. |
| 1 | 12 | 30 | 0.0 | 0.0 | 62. |
| 1 | 12 | 60 | 0.0 | 0.0 | 29. |
| 1 | 13 | 30 | 0.0 | 0.0 | 12. |
| 1 | 13 | 60 | 0.0 | 0.0 | 0. |
| 1 | 14 | 30 | 0.0 | 0.0 | 0. |
| 1 | 14 | 60 | 0.0 | 0.0 | 0. |
| 1 | 15 | 30 | 0.0 | 0.0 | 0. |
| 1 | 15 | 60 | 0.0 | 0.0 | 0. |
| 1 | 16 | 30 | 0.0 | 0.0 | 0. |
| 1 | 16 | 60 | 0.0 | 0.0 | 0. |
| 1 | 17 | 30 | 0.0 | 0.0 | 0. |
| 1 | 17 | 60 | 0.0 | 0.0 | 0. |
| 1 | 18 | 30 | 0.0 | 0.0 | 0. |
| 1 | 18 | 60 | 0.0 | 0.0 | 0. |
| 1 | 19 | 30 | 0.0 | 0.0 | 0. |
| 1 | 19 | 60 | 0.0 | 0.0 | 0. |
| 1 | 20 | 30 | 0.0 | 0.0 | 0. |
| 1 | 20 | 60 | 0.0 | 0.0 | 0. |
| 1 | 21 | 30 | 0.0 | 0.0 | 0. |
| 1 | 21 | 60 | 0.0 | 0.0 | 0. |
| 1 | 22 | 30 | 0.0 | 0.0 | 0. |
| 1 | 22 | 60 | 0.0 | 0.0 | 0. |
| 1 | 23 | 30 | 0.0 | 0.0 | 0. |
| 1 | 23 | 60 | 0.0 | 0.0 | 0. |
| 2 | 0 | 30 | 0.0 | 0.0 | 0. |
| 2 | 0 | 60 | 0.0 | 0.0 | 0. |
| 2 | 1 | 30 | 0.0 | 0.0 | 0. |
| 2 | 1 | 60 | 0.0 | 0.0 | 0. |
| 2 | 2 | 30 | 0.0 | 0.0 | 0. |
| 2 | 2 | 60 | 0.0 | 0.0 | 0. |
| 2 | 3 | 30 | 0.0 | 0.0 | 0. |
| 2 | 3 | 60 | 0.0 | 0.0 | 0. |
| 2 | 4 | 30 | 0.0 | 0.0 | 0. |
| 2 | 4 | 60 | 0.0 | 0.0 | 0. |
| 2 | 5 | 30 | 0.0 | 0.0 | 0. |

**Table 10-21** *(Continued)*

```
                              2  5 60   0.0    0.0          0.
                              2  6 30   0.0    0.0          0.
                              2  6 60   0.0    0.0          0.
                              2  7 30   0.0    0.0          0.
                              2  7 60   0.0    0.0          0.
                              2  8 30   0.0    0.0          0.
                              2  8 60   0.0    0.0          0.
                              2  9 30   0.0    0.0          0.
                              2  9 60   0.0    0.0          0.
                              2 10 30   0.0    0.0          0.
                              2 10 60   0.0    0.0          0.
                              2 11 30   0.0    0.0          0.
                              2 11 60   0.0    0.0          0.
                              2 12 30   0.0    0.0          0.
                              2 12 60   0.0    0.0          0.
                              2 13 30   0.0    0.0          0.

                                 SUM    4.29   4.29    148721.

                      PEAK    6-HOUR   24-HOUR   72-HOUR   TOTAL VOLUME
           CFS      17301.    11207.     3098.     1983.        148720.
        INCHES                  3.88      4.29      4.29           4.29
         AC-FT                 5560.     6149.     6149.          6149.

  **********        **********        **********        **********        **********

                              COMBINE HYDROGRAPHS

                          COMBINE 2 HYDS AT PT 2
                ISTAQ   ICOMP   IECON   ITAPE    JPLT    JPRT   INAME
                    2       2       0       0       0       0       1

                           SUM OF 2 HYDROGRAPHS AT     2
      170.      451.     1062.     2119.     4316.     7896.    12100.    16308.    19908.    22634.
    24482.    25704.    27029.    28521.    29666.    30079.    29599.    28315.    26425.    24151.
    21598.    18986.    16356.    13818.    11465.     9348.     7504.     5928.     4620.     3550.
     2693.     2018.     1494.     1093.      791.      566.      401.      281.      195.      135.
       92.       63.       43.       29.       19.       13.        8.        6.        4.        2.
        2.        1.        1.        0.        0.        0.        0.        0.        0.        0.
        0.        0.        0.        0.        0.        0.        0.        0.        0.        0.
        0.        0.        0.        0.        0.
```

```
                 PEAK    6-HOUR   24-HOUR   72-HOUR   TOTAL VOLUME
        CFS    30079.    26517.    10084.     6454.        484055.
     INCHES                4.09      6.22      6.22           6.22
      AC-FT              13156.    20012.    20013.         20013.


**********      **********      **********      **********      **********

                              HYDROGRAPH ROUTING

                         ROUTE B OUTFLOW TO PT 3
             ISTAQ   ICOMP   IECON   ITAPE    JPLT    JPRT   INAME
                 3       1       0       0       0       0       1
                               ROUTING DATA
                     QLOSS   CLOSS     AVG    IRES   ISAME
                       0.0     0.0     0.0       0       0

             NSTPS   NSTDL     LAG   AMSKK       X     TSK   STORA
                 3       0       0   1.200   0.150     0.0      0.

                            ROUTED FLOWS AT      3
   170.      170.      171.      181.      222.      340.      603.     1129.     2054.     3469.
  5381.     7701.    10273.    12918.    15486.    17898.    20123.    22133.    23875.    25275.
 26254.    26754.    26752.    26257.    25311.    23978.    22337.    20473.    18473.    16421.
 14387.    12432.    10602.     8928.     7428.     6110.     4970.     4001.     3188.     2516.
  1967.     1524.     1171.      893.      675.      507.      378.      280.      206.      151.
   109.       79.       57.       41.       29.       21.       15.       10.        7.        5.
     3.        2.        2.        1.        1.        1.        0.        0.        0.        0.
     0.        0.        0.        0.        0.

                 PEAK    6-HOUR   24-HOUR   72-HOUR   TOTAL VOLUME
        CFS    26754.    24127.    10095.     6470.        485281.
     INCHES                3.72      6.23      6.24           6.24
      AC-FT              11970.    20034.    20063.         20063.


**********      **********      **********      **********      **********
```

***Table 10-22*** *Runoff Summary of Simulated Peak and Average Flows at Points 1 through 8 for the June, 1963 Storm*

RUNOFF SUMMARY, AVERAGE FLOW

| | | PEAK | 6-HOUR | 24-HOUR | 72-HOUR | AREA |
|---|---|---|---|---|---|---|
| HYDROGRAPH AT | 1 | 34457. | 24144. | 6971. | 4461. | 33.40 |
| ROUTED TO | 2 | 25610. | 20912. | 6986. | 4471. | 33.40 |
| HYDROGRAPH AT | 2 | 17301. | 11207. | 3098. | 1983. | 26.90 |
| 2 COMBINED | 2 | 30079. | 26517. | 10084. | 6454. | 60.30 |
| ROUTED TO | 3 | 26754. | 24127. | 10095. | 6470. | 60.30 |
| HYDROGRAPH AT | 3 | 15392. | 10572. | 2993. | 1916. | 27.30 |
| 2 COMBINED | 3 | 27533. | 25878. | 13088. | 8386. | 87.60 |
| ROUTED TO | 4 | 25912. | 24551. | 13052. | 8409. | 87.60 |
| HYDROGRAPH AT | 4 | 3362. | 2382. | 689. | 441. | 9.20 |
| 2 COMBINED | 4 | 25925. | 24606. | 13702. | 8850. | 96.80 |
| ROUTED TO | 5 | 23911. | 22858. | 13523. | 8875. | 96.80 |
| HYDROGRAPH AT | 5 | 7100. | 5604. | 1816. | 1162. | 28.30 |
| 2 COMBINED | 5 | 23911. | 22874. | 14717. | 10038. | 125.10 |
| ROUTED TO | 6 | 23911. | 22874. | 14717. | 10039. | 125.10 |
| HYDROGRAPH AT | 6 | 3349. | 2478. | 750. | 480. | 28.00 |
| 2 COMBINED | 6 | 23911. | 22874. | 15268. | 10518. | 153.10 |
| ROUTED TO | 7 | 22949. | 22024. | 14984. | 10482. | 153.10 |
| HYDROGRAPH AT | 7 | 4113. | 2764. | 773. | 495. | 17.00 |
| 2 COMBINED | 7 | 22949. | 22024. | 15112. | 10977. | 170.10 |
| ROUTED TO | 8 | 22453. | 21595. | 14994. | 10908. | 170.10 |
| HYDROGRAPH AT | 8 | 1684. | 868. | 228. | 146. | 5.00 |
| 2 COMBINED | 8 | 22453. | 21595. | 14995. | 11054. | 175.10 |

boundaries both lateral and vertical, a node coordinate system, actual or net pumpage rates, percolation and recharge rates for precipitation and other applied waters, and soil types and cropping patterns.

Time-dependent data requirements for aquifer models involve principally the formulation of time schedules, using a range of time increments for such variables as pumping rates, precipitation hyetographs, canal and streamflow hydrographs, groundwater evapotranspiration rates, and development variables such as the timing of added wells or other system components. Because each temporal schedule can apply only to a particular subset of node positions, the time-dependent requirements are also spatial.

In addition to the listed input parameters, aquifer models require reliable estimates of the percentages of waters in the land phase that actually percolate to the aquifer being modeled. These estimates can be based on knowledge of the physical processes involved in unsaturated flow through porous medium but are most often obtained as judgment parameters that are modified during the calibration phase of the simulation. Simply stated, the lateral movement and the changes in piezometer or water table levels are easily modeled if the node-by-node stresses (withdrawal rates or recharge rates) are known. The latter parameters are governed by the complex movement of water in the unsaturated soil zone and by the random precipitation and consumptive use patterns of the region. The art of modeling groundwater systems lies in the ability to evaluate these parameters.

Groundwater model calibration removes some of the guesswork involved in parameter determination. Several combinations of parameters, based on available knowledge of the physical system, are tested in the model during a period for which records are available. Simulated results are then compared with historical events. After structuring the model, calibration is achieved by operating the model during the study period by imposing historical precipitation amounts, canal diversions, evaporation and evapotranspiration rates, streamflows and stream levels, pumping rates during know periods, and other stresses on the aquifer. Calibration is achieved after the flow, storage, and other parameters have been adjusted within reasonable limits to produce the best imitation of recorded events.

A typical finite difference study involving surface and groundwater modeling in central Nebraska was performed by Marlette and Lewis.[42] The region involved is shown in Fig. 10-32. In addition to the surface irrigation system represented by the several canals and laterals, over 1200 wells withdraw water from the aquifer between the Platte River and the Gothenburg and Dawson County canals. The aquifer recharge and withdrawal amounts as percentages of precipitation, snowfall, pumped water, delivered canal water, evaporation, and evapotranspiration were estimated using a mix of judgment and physical process evaluations. The resulting set which produced the best comparison with recorded events at the six observation wells shown in Fig. 10-32 is summarized in Table 10-23. Samples of the comparison between recorded and simulated water levels in the Dawson County study during a 2-yr calibration period are shown in Figs. 10-33 and 10-34.

The Prickett and Lonnquist model was applied in the Dawson County study. The storage coefficient for this unconfined aquifer was established by calibration trials as 0.25 and the adopted permeability

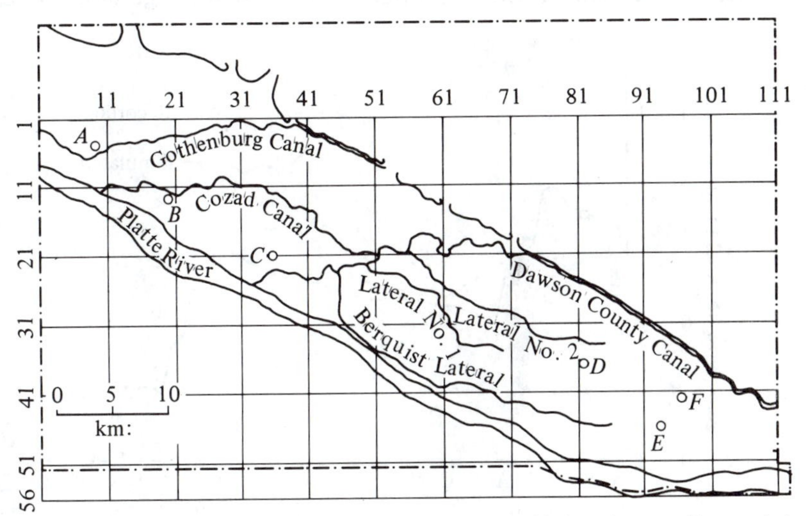


**Fig. 10-32.** *Grid coordinates for Dawson County, Nebraska, aquifer model. O = observation well.*

***Table 10-23*** *Adopted Recharge Criteria for Water Allocations over the Dawson County Aquifer*

| System Component | Allocation and Applied Amounts | Aquifer Recharge/Withdrawal as a Percentage of Applied Amount |
|---|---|---|
| Rainfall | Recorded depth if daily amount exceeded 0.25 cm at all nodes | 30 |
| Snowfall | 25% of recorded depths at all nodes | 30 |
| Pumped water | Average rate of 50 liters/sec at all well nodes during irrigation seasons | 50 |
| Delivered canal water | Recorded daily rates, applied to land surface one node laterally uphill and two nodes downhill from canal | 30 |
| Evaporation | Observed daily lake evaporation depth at all marsh and water surface nodes | 100 |
| Evapotranspiration | 125% of daily lake evaporation, applied at all alfalfa nodes | 15 |

was 61 m/day. Other trials were made using various combinations of $S$ and $K$, with $S$ ranging from 0.10 to 0.30 and with $K$ ranging between 41 and 102 m/day. As with most unconfined aquifer models, water table elevations were most sensitive to fluctuations in the storage coefficient. Figure 10-35 is a typical summary of the calibration results at a single observation well located at position $F$ in Fig. 10-32.

After verification, the Dawson County model was applied to investigate the short-term influence of several management schemes.

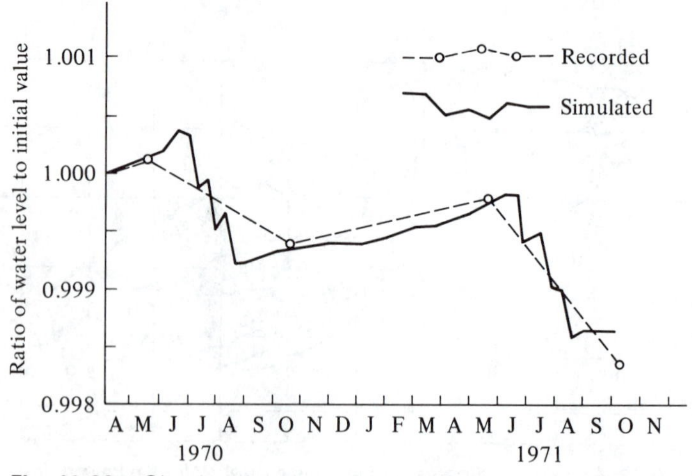


***Fig. 10-33.*** *Simulated and recorded water levels at observation well D in Fig. 10-32. I = 82; J = 37.*

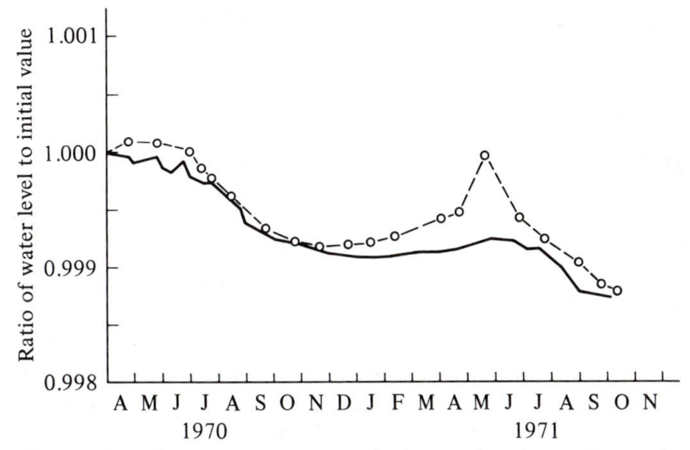


*Fig. 10-34.* Simulated and recorded water levels at observation well F in Fig. 10-32. $I = 97$; $J = 42$.

Included among the schemes were investigations involving the complete removal or shutdown of the surface water canals, and other tests in which isolated canal contributions to recharge were determined by operating the model with single canals and comparing results with water table fluctuations for identical conditions with all canals removed. Many other applications of the model are possible. This particular study revealed that recharge from the existing canal system contributes to the water balance of the aquifer but is not the dominant

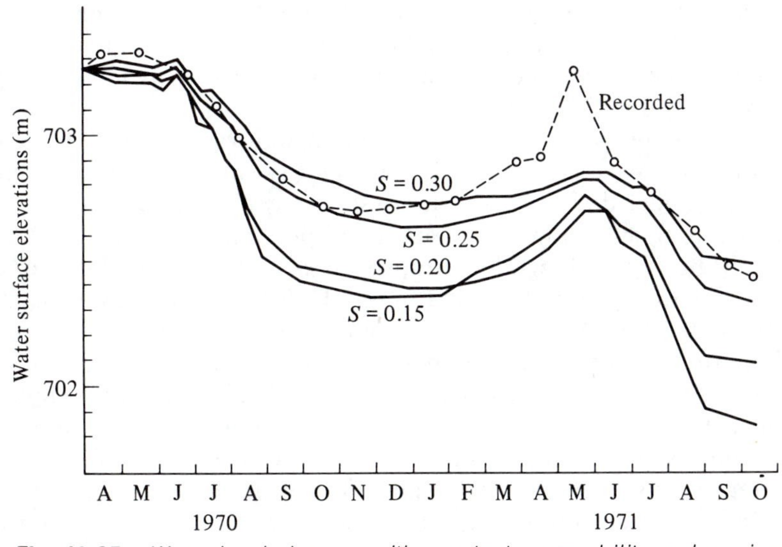


*Fig. 10-35.* Water level changes with constant permeability and varying storage coefficients. Well F, Fig. 10-32; $K = 61$ m/day; $I = 97$; $J = 42$.

factor in the short run. The natural recharge from precipitation and from the Platte River account for the long-term water table stability in the region.

## 10-7 Digital Simulation in Snow Hydrology

Anderson and Crawford,[43] and Crawford and Linsley[10] have developed special subroutines for synthesizing snowmelt runoff hydrographs. Their work is briefly described to illustrate the nature of such models.

Several models have been produced. One makes use of theoretical considerations in finding the various components of the snowmelt. The condensation melt is calculated using Eq. 9-29; the evaporation is computed by Eq. 9-31; the convection melt is based on Eq. 9-28; and the radiation melt is tied to Eq. 9-33. Data requirements of the developed subroutine are highly exacting and so have greatest utility on well-instrumented watersheds. Estimates of observed runoff compare reasonably well with observed flows for the years listed. Figure 10-36 shows the results of a computer run for 1950. A flow chart for the snowmelt routine is given in Fig. 10-37.

Many basins of hydrologic interest are not extensively or completely instrumented. Such conditions require adequate estimating techniques incorporating the data to be prepared. With this in mind, Anderson and Crawford designed a subroutine to make use of the temperature data only.[43] It was assumed that differentials measured between the air and snow-surface temperature could be used. Reduced accuracy was noted in the comparison of outflow computations obtained by this method with those found through the more basic approach already discussed. The inability of the temperature differential to adequately reflect the radiation melt was considered largely responsible.

A somewhat more sophisticated model was produced to include radiation measurements. This snowmelt subroutine utilizes maximal and minimal daily temperatures, short-wave radiation (measured or estimated), and snow evaporation and precipitation. Calculations are on an hourly basis while precipitation adds to the snowpack or contributes to the liquid water storage. Hourly net heat exchanges are calculated from temperature and radiation data. Heat exchange might be either positive or negative, the latter indicating decreases in temperature of the snowpack that can be modeled by including a negative heat storage term. As the snowpack warms, the negative heat storage decreases until the temperature of the snowpack reaches 32°F. Thereafter, positive heat initiates melt. Whenever melt alone, or melt plus rainfall, exceeds the water-holding capacity of a snowpack, the runoff can begin (see Sec. 9-2).

If the dewpoint temperature drops below freezing, evaporation

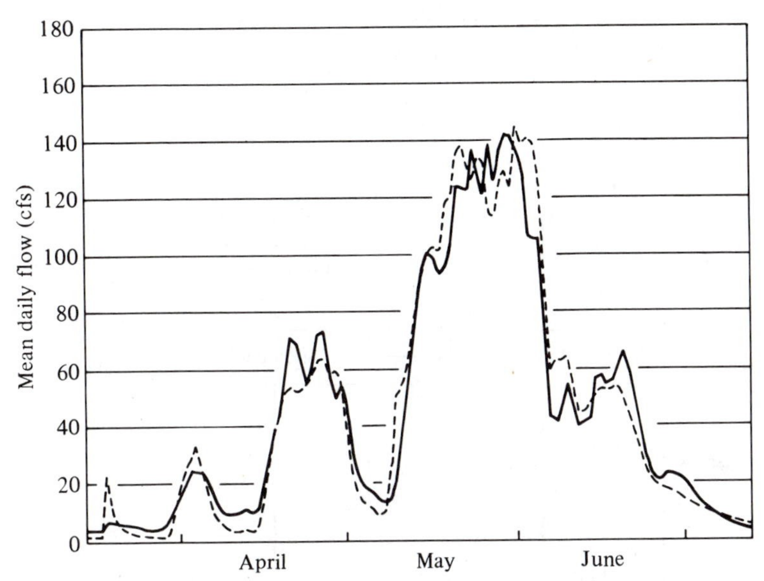


***Fig. 10-36.*** *Upper Castle Creek hydrograph.*

Mean daily flow
Water year 1950

—— Observed
--- Synthetic-theoretical subroutine

*(After E. A. Anderson and N. H. Crawford, "The Synthesis of Continuous Snowmelt Runoff Hydrographs on a Digital Computer," Department of Civil Engineering, Stanford University, Stanford, Calif., Tech. Rept. No. 36, June 1964.)*

from the snowpack can be reckoned from potential estimates. Reflected short-wave radiation is continuously calculated to get variations in the amount of absorbed solar radiation. The areal extent of snow cover in different zones is also constantly predicted; these zones are chosen on the basis of elevation and exposure.

Observed maximum and minimum temperatures are used to calculate hourly values, whereas changes with elevation are based on lapse-rate models. The amount of liquid water stored in a snowpack is limited by using an input parameter $WC$ which restricts the quantity to a function of the snowpack supply. For the usual density range of snow, this factor is considered to be constant. The density of new snow at temperatures above 0°F is presumed to be related to temperature by

$$DNS = INDS + \left(\frac{T}{100}\right)^2 \tag{10-46}$$

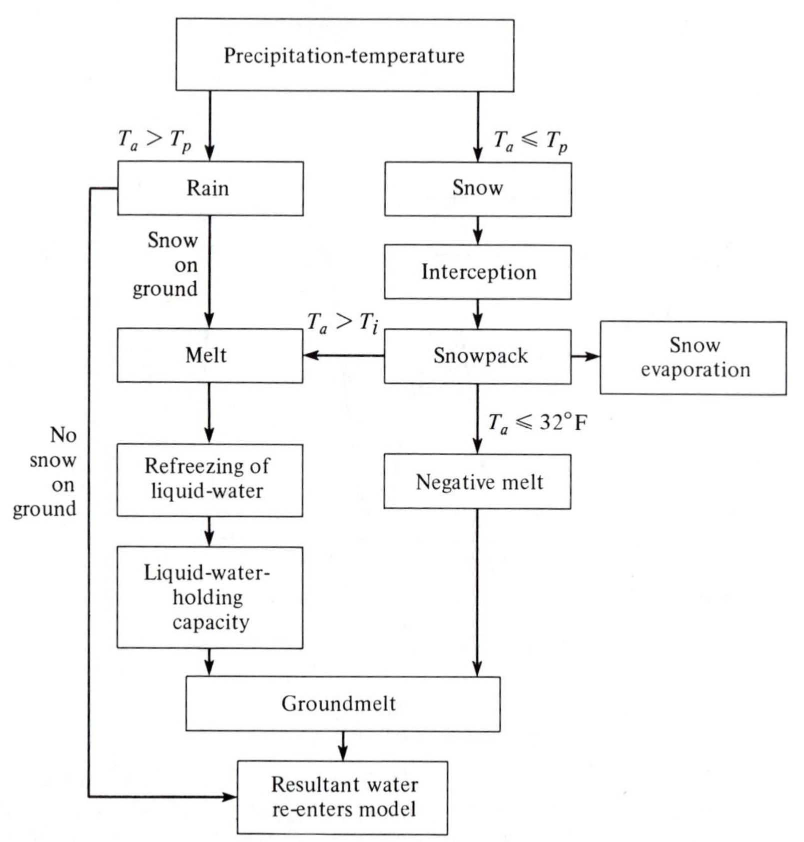


**Fig. 10-37.** *Flow chart of a snowmelt subroutine. $T_i$ = base temperature. (After E. A. Anderson and N. H. Crawford, "The Synthesis of Continuous Snowmelt Runoff Hydrographs on a Digital Computer," Department of Civil Engineering, Stanford University, Stanford, Calif., Tech. Rept. No. 36, June 1964.)*

where

$DNS$ = the new snow density
$INDS$ = the snow density at or below the datum temperature of 0°F
$T$ = the temperature

Where snow density <0.6, hourly decreases in depth are computed by multiplying the prevailing depth by a correction factor.

When the form of precipitation is not shown by records, an assumption is made that for temperatures below freezing at an elevation 750 ft above a zone, precipitation is in the form of snow. The reliability of such a supposition is affected by the method of hourly temperature determination used. Whenever temperatures are close to freezing, Linsley suggests that snow depth observations be used to check the assumption. Solar radiation data are taken from daily observations and can be estimated. The gross amount of insolation is modified to include the reflection losses, forest cover, slope, and exposure. A method of estimating net long-wave radiation exchange based on studies of the Corps of Engineers is employed.[44] The equation for net long-wave radiation in langleys/hr is

$$LW = 27.5(1.0 - F)\left[0.76\left(\frac{T_r}{273.0}\right)^4 - 1.0\right] + 27.5(F)\left[\left(\frac{T_r}{273.0}\right)^4 - 1.0\right] \quad \textbf{(10-47)}$$

where

$F$ = the fraction of the area in forest cover
$T_r$ = absolute temperature (°R)

Hourly long- and short-wave radiation are combined to produce the total net radiation. The hourly radiation melt in inches is thus

$$M_r = \frac{H_m}{203.2} \quad \textbf{(10-48)}$$

Convective heat exchange is indexed to predict the convective melt. The resulting expression is

$$M_c = CONMELT\ (T - 32) \quad \textbf{(10-49)}$$

in which

$M_c$ = convective melt (in.)
$T$ = temperature (°F)
$CONMELT$ = a convection-condensation melt parameter.

Since dewpoint data are not commonly available, the condensation melt is not estimated separately. Parameter *CONMELT* includes condensation effects, however, when forecast from simulation runs. The rainfall snowmelt in inches is appraised from

$$M_p = \frac{(T - 32)PX}{144} \quad \textbf{(10-50)}$$

where

$PX$ = the rainfall (in.)
$T$ = the temperature (°F)

Groundmelt is determined by a parameter *DGM*, which represents the water equivalent of daily melt in various zones whose elevations are reflected in the model as well as areal coverage of the snowpack. When negative values result from Eqs. 10-48 or 10-49, the negative heat storage is increased.

Snowmelt simulation indicated in the preceding paragraphs is based on the availability of only limited data and its accuracy is affected accordingly. When more complete facts are available, estimates of snowmelt can be improved and a better simulation model results. A complete discussion is given in Refs. 43, 44, and 45.

## 10-8 Hydrologic Simulation Applications

The use of hydrologic simulation as a tool in the decision-making process is not new but is of a different, more sophisticated and more encompassing form. A model is a representation of an actual or proposed system that permits the evaluation and manipulation of many years of prototype behavior. This is the feature that makes the use of these tools so attractive and holds such potential for the analysis of even the largest, most complex systems. It is also the principal feature that makes this approach so well suited to water resources system planning and analysis.

Apart from the use of conventional hand methods and some elementary models, planning has traditionally been a practice of judgment. This is changing, however, as quantitative tools are developed that permit the analysis of large numbers of alternatives and plans. Judgment, an essential element of the process, is not ruled out but is strengthened through new insights that were not available to those in the planning profession a few years ago.

Planners are continually required to anticipate the future and ask "what if?" questions. Quantitative planning techniques, such as simulation, answer these questions. In the hands of a skilled analyst, the techniques can provide detailed information about more planning alternatives for less cost than any other approach available.

An important second type of quantitative planning tool should be mentioned at this point. *Screening* models are designed to utilize limited system information to select a best plan among many alternatives for a specified objective or set of objectives. Hence screening models, or optimization models as they are often called, are oriented toward *plan formulation* in contrast to the *plan evaluation* function of simulation models. Simulation models are suited to detailed analyses of specific alternatives and yield reliable information on which to base final designs or operating policies. Screening models address the question, "If our goal is . . ., how can we best proceed?" Simulation models, on the other hand, ask, "If our plan is to . . ., will the plan work and what will the system look like after we are finished?" Used together to take advantage of the special merits of each, these two tools become a powerful adjunct to traditional planning technologies. Complete descriptions of techniques and case studies of screening, optimization, and simulation models are presented in Refs. 46, 47, 48, and 49.

Final design values should be determined by assigning the optimization results to the system elements and operating a detailed simulation model over time using a sequence of known or synthesized precipitation amounts and/or streamflows, while at the same time accumulating benefits over time for flood control, reservoir and streamside recreation, water yields, streamflow augmentation, sediment and erosion control, and any other factors not considered or approximated in the preliminary screening model. Several simulation runs with slight adjustments in reservoir capacities, and so on, should result in a plan that best meets the objectives and is a significant improvement in plans generated by conventional methods.

Results of optimization models will provide readily obtained and useful information for initiating more refined simulation analyses in order to test the most promising measures and arrive at final plans for the design, construction, and operation of a water resource system. Even though optimization models will provide information for decisions regarding both the development and management of a system, the use of postoptimization simulation is recommended, primarily because of the simplifying assumptions often required in preliminary screening. Detailed simulation is unsuited for preliminary screening of development alternatives owing to time and cost limitations. Unless a new generation of computers evolves, current time and size limitation do not allow screening by simulating all alternatives unless substantial sacrifices in realism are made. For the present, preliminary screening followed by detailed simulation appears to be the most effective means for arriving at optimal water development and management plans.

## Problems

10-1. Simulation and synthesis are treated separately in this chapter. List the most distinguishing characteristics of each method and give an example of each.

10-2. Plot cumulative inflows versus time for the 8-yr record in Example 10-2 and determine by mass curve analysis the size of reservoir needed to provide a yield of 12,000 acre-ft in each of the next 24 yr. What is the maximum yield possible?

10-3. Use the annual flows from Prob. 6-12 to generate a 10-yr sequence of synthetic annual flows for Plum Creek using Rippl's mass curve assumption.

10-4. Repeat Prob. 10-3 using random generation rather than mass curve methods. Use two-digit random numbers from Table C-3 and match these with the last two digits of the year numbers in Prob. 6-12.

10-5. Repeat Prob. 10-3 using random generation with random numbers determined by the mixed congruential method. Use $a = 41$, $b = 3$, $c = 10^6$, $p_0 = 123{,}456$, and match the last two digits in the remainder with the last two digits in each year number.

10-6. Repeat Prob. 10-3 using random generation to generate a 10-yr synthetic sequence of annual flows that has a normal CDF with a mean and standard deviation equal to that of the annual flow data from Prob. 6-12.

10-7. Repeat Prob. 10-6 assuming that the annual flows on Crooked Creek follow a Log-Pearson Type III distribution. For statistics use the mean, standard deviation, and skew of the logarithms of annual flows in Crooked Creek.

10-8. Select a gauged stream in your geographic location and prepare a quarterly model using (a) normal distribution, (b) lognormal distribution, (c) Pearson Type III distribution.

10-9. Can you convert the simulation problem in Example 10-4 to a lognormal distribution simulation? What difficulties are encountered with the data given in the example?

10-10. Review a journal article in hydrologic simulation and identify the type or combination of types of model employed. Discuss the advantages and disadvantages.

10-11. Select a month of thunderstorm activity in your region. From published NOAA hourly rainfall data, fit a distribution to the time between storms, and duration of storms, for 20 yr of recorded data covering the selected month. Prepare a computer program to randomly generate the times between storms and the times of storms.

10-12. Assume that a 30-mi$^2$ rural watershed in your locale receives a 3-in. rain in a 10-day period. Reconstruct the block diagram of Fig. 10-4 and plot approximate percentages to show, for average conditions, how the rain would be distributed: (a) initially and (b) after 30 days.

10-13. Compare, by listing traits and capabilities of each, the SWM-IV with its more sophisticated offspring HSP.

10-14. Verify Eqs. 10-36 and 10-37 by starting from Eqs. 7-4 and 7-14.

10-15. Discuss the watershed behavior that is depicted in Fig. 10-8. Is this a "typical" watershed?

10-16. A sloping, concrete parking lot experiences rain at a rate of 3.0 in./hr for 60 min. The lot is 500 ft deep and has a slope of 0.0001 ft/ft. If the water detention on the lot were zero at the start of the storm, calculate the complete overland flow hydrograph for 1 ft of width using the SWM-IV equations. Use a 5-min routing interval and continue computations until all the detained water is discharged.

10-17. Calculate the SWM-IV overland flow time-to-equilibrium for the lot of Prob. 10-16 and compare it with the Kirpich time of concentration for the lot. Should these be equal?

10-18. Synthesize a unit hydrograph for a watershed in your locale using the HYMO Model equations. Compare with corresponding unit hydrographs from Snyder's method and the SCS method in Chapter 4.

10-19. Use the HYMO Model equations to synthesize a unit hydrograph for the entire Oak Creek Watershed in Fig. 10-31.

10-20. A watershed experiences a 12-hr rainstorm having a uniform intensity of 0.1 in./hr. Using $E = 0.7$, $K_0 = 0.6$, $C = 3.0$ and $\Delta K = 0.0$, calculate the hourly loss rates $L_t$ as determined by the HEC-1 event simulation model. Determine the total and percent losses for the storm.

10-21. Repeat Prob. 10-20 using $CUML_1 = 0.5$ in.

10-22. Route the inflow hydrograph in Prob. 7-7 to the outlet of the 30-mi reach using the Straddle-Stagger method of HEC-1. Lag the averaged pairs of flows two time increments (12 hr) and compare the routed and measured outflow rates.

10-23. Route the inflow hydrograph in Prob. 7-7 through the reach using the Tatum method of HEC-1. Divide the 30-mi reach into three subreaches and treat the outflow from each as inflow to the next in line. Lag one time increment in each subreach and compare the final outflows with the measured values.

10-24. Study Table 10-21 and Fig. 10-31, and then see if you can define the following terms from Table 10-21: NQ, JOPER, TAREA, NP, STORM, TP, CP, TC, R, RAIN, EXCS, and COMP Q.

10-25. Search the HEC-1 printout in Table 10-21 to determine values (give units) of

(a) The time increment used in the model run.
(b) Snyder's $C_p$, Eq. 4-34, input for subarea *B*.
(c) The peak flow rate for the synthesized subarea-*A* unit hydrograph.
(d) The total runoff (in.) from subarea *A*.
(e) The peak outflow rate from subarea *A*.
(f) The peak-to-peak time lag in routing the outflow hydrograph from point 1 to point 2, Fig. 10-31.
(g) The percent attenuation caused by the reach between points 1 and 2.
(h) The subarea-*B* peak outflow rate if subarea *A* is neglected.
(i) The actual simulated subarea-*B* peak outflow rate.

## References


1. Rippl, W., "The Capacity of Storage Reservoirs for Water Supply," *Proc. Inst. Civil Engineers London,* 71 (1883), p. 270.
2. Hurst, H. E., "Long-Term Storage Capacities of Reservoirs," *Trans. ASCE,* 116 (1951): 776.
3. Hurst, H. E., R. P. Black, and Y. M. Sunaika, *Long-Term Storage* (London: Constable & Company Ltd., 1965).
4. Fiering, M. B., and B. B. Jackson, "Synthetic Streamflows," American Geophysical Union Water Resources Monograph No. 1, 1971.
5. Hillier, F. S., and G. J. Lieberman, *Introduction to Operations Research,* 2nd ed. (San Francisco: Holden-Day, Inc., 1974), pp. 626–628.
6. Thomas, H. A., and M. B. Fiering, "Mathematical Synthesis of Streamflow Sequences for the Analysis of River Basins by Simulation," in A. Maass, et al., *Design of Water Resources Systems.* (Cambridge, Mass.: Harvard University Press, 1962).
7. Matalas, N. C., "Mathematical Assessment of Synthetic Hydrology," *Water Resources Research,* 3, No. 4 (1967): 937–945.
8. Beard, L. R., "Simulation of Daily Streamflow," International Hydrology Symposium, Fort Collins, Colo., Sept. 1967.
9. U.S. Army Corps of Engineers, "HEC-1 Flood Hydrograph Package," Users and Programmers Manuals, HEC Program 723-X6-L2010, Jan. 1973.

10. Crawford, N. H., and R. K. Linsley, Jr., "Digital Simulation in Hydrology: Stanford Watershed Model IV," Department of Civil Engineering, Stanford University, Stanford, Calif., Tech. Rept. No. 39, July 1966.
11. U.S. Army Waterways Experiment Station, "Models and Methods Applicable to Corps of Engineers Urban Studies," Miscellaneous Paper H-74-8, National Technical Information Service, Aug. 1974.
12. Fleming, George, *Computer Simulation Techniques in Hydrology.* (New York: American Elsevier Publishing Co., Inc., 1975).
13. Sittner, W. T., C. E. Schauss, and J. C. Monro, "Continuous Hydrograph Synthesis with an API-Type Hydrologic Model," *Water Resources Research,* 5, No. 5 (1969): 1007–1022.
14. James, L. D., "An Evaluation of Relationship Between Streamflow Patterns and Watershed Characteristics Through the use of OPSET," Research Rept. No. 36, Water Resources Institute, University of Kentucky, Lexington, 1970.
15. Liou, E. Y., "OPSET: Program for Computerized Selection of Watershed Parameter Values for the Stanford Watershed Model," Research Rept. No. 34, Water Resources Institute, University of Kentucky, Lexington, 1970.
16. James, L. D., "Hydrologic Modeling, Parameter Estimation, and Watershed Characteristics," *J. Hydrology,* 17 (1972): 283–307.
17. Ricca, V. T., "The Ohio State University Version of the Stanford Streamflow Simulation Model, Part I—Technical Aspects," Ohio State University, Columbus, May 1972.
18. Holtan, H. N., and N. C. Lopez, "USDAHL-73 Revised Model of Watershed Hydrology," U.S. Department of Agriculture, Plant Physiology Institute, Rept. No. 1, 1973.
19. Claborn, B. J., and W. Moore, "Numerical Simulation in Watershed Hydrology," Hydraulic Engineering Laboratory, University of Texas Rept. No. HYD 14-7001, 1970.
20. U.S. National Weather Service Office of Hydrology, "National Weather Service River Forecast System Forecast Procedures," NOAA Tech. Mem. NWS HYDRO-14, Dec. 1972.
21. U.S. Army Corps of Engineers, "Program Description and User Manual for SSARR Model Streamflow Synthesis and Reservoir Regulation," Program 724-K5-G0010, Dec. 1972.
22. Crawford, N. H., "Studies in the Application of Digital Simulation to Urban Hydrology," Hydrocomp International, Inc., Palo Alto, Calif., Sept. 1971.
23. Clarke, D. K., "Applications of Stanford Watershed Model Concepts to Predict Flood Peaks for Small Drainage Areas," Division of Research, Kentucky Department of Highways, 1968.
24. Ross, G. A., "The Stanford Watershed Model: The Correlation of Parameter Values Selected by a Computerized Procedure with Measurable Physical Characteristics of the Watershed," Research Rept. 35, Water Resources Institute, University of Kentucky, Lexington, 1970.
25. Kirpich, Z. P., "Time of Concentration of Small Agricultural Watersheds," *Civil Engineering,* 10 (June 1940): 362.
26. Wolman, M. G., and L. B. Leopold, "River Flood Plains: Some Observa-

tions of Their Formation," U.S. Geological Survey Professional Paper 282-C, 1957.

27. Barnes, B. S., "Discussion of Analysis of Runoff Characteristics," *Trans. ASCE*, 105, 1940.
28. Shaffer, F. B., "Availability and Use of Water in Nebraska, 1970," Nebraska Water Survey Paper No. 31, Conservation and Survey Division, University of Nebraska, Lincoln, Mar. 1972.
29. Briggs, D. L., "Application of the Stanford Streamflow Simulation Model to Small Agricultural Watershed at Coshocton, Ohio," M.S. Thesis, Ohio State University, 1969.
30. Clark, C. O., "Storage and the Unit Hydrograph," *Trans. ASCE*, 110 (1945): 1419–1488.
31. Carrigan, P. H., "Calibration of U.S. Geological Survey Rainfall-Runoff Model for Peak-Flow Synthesis-Natural Basins," U.S. Geological Survey Computer Report, U.S. Department of Commerce National Technical Information Service, 1973.
32. Phillip, J. R., "Numerical Solution of Equations of the Diffusion Type with Diffusivity Concentration Dependent," *Australian J. Phys.*, 10 (1957): 29.
33. U.S. Department of Agriculture Soil Conservation Service, "Computer Program for Project Formulation Hydrology," Tech. Release No. 20, June 1973.
34. Hydrology, Supplement A to Section 4, *National Engineering Handbook.* Washington, D.C.: U.S. Department of Agriculture, Soil Conservation Service, Aug. 1972.
35. "Rainfall Frequency Atlas of the United States for Durations from 30 Minutes to 24 Hours and Return Periods from 1 to 100 Years," U.S. Weather Bureau Tech. Paper No. 40, May 1961.
36. Williams, J. R., and R. W. Hann, "HYMO:Problem-Oriented Computer Language for Hydrologic Modeling," U.S. Department of Agriculture, Agriculture Research Service, May 1973.
37. Williams, J. R., "Flood Routing with Variable Travel Time or Variable Storage Coefficients," *Amer. Soc. Agricultural Eng. Trans.*, 12, No. 1 (1969): 100–103.
38. "SWMM, Volume No. 1, Final Report," Report for U.S. Environmental Protection Agency, Water Resources Engineers, and Metcalf and Eddy Inc., July 1971.
39. Jacob, C. E., "Flow of Groundwater," in Hunter Rouse (ed.), *Engineering Hydraulics* (New York: John Wiley & Sons, Inc., 1950).
40. Peaceman, D. W., and N. H. Rachford, "The Numerical Solution of Parabolic and Elliptic Differential Equations," *J. Soc. Indust. and Appl. Math.*, 3, 1955.
41. Prickett, T. A., and C. G. Lonnquist, "Selected Digital Computer Techniques for Groundwater Resource Evaluation," Illinois State Water Survey Bull. No. 55, 1971.
42. Marlette, R. R., and G. L. Lewis, "Digital Simulation of Conjunctive-Use of Groundwater in Dawson County, Nebraska," Civil Engineering Rept., University of Nebraska, Lincoln, 1973.
43. Anderson, E. A., and N. H. Crawford, "The Synthesis of Continuous

Snowmelt Runoff Hydrographs on a Digital Computer," Department of Civil Engineering, Stanford University, Stanford, Calif., Tech. Rept. No. 36, June 1964.

44. "Runoff from Snowmelt," U.S. Army Corps of Engineers, Engineering and Design Manuals, EM 1110-2-1406, Jan. 1960.
45. Anderson, Eric A., "National Weather Service River Forecast System—Snow Accumulation and Ablation Model," NOAA Tech. Mem., NWS HYDRO-17, Silver Spring, Md., Nov. 1973.
46. Johnson, W. K., "Use of Systems Analysis in Water Resource Planning," *Proc. ASCE, J. Hyd. Div.*, Paper No. 9174, Sept. 1974.
47. deNeufville, R., and D. H. Marks, *Systems Planning and Design Case Studies in Modeling Optimization and Evaluation.* (Englewood Cliffs, N.J.: Prentice-Hall, 1974).
48. Loucks, D. P., "Stochastic Methods for Analyzing River Basin Systems," Cornell University Water Resources and Marine Sciences Center, Ithaca, N.Y., Aug. 1969.
49. Maass, A., et al., *Design of Water Resources Systems.* (Cambridge, Mass.: Harvard University Press, 1962).

# 11

# Urban and Small Watershed Hydrology

Effective disposition of storm water is essential. Drainage systems have changed from primitive ditches to complex networks of curbs, gutters, and surface and underground conduits. Along with the increasing complexity of these systems has come the need for a more thorough understanding of basic hydrologic processes. Simple rules of thumb and crude empirical formulas are generally inadequate. The approximation of maximal rates of flow to be expected with some relative frequency is not sufficient for many modern designs. Management on a day-to-day basis requires continuous time histories. Accordingly, it is often essential to account for all key hydrologic processes and combine them in composite models that yield outputs at points of interest in time and space. In addition, demands by society for better environmental control require that water quality considerations be superimposed on estimates of quantity so management of the total water resource can be effected.

## 11-1 Introduction

Methods used in estimating quantities of storm water runoff from urban drainage areas and other small watersheds may be classified as the rule-of-thumb approach, the macroscopic approach, the microscopic approach, and continuous simulation.

### The Rule-of-Thumb Approach

An early statement about urban rainfall runoff was precipitated by the storm of June 20, 1857, on the Savoy Street sewer in London. One inch of rain fell in 75 min, producing a maximum flow of 0.34 $ft^3$/sec/acre. Based upon information then available, the distinguished engineers Bidder, Hawksley, and Bazalgette concluded that 0.25 in. of rainfall would contribute about 0.125 in. to the sewer, and 0.40 in. would yield approximately 0.25 in. At this time, a general English rule of thumb was that about half of the rainfall would appear as runoff from urban surfaces. These early guidelines were forerunners of modern urban hydrologic models.

### Macroscopic Approach

Following the early rules of thumb, empirical formulas became the principal mechanism for determining quantities of runoff. Most second-generation approaches were macroscopic. They are characterized by (1) consideration of the entire drainage area as a single unit, (2) estimation of flow at only the most downstream point, and (3) the assumption that rainfall is uniformly distributed over the drainage area. The foremost example of this approach is the Rational Method which was based on 4 yr of rainfall data using nonrecording rain gauges and 1 yr of runoff data estimated from pairs of white-washed sticks. Five open ditches were used for flow determination.

The Rational Method is described by the statement $Q = CIA$, where $Q$ equals the peak runoff rate in cfs, $C$ is a runoff coefficient (the ratio of an instantaneous peak runoff rate to a rainfall rate averaged over a time of concentration), $I$ is the rainfall rate in in./hr, and $A$ is the drainage area in acres (see Sec. 11-2). The Rational Method has been used for over half a century with little change in its original form. It is the standard method of urban storm drainage design today. Persistence in the use of this formula can be attributed to its simplicity. The present analytical effort in urban hydrology should bring about some change in design concepts, but new techniques should not sacrifice the practicing professional's desire for easy-to-apply procedures. In fact, sophisticated models might serve a useful purpose in evaluating parameters for simpler procedures that can be applied to routine problems.

A second example of the macroscopic approach is the unit hydrograph method developed by Leroy K. Sherman in 1932. In Chapter 4 it was stated that the unit hydrograph is the hydrograph of 1 in. of runoff from a drainage area produced by a uniform rainfall lasting any unit of time. Once determined, the unit graph can be used to construct the hydrograph for a storm of any magnitude and duration. Originally, the unit hydrograph concept was applied mainly to river basins, but it is now used for urban and small watershed networks as well. The concept of the instantaneous unit hydrograph has been a major factor in its expansion to use on small drainage areas. The instantaneous unit hydrograph operates on an effective precipitation applied in zero time. This is theoretical, but the assumption makes the hydrograph independent of the duration of effective precipitation and eliminates one variable from hydrograph analysis. A number of models using an instantaneous unit hydrograph or an approximation of it have been reported in the literature.[1,2,3] One of the pioneers was J. E. Nash.[4]

### Microscopic Approach

The microscopic approach is characterized by an attempt to quantify all pertinent physical phenomena from the input (rainfall) to the output (runoff).[5,6,7] This usually involves the following steps: (1) determine a design storm; (2) deduct losses from the design storm to arrive at an excess rainfall rate; (3) determine the flow to a gutter or some defined channel by overland flow equations; (4) route these gutter or small channel flows to the main channel; (5) route the flow through the principal conveyance system (pipe, canal, or stream); and (6) determine the outflow hydrograph. The result obtained is affected by the accuracy of calculating losses and hydraulic phenomena and the validity of the simplifying assumptions. If errors are small and noncumulative, the prediction of the runoff is valid. Tholin's Hydrograph Method (Sec. 11-3) is an example of the microscopic approach.[8]

In the past, most microscopic procedures dealt solely with individual storm events. With the advent of modern computers, the trend has been more toward the continuous simulation of hydrologic processes.[3,9,17]

### Simulation

Analyses of many urban and other small watersheds are effected through the use of simulation models. As discussed in Chapter 10, hydrologic systems may be simulated by physical models, analog models, or digital models. Digital models such as the Sanford Watershed Model and EPA's Storm Water Management Model are widely used for small watershed studies and for the hydrologic design of storm drainage systems.

### Quantity-Quality Models

Evaluation of the water quality produced in the runoff from small watersheds is becoming increasingly more important.[10,11] Unfortunately, data on this aspect are very limited and usually not adequate for continuous simulation. Variations in runoff quality result from changes in season and geographical area. Urban and rural practices such as lawn watering, irrigation, car washing, and others affect the ultimate destination of fertilizers and other chemicals.[12,13,14] A better understanding of the complex nature of physical, chemical, and biological processes that affect the water quality is needed, but a number of operational models have been developed.[15,16] A more complete discussion is given in Chapter 13.

The state of the art in modeling urban and small watershed runoff quantity and quality has progressed from simple rules of thumb to complex simulation models incorporating all fundamental hydrologic processes. The mechanics of all of these processes is not completely known and some empiricism remains. In fact, small watershed modeling is still part art and part science, but as more data become available and modeling technology improves, a greater degree of sophistication and reliability will result. Perhaps the greatest underlying need is for more data characterizing the water cycle on urban areas. Information on both the quantity and quality of urban runoff is in critically short supply.

## 11-2 Peak Flow Formulas for Urban and Small Watersheds

Numerous methods are available for estimating the peak rates of runoff required for design applications in small urban and rural watersheds. Some incorporate a rational analysis of the rainfall-runoff process whereas others are completely empiric or correlative in that they predict peak runoff rates by correlating the flow rates with simple drainage basin characteristics such as area or slope.

Both categories of peak flow determination are easily adapted and have had wide application; however, two relatively major difficulties are normally encountered in applying the techniques. First, the rainfall-runoff formulas, such as the *Rational Formula,* are difficult to apply unless the return periods for rainfall and runoff are assumed to be equal. Also, estimates of coefficients required by these formulas are subjective and have received considerable criticism. The empiric and correlative methods are limited in application because they are derived from localized data and are not valid when extrapolated to other regions.

The most fundamental peak flow formulas and empiric-correlative methods, due to their simplicity, persist in dominating the urban design scene, and several of the most popular forms are briefly described to acquaint the reader with methods and assumptions. Urban runoff simulation techniques are described in Sec. 11-3, followed in Sec. 11-4 by a discussion of the effects of urbanization on the hydrologic cycle.

### Rational Formula

The Rational Formula for estimating peak runoff rates was introduced in the United States by Emil Kuichling in 1889.[18] Since then it has become the most widely used method for designing drainage facilities for small urban areas and highways. Peak flow is found from

$$Q_p = CIA \tag{11-1}$$

where

$Q_p$ = the peak runoff rate (cfs)
$C$ = the runoff coefficient (assumed to be dimensionless)
$I$ = the *average* rainfall intensity (in./hr), lasting for a critical period of time, $t_c$
$t_c$ = the time of concentration
$A$ = the size of the drainage area (acres)

The runoff coefficient can be assumed to be dimensionless because 1.008 acre-in./hr is equivalent to 1.0 ft³/sec. Typical $C$ values for storms of 5- to 10-yr return periods are provided in Table 11-1.

The rationale for the method lies in the concept that application of a steady, uniform rainfall intensity will cause runoff to reach its maximum rate when all parts of the watershed are contributing to the outflow at the point of design. That condition is met after the elapsed time, $t_c$, the time of concentration, which usually is taken as the time for water to flow from the most remote part of the watershed.

Figure 11-1 graphically illustrates the relationship. The IDF curve is the rainfall intensity-duration-frequency relation for the area and the peak intensity of the runoff is $Q/A = q$ which is proportional to the value of $I$ defined at $t_c$. The constant of proportionality is thus the runoff coefficient, $C = (Q/A)/I$. Note that $Q/A$ is a point value and that the relationship, as it stands, yields nothing of the nature of the rest of the hydrograph. The RRL Method described in a later section and the time-area concepts in Chapter 7 are extensions of the Rational Method that attempt to improve upon this limitation.

### Rational Method Applications

Most applications of the Rational Formula in determining peak flow rates utilize the following steps:

1. Estimate the time of concentration of the drainage area.
2. Estimate the runoff coefficient, Table 11-1.
3. Select a return period $T_r$ and find the intensity of rain that will be equaled or exceeded, on the average, once every $T_r$ years. To produce equilibrium flows, this design storm must have a duration equal to $t_c$. The desired intensity is easily read from a locally derived IDF curve such as Fig. 5-12 or 11-3 using a rainfall duration equal to the time of concentration.

***Table 11-1*** *Typical C Coefficients for 5- to 10-yr Frequency Design*

| Description of Area | Runoff Coefficients |
|---|---|
| Business | |
| Downtown areas | 0.70–0.95 |
| Neighborhood areas | 0.50–0.70 |
| Residential | |
| Single-family areas | 0.30–0.50 |
| Multiunits, detached | 0.40–0.60 |
| Multiunits, attached | 0.60–0.75 |
| Residential (suburban) | 0.25–0.40 |
| Apartment dwelling areas | 0.50–0.70 |
| Industrial | |
| Light areas | 0.50–0.80 |
| Heavy areas | 0.60–0.90 |
| Parks, cemeteries | 0.10–0.25 |
| Playgrounds | 0.20–0.35 |
| Railroad yard areas | 0.20–0.40 |
| Unimproved areas | 0.10–0.30 |
| Streets | |
| Asphaltic | 0.70–0.95 |
| Concrete | 0.80–0.95 |
| Brick | 0.70–0.85 |
| Drives and walks | 0.75–0.85 |
| Roofs | 0.75–0.95 |
| Lawns; Sandy Soil: | |
| Flat, 2% | 0.05–0.10 |
| Average, 2–7% | 0.10–0.15 |
| Steep, 7% | 0.15–0.20 |
| Lawns; Heavy Soil: | |
| Flat, 2% | 0.13–0.17 |
| Average, 2–7% | 0.18–0.22 |
| Steep, 7% | 0.25–0.35 |

4. Determine the desired peak flow $Q_p$ from Eq. 11-1.
5. Some design situations produce larger peak flows if design storm intensities for durations less than $t_c$ are used. Substituting intensities for durations less than $t_c$ is justified only if the contributing area term in Eq. 11-1 is also reduced to accommodate the shortened storm duration.

One of the principal assumptions of the Rational Method is that the predicted peak discharge has the same return period as the rainfall IDF relationship used in the prediction. Another assumption, and one that has received close scrutiny by investigators,[19,20] is the constancy of the runoff coefficient during the progress of individual

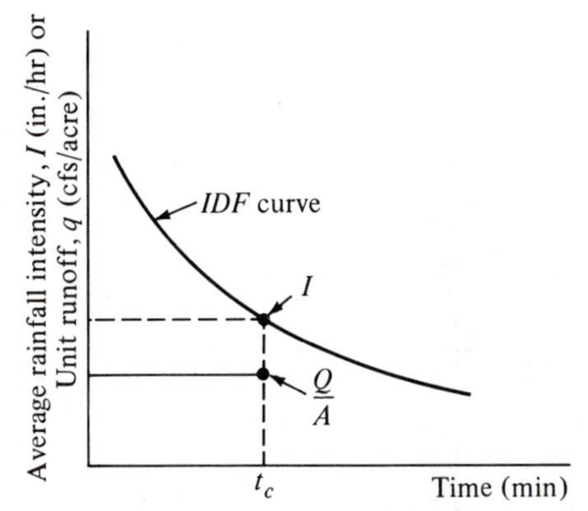


***Fig. 11-1.*** *Rainfall runoff relation for the Rational Method.*

storms and also from storm to storm. The coefficient is usually selected from a list based upon the degree of imperviousness and infiltration capacity of the drainage surface. But, as should be apparent, the coefficient must vary if it is to account for antecedent moisture, non-uniform rainfall, and the numerous conditions that cause abstractions and attenuation of flood-producing rainfalls. In practice, a composite, weighted average runoff coefficient is computed for the various surface conditions. Times of concentration are determined from the hydraulic characteristics of the principal flow path which typically is divided into two parts, overland flow and flow in defined channels; the times of flow in each segment are added to obtain $t_c$.

***Example 11-1*** Use the Rational Method to find the 10-yr design runoff for the area shown in Fig. 11-2. The IDF rainfall curves shown in Fig. 11-3 are applicable.

*Solution:*

1. Time of concentration:

$$t_c = t_1 + t_2 = 15 + 5 = 20 \text{ min}$$

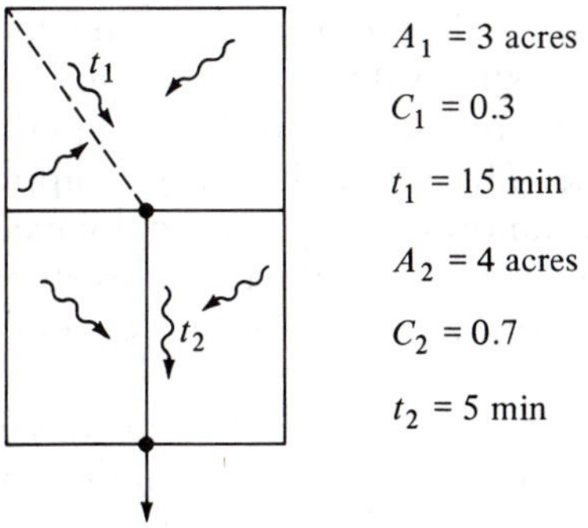

$t_1$

$t_2$

$A_1 = 3$ acres

$C_1 = 0.3$

$t_1 = 15$ min

$A_2 = 4$ acres

$C_2 = 0.7$

$t_2 = 5$ min

***Fig. 11-2.*** *Hypothetical drainage system for Example 11-1.*

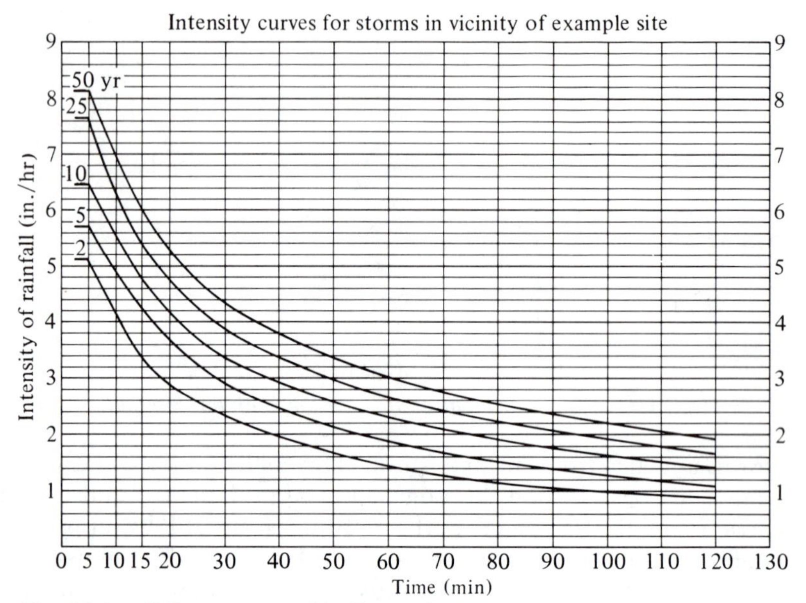


***Fig. 11-3.*** *IDF curves used in Example 11-1.*

2. Runoff coefficient:

$$\bar{C} = [(3 \times 0.3) + (4 \times 0.7)] \div (3 + 4) = 0.53$$

3. Rainfall intensity: From Fig. 11-3,

$$I = 4.2 \text{ in./hr}$$

4. Design peak runoff:

$$Q_{10} = CIA = 0.53 \times 4.2 \times 7 = 16 \text{ cfs}$$

Attempts to verify the Rational Method have not produced encouraging results. It is impossible to observe or measure a uniform, average rainfall; times of concentration do not seem to accord with the time to peak; and the ratio of runoff to rainfall produces anything but a constant coefficient. The reasons are inherent in adopting a simple model to express a complex hydrologic system. Yet the method continues to be used in practice with results implying acceptance by designers, officials, and the public. The method is easy to apply and gives consistent results. From the standpoint of planning, for example, the method demonstrates in clear terms the effects of development: Runoff from developed surfaces increases because times of concentration decrease and runoff coefficients increase.

Part of the success of the method can be explained by viewing $Q = CIA$ as a probability model rather than a physical model. Studies at the Johns Hopkins University have explored this view.[21] Figure 11-4 shows cumulative lognormal probability functions fitted to observations of rainfall and runoff on a 47-acre area in Baltimore, Maryland, with an average surface imperviousness of 0.44. The data are partial series fitted independently to the observed rainfall sequence and the observed runoff sequence. Thus the largest runoff does not necessarily correspond to the largest ranked rainfall, and a similar lack of correspondence between any runoff and the rainfall that produced it holds for the ranked position of the observations in the arrays of the two separate sequences. The design problem is to select from the rainfall history an average intensity and to determine the fraction that corresponds to a runoff intensity. In Fig. 11-4, the 5-yr rainfall frequency of 6.5 in./hr corresponds to a runoff frequency of 4.0 cfs/acre; the ratio indicates a runoff coefficient of approximately 0.6. Although the two sequences are each closely lognormal, they tend to converge, which suggests that the runoff coefficient increases slightly with more intense, less frequent storms. In the design range, however, the results tend to support the assumption of the Rational Method that the recurrence interval of the runoff equals the recurrence interval of the rainfall. It should be noted that the rainfall distributions in Figs. 11-1 and 11-4 have similar properties. All IDF curves are drawn through the average rainfall intensities

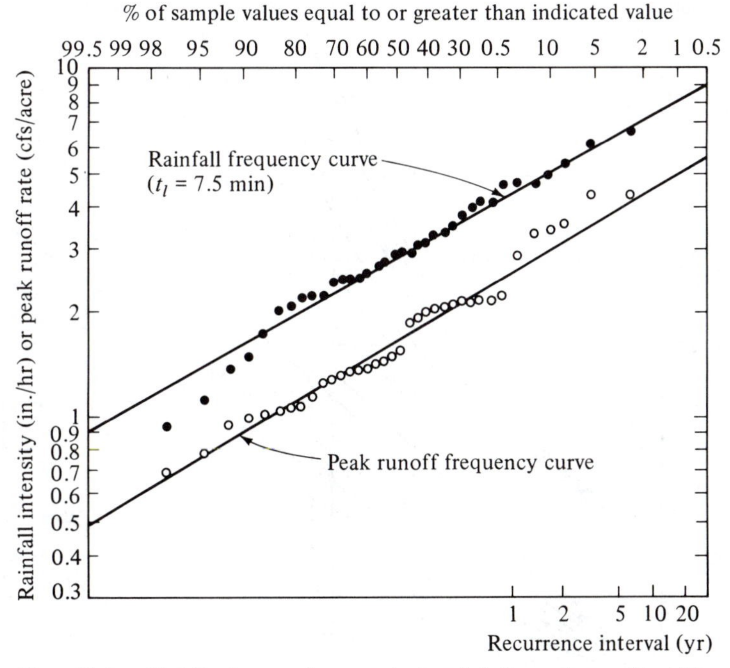


***Fig. 11-4.*** *Distributions of recorded rainfall and runoff. (After J. C. Schaake, Jr., J. C. Geyer, and J. W. Knapp, "Experimental Examination of the Rational Method,"* Proc. ASCE, J. Hyd. Div., *93, No. HY6 (Nov. 1967).)*

derived from many different storms of record; any single IDF curve does not represent the progress of a single storm. For lack of historical runoff data, the designer turns to the Rational Method to construct from the rainfall history what amounts to a *runoff* intensity-duration-frequency relationship.

The Rational Method is used for the design of most urban storm drainage systems serving areas up to several hundred acres in size. For areas larger than 1 mi², hydrograph techniques are generally warranted. Considerable judgment is required in selecting both the runoff coefficients and times of concentration. A common procedure is to select coefficients and assume that they remain constant throughout the storm. As the design proceeds from point to point downstream, a composite weighted $C$-factor is computed for the drainage area above each point. Table 11-1 is a typical list of $C$-factors. The time of concentration is composed of an inlet time (the overland and any

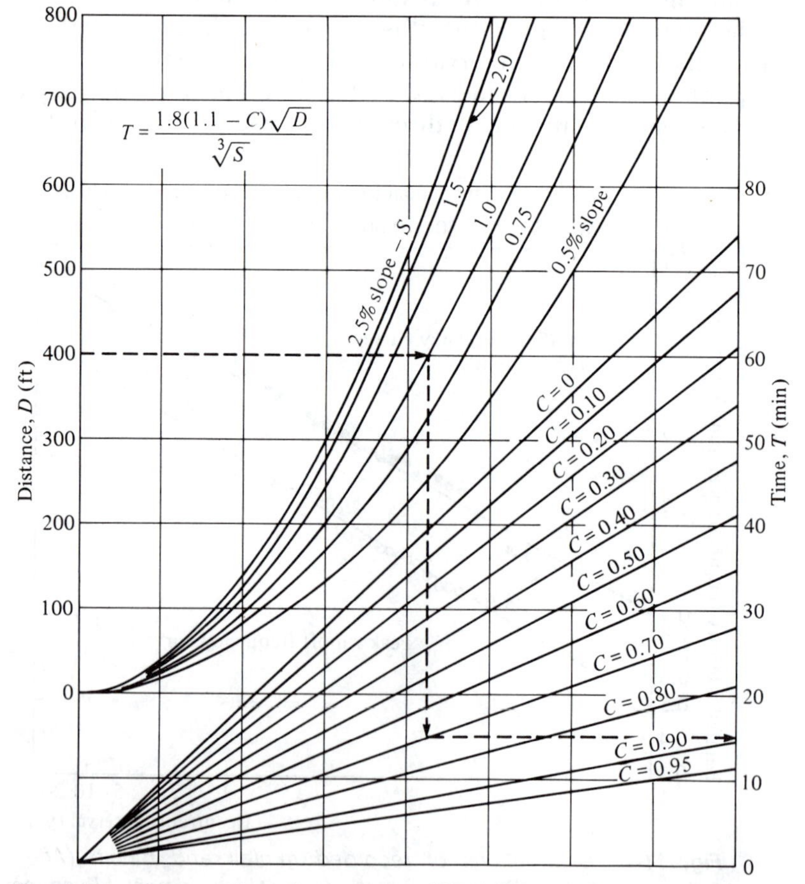


**Fig. 11-5.** *Surface flow time curves. (After "Airport Drainage," Federal Aviation Agency, Department of Transportation, Advisory Circular, A/C 150-5320-5B, Washington, D.C., 1970.)*

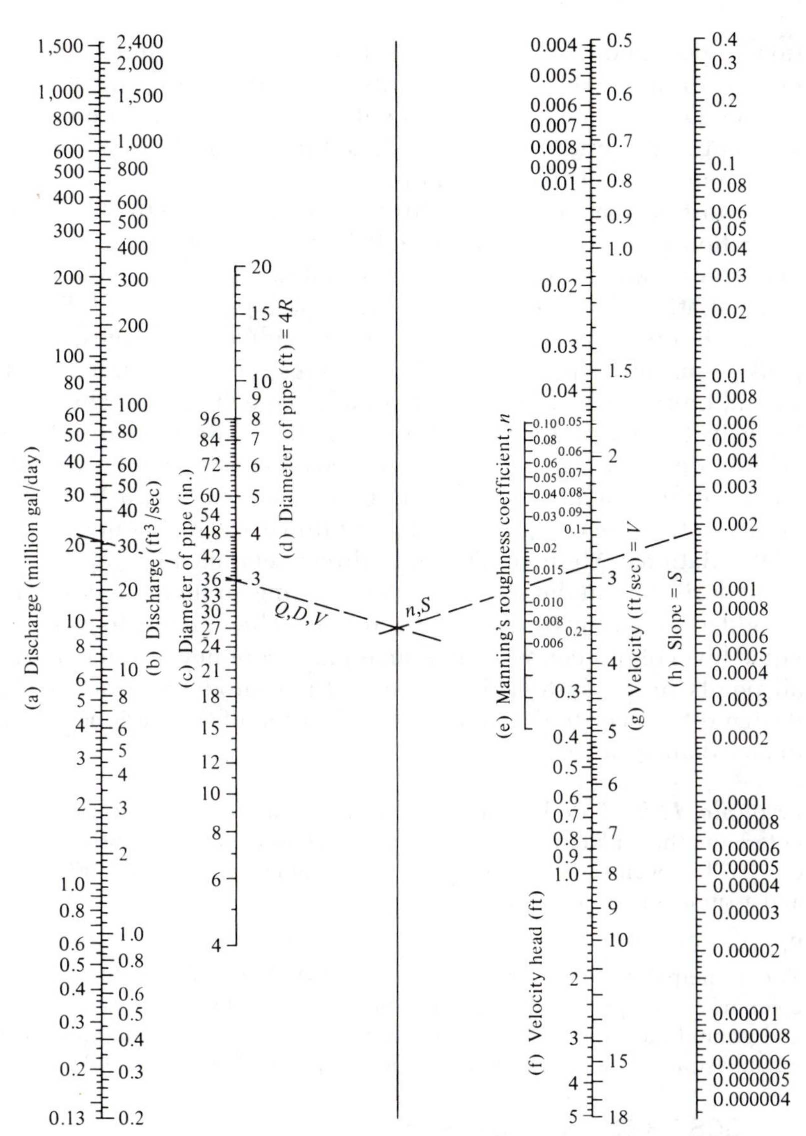


***Fig. 11-6.*** *Flow in pipes (Manning's formula). (After "Design and Construction of Sanitary Storm Sewers," ASCE Manuals and Reports on Engineering Practice, No. 37, 1970.)*

channel flow times to the first inlet) plus the accumulated time of flow in the system to the point of design.

Figure 11-5 is an example of a design aid for predicting overland flow times. Calculation of flow time in storm drains can readily be estimated knowing the type of pipe, slope, size, and discharge.[22] Generally, the pipe is assumed to flow full for this calculation. (See Fig. 11-6.) Nomographs also are available to solve the Manning equation for flow in ditches and gutters. The estimation of inlet time is frequently based solely on judgment; reported values vary from 5 to 30 min. Densely developed areas with impervious tracts immediately adjacent to the inlet might be assigned inlet periods of 5 min, but a minimum value of 10 to 20 min is more usual.

Most designers applying this method do not use the time of concentration in its strictest sense; rather, the largest sum of inlet time plus travel time in the storm drain system is taken as the time of concentration. Caution is required in applying the method. Peak discharge is not the summation of the individual discharges, because peaks from subareas occur at different times. The runoff from subareas should be rechecked for each area under consideration. The average intensity $I$ is that for the time of concentration of the total area drained. While $I$ decreases as the design proceeds downstream, the size of the contributing area increases to make $Q$ increase continuously. It should be noted that the design at each point downstream is a new solution of the Rational Method. The only direct relationship from point to point derives from the means for determining an increment of time to be added for a new time of concentration. The effect is to provide an equal level of protection (that is, an equal frequency of surcharging) at all points in the system. Example 11-2 is reproduced from standard design references to illustrate the application of the Rational Method to an urban system.[23]

***Example 11-2*** Based on the storm sewer arrangement of Fig. 11-7(a), determine the outfall discharge. Assume that $C = 0.3$ for residential areas and $C = 0.6$ for business tracts. Use a 5-yr frequency rainfall from Fig. 11-7(b) and assume a minimum 20-min inlet time.

*Solution:*

The principal factors in the design are listed in Table 11-2. Additional columns can be provided to list elevations of manhole inverts, sewer inverts, and ground elevations. This information is helpful in checking designs and for subsequent use in drawing final design plans. (See Table 11-3.)

### SCS TR-55 Peak Flow Graph

A 1975 SCS report on Urban Hydrology for Small Watersheds[24] provides a simple rainfall-runoff method for determining the peak flow rate from the 24-hr net rain depth and the time of concentration. The graphical method was developed for homogeneous watersheds on which the land use and soil type may be represented by a single parameter called the *runoff curve number.* As shown in Chapter 12, the runoff curve number is simply a third variable in a graph of rainfall versus runoff.

The SCS peak discharge graph shown as Fig. 11-8 is limited to applications where only the peak flow rate is desired for 24 hr; type-II

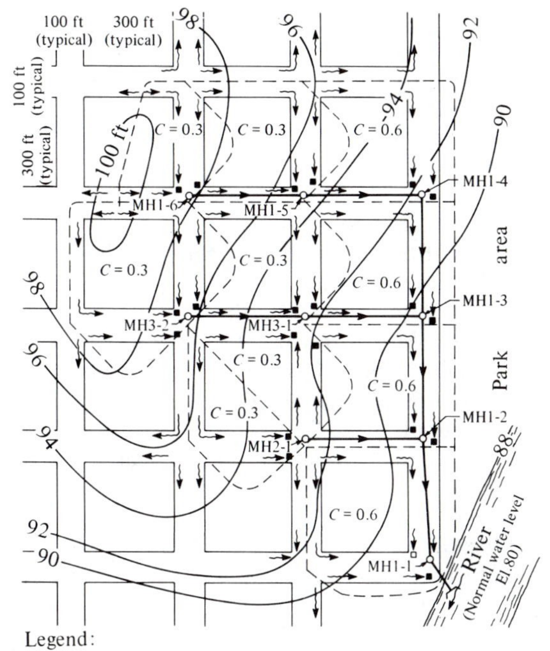


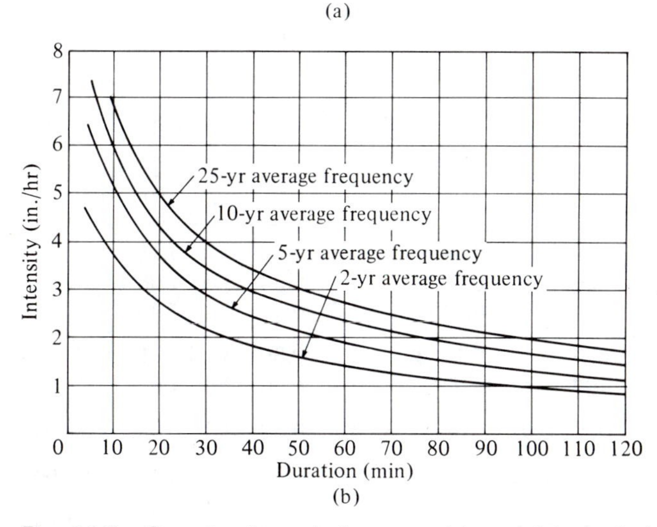


***Fig. 11-7.*** *Sample storm drainage problem: (a) typical stormsewer design plan; and (b) intensity-duration rainfall curves for Davenport, Iowa. (After "Design and Construction of Sanitary Storm Sewers," ASCE Manuals and Reports on Engineering Practice, No. 37, 1970.)*

**Table 11-2** *Principal Factors in Table 11-3*

| Column | Comment |
|---|---|
| 1 | Line being investigated |
| 2, 3 | Inlet or manhole being investigated |
| 4 | Length of the line |
| 5 | Subarea for the inlet |
| 6 | Accumulated subareas |
| 7 | Value of the concentration time for the area draining into the inlet |
| 8 | Travel time in the pipe line |
| 9 | Weighted $C$ for the area being drained |
| 10 | Rainfall intensity based on time of concentration and a 5-yr frequency curve |
| 11 | Unit runoff $q = CI$ |
| 12 | Accumulated runoff that must be carried by line |
| 13 | Slope of line |
| 14 | Size of pipe |
| 15 | Pipe capacity |
| 16 | Velocity in full pipe |
| 17 | Actual velocity in pipe |

**Table 11-3** *Typical Storm Sewer Computations for the Rational Method for a System*

| | Manhole No. | | | Area (acres) | | Flow Time (min) | | | | | | | |
|---|---|---|---|---|---|---|---|---|---|---|---|---|---|
| *Line* (1) | *From* (2) | *To* (3) | *Length (ft)* (4) | *Increment* (5) | *Total* (6) | *To Upper End* (7) | *In Section* (8) | *Average Runoff Coefficient* (9) | *Rainfall (in./hr)* (10) | *Runoff (cfs/acre)* (11) | *Total Runoff (cfs)* (12) | *Slope of Sewer (%)* (13) | *Diameter (in.)* (14) |
| 1 | 1–6 | 1–5 | 400 | 2.64 | 2.64 | 20.0 | 1.4 | 0.30 | 3.7 | 1.11 | 2.93 | 0.85 | 12 |
| 1 | 1–5 | 1–4 | 400 | 3.61 | 6.25 | 21.4 | 1.2 | 0.30 | 3.6 | 1.08 | 6.75 | 0.75 | 18 |
| 1 | 1–4 | 1–3 | 400 | 3.88 | 10.13 | 22.6 | 1.2 | 0.42 | 3.4 | 1.43 | 14.50 | 0.45 | 24 |
| 3 | 3–2 | 3–1 | 400 | 5.55 | 5.55 | 20 | 1.1 | 0.30 | 3.7 | 1.11 | 6.16 | 1.00 | 15 |
| 3 | 3–1 | 1–3 | 400 | 6.43 | 11.98 | 21.1 | 1.1 | 0.30 | 3.6 | 1.08 | 12.92 | 0.60 | 24 |
| 1 | 1–3 | 1–2 | 400 | 3.92 | 26.03 | 23.8 | 1.1 | 0.39 | 3.3 | 1.29 | 33.60 | 0.30 | 36 |
| 2 | 2–1 | 1–2 | 400 | 2.52 | 2.52 | 20 | 1.4 | 0.30 | 3.7 | 1.11 | 2.80 | 0.90 | 12 |
| 1 | 1–2 | 1–1 | 400 | 3.86 | 32.41 | 24.9 | 1.1 | 0.41 | 3.2 | 1.31 | 42.50 | 0.24 | 42 |
| 1 | 1–1 | Out-fall | 125 | 5.44 | 37.85 | 26.0 | . . . | 0.44 | 3.2 | 1.41 | 53.20 | 0.30 | 42 |

storm distributions. A type II storm distribution is typical of the 24-hr thunderstorm experienced in all states except the Pacific Coast states. Figure 11-8 was developed from numerous applications of the SCS TR-20 event simulation model described in Sec. 10-5. To apply Fig. 11-8, one need only estimate the watershed time of concentration in hours, which is entered into the graph to produce the peak discharge rate in cfs/mi² of watershed per inch of net rain during the 24-hr period. The 24-hr net rain is estimated from the 24-hr gross amount using the SCS curve number approach described in Chapter 12.

The effect of urbanization of a previously pervious watershed can be estimated using Fig. 11-9. Once the composite curve number (*CN*) has been estimated for the pervious area, a modified curve number is determined by entering Fig. 11-9 along the abscissa with the value of the percent of impervious area on the modified watershed, reading vertically to the curve corresponding to the *CN* for the pervious watershed, and then reading horizontally to determine the modified composite runoff curve number which would then be used in determining the net rain depth for the urbanized watershed.

| | | Design Flow | | | | | | | Sewer Invert | | Ground Elevation | |
|---|---|---|---|---|---|---|---|---|---|---|---|---|
| Capacity, Full | Velocity, Full | Velocity (fps) | Velocity Head (ft) | Depth of Flow (in.) | Total Energy (ft) | Manhole Losses (ft) | Manhole Invert Drop (ft) | Fall in Sewer (ft) | Upper End | Lower End | Upper End | Lower End |
| (15) | (16) | (17) | (18) | (19) | (20) | (21) | (22) | (23) | (24) | (25) | (26) | (27) |
| 3.3 | 4.0 | 4.6 | | 9 | | | ... | 3.40 | 93.00 | 89.60 | 98.4 | 94.9 |
| 9.2 | 5.1 | 5.6 | | 11 | | | 0.40 | 3.00 | 89.20 | 86.20 | 94.9 | 91.8 |
| 15.2 | 4.8 | 5.6 | | 18 | | | 0.40 | 1.80 | 85.80 | 84.00 | 91.8 | 89.7 |
| 6.4 | 5.1 | 5.9 | | 12 | | | ... | 4.00 | 91.00 | 87.00 | 96.2 | 92.3 |
| 17.5 | 5.5 | 6.1 | | 15 | | | 0.60 | 2.40 | 86.40 | 84.00 | 92.3 | 89.7 |
| 37.0 | 5.1 | 5.9 | | 26 | | | 0.80 | 1.20 | 83.20 | 82.00 | 89.7 | 89.5 |
| 3.2 | 4.1 | 4.7 | | 9 | | | ... | 3.60 | 87.50 | 83.90 | 92.7 | 89.5 |
| 50.0 | 5.2 | 5.9 | | 29 | | | 0.40 | 0.96 | 81.60 | 80.64 | 89.5 | 88.5 |
| 56.0 | 5.7 | 6.6 | | 33 | | | 0.10 | 0.38 | 80.54 | 80.16 | 88.5 | ... |

*Source:* Design and Construction of Sanitary Storm Sewers, ASCE Manuals and Reports on Engineering Practice, No. 37, 1970.

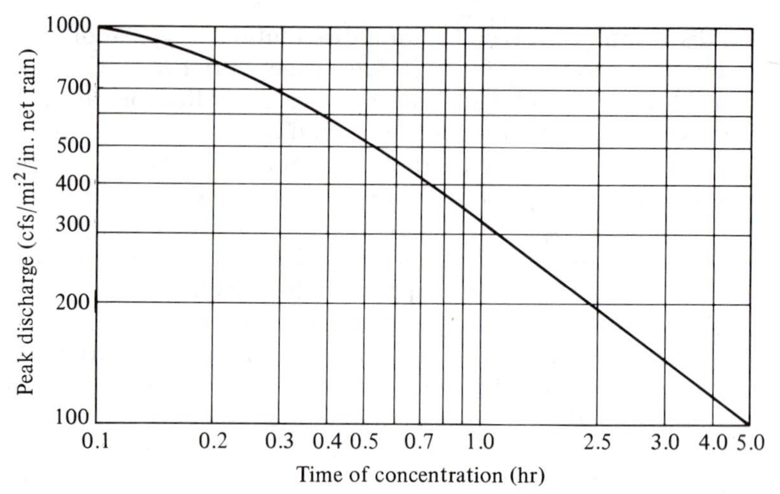


***Fig. 11-8.*** *Peak discharge (cfs/(mi²)(in.)) of runoff versus time of concentration ($T_c$) for 24-hr, type-II storm distribution. (After "Urban Hydrology for Small Watersheds," U.S. Soil Conservation Service, Tech. Release No. 55, Jan. 1975.)*

## Routing Method

Larger cities are developing rainfall-runoff methods to determine peak flow rates needed in designing urban storm drain systems. Typically, the approach is to assume an average set of conditions that apply to a typical block in the urban area, establish the degree of imperviousness, estimate depression storage, and route the flow in a

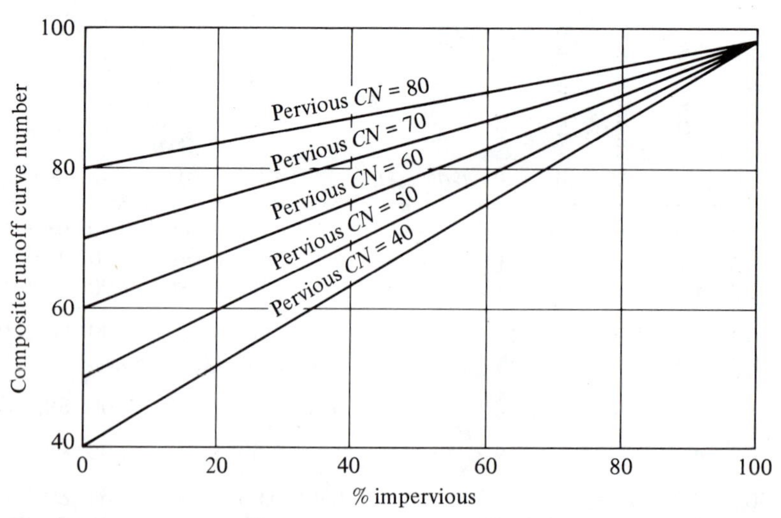


***Fig. 11-9.*** *Percentage of impervious areas versus composite CN's for given pervious area CN's. (After "Urban Hydrology for Small Watersheds," U.S. Soil Conservation Service, Tech. Release No. 55, Jan. 1975.)*

storm sewer to obtain a discharge hydrograph based upon a design storm typical of the area.

Keifer utilized this procedure to prepare peak flow design curves for the city of Chicago (Fig. 11-10).[25] A 4.01-in. average-intensity 5-yr frequency rainfall was used to develop a rainfall hyetograph of 3-hr duration with a peak intensity of 8.1 in./hr at the $\frac{3}{8}$ point of the period. Infiltration losses were assumed according to earlier work by Horner and Jens as shown in Fig. 11-11. Runoff was presumed to occur when the mass curve of rainfall exceeded that of infiltration. Depression storage was also subtracted from the rainfall before overland flow equations were used to compute the discharge from elemental strips in the typical block. Routing these strip inflows through street gutters to catch basins or inlets was done by hydrologic routing procedures. Additional input came from the roof and street runoff. Design charts giving the peak discharge to be expected under various conditions of imperviousness, depression storage, and slope of land are available.[25] A sample is shown in Fig. 11-10 and an example problem presented in Example 11-3.[25]

***Example 11-3*** A layout of a proposed storm sewer system is shown in Fig. 11-12. Determine the peak discharge at the outfall.

*Solution:*
A tabular arrangement of the calculations is shown in Table 11-4.

Data for columns 1 to 4 are taken from Fig. 11-12. Computations then proceed as follows:

Columns 8 to 14 show the size of areas and derivations of the percentage of imperviousness for areas tributary to the street intersections indicated in columns 1 and 2. Multiplying the acreage of each subdrainage tract by its percentage of imperviousness, adding the products, and then dividing the cumulative total by its area, we obtain the percentage of imperviousness of the cumulative area and list it in column 14.

Taking the value of imperviousness from column 14 and the time of travel in column 7, we obtain the unit rate of flow from Fig. 11-10(b) and enter it in column 15.

We can then compute the total flow by multiplying the unit rate by the cumulative area to the point in question.

Using the total flow from column 16 and the slope of the hydraulic gradient from column 4, we can select the velocity of flow and the time of travel from conventional hydraulic charts in the customary manner.

We obtain the time of flow shown in column 6 by dividing the length of the reach by the velocity of flow in column 5; we then record the total at each junction point, column 7.

Where additional branches enter the main sewer as at the corner of Sixth and Main Street, separate drainage areas are individually listed in lines 9 and 10, and the summation of all areas tributary to that point is used to calculate the flow rates and time of travel below such a junction. For example, the value of 160 acres for column 10, line 10, is found by adding the 80 acres

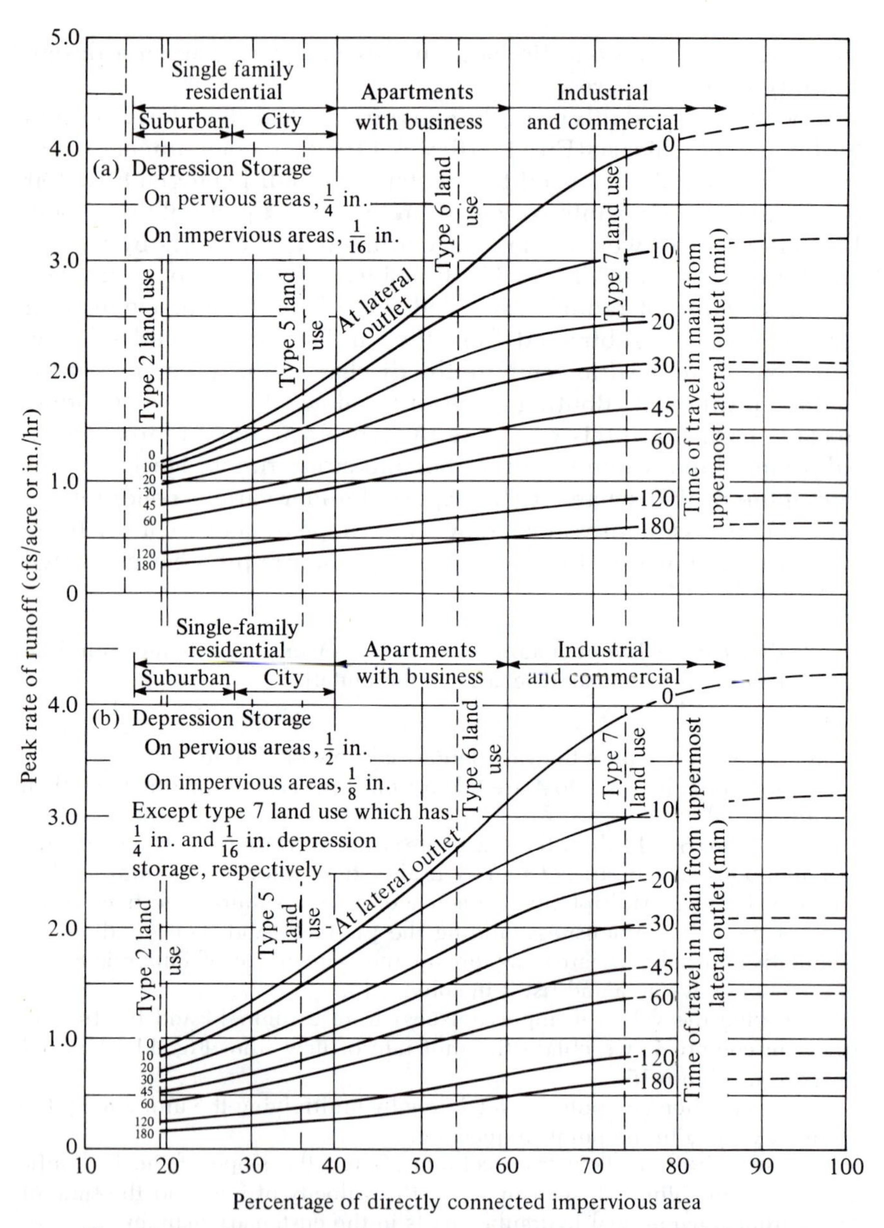


***Fig. 11-10.*** *Runoff rates in sewer system versus the travel time and percentage of directly connected impervious areas; the average ground slope =0.01. (After A. L. Tholin and C. J. Keifer, "Hydrology of Urban Runoff,"* Trans. ASCE, *Paper No. 3061, Vol. 125 (1960).)*

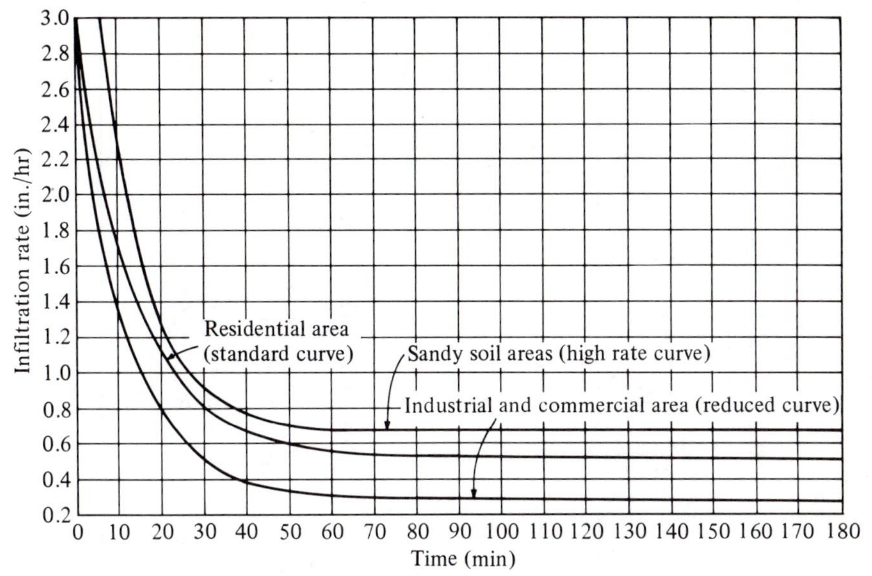


***Fig. 11-11.*** *Standard infiltration capacity curves for pervious surface. (After A. L. Tholin and C. J. Keifer, "Hydrology of Urban Runoff,"* Trans. ASCE, *Paper No. 3061, Vol. 125 (1960).)*

tabulated in column 10, line 7, to the 20 acres from column 9, lines 9 and 10. Similarly, the entry in column 13, line 10, is the summation of the figures in column 13, line 7, and column 12, lines 9 and 10.

### Synthetic Unit Hydrograph Peak Rate Formulas

Peak flow rates from urban and small watersheds can also be determined from critical or design storm information using the synthetic unit hydrograph techniques described in Chapters 4 and 10. A storm having a duration defined by Eq. 4-32 will produce, according to Snyder's method of synthesizing unit hydrographs, a peak discharge for 1.0 in. of net rain given by Eq. 11-2, or

$$Q_p = \frac{640 C_p A}{t_l} \tag{11-2}$$

Similarly, the peak flow rate resulting from a storm with any duration $D$ is, according to the SCS method for constructing synthetic unit hydrographs, equal to

$$Q_p = \frac{484A}{t_P} \tag{11-3}$$

where $t_P$ is the time from the beginning of the effective rain to the time of the peak runoff rate, which by definition is the watershed lag time plus half the storm duration. Both of Eqs. 11-2 and 11-3 apply

**Table 11-4** *Computation of Runoff by Hydrograph Method*

| | Location | | | | | Time of Travel | | Area of Addition | | | Imperviousness | | | | Peak Flows | |
|---|---|---|---|---|---|---|---|---|---|---|---|---|---|---|---|---|
| | In (1) | At (2) | Length of Reach (ft) (3) | Slope of Hydraulic Gradient (ft/ft) (4) | Velocity in Reach (ft/sec) (5) | In Reach (min) (6) | To (min) (7) | No. (8) | A (acres) (9) | $\sum A$ (acres) (10) | $I_p$ (%) (11) | $A \times I_p$ (acres) (12) | $\sum AI_p$ (acres) (13) | $I_p$ for $\sum A$ (%) (14) | Unit rate (cfs/acre) (15) | Total (cfs) (16) |
| 1 | | Cornelia Avenue | | | | | 0 | (1) | 20 | 20 | 38.6 | 772 | 772 | 38.6 | 1.82 | 36 |
| 2 | Tenth Street | | 330 | 0.0020 | 4.5 | 1.2 | | | | | | | | | | |
| 3 | | Blackhawk | | | | | 1.2 | (2) | 20 | 40 | 34.5 | 690 | 1462 | 36.5 | 1.69 | 68 |
| 4 | Tenth Street | | 330 | 0.0017 | 4.9 | 1.1 | | | | | | | | | | |
| 5 | | Albany Avenue | | | | | 2.3 | (3) | 20 | 60 | 36.0 | 720 | 2182 | 36.4 | 1.67 | 100 |
| 6 | Tenth Street | | 330 | 0.0014 | 5.0 | 1.1 | | | | | | | | | | |
| 7 | | Main Street | | | | | 3.4 | (4) | 20 | 80 | 45.5 | 910 | 3092 | 38.6 | 1.75 | 140 |
| 8 | Main Street | | 2640 | 0.0012 | 5.2 | 8.5 | | | | | | | | | | |
| 9 | | Sixth Street | | | | | | (5) | 20 | | 42.8 | 856 | | | | |
| 10 | | Sixth Street | | | | | 11.9 | Br.[a] A | 60 | 160 | 35.2 | 2112 | 6060 | 37.9 | 1.51 | 242 |
| 11 | Main Street | | 2640 | 0.0010 | 5.5 | 8.0 | | | | | | | | | | |
| 12 | | Second Street | | | | | | (6) | 23 | | 50.1 | 1152 | | | | |
| 13 | | Second Street | | | | | 19.9 | Br.[a] B | 34 | 217 | 37.2 | 1268 | 8480 | 39.0 | 1.30 | 282 |
| 14 | Main Street | | 1650 | 0.0009 | 5.6 | 4.9 | | | | | | | | | | |
| 15 | | River | | | | | 24.8 | | | | | | | | | 282 |

[a] Branch

*Source:* A. L. Tholin and C. J. Keifer, "Hydrology of Urban Runoff," *Trans. ASCE*, Paper No. 3061, Vol. 125 (1960).

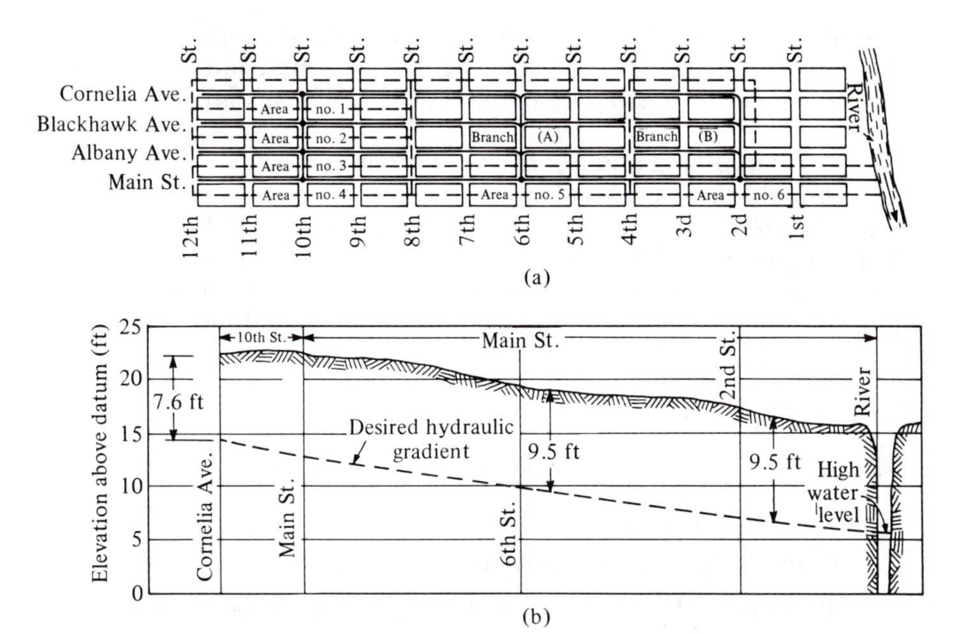


**Fig. 11-12.** *Layout of proposed sewer system by hydrograph method: (a) plan; and (b) profile. (After A. L. Tholin and C. J. Keifer, "Hydrology of Urban Runoff,"* Trans. ASCE, *Paper No. 3061, Vol. 125 (1960).)*

to 1.0 in. of net rain occurring in the specified duration $D$. Either can be multiplied by $P_{net}$ for other storms with equal durations but different depths. Peak flows for storms with durations other than $D$ could be determined by unit hydrograph methods.

### Flood Frequency Method

Thus far, the methods used to determine peak discharge rates have been based upon a rainfall frequency. The frequency of the storm runoff derived by this method is not necessarily the same as that of the rainfall due to assumptions regarding antecedent conditions and the estimate of abstractions from the initial precipitation. A method of dealing with the runoff directly is called the *flood frequency method.* Unfortunately, this method does not provide a hydrograph shape but gives only a peak discharge of known frequency.

The total direct runoff in this method must be approximated for most cases by using precipitation values from Ref. 26 and a direct runoff computation such as the $CN$ approach in Chapter 12. Storm-volume-frequency curves can also be developed for regional areas to aid in finding the direct runoff to be expected for a selected design frequency. The shape of the hydrograph can be sketched to produce the desired peak discharge and direct runoff volume. This procedure is trial and error but is accomplished quickly. For sketching purposes, the time base of the hydrograph is generally a 2.67 multiple of the period to peak (estimated by the SCS as 0.70 of concentration time).

### Discharge-Area and Regression Formulas

A multiple of peak flow formulas relating the discharge rate to drainage area have been proposed and applied. Gray[27] lists 35 such formulas, and Chow[28] compares many others. Most of these empiric equations are derived using pairs of measurements of drainage area and peak flow rates in a regression equation having the form

$$Q = CA^m \tag{11-4}$$

where

$Q$ = the peak discharge associated with a given return period
$A$ = the drainage area
$C$ and $m$ = regression constants

Popular discharge-area formulas in the form of Eq. 11-4 include the Meyers equation[29]

$$Q = 10{,}000A^{0.5} \tag{11-5}$$

where

$A$ = the drainage area, which must be 4 mi$^2$ or more
$Q$ = the ultimate maximum flood flow (cfs)

This example gives only one flow rate of unknown frequency and is chosen only to illustrate the form of flood flow equations. A recent federal program of determining flood magnitudes for a range of frequencies on a state-by-state basis has been completed by the USGS using the multiple regression techniques discussed in Sec. 6-6 and illustrated for Virginia in Prob. 6-14. Similar formulas are available from the USGS for other states.

### U.S. Geological Survey Index-Flood Method

The U.S. Geological Survey *Index-Flood* method described in Sec. 6-5 is a graphical regional correlation of the recurrence interval with peak discharge rates. The steps involved in the derivation of a regional flood index curve were outlined in Chapter 6, Sec. 6-5. The first step in applying the technique to a watershed is to determine the mean annual flood, defined as the flood magnitude having a return period of 2.33 yr. Mean annual floods for ungauged watersheds are found from regression equations similar in form to Eq. 11-4. For example, the USGS report[30] on flood magnitudes and frequencies in Nebraska gives, in cfs,

$$Q_{2.33} = CA^{0.7} \tag{11-6}$$

where

$A$ = the contributing drainage area in mi²
$C$ = a regional coefficient obtained from Fig. 11-13

Once the mean annual flood magnitude is obtained, other annual flood magnitudes can easily be determined from the appropriate index flood curve (see Fig. 6-5). The use of such curves in urban hydrology is limited because the USGS data network seldom includes watersheds smaller than 10 mi².

### Soil Conservation Service "Cook" Method

The Soil Conservation Service 1940 equation, called the Cook equation, for estimating the peak flow $Q$ for watersheds not more than 2500 acres in size is, in cfs units,[31]

$$Q = TRF \tag{11-7}$$

where $T$ and $R$ are determined from Fig. 11-14. The $\Sigma W$ term in Fig. 11-14 is a function of certain watershed characteristics that are described in Table 11-5. The rainfall factor $R$, determined from the map in Fig. 11-14, is a geographical factor; $F$ is a frequency factor equal to 1.00 for a 50-yr frequency, 0.83 for a 25-yr frequency, and 0.71 for a 10-yr frequency.

### U.S. Bureau of Public Roads "Potter" Method

Another rainfall-runoff method for determining peak flows for storms with selected recurrence intervals was devised for the U.S. Bureau of Public Roads by William D. Potter.[32] The graphical method is the result of a research study of 96 gauged watersheds in the United States with areas less than 25 mi² all located east of the 105th meridian.

Potter's method groups all watersheds in the study region into four physiographic zones based on large regions that are underlaid by similar rock formations. Figure 11-15 is the zone classification map for the North Central States.

To apply Potter's method in estimating the 10-yr and 50-yr peak flow rates, a topographic index $T$ is first determined as

$$T = \frac{0.3L}{\sqrt{S_1}} + \frac{0.7L}{\sqrt{S_2}} \tag{11-8}$$

where

$L$ = the total length of the stream channel (miles)
$S_1$ and $S_2$ = the average slopes (ft/mi) of the upper third and lower two-thirds of the main stream channel, respectively

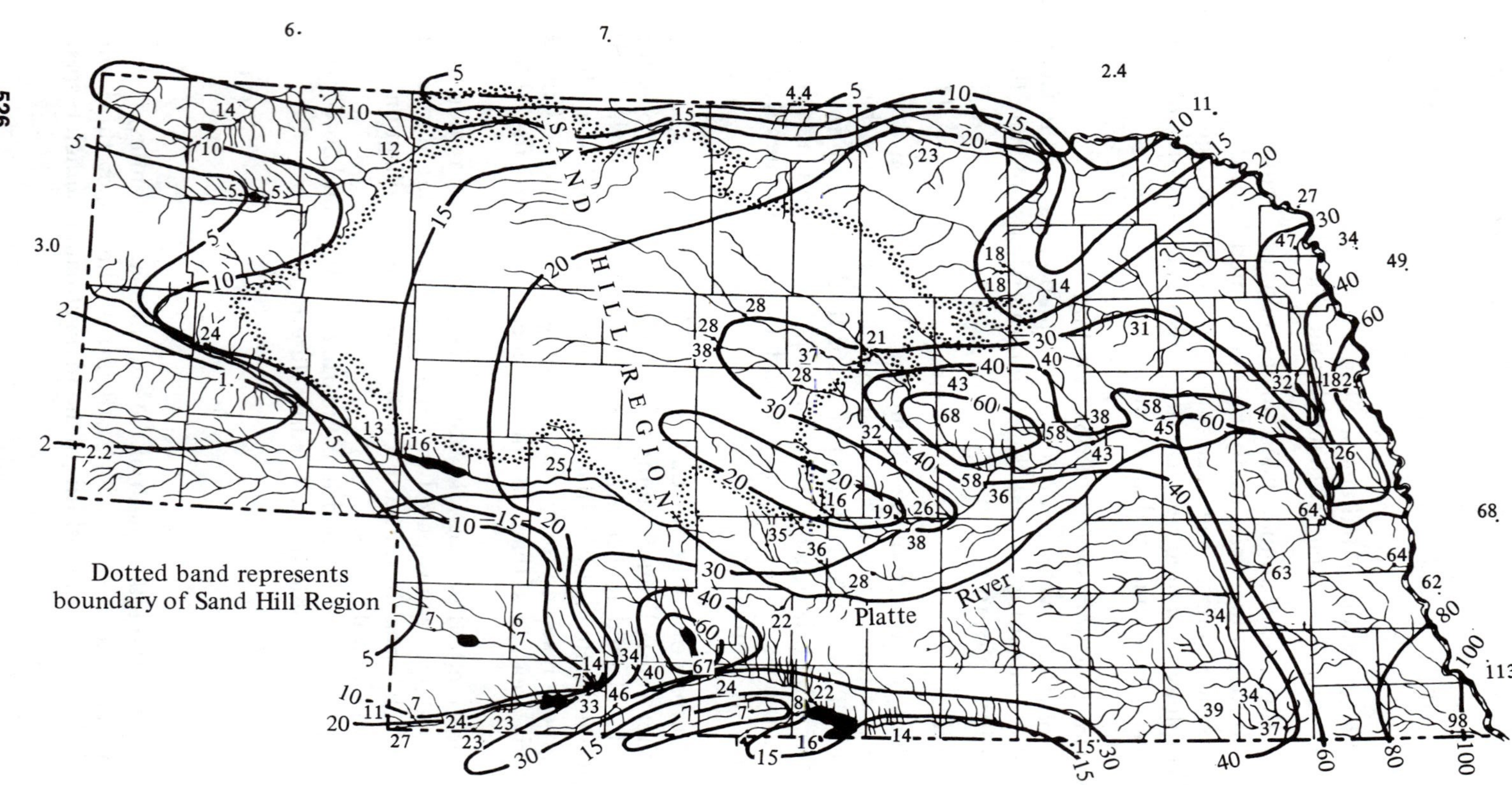


**Fig. 11-13.** *Variation of the mean annual flood coefficient, C, in Nebraska, exclusive of the main stems of the Platte Rivers. Dotted band represents the boundary of the Sand Hill Region. (After L. W. Furness, "Floods in Nebraska, Magnitude and Frequency," U.S. Geological Survey Rept. to Nebraska Department of Roads and Irrigation, Apr. 1955.)*

**Table 11-5** *Runoff Producing Characteristics of Drainage Basins with Corresponding Weights W (The weights are shown in brackets.)*

| Designation of Basin Characteristics | Runoff Producing Characteristics | | | |
|---|---|---|---|---|
| | *(100) Extreme* | *(75) High* | *(50) Normal* | *(25) Low* |
| Relief | (40) Steep, rugged, terrain, with average slopes generally above 30% | (30) Hilly, with average slopes of 10–30% | (20) Rolling with average slopes of 5–10% | (10) Relatively flat land, with average slopes of 0–5% |
| Soil infiltration | (20) No effective soil cover; either rock or thin soil mantle of negligible infiltration capacity | (15) Slow to take up water; clay or other soil of low infiltration capacity, such as heavy gumbo | (10) Normal, deep loam with infiltration about equal to that of typical prairie soil | (5) High; deep sand or other soil that takes up water readily and rapidly |
| Vegetal cover | (20) No effective plant cover; bare except for very sparse cover | (15) Poor to fair; clean-cultivated crops or poor natural cover; less than 10% of drainage area under good cover | (10) Fair to good; about 50% of drainage area in good grassland, woodland, or equivalent cover; not more than 50% of area in clean-cultivated crops | (5) Good to excellent, about 90% of drainage area in good grassland, woodland, or equivalent cover |
| Surface storage | (20) Negligible; surface depressions are few and shallow; drainage-ways steep and small; no ponds or marshes | (15) Low; well-defined system of small drainage-ways; no ponds or marshes | (10) Normal; considerable surface depression storage; drainage system similar to that of typical prairie lands; lakes, ponds and marshes less than 2% of drainage area | (5) High; surface-depression storage high; drainage system not sharply defined; large flood plain storage or a large number of lakes, ponds or marshes |

*Source:* After C. L. Hamilton and H. G. Jepson, "Stock-Water Developments: Wells, Springs, and Ponds," U.S. Department of Agriculture, Farmers' Bulletin No. 1859, 1940.

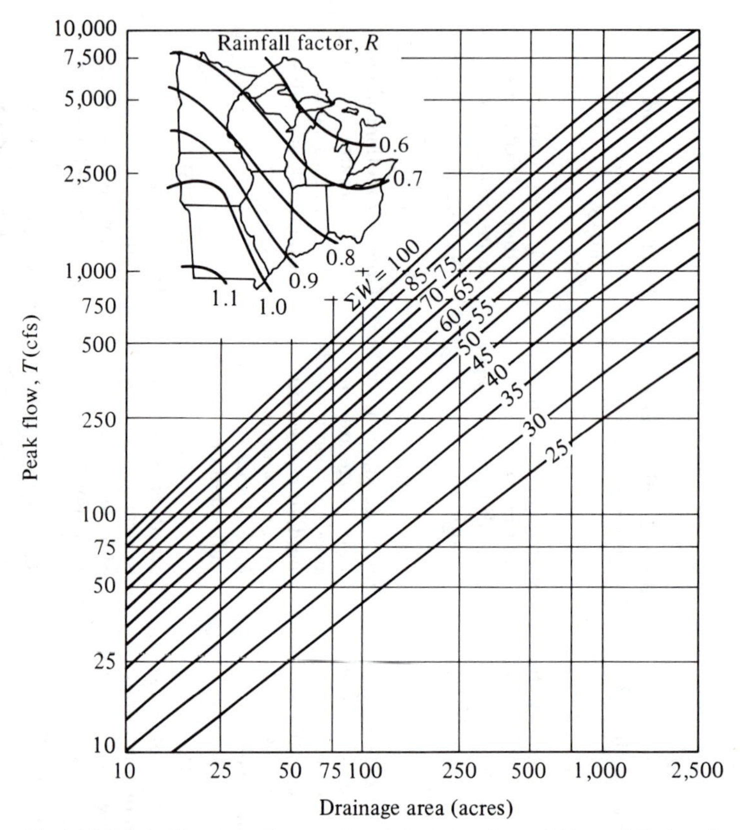


***Fig. 11-14.*** *Chart and map for determination of P and R for the Cook equation. (After C. L. Hamilton and H. G. Jepson, "Stock-Water Developments: Wells, Springs, and Ponds," U.S. Department of Agriculture, Farmers Bulletin No. 1859, 1940.)*

A precipitation index $P$ is next determined as the amount of rain (in.) which may be expected to be equaled or exceeded on an average once in 10 yr for a 60-min duration. This depth is available from applicable IDF curves such as those shown in Figs. 5-12 and 11-3, or may be determined from Fig. 11-16.

An initial trial value for the 10-yr peak flow $\hat{Q}_{10(ATP)}$ is obtained graphically from Fig. 11-17 or 11-18 by reading vertically from the area to the proper $T$-line, then horizontally to the precipitation intensity, then vertically. Next, Fig. 11-19 for Zones I or II is used to determine if the watershed is topographically similar to the basins studied by Potter. Having a value for the topographical index, $\hat{T}_{AP}$, the percent difference between the actual and ideal topographic indexes is found from

$$D = \left[\frac{\hat{T}_{AP} - T}{\hat{T}_{AP}}\right](100) \qquad \textbf{(11-9)}$$

If $D$ is less than 30%, trial $\hat{Q}_{10(ATP)}$ is satisfactory and this is accepted as the 10-yr peak discharge. If $D$ is equal to or greater than 30%, the ratio $T/\hat{T}_{AP}$ is determined and used in Fig. 11-20 to obtain a compensating multiplier, $C$. The trial $\hat{Q}_{10(ATP)}$ is multiplied by $C$ to obtain the 10-yr discharge. The 50-yr flood is then determined from Fig. 11-21. Potter suggests that peak discharges for other recurrence intervals be obtained by plotting $Q_{10}$ and $Q_{50}$ on extremal probability paper and drawing a straight line through the plotted points.

## 11-3 Urban Runoff Models

The digital simulation of urban runoff is characterized by an attempt to quantify all pertinent physical phenomena from the input (rainfall) to the output (runoff). This usually involves the following steps: (1) determine a design storm; (2) deduct losses from the design storm to arrive at an excess rainfall rate; (3) determine the flow to the gutter by overland flow equations; (4) route the gutter flow; (5) route the

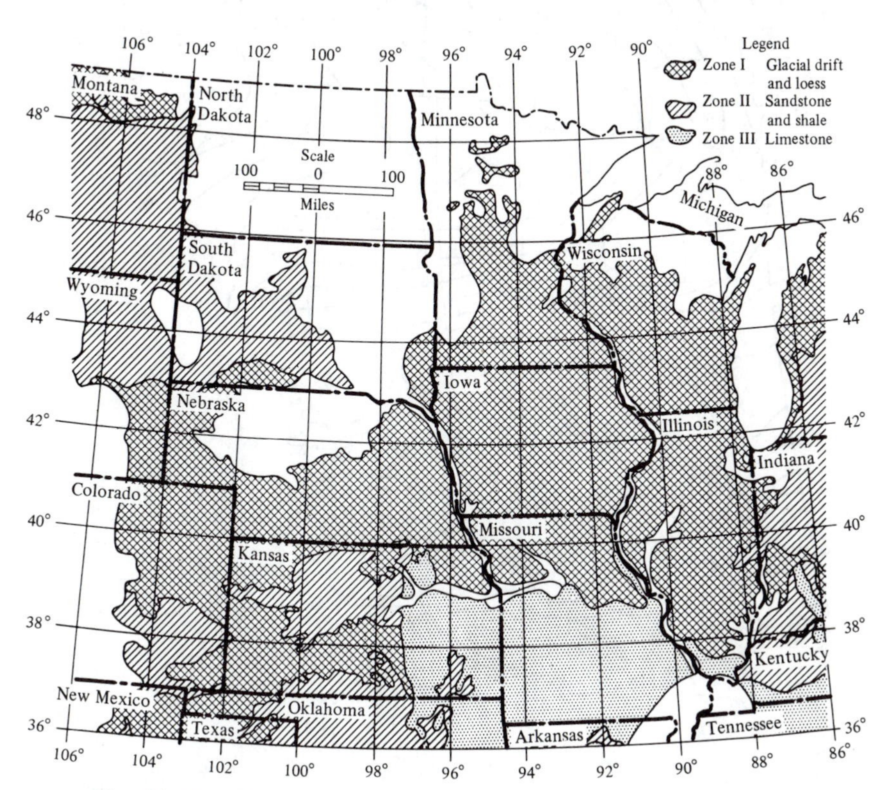


***Fig. 11-15.*** *Classification by zones for the Potter method in the North Central states (After W. D. Potter, "Peak Rates of Runoff from Small Watersheds," Hydraulic Design Series No. 2, Bureau of Public Roads (washington, D.C.: Government Printing Office, Apr. 1961).)*

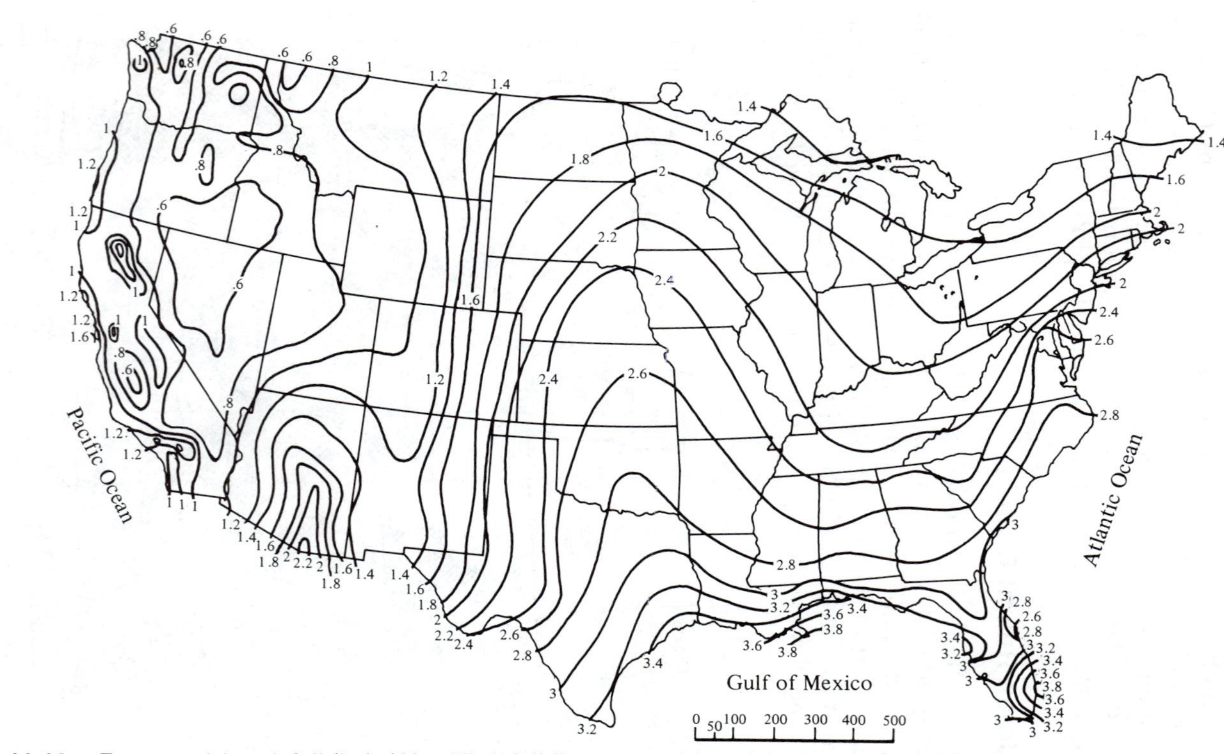


***Fig. 11-16.*** *Ten-year 1-hr rainfall (in.). (After "Rainfall Frequency Atlas of the United States for Durations from 30 Minutes to 24 Hours and Return Periods from 1 to 100 Years," U.S. Weather Bureau Tech. Paper 40, May 1961.)*

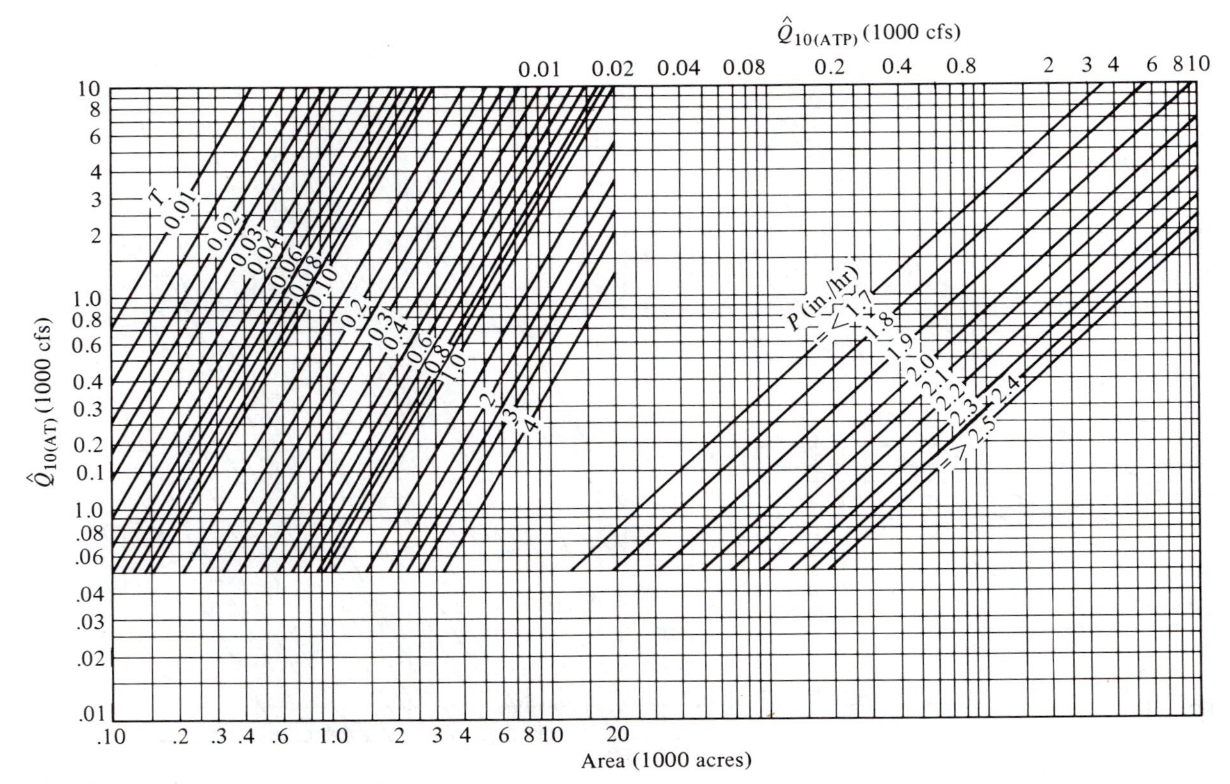


**Fig. 11-17.** *Relations between $Q_{10}$ A, T, and P for Zone I, Potter's method.*

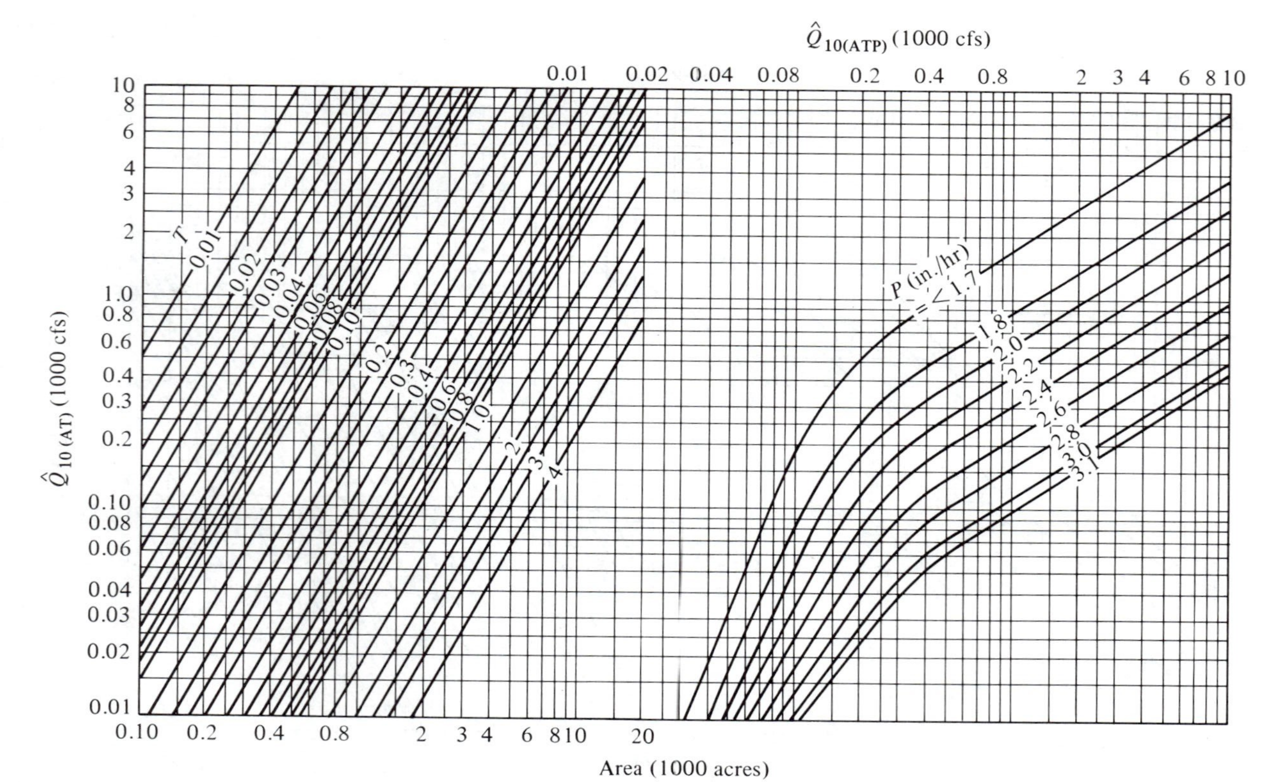


**Fig. 11-18.** *Relations between $Q_{10}$, A, T, and P for Zone II, Potter's method.*

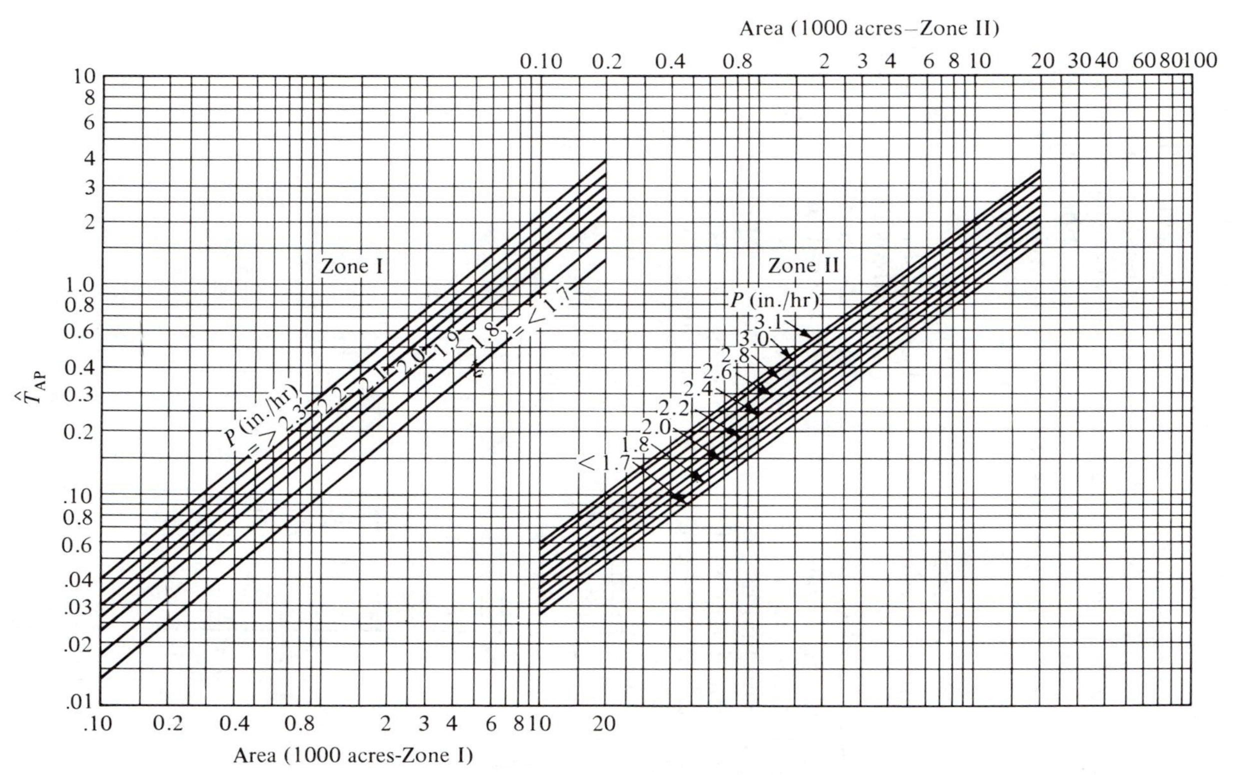


*Fig. 11-19.* *Relations between T, A, and P for Zones I and II.*

flow in pipes; and (6) determine the outflow hydrograph. The result obtained is affected by the accuracy of the determination of losses and by the validity of simplifying assumptions. If errors are small and noncumulative, the predicted runoff is valid.

Most urban runoff models deal solely with individual storm events. With the advent of modern computers, the trend has been more toward the continuous time simulation of many storm and dry periods using the hydrologic processes described in Chapter 10. The following urban watershed models are typical of many in use. Comparisons are included at the end of this section.

### Chicago Hydrograph Method (CHM)

Tholin's[33] Hydrograph Method, known as the *Chicago Hydrograph Method,* is an example of early urban runoff models. The procedure programmed is (1) develop a design storm pattern from local intensity-duration frequency curves and an average chronological storm pattern; (2) compute the overland flow using selected Horton-type infiltration capacity curves, the estimated depth of the rainfall retained in surface depressions, and Izzard's[34] overland flow equations; (3) route overland flow through gutters using the storage

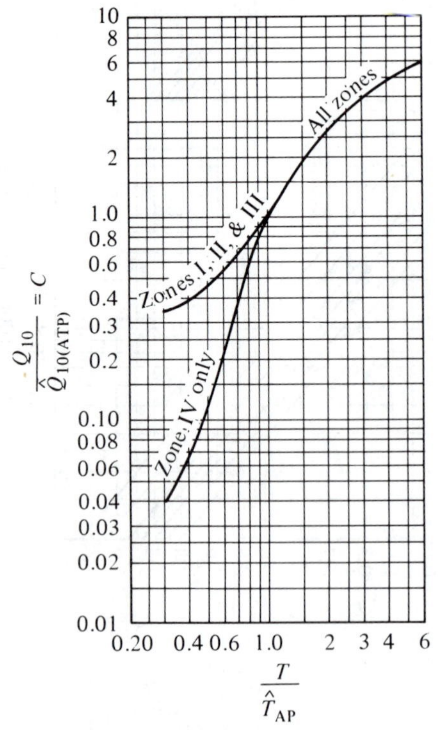


***Fig. 11-20.*** *Coefficient C as a function of $T/T_{AP}$.*

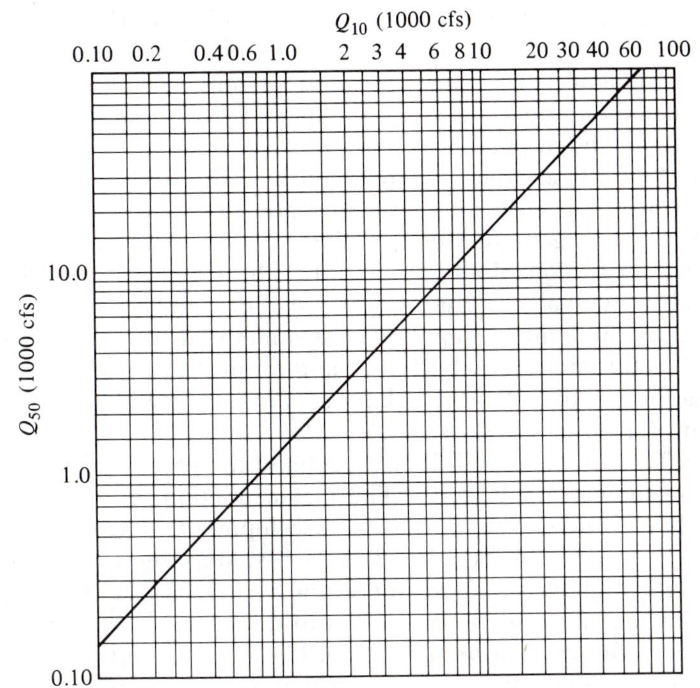


*Fig. 11-21. Relation between $Q_{10}$ and $Q_{50}$ for all zones.*

equation to obtain the runoff into catchbasins; (4) route sewer supply hydrographs from roofs and street inlets along a typical headwater sewer lateral to produce a lateral outflow hydrograph; and (5) route the lateral outflow hydrograph by a time-offset method along submains and the main sewer to a point of discharge. The method originally[33] involved a graphical hand computation but was later programmed for digital computer solution in 1970 by Keifer.[35]

### Road Research Laboratory Method (RRL) and Illinois Simulator

An urban runoff model that utilizes the time-area runoff routing method described in Sec. 7-3 was developed in England and described by Watkins.[36] The technique was developed specifically for the analysis of urban runoff and ignores completely all pervious areas and all impervious areas that are not directly connected to the storm drain system; hence the estimates of peak flow rates and runoff volumes are likely to be low.

The RRL Model could be used for continuous streamflow simulation but tends to be used as an event simulation model. It has been extensively applied in Great Britain, and moderate success has been reported by Terstriep and Stall[37] for North American applications in the Chicago, Baltimore, and Champaign, Illinois, areas. Other applications are reported in Refs. 38, 39, 40, 41, and 42. The Illinois Urban Drainage Area Simulator (ILLUDAS) described in Ref. 42 is an improved version of RRL that has a wider range of capabilities. It incorporates the impervious areas neglected by RRL and is a demonstrated improvement over RRL.

The hydrologic processes modeled by RRL are summarized in column 4 of Table 11-6. Also tabulated are procedures used by the Storm Water Management Model (SWMM) and the University of Cincinnati Urban Runoff Model (UCURM). The latter two models are described individually in the next two sections, followed by a comparison of all three models applied to several urban areas. Table 11-6 is reproduced from a study by Heeps and Mein.[40] The results of their investigation are discussed in a later section along with results of a similar investigation by Marsalek, et al.[41]

The Terstriep and Stall flow diagram of the processes simulated by RRL is shown in Fig. 11-22. The major functions of the program involve five principal steps in the development of runoff hydrographs as illustrated in Fig. 11-23. As a first step, the total basin is divided into subbasins similar to the one in Fig. 11-23a, and impervious areas that are directly connected to the storm drain system are identified as shown. The remainder of the basin including surfaces such as lawns, floodways, parks, roofs that are not connected to the storm drainage system, and impervious areas that drain into pervious areas are all ignored by the RRL model but accounted for in the ILLUDAS model.

After hydraulic characteristics such as lengths, slopes, and roughness coefficients are estimated, the second step is the calculation of flow velocities for all segments. These velocities are then used to construct lines of equal travel time to the outlet of the basin, called isochrones, on the basin map. The areas between isochrones are then determined and plotted against travel time as shown in Fig. 11-23b.

The third step is to apply the specified rainfall pattern to the directly connected impervious area, and then apply Eq. 7-32 to determine the translated hydrograph at the sub-basin outlet. Excess rainfall hyetograph ordinates in Eq. 7-32 are obtained by subtracting the losses from rainfall to give the net supply rate shown in Fig. 11-23c, d, and e. The application of Eq. 7-32 is shown in Fig. 11-23f.

Because the routed time-area hydrograph in Fig. 11-23f represents translation effects only, the hydrograph must now be routed through reservoir-type storage to account for the effects of storage within the basin. This is accomplished by routing the hydrograph of Fig. 11-23f through a reservoir using the storage-indication method described in Chapter 7 and illustrated in Example 7-3.

The fifth and final step in the RRL Method is the routing of the sub-basin outflow hydrograph to the next confluence or the next input

**Table 11-6** *Comparison of Urban Runoff Model Simulation Procedures*

| Process (1) | SWMM (2) | UCURM (3) | RRL (4) |
|---|---|---|---|
| Simulation | Noncontinuous | Noncontinuous | Noncontinuous |
| Interception | Neglects | Neglects | Neglects |
| Evaporation | Neglects | Neglects | Neglects |
| Transpiration | Neglects | Neglects | Neglects |
| Depression storage | Fills before overland flow begins—part of impervious area assigned zero depression storage<br>Depleted by infiltration | Exponential filling rate<br>No allowance to be depleted by infiltration on pervious areas | Neglects |
| Infiltration | Horton equation<br>No time offset<br>Satisfied by water on ground surface and rainfall | Horton equation<br>Time offset<br>Satisfied by rainfall only | Impermeability assumed constant (equal to fraction of impervious area directly connected) |
| Overland flow | Uniform depth of detention | Profile with increasing depth | Linear time-area routing (only directly connected impervious areas) |
| | Storage routing using Manning turbulent flow equation and continuity equation | Solved using an empirical relation, continuity equation and Manning turbulent flow equation | Time of entry required as input data |
| | Quasisteady state | Quasisteady state | |
| Gutter flow | Uniform flow storage routing | Outflow = sum of inflows | Neglects |
| Inlet pits and junctions | Outflow = sum of inflows | Outflow = sum of inflows | Outflow = sum of inflows |
| Pipe flow | Storage routing (Manning equation based on the slope of energy line) | No storage routing | Storage routing (Manning equation for uniform flow) |
| | Quasisteady state | Lagged using weighted average velocity | Lagged using full bore velocity |
| | | Quasisteady state | Quasisteady state |
| Surcharge | Stores (preserves volume continuity) | Neglects | Increases pipe diameter |

*Source:* D. P. Heeps and R. G. Mein, "Independent Comparison of Three Urban Runoff Models," *Proc. ASCE, J. Hyd. Div.*, 100, No. HY7 (July 1974): 995–1010.

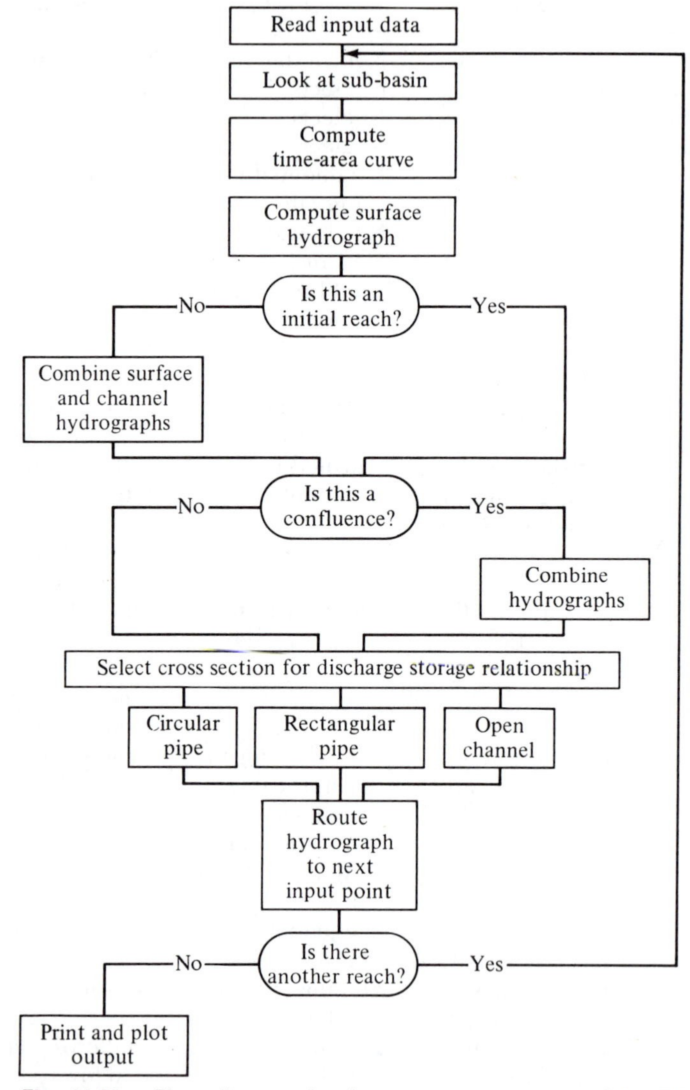


**Fig. 11-22.** *Flow diagram for the computer program of the RRL method. (After John B. Stall and Michael L. Terstriep, "Storm Sewer Design–An Evaluation of the RRL Method," EPA Technology Series EPA-R2-72-068, Oct. 1972.)*

point by a simple storage routing technique similar to the storage indication method of Chapter 7. The final result is a total basin runoff hydrograph that would result from the specified storm rainfall.

Stall and Terstriep evaluated the merits of the RRL method by applying it to 10 urban watersheds located largely in the east, south, and midwest regions of the United States.[38] The criteria followed in

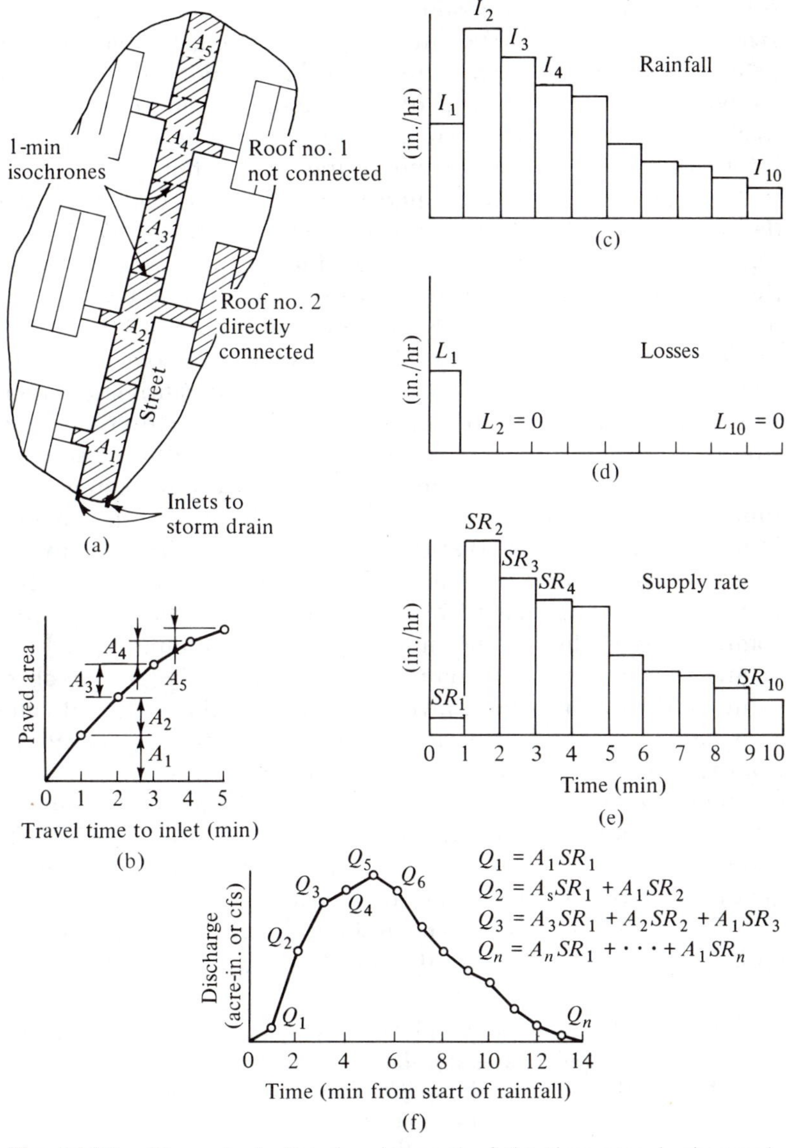


***Fig. 11-23.*** *Elements in the development of the time-area hydrograph: (a) subbasin map (directly connected paved area shaded); (b) time-area curve; (c) rainfall; (d) losses; (e) supply rate; and (f) the hydrograph. (After John B. Stall and Michael L. Terstriep, "Storm Sewer Design–An Evaluation of the RRL Method," EPA Technology Series EPA-R2-72-068, Oct. 1972.)*

selecting basins for the evaluation were (1) basins less than 5 mi² in size, (2) basins that were intensely urbanized, (3) basins that had extensive storm drainage systems, (4) basins with a high amount of paved area, (5) long records of rainfall and runoff, (6) the degree of quality of the data on storm rainfall and runoff, (7) the degree of information available on the storm drainage system, and (8) data that had not already been published in one form or another. Pertinent data for the selected basins are provided in Table 11-7. The basins represent a variety of hydrologic regimes in the United States, and the storms selected are characteristic of the variable storm rainfall occurring within the United States. As shown in Table 11-7, the basins ranged in size from 14.7 to 5326 acres; the percent areas directly connected ranged from 14.4 to 61%; a variety of hydrologic soil groups and basin slopes were represented; and the number of historical storms simulated for each watershed varied from 2 to 18.

A comparison of the computed and observed peak flow rates and runoff volumes from the Woodoak Drive basin are shown in Figs. 11-24a and b. The observed and computed hydrographs for one of these storms are shown as Fig. 11-24c. As indicated in Table 11-7, this watershed had the overall best comparison between measured and computed peak flows and runoff volumes. Columns 9 and 10 and 12 and 13 indicate that 5 computed peaks were higher than observed values and 5 runoff volumes were greater than the observed volumes. This balance was not experienced in some of the other watersheds. For example, computed peak and runoff volumes are underestimated for practically all storms simulated for the Echo Park Avenue basin. The results for the Echo Park basin are presented in Fig. 11-25. These results are not surprising, since the Echo Park basin had significant grassed area runoff for all 18 of the storms.

Stall and Terstriep arrived at the following conclusions based on their evaluation of the RRL Method:[38]

1. The RRL method provides an accurate means of computing runoff from the paved area portion of an urban basin.
2. The RRL method adequately represents the runoff from actual urban basins under the following conditions:
   a. The basin area is less than 5 mi².
   b. The directly connected paved area is equal to at least 15% of the basin area.
   c. The frequency of the storm event being considered is not greater than 20 yr.
3. The RRL method cannot be used for all urban basins in the United States; the method breaks down when significant grassed area runoff occurs, which happens if one or more of the following conditions exist:
   a. The directly connected paved area is less than 15% of the basin area.

**Table 11-7** *Watershed Data and Results of the Stall and Terstriep Evaluation*

| | | | | | | | | Computed Peaks | | | Computed Runoff | | |
|---|---|---|---|---|---|---|---|---|---|---|---|---|---|
| Basin (1) | Basin Area (acres) (2) | Total Paved Area (acres) (3) | Total Paved Area (%) (4) | Direct Connected Paved Area (acres) (5) | Direct Connected Paved Area (%) (6) | Hydro-logic Soil Group (7) | Basin Slope (8) | No. High (9) | No. Low (10) | Mean Absolute Error (%) (11) | No. High (12) | No. Low (13) | Mean Absolute Error (%) (14) |
| Woodoak Drive | 14.7 | 4.9 | 33.9 | 2.8 | 19.4 | B | Flat | 5 | 5 | 28.2 | 5 | 5 | 18.3 |
| Ross-Ade (Upper) | 54.0 | 13.3 | 24.7 | 7.8 | 14.4 | B-C | Steep | 0 | 2 | 50.5 | 0 | 2 | 62.8 |
| Sewer Dist No. 8 | 206 | 43 | 21.0 | 37.5 | 18.2 | C-D | Flat | 9 | 1 | 68.0 | 10 | 0 | 21.4 |
| Echo Park Avenue | 252 | 136 | 53.8 | 97.7 | 38.8 | B-C | Steep | 0 | 18 | 47.2 | 2 | 16 | 30.3 |
| Crane Creek | 273 | 65.5 | 23.9 | 39.7 | 14.5 | C-D | Moderate | 5 | 12 | 41.5 | 1 | 16 | 45.9 |
| Tripps Run Trib. | 322 | 100 | 31.0 | 56.9 | 17.7 | B-C | Moderate | 3 | 6 | 37.9 | 1 | 8 | 44.1 |
| Tar Branch | 384 | 227 | 59.0 | 195 | 51.0 | B | Moderate | 12 | 5 | 33.4 | 13 | 4 | 57.3 |
| Third Fork | 1075 | 397 | 37.0 | 293 | 27.0 | B-D | Moderate | 3 | 12 | 31.2 | 0 | 15 | 36.7 |
| Dry Creek | 1882 | 583 | 31.0 | 365 | 19.0 | B-C | Flat | 2 | 6 | 20.3 | 2 | 6 | 27.8 |
| Wingohocking | 5326 | 3246 | 61.0 | 3246 | 61.0 | B-D | Moderate | 14 | 2 | 49.9 | 13 | 3 | 72.5 |

*Source:* After John B. Stall and Michael L. Terstreip, "Storm Sewer Design—An Evaluation of the RRL Method," EPA Technology Series EPA-R2-72-068, Oct. 1972.

b. The frequency of the event being considered is greater than 20 yr.
c. The grassed area of the basin has steep slopes and tight soils, regardless of the antecedent moisture condition.
d. The grassed area of the basin has steep slopes, moderately tight soils, and an antecedent moisture condition of 3 or 4.
e. The grassed area of the basin has moderate slopes, moderately tight soils, and an antecedent moisture condition of 4.

4. The principal strength of the RRL method is that, by confining runoff calculations to the paved area of a basin, it utilizes hydraulic functions that are largely determinate such as gravity flow from plain

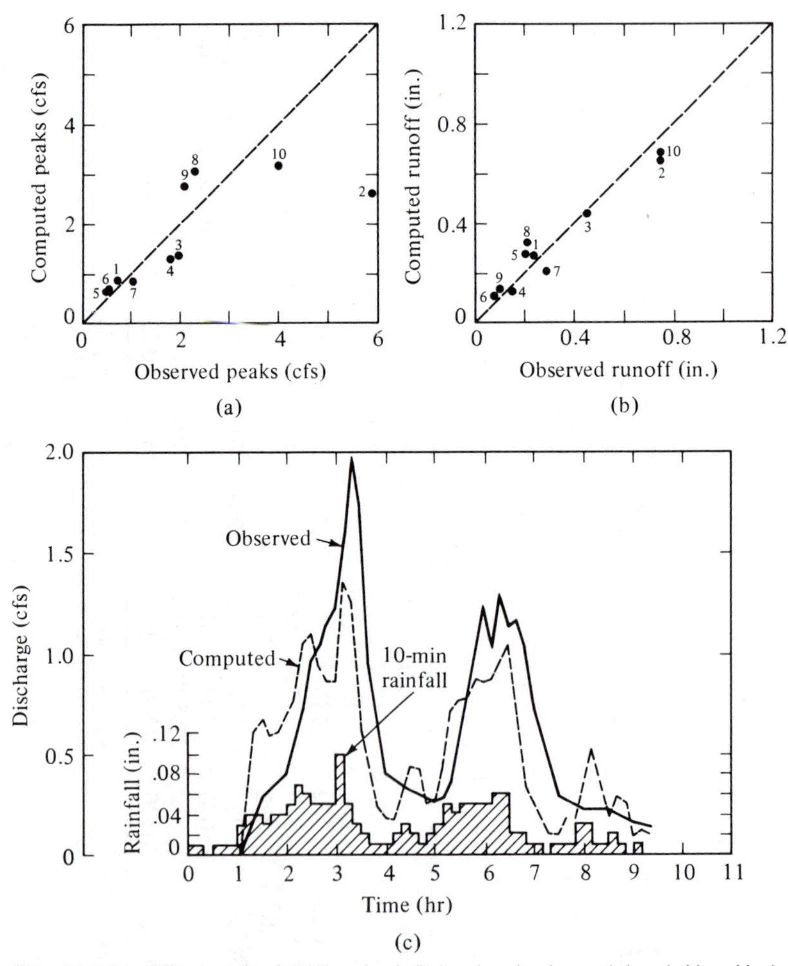


***Fig. 11-24.*** *RRL results for Woodoak Drive basin, Long Island, New York, storm of Oct. 19, 1966; (a) peaks; (b) volumes; and (c) the hydrographs. (After John B. Stall and Michael L. Terstriep, "Storm Sewer Design–An Evaluation of the RRL Method," EPA Technology Series EPA-R2-72-068, Oct. 1972.)*

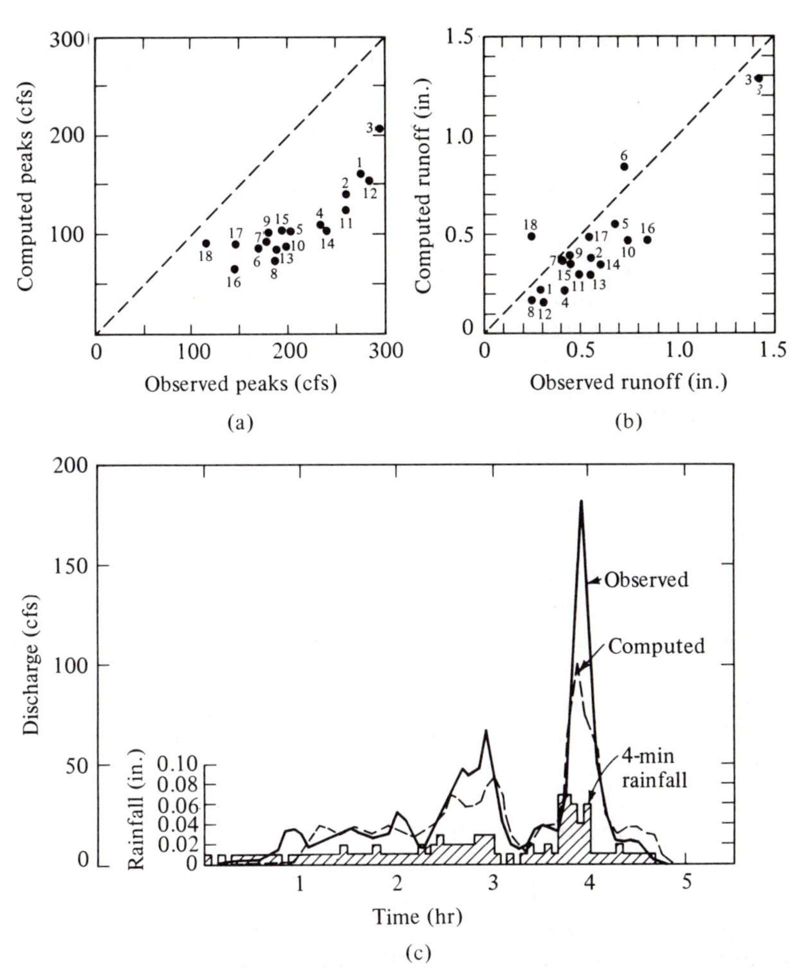


***Fig. 11-25.*** *RRL results for Echo Park Avenue basin, Los Angeles, California, storm of April 18, 1965: (a) peaks; (b) volumes; (c) the hydrographs. (After John B. Stall and Michael L. Terstriep, "Storm Sewer Design–An Evaluation of the RRL Method," EPA Technology Series EPA-R2-72-068, Oct. 1972.)*

sloping concrete surfaces, gutters, pipes, and open channels. Physical understanding of these flow phenomena is much greater than the present understanding of the many complex phenomena governing runoff from rural areas such as antecedent moisture conditions, infiltration, soil moisture movement, transpiration, evaporation, and so forth.

5. A modification of the RRL method that would provide a function for grassed area contributions to runoff could be developed into a valuable design tool for urban drainage. This is believed to be possible

in spite of the many complexities involved. Further flexibility could be offered by the additional provision for routing surface runoff through surface storage.

6. The input data requirements for use of the RRL method on an urban basin are reasonable for the engineering evaluation of a basin for storm drainage design. The necessary data are no more complex or elaborate than the data usually compiled for a traditional storm drainage design.
7. It appears that rainfall occurs in greater amounts in the United States than in Great Britain. This may account for the fact that the RRL method is successful and widely used in Great Britain and yet suffers the above-described breakdowns for some of the basins studied in the United States.
8. Better urban rainfall and runoff data are required for the proper testing of all mathematical models. Research basins that do not have hydraulic problems, such as undersized drains or inadequate inlets, should be selected and instrumented.

### Storm Water Management Model (SWMM)

A very widely accepted and applied storm runoff simulation model was jointly prepared by Metcalf and Eddy, Inc.; the University of Florida; and Water Resources Engineers[43] for use by the U.S. Environmental Protection Agency (EPA). This model is designed to simulate the runoff of a drainage basin for any predescribed rainfall pattern. The total watershed is broken into a finite number of smaller units or subcatchments that can be readily described by their hydraulic or geometric properties. A flow chart for the process is shown in Fig. 11-26.

The SWMM model has the capability of determining, for short-duration storms of given intensity, the locations and magnitudes of local floods as well as the quantity and quality of storm water runoff at several locations both in the system and in the receiving waters. The SWMM is an event simulation model and does not keep track of long-term water budgets.

The fine detail in the design of the model allows the simulation of both water quantity and quality aspects associated with urban runoff and combined sewer systems. The water quality aspects are described in Chapter 13. Information obtained from SWMM would be used to design storm sewer systems for storm water runoff control. Use of the model is limited to relatively small urban watersheds in regions where seasonal differences in the quality aspects of water are adequately documented.

The simulation is facilitated by five main subroutine blocks. Each block has a specific function, and the results of each block are entered on working storage devices to be used as part of the input to other blocks.

The main calling program of the model is called the Executive

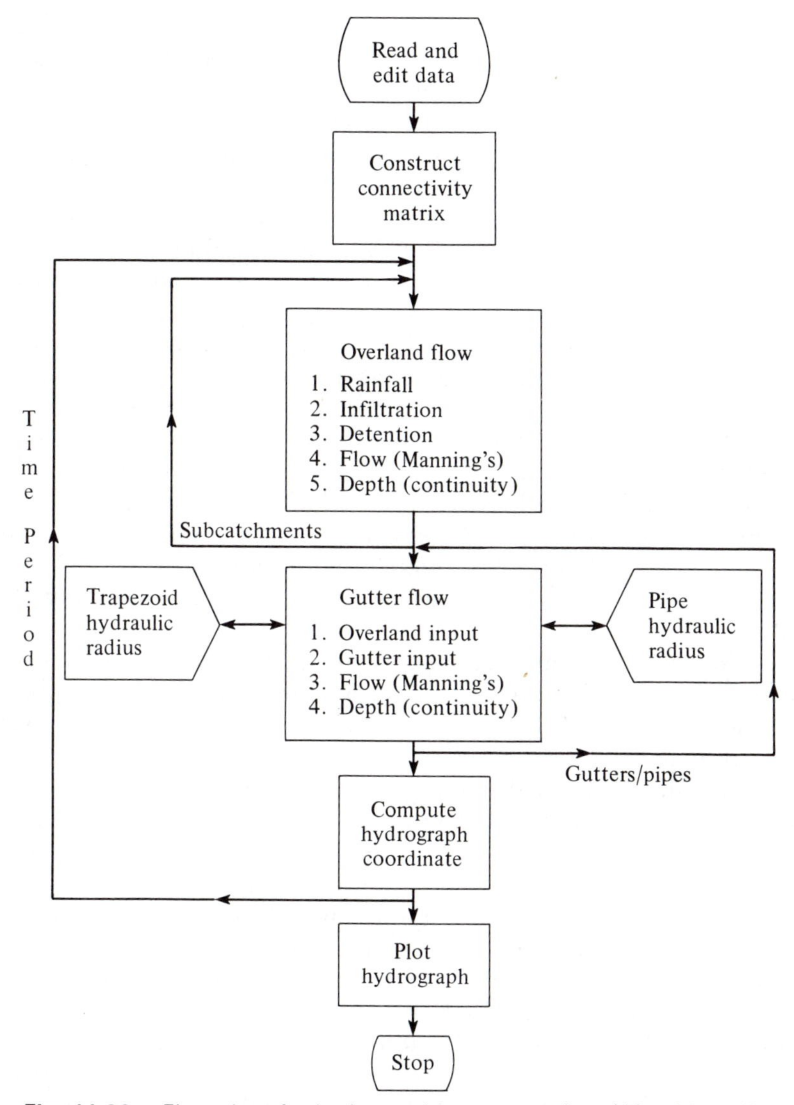


**Fig. 11-26.** *Flow chart for hydrographic computation. (After Metcalf and Eddy, Inc., University of Florida, Gainesville, Fla., Water Resources Engineers, Inc., "Storm Water Management Model," Environmental Protection Agency, Vol. 1, 1971.*

Block. This block is the first and last to be used, and performs all the necessary interfacing between the other blocks.

The Runoff Block uses Manning's equation to route the uniform rainfall intensity over the overland flow surfaces, through the small gutters and pipes of the sewer system into the main sewer pipes, and out of the sewer pipes into the receiving streams. This block also provides time-dependent pollutional graphs (pollutographs).

A third package of subroutines, the Transport Block, determines the quality and quantity of dry weather flow; calculates the system infiltration; calculates the water quality of the flows in the system; and will also calculate the land, capital, and operation and maintenance costs of two optional internal storage tanks through which the combined dry weather and infiltrated flows are routed.

A useful package of subroutines for water quality determination is contained in the Storage Block. The Storage Block allows the user to specify or have the model select sizes of several treatment processes in an optional wastewater treatment facility that receives a user-selected percentage of the peak flow. If used, this block simulates the changes in the hydrographs and pollutographs of the sewage as the sewage passes through the selected sequence of unit processes.

The hydraulic and water quality effects of the effluent from the modeled sewer system on the receiving water body are modeled in the Receiving Water Block. This fifth block of subroutines models the receiving body of water as a network of nodes connected by channel segments. The hydraulics (which determine the resulting water quality) of the flow network are simulated by the Receiving Water Block.

Subcatchment areas, slopes, widths, and linkages must be specified by the user. Manning's roughness coefficients can be supplied for pervious and impervious parts of each subcatchment, or respective default values of 0.250 and 0.013 are assigned by the model.

As indicated in Table 10-19, SWMM is the only event simulation model listed that utilizes Horton's equation for calculating watershed infiltration losses. If parameters for Horton's equation are unavailable, the user can specify the ASCE standard infiltration capacity curves shown in Fig. 11-11. Infiltration amounts thus determined for each time step are compared with instantaneous amounts of water existing on the subcatchment surface plus any rainfall that occurred during the time step, and if the infiltration loss is larger, it is set equal to the amount available. Input for Horton's equation consists of the maximal and minimal infiltration rates and the recession constant $k$ in Eq. 3-31. Respective default values in SWMM are 3.00 in./hr, 0.52 in./hr, and 0.00115 in./sec.

Urban storm drainage components are modeled using Manning's equation and the continuity equation. The hydraulic radius of the trapezoidal gutters and circular pipes is calculated from component dimensions and flow depths. A pipe surcharges if it is full, provided that the inflow is greater than the outflow capacity. In this case, the surcharged amount will be computed and stored at the head end of the pipe. The pipe will remain full until the stored water is completely drained.

Necessary inputs in the model are the surface area, width of subcatchment, ground slope, Manning's roughness coefficient, infiltration rate, and detention depth. Gutter descriptions are the length, Manning's roughness coefficient, invert slope, diameter for pipes, and cross-sectional dimensions of the gutter. General data requirements are summarized in Table 11-8. A step-by-step process accounts for all inflow, infiltration losses, and flow from upstream subcatchment areas, providing a calculated discharge hydrograph at the drainage basin

***Table 11-8*** *General Data Requirements Storm Water Management Model (SWMM)*

Item 1. Define the Study Area.

Land use, topography, population distribution, census tract data, aerial photos, and area boundaries.

Item 2. Define the System.

Plans of the collection system to define branching, sizes, and slopes. Types and general locations of inlet structures.

Item 3. Define the System Specialties.

Flow diversions, regulators, and storage basins.

Item 4. Define the System Maintenance.

Street sweeping (description and frequency), catchbasin cleaning. Trouble spots (flooding).

Item 5. Define the Receiving Waters.

General description (estuary, river, or lake), measured data (flow, tides, topography, and water quality).

Item 6. Define the Base Flow (DWF).

Measured directly or through sewerage facility operating data. Hourly variation and weekday versus weekend. The DWF characteristics (composited BOD and SS results). Industrial flows (locations, average quantities, and quality).

Item 7. Define the Storm Flow.

Daily rainfall totals over an extended period (6 months or longer) encompassing the study events. Continuous rainfall hyetographs, continuous runoff hydrographs, and combined flow quality measurements (BOD and SS) for the study events. Discrete or composited samples as available (describe fully when and how taken).

outlet. The following description of the simulation process will aid in understanding the logic of the model.[44]

1. Rainfall is added to the subcatchment according to the specified hyetograph:

$$D_1 = D_t + R_t \, \Delta t \tag{11-10}$$

where

$D_1$ = the water depth after rainfall
$D_t$ = the water depth of the subcatchment at time $t$
$R_t$ = the intensity of rainfall in time interval $\Delta t$

2. Infiltration $I_t$ is computed by Horton's exponential function, $I_t = f_c + (f_0 - f_c)e^{-kt}$, and subtracted from the water depth existing on the subcatchment

$$D_2 = D_1 - I_t \, \Delta t \tag{11-11}$$

where

$f_c$, $f_0$, and $k$ = coefficients in Horton's equations (see Eq. 3-31)
$D_2$ = the intermediate water depth after accounting for infiltration

3. If the resulting water depth of subcatchment $D_2$ is larger than the specified detention depth $D_d$, an outflow rate is computed using Manning's equation.

$$V = \frac{1.49}{n} (D_2 - D_d)^{2/3} S^{1/2} \tag{11-12}$$

and

$$Q_w = VW(D_2 - D_d) \tag{11-13}$$

where

$V$ = the velocity
$n$ = Manning's coefficient
$S$ = the ground slope
$W$ = the width
$Q_w$ = the outflow rate

4. The continuity equation is solved to determine water depth of the subcatchments resulting from rainfall, infiltration, and outflow. Thus

$$D_{t+\Delta t} = D_2 - \frac{Q_w}{A} \Delta t \tag{11-14}$$

where $A$ is the surface area of the subcatchment.

5. Steps 1 to 4 are repeated until computations for all subcatchments are completed.

6. Inflow ($Q_{in}$) to a gutter is computed as a summation of outflow from tributary subcatchments ($Q_{w,i}$) and flow rate of immediate upstream gutters ($Q_{g,i}$)

$$Q_{in} = \sum Q_{w,i} + \sum Q_{g,i} \tag{11-15}$$

7. The inflow is added to raise the existing water depth of the gutter according to its geometry. Thus

$$Y_1 = Y_t + \frac{Q_{in}}{A_s} \Delta t \tag{11-16}$$

where

$Y_1$ and $Y_t$ = the water depth of the gutter
$A_s$ = the mean water surface area between $Y_1$ and $Y_t$

8. The outflow is calculated for the gutter using Manning's equation:

$$V = \frac{1.49}{n} R^{2/3} S_i^{1/2} \tag{11-17}$$

and

$$Q_g = VA_c \tag{11-18}$$

where

$R$ = the hydraulic radius
$S_i$ = the invert slope
$A_c$ = the cross-sectional area at $Y_1$

9. The continuity equation is solved to determine the water depth of the gutter resulting from the inflow and outflow. Thus

$$Y_{t+\Delta t} = Y_1 + (Q_{in} - Q_g)\frac{\Delta t}{A_s} \tag{11-19}$$

10. Steps 6 to 9 are repeated until all the gutters are finished.
11. The flows reaching the point concerned are added to produce a hydrograph coordinate along the time axis.
12. The processes from 1 to 11 are repeated in succeeding time periods until the complete hydrograph is computed.

The model is presently functioning to provide storm hydrographs for areas having lateral sewers up to 30 in. in diameter.

Three general types of output are provided by SWMM. If waste treatment processes are simulated or proposed, the capital, land, and operation and maintenance costs are printed. Plots of water quality constituents versus time form the second type of output. These pollutographs are produced for several locations in the system and in the receiving waters. Quality constituents handled by SWMM include suspended solids, settleable solids, BOD, nitrogen, phosphorus, and grease. The third type of output is hydrologic. Hydrographs at any point, for example, the end of a gutter or inlet, are printed for designated time periods.

### University of Cincinnati Urban Runoff Model (UCURM)

The University of Cincinnati Urban Runoff Model (UCURM) was developed by the Division of Water Resources, the Department of Civil Engineering, of the University of Cincinnati.[39] A flow chart is reproduced in Fig. 11-27. The program consists of three sections: (1) MAIN—infiltration and depression storage and two subroutines, (2) GUTFL—gutter flow, and (3) PIROU—pipe routing. It is similar to the EPA model and divides the drainage basin into subcatchments whose flows are routed overland into gutters and sewer pipes. The rainfall is read in as a hyetograph. The infiltration and depression storage are summed and subtracted from the rainfall to give overland flow. This is routed through the gutter system to storm water inlets and the pipe network. Starting at the upstream inlet, the flows are calculated in successive segments of the sewer system, including discharges from inlets, to produce the total outflow.

The drainage area is divided into small subcatchments with closely matched characteristics. The rainfall data are introduced and the infiltration is computed for each subcatchment. Principal elements of the modeling process follow.

1. It is assumed that runoff begins whenever the rainfall rate equals the infiltration rate and the mass of precipitation balances infiltration. The equations representing these conditions are

$$t = -\frac{1}{k}\ln\left[\frac{i(I) + x\,(i\,(I+1) - i\,(I))\,/DT - f_c}{f_o - f_c}\right] \tag{11-20}$$

and

$$\frac{f_c t}{60} + \frac{f_0 - f_c}{60k}(1 - e^{-kt}) = mi(I) + \left(i(I) + \frac{x}{2DT}(i(I+1) - i(I))\right)\frac{x}{60} \tag{11-21}$$

where

$mi(I)$ = the mass precipitated until time $t$ (in.)
$i(I)$ = the ordinates of rainfall intensity curve
$k$ = the decay rate of infiltration (units/min)
$f_0$ = the initial infiltration capacity (in./hr.)
$f_c$ = the ultimate infiltration capacity (in./hr.)
$DT$ = the time increment of rainfall intensity curve
$t$ = the time to intersection of rainfall curve and infiltration curve
$x$ = an increment of $DT$

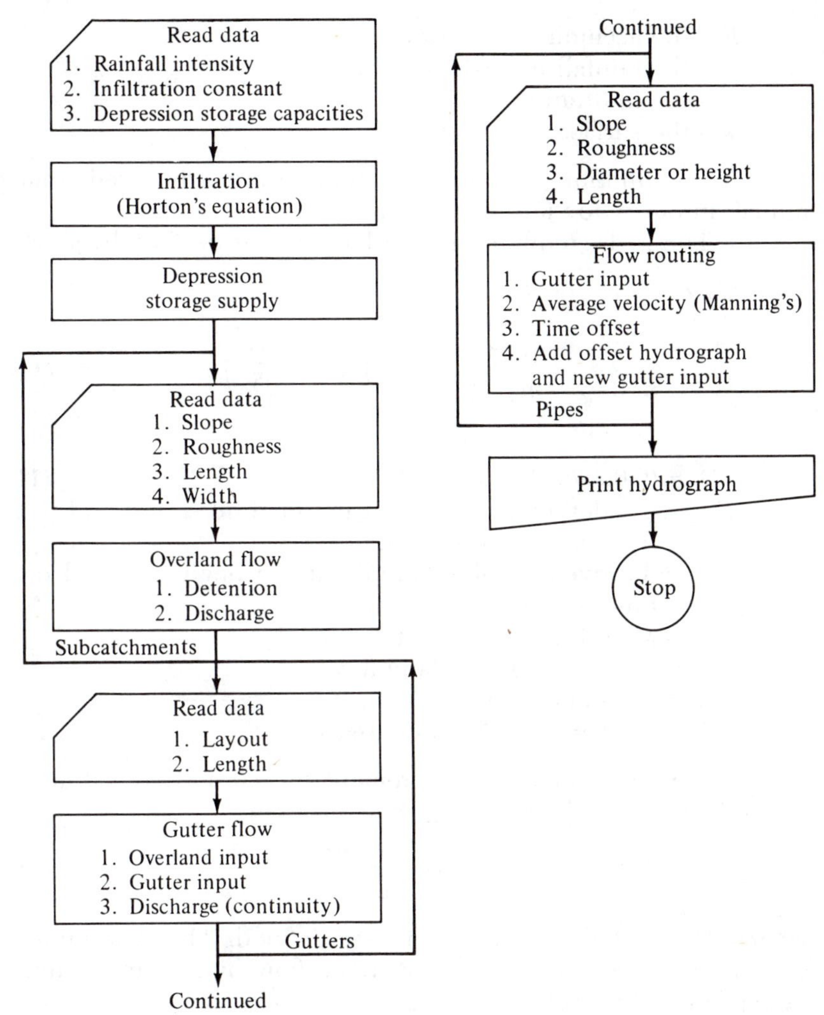


***Fig. 11-27.*** *UCURM model flow chart. (After C. N. Papadakis and H. C. Preul, "University of Cincinnati, Urban Runoff Model,"* Proc. ASCE, J. Hyd. Div., *98, No. HY10 (Oct. 1972): 1789–1804.)*

The infiltration curve is computed from the equations and $t$, $I$, and $x$ are stored.

2. Surface retention is related to depression storage by an equation derived by Linsley, Kohler, and Paulhus,[44]

$$s = (i - f)e^{-(P-F)/S_d} \qquad \textbf{(11-22)}$$

.where

$S_d$ = the total depression storage (in.)
$P$ = the accumulated rainfall in storage (in.)

$F$ = the accumulated infiltration (in.)
$i$ = the rainfall intensity (in./hr)
$f$ = the infiltration (in./hr)
$s$ = the surface retention (in./hr)

The infiltration and surface retention are subtracted from the rainfall intensity to yield the runoff.

3. The hydrograph of the overland flow is derived by solving

$$\frac{r_1 + r_2}{2} - \frac{q_1}{2} + \frac{60D_1}{\Delta t} = \frac{510.35}{nl} s^{1/2} D_2^{5/3} \left(1 + 0.6\left(\frac{D_2}{D_{e_2}}\right)^3\right)^{5/3} + \frac{60D_2}{\Delta t} \quad \textbf{(11-23)}$$

where

$$D_e = (0.00979 n^{0.6} r^{0.6} l^{0.6}) / s^{0.3} \quad \textbf{(11-24)}$$

$D_{1,2}$ = the detention storage at the beginning and end of time interval $t$ (in./unit area)
$r_1, r_2$ = the overland flow supply at the beginning and end of time interval $t$ (cfs/min)
$n$ = Manning's coefficient
$l$ = the length of overland flow
$s$ = the slope (ft/ft)
$q$ = discharge (in./hr/unit area)

4. For the initial time increment, $q_1 = 0$ and $D_1 = 0$ are substituted, $D_2$ is calculated, and $q_2$ is found from

$$q = \frac{1020.7}{nl} s^{1/2} D^{5/3} \left(1 + 0.6\left(\frac{D}{D_e}\right)^3\right)^{5/3} \quad \textbf{(11-25)}$$

where the symbols are as previously defined. The determined $D_2$ and $q_2$ become $D_1$ and $q_1$. The overland flow hydrograph is derived by repeating this cycle.

5. The gutter flow is computed using the continuity equation

$$\frac{\partial Q}{\partial x} + \frac{\partial y}{\partial t} T = q_L \quad \textbf{(11-26)}$$

where $T$ is the width of the water surface.

The term $(\partial y / \partial t)T$ is neglected because the change in the depth of the gutter flow is very small with respect to time. After integration, the equation becomes

$$Q = q_L L + Q_0 \quad \textbf{(11-27)}$$

where

$Q_0$ = upstream gutter contributions
$L$ = the length of the gutter (ft)

$q_L$ = the overland flow from the hydrograph
$Q$ = the flow from the gutter system

Inlet flows are routed through the pipe network by delaying the hydrographs by the flow time required to reach the next inlet and summing at a terminal point in the network. Manning's equation is used to find the velocity of flow in the sewer and the corresponding time delay. Provision is made for sewers of varying cross section. The model has been found to closely simulate gauged flows.[39] The calculation of pollutant concentrations in the runoff is described in Chapter 13.

### Urban Runoff Models Compared

Several good comparisons of the RRL, SWMM, and the UCURM models have been reported in the literature. One of the first was an application by Heeps and Mein[40] of all three models to two urban catchments in Australia for a total of 20 storm events. A similar statistical comparison of the same three models applied to 12 storms over each of three urban watersheds was performed by Marsalek, et al.[41] Significant results from these independent evaluations are presented here.

The Heeps and Mein comparisons of model performance for 4 of the 20 storms simulated for the two watersheds are summarized in Table 11-9. The Vine Street catchment has an area of 172 acres with 55 acres directly connected to the storm drain system. The Yarralumla Creek catchment is 7.2 times larger at 1240 acres with 250 acres directly connected. Conclusions drawn by Heeps and Mein based on the information in Table 11-9 are as follows:[40]

1. The degree of subdivision of the catchment has a significant influence on the peak discharge predicted by each of the models. The RRL and SWMM methods give lower peaks and the UCURM gives higher peaks for finer subdivision.
2. The UCURM contains several deficiencies. The major ones, the effects of which can be seen in the predicted hydrographs, are that depression storages are assigned full when the rainfall intensity falls below the infiltration capacity, and that depression storages are not depleted by infiltration. The use of instantaneous values of the rainfall intensity (difficult to obtain from recorder charts) can cause volume errors.
3. The SWMM was the model with the best overall performance but at the expense of large computer storage and time requirements.
4. The RRL model predicted poorly for storms in which pervious runoff was significant but performed reasonably well for many other types of storms. The results, in general, support those of Stall and Terstriep.[38]
5. A major problem with using noncontinuous models is the prediction of antecedent conditions. This problem is further aggravated by use

**Table 11-9** *Comparisons of Model Performance*

| | Storm 1 | | | Storm 2 | | | Storm 3 | | | Storm 4 | | |
|---|---|---|---|---|---|---|---|---|---|---|---|---|
| | *Nov. 6, 1971 Vine Street* | | | *Dec. 24, 1971 Vine Street* | | | *Dec. 20, 1971 Yarralumla Creek* | | | *Mar. 3, 1972 Yarralumla Creek* | | |
| *Objective Function (1)* | *RRL (2)* | *SWMM (3)* | *UCURM (4)* | *RRL (5)* | *SWMM (6)* | *UCURM (7)* | *RRL (8)* | *SWMM (9)* | *UCURM (10)* | *RRL (11)* | *SWMM (12)* | *UCURM (13)* |
| Percentage of observed runoff volume | 44 | 102 | 139 | 176 | 175 | 307 | 60 | 126 | 193 | 77 | 104 | 234 |
| Percentage of observed runoff peak | 91 | 106 | 189 | 175 | 207 | 266 | 45 | 101 | 139 | 36 | 47 | 64 |
| Sum of squares of errors, in $10^3$ $ft^6/sec^2$ ($m^6/s^2$) | 39 (31.2) | 4 (3.2) | 33 (26.2) | 2.2 (1.8) | 3.6 (2.9) | 9.2 (7.4) | 200 (161) | 157 (126) | 690 (554) | 88 (71) | 51 (41) | 381 (306) |
| B5500 Processor time, in minutes | 1.87 | 34.23 | 31.05 | 0.63 | 12.97 | 4.80 | 0.08[a] | 2.32[a] | 0.87[a] | 0.11[a] | 2.95[a] | 1.28[a] |

[a] B6700 Computer.

*Source:* D. P. Heeps and R. G. Mein, "Independent Comparison of Three Urban Runoff Models," *Proc. ASCE, J. Hyd. Div.*, 100, No. HY7 (July 1974): 995–1010.

of the Horton infiltration equation for which prediction of the parameters is virtually impossible.

The Marsalek study results,[41] using the same three models for three watersheds in Illinois, Ontario, and Maryland, indicated that the SWMM model performed slightly better than the RRL model and both these models were more accurate than the UCURM model for the small watersheds studied.

Table 11-10 provides descriptions of the urban watersheds used by Marsalek in the runoff model evaluation. Postcalibration values of some of the model parameters used are presented in Table 11-11. Typical comparisons of observed and calculated times to peak, peak flows, and runoff volumes for the three models are provided in Fig. 11-28 for the Calvin Park watershed in Kingston, Ontario.

Marsalek et al. used the information described to develop the qualitative comparison in Table 11-12 and arrived at the following conclusions:[41]

1. Uncalibrated deterministic models for urban runoff, such as RRL, SWMM, and UCURM, yielded a fairly good agreement between the simulated and measured runoff events on typical urban catchments of small size.
2. On the average, about 60% of the simulated peak flows were within 20% of the measured values. About the same scatter was found for the simulated times to peak and runoff volumes. The agreement between the measured and simulated values could be further improved by model calibration.
3. When comparing the entire simulated and measured hydrographs using statistical measures, the agreement was found good for SWMM, good to fair in the case of RLL, and fair in the case of UCURM.
4. Out of the three studied models, the SWMM simulations were marginally better than those by RRL, and both these models were more accurate than UCURM, with all models applied in an uncali-

***Table 11-10*** *Description of Test Urban Watersheds Used for the Evaluation of Urban Runoff Models*

| Test Urban Drainage Basin | Location | Area (acres) | Imperviousness | Land Use |
|---|---|---|---|---|
| Oakdale | Chicago, Ill. | 13.0 | 45.8 | Residential |
| Calvin Park | Kingston, Ontario | 89.4 | 27 | Residential and institutional |
| Gray Haven | Baltimore, Md. | 23.3 | 52 | Residential |

*Source:* J. Marsalek, D. T. M. Dick, P. E. Wisner, and W. G. Clarke, "Comparative Evaluation of Three Urban Runoff Models," *Water Resources Bulletin*, AWRA, 11, No. 2 (Apr. 1975): 306–328.

**Table 11-11** *Model Parameters and Catchment Discretization*

| Model Parameters | Model | | |
|---|---|---|---|
| | RRL | SWMM | UCURM |
| Infiltration | | | |
| Maximum (in./hr) | | 3.0 | 3.0 |
| Minimum (in./hr) | NA | 0.52 | 0.5 |
| Decay (l/sec) | | 0.00115 | 0.00116 |
| Retention Storage | | | |
| Impervious (in.) | NA | 0.062 | 0.0625 |
| Pervious (in.) | | 0.184 | 0.25 |
| Manning's $n$ | | | |
| Impervious | | 0.013 | 0.012–0.013 |
| Pervious | NA | 0.250 | 0.20–0.35 |
| Pipes | | 0.013 | 0.013 |

| Catchment Dis-cretization | RRL Min. CPU | RRL No. of Time Steps | RRL No. of Surface Elements | RRL No. of Pipes | SWMM Min. CPU | SWMM No. of Time Steps | SWMM No. of Surface Elements | SWMM No. of Pipes | UCURM Min. CPU | UCURM No. of Time Steps | UCURM No. of Surface Elements | UCURM No. of Pipes |
|---|---|---|---|---|---|---|---|---|---|---|---|---|
| Oakdale | 0.013 | 180 | NA | 14 | 0.244 | 150 | 27 | 31 | 0.839[a] | 87 | 92 | 14 |
| Gray Haven | 0.01 | 72 | NA | 11 | 0.097 | 72 | 17 | 26 | — | — | — | — |
| Calvin Park | 0.021 | 67 | NA | 53 | 0.243 | 69 | 53 | 58 | 0.190[b] | 54 | 225 | 53 |

[a] Before programming improvements.
[b] After programming improvements.

*Source:* After J. Marsalek, T. M. Dick, P. E. Wisner, and W. G. Clarke, "Comparative Evaluation of Three Urban Runoff Models," ***Water Resources Bulletin***, AWRA, 11, No. 2 (Apr. 1975): 306–328.

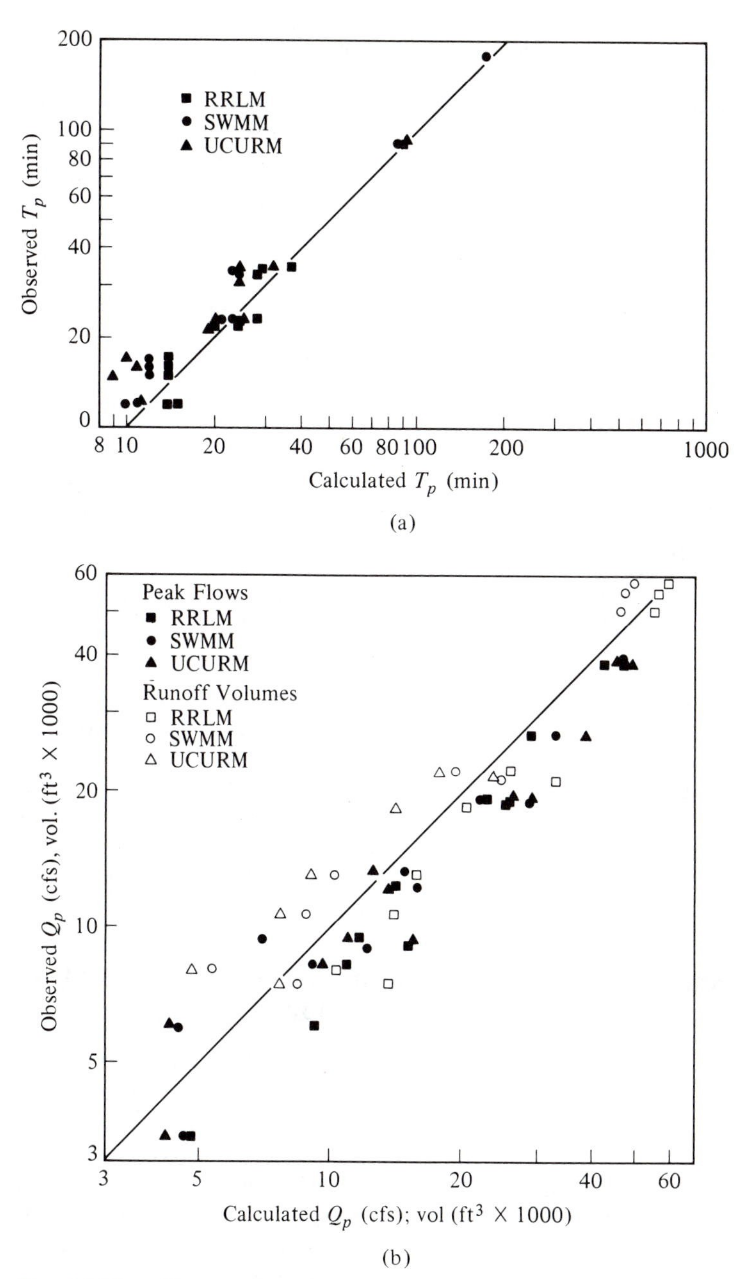


***Fig. 11-28.*** *Observed and computed values compared for the Calvin Park subcatchment.*

$T_p$ = Time to Peak
$Q_p$ = Runoff Peak Flow
*Vol.* = Total Runoff Volume

*(After J. Marsalek, T. M. Dick, P.E. Wisner, and W. G. Clarke, "Comparative Evaluation of Three Urban Runoff Models," Water Resources Bulletin, AWRA, 11, No. 2 (Apr. 1975): 306–328.)*

*Table 11-12* Subjective Evaluation of Three Urban Models

| | Model | | |
|---|---|---|---|
| | *RRL* | *SWMM* | *UCURM* |
| Effort for input data preparation | Low | Medium | High |
| Flexibility of schematization | Good | Good | Fair |
| Accuracy of employed routing scheme | Low | Medium | Low |
| Model and computer program availability | Good | Excellent | Limited |
| Availability of a runoff quality submodel | No | Yes | Yes |
| Computer time required | Low | Moderate | Moderate |
| Continuous refinement by the corresponding agency | No | Yes | No |

*Source:* J. Marsalek, T. M. Dick, P. E. Wisner, and W. G. Clarke, "Comparative Evaluation of Three Urban Runoff Models," *Water Resources Bulletin*, AWRA, 11, No. 2 (Apr. 1975): 306–328.

brated version. The main advantage of the RRL is its simplicity, since it can be applied without a computer. The SWMM, on the other hand, is much more general and versatile, and is being continuously refined and improved.

5. The SWMM model has the most advanced routing scheme (Transport Block) among the considered methods. The accuracy of flow routing becomes particularly important when studying large watersheds.

## 11-4 Small Watershed Simulation Using Equations of Gradually Varied Unsteady Flow

Equations describing gradually varied unsteady flow in open channels can be used to calculate the runoff from the component parts of a drainage area. A procedure developed by Schaake illustrates this approach.[17]

Consider the parking lot illustrated in Fig. 11-29. A swale runs longitudinally through the area and drains to a storm water inlet. The area is partitioned into smaller components as shown. For convenience, each component part is numbered and its physical characteristics tabulated (Fig. 11-29d). The outflow from component 6 discharges into a storm water inlet.

Beginning at the upstream edge of the area, components 1 and 2 contribute overland flow to component 4, the upstream part of the swale. Since the overland flow to the swale can be measured in discharge per foot of swale length, only the outflow from a 1-ft wide strip of overland flow need be computed. The rectangular overland

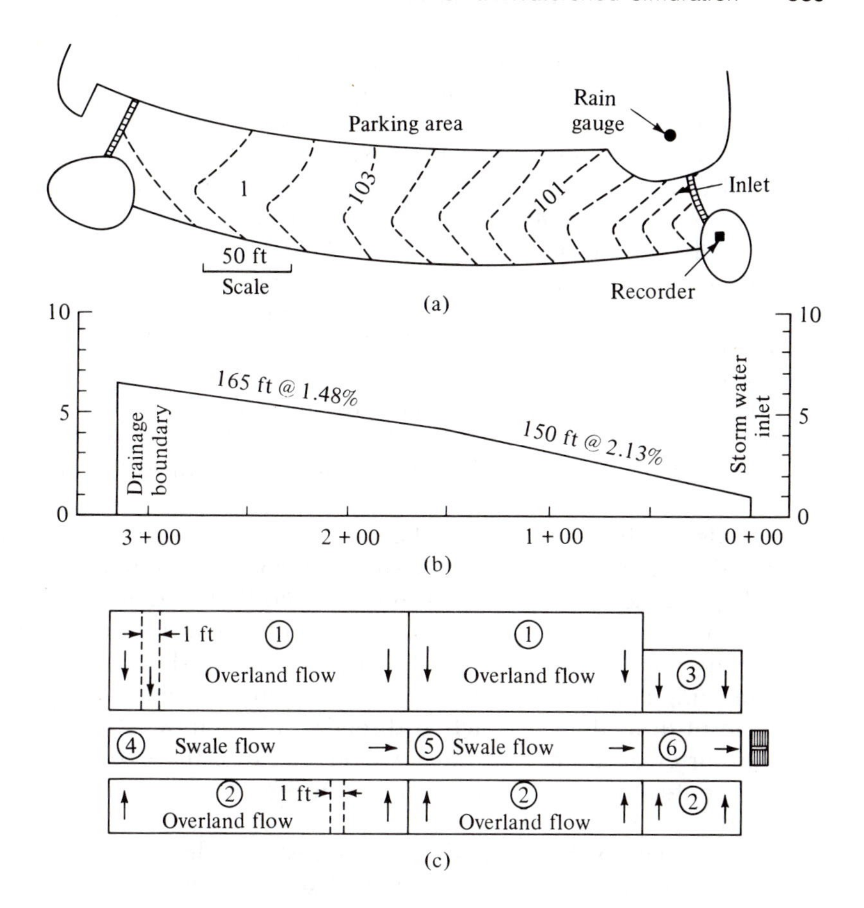


| Component | Length (ft) | Width or diameter (ft) | Slope (ft/ft) | Side slope | Inflow to component | |
|---|---|---|---|---|---|---|
| | | | | | Lateral inflow | Upstream inflow |
| ① | 36 | 1 | 0.019 | – | Rainfall | – |
| ② | 20 | 1 | 0.0167 | – | Rainfall | – |
| ③ | 25 | 1 | 0.019 | – | Rainfall | – |
| ④ | 165 | – | 0.0148 | 113:1 | ① and ② | – |
| ⑤ | 100 | – | 0.0213 | 113:1 | ① and ② | ④ |
| ⑥ | 50 | – | 0.0213 | 113:1 | ② and ③ | ⑤ |

(d)

***Fig. 11-29.*** *Physical Characteristics of inlet area SPL1. (After John C. Schaake, Jr., "Synthesis of the Inlet Hydrograph," Tech. Rept. 3, Storm Drainage Research Project, Johns Hopkins University, Baltimore, Md., June 1965.)*

flow component is thus composed of a sequence of 1-ft wide strips. The sum of the outflows from these strips in components 1 and 2 represents uniform lateral inflow along the length of component 4. The outflow from component 4 is the upstream inflow to component 5 which also receives uniform lateral inflow from components 1 and 2. Component 4 terminates where there is a change in slope of the swale as shown on the profile (Fig. 11-29b). The outflow from component 5 becomes upstream inflow to component 6 which also receives uniform lateral inflow from components 2 and 3. Component 5 terminates at the point where the length of overland flow on one side is reduced from 36 to 25 ft.

The schematic representation of the drainage area given in Fig. 11-29c illustrates the manner in which an area can be divided into smaller components representing the principal features of the total area.

In describing the overland flow contribution to the swales and gutters, the direction of flow was assumed perpendicular to the gutter. In reality, the actual direction of the overland flow is determined by the gutter slope as well as the cross-slope of the surface and is not exactly the same as the assumed direction. It can be shown, however, that the computed time of flow along the assumed direction is equal to the time of flow along the actual direction if the flow is laminar. A very minor error is introduced, however, because the overland flow actually enters the gutter at a point further downstream than is assumed.

The equations for gradually varied unsteady flow in open channels can be applied to areas such as that shown in Fig. 11-29. These equations describe the relationship between the rainfall (more precisely, the surface runoff supply rate) and the area, depth, velocity, and rate of flow of the surface runoff in the drainage area. They also account for the conservation of mass and momentum of the surface runoff at any point in space and time. Because of the complexity of the equations, analytic solutions cannot be obtained for synthesizing hydrographs and numerical methods must be used.

The first descriptive equation is a continuity equation written as[17]

$$\frac{\partial A}{\partial t} + \frac{\partial Q}{\partial x} = q \qquad \textbf{(11-28)}$$

This equation is derived by considering the water entering and leaving an infinitesimal section of the channel (Fig. 11-30). The term $\partial A/\partial t$ accounts for the change in storage with time in the infinitesimal section; the term $\partial Q/\partial x$ accounts for the difference between the outflow and the inflow to the infinitesimal section; and the term $q$ is the lateral inflow in cfs/ft along the channel.

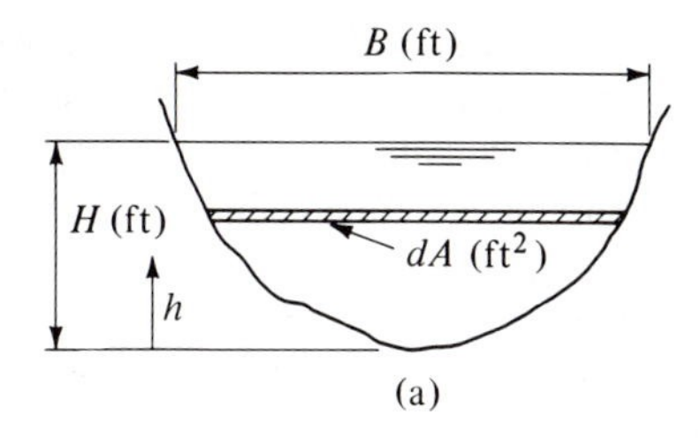


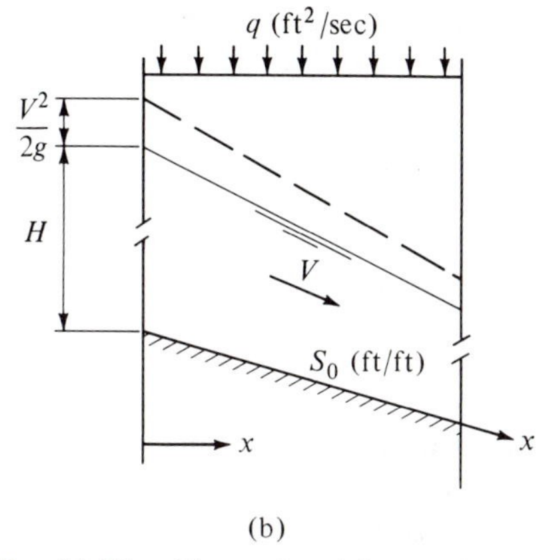


***Fig. 11-30.*** *Elements of open channel flow: (a) channel cross section; (b) longitudinal profile.*

$$dA = Bdh$$

$$A = \int_0^H Bdh \text{ ft}^2$$

$$V = \frac{\int_A vdA}{A} \text{ (ft/sec)}$$

$$Q = AV \text{ (cfs)}$$

*(After John C. Schaake, Jr., "Synthesis of the Inlet Hydrograph," Tech. Rept. 3, Storm Drainage Research Project, Johns Hopkins University, Baltimore, Md., June 1965.)*

The second equation (the momentum equation) refers to the dynamic behavior of the flow. It is written

$$\frac{\partial V}{\partial t} + V\frac{\partial V}{\partial x} + g\frac{\partial H}{\partial x} + g(S_f - S_0) + \frac{Vq}{A} = 0 \qquad \textbf{(11-29)}$$

and is derived by considering all of the forces acting on a fluid element. Neglecting the first and last terms, the remaining terms are commonly used to compute backwater profiles for steady flow in reservoirs and stream channels. The first term, $\partial V/\partial t$, accounts for the local acceleration of the fluid. The convective terms, $V(\partial V/\partial x)$ and $g(\partial H/\partial x)$, relate to changes in the kinetic and potential energy, respectively. The terms $gS_f$ and $gS_0$ account for friction along the channel and the component of gravitational force in the direction of flow, respectively. The last term, $qV/A$, accounts for momentum that must be imparted to the lateral inflow by the water flowing in the channel.

The most important term in the dynamic equation is the friction term $S_f$. Since its magnitude is usually larger than any of the other terms except $S_0$, the method of evaluating $S_f$ is important. A uniform flow formula is generally used to determine $S_f$. Where flows are laminar, the Darcy-Weisbach formula is suitable.

$$S_f = \frac{f}{H}\frac{V^2}{2g} \tag{11-30}$$

Manning's equation is used for turbulent flows.

$$S_f = \frac{(nV)^2}{1.486R^{2/3}} \tag{11-31}$$

It is generally assumed that the value of $S_f$ occurring for gradually varied unsteady flow anywhere in a drainage basin is the same as the value of $S_f$ at the same velocity and hydraulic radius for uniform steady flow at that point. The error introduced by this assumption is probably less than the error involved in selecting proper values for the coefficients in the foregoing equations for $S_f$.[17]

Assumptions used in deriving the equations of flow normally include

1. Accelerations normal to the direction of flow have been neglected.
2. Velocities normal to the direction of flow have been neglected.
3. Velocities are assumed uniform throughout a section normal to the direction of flow. This assumption is more restrictive than necessary, since coefficients can be introduced into the dynamic equation to account for a variable velocity profile.
4. Frictional resistance is assumed to be the same as for steady uniform flow at the same velocity and depth of flow.

The importance of these assumptions has been thoroughly studied.[9,17]

The equations of flow written in finite difference form become

$$\frac{\Delta A}{\Delta t} + \frac{\Delta Q}{\Delta x} = q \tag{11-32}$$

$$\frac{\Delta V}{\Delta t} + \frac{\Delta V}{\Delta x} + g\frac{\Delta H}{\Delta x} + g(S_f - S_0) + \frac{qV}{A} = 0 \tag{11-33}$$

To be computationally useful, the solution of the finite difference equations should approach the solution of the partial differential equations as $\Delta x \to 0$ and $\Delta t \to 0$. The strategy followed is to divide the channel into a number of intervals of length, $\Delta x$, and then solve the difference equations for $A(x, t + \Delta t)$ and $V(x, t + \Delta t)$ at successive intervals of time. Schaake and others have applied such equations to modeling watershed runoff.[9,17]

## 11-5 Land Use Effects

In Sec. 4-1, basin characteristics affecting runoff were discussed. The effects of slope, area size, soil and rock structure, and other factors were illustrated. From these discussions it is easy to understand that modifications of the land surface have varying effects on the runoff characteristics of a given drainage area.

If a heavily forested area with its thick layer of mulch is converted to cropland or pasture, the soil is disturbed and the overlying absorptive cover is destroyed. The result is increased runoff volume and a change in the timing of flows. When lowlands or marshes are surface-drained, the flooding characteristics of these areas are modified. The drains serve to remove the water at an accelerated rate, thus increasing the peak flows and runoff volumes. Inasmuch as there is usually a significant linkage between low, swampy areas and the underlying underground system, this relationship is changed as well. The rapid removal of water from the drained area decreases the time; consequently, the opportunity for infiltration and the net effect is usually a lowering of the underlying water table. Changes in the vegetal cover affect the infiltration capacities of soils and land use changes that modify the nature of vegetation can have significant impact on the timing and volume of flows.

Urbanization of the land usually results in the highly accelerated removal of storm water with corresponding increases in the volume and peak rate of runoff. In many cases, infiltration might be all but eliminated and a very high percentage of the storm rainfall becomes runoff. On the other hand, by increasing an area's storage capacity and delaying the outflow, it is possible to actually increase the timing and delay the peak rate of runoff. For example, a shopping center parking lot can be graded and its drains sized to permit several inches of ponding during intense storms. This delays the downstream arrival of flows from the area and significantly reduces the hydrograph peaks. By understanding the effects of land use change on the hydrology of an area, it is possible to put this knowledge to beneficial use. Several aspects of this are discussed in Refs. 47 to 55.

The principal effects of land use change have been classified by Leopold as follows:[45] (1) changes in peak flow characteristics, (2) changes in total runoff, (3) changes in water quality, and (4) changes in hydrologic amenities (the appearance or impression a watercourse and its environment leaves with the observer).

### Change in Runoff Characteristics

Land use changes can increase or decrease the volume of runoff and the maximal rate and timing of flow from a given area. The most influential factors affecting flow volume are the infiltration rate and surface storage. Changes in interception and other factors are usually of negligible importance.[46] The peak rate of flow is related to the flow volume. This can be exemplified by the unit hydrograph principle which states that with other things constant, the peak flow rate varies directly with the volume of flow. This relationship is illustrated by Fig. 11-31. From this, it can be seen that land use practices which decrease flow volume also decrease the peak rate of flow, and vice versa.

Some effects of land use treatment on the direct runoff are shown in Table 11-13. The overall effect of any measure is a function of the extent to which that measure can be or is applied in a watershed or the way in which the measure is used. For example, contour furrows can be made to have a large or small effect on the direct runoff depending

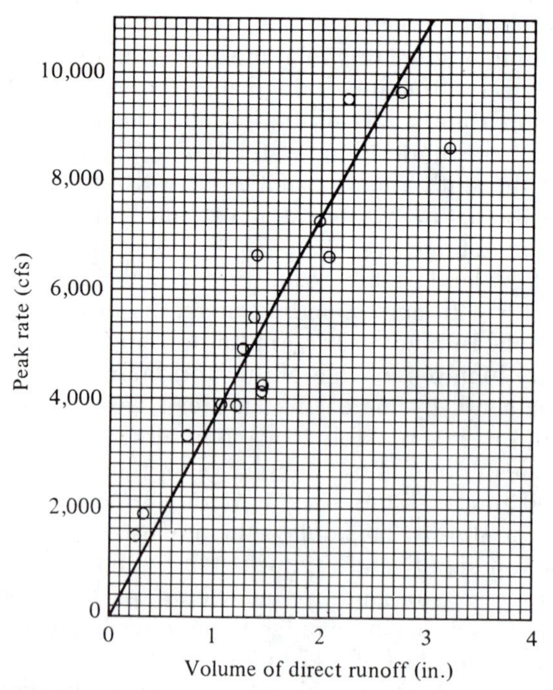


***Fig. 11-31.*** *Typical peak-volume relationship of annual floods at Eagle Creek, Indianapolis, Indiana. $A = 170$ mi² (After Victor Mockus, "Hydrologic Effects of Land Use and Treatment," Chap. 12, SCS National Engineering Handbook, Sec. 4, "Hydrology," U.S. Department of Agriculture, Soil Conservation Service, Washington, D.C., 1972.)*

***Table 11-13*** *Effects of Some Land Use and Treatment Measures on the Direct Runoff*

| Measure | Reduction in Direct Runoff Volume Is Due to | |
|---|---|---|
| | *Increasing Infiltration Rates*[a] | *Increasing Surface Storage* |
| Land use change that increases plant or root density[b] | X | |
| Increasing mulch or litter | X | |
| Contouring | | X |
| Contour furrowing | | X |
| Level terracing | | X |
| Graded terracing | | X |

[a] Assuming soils not frozen.
[b] Example: Row crop to grass for hay; poor pasture to good pasture.

*Source:* After Victor Mockus, "Hydrologic Effects of Land Use and Treatment," Chap. 12, *SCS National Engineering Handbook*, Sec. 4, "Hydrology," U.S. Department of Agriculture, Soil Conservation Service, Washington, D.C., 1972.

on the dimensions of the furrows. Changing the vegetative cover from spring oats to spring wheat would probably have little effect on runoff produced while a change from oats to a permanent meadow could have a significant effect.[46]

In addition to land use effects on the volume and rate of flow, lag effects (delay between upstream production of flow and its arrival at a downstream location) are also noticed. Land use and treatment measures can modify lag by (1) increasing or decreasing the infiltration (reducing or increasing the surface runoff), or (2) by increasing or decreasing the flow distance or flow velocity. Some of these effects are shown in Table 11-14.

The effects of urbanization require special mention, as they often have a pronounced impact on the characteristics of an area's hydrology. Urbanization generally increases the volume of the runoff and peak rate of flow and decreases the watershed's time lag. Figure 11-32 illustrates the effects on lag time and hydrograph peak for hypothetical unit hydrographs.[45] The runoff volume is determined mainly by infiltration and the nature of surface storage. The land slope, the soil type, the nature of the vegetative cover, and the degree of imperviousness of the watershed are all important factors. Figure 11-33 illustrates the combined effects of increased imperviousness and sewerage on the mean annual flood for a 1-$mi^2$ drainage area. An often overlooked but potentially important effect of increased runoff is the accompanying reduction in groundwater recharge. Where urban areas are expansive, local groundwater supplies can be seriously reduced.

***Table 11-14*** *Relative Effects of Land Use and Treatment Measures on Types of Lag*

| | Increasing Infiltration with Resultant Effect on Subsurface Flow[a] | | Effect of Increasing Surface Flow Distance or Decreasing Velocity | |
|---|---|---|---|---|
| *Measure* | *Small Watersheds* | *Large Watersheds* | *Small Watersheds* | *Large Watersheds* |
| Land use changes that increase plant or root density[b] | Can be large | Can be large | Not usually considered | |
| Increasing mulch or litter | Can be large | Can be large | Not usually considered | |
| Contouring | Can be large | Usually negligible | Can be large | Negligible |
| Contour furrowing | Can be large | Can be large | Not usually considered | |
| Level terracing | Can be large | Can be large | Not usually considered | |
| Graded terracing | Usually negligible | Usually negligible | Can be large | Negligible |

[a] Assuming soils not frozen.
[b] Examples: Row crop to grass; poor pasture to good pasture.

*Source:* Victor Mockus, "Hydraulic Effects of Land Use and Treatment," Chap. 12, *SCS National Engineering Handbook,* Sec. 4, "Hydrology," U.S. Department of Agriculture, Soil Conservation Service, Washington, D.C., 1972.

The peak rate of runoff will, in general, increase more rapidly than the volume of runoff as urbanization occurs. This is because of the increase in the rate of overland flow to stream channels and the resultant decrease in concentration time of the basin. Water flows more quickly from streets and roofs than from naturally vegetated areas and conveyances, such as storm sewers and lined open channels, increase the flow velocities and thus decrease the lag time. The reduction in time lag (or concentration time) of the basin is extremely important as it affects the frequency or return period for a given level of flow. For example, the storm that was found to be the "50-yr" storm on a basin having a 6-hr lag, will no longer be the "50-yr" storm if the lag is reduced to 3 hr by urbanization. A study of Fig. 11-34 illustrates this point.

### Water Quality

Changes in water quality due to land use practices can be either positive or negative. The principal effect of land use change on water quality is the introduction of waste materials such as nutrients, road salts, various chemicals, oil and gasoline products, and other materials. An especially important water quality problem is the rapid increase in sediment load in streams owing to the exposure of bare soil to storm runoff during and after periods of development. Urbanization has caused increases in sediment yield on the order of 100 to 250 times that of rural areas. Such increases result from the denuding of sites and the upsetting of balances of natural drainage networks to the flows they must carry. Streams tend to construct and maintain channels that exceed the bank-full stage at a recurrence interval of about 2 yr. If the number of flows above bank-full stage is increased due to urbanization, or other causes, the banks and bed of erodible channels will not remain stable but will be enlarged through erosion.

### Application

A knowledge of the manner in which land use changes and land treatments can modify the runoff process is extremely important. Various proposed changes can be simulated and their effect evalu-

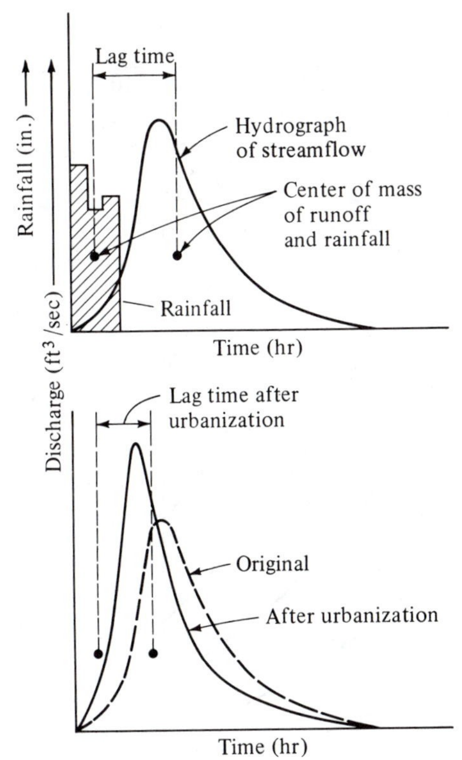


*Fig. 11-32.* *Hypothetical unit hydrographs relating the runoff to rainfall, with definitions of significant parameters. (U.S. Geological Survey Circular 554.)*

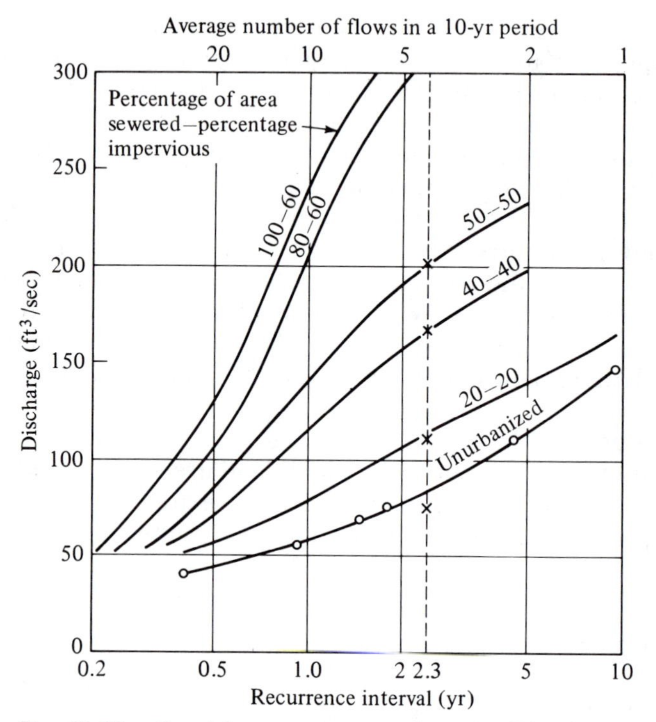


***Fig. 11-33.*** *Flood frequency curves for a 1-mi² basin in various states of urbanization. (U.S. Geological Survey Circular 554.)*

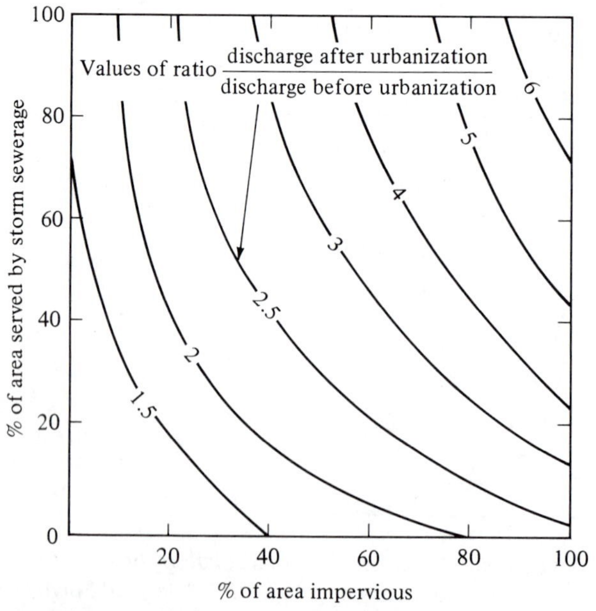


***Fig. 11-34.*** *Effect of urbanization on the mean annual flood for a 1-mi² drainage area. (U.S. Geological Survey Circular 554.)*

ated before decisions to implement these practices are made. Designs can be improved and features incorporated into traditional design practices that will save funds, reduce adverse environmental impacts, and even enhance the quality of life. New uses for excess flows such as recreational ponds, artificial recharge, urban irrigation, and others can be found. By considering the total water management instead of only the fast removal of storm water runoff, many positive impacts are obtainable. Table 11-15 summarizes some measures for modifying the runoff process.

**Table 11-15** *Measures for Reducing and Delaying Urban Storm Runoff*

| Area | Reducing Runoff | Delaying Runoff |
|---|---|---|
| Large flat roof | Cistern storage<br>Rooftop gardens<br>Pool storage or fountain storage<br>Sod roof cover | Ponding on roof by constricted downspouts<br>Increasing roof roughness<br>1. Rippled roof<br>2. Graveled roof |
| Parking lots | Porous pavement<br>1. Gravel parking lots<br>2. Porous or punctured asphalt<br>Concrete vaults and cisterns beneath parking lots in high value areas<br>Vegetated ponding areas around parking lots<br>Gravel trenches | Grassy strips on parking lots<br>Grassed waterways draining parking lot<br>Ponding and detention measures for impervious areas<br>1. Rippled pavement<br>2. Depressions<br>3. Basins |
| Residential | Cisterns for individual homes or groups of homes<br>Gravel driveways (porous)<br>Contoured landscape<br>Groundwater recharge<br>1. Perforated pipe<br>2. Gravel (sand)<br>3. Trench<br>4. Porous pipe<br>5. Dry wells<br>Vegetated depressions | Reservoir or detention basin<br>Planting a high delaying grass (high roughness)<br>Gravel driveways<br>Grassy gutters or channels<br>Increased length of travel of runoff by means of gutters, diversions, and so on |
| General | Gravel alleys<br>Porous sidewalks<br>Mulched planters | Gravel alleys |

*Source:* After U.S. Department of Agriculture, Soil Conservation Service, 1972.

## Problems

11-1. In using the rational formula, $Q = CIA$, for the design of any structure, what do the terms $Q$, $C$, $I$, and $A$, represent? In selecting a $T$-yr design storm intensity, why are the rainfall duration and time of concentration equated? Answer by noting the effect of selecting duration less than, and greater than, the time of concentration.

11-2. A 53,200-acre area has a $\phi$-index of 0.10 in./hr. A storm with a constant rainfall rate of 0.7 in./hr lasts for 6 hr.

(a) What is the Rational Formula peak discharge in cfs if the time of concentration is 4 hr?

(b) What is the runoff rate (cfs) at the end of the fifth hour after the rainfall begins?

11-3. The time of concentration for a 6-acre parking lot is 20 min. Which of the following storms gives the greatest peak rate of runoff by the Rational Formula ($Q = CIA$) if $C = 0.6$? State any assumptions.

(a) 4 in./hr for 10-min duration.

(b) 1 in./hr for 40-min duration.

11-4. A 10.0-mi² drainage basin has a time of concentration of 100 min and a constant $\phi$-index of 0.25 in./hr. If rain falls uniformly over the basin at a rate of 2.75 in./hr for a duration of 200 min, sketch the approximate hydrograph and determine the maximum discharge rate (cfs) at the basin outlet.

11-5. A 10.0-mi² drainage basin with a time of concentration of 100 min receives rainfall at a rate of 2.75 in./hr for a period of 200 min. If the runoff coefficient is 0.8, determine the discharge rate (cfs) from the basin 130 min after the rain began.

11-6. In using the Rational Formula ($Q = CIA$) for peak discharge rates, what does the product $CI$ represent? What is the meaning of $I$ in terms of the design discharge and rainfall duration?

11-7. The 4-hr unit hydrograph for a 5600-acre watershed is

| Time (hr) | 0 | 2 | 4 | 6 | 8 | 10 | 12 |
|---|---|---|---|---|---|---|---|
| $Q$ (cfs) | 0 | 400 | 1000 | 800 | 400 | 200 | 0 |

The local 10-yr IDF curve is linear with the equation $I = 5.6 - 0.2D$, where $I$ is rain intensity in in./hr and $D$ is the rain duration in hours. Use the unit hydrograph to determine the peak 10-yr flow rate for the watershed. Compare this with the Rational Formula estimate of the 10-yr peak. ($\phi = 1.0$ in./hr.) Note that $t_c$ is the time from the cessation of rain to the cessation of runoff.

11-8. Rework Example 11-2 based on a $C = 0.2$ and $C = 0.4$. Compare and discuss the effect of $C$ on the discharge at the outfall.

11-9. A watershed has area $A$. Starting with a triangular-shaped unit hydrograph with a base length of $2.67t_p$ and a height of $Q_p$, derive Eq. 4-37, $Q_p = 484A/t_p$. State and carry units of each term used in the derivation.

11-10. Use both the Rational and the SCS peak flow Eq. 11-3 to determine the 5-yr design discharge (cfs) for a storm drain that receives runoff from a 300-acre area having a length of 1200 ft, a Manning $n$ value of 0.050, a slope of 0.001 ft/ft, and a runoff coefficient of 0.60. Use a lag time of $0.6t_c$

and the Kirpich formula for $t_c$; also, use the Fig. 11-7 IDF curves for predicting needed precipitation depths.

11-11. Using the SCS dimensionless unit hydrograph described in Chapter 4, determine the peak discharge for a net storm of 10 in. in 2 hr on a 400-acre basin with a time to peak of 4 hr and a lag time of 3 hr. Compare with Eq. 4-34.

11-12. A 10.00-mi² watershed with a 100-min time of concentration receives rainfall at a rate of 2.75 in./hr for a period of 200 min.

(a) Determine the peak discharge (cfs) from the watershed if $C = 0.4$.
(b) Determine the discharge rate (cfs) 150 min after the beginning of rainfall.
(c) Using the sketch, compute the discharge rate from the watershed 40 min after the beginning of rainfall.

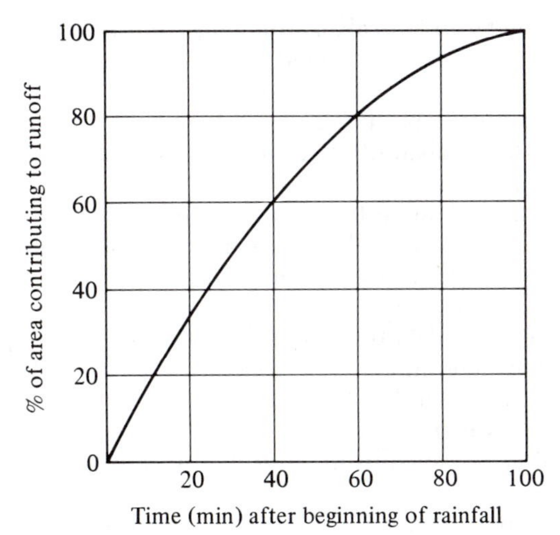


***Fig. 11-A.*** *Time-area curve for problem 11-12.*

11-13. A storm gutter receives drainage from both sides. On the left it drains a rectangular 600-acre area of $t_c = 60$ min. On the right it drains a relatively steep 300-acre area of $t_c = 10$ min. The $\phi$-index on both sides is 0.5 in./hr. Use the Fig. 11-7 intensity-duration-frequency curves to determine the peak discharge (cfs) with a 25-yr recurrence interval for (a) the 600-acre area alone, (b) the 300-acre area alone, and (c) the combined area assuming that the proportion of the 600-acre area contributing to runoff at any time $t$ after rain begins is $t/60$.

11-14. A drainage basin has a time of concentration of 8 hr and produces a peak $Q$ of 4032 cfs for a 10-hr storm with a net intensity of 2 in./hr. Determine the peak flow rate and the time base (duration) of the direct surface runoff for a net rain of 4 in./hr lasting (a) 12 hr, (b) 8 hr, and (c) 4 hr. State any assumptions used.

11-15. A 1.0-mi² parking lot has a runoff coefficient of 0.8 and a time of concentration of 40 min. For the following three rainstorms, determine the peak discharge (cfs) by the Rational Method: (a) 4.0 in./hr for 10 min, (b)

1.0 in./hr for 40 min, and (c) 0.5 in./hr for 60 min. State any assumptions regarding area contributing after various rainfall durations.

11-16. The concentration time varies with discharge but is relatively constant for large discharges. From this statement, why do engineers feel confident in using the Rational Formula?

11-17. Determine the 50-yr flood for a 20-$mi^2$ basin at the northwest corner of Nebraska. Use the index flood method and assume that Fig. 6-5c applies.

11-18. Determine and plot the entire frequency curve for the basin in Prob. 11-17 on Gumbel plotting paper.

11-19. Use the index flood method to determine the 10-yr and 50-yr peaks for a 6400-acre drainage basin near Lincoln, Neb. Assume that Fig. 6-5c applies.

11-20. For the drainage basin in Prob. 11-19 determine the probability that the 20-yr peak will be equaled or exceeded at least once (a) next year, and (b) in a 4-yr period.

11-21. For a 100-$mi^2$ drainage basin near Lincoln, Neb., use the index flood method to determine the probability that next year's flood will equal or exceed 3000 cfs.

11-22. Use Fig. 6-5c to determine the return period (yr) of the mean annual flood for that region. How does this compare with the theoretical value for a Gumbel distribution? How does it compare with a normal distribution?

11-23. Use the Potter method to determine the 50-yr peak (cfs) for a watershed if the 10-yr peak is determined to be 4000 cfs.

11-24. You are asked to determine the magnitude of the 50-yr flood for a small, rural drainage basin (near your town) that has no streamflow records. State the names of at least two techniques that would provide estimates of the desired value.

11-25. The drainage areas, channel lengths, and relevant elevations (underlined) for several subbasins of the Oak Creek Watershed at Lincoln, Neb., are shown in Fig. 10-31. The watershed has a SCS curve number of $CN = 75$ which may be used in Fig. 12-25 to determine the direct runoff for any storm. The IDF curves in Fig. 5-12 apply at Lincoln. Treat the entire watershed as a single basin and determine the 50-yr flood magnitude at point 8 using:

(a) The Rational Method.
(b) The SCS peak flow graph, Fig. 11-8.
(c) Snyder's method of synthetic unit hydrographs, Eq. 4-34.
(d) SCS method of synthetic unit hydrographs, Eq. 4-37.
(e) The USGS index flood method. Figure 6-5c applies.
(f) The SCS Cook Method.
(g) The USBPR Potter Method (extrapolation of Fig. 11-14 will be necessary).

11-26. Repeat Prob. 11-25 with subarea I excluded. Compare the results with Prob. 11-25 and comment on the effectiveness at point 8 for the 50-yr event of the Branched Oak reservoir at point 9. (This reservoir will easily store the 100-yr flood from area I).

11-27. How could the elongated shape of the watershed of Prob. 11-26 affect the results at point 8 if techniques a through g were all developed for ovoid (pear)-shaped watersheds?

11-28. How could synthesis and routing techniques be incorporated to account for the elongated shape? Hint: See Chapter 10.

11-29. Repeat Prob. 11-25 for subarea A.

11-30. Repeat Prob. 11-25 for subarea I.

11-31. Determine the maximum and minimum infiltration rates and the recession constant $k$ (give units) for the industrial-and-commercial-area curve in Fig. 11-11. Compare these with default values used in SWMM.

11-32. Do the values compared in Table 11-9 support the conclusions drawn by Heeps and Mein?

11-33. Do the Heeps and Mein conclusions regarding RRL, SWMM, and UCURM agrea with the conclusions drawn by Marsalek et al?

11-34. Prepare a table similar to Table 11-12 but use the subjective terms "neglects," "adequate," "too complex," or "poorly modeled" to evaluate how adequately each of the three models handles each of the processes in column 1 of Table 11-6.

## References

1. Amorocho, J., and W. E. Hart, "A Critique of Current Methods in Hydrologic Systems Investigations," *Trans. Amer. Geophys. Union,* 45, No. 2 (June 1964): 307–321.
2. Singh, K. P., "Nonlinear Instantaneous Unit-Hydrograph Theory," *ASCE, J. Hyd. Div.,* 90, No. HY2, Part I (Mar. 1964): 313–347.
3. Sittner, W. T., C. E. Schauss, and J. C. Monro, "Continuous Hydrograph Synthesis with an API-Type Hydrologic Model," Water Resources Research, 5, No. 5 (1969): 1007–1022.
4. Nash, J. E., "The Form of the Instantaneous Unit Hydrograph," *International Association of Science Hydrology,* 3, No. 45 (1957): 114–121.
5. Dawdy, D. R., and T. O'Donnel, "Mathematical Models of Catchment Behavior," *ASCE, J. Hyd. Div.,* 91, No. HY 4, Proc. Paper 4410 (July 1965): 124–127.
6. Jacoby, S. L. S., "A Mathematical Model for Nonlinear Hydrologic Systems," *J. Geophy. Res.,* 71, No. 20 (Oct. 1966): 4811–4824.
7. Prasad, R., "A Nonlinear Hydrologic System Response Model," *ASCE, J. Hyd. Div.,* 93, No. HY4 (1967).
8. Tholin, A. L., and C. T. Keifer, "Hydrology of Urban Runoff," *J. ASCE,* 85 (Mar. 1959): 47–106.
9. Crawford, N. H., and R. K. Linsley, Jr., "Digital Simulation in Hydrology: Stanford Watershed Model IV," Department of Civil Engineering, Stanford University, Stanford, Calif., Tech. Rept. No. 39, July 1966.
10. American Public Works Association, "Water Pollution Aspects of Urban Runoff," Federal Water Pollution Control Administration, 1969.
11. American Society of Civil Engineers, First Year Report, "Urban Water Resources Research," Sept. 1968.
12. Viessman, W., Jr., "Modeling of Water Quality Inputs from Urbanized Areas," Urban Water Resources Research, Study by ASCE Urban Hydrology Research Council, Sept. 1968, pp. A79–A103.
13. Weible, S. R., R. B. Weidner, A. G. Christianson, and R. J. Anderson,

"Characterization, Treatment, and Disposal of Urban Storm Water," Proceedings of the Third International Conference, International Association on Water Pollution Research, S. H. Jenkins, ed. (Elmsford, N.Y.: Pergamon Press, 1969).

14. Weible, S. R., R. B. Weidner, J. M. Cohan, and A. G. Christianson, "Pesticides and Other Contaminants in Rainfall and Runoff," *Journal of the American Water Works Association,* 58, No. 8 (Aug. 1966): 1675.
15. Division of Water Resources, Department of Civil Engineering, University of Cincinnati, Cincinnati, Ohio, "Urban Runoff Characteristics," Water Pollution Control Research Series, EPA, 1970.
16. Metcalf and Eddy, Inc., University of Florida, Gainesville, Fla., Water Resources Engineers, Inc., "Storm Water Management Model," Environmental Protection Agency, Vol. 1, 1971.
17. Schaake, John C., Jr., "Synthesis of the Inlet Hydrograph," Tech. Rept. 3, Storm Drainage Research Project, Johns Hopkins University, Baltimore, Md, June 1965.
18. Kuichling, E., "The Relation Between the Rainfall and the Discharge of Sewers in Populous Districts," *Trans. ASCE,* 20 (1889).
19. Horner, W. W., "Modern Procedure in District Sewer Design," *Eng. News,* 64 (1910): 326.
20. Horner, W. W., and F. L. Flynt, "Relation Between Rainfall and Runoff from Small Urban Areas," *Trans. ASCE,* 20, No. 140 (1936).
21. Schaake, J. C., Jr., J. C. Geyer, and J. W. Knapp, "Experimental Examination of the Rational Method," *Proc. ASCE, J. Hyd. Div.,* 93, No. HY6 (Nov. 1967).
22. "Airport Drainage," Federal Aviation Agency, Department of Transportation, Advisory Circular, A/C 150-5320-5B, Government Printing Office: Washington, D.C., 1970.
23. "Design and Construction of Sanitary Storm Sewers," *ASCE Manuals and Reports on Engineering Practice,* No. 37, 1970.
24. "Urban Hydrology for Small Watersheds," U.S. Soil Conservation Service, Tech. Release No. 55, Jan. 1975.
25. Tholin, A. L., and C. J. Keifer, "Hydrology of Urban Runoff," *Trans. ASCE,* Paper No. 3061, Vol. 125 (1960).
26. "Rainfall Frequency Atlas of the United States for Durations from 30 Minutes to 24 Hours and Return Periods from 1 to 100 Years," U.S. Weather Bureau Tech. Paper 40, May 1961.
27. Gray, D. M. (ed.), *Handbook on the Principles of Hydrology* (Port Washington, N.Y., Water Information Center, Inc., 1973).
28. Chow, Ven Te (ed.), *Handbook of Applied Hydrology* (New York: McGraw-Hill Book Company, 1964).
29. Jarvis, C. S., "Floods," Chap. 11-G, in O. E. Meinzer (ed.), *Hydrology* (New York: McGraw-Hill Book Company, Inc., 1942).
30. Furness, L. W., "Floods in Nebraska, Magnitude and Frequency," U.S. Geological Survey Rept. to Nebraska Department of Roads and Irrigation, Apr. 1955.
31. Hamilton, C. L., and H. G. Jepson, "Stock-Water Developments: Wells, Springs, and Ponds," U.S. Department of Agriculture, Farmers' Bulletin No. 1859, 1940.

32. Potter, W. D., "Peak Rates of Runoff from Small Watersheds," Hydraulic Design Series No. 2, Bureau of Public Roads (Washington, D.C.: Government Printing Office, Apr. 1961).
33. Tholin, A. L., and C. T. Keifer, "Hydrology of Urban Runoff," *ASCE, J. San. Engr. Div.*, 85, No. SA2 (Mar. 1959): 47–106.
34. Izzard, C. F., "Hydraulics of Runoff from Developed Surfaces," Proceedings 26th Annual Meeting Highway Research Board, 26 (1946): 129–146.
35. Keifer, C. J., J. P. Harrison, and T. O. Hixson, "Chicago Hydrograph Method Network Analysis of Runoff Computations," Preliminary Report, City of Chicago, Bureau of Engineering, July 1970.
36. Watkins, L. H., *The Design of Urban Sewer Systems*, Road Research Tech. Paper No. 55, Department of Scientific and Industrial Research (London: Her Majesty's Stationery Office, 1962).
37. Terstriep, Michael L., and John B. Stall, "Urban Runoff by the Road Research Laboratory Method," *ASCE,J. Hyd. Div.*, 95, No. HY6 (Nov. 1969): 1809–1834.
38. Stall, John B., and Michael L. Terstriep, "Storm Sewer Design—An Evaluation of the RRL Method," EPA Technology Series EPA-R2-72-068, Oct. 1972.
39. Papadakis, C. N., and H. C. Preul, "University of Cincinnati Urban Runoff Model," *Proc. ASCE, J. Hyd. Div.*, 98, No. HY10 (Oct. 1972): 1789–1804.
40. Heeps, D. P., and R. G. Mein, "Independent Comparison of Three Urban Runoff Models," *Proc. ASCE, J. Hyd. Div.*, 100, No. HY7 (July 1974): 995–1010.
41. Marsalek, J., T. M. Dick, P. E. Wisner, and W. G. Clarke, "Comparative Evaluation of Three Urban Runoff Models," Water Resources Bulletin, AWRA, 11, No. 2 (Apr. 1975): 306–328.
42. Terstriep, M. L., and J. B. Stall, "The Illinois Urban Drainage Area Simulator, ILLUDAS," Illinois State Water Survey Bulletin 58, 1974.
43. Metcalf and Eddy, Inc., University of Florida, Gainesville, Fla., Water Resources Engineers, Inc., "Storm Water Management Model," Environmental Protection Agency, Vol. 1, 1971.
44. Linsley, R. K., Jr., M. A. Kohler, and J. A. H. Paulhus, *Applied Hydrology* (New York: McGraw-Hill Book Company, 1949).
45. Leopold, Luna B., "Hydrology for Urban Land Planning," USGS Circular No. 554, U.S. Government Printing Office, Washington, D.C., 1968.
46. Mockus, Victor, "Hydrologic Effect of Land Use and Treatment," Chap. 12, *SCS National Engineering Handbook*, Sec. 4, "Hydrology," U.S. Department of Agriculture, Soil Conservation Service, Washington, D.C., 1972.
47. Anderson, D. G., "Effects of Urban Development on Floods in Northern Virginia," U.S. Geol. Survey open-file report, 1968.
48. Espey, W. H., C. W. Morgan, and F. D. Masch, "Study of Some Effects of Urbanization on Storm Runoff from a Small Watershed," Texas Water Development Board, Rept. No. 23, 1966.
49. Felton, P. N., and H. W. Lull, "Surburban Hydrology Can Improve Watershed Conditions," *Public Works*, 94, 1963.
50. Harris, E. E., and S. E. Tantz, "Effect of Urban Growth on Streamflow

Regimen of Permanente Creek, Santa Clara County, California," U.S. Geological Survey Water Supply Paper 1591-B, 1964.

51. James, L. D., "Using a Computer to Estimate the Effects of Urban Development of Flood Peaks," *Water Resources Research*, 1, No. 2, 1965.
52. Keller, F. J., "The Effect of Urban Growth on Sediment Discharge," Northwest Branch Anacostia River Basin, Maryland, in "Short Papers in Geology and Hydrology," U.S. Geological Survey Professional Paper 450-C, 1962.
53. Lull, H. W., and W. E. Sopper, "Hydrologic Effects from Urbanization of Forested Watersheds in the Northeast," Upper Darby, Pa., Northeastern Forest Experimental Station, 1966.
54. Wolman, M. G., and P. A. Schick, "Effects of Construction on Fluvial Sediment, Urban and Suburban Areas of Maryland," Water Resources Research, Vol. 3, No. 2, 1967.
55. Bras, R. L., and F. E. Perkins, "Effects of Urbanization on Catchment Response," *ASCE, J. Hyd. Div.*, 101, No. HY3 (Mar. 1975).

# 12

# Hydrologic Design

Predicting peak discharge rates or synthesizing complete discharge hydrographs for use in designing minor and major structures are two of the more challenging aspects of engineering hydrology. Minor types of hydraulic structures range from small crossroad culverts, levees, drainage ditches, urban storm drain systems, and airport drainage structures to the spillway appurtenances of small dams. When lumped together with major structures, all require varying amounts of hydrologic design information. Generally, a hydrologist is required to provide peak rates of discharge for a design frequency, a stage height at a design frequency, or synthesize a complete discharge hydrograph for a design storm. Other information such as sedimentation rates, low flow frequency analysis, abstractions, and reservoir yield studies are also a part of the hydrologist's notebook of skills.

Most designs involving hydrologic analyses utilize a design or critical flood that imitates some severe future or historical event. If stream flow records are unavailable, design flood hydrographs are synthesized from available storm records using the rainfall-runoff procedures of Chapters 3, 10, and 11. Only in rare cases are streamflow records adequate for complete designs, particularly in small watersheds. Regional analyses and the empiric-correlative methods discussed in Chapter 11 are useful for determining peak flow rates at ungauged sites. Methods presented in Chapter 4 and in this chapter are used for developing entire hydrographs necessary for many engineering designs.

Hydrologic methods for designing minor and major structures are described in this chapter. Included are discussions of data needs, frequency levels, and methods for synthesizing design storms.

Several steps in the hydrologic approach to *minor* structure design are common to most design handbooks and adopted techniques. The general steps (each is illustrated subsequently) are

1. Determine the duration of the critical storm, usually equated to the time of concentration of the watershed.
2. Choose the design frequency.
3. Obtain the storm depth based upon the selected frequency and duration.
4. Compute the direct runoff (several methods were presented in Chapter 2, and one more is included here).
5. Select the time distribution of the rainfall excess.
6. Synthesize the unit hydrograph for the watershed (see Chapter 3).
7. Pick the probability level of the storm pattern.
8. Apply the derived rainfall excess pattern to the synthetic unit hydrograph to get the storm hydrograph.

Hydrologic design aspects of a *major* structure are considerably more complex than those of a small dam, crossroad culvert, or urban drainage system. A design storm hydrograph for a large dam still is required but it is put to greater use. The design storm hydrograph is routed to determine the adequacy of spillways and outlets operated in conjunction with reservoir storage. The economic selection of the spillway size from the various possibilities dictates the final design and is a function of the degree of protection provided for downstream life and property, project economy, agency policy and construction standards, and reservoir operational requirements.

Most information and techniques presented in this chapter are directed toward the flood protection aspect of small and large structures. Needless to say, a major structure is designed for more than just flood protection; they are multipurpose and may provide storage for irrigation, power, water supply, navigation, and low flow augmentation. The proper allocation of storage to these uses requires an understanding of the entire streamflow history in terms of the frequency of occurrence of low flows, average monthly, seasonal, and yearly flows, as well as the historical and design floods. Material was presented in Chapter 10 to provide a hydrologist with the tools to develop complete streamflow histories for a complete multiple-purpose system involving various combinations of minor and major structures, water development projects, and management practices.

## 12-1 Data for Hydrologic Design

The design of any structure requires a certain degree of data if only a field estimate of the drainage area and a description of terrain type and cover. The following material identifies some general data types and sources.

### Physiographic Data

The hydrologic study for any structure always requires a reliable topographic map. United States Geological Survey topographic maps are usually available. The mapping of the United States is almost complete with 15-min quadrangles and many of these areas are mapped by $7\frac{1}{2}$-min quadrangles. County maps and aerial photos can also be utilized to advantage in making preliminary studies of the watershed.

Based on an area map, a careful investigation of the watershed's drainage behavior must be made. Additional information can be obtained from Geological Survey maps which depict predominant rock formations. Soil types and the infiltration and erosive characteristics of soils can be secured from U.S. Soil Conservation districts or University Extension divisions.

The drainage areas contributing to large dams require stricter analysis of an area's hydrology than is necessary in designing minor structures. The possibility of a uniformly intense rainfall over the entire basin is an unrealistic assumption for large watersheds. The influence of temporal and spatial variations of the rainfall should thus be considered. For major dams, the estimated "worst possible" rainfall values are generally converted to a design discharge hydrograph which is then used in reservoir routing calculations to proportion reservoir and spillway size, surcharge storage, and any additional outlets needed to maintain power requirements or sustained downstream flow for navigation, irrigation, or water supply. The basic concern in hydrologic design of a large dam is to preserve downstream interests and have a realistic estimate for the design storm hydrograph.

Topographic map detail necessarily shifts with the type and purpose of the structure being designed. Careful field reconnaissance always increases the understanding of an area's hydrology no matter how insignificant the structure might be.

### Hydrologic Data

One difficulty in hydrologic design is that of getting adequate data for the region under study. Considerable data can be acquired from previously published reports issued by governmental agencies and/or universities. The following is a list of federal agencies that gather hydrologic data in a usable form:

1. Agricultural Research Service
2. Soil Conservation Service
3. Forest Service
4. U.S. Army Corps of Engineers
5. National Oceanic and Atmospheric Administration
6. Bureau of Reclamation

7. Department of Transportation
8. U.S. Geological Survey, Topographic Division
9. U.S. Geological Survey, Water Resources Division

Additional data can often be procured from departments of state governments, interstate compact commissions, and regional and local agencies.

Hydrologic data necessary to design large dams stem from the need for a more intensive examination of rainfall, storm pattern (both temporal and spatial), and runoff records. Drainage areas contributing to the runoff can vary from 1000 to as many as 200,000 mi². A sample of reservoir sizes in scattered locations throughout the United States is given in Table 12-1.

***Table 12-1*** *Largest Reservoirs of 10 mi² or More in the United States*

| State | Reservoir and Stream | Surface Area (mi²) | Reservoir Capacity (acre-ft) |
|---|---|---|---|
| Alabama | Guntersville (Tennessee River) | 110 | 1,018,700 |
| Arizona | Lake Mead (Colorado River)[a] | 229 | 29,040,000 |
| Arkansas | Bull Shoals (White River) | 111 | 5,408,000 |
| California | Shasta (Sacramento River) | 46 | 4,500,000 |
| Colorado | John Martin (Arkansas River) | 29 | N.A.[i] |
| Florida | Conservation area No. 1 (Everglades) | 216 | N.A. |
| Georgia | Clark Hill (Savannah River)[b] | 123 | 2,900,000 |
| Idaho | American Falls (Snake River) | 88 | 1,700,000 |
| Illinois | Crab Orchard (Crab Orchard Creek)[c] | 11 | N.A. |
| Iowa | Coralville (Iowa River)[c] | 39 | N.A. |
| Kansas | Tuttle Creek (Big Blue River) | 83 | 2,280,000 |
| Kentucky | Kentucky (Tennessee River) | 407 | 6,002,600 |
| Louisiana | Bayou Bodcau (Bayou Bodcau) | 33 | N.A. |
| Maryland | Conowingo (Susquehanna River) | 13 | N.A. |
| Massachusetts | Quabbin (Swift River) | 39 | 1,265,000 |
| Michigan | Fletcher Pond, Upper South Branch (Thunder Bay River) | 13 | N.A. |
| Mississippi | Grenada (Yalobusha River) | 101 | 1,337,400 |
| Missouri | Lakes of the Ozarks (Bagnall), (Osage River) | 93 | 1,973,000 |
| Montana | Fort Peck (Missouri River) | 382 | 19,400,000 |
| Nebraska | McConaughy (Kingsley), (North Platte River) | 50 | 2,000,000 |
| Nevada | Lake Mead (Colorado River)[d] | 229 | 29,827,000 |
| New Mexico | Elephant Butte (Rio Grande) | 57 | 2,185,400 |
| New York | Sacandaga (Sacandaga River) | 42 | 690,000 |
| North Carolina | High Rock (Yadkin River) | 24 | N.A. |
| North Dakota | Garrison (Missouri River) | 610 | 23,000,000 |

## 12-2 Frequency Levels in Hydrologic Design

### Minor Structures

The design frequencies shown in Table 12-2 are typical of levels generally encountered in minor structure design. An example of variations that do occur is the design frequency of a culvert which under cases of excessive backwater could effectively halt interstate traffic. Because of the economic consequences of such an event, these structures are generally designed for higher frequencies.

***Table 12-1*** *(Cont.)*

| State | Reservoir and Stream | Surface Area (mi²) | Reservoir Capacity (acre-ft) |
|---|---|---|---|
| Ohio | St. Marys (Grand), (Beaver and Jennings Creeks) | 21 | N.A. |
| Oklahoma | Lake Texoma (Denison), (Red River)[e] | 223 | 5,530,000 |
| Oregon | McNary (Columbia River)[f] | 59 | 1,345,000 |
| Pennsylvania | Pymatuning (Shenango River) | 26 | N.A. |
| South Carolina | Lake Marion (Santee River) | 172 | 1,350,000 |
| South Dakota | Oahe (Missouri River) | 587 | 23,600,000 |
| Tennessee | Pickwick Landing (Tennessee River) | 73 | 1,091,400 |
| Texas | Lake Texoma (Denison), (Red River)[g] | 223 | 5,530,000 |
| Utah | Sevier Bridge (Sevier River)[h] | 17 | N.A. |
| Virginia | John H. Kerr (Roanoke River) | 149 | 2,808,400 |
| Washington | Franklin D. Roosevelt (Grand Coulee), (Columbia River) | 130 | 9,402,000 |
| West Virginia | Blue Stone (New River) | 14 | N.A. |
| Wisconsin | Petenwell (Wisconsin River) | 36 | N.A. |
| Wyoming | Boysen (Big Horn River) | 35 | 970,000 |

[a] Arizona and Nevada. The largest reservoir wholly within Arizona is San Carlos (Coolidge) Reservoir on the Gila River, surface area, 31 mi².
[b] Georgia and South Carolina. The largest reservoir wholly within Georgia is Buford Reservoir (Lake Sidney Lanier) on Chattahoochee River, surface area, 59 mi².
[c] Keokuk Lock and Dam No. 19 on the Mississippi River, common to Illinois and Iowa, has a surface area of 44 mi².
[d] Nevada and Arizona. The largest reservoir wholly within Nevada is Rye Patch Reservoir on the Humboldt River, surface area, 17 mi².
[e] Oklahoma and Texas. Eufaula Reservoir under construction on the Canadian River, and wholly within Oklahoma, may have a surface area as large as that of Lake Texoma, or slightly larger.
[f] Oregon and Washington. The largest reservoir wholly within Oregon is Owyhee Reservoir on Owyhee River, surface area, 20 mi².
[g] Texas and Oklahoma. McGhee Bend Reservoir under construction on Angelina River and wholly within Texas, may have a larger surface area than Lake Texoma.
[h] Lake Powell under construction on the Colorado River (dam is in Arizona, but most of reservoir is in Utah) will have a surface area of 250 mi².
[i] N.A. = not available.

*Source:* U.S. Geological Survey, 1963.

**Table 12-2** *Minor Structure Design Frequencies*

| Type of Minor Structure | | Return Period, $T_r$ | Frequency $= 1/T_r$ |
|---|---|---|---|
| Highway crossroad drainage | | | |
| 0–400 | ADT | 10 yr | 0.10 |
| 400–1,700 | ADT | 10–25 yr | 0.10–0.04 |
| 1,700–5,000 | ADT | 25 yr | 0.04 |
| 5,000– | ADT | 50 yr | 0.02 |
| Airfields | | 5 yr | 0.20 |
| Storm drainage | | 2–10 yr | 0.50–0.10 |
| Levees | | 2–50 yr | 0.50–0.02 |
| Drainage ditches | | 5–50 yr | 0.20–0.02 |

### Dams

Critical in the design of a structure such as a dam is the possibility of failure. The sudden release of large volumes of impounded water can create damages even far greater than those experienced prior to construction. Damages from several dam failures are summarized in Table 12-3. Projected damages are illustrated in Table 12-4.

Initial heights of retarded water behind the dam, disregarding the total volume of stored water, can produce destructive flood waves for a considerable distance downstream. Based upon these two criteria, the Task Force on Spillway Design Floods[1] recommended the classification of dams as listed in Table 12-5. The type of construction has not been included in this grouping although it governs the extent of failure resulting from overtopping.

### Small Dams

Small dams generally receive special attention if they are constructed in populated areas where dam failure could cause the loss of life. As shown in Table 12-6, many flood deaths have been caused by dam or levee failure. When this possibility exists, the design storm for portions of small dams is established by use of the probable maximum precipitation or PMP. The PMP is defined as the reasonable maximization of the meteorological factors that operate to produce a maximum storm.

Estimates of PMP are based on an investigation by the Environmental Science Service Administration conducted to establish the maximum possible amount of precipitable water that could be achieved throughout the United States.[2,3] Point values of PMP for a 6-hr storm are shown in Fig. 12-1. This chart is helpful in selecting the worst possible conditions to be expected for a given region in the United States.

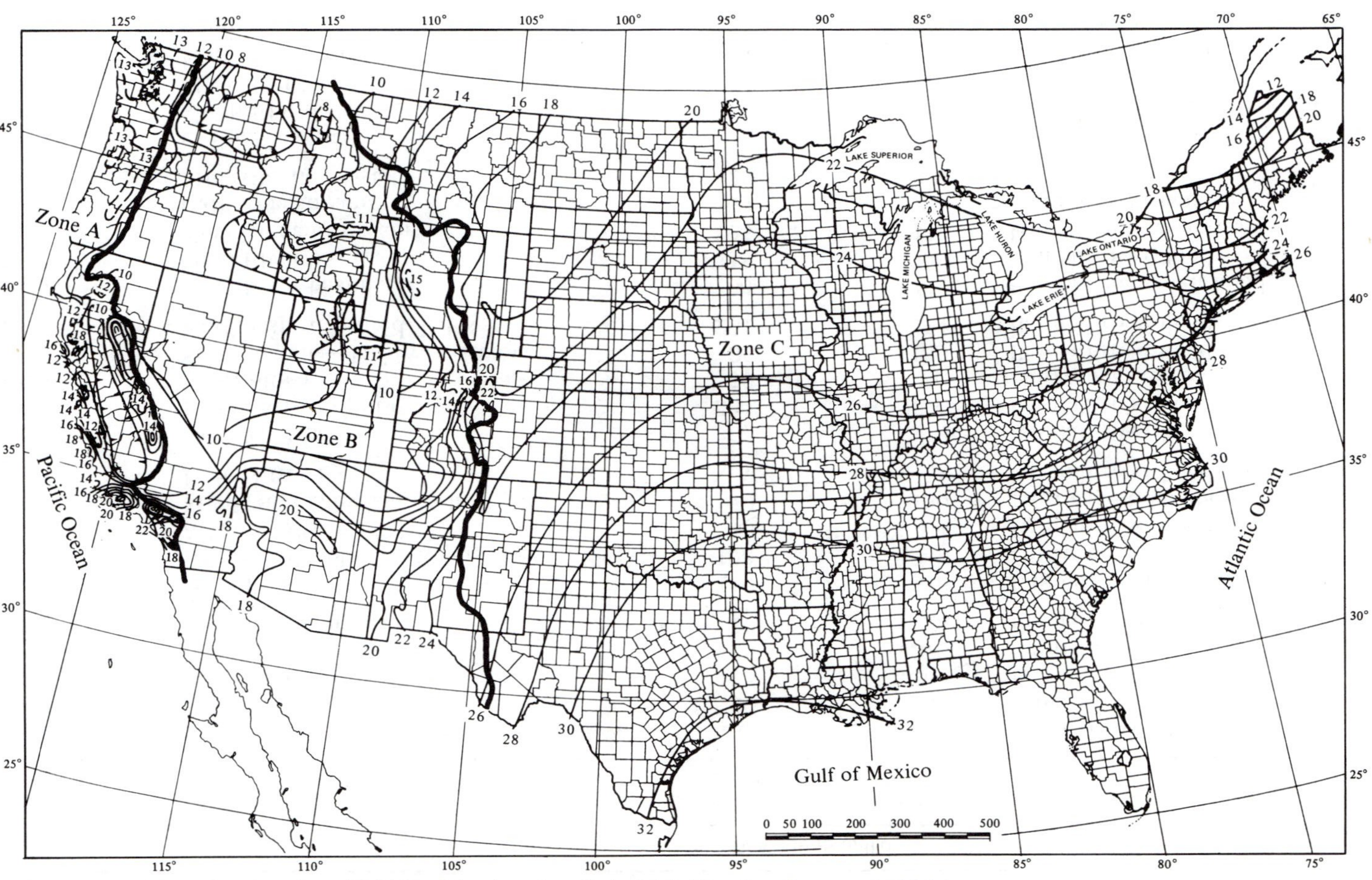


**Fig. 12-1.** *The 10-mi² or less PMP for 6-hr duration (in.). (U.S. Weather Bureau, NOAA.)*

**Table 12-3** *Floods Resulting in Damages Exceeding $50 Million in the United States*

| Year | Stream or Place | Damage ($ millions) | | Cause |
|---|---|---|---|---|
| | | *Contemporary Dollars* | *1966 Dollars* | |
| 1844 | Upper Mississippi River | N.A. | 1,161 | Rainfall-river flood |
| 1889 | Johnstown, Pa. | 20 | 84 | Dam failure |
| 1900 | Galveston, Tex. | 25 | 100 | Hurricane tidal floods |
| 1903 | Passaic and Delaware Rivers | 25 | 273 | Rainfall and dam failure |
| 1903 | Missouri River basin | 50 | N.A. | Rainfall-river flood |
| 1913 | Ohio River basin | 150 | 516 | Rainfall-river flood |
| 1913 | Brazos and Colorado Rivers, Tex. | 128 | 349 | Hurricane rainfall-river floods |
| 1921 | Arkansas River | 13 | 64 | Rainfall-river flood |
| 1926 | Miami and Clewiston, Fla. | 70 | 130 | Hurricane-tidal and river floods |
| 1926 | Illinois River | N.A. | 51 | Rainfall-river floods |
| 1927 | New England | 50 | 178 | Rainfall-river flood |
| 1927 | Lower Mississippi | 284 | N.A. | Rainfall-river flood |
| 1928 | Puerto Rico | 50 | 90 | Hurricane tide and waves |
| 1935 | Susquehanna—Delaware Rivers | 36 | 185 | Rainfall-river flood |
| 1936 | Northeastern United States | 221 | 374 | Rainfall-river flood |
| 1936 | Ohio River basin | 150 | 371 | Rainfall snowmelt flood |
| 1937 | Ohio River basin | 418 | 996 | Rainfall-river flood |
| 1938 | New England streams | 125 | 376 | Hurricane tidal and river floods |
| 1938 | California streams | 100 | 294 | Rainfall-river floods |
| 1942 | Mid-Atlantic coastal streams | 28 | 103 | Rainfall-river floods |
| 1943 | Central States | 172 | N.A. | Rainfall-river floods |
| 1944 | South Florida | 63 | 117 | Hurricane tidal and river floods |
| 1944 | Missouri River basin | 52 | N.A. | Rainfall-river floods |
| 1945 | Hudson River basin | 24 | 75 | Rainfall-river floods |
| 1945 | South Florida | 54 | 98 | Hurricane tidal and river floods |

| | | | | |
|---|---|---|---|---|
| 1945 | Ohio River basin | 34 | 61 | Rainfall-river floods |
| 1947 | South Florida | 60 | 88 | Hurricane tidal and river floods |
| 1947 | Missouri River basin | 178 | N.A. | Rainfall-river floods |
| 1948 | Columbia River basin | 102 | 226 | Rainfall-river floods |
| 1950 | San Joaquin River, Calif. | 32 | 57 | Rainfall-river floods |
| 1951 | Kansas River basin | 883 | N.A. | Rainfall-river floods |
| 1952 | Missouri River basin | 180 | N.A. | Snowmelt floods |
| 1952 | Upper Mississippi River | 198 | N.A. | Rainfall-river floods |
| 1954 | New England streams | 180 | 216 | Hurricane tidal floods |
| 1955 | Northeastern United States | 684 | 879 | Hurricane tidal and river floods |
| 1955 | California and Oregon streams | 271 | 405 | Rainfall-river floods |
| 1957 | Ohio River basin | 65 | 72 | Rainfall-river floods |
| 1957 | Texas rivers | 144 | 188 | Rainfall-river floods |
| 1959 | Ohio River basin | 114 | 120 | Rainfall-river floods |
| 1960 | South Florida | 78 | 86 | Hurricane tidal and river floods |
| 1961 | Texas coast | 300 | 336 | Hurricane tidal floods |
| 1964 | Florida | 325 | 342 | Hurricane tidal and river floods |
| 1964 | Ohio River basin | 106 | 112 | Rainfall-river floods |
| 1964 | California streams | 173 | 183 | Rainfall-river floods |
| 1964 | Columbia River—N. Pacific | 289 | 311 | Rainfall-river floods |
| 1965 | South Florida | 139 | 144 | Hurricane tidal and river floods |
| 1965 | Upper Mississippi River | 158 | 162 | Rainfall snowmelt river flood |
| 1965 | Platte River, Colo.—Nebr. | 191 | N.A. | Rainfall-river flood |
| 1965 | Arkansas River, Colo.—Kans. | 61 | 65 | Rainfall-river flood |
| 1965 | New Orleans and vicinity | 322 | 338 | Hurricane tidal flood |

N.A. = not available.

*Source:* U.S. Water Resources Council, 1968.

**Table 12-4** *Projected Annual Flood Damages in the United States*

(Values in $ millions. Upstream refers to those streams above a point where the total area drained is 250,000 acres or less; downstream refers to the stream pattern below that point.)

| Region | 1957 | | 1966 | | 1980[a] | | 2000[a] | | 2020[a] | |
|---|---|---|---|---|---|---|---|---|---|---|
| | *Downstream* | *Upstream* | *Downstream* | *Upstream* | *Downstream* | *Upstream* | *Downstream* | *Upstream* | *Downstream* | *Upstream* |
| North Atlantic | 64.3 | 62.6 | 63.1 | 70.7 | 75.6 | 91.2 | 89.8 | 120.9 | 116.2 | 163.3 |
| South Atlantic—Gulf | 46.7 | 109.6 | 44.1 | 123.8 | 55.8 | 183.2 | 74.8 | 267.3 | 90.4 | 383.7 |
| Great Lakes | 12.3 | 29.8 | 13.0 | 33.7 | 15.8 | 43.8 | 21.0 | 57.2 | 27.7 | 76.1 |
| Ohio | 78.7 | 49.2 | 73.9 | 55.6 | 99.5 | 68.3 | 151.0 | 90.9 | 237.0 | 116.6 |
| Tennessee | 3.5 | 27.3 | 4.9 | 30.9 | 7.6 | 42.6 | 8.3 | 58.3 | [b] | 80.2 |
| Upper Mississippi | 60.0 | 52.0 | 64.5 | 68.5 | 96.0 | 101.9 | 151.0 | 143.2 | 218.0 | 197.2 |
| Lower Mississippi | 66.0 | 38.5 | 86.8 | 43.5 | 117.2 | 55.3 | 164.2 | 73.5 | 224.5 | 100.1 |
| Souris—Red—Rainy | 5.8 | 13.5 | 5.6 | 15.3 | 6.4 | 18.4 | 7.5 | 24.1 | 8.6 | 32.2 |
| Missouri | 101.4 | 148.1 | 44.0 | 167.3 | 69.0 | 222.3 | 118.0 | 302.7 | 221.0 | 430.1 |
| Arkansas—White—Red | 48.6 | 129.2 | 50.0 | 146.0 | 61.6 | 184.0 | 90.6 | 245.3 | 127.0 | 330.0 |
| Texas—Gulf | 32.5 | 49.5 | 28.2 | 55.9 | 39.5 | 86.1 | 59.3 | 125.3 | 86.4 | 178.4 |
| Rio Grande | 12.2 | 10.4 | 14.7 | 11.8 | 14.8 | 19.5 | 15.8 | 30.9 | 18.8 | 44.9 |
| Upper Colorado | 0.9 | 16.9 | 13.0 | 19.1 | 19.0 | 27.4 | 30.3 | 42.1 | 57.0 | 62.1 |
| Lower Colorado | 5.4 | 25.8 | 10.0 | 29.1 | 20.2 | 59.3 | 42.2 | 93.3 | 96.7 | 141.3 |
| Great Basin | 3.0 | 8.4 | 4.1 | 9.5 | 6.7 | 17.5 | 10.2 | 27.5 | 14.1 | 42.0 |
| Columbia—North Pacific | 52.2 | 106.7 | 52.1 | 120.6 | 73.6 | 170.1 | 120.6 | 235.3 | 197.7 | 325.8 |
| California | 36.9 | 67.2 | 61.6 | 75.9 | 102.1 | 134.3 | 185.9 | 211.0 | 262.6 | 311.2 |
| Alaska | 3.2 | [c] | 4.3 | [c] | 5.6 | [c] | 8.4 | [c] | 12.4 | [c] |
| Hawaii | 1.2 | 10.6 | 1.8 | 12.2 | 2.2 | 16.8 | 2.8 | 23.7 | 3.6 | 34.0 |
| Puerto Rico—Virgin Islands | 2.6 | 3.8 | 2.9 | 4.3 | 3.2 | 6.0 | 3.5 | 8.5 | 4.0 | 12.0 |
| TOTAL[d] | 637 | 959 | 643 | 1094 | 891 | 1548 | 1355 | 2181 | 2024 | 3061 |

[a] Projected damages based on 1968 flood control works.
[b] Not reported.
[c] Not available.
[d] Rounded.

*Source:* U.S. Water Resources Council, 1968.

**Table 12-5** *Classification of Dams*

| Category (1) | Impoundment Danger Potential: *Storage (acre-ft)*[b] (2) | Impoundment Danger Potential: *Height (ft)* (3) | Failure Damage Potential[a]: *Loss of Life* (4) | Failure Damage Potential[a]: *Damage* (5) | Spillway Design Flood (6) |
|---|---|---|---|---|---|
| Major; failure cannot be tolerated | >50,000 | >60 | Considerable | Excessive or as matter of policy | Probable maximum; most severe flood considered reasonably possible on the basin |
| Intermediate | 1000 to 50,000 | 40 to 100 | Possible but small | Within financial capability of owner | Standard project; based on most severe storm or meteorological conditions considered reasonably characteristic of the specific region |
| Minor | <1000 | <50 | None | Of same magnitude as cost of the dam | Frequency basis; 50–100-yr recurrence interval |

[a] Based on consideration of height of dam above tailwater, storage volume, and length of damage reach, present and future potential population and economic development of the floodplain.
[b] Storage at design spillway pool level.

*Source:* After F. F. Snyder, "Hydrology of Spillway Design: Large Structures—Adequate Data," *ASCE, J. Hyd. Div.* 90, No. HY3 (May 1964): 239–259.

**Table 12-6** *Floods Causing 100 or More Deaths in the United States*

| Year | Stream or Place | Lives Lost | Cause |
|---|---|---|---|
| 1831 | Barataria Isle, La. | 150 | Hurricane tidal flood |
| 1856 | Isle Derniere, La. | 320 | Hurricane tidal flood |
| 1874 | Connecticut River tributary | 143 | Dam failure |
| 1875 | Indianola, Tex. | 176 | Hurricane tidal flood |
| 1886 | Sabine, Tex. | 150 | Hurricane tidal flood |
| 1889 | Johnstown, Pa. | 2100 | Dam failure |
| 1893 | Vic. Grand Isle, La. | 2000 | Hurricane tidal flood |
| 1899 | Puerto Rico | 3000 | Hurricane tide and waves |
| 1900 | Galveston, Tex. | 6000+ | Hurricane tidal flood |
| 1903 | Central States | 100+ | Rainfall-river floods |
| 1903 | Heppner, Ore. | 247 | Rainfall-river floods |
| 1906 | Gulf coast | 151 | Hurricane tidal flood |
| 1909 | Gulf coast—New Orleans | 700 | Hurricane tidal flood |
| 1913 | Miami, Muskingham, and Ohio Rivers | 467 | Rainfall-river floods |
| 1913 | Brazos River, Tex. | 177 | Rainfall-river floods |
| 1915 | Louisiana and Texas Gulf coast | 550 | Hurricane tidal flood |
| 1919 | Louisiana and Texas Gulf coast | 284 | Hurricane tidal flood |
| 1921 | Upper Arkansas River | 120 | Rainfall-river flood |
| 1926 | Miami and Clewiston, Fla. | 350 | Hurricane tidal and river flood |
| 1927 | Lower Mississippi River | 100+ | Rainfall-river flood |
| 1927 | Vermont | 120 | Rainfall-river flood |
| 1928 | Puerto Rico | 300 | Hurricane tide and waves |
| 1928 | Lake Okeechobee, Fla. | 2400 | Hurricane tidal flood |
| 1928 | San Francisco, Calif. | 350 | Dam failure |
| 1932 | Puerto Rico | 225 | Hurricane tide and waves |
| 1935 | Florida Keys | 400 | Hurricane tidal flood |
| 1935 | Republican R., Kans., Nebr. | 110 | Rainfall-river flood |
| 1936 | Northeastern United States | 107 | Rainfall, snow melt-river floods |
| 1937 | Ohio River | 137 | Rainfall-river flood |
| 1938 | New England coast | 200 | Hurricane tidal and river flood |
| 1955 | Northeastern United States | 115 | Hurricane rainfall-river floods |
| 1957 | West coast, La. | 556 | Hurricane tide and river floods |
| 1960 | Puerto Rico | 107 | Hurricane rainfall-river floods |

*Source:* U.S. Water Resources Council, 1968.

Small dams customarily are designed using two or more levels of frequency to provide an emergency spillway and insure an adequate allowable freeboard. Figure 12-2 shows a typical small dam with normal freeboard, NF, and minimal freeboard, MF. The freeboard values, for earth dams with riprap protection on the upstream slope, are based upon wave runup caused by storm winds with 100-mph wind velocities. Minimal freeboard pertains to wind velocities of 50 mph. The fetch is defined as the perpendicular distance from the structure to the

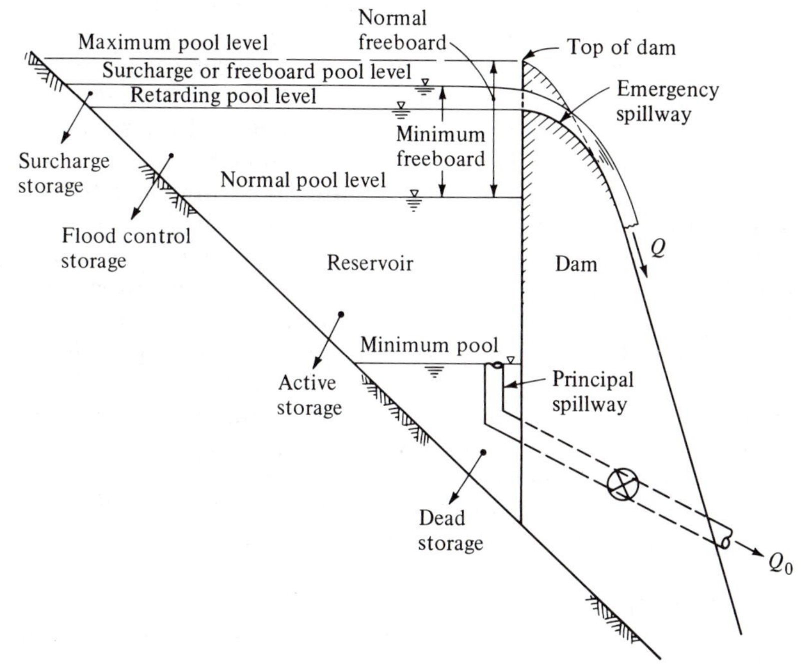


***Fig. 12-2.*** *Multipurpose reservoir pool levels and storage zones.*

*Typical Requirements for Normal and Minimum Feedboard (ft)*

| Fetch, (mi) | NF[a] | MF[a] |
|---|---|---|
| <1 | 4 | 3 |
| 1 | 5 | 4 |
| 2.5 | 6 | 5 |
| 5 | 8 | 6 |
| 10 | 10 | 7 |

[a] After "Designs of Small Dams," Bureau of Reclamation, U.S. Department of the Interior (Washington, D.C.: Government Printing Office, 1965).

windward shore. If smooth concrete rather than riprap is used on the upstream face, the freeboard values shown should be increased 50%.[4]

Design frequency requirements are altered to fit the planned or foreseeable use of the structures. The SCS classifies structures into three groups:[5]

Class a. Structures located in rural or agricultural areas where failure might damage farm buildings, agricultural land, or township or country roads.

Class b. Structures located in predominately rural or agricultural areas where failure might damage isolated homes, main highways or minor railroads, or cause interruption of use or service of relatively important public utilities.

Class c. Structures located where failure might cause loss of life, serious damage of homes, industrial and commercial buildings, important public utilities, main highways, or railroads.

The physical size of a small dam can range to over 100 ft in height but generally is restricted to structures retarding less than 25,000 acre-ft of storage at the emergency spillway crest.

The rainfall frequency levels for use in synthesizing a design storm hydrograph of both the emergency spillway dimensions and freeboard pool level are given in Table 12-7 for the three classifications noted. These are based upon 6-hr rainfall depths for (a) the 100-yr frequency (Fig. 12-3), and (b) the PMP (Fig. 12-1). Design storm depths for all watersheds having a time of concentration less than 6-hr are established in Table 12-7. For those watersheds with greater time of concentration, adjustments are made to the 6-hr storm depth to account for the greater amounts of direct runoff in a longer period of time. These adjustments are discussed in Sec. 12-3.

### Major Structures

Three general terms described in Table 12-5 are employed to designate design floods for major structures: (a) the probable maximum flood, (b) the standard project flood, and (c) the frequency-based flood. The concept of *flood* in this section is described by an entire discharge hydrograph that is generally synthesized from rainfall estimates. Then, corresponding to the three flood designations are the

**Table 12-7** *Small-Dam Storm Frequency Requirements*

| Classification | Emergency Spillway Design[a] | Freeboard Pool Design |
|---|---|---|
| a | $P_{100}$ | $P_{100} + 0.12(\text{PMP} - P_{100})$ |
| b | $P_{100} + 0.12(\text{PMP} - P_{100})$ | $P_{100} + 0.40(\text{PMP} - P_{100})$ |
| c | $P_{100} + 0.26(\text{PMP} - P_{100})$ | PMP |

[a] $P_{100}$ represents a 6-hr rainfall depth for the 100-yr frequency event. (After the Soil Conservation Service.)

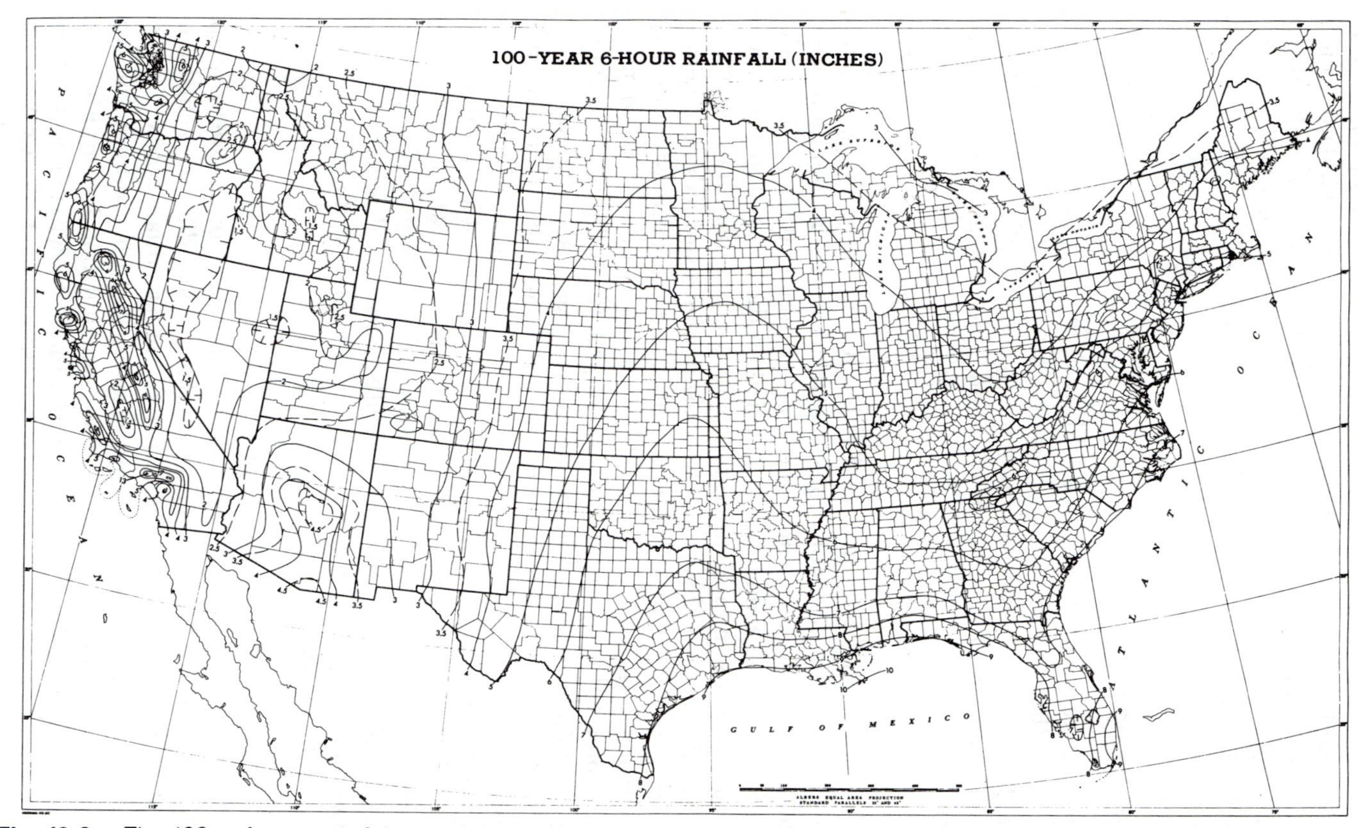


**Fig. 12-3.** *The 100-yr frequency 6-hr precipitation (in.) for 10-mi² or less. (U.S. Weather Bureau NOAA.)*

storm values, that is, the depth of rainfall, referred to in terms of (a) the probable maximum precipitation, PMP; (b) the standard project storm, SPS; and (c) the frequency-based storm. Design of the reservoir system components is commonly based upon one of these representative terms. Descriptions of each are given at the end of the next section on design storms.

## 12-3 Design Storms

Once the frequency has been established, the next step in a structure design is the determination of the storm parameters, which include the storm duration, the point depth, any areal depth adjustment, the storm intensity and time distribution, and the areal distribution pattern.

### Duration

The length of storm used by the SCS in designing emergency and freeboard hydrographs for small dams is of 6-hr duration or *tc*, whichever is greater. Often, the type of minor structure being designed cannot be economically justified on the basis of this length of storm. For many minor structures, particularly urban drainage structures, a design flood hydrograph is based upon a storm duration equal to the time of concentration of the watershed. This procedure utilizes the Rational Method of Chapter 11 or the synthetic unit hydrographs of Chapter 4 along with a critical storm pattern produced by rearranging the rainfall excess pattern into the most critical sequence.

Durations of approximately 6 hr or less are satisfactory for small watersheds, but the length of storms in large areas requires storm depths for periods of up to 10 days. Frequency-based values are available for durations of from 2 to 10 days for locations within the United States.[6] Similar data are also available for other selected areas outside of the United States. Generally, however, design criteria for large dams require estimates of storm depths that do not have frequency levels assigned.

### Duration of Rainfall Excess

Initial rainfall during most storms infiltrates or is otherwise abstracted, and the duration of excess rain $T_0$ is less than the actual rain duration by an amount equal to the time that initial abstractions occur. Excess rain duration $T_0$ can be estimated for a 6-hr storm as a function of the curve number $CN$ and precipitation $P$ from Fig. 12-4. This family of curves was developed by the SCS,[9] where $P$ is the storm depth and $CN$ is a loss parameter defined in Sec. 12-5. A $CN$ of 100 represents zero losses so that $T_0 = 6$ hr for $CN = 100$. Table 12-8 is used to find the duration of excess rain for any storm duration greater

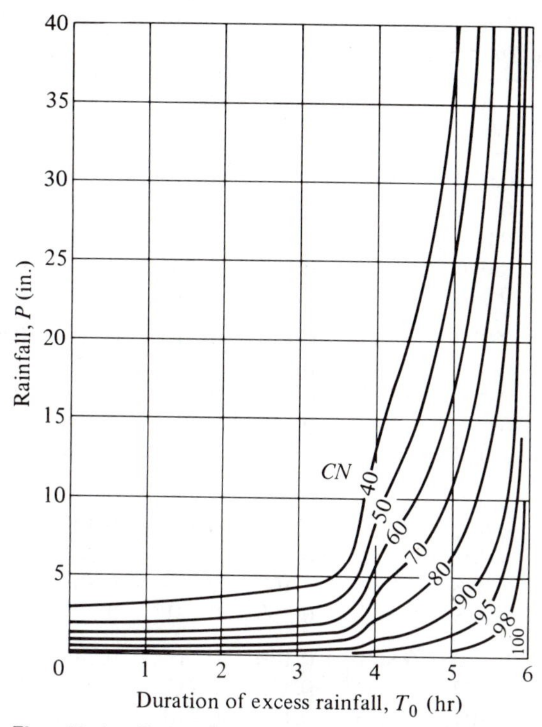


***Fig. 12-4.*** *Duration of excess rainfall. (After "Hydrology," Suppl. A to Sec. 4,* Engineering Handbook, *U.S. Department of Agriculture, Soil Conservation Service, 1968.*

than 6 hr. The rainfall ratio is the abstraction $P^*$ lost before runoff, Table 12-9, divided by the total precipitation amount $P$. The time ratio from Table 12-8 is multiplied by the rainfall duration to obtain $T_0$.

### Depth

The probable maximum precipitation or frequency-based 100-yr 6-hr storm depths at any point can be determined from Figs. 12-1 and 12-3. A convenient means of obtaining storm depths for durations other than 6 hr is to use a table or graph of multipliers for various durations. The U.S. Bureau of Reclamation[4] applies the multipliers in Table 12-10 to the PMP from Fig. 12-1 to determine other duration PMP depths for areas west of the 105° meridian. Similar USBR data east of the meridian are not available.

The U.S. Soil Conservation Service curve[7] of Fig. 12-5 is available for use in adjusting the PMP and 100-yr 6-hr point rainfall depths from Figs. 12-1 and 12-3. Taken together, Figs. 12-1, 12-3, and 12-5 allow the determination for minor structure design of all values in Table 12-7 for any storm duration.

***Table 12-8*** *Rainfall and Time Ratios for Determining $T_0$ When Storm Duration Is Greater Than 6 hr*

| Rainfall Ratio | Time Ratio | Rainfall Ratio | Time Ratio | Rainfall Ratio | Time Ratio | Rainfall Ratio | Time Ratio |
|---|---|---|---|---|---|---|---|
| 0 | 1.00 | 0.070 | 0.852 | 0.140 | 0.746 | 0.210 | 0.684 |
| 0.002 | 0.995 | 0.072 | 0.848 | 0.142 | 0.744 | 0.212 | 0.682 |
| 0.004 | 0.990 | 0.074 | 0.844 | 0.144 | 0.742 | 0.214 | 0.680 |
| 0.006 | 0.985 | 0.076 | 0.841 | 0.146 | 0.740 | 0.216 | 0.679 |
| 0.008 | 0.981 | 0.078 | 0.837 | 0.148 | 0.739 | 0.218 | 0.677 |
| 0.010 | 0.976 | 0.080 | 0.833 | 0.150 | 0.737 | 0.220 | 0.675 |
| 0.012 | 0.971 | 0.082 | 0.830 | 0.152 | 0.735 | 0.222 | 0.673 |
| 0.014 | 0.967 | 0.084 | 0.827 | 0.154 | 0.733 | 0.224 | 0.672 |
| 0.016 | 0.962 | 0.086 | 0.824 | 0.156 | 0.732 | 0.226 | 0.670 |
| 0.018 | 0.957 | 0.088 | 0.821 | 0.158 | 0.730 | 0.228 | 0.668 |
| 0.020 | 0.952 | 0.090 | 0.818 | 0.160 | 0.728 | 0.230 | 0.667 |
| 0.022 | 0.948 | 0.092 | 0.815 | 0.162 | 0.726 | 0.232 | 0.666 |
| 0.024 | 0.943 | 0.094 | 0.812 | 0.164 | 0.724 | 0.234 | 0.666 |
| 0.026 | 0.938 | 0.096 | 0.809 | 0.166 | 0.723 | 0.236 | 0.665 |
| 0.028 | 0.933 | 0.098 | 0.806 | 0.168 | 0.721 | 0.238 | 0.665 |
| 0.030 | 0.929 | 0.100 | 0.803 | 0.170 | 0.719 | 0.240 | 0.664 |
| 0.032 | 0.924 | 0.102 | 0.800 | 0.172 | 0.717 | | |
| 0.034 | 0.919 | 0.104 | 0.797 | 0.174 | 0.716 | (Change in | |
| 0.036 | 0.915 | 0.106 | 0.794 | 0.176 | 0.714 | tabulation | |
| 0.038 | 0.911 | 0.108 | 0.791 | 0.178 | 0.712 | increment) | |
| 0.040 | 0.908 | 0.110 | 0.788 | 0.180 | 0.710 | 0.250 | 0.662 |
| 0.042 | 0.904 | 0.112 | 0.785 | 0.182 | 0.709 | 0.300 | 0.651 |
| 0.044 | 0.900 | 0.114 | 0.782 | 0.184 | 0.707 | 0.350 | 0.640 |
| 0.046 | 0.896 | 0.116 | 0.779 | 0.186 | 0.705 | 0.400 | 0.628 |
| 0.048 | 0.893 | 0.118 | 0.776 | 0.188 | 0.703 | 0.450 | 0.617 |
| 0.050 | 0.889 | 0.120 | 0.773 | 0.190 | 0.702 | 0.500 | 0.606 |
| 0.052 | 0.885 | 0.122 | 0.770 | 0.192 | 0.700 | 0.550 | 0.595 |
| 0.054 | 0.882 | 0.124 | 0.767 | 0.194 | 0.698 | 0.600 | 0.583 |
| 0.056 | 0.878 | 0.126 | 0.764 | 0.196 | 0.696 | 0.650 | 0.542 |
| 0.058 | 0.874 | 0.128 | 0.761 | 0.198 | 0.695 | 0.700 | 0.500 |
| 0.060 | 0.870 | 0.130 | 0.758 | 0.200 | 0.693 | 0.750 | 0.447 |
| 0.062 | 0.867 | 0.132 | 0.755 | 0.202 | 0.691 | 0.800 | 0.386 |
| 0.064 | 0.863 | 0.134 | 0.751 | 0.304 | 0.689 | 0.850 | 0.310 |
| 0.066 | 0.859 | 0.136 | 0.749 | 0.206 | 0.687 | 0.900 | 0.220 |
| 0.068 | 0.856 | 0.138 | 0.747 | 0.208 | 0.686 | 0.950 | 0.116 |

*Source:* After "Hydrology," Suppl. A to Sec. 4, *Engineering Handbook*, U.S. Department of Agriculture, Soil Conservation Service, 1968.

**Table 12-9** *Rainfall Prior to Excess Rainfall*

| *CN* | *P**(in.) | *CN* | *P**(in.) | *CN* | *P**(in.) | *CN* | *P**(in.) | *CN* | *P**(in.) |
|---|---|---|---|---|---|---|---|---|---|
| 100 | 0 | 86 | 0.33 | 72 | 0.78 | 58 | 1.45 | 44 | 2.54 |
| 99 | 0.02 | 85 | 0.35 | 71 | 0.82 | 57 | 1.51 | 43 | 2.64 |
| 98 | 0.04 | 84 | 0.38 | 70 | 0.86 | 56 | 1.57 | 42 | 2.76 |
| 97 | 0.06 | 83 | 0.41 | 69 | 0.90 | 55 | 1.64 | 41 | 2.88 |
| 96 | 0.08 | 82 | 0.44 | 68 | 0.94 | 54 | 1.70 | 40 | 3.00 |
| 95 | 0.11 | 81 | 0.47 | 67 | 0.98 | 53 | 1.77 | 39 | 3.12 |
| 94 | 0.13 | 80 | 0.50 | 66 | 1.03 | 52 | 1.85 | 38 | 3.26 |
| 93 | 0.15 | 79 | 0.53 | 65 | 1.08 | 51 | 1.92 | 37 | 3.40 |
| 92 | 0.17 | 78 | 0.56 | 64 | 1.12 | 50 | 2.00 | 36 | 3.56 |
| 91 | 0.20 | 77 | 0.60 | 63 | 1.17 | 49 | 2.08 | 35 | 3.72 |
| 90 | 0.22 | 76 | 0.63 | 62 | 1.23 | 48 | 2.16 | 34 | 3.88 |
| 89 | 0.25 | 75 | 0.67 | 61 | 1.28 | 47 | 2.26 | 33 | 4.06 |
| 88 | 0.27 | 74 | 0.70 | 60 | 1.33 | 46 | 2.34 | 32 | 4.24 |
| 87 | 0.30 | 73 | 0.74 | 59 | 1.39 | 45 | 2.44 | 31 | 4.44 |

*Source:* After "Hydrology," Suppl. A to Sec. 4, *Engineering Handbook*, U.S. Department of Agriculture, Soil Conservation Service, 1968.

**Table 12-10** *Constants for Extending 6-hr PMP Design Storms in Areas West of the 105° Meridian to Longer Duration Periods*

| Duration (hr) | Constant[a] |
|---|---|
| 8 | 1.16 |
| 10 | 1.31 |
| 12 | 1.43 |
| 14 | 1.50 |
| 16 | 1.56 |
| 18 | 1.62 |
| 20 | 1.68 |
| 22 | 1.74 |
| 24 | 1.80 |
| 30 | 1.95 |
| 36 | 2.10 |
| 42 | 2.25 |
| 48 | 2.38 |

[a] Multiply 6-hr point rainfall from Fig. 12-1 by the indicated constant.

*Source:* After "Design of Small Dams," Bureau of Reclamation U.S. Department of the Interior, Washington, D.C.: Government Printing Office, 1965.

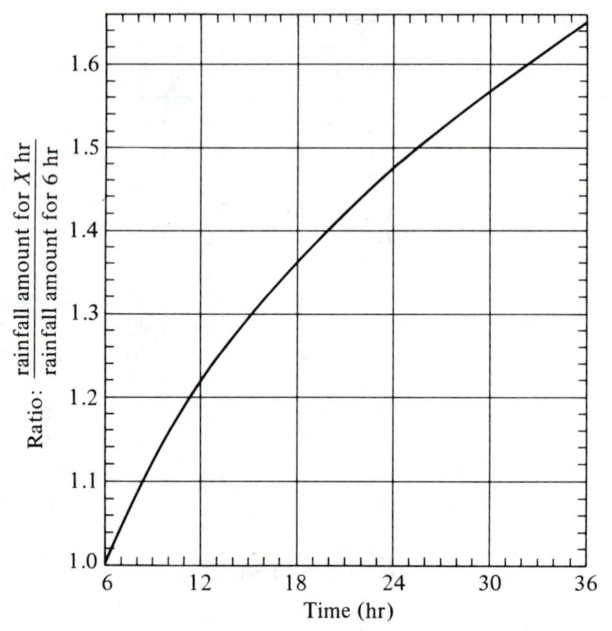


***Fig. 12-5.*** *Relative increase in rainfall amount for storm durations over 6 hr. (After "Hydrology," Suppl. A to Sec. 4,* Engineering Handbook, *U.S. Department of Agriculture, Soil Conservation Service, 1968.)*

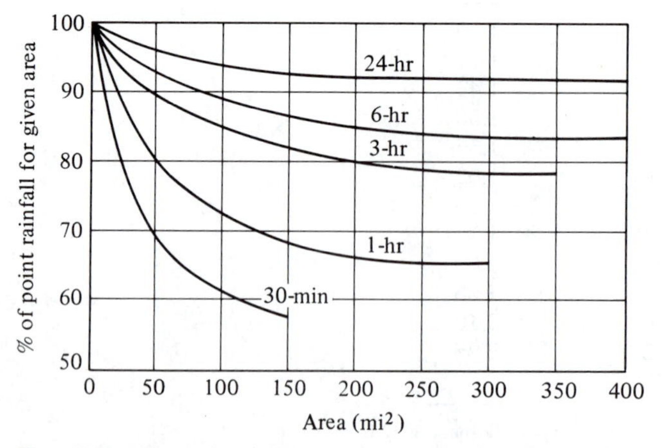


***Fig. 12-6.*** *Area-depth curves for use with duration frequency values. (After Rainfall Frequency Atlas of the United States for Durations from 30 Minutes to 24 Hours and Return Periods from 1 to 100 Years, U.S. Weather Bureau Tech. Paper 40, 1963.)*

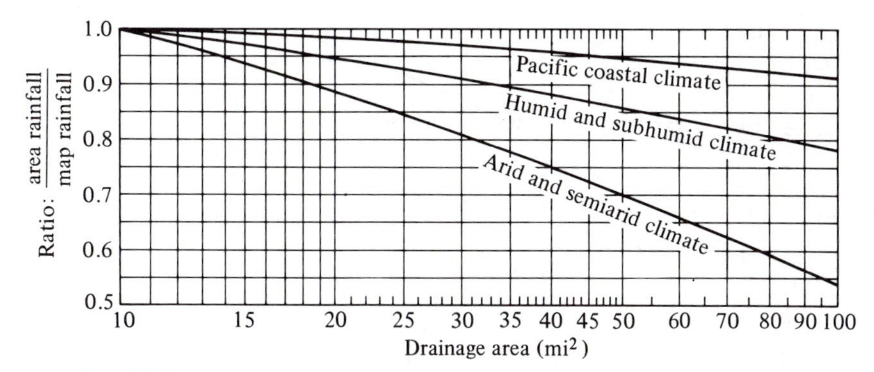


***Fig. 12-7.*** *Rainfall ratios for 10 to 100-mi². (After "Hydrology," Suppl. A to Sec. 4,* Engineering Handbook, *U.S. Department of Agriculture, Soil Conservation Service, 1968.*

## Areal Adjustment

The rainfall depths shown in Figs. 12-1 and 12-3 were derived from frequency analyses (Chapter 6) of *point* measurements and are considered to be applicable only for areas up to 10 $mi^2$. For larger watersheds the areal depths are less; adjustment must be made to account for smaller rainfall depths over larger areas.

The U.S. Weather Bureau[8] developed Fig. 12-6 as a guide in reducing point depths to areal depths for areas up to 400 $mi^2$. For small watersheds, the SCS applies the ratios from Fig. 12-7 to 6-hr map values from Figs. 12-1 and 12-3. Any PMP value from Fig. 12-1 for major designs is modified according to Fig. 12-8. This curve is used by the U.S. Bureau of Reclamation in areas west of the 105° meridian.

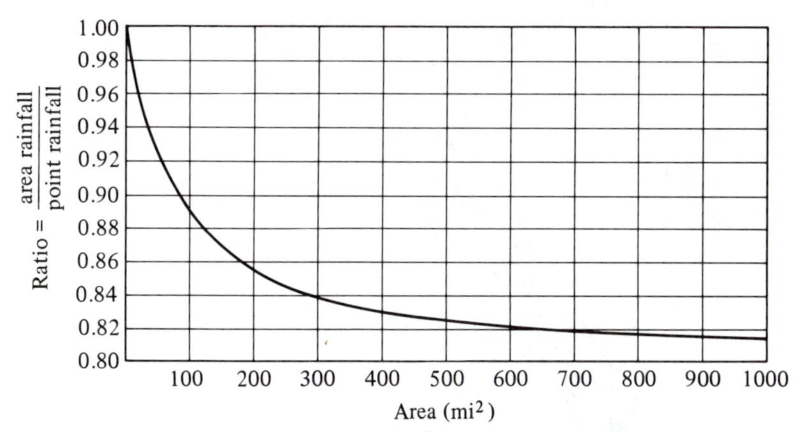


***Fig. 12-8.*** *Conversion ratio from 6-hr point PMP rainfall to 6-hr area rainfall for area west of 105° meridian. (After "Design of Small Dams," Bureau of Reclamation, U.S. Department of the Interior (Washington, D.C.: Government Printing Office, 1965).)*

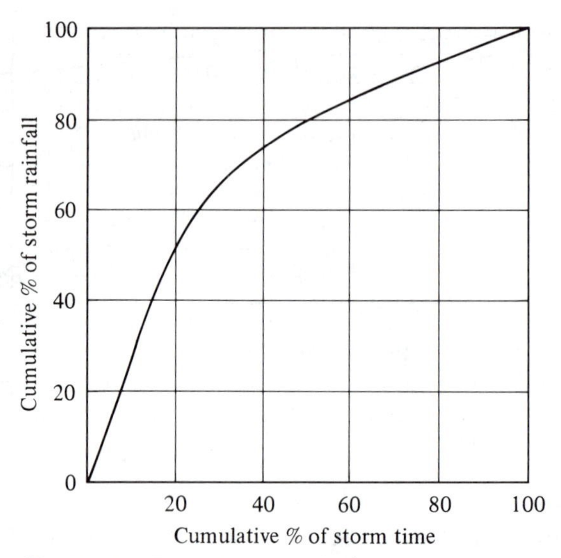


***Fig. 12-9.*** *Time distribution of storm rainfall, median first quartile curve for point rainfall. (After F. A. Huff, "Time Distribution of Rainfall in Heavy Storms,"* Water Resources Research, *3, No. 4 (1967): 1007–1019.)*

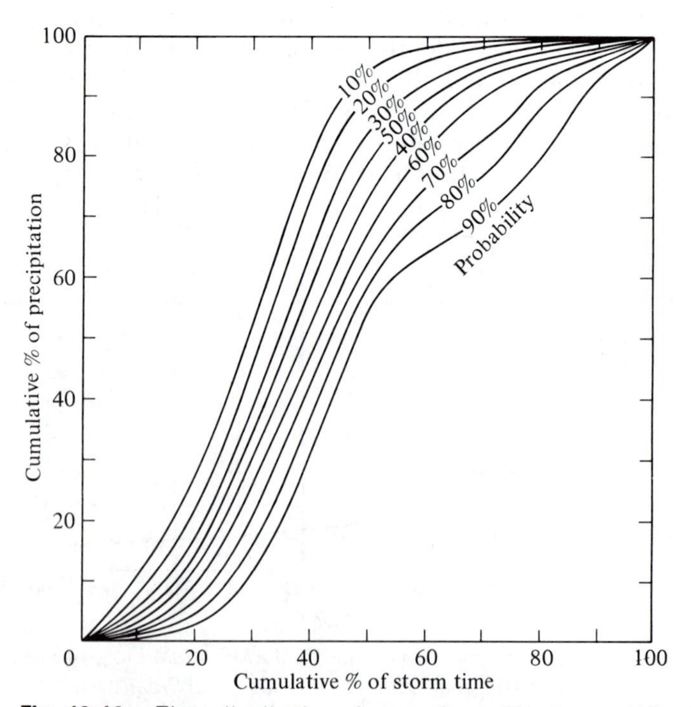


***Fig. 12-10.*** *Time distribution of second quartile storms. (After F. A. Huff, "Time Distribution of Rainfall in Heavy Storms,"* Water Resources Research, *3, No. 4 (1967): 1007–1019.)*

## Time Distribution

After the storm depth and duration have been established, the designer must select a representative hyetograph. His choice will significantly affect the shape and peak value of the resulting runoff hydrograph. Any decision must be based on either the worst-possible storm pattern or on an analysis of recorded storm distribution patterns.

Huff[9] divided recorded storm distribution patterns from small midwestern watersheds into four equal probability groups from the most severe (first quartile) to the mildest (fourth quartile). The median curve for first quartile storms is given in Fig. 12-9 which is used, for example, in the RRL simulation model of Chapter 11.

The two rainfall patterns normally investigated are first quartile and second quartile storms. A first quartile distribution has greater portions of rainfall occurring during the early minutes of the storm. Additional information on the most probable storm pattern over areas up to 400 mi² has been provided by Huff. Taken together, the two types make up 66% of the total number of storms registered on small watersheds in the midwest. Each has an almost equal chance of occurrence.

Curves for each probability level drawn in Figs. 12-10 and 12-11

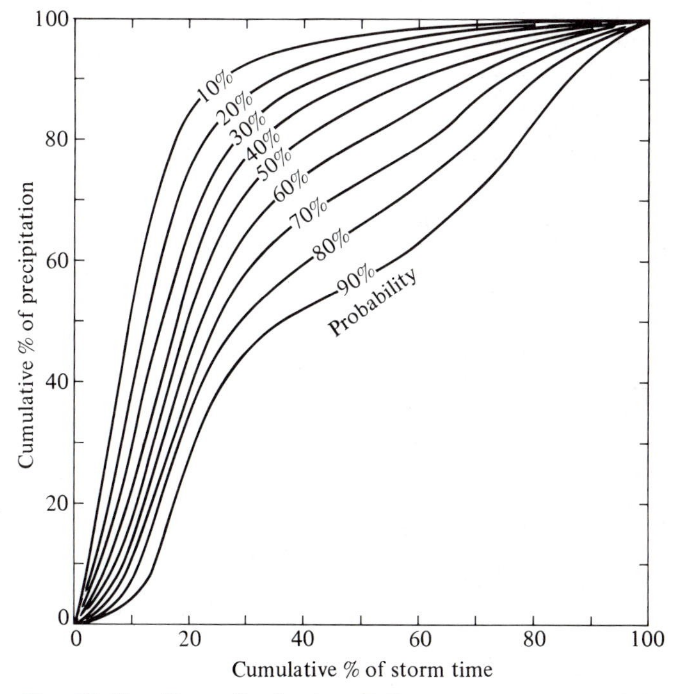


***Fig. 12-11.*** *Time distribution of first quartile storms. (After F. A. Huff, "Time Distribution of Rainfall in Heavy Storms,"* Water Resources Research, *3, No. 4 (1967): 1007–1019.)*

can be used to design for several levels of severity. A 90% level is the distribution occurring in 10% or less of the storms. Eighty percent of the total rainfall occurs in the first 20% of storm time for the 10% level in the first quartile storm. The passage of an intense, prefrontal squall lines, typical of thunderstorms, will produce this particular rainfall distribution. On the other hand, the 90% level is more indicative of a steady rain or a series of rain showers. The 50% or median curve is recommended for most applications.

Figure 12-12 shows the 10, 50, and 90% histograms for first quartile storms. Using these storm distributions permits the construction of rainfall hyetographs for the design rainfall.

Time distributions for critical storms for small-dam or other minor structure designs are usually assumed to be uniform. The SCS uses a uniform distribution for short-duration storms. Alternatively, Fig. 12-13 is the SCS distribution of the 6-hr storm used in developing

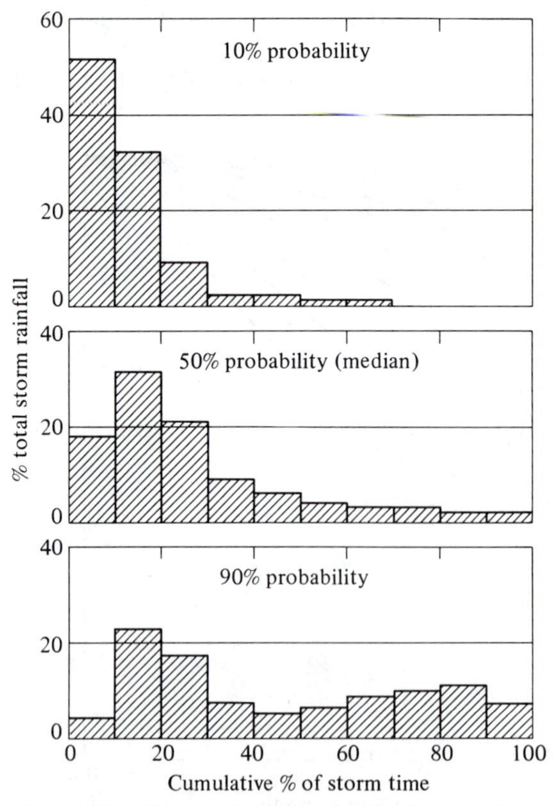


***Fig. 12-12.*** *Selected histograms for first quartile storms. (After F. A. Huff, "Time Distribution of Rainfall in Heavy Storms,"* Water Resources Research, *3, No. 4 (1967): 1007–1019.)*

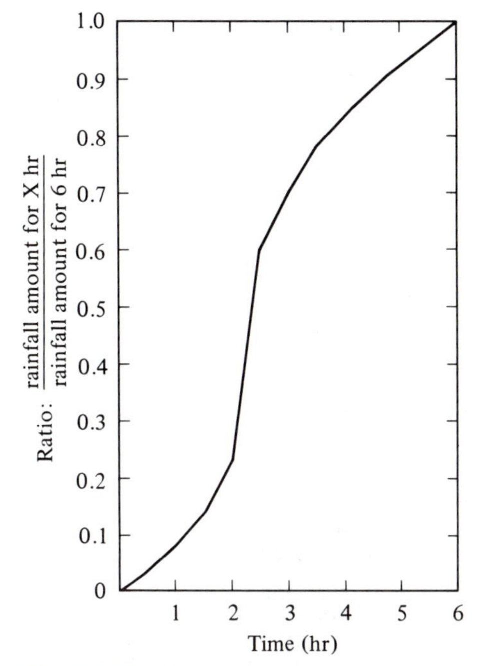


***Fig. 12-13.*** *A 6-hr design storm distribution. (After "Hydrology," Suppl. A to Sec. 4,* Engineering Handbook, *U.S. Department of Agriculture, Soil Conservation Service, 1968.)*

emergency spillway and freeboard hydrographs.[7] This curve is very similar to the 50% (median) second quartile curve in Fig. 12-10.

Time distributions for PMP and other storms used in major structure design can be constructed from Fig. 12-14. This family of curves is used by the U.S. Bureau of Reclamation[4] in three geographical zones shown in Fig. 12-1. The Corps of Engineers uses a distribution curve similar to Fig. 12-13 for 6-hr SPS analyses.

### Areal Distribution

Precipitation depths can and do vary from point to point during a storm (see Fig. 2-14). Areal variation in design storm depth is normally disregarded except in major structure designs. The usual approach in major structure analysis is to select a design (usually elliptical) or historic (transposed) isohyetal pattern for the PMP or SPS depth and assign precipitation depths to the isohyets in a fashion that gives the desired average depth over the basin. The average depth is determined by the isohyetal method illustrated in Fig. 6-7b in Chapter 6.

Four major types of storm patterns are shown in Fig. 12-15 for areas up to 400 mi². These were identified by Huff in his analysis of midwestern storm patterns.[9] The letters $H$ and $L$ represent areas with

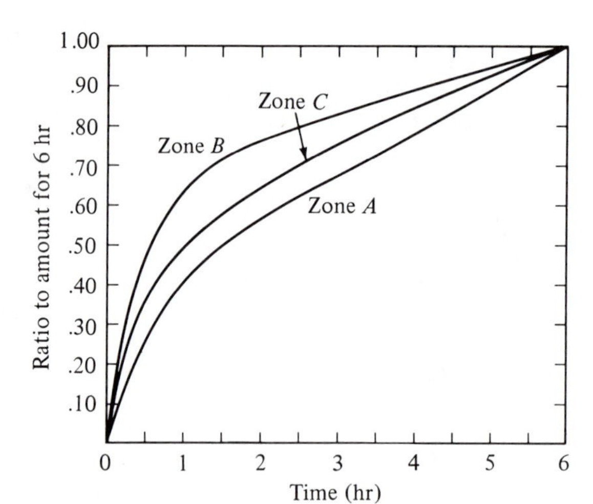


**Fig. 12-14.** *Distribution of 6-hr PMP for an area west of the 105° meridian. (After "Design of Small Dams," Bureau of Reclamation, U.S. Department of the Interior (Washington, D.C.: Government Printing Office, 1965).)*

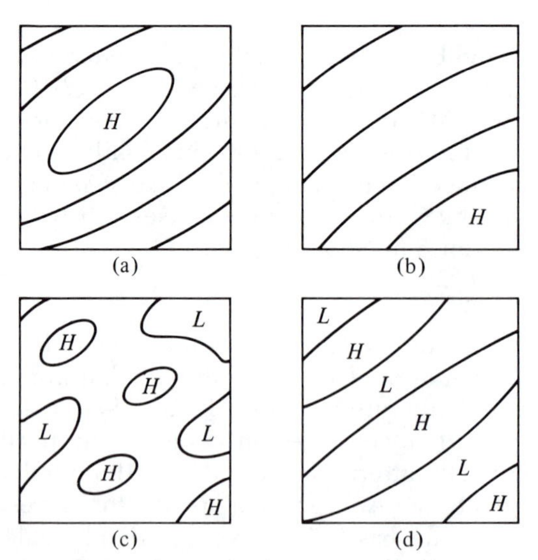


**Fig. 12-15.** *Major types of storm patterns. (a) closed elliptical; (b) open elliptical; (c) multicellular; (d) banded. (After F. A. Huff, "Time Distribution of Rainfall in Heavy Storms,"* Water Resources Research, *3, No. 4 (1967): 1007–1019.)*

high and low precipitation depths, respectively. The typical isohyetal pattern for SPS storms has been established as generally elliptical in shape as shown in Fig. 12-16. This pattern is used by the Tennessee Valley Authority (TVA)[10] for areas up to 3000 mi². Variations in the rainfall depth found in a standard project storm will diverge from a maximum at the storm center to a value considerably less than the average depth at the edges of the watershed boundaries. This variation can be determined[10] and incorporated in the design storm.

A slightly modified isohyetal pattern for SPS storms is used by the Corps of Engineers[11] as shown in Fig. 12-17. The percentages shown for isohyets *A*, *B*, . . ., *G* are multiplied by the 96-hr SPS depth to give an elliptical pattern with the desired average depth. Similar maps for 24-, 48-, or 72-hr storms can be obtained simply by modifying the 96-hr percentages of Fig. 12-17. This is accomplished using

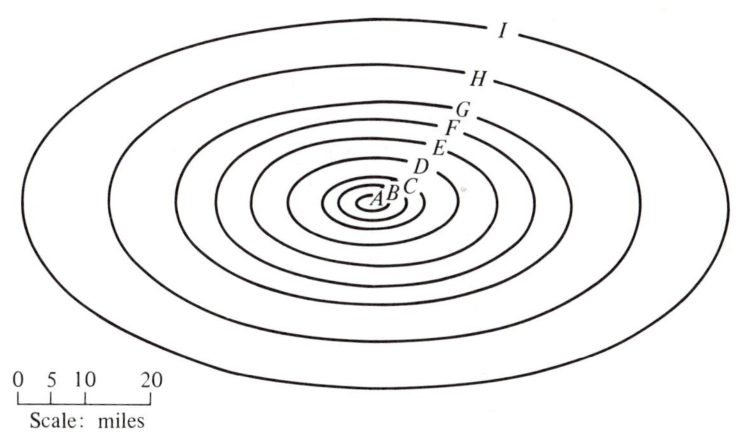


***Fig. 12-16.*** *Generalized pattern storm.*

| Isohyet | Area Enclosed (mi²) |
|---|---|
| *A* | 11 |
| *B* | 45 |
| *C* | 114 |
| *D* | 279 |
| *E* | 546 |
| *F* | 903 |
| *G* | 1349 |
| *H* | 2508 |
| *I* | 4458 |

*(After "Probable Maximum TVA Precipitation for Tennessee River Basins Up to 3000 Square Miles in Area and Durations to 72 Hours," U.S. Department of Commerce, ESSA, Weather Bureau Hydrometeorological Rept. No. 45, 1969.)*

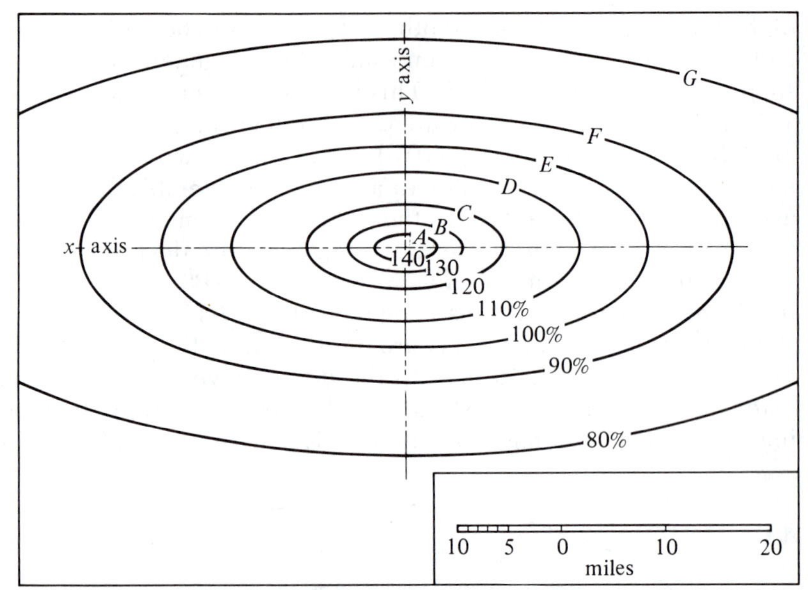


***Fig. 12-17.*** *Generalized PMP isohyetal pattern for a 96-hr storm. Notes: 1. The pattern may be oriented in any direction. 2. The pattern corresponds to the depth-area relation represented by a 96-hr storm.*

| Isohyet No. | Area ($mi^2$) |
|---|---|
| *A* | 16 |
| *B* | 100 |
| *C* | 320 |
| *D* | 800 |
| *E* | 1800 |
| *F* | 3700 |
| *G* | 7100 |

*(After "Standard Project Flood Determinations," Civil Engineer Bulletin No. 52-8B, EM 1110-2-1411, Department of the Army, Office of the Chief of Engineers, Washington, D.C., revised 1965, pp. 1–19.)*

the depth-area-duration curves in Fig. 12-18. For example, if a 24-hr storm is used, first note that the A isohyet of Fig. 12-17 encloses an area of 16 $mi^2$. From Fig. 12-18 the corresponding SPS percentage for a 24-hr storm is 116% rather than the 140% value used with a 96-hr storm. Hence the pattern percentages vary with the selected design storm duration.

Additional aid for constructing design storm distributions over smaller midwestern[9] watersheds (up to 400 $mi^2$) is presented in Table 12-11. The ratio of maximum point rainfall to mean rainfall over the basin is provided and can be used to estimate the maximum depth occurring at a storm center if the mean areal depth is known. Ratios for 50-, 100-, and 200-mi² areas are equal to those in Table 12-11 multiplied by 0.91, 0.94, and 0.97, respectively. For uniform rainfall the 95% ratios of the table are recommended. With extreme variability the 5% ratio applies. The 50% ratios approximate average conditions.

### Probable Maximum Precipitation

Typical storm depths required for major structure design are the probable maximum precipitation, PMP, or the standard project storm,

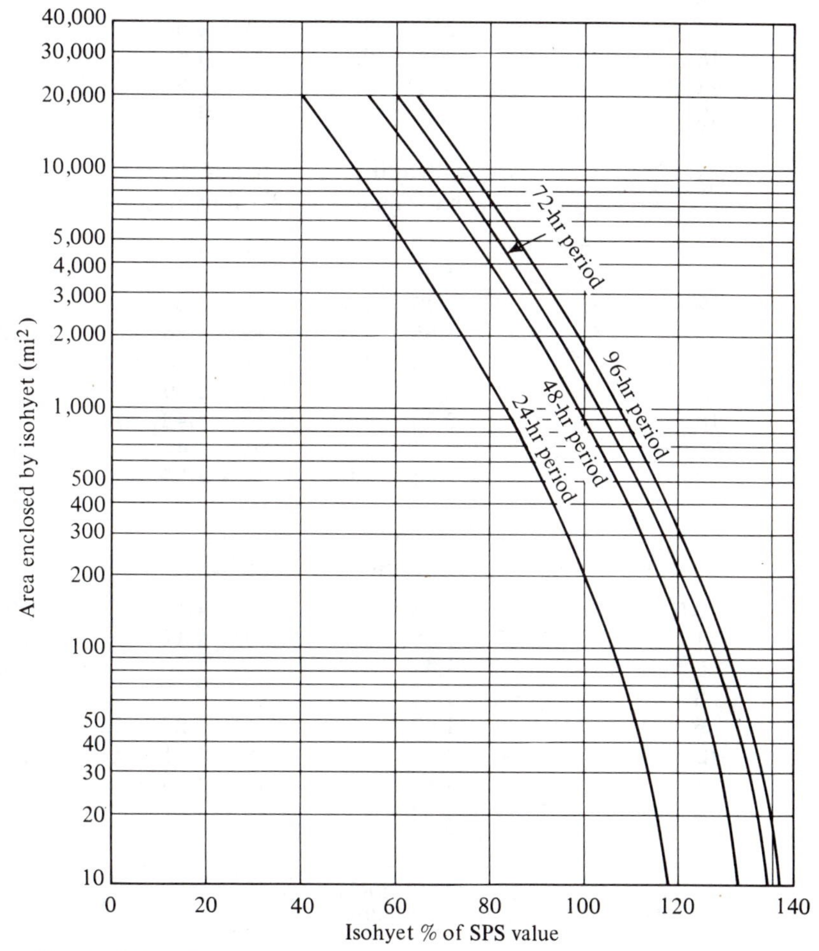


***Fig. 12-18.*** *SPS Depth-area-duration curves by 24-hr storm increments. (After "Standard Project Flood Determinations," Civil Engineers Bulletin No. 52-8, EM 1110-2-1411, U.S. Department of the Army, Office of the Chief of Engineers, Washington, D.C., revised 1965, pp. 1–19.)*

**Table 12-11** *Ratio of Maximum Point to Mean Rainfall on 400 mi²*

| Rainfall Period (hr) | Mean Rainfall (in.) | | | | | | | |
|---|---|---|---|---|---|---|---|---|
| | 0.5 | 1.0 | 1.5 | 2.0 | 2.5 | 3.0 | 4.0 | 5.0 |
| | 5% Probability Level Ratios (Storms with extreme variation in intensity) | | | | | | | |
| 0.5 | 5.20 | 3.00 | 2.18 | 1.70 | 1.41 | 1.30 | 1.26 | 1.22 |
| 1 | 5.50 | 3.21 | 2.29 | 1.80 | 1.48 | 1.35 | 1.30 | 1.25 |
| 2 | 5.80 | 3.38 | 2.44 | 1.90 | 1.55 | 1.41 | 1.33 | 1.28 |
| 3 | 6.05 | 3.54 | 2.53 | 1.99 | 1.61 | 1.46 | 1.36 | 1.30 |
| 6 | | 3.77 | 2.69 | 2.12 | 1.72 | 1.52 | 1.43 | 1.35 |
| 12 | | 4.01 | 2.86 | 2.25 | 1.83 | 1.60 | 1.50 | 1.40 |
| 18 | | 4.14 | 2.96 | 2.33 | 1.90 | 1.65 | 1.54 | 1.43 |
| 24 | | 4.27 | 3.05 | 2.40 | 1.96 | 1.69 | 1.57 | 1.45 |
| 48 | | 4.55 | 3.25 | 2.55 | 2.08 | 1.77 | 1.63 | 1.50 |
| | 50% Probability Level Ratios (Storms with average time distributions) | | | | | | | |
| 0.5 | 2.66 | 2.02 | 1.57 | 1.32 | 1.22 | 1.16 | 1.14 | 1.12 |
| 1 | 3.03 | 2.15 | 1.65 | 1.39 | 1.27 | 1.20 | 1.18 | 1.16 |
| 2 | 3.46 | 2.29 | 1.75 | 1.46 | 1.32 | 1.24 | 1.21 | 1.19 |
| 3 | 3.77 | 2.42 | 1.85 | 1.52 | 1.38 | 1.28 | 1.23 | 1.22 |
| 6 | | 2.59 | 1.98 | 1.63 | 1.43 | 1.33 | 1.28 | 1.26 |
| 12 | | 2.78 | 2.12 | 1.75 | 1.50 | 1.39 | 1.32 | 1.30 |
| 18 | | 2.89 | 2.20 | 1.81 | 1.57 | 1.43 | 1.35 | 1.32 |
| 24 | | 3.00 | 2.28 | 1.87 | 1.60 | 1.47 | 1.38 | 1.33 |
| 48 | | 3.17 | 2.44 | 1.99 | 1.68 | 1.53 | 1.46 | 1.38 |
| | 95% Probability Level Ratios (Storms with uniform intensities) | | | | | | | |
| 0.5 | 2.38 | 1.53 | 1.28 | 1.18 | 1.16 | 1.13 | 1.11 | 1.10 |
| 1 | 2.75 | 1.72 | 1.38 | 1.23 | 1.20 | 1.17 | 1.15 | 1.14 |
| 2 | 3.15 | 1.90 | 1.47 | 1.28 | 1.24 | 1.20 | 1.18 | 1.16 |
| 3 | 3.46 | 2.02 | 1.53 | 1.33 | 1.27 | 1.22 | 1.20 | 1.18 |
| 6 | | 2.24 | 1.67 | 1.43 | 1.31 | 1.27 | 1.24 | 1.21 |
| 12 | | 2.50 | 1.78 | 1.50 | 1.38 | 1.31 | 1.28 | 1.25 |
| 18 | | 2.67 | 1.89 | 1.53 | 1.41 | 1.33 | 1.30 | 1.27 |
| 24 | | 2.77 | 1.92 | 1.58 | 1.43 | 1.35 | 1.32 | 1.29 |
| 48 | | 3.07 | 2.07 | 1.64 | 1.47 | 1.40 | 1.36 | 1.33 |

*Source:* After F. A. Huff, "Time Distribution of Rainfall in Heavy Storms," Water Resources Research, 3, No. 4 (1967): 1007–1019.

SPS. Both were defined in Sec. 12-2; additional information is provided here.

Examples of extreme rainfall occurrences are listed in Tables 12-12 and 12-13. Numbers in the former table are for selected point events exceeding 10 in. in 6 hr whereas those in Table 12-13 are for maximum average rainfall depths over an area of 2000 $mi^2$.

The probable maximum storm is defined as the most severe storm considered reasonably possible to occur. The resulting probable maximum flood is customarily obtained by using unit hydrographs and rainfall estimates of the PMP prepared by the National Weather Service[10] (see Figs. 12-1 and 12-19). In areas of sparse data it can be obtained by drawing envelope curves of the maximal floods of record in the region under study; see Fig. 12-20. In cases where estimates of PMP have not been made, volumes of rainfall to be expected can also be approximated from envelope curves of the world record rainfalls depicted in Fig. 12-21.

A proposed method to estimate PMP advocated by Hershfield[12] suggests that the 24-hr PMP at a point be computed by the equation

$$\mathrm{PMP}_{24} = \bar{P} + KS_n$$

where

$\mathrm{PMP}_{24}$ = the 24-hr probable maximum precipitation
$\bar{P}$ = the mean of the 24-hr annual maximums over the period of record
$S_n$ = the standard deviation of the 24-hr annual maximums
$K$ = a constant equal to 15

Adjustments to the value of $\bar{P}$ and $S_n$ for the record length are noted by Hershfield. However, for appraisal purposes these adjustments probably will not significantly alter results more than 5 to 10%.

### Standard Project Storm

The standard project storm is another rainfall depth that is crucial in the design of large dams. This value is usually obtained from a survey of severe storms in the general vicinity of the drainage basin (Tables 12-12 and 12-13). The storm selected as the SPS may be oriented to produce the maximum amount of runoff for the SPF. Alternatively, severe storms experienced in meteorologically "similar" areas can be transposed over the study area.

Transposition limits are based upon climatological similarity, synoptic weather patterns, and a knowledge of atmospheric processes. A number of severe storms are selected from the storms of record. The depth-area-duration curves for these are developed and presented in graphical form similar to Fig. 12-18, usually for a particular point location within the basin (referred to as an index point). This

**Table 12-12** *U.S. Rainfall Occurrences Equaling or Exceeding 10 in. in 6 hr*[a]

| Date | Amount[b] (in.) | Location | Date | Amount (in.) | Location |
|---|---|---|---|---|---|
| Jun. 13–17, 1886 | 11.5 | Alexandria, La. | May 30–31, 1938 | 10.0 | Sharon Springs, Kans. |
| Jun. 23–27, 1891 | 10.4 | Larabee, Iowa | July 19–25, 1938 | 11.5 | Eldorado, Tex. |
| Jun. 4–7, 1896 | 12.0 | Greenley, Nebr. | Aug. 12–15, 1938 | 10.9 | Koll, La. |
| July 26–29, 1897 | 13.0 | Jewell, Md. | May 25, 1939 | 8.2 | Lebanon, Va. |
| June 12–13, 1907 | 6.2 (3 hr) | Fort Meade, S. Dak. | June 19–20, 1939 | 18.8 | Synder, Tex. |
| July 18–23, 1909 | 10.5 | Ironwood, Mich. | July 4–5, 1939 | 18.6 (3 hr) | Simpson P.O., Ky. |
| July 18–23, 1909 | 10.5 | Beaulieu, Mont. | July 4–5, 1939 | 20.0 | Simpson P.O., Ky. |
| Aug. 28–31, 1911 | 14.9 | St. George, Ga. | Aug. 21, 1939 | 9.5 | Baldwin, Me. |
| Aug. 31–Sept. 1, 1914 | 12.6 | Cooper, Mich. | June 3–4, 1940 | 13.0 | Grant Township, Nebr. |
| Aug. 1–3, 1915 | 12.9 | St. Petersburg, Fla. | June 28–30, 1940 | 11.0 | Engle, Tex. |
| Sept. 28–30, 1915 | 10.1 | Franklinton, La. | Sept. 1, 1940 | 20.1 | Ewan, N.J. |
| July 5–10, 1916 | 15.9 | Bonifay, Fla. | Sept. 2–6, 1940 | 18.4 | Hallett, Okla. |
| June 2–6, 1921 | 10.4 | Pueblo, Colo. | May 22, 1941 | 6.5 (3 hr) | Plainville, Ill. |
| June 17–21, 1921 | 10.5 | Springbrook, Mont. | Oct. 17–22, 1941 | 12.9 | Trenton, Fla. |
| Sept. 8–10, 1921 | 22.4 | Thrall (Taylor), Tex. | Apr. 14–17, 1942 | 13.1 | Green Acres City, Fla. |

| | | | | | |
|---|---|---|---|---|---|
| July 9–12, 1922 | 10.8 | Grant City, Mo. | July 17–18, 1942 | 24.7 | Smethport, Pa. |
| Oct. 4–11, 1924 | 13.6 | New Smyrna, Fla. | May 12–20, 1943 | 15.9 | Near Mounds, Okla. |
| Sept. 11–16, 1926 | 13.4 | Neosho Falls, Kans. | June 5–7, 1943 | 14.2 | Silver Lake, Tex. |
| Sept. 17–19, 1926 | 18.4 | Boyden, Iowa | July 27–29, 1943 | 10.7 | Devers, Tex. |
| Apr. 12–16, 1927 | 13.8 | Jeff-Plaq Drainage District, La. | Aug. 4–5, 1943 | 11.1 | Glenville, W. Va. |
| Mar. 11–16, 1929 | 14.0 | Elba, Ala. | June 10–13, 1944 | 13.4 | Stanton, Nebr. |
| May 25–30, 1929 | 11.3 | Henly, Tex. | July 9, 1945 | 9.1 (4 hr) | Easton, Pa. |
| June 30–July 2, 1932 | 13.3 | State Fish Hatchery, Tex. | Aug. 26–29, 1945 | 10.1 | Hockley, Tex. |
| Aug. 30–Sept. 5, 1932 | 10.0 | Fairfield, Tex. | Aug. 12–15, 1946 | 10.6 | Cole Camp, Mo. |
| Apr. 3–4, 1934 | 17.3 | Cheyenne, Okla. | Sept. 26–27, 1946 | 15.8 | San Antonio, Tex. |
| May 2–7, 1935 | 10.6 | Melville, La. | June 18–23, 1947 | 11.5 | Holt, Mo. |
| May 16–20, 1935 | 13.8 | Simmesport, La. | Aug. 27–28, 1947 | 13.8 | Wickes, Ark. |
| May 30–31, 1935 | 20.6 | NE of Colorado Springs, Colo. | Aug. 24–27, 1947 | 10.9 | Dallas, Tex. |
| June 27–July 4, 1936 | 14.0 | Bebe, Tex. | June 23–24, 1948 | 13.2 | Del Rio, Tex. |
| Sept. 14–18, 1936 | 16.0 | Broome, Tex. | Sept. 3–7, 1950 | 16.0 | Yankeetown, Fla. |

[a] A few cases of storms less than 6-hr duration are included.
[b] 6-hr duration on 10 $mi^2$.

*Source:* "Probable Maximum TVA Precipitation for Tennessee River Basins up to 3000 Square Miles in Area and Durations to 72 Hours," U.S. Department of Commerce, ESSA, Weather Bureau Hydrometeorological Rept. No. 45, 1969.

***Table 12-13*** *Maximum Observed and Moisture-Maximized Storm Rainfall for 24 hr over 2000 mi²*

| Date | Location | Observed amount (in.) |
|---|---|---|
| 9/10–13/1878 | Jefferson, Ohio | 10.4 |
| 6/13–17/1886 | Alexandria, La. | 17.3 |
| 6/27–7/1/1899 | Hearne, Tex. | 19.0 |
| 4/15–18/1900 | Eautaw, Ala. | 10.8 |
| 10/7–11/1903 | Cortland, N.Y. | 10.2 |
| 8/28–31/1911 | St. George, Ga. | 11.3 |
| 3/24–28/1914 | Merryville, La. | 10.1 |
| 9/28–30/1915 | Franklinton, La. | 11.4 |
| 7/5–10/1916 | Bonifay, Fla. | 14.6 |
| 7/13–17/1916 | Altapass, N.C. | 13.3 |
| 9/8–10/1921 | Thrall, Tex. | 20.6 |
| 9/13–17/1924 | Beaufort, N.C. | 10.7 |
| 10/4–11/1924 | New Smyrna, Fla. | 11.9 |
| 4/12–16/1927 | Louisiana | 13.3 |
| 6/1–5/1928 | Thomasville, Ala. | 10.9 |
| 9/16–19/1928 | Darlington, S.C. | 10.3 |
| 3/11–16/1929 | Elba, Ala. | 15.0 |
| 9/23–28/1929 | Washington, Ga. | 12.1 |
| 6/30–7/2/1932 | State Fish Hatchery, Tex. | 16.9 |
| 8/30–9/5/1932 | Fairfield, Tex. | 12.8 |
| 7/22–27/1933 | Logansport, La. | 13.0 |
| 12/5–8/1935 | Satsuma, Tex. | 11.9 |
| 6/27–7/4/1936 | Bebe, Tex. | 12.2 |
| 9/14–18/1936 | Broome, Tex. | 11.6 |
| 8/6–9/1940 | Miller Is., La. | 16.7 |
| 9/2–6/1940 | Hallett, Okla. | 10.7 |
| 10/17–22/1941 | Trenton, Fla. | 15.2 |
| 7/17–18/1942 | Smethport, Pa. | 10.2 |
| 9/3–7/1950 | Yankeetown, Fla. | 24.8 |
| 6/23–28/1954 | Pierce, Tex. | 14.7 |

*Source:* "Probable Maximum TVA Precipitation for Tennessee River Basins Up to 3000 Square Miles in Area and Durations to 72 Hours," U.S. Department of Commerce, ESSA, Weather Bureau Hydrometeorological Rept. No. 45, 1969.

position is used as a reference location for further calculations regarding the SPS.

The SPS differs from a PMP estimate and is patterned after a storm of record that causes the most severe rainfall depth-area-duration relationship. Appropriate allowances should be made for inclusion of snow melt in calculating design-storm hydrographs from the standard project storm. Generally, the standard project storm rainfall is approximately 50% of the PMP. Records of the four or five largest

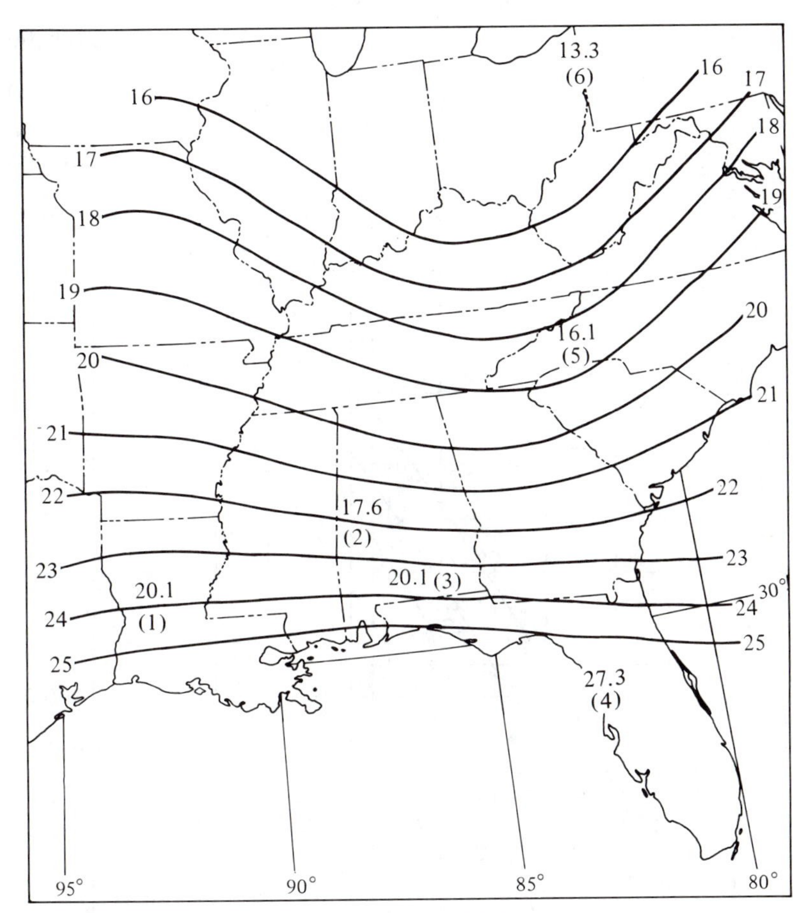


**Fig. 12-19.** *Twenty-four-hour 2000-mi² PMP (in.).*

*Legend:*

1. Alexandria, La., June 13–17, 1886.
2. Eautaw, Ala., April 15–18, 1900.
3. Elba, Ala., March 11–16, 1929.
4. Yankeetown, Fla., September 3–7, 1950.
5. Altapass, N.C., July 13–17, 1916.
6. Jefferson, Ohio, September 10–13, 1878.

*(After "Probable Maximum TVA Precipitation for Tennessee River Basins Up to 3000 Square Miles in Area and Durations to 72 Hours," U.S. Department of Commerce, ESSA, Weather Bureau Hydrometeorological Rept. No. 45, 1969.)*

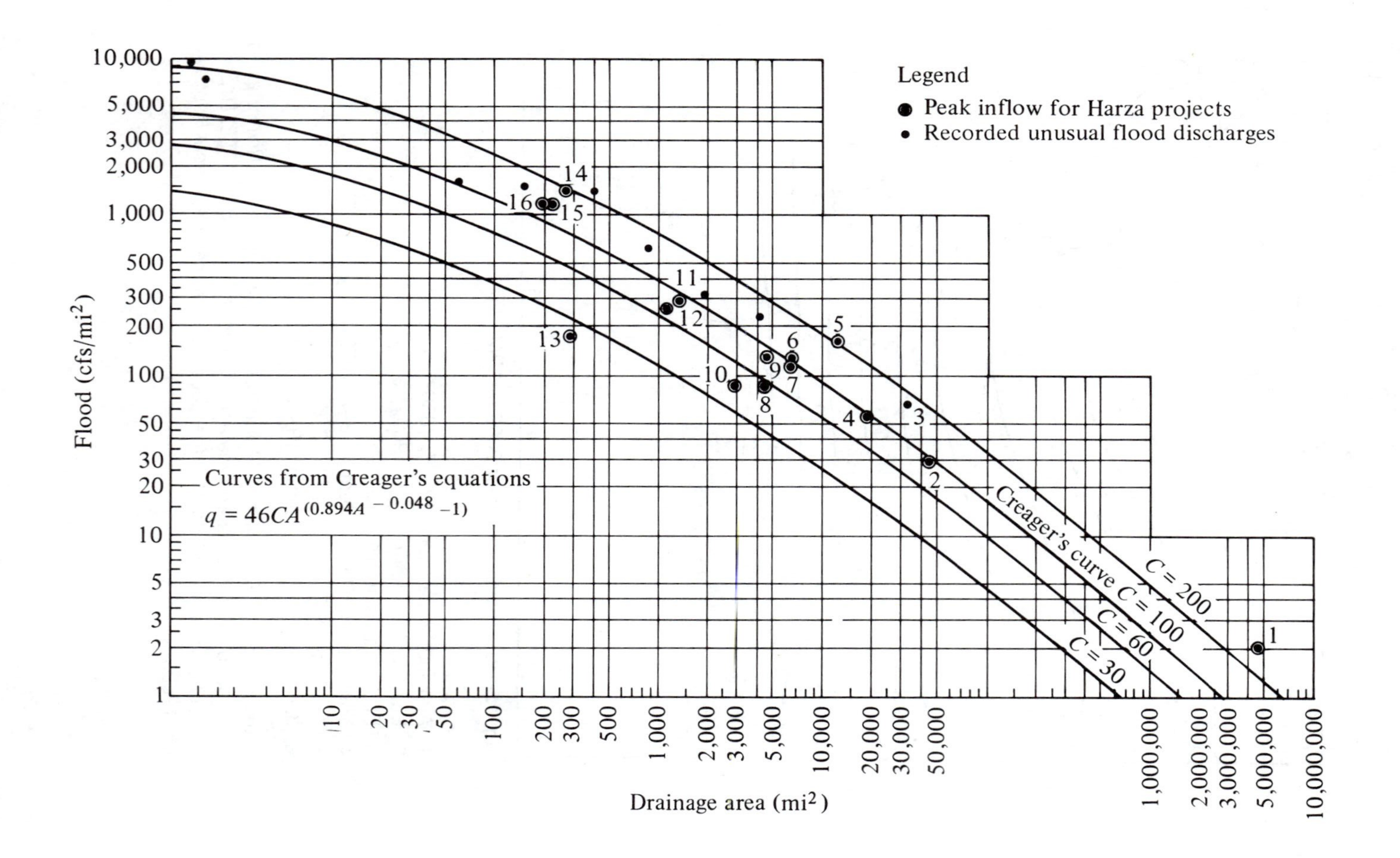

Legend
Peak inflow for Harza projects
Recorded unusual flood discharges
Curves from Creager's equations
$q = 46CA^{(0.894A^{-0.048} - 1)}$
Creager's curve
C = 200
C = 100
C = 60
C = 30
Flood (cfs/mi²)
Drainage area (mi²)
10,000
5,000
3,000
2,000
1,000
500
300
200
100
50
30
20
10
5
3
2
1
10
20
30
50
100
200
300
500
1,000
2,000
3,000
5,000
10,000
20,000
30,000
50,000
1,000,000
2,000,000
3,000,000
5,000,000
10,000,000
1
2
3
4
5
6
7
8
9
10
11
12
13
14
15
16

**Fig. 12-20.** *Creager envelope curves.*

*Legend:*
○ Peak inflow for Harza projects.
• Recorded unusual flood discharges.

Note: Curves taken from *Hydroelectric Handbook,* Creager and Justin (New York: John Wiley & Sons, Inc., 1950).

1. Congo at Inga, Congo.
2. Tigris at Samarra, Iraq.
3. Caroni at Guri, Venezuela.
4. Tigris at Eski Mosul, Iraq.
5. Jhelum at Mangla, Pakistan.
6. Diyala at Derbendi Khan, Iraq.
7. Greater Zab at Bekhme, Iraq.
8. Suriname at Brokopondo, Suriname.
9. Lesser Zab at Doken Dam, Iraq.
10. Pearl River, U.S.A.
11. Cowlitz at Mayfield, U.S.A.
12. Cowlitz at Mossyrock, U.S.A.
13. Karadj, Iran.
14. Agno at Ambuklao, Philippines.
15. Angat, Philippines.
16. Tachien, Formosa.

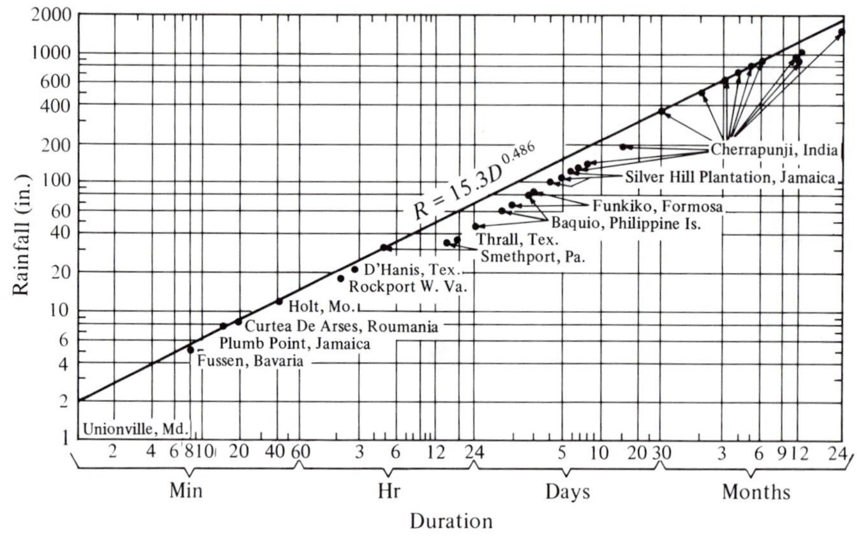


***Fig. 12-21.*** *World's greatest rainfalls. (After "Probable Maximum TVA Precipitation for Tennessee River Basins Up to 3000 Square Miles in Area and Durations to 72 Hours," U.S. Department of Commerce, ESSA, Weather Bureau Hydrometeorological Rept. No. 45, 1969.)*

storms should be critically examined to find a suitable composite for use in calculating the SPS. When these data are not available, a reasonable percentage of the PMP can be substituted. For example, when the loss of life is not a problem, the U.S. Bureau of Reclamation[4] divides the 6-hr PMP from Fig. 12-1 by the factors from Figs. 12-22 and 12-23 to obtain a reasonable design storm depth.

### Frequency-Based Flood

Frequency curves can be plotted and used in major and minor structure design for streams for which lengthy records are available. Most often, however, the location of the dam is not the gauging site, but stream routing techniques can effectively transfer the flood peaks. Regionalized flood frequency data may also be employed to advantage for small structures. For example, Log-Pearson Type III estimates of peak flows for assigned return periods are readily found from Eq. 5-47 if the mean, standard deviation, and skew of logarithms of annual peaks can be estimated. The regional mean and standard deviation of logarithms often correlate well with drainage area and can be determined from nearby gauged stations. Because large samples are required for the determination of skew coefficients, regional skews such as those in Fig. 12-24 are preferred. Customarily, frequency-based floods are not a part of the design criteria for major structures.

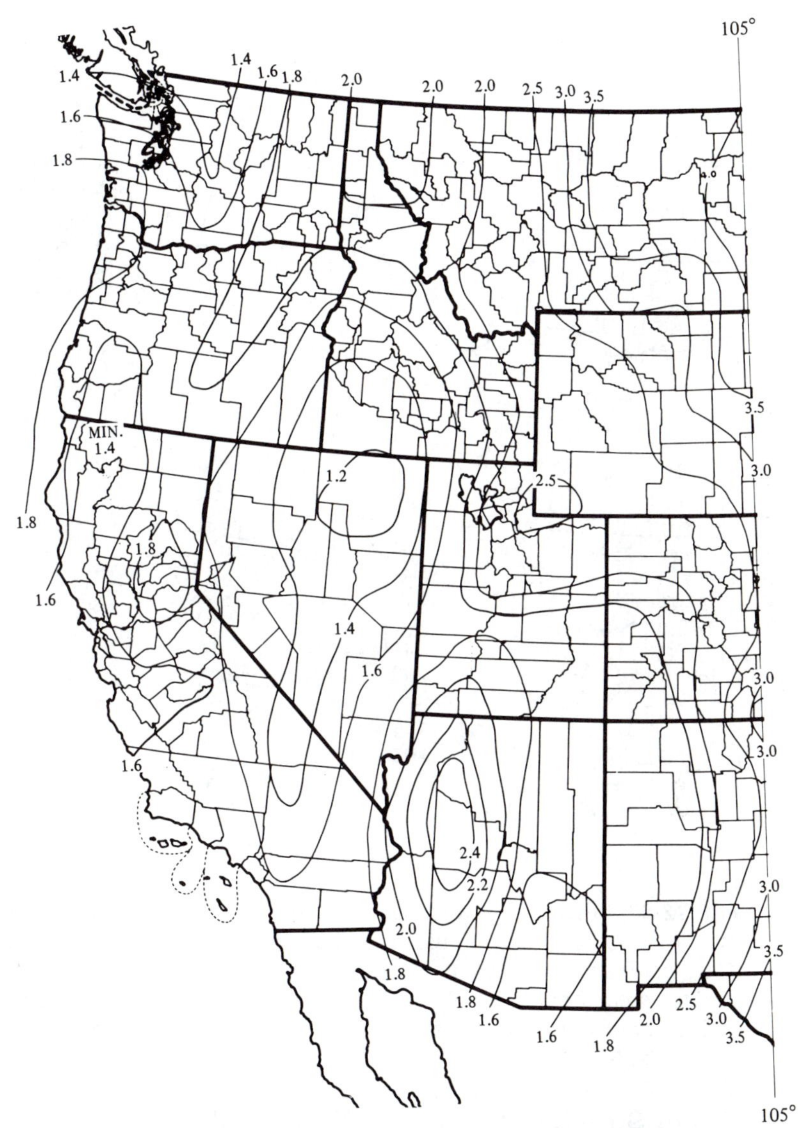


**Fig. 12-22.** *Ratio for determining the rainfall applicable for computing the inflow design flood less than the maximum probable for the area west of the 105° meridian. Note: Divide the probable maximum storm values by the indicated number. (After "Design of Small Dams," Bureau of Reclamation, U.S. Department of the Interior (Washington, D.C.: Government Printing Office, 1965).)*

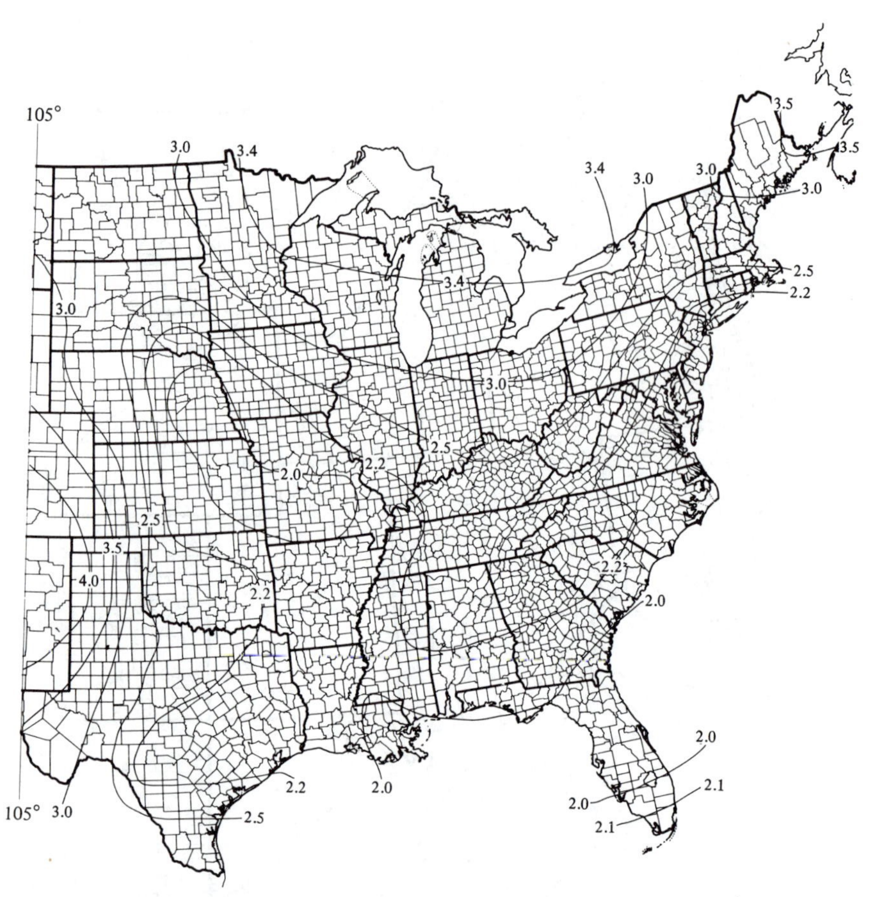


***Fig. 12-23.*** *Ratio for determining the rainfall applicable for computing the inflow design flood less than maximum probable for the area east of the 105° meridian. Note: Divide the probable maximum precipitation by the indicated number. (After "Design of Small Dams," Bureau of Reclamation, U.S. Department of the Interior (Washington, D.C.: Government Printing Office, 1965).)*

## 12-4 Minor Structure Design—SCS Method

A small dam probably presents the best example of structure design because it involves development of a design storm, a design hydrograph, and the use of hydrologic or hydraulic routing techniques to size the spillway. Design storms were just covered, routing techniques have been covered in Chapter 7 and the unit hydrographs in Chapter 4; now application of the various methods in preparing a design storm hydrograph is presented. Routing procedures to be used with the developed design-storm hydrograph are the same as those covered in Chapter 7.

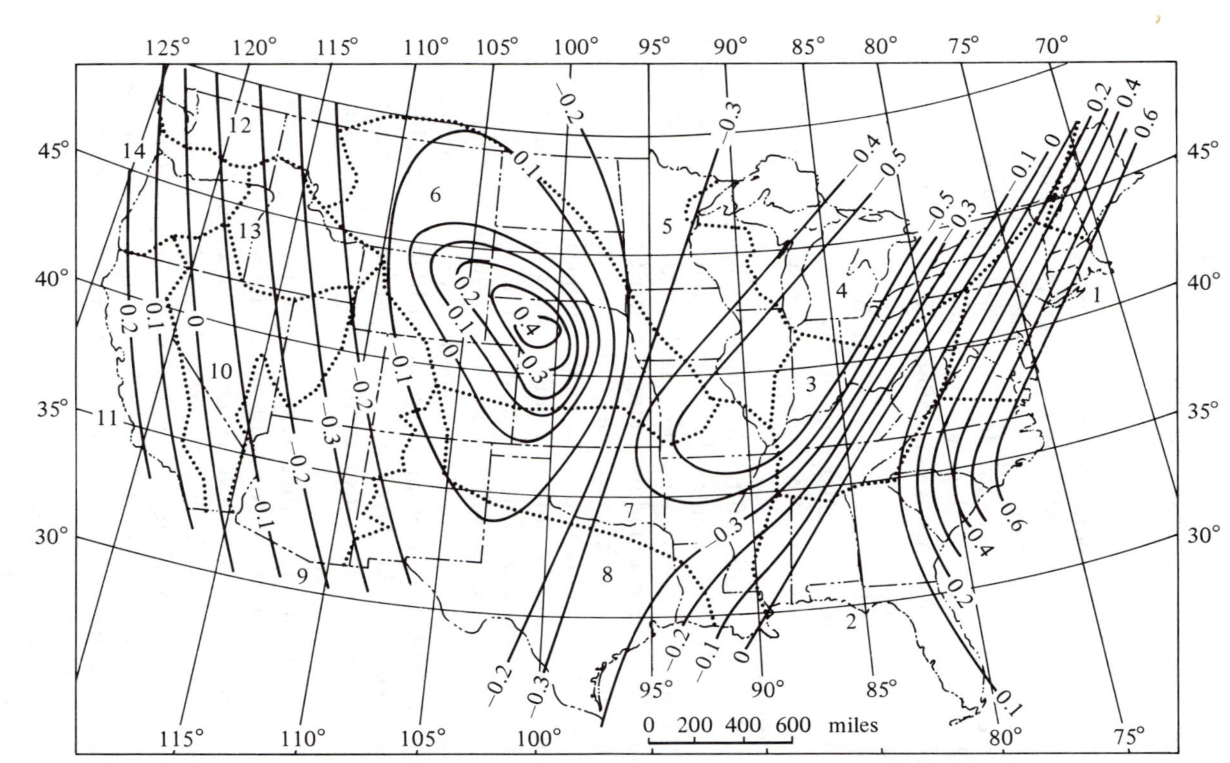


**Fig. 12-24.** *Generalized skew coefficients of annual maximum streamflow logarithms. Prepared by U.S. Geological Survey. Note: Zone numbers added. (After "Guidelines for Determining Flood Flow Frequencies," U.S. Water Resources Council, Washington, D.C., 1976, pp. 14–20.)*

A design hydrograph for both the emergency spillway design and freeboard pool design can be found through a procedure developed by the U.S. Soil Conservation Service. The solution provides a design inflow hydrograph which can be used in selecting the spillway proportions and sizing the reservoir storage for flood detention.

### Runoff Curve Number

The SCS procedure consists of selecting a storm and computing the direct runoff by the use of curves founded on field studies of the amount of measured runoff from numerous soil cover combinations. A runoff curve number (*CN*) is extracted from Table 12-14. Selection of the runoff curve number is dependent upon antecedent conditions and the types of cover. Soils are classified A, B, C, or D according to the following criteria:

A. (Low runoff potential.) Soils having high infiltration rates even if thoroughly wetted and consisting chiefly of deep well to excessively drained sands or gravels. They have a high rate of water transmission.

B. Soils having moderate infiltration rates if thoroughly wetted and consisting chiefly of moderately deep to deep, moderately well to well-drained soils with moderately fine to moderately coarse textures. They have a moderate rate of water transmission.

C. Soils having slow infiltration rates if thoroughly wetted and consisting chiefly of soils with a layer that impedes the downward movement of water, or soils with moderately fine to fine texture. They have a slow rate of water transmission.

D. (High runoff potential.) Soils having very slow infiltration rates if thoroughly wetted and consisting chiefly of clay soils with a high swelling potential, soils with a permanent high water table, soils with a claypan or clay layer at or near the surface, and shallow soils over nearly impervious material. They have a very slow rate of water transmission.

A composite curve number (*CN*) for a watershed having more than one land use, treatment, or soil type can be found by weighting each curve number according to its area. If, for example, 80% of a watershed has a *CN* of 75 and the remaining 20% is impervious ($CN = 100$), then the weighted $CN = 0.80 \times 75 + 0.20 \times 100 = 80$.

The curve numbers in Table 12-14 are applicable to average antecedent moisture conditions. Other antecedent moisture conditions (AMC) are

AMC I. A condition of watershed soils where the soils are dry but not to the wilting point, and when satisfactory plowing or cultivation takes place. (This condition is not considered applicable to the design flood computation methods presented in this text.)

AMC II. The average case for annual floods, that is, an average of the conditions that have preceded the occurrence of the maximum annual flood on numerous watersheds.

AMC III. If heavy rainfall or light rainfall and low temperatures have occurred during the 5 days previous to the given storm and the soil is nearly saturated.

The corresponding curve numbers for condition I and condition III can be obtained from Table 12-15 if the *CN* for AMC II is known.

### Design Storm

Rainfall can be obtained from the rainfall atlas[8] for a selected design frequency (Table 12-7) and duration. The direct runoff resulting from this precipitation is estimated from Figs. 12-25 or 12-26, which are applicable to basin areas up to 2000 acres. The antecedent condition II for selection of *CN* is typical of midwestern small watersheds for a design storm precipitation greater than 1 in. Note that $Q$ is the net rain depth in inches and should not be mistaken as a discharge rate.

### Design-Storm Hydrograph

A design-storm hydrograph is next synthesized by selecting a unit hydrograph from Table 12-17 using the following equations:

$$T_p = 0.7T_c \tag{12-1}$$

$$\text{rev. } T_p = \frac{T_o}{(T_o/T_p)\text{ rev.}} \tag{12-2}$$

$$q_p = \frac{484A}{\text{rev. } T_p} \tag{12-3}$$

$$t = \left(\frac{t}{T_p}\right)(\text{rev. } T_p) \tag{12-4}$$

$$q = (q_c/q^p)(Qq^p) \tag{12-5}$$

where

$A$ = the drainage area (mi²)
$q$ = the synthesized hydrograph rate (cfs)
$q_c$ = the hydrograph rate for $Q = 1$ in.
$q_p$ = the hydrograph peak rate for $Q = 1$ in.
$Q$ = the synthesized hydrograph runoff (in.)
rev. $T_p$ = the revised time to peak (hr)
$t$ = the time (hr) at which the synthesized hydrograph rate is computed
$T_c$ = the time of concentration (hr)
$(T_o/T_p)_{rev}$ = the revised ratio from Table 12-16
$T_p$ = the time to peak (hr) for the synthesized hydrograph

**Table 12-14** *Runoff Curve Numbers for Hydrologic Soil-Cover Complexes*
(Antecedent moisture condition II, and $I_a = 0.2S$)

| Cover | | | Hydrologic Soil Group | | | |
|---|---|---|---|---|---|---|
| *Land Use or Cover* | *Treatment or Practice* | *Hydrologic Condition* | *A* | *B* | *C* | *D* |
| Fallow | Straight row | — | 77 | 86 | 91 | 94 |
| Row crops | Straight row | Poor | 72 | 81 | 88 | 91 |
| | Straight row | Good | 67 | 78 | 85 | 89 |
| | Contoured | Poor | 70 | 79 | 84 | 88 |
| | Contoured | Good | 65 | 75 | 82 | 86 |
| | Contoured and terraced | Poor | 66 | 74 | 80 | 82 |
| | Contoured and terraced | Good | 62 | 71 | 78 | 81 |
| Small grain | Straight row | Poor | 65 | 76 | 84 | 88 |
| | | Good | 63 | 75 | 83 | 87 |
| | Contoured | Poor | 63 | 74 | 82 | 85 |
| | | Good | 61 | 73 | 81 | 84 |
| | Contoured and terraced | Poor | 61 | 72 | 79 | 82 |
| | | Good | 59 | 70 | 78 | 81 |
| Close-seeded legumes[a] or rotation meadow | Straight row | Poor | 66 | 77 | 85 | 89 |
| | Straight row | Good | 58 | 72 | 81 | 85 |
| | Contoured | Poor | 64 | 75 | 83 | 85 |
| | Contoured | Good | 55 | 69 | 78 | 83 |
| | Contoured and terraced | Poor | 63 | 73 | 80 | 83 |
| | Contoured and terraced | Good | 51 | 67 | 76 | 80 |

| | | | | | | |
|---|---|---|---|---|---|---|
| Pasture or range | | Poor | 68 | 79 | 86 | 89 |
| | | Fair | 49 | 69 | 79 | 84 |
| | | Good | 39 | 61 | 74 | 80 |
| | Contoured | Poor | 47 | 67 | 81 | 88 |
| | Contoured | Fair | 25 | 59 | 75 | 83 |
| | Contoured | Good | 6 | 35 | 70 | 79 |
| Meadow | | Good | 30 | 58 | 71 | 78 |
| Woods | | Poor | 45 | 66 | 77 | 83 |
| | | Fair | 36 | 60 | 73 | 79 |
| | | Good | 25 | 55 | 70 | 77 |
| Farmsteads | | — | 59 | 74 | 82 | 86 |
| Roads (dirt)[b] | | — | 72 | 82 | 87 | 89 |
| (hard surface)[b] | | — | 74 | 84 | 90 | 92 |

[a] Close-drilled or broadcast.
[b] Including right of way.

*Source:* After "Hydrology," Suppl. A to Sec. 4, *Engineering Handbook*, U.S. Department of Agriculture, Soil Conservation Service, 1968.

***Table 12-15*** *Curve Numbers (CN) for Wet (AMC III) and Dry (AMC I) Antecedent Moisture Conditions Corresponding to an Average Antecedent Moisture Condition*

| *CN* for AMC II | Corresponding *CN*'s | |
|---|---|---|
| | *AMC I* | *AMC III* |
| 100 | 100 | 100 |
| 95 | 87 | 98 |
| 90 | 78 | 96 |
| 85 | 70 | 94 |
| 80 | 63 | 91 |
| 75 | 57 | 88 |
| 70 | 51 | 85 |
| 65 | 45 | 82 |
| 60 | 40 | 78 |
| 55 | 35 | 74 |
| 50 | 31 | 70 |
| 45 | 26 | 65 |
| 40 | 22 | 60 |
| 35 | 18 | 55 |
| 30 | 15 | 50 |
| 25 | 12 | 43 |
| 20 | 9 | 37 |
| 15 | 6 | 30 |
| 10 | 4 | 22 |
| 5 | 2 | 13 |

AMC I. Lowest runoff potential. Soils in the watershed are dry enough for satisfactory plowing or cultivation.
AMC II. The average condition.
AMC III. Highest runoff potential. Soils in the watershed are practically saturated from antecedent rains.

*Source:* After "Hydrology," Suppl. A to Sec. 4, *Engineering Handbook,* U.S. Department of Agriculture, Soil Conservation Service, 1968.

Five families of unit hydrographs are listed, Table 12-17 for different ratios of storm duration, $T_0$, to time to peak, $T_p$. Selection of family is based on the rainfall and composite *CN* according to Fig. 12-27. A summary of these synthesized hydrographs is shown in Table 12-16. After the inflow hydrograph has been synthesized, the flood is routed through the reservoir and the spillway dimensions can be determined on the basis of allowable velocities or maximum discharge rates.

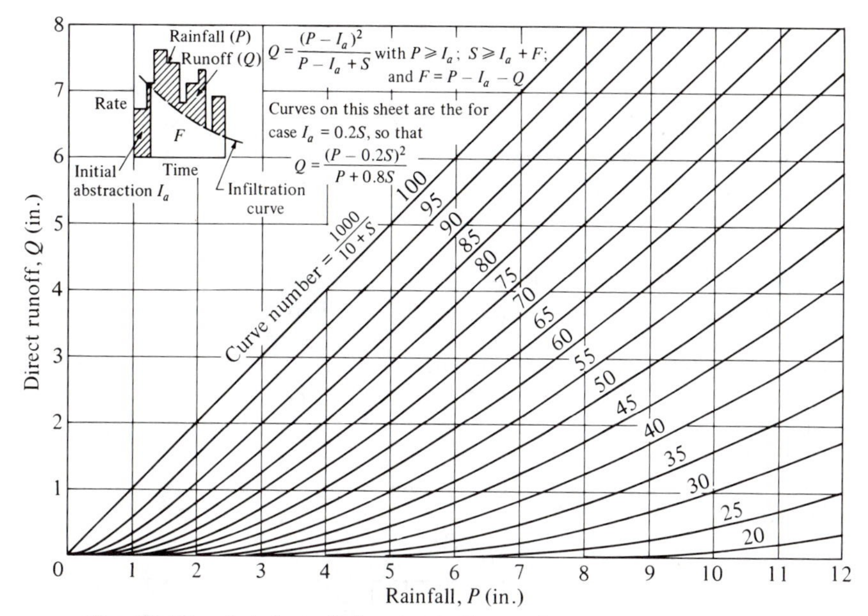


***Fig. 12-25.*** *Solution of direct runoff equation.*

$$Q = \frac{(P - I_a)^2}{P - I_a + S} \text{ with } P \geqslant I_a;\ S \geqslant I_a + F;\ \text{and } F = P - I_a - Q$$

Curves on this sheet are for the case $I_a = 0.2S$, so that

$$Q = \frac{(P - 0.2S)^2}{P + 0.8S}$$

*(After "Hydrology," Suppl. A to Sec. 4,* Engineering Handbook, *U.S. Department of Agriculture, Soil Conservation Service, 1968.)*

***Table 12-16*** *Hydrograph Families and $T_o/T_p$ Ratios for Which Dimensionless Hydrograph Ratios Are Given in Table 12-17*

| Hydrograph Family | $T_o/T_p$ 1 | 1.5 | 2 | 3 | 4 | 6 | 10 | 16 | 25 | 36 | 50 | 75 |
|---|---|---|---|---|---|---|---|---|---|---|---|---|
| 1 | * | * | * | * | * | * | * | * | * | * | * | * |
| 2 | * | * | * | * | * | * | * | * | * | * | * | * |
| 3 | * | * | * | * | * | * | * | * | * | * | * | * |
| 4 | * | * | * | * | * | * | * | * | * | * | * | |
| 5 | * | * | * | * | * | * | * | * | * | * | * | |

* Asterisks signify that dimensionless hydrograph tabulations are given in Table 12-17.

*Source:* "Hydrology," Suppl. A to Sec. 4, *Engineering Handbook,* U.S. Department of Agriculture, Soil Conservation Service, 1968.

### Examples Using SCS Method

The SCS procedure for determining the design storm hydrograph for a watershed with a time of concentration less than 6 hr is shown as Example 12-1. The determination of the design storm hydrograph for a watershed with a time concentration greater than 6 hr is given in Example 12-2. Both examples assume that the design storm is uniformly distributed in time and space.

***Example 12-1*** Determine the emergency spillway hydrograph for a class (b) structure, Table 12-7, with a drainage area of 1.86 mi², time of concentration = 1.25 hr, runoff curve number = 82, which is geographically located to provide a 6-hr design rainfall depth of 9.4 in. If not given, this depth would be determined from Table 12-7 using Figs. 12-1 and 12-3.

1. Duration adjustment. No adjustment necessary for $T < 6$ hr; by the SCS method $D = 6$ hr.
2. Determine the areal factor. No adjustment is necessary, since the area is less than 10 mi².

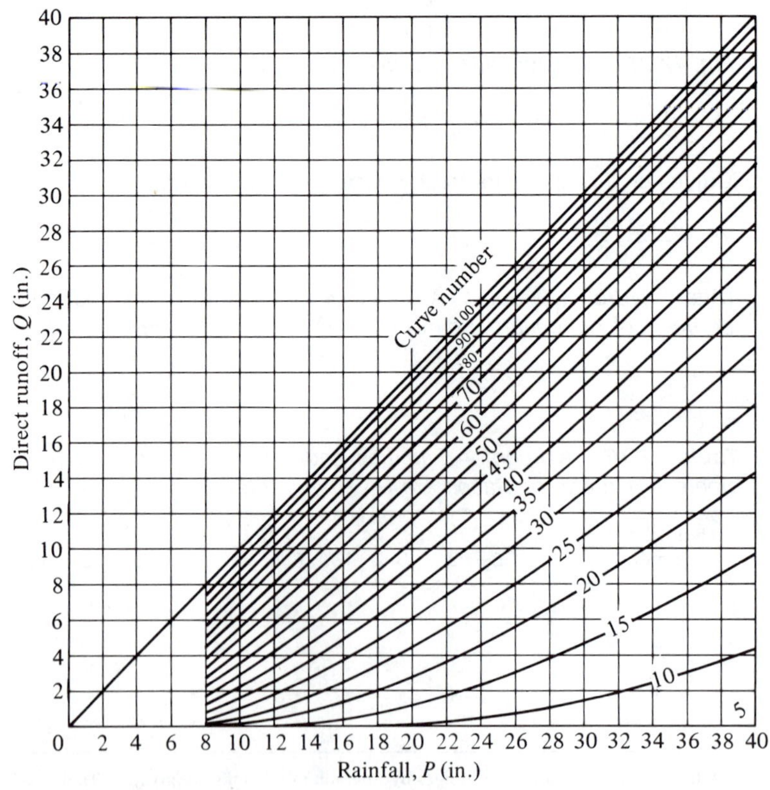


**Fig. 12-26.** *Solution of direct runoff equation. (After "Hydology," Suppl. A to Sec. 4,* Engineering Handbook, *U.S. Department of Agriculture, Soil Conservation Service, 1968.)*

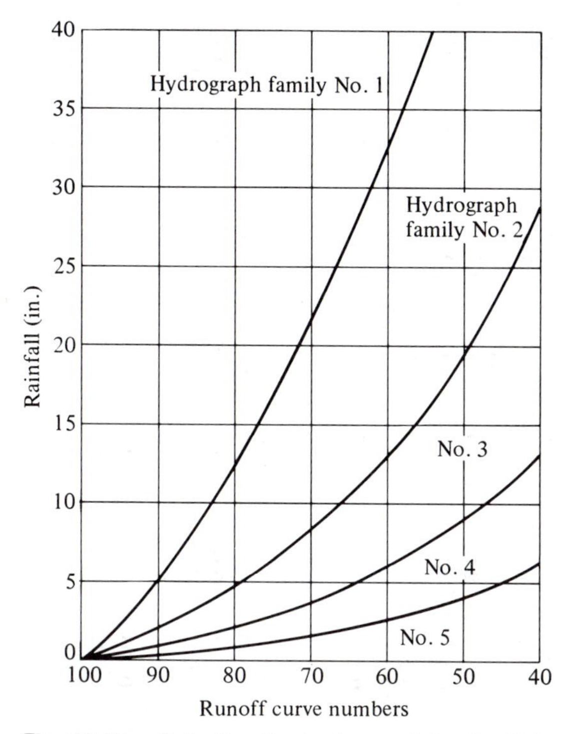


***Fig. 12-27.*** *Selecting the hydrograph family. (After "Hydrology," Suppl. A to Sec. 4, Engineering Handbook, U.S. Department of Agriculture, Soil Conservation Service, 1968.)*

3. Direct runoff $Q$. Enter Fig. 12-25 and find $Q = 7.21$ in. for $P = 9.4$ in. and $CN = 82$.
4. Hydrograph family. Enter Fig. 12-27 with $CN = 82$ and $P = 9.4$, and read hydrograph family No. 2.
5. Duration of rainfall excess: Enter Fig. 12-4 with $P = 9.4$ in. at $CN = 82$ and read $T_0 = 5.37$ hr.
6. Compute $T_p$; $T_p = 0.7\ T_c = 0.88$ hr.
7. $T_0/T_p$ ratio; $5.37/0.88 = 6.10$.
8. Revised $T_0/T_p$ based on available curves from Table 12-16: $(T_0/T_p)_{\text{rev}} = 6$.
9. Rev $T_p$: rev $T_p = 5.37/6 = 0.895$ hr.
10. $q_p$:$q_p = 484(A) \div$ rev $T_p = 1006$ cfs.
11. $Qq_p$:$(Q)(q_p) = (7.21)(1006) = 7250$ cfs.
12. From the selected hydrograph family in Table 12-17, multiply each value of the $t/T_p$ column by rev $T_p$ to obtain $t$ in hours.
13. From the selected hydrograph family in Table 12-17, multiply each value of the $q_c/q_p$ column by $(Q)(q_p)$ to obtain values of $q$ in cfs.

The solution for Example 12-1 is given in Table 12-18.

**Table 12-17** *Hydrograph Family No. 1*

| Line No. | $T_0/T_p = 1$ | | $T_0/T_p = 1.5$ | | $T_0/T_p = 2$ | | $T_0/T_p = 3$ | | $T_0/T_p = 4$ | | $T_0/T_p = 6$ | |
|---|---|---|---|---|---|---|---|---|---|---|---|---|
| | $t/T_p$ | $q_c/q_p$ | $t/T_p$ | $q_c/q_p$ | $t/T_p$ | $q_c/q_p$ | $t/T_p$ | $q_c/q_p$ | $t/T_p$ | $q_c/q_p$ | $t/T_p$ | $q_c/q_p$ |
| 1 | 0 | 0 | 0 | 0 | 0 | 0 | 0 | 0 | 0 | 0 | 0 | 0 |
| 2 | 0.28 | 0.029 | 0.32 | 0.012 | 0.29 | 0.007 | 0.35 | 0.005 | 0.35 | 0.003 | 0.44 | 0.003 |
| 3 | 0.56 | 0.150 | 0.64 | 0.118 | 0.58 | 0.035 | 0.70 | 0.027 | 0.70 | 0.015 | 0.98 | 0.018 |
| 4 | 0.84 | 0.472 | 0.96 | 0.377 | 0.87 | 0.164 | 1.05 | 0.101 | 1.05 | 0.049 | 1.32 | 0.041 |
| 5 | 1.12 | 0.798 | 1.28 | 0.711 | 1.16 | 0.432 | 1.40 | 0.302 | 1.40 | 0.122 | 1.76 | 0.084 |
| 6 | 1.40 | 0.901 | 1.60 | 0.815 | 1.45 | 0.669 | 1.75 | 0.563 | 1.75 | 0.298 | 2.20 | 0.176 |
| 7 | 1.68 | 0.776 | 1.92 | 0.719 | 1.74 | 0.740 | 2.10 | 0.650 | 2.10 | 0.528 | 2.64 | 0.386 |
| 8 | 1.96 | 0.568 | 2.24 | 0.526 | 2.03 | 0.680 | 2.45 | 0.576 | 2.45 | 0.585 | 3.08 | 0.497 |
| 9 | 2.24 | 0.389 | 2.56 | 0.352 | 2.32 | 0.561 | 2.80 | 0.460 | 2.80 | 0.518 | 3.52 | 0.430 |
| 10 | 2.52 | 0.258 | 2.88 | 0.225 | 2.61 | 0.441 | 3.51 | 0.374 | 3.15 | 0.413 | 3.96 | 0.335 |
| 11 | 2.80 | 0.173 | 3.20 | 0.143 | 2.90 | 0.319 | 3.60 | 0.290 | 3.50 | 0.334 | 4.40 | .258 |
| 12 | 3.08 | 0.115 | 3.52 | 0.090 | 3.19 | 0.212 | 3.85 | 0.201 | 3.85 | 0.273 | 4.84 | 0.202 |
| 13 | 3.36 | 0.078 | 3.84 | 0.057 | 3.48 | 0.140 | 4.20 | 0.127 | 4.20 | 0.231 | 5.28 | 0.164 |
| 14 | 3.64 | 0.052 | 4.16 | 0.037 | 3.77 | 0.094 | 4.55 | 0.078 | 4.55 | 0.189 | 5.72 | 0.139 |
| 15 | 3.92 | 0.036 | 4.48 | 0.024 | 4.06 | 0.063 | 4.90 | 0.047 | 4.90 | 0.128 | 6.16 | 0.124 |
| 16 | 4.20 | 0.024 | 4.80 | 0.015 | 4.35 | 0.042 | 5.25 | 0.028 | 5.25 | 0.080 | 6.60 | 0.100 |
| 17 | 4.48 | 0.016 | 5.12 | 0.008 | 4.64 | 0.028 | 5.60 | 0.016 | 5.60 | 0.047 | 7.04 | 0.060 |
| 18 | 4.76 | 0.009 | 5.44 | 0.004 | 4.93 | 0.017 | 5.95 | 0.009 | 5.95 | 0.028 | 7.48 | 0.033 |
| 19 | 5.04 | 0.005 | 5.76 | 0.002 | 5.22 | 0.011 | 6.30 | 0.005 | 6.30 | 0.017 | 7.92 | 0.018 |
| 20 | 5.32 | 0.002 | 6.08 | 0.001 | 5.51 | 0.007 | 6.65 | 0.003 | 6.65 | 0.010 | 8.36 | 0.009 |
| 21 | 5.60 | 0.001 | 6.40 | 0 | 5.80 | 0.004 | 7.00 | 0.002 | 7.00 | 0.006 | 8.80 | 0.005 |
| 22 | 5.88 | 0 | | | 6.09 | 0.002 | 7.35 | 0.001 | 7.35 | 0.004 | 9.24 | 0.003 |
| 23 | | | | | 6.38 | 0.001 | 7.70 | 0 | 7.70 | 0.003 | 9.68 | 0.002 |
| 24 | | | | | 6.67 | 0 | | | 8.05 | 0.002 | 10.12 | 0.001 |
| 25 | | | | | | | | | 8.40 | 0.001 | 10.56 | 0 |
| 26 | | | | | | | | | 8.75 | 0 | | |

*Source:* After "Hydrology," Suppl. A to Sec. 4, *Engineering Handbook*, U.S. Department of Agriculture, Soil Conservation Service, 1968.

| Line No. | $T_0/T_p = 10$ | | $T_0/T_p = 16$ | | $T_0/T_p = 25$ | | $T_0/T_p = 36$ | | $T_0/T_p = 50$ | | $T_0/T_p = 75$ | |
|---|---|---|---|---|---|---|---|---|---|---|---|---|
| | $t/T_p$ | $q_c/q_p$ | $t/T_p$ | $q_c/q_p$ | $t/T_p$ | $q_c/q_p$ | $t/T_p$ | $q_c/q_p$ | $t/T_p$ | $q_c/q_p$ | $t/T_p$ | $q_c/q_p$ |
| 1 | 0 | 0 | 0 | 0 | 0 | 0 | 0 | 0 | 0 | 0 | 0 | 0 |
| 2 | 0.56 | 0.002 | 0.66 | 0.001 | 1.22 | 0.002 | 1.70 | 0.002 | 2.00 | 0.0019 | 3.00 | 0.0017 |
| 3 | 1.12 | 0.013 | 1.32 | 0.006 | 2.44 | 0.009 | 3.40 | 0.008 | 4.00 | 0.0052 | 6.00 | 0.0039 |
| 4 | 1.68 | 0.027 | 1.98 | 0.015 | 3.66 | 0.018 | 5.10 | 0.014 | 6.00 | 0.0085 | 9.00 | 0.0054 |
| 5 | 2.24 | 0.047 | 2.64 | 0.027 | 4.88 | 0.027 | 6.80 | 0.020 | 8.00 | 0.0118 | 12.00 | 0.0084 |
| 6 | 2.80 | 0.071 | 3.30 | 0.037 | 6.10 | 0.036 | 8.50 | 0.026 | 10.00 | 0.0151 | 15.00 | 0.0106 |
| 7 | 3.36 | 0.115 | 3.96 | 0.047 | 7.32 | 0.046 | 10.20 | 0.033 | 12.00 | 0.0192 | 18.00 | 0.0137 |
| 8 | 3.92 | 0.278 | 4.62 | 0.062 | 8.54 | 0.116 | 11.90 | 0.077 | 14.00 | 0.0259 | 21.00 | 0.0197 |
| 9 | 4.48 | 0.394 | 5.28 | 0.092 | 9.76 | 0.232 | 13.60 | 0.177 | 16.00 | 0.0578 | 24.00 | 0.0516 |
| 10 | 5.04 | 0.322 | 5.94 | 0.223 | 10.98 | 0.146 | 15.30 | 0.101 | 18.00 | 0.1330 | 27.00 | 0.0900 |
| 11 | 5.60 | 0.235 | 6.60 | 0.309 | 12.20 | 0.088 | 17.00 | 0.058 | 20.00 | 0.0941 | 30.00 | 0.0593 |
| 12 | 6.16 | 0.174 | 7.26 | 0.243 | 13.42 | 0.062 | 18.70 | 0.044 | 22.00 | 0.0506 | 33.00 | 0.0321 |
| 13 | 6.72 | 0.136 | 7.92 | 0.171 | 14.64 | 0.051 | 20.40 | 0.036 | 24.00 | 0.0357 | 36.00 | 0.0226 |
| 14 | 7.28 | 0.110 | 8.58 | 0.124 | 15.86 | 0.045 | 22.10 | 0.030 | 26.00 | 0.0297 | 39.00 | 0.0188 |
| 15 | 7.84 | 0.092 | 9.24 | 0.097 | 17.08 | 0.039 | 23.80 | 0.027 | 28.00 | 0.0254 | 42.00 | 0.0161 |
| 16 | 8.40 | 0.079 | 9.90 | 0.081 | 18.30 | 0.035 | 25.50 | 0.024 | 30.00 | 0.0219 | 45.00 | 0.0142 |
| 17 | 8.96 | 0.073 | 10.56 | 0.070 | 19.52 | 0.031 | 27.20 | 0.022 | 32.00 | 0.0192 | 48.00 | 0.0125 |
| 18 | 9.52 | 0.068 | 11.22 | 0.061 | 20.74 | 0.027 | 28.90 | 0.020 | 34.00 | 0.0172 | 51.00 | 0.0112 |
| 19 | 10.08 | 0.065 | 11.88 | 0.055 | 21.96 | 0.025 | 30.60 | 0.018 | 36.00 | 0.0159 | 54.00 | 0.0105 |
| 20 | 10.64 | 0.053 | 12.54 | 0.050 | 23.18 | 0.025 | 32.30 | 0.017 | 38.00 | 0.0150 | 57.00 | 0.0100 |
| 21 | 11.20 | 0.027 | 13.20 | 0.047 | 24.40 | 0.025 | 34.00 | 0.017 | 40.00 | 0.0145 | 60.00 | 0.0097 |
| 22 | 11.76 | 0.012 | 13.86 | 0.045 | 25.62 | 0.020 | 35.70 | 0.017 | 42.00 | 0.0140 | 63.00 | 0.0094 |
| 23 | 12.32 | 0.006 | 14.52 | 0.044 | 26.84 | 0.005 | 37.40 | 0.004 | 44.00 | 0.0136 | 66.00 | 0.0090 |
| 24 | 12.88 | 0.003 | 15.18 | 0.043 | 28.06 | 0.002 | 39.10 | 0.002 | 46.00 | 0.0131 | 69.00 | 0.0087 |
| 25 | 13.44 | 0.002 | 15.84 | 0.040 | 29.28 | 0 | 40.80 | 0 | 48.00 | 0.0125 | 72.00 | 0.0084 |
| 26 | 14.00 | 0.001 | 16.50 | 0.034 | | | | | 50.00 | 0.0123 | 75.00 | 0.0081 |
| 27 | 14.56 | 0 | 17.16 | 0.020 | | | | | 52.00 | 0.0016 | 78.00 | 0.0002 |
| 28 | | | 17.82 | 0.008 | | | | | 54.00 | 0 | 81.00 | 0 |
| 29 | | | 18.48 | 0.004 | | | | | | | | |
| 30 | | | 19.14 | 0.002 | | | | | | | | |
| 31 | | | 19.80 | 0.001 | | | | | | | | |
| 32 | | | 20.46 | 0 | | | | | | | | |

**Table 12-17** *Hydrograph Family No. 2*

| Line No. | $T_0/T_p = 1$ | | $T_0/T_p = 1.5$ | | $T_0/T_p = 2$ | | $T_0/T_p = 3$ | | $T_0/T_p = 4$ | | $T_0/T_p = 6$ | |
|---|---|---|---|---|---|---|---|---|---|---|---|---|
| | $t/T_p$ | $q_c/q_p$ | $t/T_p$ | $q_c/q_p$ | $t/T_p$ | $q_c/q_p$ | $t/T_p$ | $q_c/q_p$ | $t/T_p$ | $q_c/q_p$ | $t/T_p$ | $q_c/q_p$ |
| 1 | 0 | 0 | 0 | 0 | 0 | 0 | 0 | 0 | 0 | 0 | 0 | 0 |
| 2 | 0.28 | 0.026 | 0.22 | 0.003 | 0.28 | 0.004 | 0.32 | 0.003 | 0.32 | 0.002 | 0.34 | 0.001 |
| 3 | 0.56 | 0.170 | 0.44 | 0.041 | 0.56 | 0.040 | 0.64 | 0.017 | 0.64 | 0.009 | 0.68 | 0.005 |
| 4 | 0.84 | 0.480 | 0.66 | 0.161 | 0.84 | 0.170 | 0.96 | 0.093 | 0.96 | 0.036 | 1.02 | 0.015 |
| 5 | 1.12 | 0.802 | 0.88 | 0.362 | 1.12 | 0.428 | 1.28 | 0.311 | 1.28 | 0.129 | 1.36 | 0.037 |
| 6 | 1.40 | 0.885 | 1.10 | 0.604 | 1.40 | 0.645 | 1.60 | 0.530 | 1.60 | 0.332 | 1.70 | 0.098 |
| 7 | 1.68 | 0.770 | 1.32 | 0.740 | 1.68 | 0.715 | 1.92 | 0.615 | 1.92 | 0.501 | 2.04 | 0.244 |
| 8 | 1.96 | 0.550 | 1.54 | 0.790 | 1.96 | 0.677 | 2.24 | 0.575 | 2.24 | 0.550 | 2.38 | 0.407 |
| 9 | 2.24 | 0.380 | 1.76 | 0.746 | 2.24 | 0.574 | 2.56 | 0.487 | 2.56 | 0.500 | 2.72 | 0.464 |
| 10 | 2.52 | 0.257 | 1.98 | 0.640 | 2.52 | 0.472 | 2.88 | 0.409 | 2.88 | 0.422 | 3.06 | 0.429 |
| 11 | 2.80 | 0.166 | 2.20 | 0.536 | 2.80 | 0.369 | 3.20 | 0.344 | 3.20 | 0.358 | 3.40 | 0.367 |
| 12 | 3.08 | 0.113 | 2.42 | 0.414 | 3.08 | 0.247 | 3.52 | 0.279 | 3.52 | 0.302 | 3.74 | 0.309 |
| 13 | 3.36 | 0.078 | 2.64 | 0.303 | 3.36 | 0.168 | 3.84 | 0.206 | 3.84 | 0.274 | 4.08 | 0.261 |
| 14 | 3.64 | 0.052 | 2.86 | 0.219 | 3.64 | 0.113 | 4.16 | 0.135 | 4.16 | 0.230 | 4.42 | 0.224 |
| 15 | 3.92 | 0.034 | 3.08 | 0.160 | 3.92 | 0.075 | 4.48 | 0.087 | 4.48 | 0.195 | 4.76 | 0.193 |
| 16 | 4.20 | 0.023 | 3.30 | 0.117 | 4.20 | 0.050 | 4.80 | 0.054 | 4.80 | 0.147 | 5.10 | 0.169 |
| 17 | 4.48 | 0.015 | 3.52 | 0.088 | 4.48 | 0.034 | 5.12 | 0.032 | 5.12 | 0.099 | 5.44 | 0.152 |
| 18 | 4.76 | 0.009 | 3.74 | 0.064 | 4.76 | 0.021 | 5.44 | 0.019 | 5.44 | 0.061 | 5.78 | 0.139 |
| 19 | 5.04 | 0.004 | 3.96 | 0.047 | 5.04 | 0.014 | 5.76 | 0.012 | 5.76 | 0.037 | 6.12 | 0.129 |
| 20 | 5.32 | 0.002 | 4.18 | 0.035 | 5.32 | 0.008 | 6.08 | 0.008 | 6.08 | 0.023 | 6.46 | 0.113 |
| 21 | 5.60 | 0.001 | 4.40 | 0.025 | 5.60 | 0.004 | 6.40 | 0.005 | 6.40 | 0.013 | 6.80 | 0.085 |
| 22 | 5.88 | 0 | 4.62 | 0.018 | 5.88 | 0.003 | 6.72 | 0.003 | 6.72 | 0.008 | 7.14 | 0.055 |
| 23 | | | 4.84 | 0.012 | 6.16 | 0.002 | 7.04 | 0.002 | 7.04 | 0.005 | 7.48 | 0.035 |
| 24 | | | 5.06 | 0.007 | 6.44 | 0.001 | 7.36 | 0.001 | 7.36 | 0.004 | 7.82 | 0.020 |
| 25 | | | 5.28 | 0.004 | 6.72 | 0 | 7.68 | 0 | 7.68 | 0.003 | 8.16 | 0.012 |
| 26 | | | 5.50 | 0.003 | | | | | 8.00 | 0.002 | 8.50 | 0.008 |
| 27 | | | 5.72 | 0.002 | | | | | 8.32 | 0.001 | 8.84 | 0.005 |
| 28 | | | 5.94 | 0.001 | | | | | 8.64 | 0 | 9.18 | 0.004 |
| 29 | | | 6.16 | 0 | | | | | | | 9.52 | 0.003 |
| 30 | | | | | | | | | | | 9.86 | 0.002 |
| 31 | | | | | | | | | | | 10.20 | 0.001 |
| 32 | | | | | | | | | | | 10.54 | 0 |

| Line No. | $T_0/T_p = 10$ | | $T_0/T_p = 16$ | | $T_0/T_p = 25$ | | $T_0/T_p = 36$ | | $T_0/T_p = 50$ | | $T_0/T_p = 75$ | |
|---|---|---|---|---|---|---|---|---|---|---|---|---|
| | $t/T_p$ | $q_c/q_p$ | $t/T_p$ | $q_c/q_p$ | $t/T_p$ | $q_c/q_p$ | $t/T_p$ | $q_c/q_p$ | $t/T_p$ | $q_c/q_p$ | $t/T_p$ | $q_c/q_p$ |
| 1 | 0 | 0 | 0 | 0 | 0 | 0 | 0 | 0 | 0 | 0 | 0 | 0 |
| 2 | 0.63 | 0.002 | 0.90 | 0.002 | 1.30 | 0.002 | 1.79 | 0.002 | 2.50 | 0.0018 | 3.00 | 0.0012 |
| 3 | 1.26 | 0.009 | 1.80 | 0.007 | 2.60 | 0.006 | 3.58 | 0.006 | 5.00 | 0.0047 | 6.00 | 0.0027 |
| 4 | 1.89 | 0.027 | 2.70 | 0.020 | 3.90 | 0.014 | 5.37 | 0.012 | 7.50 | 0.0087 | 9.00 | 0.0044 |
| 5 | 2.52 | 0.063 | 3.60 | 0.037 | 5.20 | 0.024 | 7.16 | 0.019 | 10.00 | 0.0145 | 12.00 | 0.0067 |
| 6 | 3.15 | 0.236 | 4.50 | 0.148 | 6.50 | 0.088 | 8.95 | 0.057 | 12.50 | 0.0615 | 15.00 | 0.0108 |
| 7 | 3.78 | 0.364 | 5.40 | 0.277 | 7.80 | 0.210 | 10.74 | 0.157 | 15.00 | 0.1184 | 18.00 | 0.0309 |
| 8 | 4.41 | 0.307 | 6.30 | 0.214 | 9.10 | 0.146 | 12.53 | 0.104 | 17.50 | 0.0621 | 21.00 | 0.0790 |
| 9 | 5.04 | 0.226 | 7.20 | 0.149 | 10.40 | 0.097 | 14.32 | 0.068 | 20.00 | 0.0433 | 24.00 | 0.0624 |
| 10 | 5.67 | 0.172 | 8.10 | 0.112 | 11.70 | 0.072 | 16.11 | 0.047 | 22.50 | 0.0342 | 27.00 | 0.0357 |
| 11 | 6.30 | 0.136 | 9.00 | 0.088 | 13.00 | 0.057 | 17.90 | 0.040 | 25.00 | 0.0274 | 30.00 | 0.0283 |
| 12 | 6.93 | 0.113 | 9.90 | 0.073 | 14.30 | 0.049 | 19.69 | 0.034 | 27.50 | 0.0234 | 33.00 | 0.0234 |
| 13 | 7.56 | 0.097 | 10.80 | 0.063 | 15.60 | 0.044 | 21.48 | 0.030 | 30.00 | 0.0209 | 36.00 | 0.0196 |
| 14 | 8.19 | 0.085 | 11.70 | 0.056 | 16.90 | 0.039 | 23.27 | 0.026 | 32.50 | 0.0187 | 39.00 | 0.0167 |
| 15 | 8.82 | 0.078 | 12.60 | 0.052 | 18.20 | 0.035 | 25.06 | 0.025 | 35.00 | 0.0167 | 42.00 | 0.0150 |
| 16 | 9.45 | 0.074 | 13.50 | 0.048 | 19.50 | 0.033 | 26.85 | 0.023 | 37.50 | 0.0159 | 45.00 | 0.0137 |
| 17 | 10.08 | 0.069 | 14.40 | 0.045 | 20.80 | 0.031 | 28.64 | 0.021 | 40.00 | 0.0153 | 48.00 | 0.0126 |
| 18 | 10.71 | 0.053 | 15.30 | 0.044 | 22.10 | 0.029 | 30.43 | 0.020 | 42.50 | 0.0147 | 51.00 | 0.0115 |
| 19 | 11.34 | 0.025 | 16.20 | 0.042 | 23.40 | 0.028 | 32.22 | 0.019 | 45.00 | 0.0142 | 54.00 | 0.0108 |
| 20 | 11.97 | 0.009 | 17.10 | 0.023 | 24.70 | 0.027 | 34.01 | 0.018 | 47.50 | 0.0136 | 57.00 | 0.0104 |
| 21 | 12.60 | 0.004 | 18.00 | 0.006 | 26.00 | 0.014 | 35.80 | 0.017 | 50.00 | 0.0131 | 60.00 | 0.0101 |
| 22 | 13.23 | 0.002 | 18.90 | 0.003 | 27.30 | 0.004 | 37.59 | 0.007 | 52.50 | 0.0008 | 63.00 | 0.0098 |
| 23 | 13.86 | 0.001 | 19.80 | 0.001 | 28.60 | 0.001 | 39.38 | 0.001 | 55.00 | 0 | 66.00 | 0.0095 |
| 24 | 14.49 | 0 | 20.70 | 0 | 29.90 | 0 | 41.17 | 0 | | | 69.00 | 0.0092 |
| 25 | | | | | | 0 | | 0 | | | 72.00 | 0.0089 |
| 26 | | | | | | | | | | | 75.00 | 0.0086 |
| 27 | | | | | | | | | | | 78.00 | 0.0003 |
| 28 | | | | | | | | | | | 81.00 | 0 |

**Table 12-17** *Hydrograph Family No. 3*

| Line No. | $T_0/T_p = 1$ | | $T_0/T_p = 1.5$ | | $T_0/T_p = 2$ | | $T_0/T_p = 3$ | | $T_0/T_p = 4$ | | $T_0/T_p = 6$ | |
|---|---|---|---|---|---|---|---|---|---|---|---|---|
| | $t/T_p$ | $q_c/q_p$ | $t/T_p$ | $q_c/q_p$ | $t/T_p$ | $q_c/q_p$ | $t/T_p$ | $q_c/q_p$ | $t/T_p$ | $q_c/q_p$ | $t/T_p$ | $q_c/q_p$ |
| 1 | 0 | 0 | 0 | 0 | 0 | 0 | 0 | 0 | 0 | 0 | 0 | 0 |
| 2 | 0.26 | 0.048 | 0.29 | 0.028 | 0.30 | 0.012 | 0.34 | 0.004 | 0.36 | 0.003 | 0.42 | 0.002 |
| 3 | 0.52 | 0.219 | 0.58 | 0.190 | 0.60 | 0.123 | 0.68 | 0.088 | 0.72 | 0.044 | 0.84 | 0.021 |
| 4 | 0.78 | 0.521 | 0.87 | 0.450 | 0.90 | 0.343 | 1.02 | 0.289 | 1.08 | 0.203 | 1.26 | 0.138 |
| 5 | 1.04 | 0.762 | 1.16 | 0.656 | 1.20 | 0.570 | 1.36 | 0.489 | 1.44 | 0.400 | 1.68 | 0.320 |
| 6 | 1.30 | 0.844 | 1.45 | 0.734 | 1.50 | 0.657 | 1.70 | 0.543 | 1.80 | 0.478 | 2.10 | 0.390 |
| 7 | 1.56 | 0.778 | 1.74 | 0.685 | 1.80 | 0.630 | 2.04 | 0.507 | 2.16 | 0.450 | 2.52 | 0.363 |
| 8 | 1.82 | 0.621 | 2.03 | 0.585 | 2.10 | 0.562 | 2.38 | 0.445 | 2.52 | 0.397 | 2.94 | 0.314 |
| 9 | 2.08 | 0.441 | 2.32 | 0.445 | 2.40 | 0.484 | 2.72 | 0.385 | 2.88 | 0.342 | 3.36 | 0.270 |
| 10 | 2.34 | 0.305 | 2.61 | 0.350 | 2.70 | 0.379 | 3.06 | 0.340 | 3.24 | 0.296 | 3.78 | 0.232 |
| 11 | 2.60 | 0.214 | 2.90 | 0.199 | 3.00 | 0.267 | 3.40 | 0.294 | 3.60 | 0.257 | 4.20 | 0.199 |
| 12 | 2.86 | 0.149 | 3.19 | 0.132 | 3.30 | 0.177 | 3.74 | 0.223 | 3.96 | 0.234 | 4.62 | 0.174 |
| 13 | 3.12 | 0.103 | 3.48 | 0.089 | 3.60 | 0.116 | 4.08 | 0.149 | 4.32 | 0.210 | 5.04 | 0.155 |
| 14 | 3.38 | 0.070 | 3.77 | 0.057 | 3.90 | 0.076 | 4.42 | 0.096 | 4.68 | 0.169 | 5.46 | 0.144 |
| 15 | 3.64 | 0.048 | 4.06 | 0.038 | 4.20 | 0.050 | 4.76 | 0.056 | 5.04 | 0.111 | 5.88 | 0.137 |
| 16 | 3.90 | 0.034 | 4.35 | 0.025 | 4.50 | 0.033 | 5.10 | 0.033 | 5.40 | 0.067 | 6.30 | 0.127 |
| 17 | 4.16 | 0.024 | 4.64 | 0.015 | 4.80 | 0.020 | 5.44 | 0.019 | 5.76 | 0.037 | 6.72 | 0.101 |
| 18 | 4.42 | 0.016 | 4.93 | 0.008 | 5.10 | 0.011 | 5.78 | 0.013 | 6.12 | 0.022 | 7.14 | 0.063 |
| 19 | 4.68 | 0.010 | 5.22 | 0.005 | 5.40 | 0.006 | 6.12 | 0.008 | 6.48 | 0.014 | 7.56 | 0.033 |
| 20 | 4.94 | 0.006 | 5.51 | 0.003 | 5.70 | 0.004 | 6.46 | 0.004 | 6.84 | 0.008 | 7.98 | 0.018 |
| 21 | 5.20 | 0.003 | 5.80 | 0.002 | 6.00 | 0.002 | 6.80 | 0.003 | 7.20 | 0.006 | 8.40 | 0.010 |
| 22 | 5.46 | 0.001 | 6.09 | 0.001 | 6.30 | 0.001 | 7.14 | 0.002 | 7.56 | 0.004 | 8.82 | 0.005 |
| 23 | 5.72 | 0 | 6.38 | 0 | 6.60 | 0 | 7.48 | 0.001 | 7.92 | 0.002 | 9.24 | 0.003 |
| 24 | | | | | | | 7.82 | 0 | 8.28 | 0.001 | 9.66 | 0.002 |
| 25 | | | | | | | | | 8.64 | 0 | 10.08 | 0.001 |
| 26 | | | | | | | | | | | 10.50 | |
| 27 | | | | | | | | | | | 10.92 | 0 |

| Line No. | $T_0/T_p = 10$ | | $T_0/T_p = 16$ | | $T_0/T_p = 25$ | | $T_0/T_p = 36$ | | $T_0/T_p = 50$ | | $T_0/T_p = 75$ | |
|---|---|---|---|---|---|---|---|---|---|---|---|---|
| | $t/T_p$ | $q_c/q_p$ | $t/T_p$ | $q_c/q_p$ | $t/T_p$ | $q_c/q_p$ | $t/T_p$ | $q_c/q_p$ | $t/T_p$ | $q_c/q_p$ | $t/T_p$ | $q_c/q_p$ |
| 1 | 0 | 0 | 0 | 0 | 0 | 0 | 0 | 0 | 0 | 0 | 0 | 0 |
| 2 | 0.54 | 0.001 | 0.90 | 0.002 | 1.23 | 0.002 | 1.62 | 0.002 | 2.25 | 0.0008 | 3.25 | 0.0009 |
| 3 | 1.08 | 0.008 | 1.80 | 0.016 | 2.46 | 0.009 | 3.24 | 0.006 | 4.50 | 0.0070 | 6.50 | 0.0057 |
| 4 | 1.62 | 0.069 | 2.70 | 0.122 | 3.69 | 0.073 | 4.86 | 0.047 | 6.75 | 0.0474 | 9.75 | 0.0289 |
| 5 | 2.16 | 0.231 | 3.60 | 0.230 | 4.92 | 0.173 | 6.48 | 0.130 | 9.00 | 0.0972 | 13.00 | 0.0667 |
| 6 | 2.70 | 0.303 | 4.50 | 0.185 | 6.15 | 0.132 | 8.10 | 0.097 | 11.25 | 0.0642 | 16.25 | 0.0445 |
| 7 | 3.24 | 0.269 | 5.40 | 0.139 | 7.38 | 0.096 | 9.72 | 0.069 | 13.50 | 0.0460 | 19.50 | 0.0317 |
| 8 | 3.78 | 0.223 | 6.30 | 0.113 | 8.61 | 0.076 | 11.34 | 0.052 | 15.75 | 0.0375 | 22.75 | 0.0257 |
| 9 | 4.32 | 0.188 | 7.20 | 0.094 | 9.84 | 0.064 | 12.96 | 0.045 | 18.00 | 0.0322 | 26.00 | 0.0219 |
| 10 | 4.86 | 0.159 | 8.10 | 0.081 | 11.07 | 0.055 | 14.58 | 0.041 | 20.25 | 0.0285 | 29.25 | 0.0195 |
| 11 | 5.40 | 0.139 | 9.00 | 0.072 | 12.30 | 0.050 | 16.20 | 0.037 | 22.50 | 0.0258 | 32.50 | 0.0176 |
| 12 | 5.94 | 0.122 | 9.90 | 0.064 | 13.53 | 0.046 | 17.82 | 0.034 | 24.75 | 0.0239 | 35.75 | 0.0160 |
| 13 | 6.48 | 0.108 | 10.80 | 0.057 | 14.76 | 0.042 | 19.44 | 0.031 | 27.00 | 0.0219 | 39.00 | 0.0147 |
| 14 | 7.02 | 0.097 | 11.70 | 0.053 | 15.99 | 0.038 | 21.06 | 0.028 | 29.25 | 0.0201 | 42.25 | 0.0136 |
| 15 | 7.56 | 0.089 | 12.60 | 0.050 | 17.22 | 0.035 | 22.68 | 0.025 | 31.50 | 0.0185 | 45.50 | 0.0127 |
| 16 | 8.10 | 0.081 | 13.50 | 0.049 | 18.45 | 0.033 | 24.30 | 0.024 | 33.75 | 0.0173 | 48.75 | 0.0118 |
| 17 | 8.64 | 0.078 | 14.40 | 0.048 | 19.68 | 0.032 | 25.92 | 0.024 | 36.00 | 0.0165 | 52.00 | 0.0113 |
| 18 | 9.18 | 0.077 | 15.30 | 0.047 | 20.91 | 0.031 | 27.54 | 0.024 | 38.25 | 0.0162 | 55.25 | 0.0109 |
| 19 | 9.72 | 0.077 | 16.20 | 0.046 | 22.14 | 0.031 | 29.16 | 0.024 | 40.50 | 0.0159 | 58.50 | 0.0107 |
| 20 | 10.26 | 0.075 | 17.10 | 0.024 | 23.37 | 0.031 | 30.78 | 0.023 | 42.75 | 0.0156 | 61.75 | 0.0105 |
| 21 | 10.80 | 0.055 | 18.00 | 0.006 | 24.60 | 0.031 | 32.40 | 0.023 | 45.00 | 0.0153 | 65.00 | 0.0103 |
| 22 | 11.34 | 0.030 | 18.90 | 0.004 | 25.83 | 0.025 | 34.02 | 0.023 | 47.25 | 0.0150 | 68.25 | 0.0101 |
| 23 | 11.88 | 0.012 | 19.80 | 0.002 | 27.06 | 0.004 | 35.64 | 0.023 | 49.50 | 0.0147 | 71.50 | 0.0099 |
| 24 | 12.42 | 0.006 | 20.70 | 0 | 28.29 | 0.001 | 37.26 | 0.007 | 51.75 | 0.0028 | 74.75 | 0.0097 |
| 25 | 12.96 | 0.004 | | | 29.52 | 0 | 38.88 | 0.003 | 54.00 | 0 | 78.00 | 0.0003 |
| 26 | 13.50 | 0.002 | | | | | 40.50 | 0 | | | 81.25 | 0 |
| 27 | 14.04 | 0.001 | | | | | | | | | | |
| 28 | 14.58 | 0 | | | | | | | | | | |

**Table 12-17** *Hydrograph Family No. 4*

| Line No. | $T_0/T_p = 1$ | | $T_0/T_p = 1.5$ | | $T_0/T_p = 2$ | | $T_0/T_p = 3$ | | $T_0/T_p = 4$ | | $T_0/T_p = 6$ | |
|---|---|---|---|---|---|---|---|---|---|---|---|---|
| | $t/T_p$ | $q_c/q_p$ | $t/T_p$ | $q_c/q_p$ | $t/T_p$ | $q_c/q_p$ | $t/T_p$ | $q_c/q_p$ | $t/T_p$ | $q_c/q_p$ | $t/T_p$ | $q_c/q_p$ |
| 1 | 0 | 0 | 0 | 0 | 0 | 0 | 0 | 0 | 0 | 0 | 0 | 0 |
| 2 | 0.28 | 0.051 | 0.28 | 0.038 | 0.32 | 0.031 | 0.28 | 0.018 | 0.40 | 0.023 | 0.40 | 0.014 |
| 3 | 0.56 | 0.220 | 0.56 | 0.166 | 0.64 | 0.173 | 0.56 | 0.086 | 0.80 | 0.143 | 0.80 | 0.088 |
| 4 | 0.84 | 0.490 | 0.84 | 0.360 | 0.96 | 0.360 | 0.84 | 0.200 | 1.20 | 0.272 | 1.20 | 0.191 |
| 5 | 1.12 | 0.738 | 1.12 | 0.551 | 1.28 | 0.494 | 1.12 | 0.311 | 1.60 | 0.326 | 1.60 | 0.244 |
| 6 | 1.40 | 0.830 | 1.40 | 0.651 | 1.60 | 0.555 | 1.40 | 0.386 | 2.00 | 0.340 | 2.00 | 0.250 |
| 7 | 1.68 | 0.751 | 1.68 | 0.686 | 1.92 | 0.567 | 1.68 | 0.415 | 2.40 | 0.337 | 2.40 | 0.246 |
| 8 | 1.96 | 0.573 | 1.96 | 0.650 | 2.24 | 0.555 | 1.96 | 0.422 | 2.80 | 0.323 | 2.80 | 0.240 |
| 9 | 2.24 | 0.392 | 2.24 | 0.543 | 2.56 | 0.490 | 2.24 | 0.417 | 3.20 | 0.306 | 3.20 | 0.233 |
| 10 | 2.52 | 0.259 | 2.52 | 0.392 | 2.88 | 0.370 | 2.52 | 0.402 | 3.60 | 0.293 | 3.60 | 0.223 |
| 11 | 2.80 | 0.174 | 2.80 | 0.267 | 3.20 | 0.242 | 2.80 | 0.394 | 4.00 | 0.286 | 4.00 | 0.212 |
| 12 | 3.08 | 0.118 | 3.08 | 0.180 | 3.52 | 0.150 | 3.08 | 0.387 | 4.40 | 0.266 | 4.40 | 0.202 |
| 13 | 3.36 | 0.079 | 3.36 | 0.120 | 3.84 | 0.098 | 3.36 | 0.363 | 4.80 | 0.197 | 4.80 | 0.194 |
| 14 | 3.64 | 0.053 | 3.64 | 0.081 | 4.16 | 0.063 | 3.64 | 0.316 | 5.20 | 0.122 | 5.20 | 0.189 |
| 15 | 3.92 | 0.036 | 3.92 | 0.055 | 4.48 | 0.038 | 3.92 | 0.236 | 5.60 | 0.067 | 5.60 | 0.187 |
| 16 | 4.20 | 0.025 | 4.20 | 0.036 | 4.80 | 0.024 | 4.20 | 0.164 | 6.00 | 0.036 | 6.00 | 0.185 |
| 17 | 4.48 | 0.017 | 4.48 | 0.024 | 5.12 | 0.013 | 4.48 | 0.108 | 6.40 | 0.021 | 6.40 | 0.175 |
| 18 | 4.76 | 0.011 | 4.76 | 0.015 | 5.44 | 0.008 | 4.76 | 0.073 | 6.80 | 0.013 | 6.80 | 0.131 |
| 19 | 5.04 | 0.006 | 5.04 | 0.009 | 5.76 | 0.004 | 5.04 | 0.047 | 7.20 | 0.008 | 7.20 | 0.080 |
| 20 | 5.32 | 0.003 | 5.32 | 0.005 | 6.08 | 0.002 | 5.32 | 0.030 | 7.60 | 0.005 | 7.60 | 0.046 |
| 21 | 5.60 | 0.001 | 5.60 | 0.003 | 6.40 | 0.001 | 5.60 | 0.020 | 8.00 | 0.002 | 8.00 | 0.027 |
| 22 | 5.88 | 0 | 5.88 | 0.001 | 6.72 | 0 | 5.88 | 0.013 | 8.40 | 0.001 | 8.40 | 0.016 |
| 23 | | | 6.16 | 0 | | | 6.16 | 0.008 | 8.80 | 0 | 8.80 | 0.009 |
| 24 | | | | | | | 6.44 | 0.005 | | | 9.20 | 0.005 |
| 25 | | | | | | | 6.72 | 0.003 | | | 9.60 | 0.003 |
| 26 | | | | | | | 7.00 | 0.002 | | | 10.00 | 0.002 |
| 27 | | | | | | | 7.28 | 0.001 | | | 10.40 | 0.001 |
| 28 | | | | | | | 7.56 | | | | 10.80 | 0 |
| 29 | | | | | | | 7.84 | 0 | | | | |

| Line No. | $T_0/T_p = 10$ $t/T_p$ | $q_c/q_p$ | $T_0/T_p = 16$ $t/T_p$ | $q_c/q_p$ | $T_0/T_p = 25$ $t/T_p$ | $q_c/q_p$ | $T_0/T_p = 36$ $t/T_p$ | $q_c/q_p$ | $T_0/T_p = 50$ $t/T_p$ | $q_c/q_p$ |
|---|---|---|---|---|---|---|---|---|---|---|
| 1 | 0 | 0 | 0 | 0 | 0 | 0 | 0 | 0 | 0 | 0 |
| 2 | 0.50 | 0.015 | 0.62 | 0.015 | 1.02 | 0.025 | 1.50 | 0.0306 | 2.00 | 0.0277 |
| 3 | 1.00 | 0.079 | 1.24 | 0.064 | 2.04 | 0.070 | 3.00 | 0.0575 | 4.00 | 0.0464 |
| 4 | 1.50 | 0.151 | 1.86 | 0.112 | 3.06 | 0.092 | 4.50 | 0.0672 | 6.00 | 0.0435 |
| 5 | 2.00 | 0.177 | 2.48 | 0.128 | 4.08 | 0.082 | 6.00 | 0.0492 | 8.00 | 0.0378 |
| 6 | 2.50 | 0.170 | 3.10 | 0.119 | 5.10 | 0.068 | 7.50 | 0.0433 | 10.00 | 0.0335 |
| 7 | 3.00 | 0.159 | 3.72 | 0.105 | 6.12 | 0.062 | 9.00 | 0.0418 | 12.00 | 0.0307 |
| 8 | 3.50 | 0.152 | 4.34 | 0.097 | 7.14 | 0.059 | 10.50 | 0.0408 | 14.00 | 0.0291 |
| 9 | 4.00 | 0.146 | 4.96 | 0.094 | 8.16 | 0.056 | 12.00 | 0.0400 | 16.00 | 0.0282 |
| 10 | 4.50 | 0.141 | 5.58 | 0.091 | 9.18 | 0.055 | 13.50 | 0.0391 | 18.00 | 0.0274 |
| 11 | 5.00 | 0.136 | 6.20 | 0.089 | 10.20 | 0.054 | 15.00 | 0.0382 | 20.00 | 0.0266 |
| 12 | 5.50 | 0.131 | 6.82 | 0.087 | 11.22 | 0.053 | 16.50 | 0.0371 | 22.00 | 0.0258 |
| 13 | 6.00 | 0.126 | 7.44 | 0.085 | 12.24 | 0.052 | 18.00 | 0.0358 | 24.00 | 0.0250 |
| 14 | 6.50 | 0.121 | 8.06 | 0.082 | 13.26 | 0.050 | 19.50 | 0.0341 | 26.00 | 0.0242 |
| 15 | 7.00 | 0.166 | 8.68 | 0.079 | 14.28 | 0.049 | 21.00 | 0.0319 | 28.00 | 0.0234 |
| 16 | 7.50 | 0.112 | 9.30 | 0.076 | 15.30 | 0.047 | 22.50 | 0.0382 | 30.00 | 0.0230 |
| 17 | 8.00 | 0.112 | 9.92 | 0.074 | 16.32 | 0.046 | 24.00 | 0.0306 | 32.00 | 0.0229 |
| 18 | 8.50 | 0.111 | 10.54 | 0.072 | 17.34 | 0.045 | 25.50 | 0.0306 | 34.00 | 0.0227 |
| 19 | 9.00 | 0.111 | 11.16 | 0.071 | 18.36 | 0.044 | 27.00 | 0.0306 | 36.00 | 0.0226 |
| 20 | 9.50 | 0.110 | 11.78 | 0.070 | 19.38 | 0.044 | 28.50 | 0.0306 | 38.00 | 0.0225 |
| 21 | 10.00 | 0.110 | 12.40 | 0.069 | 20.40 | 0.044 | 30.00 | 0.0306 | 40.00 | 0.0224 |
| 22 | 10.50 | 0.100 | 13.02 | 0.069 | 21.42 | 0.044 | 31.50 | 0.0306 | 42.00 | 0.0222 |
| 23 | 11.00 | 0.065 | 13.64 | 0.069 | 22.44 | 0.044 | 33.00 | 0.0306 | 44.00 | 0.0221 |
| 24 | 11.50 | 0.033 | 14.26 | 0.069 | 23.46 | 0.044 | 34.50 | 0.0306 | 46.00 | 0.0219 |
| 25 | 12.00 | 0.025 | 14.88 | 0.069 | 24.48 | 0.044 | 36.00 | 0.0306 | 48.00 | 0.0219 |
| 26 | 12.50 | 0.007 | 15.50 | 0.069 | 25.50 | 0.039 | 37.50 | 0.0085 | 50.00 | 0.0217 |
| 27 | 13.00 | 0.004 | 16.12 | 0.068 | 26.52 | 0.012 | 39.00 | 0.0009 | 52.00 | 0.0029 |
| 28 | 13.50 | 0.002 | 16.74 | 0.053 | 27.54 | 0.004 | 40.50 | 0 | 54.00 | 0 |
| 29 | 14.00 | 0.001 | 17.36 | 0.023 | 28.56 | 0.001 | | | | |
| 30 | 14.50 | 0 | 17.98 | 0.009 | 29.58 | 0 | | | | |
| 31 | | | | | | | | | 18.60 | 0.004 |
| 32 | | | | | | | | | 19.22 | 0.002 |
| 33 | | | | | | | | | 19.84 | 0.001 |
| 34 | | | | | | | | | 20.46 | 0 |

**Table 12-17** *Hydrograph Family No. 5*

| Line | $T_0/T_p = 1$ | | $T_0/T_p = 1.5$ | | $T_0/T_p = 2$ | | $T_0/T_p = 3$ | | $T_0/T_p = 4$ | | $T_0/T_p = 6$ | |
|---|---|---|---|---|---|---|---|---|---|---|---|---|
| No. | $t/T_p$ | $q_c/q_p$ | $t/T_p$ | $q_c/q_p$ | $t/T_p$ | $q_c/q_p$ | $t/T_p$ | $q_c/q_p$ | $t/T_p$ | $q_c/q_p$ | $t/T_p$ | $q_c/q_p$ |
| 1 | 0 | 0 | 0 | 0 | 0 | 0 | 0 | 0 | 0 | 0 | 0 | 0 |
| 2 | 0.26 | 0.021 | 0.25 | 0.013 | 0.25 | 0.010 | 0.34 | 0.010 | 0.36 | 0.010 | 0.52 | 0.015 |
| 3 | 0.52 | 0.106 | 0.50 | 0.065 | 0.50 | 0.048 | 0.68 | 0.068 | 0.72 | 0.053 | 1.04 | 0.070 |
| 4 | 0.78 | 0.289 | 0.75 | 0.173 | 0.75 | 0.127 | 1.02 | 0.150 | 1.08 | 0.124 | 1.56 | 0.130 |
| 5 | 1.04 | 0.530 | 1.00 | 0.306 | 1.00 | 0.227 | 1.36 | 0.229 | 1.44 | 0.181 | 2.08 | 0.159 |
| 6 | 1.30 | 0.740 | 1.25 | 0.434 | 1.25 | 0.318 | 1.70 | 0.283 | 1.80 | 0.220 | 2.60 | 0.172 |
| 7 | 1.56 | 0.848 | 1.50 | 0.562 | 1.50 | 0.389 | 2.04 | 0.315 | 2.16 | 0.243 | 3.12 | 0.178 |
| 8 | 1.82 | 0.767 | 1.75 | 0.680 | 1.75 | 0.448 | 2.38 | 0.339 | 2.52 | 0.256 | 3.64 | 0.182 |
| 9 | 2.08 | 0.590 | 2.00 | 0.737 | 2.00 | 0.523 | 2.72 | 0.378 | 2.88 | 0.263 | 4.16 | 0.183 |
| 10 | 2.34 | 0.406 | 2.25 | 0.673 | 2.25 | 0.609 | 3.06 | 0.459 | 3.24 | 0.273 | 4.68 | 0.184 |
| 11 | 2.60 | 0.279 | 2.50 | 0.530 | 2.50 | 0.642 | 3.40 | 0.509 | 3.60 | 0.308 | 5.20 | 0.218 |
| 12 | 2.86 | 0.193 | 2.75 | 0.381 | 2.75 | 0.576 | 3.74 | 0.446 | 3.96 | 0.380 | 5.72 | 0.285 |
| 13 | 3.12 | 0.134 | 3.00 | 0.262 | 3.00 | 0.450 | 4.08 | 0.310 | 4.32 | 0.427 | 6.24 | 0.324 |
| 14 | 3.38 | 0.092 | 3.25 | 0.185 | 3.25 | 0.322 | 4.42 | 0.190 | 4.68 | 0.377 | 6.76 | 0.267 |
| 15 | 3.64 | 0.065 | 3.50 | 0.129 | 3.50 | 0.222 | 4.76 | 0.117 | 5.04 | 0.260 | 7.28 | 0.133 |
| 16 | 3.90 | 0.044 | 3.75 | 0.090 | 3.75 | 0.156 | 5.10 | 0.069 | 5.40 | 0.155 | 7.80 | 0.064 |
| 17 | 4.16 | 0.030 | 4.00 | 0.063 | 4.00 | 0.109 | 5.44 | 0.040 | 5.76 | 0.094 | 8.32 | 0.029 |
| 18 | 4.42 | 0.021 | 4.25 | 0.045 | 4.25 | 0.075 | 5.78 | 0.025 | 6.12 | 0.055 | 8.84 | 0.016 |
| 19 | 4.68 | 0.015 | 4.50 | 0.031 | 4.50 | 0.053 | 6.12 | 0.016 | 6.48 | 0.032 | 9.36 | 0.007 |
| 20 | 4.94 | 0.009 | 4.75 | 0.022 | 4.75 | 0.037 | 6.46 | 0.009 | 6.84 | 0.019 | 9.88 | 0.003 |
| 21 | 5.20 | 0.005 | 5.00 | 0.014 | 5.00 | 0.025 | 6.80 | 0.005 | 7.20 | 0.012 | 10.40 | 0.001 |
| 22 | 5.46 | 0.002 | 5.25 | 0.009 | 5.25 | 0.017 | 7.14 | 0.003 | 7.56 | 0.007 | 10.92 | 0 |
| 23 | 5.72 | 0 | 5.50 | 0.005 | 5.50 | 0.011 | 7.48 | 0.001 | 7.92 | 0.004 | | 0 |
| 24 | | | 5.75 | 0.003 | 5.75 | 0.007 | 7.82 | 0 | 8.28 | 0.002 | | |
| 25 | | | 6.00 | 0.001 | 6.00 | 0.004 | | | 8.64 | 0 | | |
| 26 | | | 6.25 | 0 | 6.25 | 0.002 | | | | | | |
| 27 | | | | | 6.50 | 0.001 | | | | | | |
| 28 | | | | | 6.75 | 0 | | | | | | |

| Line No. | $T_0/T_p = 10$ | | $T_0/T_p = 16$ | | $T_0/T_p = 25$ | | $T_0/T_p = 36$ | | $T_0/T_p = 50$ | |
|---|---|---|---|---|---|---|---|---|---|---|
| | $t/T_p$ | $q_c/q_p$ | $t/T_p$ | $q_c/q_p$ | $t/T_p$ | $q_c/q_p$ | $t/T_p$ | $q_c/q_p$ | $t/T_p$ | $q_c/q_p$ |
| 1 | 0 | 0 | 0 | 0 | 0 | 0 | 0 | 0 | 0 | 0 |
| 2 | 0.67 | 0.013 | 0.80 | 0.008 | 1.25 | 0.015 | 1.50 | 0.0195 | 2.00 | 0.0167 |
| 3 | 1.34 | 0.061 | 1.60 | 0.046 | 2.50 | 0.039 | 3.00 | 0.0275 | 4.00 | 0.0204 |
| 4 | 2.01 | 0.091 | 2.40 | 0.060 | 3.75 | 0.043 | 4.50 | 0.0294 | 6.00 | 0.0214 |
| 5 | 2.68 | 0.102 | 3.20 | 0.065 | 5.00 | 0.044 | 6.00 | 0.0300 | 8.00 | 0.0216 |
| 6 | 3.35 | 0.107 | 4.00 | 0.067 | 6.25 | 0.044 | 7.50 | 0.0301 | 10.00 | 0.0216 |
| 7 | 4.02 | 0.110 | 4.80 | 0.067 | 7.50 | 0.044 | 9.00 | 0.0301 | 12.00 | 0.0216 |
| 8 | 4.69 | 0.111 | 5.60 | 0.068 | 8.75 | 0.044 | 10.50 | 0.0301 | 14.00 | 0.0216 |
| 9 | 5.36 | 0.111 | 6.40 | 0.068 | 10.00 | 0.044 | 12.00 | 0.0301 | 16.00 | 0.0216 |
| 10 | 6.03 | 0.112 | 7.20 | 0.068 | 11.25 | 0.044 | 13.50 | 0.0301 | 18.00 | 0.0216 |
| 11 | 6.70 | 0.112 | 8.00 | 0.068 | 12.50 | 0.044 | 15.00 | 0.0301 | 20.00 | 0.0216 |
| 12 | 7.37 | 0.112 | 8.80 | 0.068 | 13.75 | 0.044 | 16.50 | 0.0301 | 22.00 | 0.0216 |
| 13 | 8.04 | 0.116 | 9.60 | 0.068 | 15.00 | 0.044 | 18.00 | 0.0301 | 24.00 | 0.0216 |
| 14 | 8.71 | 0.160 | 10.40 | 0.068 | 16.25 | 0.044 | 19.00 | 0.0301 | 26.00 | 0.0216 |
| 15 | 9.38 | 0.198 | 11.20 | 0.068 | 17.50 | 0.044 | 21.00 | 0.0301 | 28.00 | 0.0216 |
| 16 | 10.05 | 0.212 | 12.00 | 0.068 | 18.75 | 0.045 | 22.50 | 0.0301 | 30.00 | 0.0216 |
| 17 | 10.72 | 0.168 | 12.80 | 0.086 | 20.00 | 0.067 | 24.00 | 0.0311 | 32.00 | 0.0217 |
| 18 | 11.39 | 0.074 | 13.60 | 0.121 | 21.25 | 0.083 | 25.50 | 0.0364 | 34.00 | 0.0243 |
| 19 | 12.06 | 0.027 | 14.40 | 0.133 | 22.50 | 0.087 | 27.00 | 0.0425 | 36.00 | 0.0287 |
| 20 | 12.73 | 0.010 | 15.20 | 0.136 | 23.75 | 0.087 | 28.50 | 0.0480 | 38.00 | 0.0329 |
| 21 | 13.40 | 0.005 | 16.00 | 0.137 | 25.00 | 0.088 | 30.00 | 0.0525 | 40.00 | 0.0363 |
| 22 | 14.07 | 0.002 | 16.80 | 0.098 | 26.25 | 0.035 | 31.50 | 0.0561 | 42.00 | 0.0391 |
| 23 | 14.74 | 0 | 17.60 | 0.033 | 27.50 | 0.006 | 33.00 | 0.0584 | 44.00 | 0.0411 |
| 24 | | | 18.40 | 0.012 | 28.75 | 0.002 | 34.50 | 0.0598 | 46.00 | 0.0423 |
| 25 | | | 19.20 | 0.004 | 30.00 | 0 | 36.00 | 0.0603 | 48.00 | 0.0430 |
| 26 | | | 20.00 | 0.001 | | | 37.50 | 0.0167 | 50.00 | 0.0433 |
| 27 | | | 20.80 | 0 | | | 39.00 | 0.0018 | 52.00 | 0.0058 |
| 28 | | | | | | | 40.50 | 0 | 54.00 | 0.0002 |
| 29 | | | | | | | | | 56.00 | 0 |

**Table 12-18** *Solution to Example 12-1*

| Line No. | $t$ (hr) | $q$ (cfs) | Line No. | $t$ (hr) | $q$ (cfs) |
|---|---|---|---|---|---|
| 1 | 0 | 0 | 21 | 6.09 | 616 |
| 2 | 0.30 | 7 | 22 | 6.39 | 399 |
| 3 | 0.61 | 36 | 23 | 6.69 | 254 |
| 4 | 0.91 | 109 | 24 | 7.00 | 145 |
| 5 | 1.22 | 268 | 25 | 7.30 | 87 |
| 6 | 1.52 | 710 | 26 | 7.61 | 58 |
| 7 | 1.82 | 1769 | 27 | 7.91 | 36 |
| 8 | 2.13 | 2951 | 28 | 8.22 | 29 |
| 9 | 2.43 | 3364 | 29 | 8.52 | 22 |
| 10 | 2.74 | 3110 | 30 | 8.82 | 14 |
| 11 | 3.04 | 2661 | 31 | 9.13 | 7 |
| 12 | 3.35 | 2240 | 32 | 9.43 | 0 |
| 13 | 3.65 | 1892 | 33 | | |
| 14 | 3.96 | 1624 | 34 | | |
| 15 | 4.26 | 1399 | 35 | | |
| 16 | 4.56 | 1225 | 36 | | |
| 17 | 4.87 | 1102 | 37 | | |
| 18 | 5.17 | 1008 | 38 | | |
| 19 | 5.48 | 935 | 39 | | |
| 20 | 5.78 | 819 | 40 | | |

***Example 12-2*** Determine the freeboard design hydrograph for a class (c) structure, Table 12-7, with a drainage area = 23.0 mi², time of concentration = 10.8 hr, and $CN$ of 77, which is geographically located to produce a $P = 25.5$ in.

1. Determine the areal factor. Use the appropriate curve of Fig. 12-7. As an example, the humid and subhumid climate curve factor is 0.93; $P$ adjusted = 25.5 × 0.93 = 23.72 in.
2. Duration adjustment. The duration is made equal to the time of concentration by adjusting the rainfall amount. Enter Fig. 12-5 to find the adjustment factor of 1.18; $P$ adjusted = 23.72 × 1.18 = 27.99 in. or 28 in.
3. Direct runoff $Q$. Enter Fig. 12-26 and find $Q = 24.7$ in. for $P = 28$ in. and $CN = 77$.
4. Hydrograph family. Enter Fig. 12-27 with $CN = 77$ and $P = 28$, and read hydrograph family No. 1.
5. Duration of rainfall excess. Enter Table 12-9 with $CN = 77$ and find that $P^*$, the rainfall prior to $P$, is 0.60 in. Enter Table 12-8 with the ratio $P^*/P = 0.60/28.0 = 0.0214$ and read a time ratio of 0.950. $T_0$ = (time ratio) × (storm duration) = 0.950(10.8) = 10.26 hr. The excess rainfall is assumed to be uniformly distributed in this example.
6. Compute $T_p$: $T_p = 0.7T_c = 7.56$ hr.
7. $T_o/T_p$ ratio=10.26/7.56 = 1.357.
8. Revised $T_o/T_p$ based on available curves from Table 12-16. $(T_o/T_p)_{rev} =$ 1.5

9. Rev. $T_p$:rev $T_p = 10.26/1.5 = 6.84$ hr.
10. $q_p$:$q_p = 484(23.0)/6.84 = 1628$ cfs.
11. $Qq_p$:$(Q)(q_p) = (24.7)(1628) = 40{,}212$ cfs.
12. From selected hydrograph family in Table 12-17, multiply each value of the $t/T_p$ column by rev $T_p$ to obtain $t$ in hours.
13. From the selected hydrograph family in Table 12-17, multiply each value of the $q_c/q_p$ column by $(Q)(q_p)$ to obtain values of $q$ in cfs.

The solution for Example 12-2 is given in Table 12-19.

**Table 12-19** *Solution to Example 12-2*

| Line No. | $t$ (hr) | $q$ (cfs) | Line No. | $t$ (hr) | $q$ (cfs) |
|---|---|---|---|---|---|
| 1 | 0 | 0 | 21 | 43.78 | 0 |
| 2 | 2.19 | 482 | 22 | | |
| 3 | 4.38 | 4745 | 23 | | |
| 4 | 6.57 | 15160 | 24 | | |
| 5 | 8.76 | 28591 | 25 | | |
| 6 | 10.94 | 32773 | 26 | | |
| 7 | 13.13 | 28912 | 27 | | |
| 8 | 15.32 | 21152 | 28 | | |
| 9 | 17.51 | 14155 | 29 | | |
| 10 | 19.70 | 9048 | 30 | | |
| 11 | 21.89 | 5750 | 31 | | |
| 12 | 24.08 | 3619 | 32 | | |
| 13 | 26.26 | 2292 | 33 | | |
| 14 | 28.45 | 1488 | 34 | | |
| 15 | 30.64 | 965 | 35 | | |
| 16 | 32.83 | 603 | 36 | | |
| 17 | 35.02 | 322 | 37 | | |
| 18 | 37.21 | 161 | 38 | | |
| 19 | 39.40 | 80 | 39 | | |
| 20 | 41.59 | 40 | 40 | | |

## 12-5 Airport Drainage Design

Airport drainage is required to dispose of surface water and minimize the interruption of traffic into and out of the area. The total drainage system has several functions:[14] (1) collect and carry off surface water, (2) remove excess groundwater, (3) lower the water table, and (4) protect all slopes from erosion. Only the problem of collecting and removing the surface water will be discussed here.

The design of a surface water disposal system relies upon the Rational Method although some efforts are being made to evaluate these systems by means of the spatially varied unsteady flow equations of Chapter 7. Drainage calculations are usually based upon the collected rainfall data[8] and a 1-ft contour topographic map of the proposed finished site. The entire system of drainage is outlined on this map with the proper identification of subareas, main and lateral storm drains, direction of flow, gradients, inlets, and surface channels. The final design is attained by calculating the most reasonable cost to provide satisfactory drainage.

A step-by-step procedure to design a portion of the surface drainage facilities of an airport is outlined in Example 12-3 as a straightforward application of the Rational Method. Each subarea size is outlined and a weighted $C$ adopted, based upon $C = 0.90$ for the pavement and $C = 0.30$ for the turf areas. The time of concentration in the system is composed of inlet time and duration of travel in the conduit. Figures 11-1 and 11-2 provide estimates for these times based upon field research by the Corps of Engineers.[14] The design of the drainage system should be adequate to insure that ponding will not be excessive in areas adjacent to the runways. The general criteria is that these areas should be at least 75 ft from the bordering pavement. To prevent saturation of the nearby ground, rough calculations to insure adequate ponding volumes for the design rainfall are desirable. If critical places occur, routing techniques such as those described in Chapter 7 facilitate calculations of ponding depth from the known storage and outflow characteristics of the system.

***Example 12-3*** Prepare a surface drainage design for the portion of an airfield shown in Fig. 12-28 for the 5-yr frequency rainfall in Fig. 12-29.

*Solution*

The solution of Example 12-3 is given in Table 12-20.

Computation of subarea sizes, values of weighted $C$, and inlet times are listed in Table 12-21. Calculation of final pipe sizes necessary to drain the system will vary with slope, and type of pipe selected. The slope of the pipe is usually controlled by the outlet elevation which must be maintained to allow the system to drain freely. Designs in this example are based upon use of a concrete pipe with $n = 0.015$. A nomograph for solution of the discharge as a function of the size and slope is provided in Fig. 12-30. The minimum velocity allowed in the pipe is 2.5 fps to prevent excessive settling of sediment. The final drainage system design is shown in Table 12-22.

## 12-6 Design of Urban Storm Drain Systems

Problems encountered in hydrologic design of urban storm drain systems are more complex than those associated with airfield projects. The ideal approach is to (a) determine the flow to each inlet area, (b) attenuate the peak flow from each subarea as it moves down the pipe, and (c) sum the attenuated peaks to get the total discharge at the design point. The calculation of individual hydrographs to the inlet of each section is influenced by the infiltration capacity of the pervious

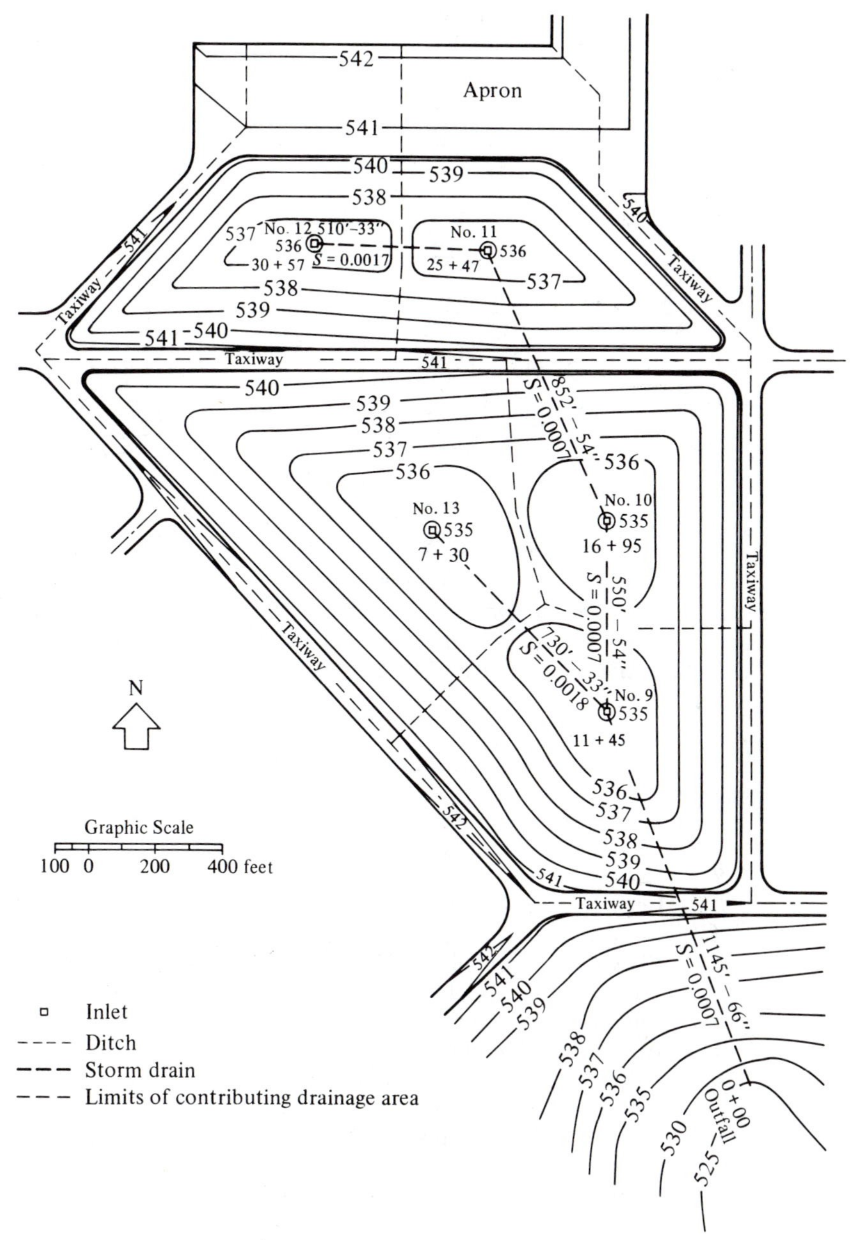


***Fig. 12-28.*** *Portion of airport showing drainage design. (After "Airport Drainage," Federal Aviation Agency, Circular, AC No. AC150/5320-5A (Washington, D.C.: Government Printing Office, 1966).)*

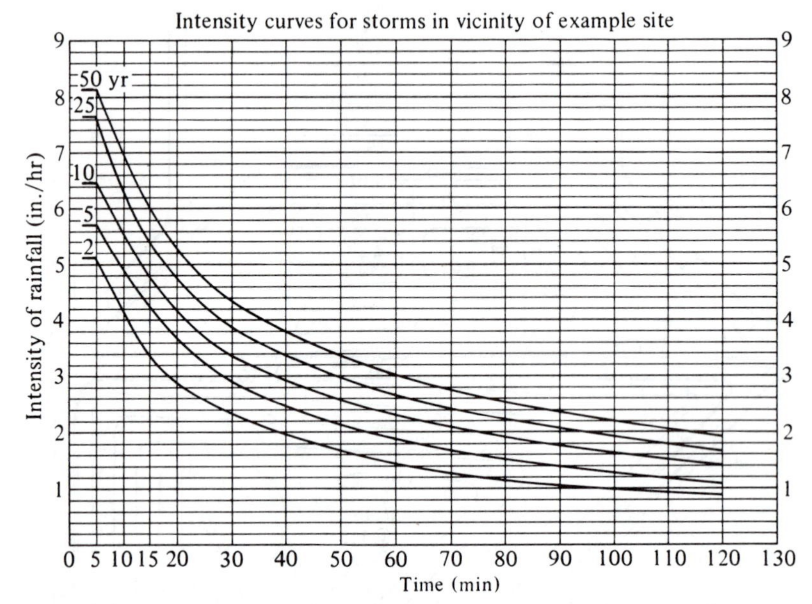


**Fig. 12-29.** *Duration of rainfall (min). (After "Airport Drainage," Federal Aviation Agency, Advisory Circular, AC No. AC150/5320-5A (Washington, D.C.: Government Printing Office, 1966).)*

**Table 12-20** *Descriptions of the Computations in Table 12-22*

| Column | Comment |
|---|---|
| 1 | Inlet being investigated |
| 2 | Line segment |
| 3 | Length of line |
| 4 | Inlet time |
| 5 | Flow time in line obtained by dividing the length of the line by the velocity of the drain |
| 6 | Time of concentration |
| 7 | Weighted value of $C$ |
| 8 | Rainfall intensity based on the time of concentration as duration and 5 yr as the frequency of design |
| 9 | Acreage of subarea immediately tributary to the inlet |
| 10 | $Q = CIA$ |
| 11 | Accumulated runoff that must be carried by the next line being computed |
| 12 | Velocity of flow through line $A/A$ or the nomograph Fig. 11-2 |
| 13 | Pipe size from Fig. 12-30 |
| 14 | Slope of the line |
| 15 | Pipe capacity that must be larger than the estimated flow |
| 16 | Invert elevation of the pipeline |
| 17 | Pertinent remarks relative to the design |

**Table 12-21** *Design Data for Drainage Example in Table 12-22*

| | Tributary Area to Inlets (acres) | | | | | Distance Remote Point to Inlet (ft) | | | Time for Overland Flow (min) | | |
|---|---|---|---|---|---|---|---|---|---|---|---|
| Inlet No. | *Pavement* | *Turf* | *Both* | *Subtotal* | $C^a$ | *Pavement* | *Turf* | *Total* | *Pavement* | *Turf* | *Total* |
| 12 | 4.78 | 9.91 | 14.69 | 14.69 | 0.49 | 100 | 790 | 890 | 4 | 37 | 41 |
| 11 | 5.48 | 9.24 | 14.72 | 29.41 | 0.53 | 90 | 750 | 840 | 4 | 36 | 40 |
| 10 | 1.02 | 10.95 | 11.97 | 41.38 | 0.35 | 65 | 565 | 630 | 3.5 | 31.3 | 34.8 |
| 13 | 1.99 | 19.51 | 21.50 | 21.50 | 0.35 | 110 | 1140 | 1250 | 4.3 | 44.3 | 48.6 |
| 9 | 1.46 | 14.59 | 16.05 | 78.93 | 0.35 | 85 | 612 | 697 | 3.9 | 32.4 | 36.3 |
| Totals | 14.73 | 64.20 | 78.93 | | | | | | | | |

[a] Weighted $C$ based on $C = 0.9$, pavement; $C = 0.3$, turf.

*Source:* "Airport Drainage," Federal Aviation Agency, Advisory Circular, AC No. AC150/5320-5A. Washington, D.C.: Government Printing Office, 1966.

areas, overland flow delays, depression storage, detention in gutters, house drains, catchbasins and the storm sewer system, and interception in extensively landscaped locations.

Two major items normally accounted for in urban storm drain design are

1. *Infiltration.* The ability of the soil to infiltrate water depends upon many characteristics of the soil as noted in Chapter 2. The range of values given in the following table is typical of various bare soils after 1 hr of continuous rainfall.

*Typical Infiltration Rates*

| Soil Group | Infiltration, (in./hr) |
|---|---|
| High (sandy, open-structured) | 0.50 to 1.00 |
| Intermediate (loam) | 0.10 to 0.50 |
| Low (clay, close-structured) | 0.01 to 0.10 |

The influence of grass cover increases these values 3 to 7.5 times.

2. *Retention.* Usually assumed to be 0.10 in. for pervious surfaces such as lawns and normal urban pervious surfaces.

Techniques employed to design urban storm drain systems vary but normally are based upon the Rational Method or the routing method of Chapter 11. The Road Research Laboratory Method of Chapter 11 is one approach of the routing method but utilizes only the runoff from the impervious areas. As a result, a less densely populated subdivision or urban area is not accurately analyzed.

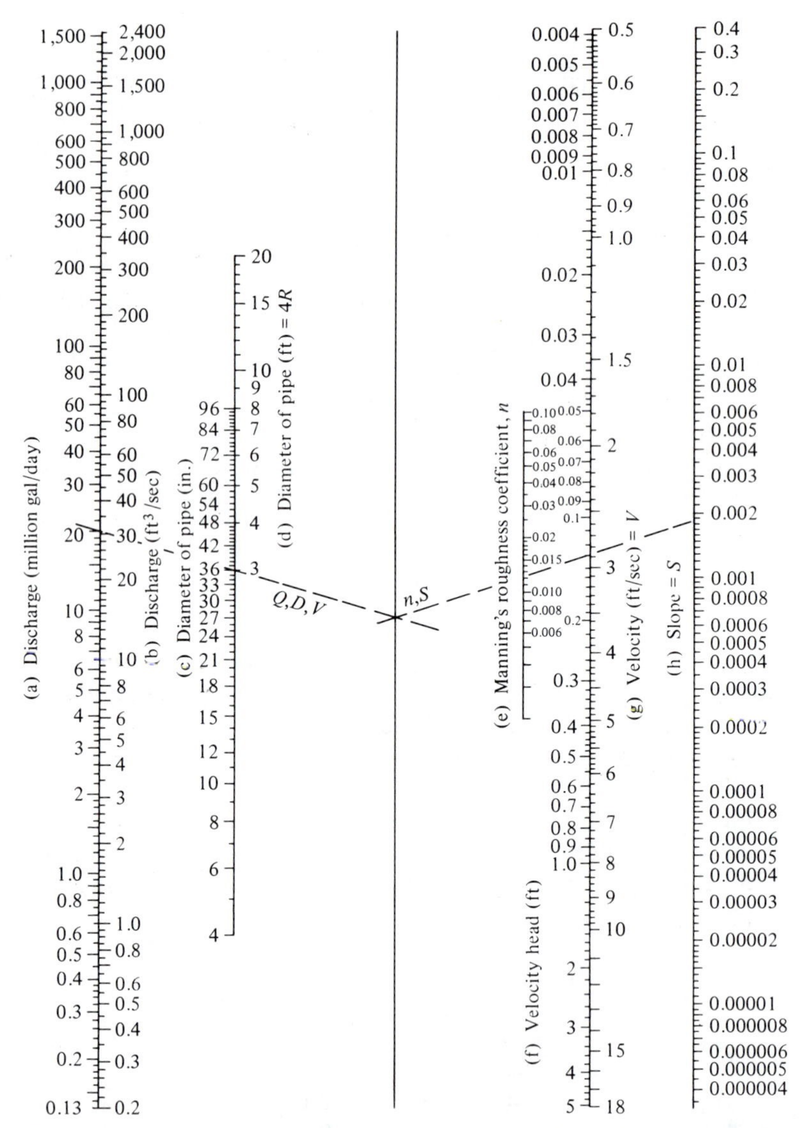


***Fig. 12-30.*** *Flow in pipes (Manning's formula). (After "Design and Construction of Sanitary and Storm Sewers,"* ASCE, *Manuals and Reports on Engineering Practice, No. 37, 1969.)*

**Table 12-22** *Drainage System Design Data*

| Inlet (1) | Line Segment (2) | Length of Segment (ft) (3) | Inlet Time (min) (4) | Flow Time (min) (5) | Time of Concentration (min) (6) | Runof Coefficient $C$ (7) | Rainfall Intensity $I$ (in./hr) (8) | Tributary Area $A$ (acres) (9) | Runoff $Q$ (cfs) (10) | Accumulated Runoff (cfs) (11) | Velocity of Drain (ft/sec)[a] (12) | Size of Pipe (in.)[b] (13) | Slope of Pipe (ft/ft) (14) | Capacity of Pipe (cfs) (15) | Invert Elevation (16) | Remarks (17) |
|---|---|---|---|---|---|---|---|---|---|---|---|---|---|---|---|---|
| 12 | 12–11 | 510 | 41 | 2.7 | 41 | 0.49 | 2.40 | 14.69 | 17.28 | 17.28 | 3.18 | 33 | 0.0017 | 18.90 | 530.65 | ($n = 0.015$) |
| 11 | 11–10 | 852 | 40 | 5.0 | 43.7 | 0.53 | 2.31 | 14.72 | 18.02 | 35.30 | 2.84 | 54 | 0.0007 | 45.00 | 528.03 | See note below |
| 10 | 10–9 | 550 | 34.8 | 3.3 | 48.7 | 0.35 | 2.15 | 11.97 | 9.01 | 44.31 | 2.84 | 54 | 0.0007 | 45.00 | 527.44 | See note below |
| 13 | 13–9 | 730 | 48.6 | 3.7 | 48.6 | 0.35 | 2.16 | 21.50 | 16.25 | 16.25 | 3.27 | 33 | 0.0018 | 19.40 | 530.11 | |
| 9 | 9–out | 1145 | 36.3 | 5.9 | 52.3 | 0.35 | 2.03 | 16.05 | 11.40 | 71.96 | 3.24 | 66 | 0.0007 | 77.00 | 526.05 | |
| Out | | | | | | | | | | | | | | | 525.25 | |

[a] Minimum velocity is 2.5 fps.
[b] Minimum pipe size is 12 in. diameter for maintenance purposes.
*Note:* The time of concentration for Inlet No. 11 is 43.7 min (41 + 2.7 = 43.7) which is the most time remote point for this inlet. Also the time of concentration for Inlet No. 10 is 48.8 min (41 + 2.7 + 5.0 = 48.7).

*Source:* "Airport Drainage," Federal Aviation Agency Advisory Circular, AC No. AC150/5320-5A. Washington, D.C.: Government Printing Office, 1966.

## Problems

12-1. Calculate a 100-yr minor structure design storm hyetograph for subarea I in the Oak Creek watershed, Fig. 10-31.

12-2. Calculate a design storm hyetograph for a 100-yr frequency level that has a 6-hr duration. Assume that the storm occurred near Springfield, Ill., and compute the rainfall excess distribution likely to occur only 10% of the time. Compare this with the median (50%) distribution.

12-3. Determine the U.S. Bureau of Reclamation minimum (50 mph) and normal (100 mph) freeboard values for a small dam with a smooth concrete upstream face. The fetch for the reservoir is 5 mi.

12-4. Needed is a design inflow hydrograph to a reservoir at a site where no records of streamflow are available. List the general steps you would take as a hydrologist in developing the entire design inflow hydrograph.

12-5. Calculate a major structure design storm for subarea I in the Oak Creek watershed, Fig. 10-31. Assume that the storm will have a uniform areal distribution.

12-6. Construct a major structure design storm pattern for the entire Oak Creek watershed, Fig. 10-31, using Figs. 12-17 and 12-18. Describe how you would orient the pattern to create the most severe conditions at point 8.

12-7. Show how the histograms in Fig. 12-12 were developed from the curves in Fig. 12-11. Also, construct a 50% probability histogram similar to the one in Fig. 12-12 for second quartile storms.

12-8. Except for soil type, two drainage basins are otherwise identical. For the same storm, which basin would produce the most direct runoff, one containing SCS group A soils, or one containing group D soils?

12-9. The SCS curve number method of estimating rainfall excess (net rain) is based on the assumption that the curve number for a watershed depends on several factors. Name or describe the factors that are considered when a curve number is determined.

12-10. A 7-mi$^2$ drainage basin has a composite curve number, $CN$, of 50. According to the SCS, exactly how much rain must fall before the direct runoff commences?

12-11. Given the following watershed data, route a storm hydrograph through a reservoir at the watershed outlet and plot the resulting outflow hydrograph. Show all the work in a neat and logical order.

(a) Area = 3.75 mi$^2$
(b) Length = 5.80 mi
(c) Elevation of the bottom of the spillway = 1160.0 ft
(d) Spillway width $B$ = 500 ft
(e) Spillway coefficient $C$ = 3.0

Required:

(a) A 45-min unit hydrograph for this watershed by Gray's method.
(b) The storm (inflow) hydrograph resulting from a storm that produced 6.0 in. of rainfall excess during the first 45-min period, and 4.0 in. during the last 45-min interval.
(c) Route the storm hydrograph through the reservoir. Assume that at the start of rainfall, the elevation of the water surface in the reservoir is 1158.0 ft.

(d) Elevation for the top of the dam. Discuss.
(e) Spillway design adequacy. Discuss.

| Elevation (ft) | Distance (ft) | Area $\times 10^{-6}$ ft$^2$ (of respective contours) |
|---|---|---|
| 1,110 | 0 | 0.0 |
| 1,120 | 1,400 | 0.85 |
| 1,140 | 4,600 | 3.75 |
| 1,160 | 8,000 | 10.80 |
| 1,180 | 11,500 | 25.00 |
| 1,200 | 14,700 | |
| 1,220 | 17,400 | |
| 1,240 | 20,000 | |
| 1,260 | 22,600 | |
| 1,280 | 24,800 | |
| 1,300 | 26,600 | |
| 1,320 | 28,000 | |
| 1,340 | 29,200 | |
| 1,360 | 30,000 | |
| 1,376 | 30,624 | |

12-12. Repeat Prob. 12-11 for a type (b) classified structure in an area selected by the instructor.

12-13. Repeat Prob. 12-12 for a type (a) classified structure. Calculate the increase in cost assuming \$125/yd$^3$ for concrete and 70 cents/yd$^3$ for fill dirt.

12-14. Rework Example 12-1 for a 5-mi$^2$ drainage area, time of concentration 3.8 hr, and runoff number of 73 in your geographic area.

12-15. Use information from Prob. 12-14 and rework Example 12-2.

12-16. A 1000-acre watershed is composed of 40% woods in fair hydrologic condition (Type B soil), 40% native pasture in good condition (Type D soil), and 20% contoured row crops in poor condition (Type A soil). Determine the runoff if 13 in. of rain falls on the watershed.

12-17. How is the storm duration for any of the SCS hydrograph family unit hydrographs in Table 12-17 determined?

12-18. Determine a composite SCS runoff curve number for a 600-acre basin that is totally within soil group C. The land use is 40% contoured row crops in poor hydrologic condition and 60% native pasture in fair hydrologic condition.

12-19. Which SCS-classified soil would have the highest infiltration rate, A, B, C, or D?

12-20. Determine the direct runoff, $Q$ (SCS hydrograph family method) for a 7-mi$^2$ drainage basin at Lincoln, Nebr. (semiarid climate) if the time of concentration is 5 hr, $CN = 85$, and $P$ from Fig. 12-3 is 5.2 in. Provide correct units along with numerical answers.

12-21. Repeat Prob. 12-20 for $A = 7$ mi$^2$, $t_c = 20$ hr.

12-22. Repeat Prob. 12-20 for $A = 40$ mi$^2$, $t_c = 5$ hr.

12-23. Repeat Prob. 12-20 for $A = 40$ mi$^2$, $t_c = 20$ hr.

12-24. For a design by the SCS hydrograph family method, the revised $T_p$ is 10 hr, $q_p = 4000$ cfs, $P = 11$ in., $CN = 69$, $Q = 7$ in., and $T_0 = 30$ hr. Determine the runoff rate (cfs) 32 hr after the beginning of the rainfall.

12-25. How are the $P^*$-values in Table 12-9 related to the curves in Fig. 12-25?

## References


1. Snyder, F. F., "Hydrology of Spillway Design: Large Structures—Adequate Data," *ASCE, J. Hyd. Div.*, 90, No. HY3 (May 1964): 239–259.
2. "Generalized Estimates of Maximum Probable Precipitation over the United States East of the 105th Meridian," U.S. Weather Bureau Hydrometeorological Rept. No. 23, June 1947.
3. "Generalized Estimates of Probable Maximum Precipitation for the United States West of the 105th Meridian for Areas to 400 Square Miles and Durations to 24 Hours," U.S. Weather Bureau Tech. Paper 38, 1960.
4. "Designs of Small Dams," Bureau of Reclamation, U.S. Department of the Interior (Washington, D.C.: Government Printing Office, 1965).
5. Ogrosky, H. O., "Hydrology of Spillway Design: Small Structures—Limited Data," *ASCE, J. Hyd. Div.*, 90, No. HY3 (May 1964): 295–310.
6. "Two-to-Ten-Day Precipitation for Return Periods of 2 to 100 Years in the Contiguous United States," U.S. Department of Commerce, Tech. Paper No. 49, 1964.
7. "Hydrology," Suppl. A to Sec. 4, *Engineering Handbook*, U.S. Department of Agriculture, Soil Conservation Service, 1968.
8. "Rainfall Frequency Atlas of the United States for Durations from 30 Minutes to 24 Hours and Returns Periods from 1 to 100 Years," U.S. Weather Bureau Tech. Paper 40, 1963.
9. Huff, F. A., "Time Distribution of Rainfall in Heavy Storms," *Water Resources Research*, 3, No. 4 (1967): 1007–1019.
10. "Probable Maximum TVA Precipitation for Tennessee River Basins Up to 3000 Square Miles in Area and Durations to 72 Hours," U.S. Department of Commerce, ESSA, Weather Bureau Hydrometeorological Rept. No. 45, 1969.
11. "Standard Project Flood Determinations," Civil Engineer Bulletin No. 52-8, EM 1110-2-1411, U.S. Department of the Army, Office of the Chief of Engineers, Washington, D.C., revised 1965, pp. 1–19.
12. Hershfield, D. M., "Estimating The Probable Maximum Precipitation," *ASCE, J. Hyd. Div.*, 87, No. HY5 (1961), pp. 99–116.
13. "Guidelines for Determining Flood Flow Frequencies," U.S. Water Resources Council, Washington, D.C., 1976, pp. 14–20.
14. "Airport Drainage," Federal Aviation Agency, Advisory Circular, AC No. AC150/5320-5A (Washington, D.C.: Government Printing Office, 1966).
15. "Design and Construction of Sanitary Storm Sewers," *ASCE* Manuals and Reports on Engineering Practice, No. 37.

# 13

# Water Quality Models

Because water quality is inextricably linked to water quantity, it is important for the hydrologist to understand the significance of developing modeling techniques that can accommodate both features. This chapter is designed to introduce the subject of water quality modeling and to illustrate the linkage between water quality modeling components and the hydrologic processes that have been discussed in earlier chapters.

## 13-1 Water Quality Models

A water quality model is a mathematical statement or set of statements that equate water quality at a point of interest to causative factors. In general, water quality models are designed to (1) accept as input, constituent concentration versus time at points of entry to the system; (2) simulate the mixing and reaction kinetics of the system; and (3) synthesize a time-distributed output at the system outlet.

Either stochastic (containing probabilistic elements) or deterministic approaches may be taken in developing methods for predicting pollutional loads. The former technique is based upon determining the likelihood (frequency) of a particular output quality response by statistical means. This is similar to frequency analysis of floods or low flows. Water quality records should be available for at least 5 years (and preferably much longer) if estimates of return periods for infrequent events are to be reliable.

The deterministic approach (output explicitly determined for a given input) requires that a model be developed to relate water quality loading to a known or assumed hydrologic input. Such a model can range from an empirical concentration discharge relationship to a physical equation representing the hydrochemical cycle. The ultimate modeling technique is that which best defines the actual mechanism triggering the water quality response. The cause of a given state of pollution can then be specifically identified.

### Water Quality Constituents

Water quality constituents are categorized as organic, inorganic, radiological, thermal, and biological or they can be subdivided into specific forms such as biochemical oxygen demand (BOD), nitrogen, phosphorus, and so forth. Pollutants of interest include silt, pesticides, fertilizers, fecal organisms, nitrates, and phosphates.[1–4]

Hydrologic data necessary to compute pollutional loadings of various constituents must be obtained concurrently with water quality data to insure the proper calibration and verification of models. A knowledge of the frequency and time distribution of loading is often more important than a knowledge of the total loading. This is particularly true where the objective is to determine the impact of a waste flow on a receiving stream. For example, a short, high-peaked surface runoff hydrograph of suspended matter could be expected to more seriously affect a stream than a hydrograph which released the same volume of suspended matter over an extended period of time. On the other hand, when waste flows enter a lake, the annual volume of contaminants takes on special importance because they are trapped, in a sense, and accumulate. In general, water quality data should be recorded in a continuous manner so that the time rate of delivery of the constituent loading can be determined. If this is not done, only very gross estimates of the impact of water quality inputs on receiving waters can be obtained.

### Nonconservative Water Quality

Unstable pollutants such as radioactive wastes and heat that have a time-dependent decay are classified as nonconservative. Biochemical oxygen demand is another example of a nonconservative pollutant.

Several nonconservative models have been proposed for rivers in the United States.[5,6,7] The application of nonconservative BOD water quality models is clearly illustrated by Smith et al.,[5] Thayer and Krutchkoff,[6] and Loucks et al.[7] The classic dissolved oxygen depletion model developed by Streeter and Phelps is an early example of a water quality model.[40] See also Ref. 41.

Development of nonconservative models is dependent upon (1) reliable input data related to the surface runoff and other discharges, (2) representative mixing models, and (3) knowledge of the reaction kinetics involved.

### Conservative Water Quality

Many inorganic pollutants can be considered conservative. The principal features of a conservative water quality model include (1) determination of the input to the system by relating it to the study area hydrology and/or other factors, and (2) determination of a mixing model to be employed. Useful studies can be found in Refs. 5, 7, 8, 9, 10, and 11. Steele[8] gives a very good summary of previous studies designed to relate water quality to several defining parameters.

Models representing the transport of solutes from a watershed can be developed according to known physical laws or might be largely empirical in nature.

The strict physical approach attempts to evaluate the change in the chemical character of the water as it moves through phases of the hydrochemical cycle. For each phase, the quantity of runoff involved, the nature of the chemical reactions governing the changes, and the character of any natural constraints must be known or simulated. The hydrologic simulation or modeling can be carried out according to different schemes.

The problem of the chemical change is difficult to resolve. Few reliable studies of the chemical phenomena of dissolution and ion exchange under natural conditions are available. Physical conditions affecting chemical reactions occurring in the field are not subject to the simple control of the laboratory. In addition, the chemical, physical, and hydraulic character of the total hydrochemical system of a watershed defies an exact determination of the contribution to total solute load by individual phases of the cycle.[8]

Fortunately, it has been demonstrated that the complexities of natural water quality systems can be somewhat circumvented. Several studies have shown that such systems frequently exhibit consistent chemical behavior that can be explained rationally in terms of adequate field data.[8] An empirical approach can often be used to identify the more important factors controlling the water quality changes on a watershed and to develop useful working relationships for predicting water quality inputs.

Several studies have indicated that chemical quality data can be satisfactorily related to discharge by an equation or series of equations of the form[5,8,9,10]

$$C = kQ^n \tag{13-1}$$

where

$C$ = the constituent concentration
$Q$ = the streamflow
$k$ and $n$ = regression parameters

Based on somewhat limited data, Weibel et al. have published graphical correlations between rain (in./storm) and BOD (lb/storm), solids (mg/l) and flow ranges (cfs) for data on a 27-acre urban area in Cincinnati.[12]

## 13-2 A Crude Water Quality Model of Lake Flushing Rates

To illustrate the conceptualization of a water quality model, the following example relative to estimation of lake flushing rates is given.[14]

### Individuality of Lakes

Lakes have many common features but often strikingly different performance characteristics. This individuality has environmental value; but to the analyst it presents the problem of having to understand both the general nature of the system and variations due to local conditions. Of significance is the fact that the hydrologic characteristics of lakes vary considerably because of differences in the depth, length, width, surface area, basin material, surrounding ground cover, reservoir, prevailing winds, climate, surface inflows and outflows, and other factors.[13] This means that each lake requires its own model and these models will be characterized by different degrees of variance from a generalized conceptual model.

### The Water Budget

As stated in Chapter 1, the water budget is an accounting of the disposition of storage, inflow, and outflow of a system during some period of time. The budget might be static or dynamic depending upon the choice of either a single interval for investigation or of a sequence of intervals. The water budget, constituent budget (to be defined later), and mixing process of the lake jointly define the flushing rates for both water and pollutants or other constituents.

Figure 13-1 illustrates the components of the water budget. Classifying these components as inputs, outputs, and storage elements permits writing the water budget in the form of Eq. 13-2.

$$\text{input} - \text{output} = \text{change in storage} \tag{13-2}$$

or

$$(I_c + I_o + I_g + P + R) - (E + T + G_s + O_c + W) = \frac{\Delta S}{\Delta t} \tag{13-3}$$

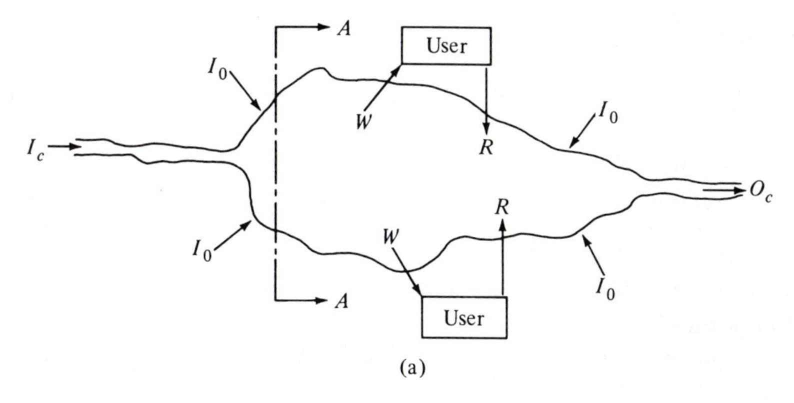


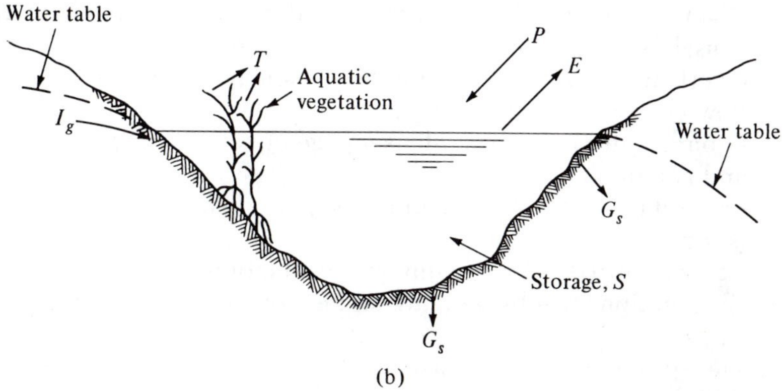


*Fig. 13-1.* *A conceptual lake illustrating variables in the water budget. (a) Plan view; (b) Section A-A.*

where

$I_c$ = the channel inflow rate
$I_o$ = the overland inflow rate
$I_g$ = the groundwater inflow rate
$E$ = the evaporation rate
$T$ = the transpiration rate
$P$ = the precipitation rate
$G_s$ = the seepage rate
$O_c$ = the channel outflow rate
$S$ = the lake storage at time $t$, volume
$W$ = the withdrawal rate
$R$ = the return flow rate
$\Delta t$ = the time interval

Rearranging Eq. 13-3, we obtain

$$(I_c + I_o + I_g + P + R) - (E + T + W) = f(\Delta S, \ldots) \qquad (13\text{-}4)$$

where

$$f(\Delta S, \ldots) = G_s(S_1 + \Delta S, \ldots) + O_c(S_1 + \Delta S, \ldots) + \frac{(S_2 - S_1)}{\Delta t} \quad \textbf{(13-5)}$$

and the subscripts on $S$ denote time periods.

Equation 13-4 can be solved for $\Delta S$ if all of the terms on the left-hand side can be evaluated and the initial storage $S_1$ is known or can be estimated. A knowledge of $\Delta S$ permits determination of the water flushing rate $(O_c + G_s)$ if the functional relationships given in Eq. 13-5 are defined.

Terms requiring quantification for solution of Eq. 13-4 may be evaluated as follows:

1. Channel inflow $(I_c)$—by gauging, deterministic relationship to causal factors, generation of stochastic records.
2. Overland inflow $(I_0)$—by rainfall-runoff relationships, physics of flow equations, and gauging.
3. Groundwater inflow $(I_g)$—by seepage equations, flow net analyses, and gauging.
4. Precipitation $(P)$—by gauging and generation of stochastic traces (synthesis).
5. Return flow $(R)$—by gauging and estimation.
6. Evaporation $(E)$—by gauging and evaporation prediction relationships.
7. Transpiration $(T)$—by consumptive use equations and gauging.
8. Withdrawal $(W)$—by gauging and estimation.

The estimation or measurement of any of these variables may be difficult, and the reliability of estimated values will vary with the accuracy of the prediction techniques, the quality of gauging, and the time period involved. As a result, the calculated flushing rates must be qualified as to expected range of errors. In practice, actual computations are not as straightforward as the previous remarks might indicate. A good deal of modeling capability is usually a requisite.

Solutions to functional relationships between water budget variables are dependent upon the quantification of parameters descriptive of the characteristics of the particular lake system to be studied. For example, seepage rates are related to the permeability of the surrounding soils and water temperatures. To solve the water budget equation, the initial storage volume of the lake is needed. Changes in the water surface level can only be determined if the geometry of the basin is known. All types of descriptive data are required. It might also be necessary to continuously monitor precipitation, streamflow, evaporation, return flows, and withdrawals. These records can serve as a historic basis for synthetic data generation or can be used directly in "real time" control systems. The system of interest must be understood and described quantitatively if projections or estimates are to be made.

### The Constituent Budget

Water quantity and quality considerations are directly linked. In order to account for the movement of pollutants in a lake, it is necessary to combine the water budget with a constituent budget and a model of the mixing mechanics of the system. Unfortunately, the complexities of the water budget are paled in comparison to those of the constituent budget. This is because the constituent budget includes conservative and nonconservative components and is highly dependent upon the circulation of the lake. The ultimate disposition of constituents and the reaction kinetics of both chemical and biological components are affected by circulation. An incomplete understanding of the mixing mechanics of large water bodies and of various aspects of the hydrochemical and hydrobiological cycles compounds the problem. Models combining these aspects are difficult to construct.

Conceptually, the constituent budget is similar to the water budget—an accounting of inputs, storages, and outputs. Basically, a separate budget must be struck for each constituent of interest. A simple conceptual model structure for a single constituent ($I$) would be

$$C_{II} - C_{IO} = \frac{\Delta C_{IC}}{\Delta t} \tag{13-6}$$

where

$C_{II}$ = the constituent input to the system per unit time
$C_{IO}$ = the constituent output from the system per unit time
$\Delta C_{IC}$ = the change in concentration of the constituent stored within the system
$\Delta t$ = the time interval of interest

Note that, in general,

$$C_{II} = \sum_{i=1}^{n} C_{IIi} \tag{13-7}$$

$$C_{IO} = \sum_{j=1}^{m} C_{IOj} \tag{13-8}$$

where $i$ and $j$ represent the total number of sources and outlets of constituent $I$, respectively. The notation $C_{II3}$ would represent the rate of input of constituent ($I$) from source 3, for example. It should be understood that the values of the $C_{IIi}$ and $C_{IOj}$ must be determined by monitoring, by the use of empirical or theoretical relationships, or by some form of forecasting. These determinations in themselves might be quite complex.

Also,

$$C_{IOj} = f(C_{IIi}, \text{mixing}, \ldots) \quad \textbf{(13-9)}$$

Data needs for constituent budgeting are extensive; they range from simple monitoring to the determination of reaction constants for functional relationships. An initial concentration of each constituent in the system must be obtained for use in the budget equation. Data on the thermal properties of the lake are vital. Unless input prediction equations can be developed, source monitoring is essential.

### Mixing

Interest in flushing rates centers around the need to determine how rapidly undesirable constituents can be removed from a lake system under varying conditions of constituent loading and hydrology. The rate of removal of a constituent is highly dependent on the mixing that occurs. To explain this, consider the following situation. Water budget calculations indicate an outflow equal to the entire volume of a lake during some time period. Does this mean that an undesirable constituent would be completely flushed during this same interval, assuming that the constituent input had ceased immediately prior to the beginning of the time period involved? No, it does not. In fact, if the system outflow were primarily from a surface current, little constituent removal would occur. The greater the mixing near the pollutants, the more rapid the flushing. Principal mixing mechanisms for lakes are the actions or interactions of wind, temperature changes, and atmospheric pressure. In shallow lakes with high volumes of boat traffic, propeller stirring may also be significant.

### Thermal Considerations

Water has its temperature of maximum density (4°C) above the freezing point, and as a result, most lakes in the temperate zone have two thermal circulation periods: one in the spring and another in the fall. During these periods, a condition of temperature homogeneity results and vertical circulation occurs.

In the fall, surface waters become cooled and these dense waters sink to become replaced by warmer water from lower elevations. As surface temperatures approach 4°C, surface layers become denser than any underlying layers and, consequently, sink to the bottom. This convective circulation finally establishes a lake of uniform density that can be mixed at all levels by the action of the wind. Continued cooling below 4°C establishes an inverse temperature structure with colder water on the top. At temperatures below 4°C and prior to freeze-up, the overturn may cease at any time once the surface density is reduced enough to preclude wind mixing to any considerable depth. After ice cover is established, the mixing potential of the wind is shielded against and wind mixing is ineffective until spring.

When springtime surface temperatures again approach 4°C, convective sinking and wind action again bring the lake to a uniform temperature and state of mixing. As the surface waters warm further, a density gradient develops which ultimately rules out complete wind mixing and again the lake tends to stagnate. This is the period of summer stratification.

The complete mixing of lake waters is, in general, limited to the periods of spring and fall overturn. During other periods, a more limited mixing due to wind-induced and flow-through currents can occur.

### Currents

Most lakes have an inlet and outlet, and the resulting flow-through generates what is known as the gradient or slope current. The relative dimensions of the inlet and outlet channels and those of the lake are usually such as to rule out any significant gradient currents. When mixing is considered, these currents are usually of negligible importance when compared to those generated by the wind.

The principal classifications of lake currents are periodic and nonperiodic.[13,15] Periodic currents are associated with seiches (water surface oscillations); although these may have important local effects on mixing, they are not often as important in dispersing constituents as nonperiodic currents. Nonperiodic currents are the product of wind stress energy transfer to the surface water. The exact mechanism of transfer is not fully known, but it is considered that wind stress energy input is a function of velocity to some power (usually considered to be from about 1.8 to 3).[13,15] A frequently used relationship is[13]

$$T = 3.2 \times 10^{-6} W^2 \tag{13-10}$$

where

$T$ = the wind stress (g/cm/sec²)
$W$ = the wind velocity (cm/sec)

The wind stress energy at the water surface produces an acceleration of water particles. Some of this acceleration generates waves while the remainder produces current. Wind produced surface currents are usually on the order of 1 to 3% of the wind velocity generating them. Along-shore and open-lake currents may be produced by wind action. The distinction is based on the point of departure from the influence of shoreline and bottom topography. Figures

13-2 and 13-3 indicate some possible wind-induced circulation patterns and the role played by multiple density strata.

### Circulation Models

Exacting analyses of circulation in stratified lakes are elusive. This is because of the highly complex nature of the circulation process. Such problems can be attacked by large computer models,[23] but even these incorporate many simplifying assumptions. Numerous models have been proposed and studied for special cases and particular solutions.[16–24] Some are research steps and others offer the potential for direct practical application to the determination of circulation in any given lake when combined with field data on velocities.[19] A thorough discussion of circulation mechanics is beyond the scope of this book, and the interested reader should consult the references cited.

### A Crude Flushing Model

Crude estimates often play a useful role in planning. They can usually be arrived at quickly and indicate the order of magnitude or range of values likely to be estimated by more sophisticated methods. Sometimes this information alone will prove adequate; in other cases, useful insights will be gained into the efforts required or the degree of refinement needed to obtain better results. In this section a simple model of the removal (flushing) of constituents from a lake is given to illustrate the water quality modeling process.

A rough estimate of both the maximum and minimum flushing rates of any constituent of interest can be obtained from some minimal knowledge of the lake characteristics and inflow and outflow hydrology.

1. Maximal Flushing (annual)
   a. Using samples, estimate the average concentration of constituent in the lake at the beginning of the year ($t = t_1$).

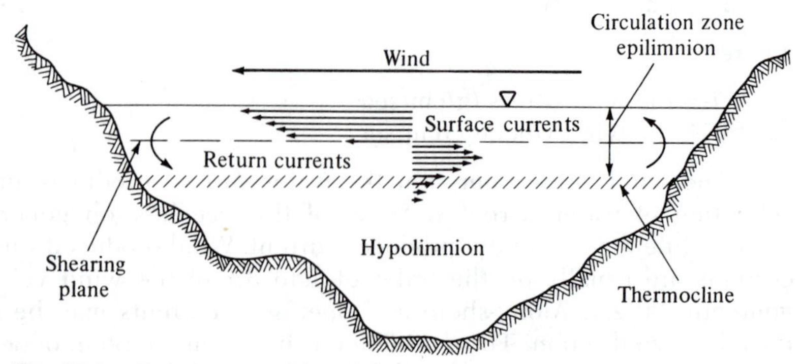


***Fig. 13-2.*** *Relative velocities of wind-induced currents.*

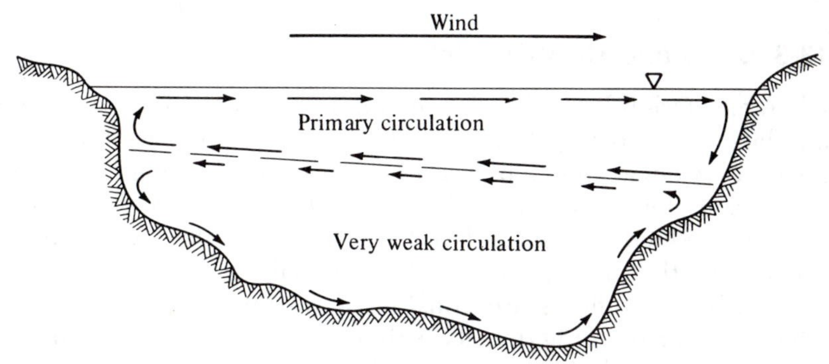


***Fig. 13-3.*** *Hypothetical circulation pattern in a stratified model.*

   b. Use projections, gauging records, or other means to estimate added inflow concentration during the year $(t_2 - t_1)$.
   c. Assume the complete mixing of the lake.
   d. From a hydrologic analysis, or based on historic records, estimate the average outlet flow over the period $(t_2 - t_1)$.
   e. Combine the mean initial constituent concentration at $t = t_1$ with the added mean concentration to obtain the total mixed mean concentration of the constituent.
   f. Multiply the total mixed mean concentration by the volume of outflow to arrive at the constituent quantity flushed.
2. Minimal Flushing (annual)
   a. Using samples, estimate the average concentration of the constituent in the lake at the start of the year $(t = t_1)$.
   b. Use projections, gauging records, or other means to estimate added inflow concentration during the year $(t_2 - t_1)$.
   c. Estimate the average length of the time span of spring and fall overturns from observations or records of similar lakes in the area.
   d. Assume complete mixing during the spring and fall overturns only.
   e. Use hydrologic analysis or historic records to estimate the mean outlet flow during overturns.
   f. Combine the mean initial constituent concentration at $t = t_1$ with the added mean concentration to obtain the total mixed mean concentration of the constituent.
   g. Multiply the total mixed mean concentration by the volume of water discharged during overturns to arrive at the constituent quantity flushed.

These are only crude approximations and do not include considerations such as fixation of constituents to bottom deposits, recycling, and others, but they could be used as preliminary planning guidelines or the first step in a more sophisticated modeling process.

## 13-3 Chen and Orlob Model

Chen and Orlob have developed water quality models for estuarine and lake (reservoir) systems.[25,26] The lake model is based on a multilayered system as shown in Fig. 13-4. Horizontal sections are assumed to be completely mixed and thus unstratified. The lake's hydrodynamic behavior is considered density-dependent with temperatures being the principal determining variable.

Tributary inflows are considered to enter the lake at a level where the surrounding density is their equivalent. Because of incompressibility, the inflow is assumed to generate an advective flow between the elements above the level of entry. Orlob and Selna have demonstrated that this approach is satisfactory for temperature simulation of lakes and reservoirs.[33] The differential equations describing the quality of the lake ecosystem are solved implicitly.[26,28]

To use the model, various input data are required. They include hydrologic and water quality parameters of tributary inflows, the amount and nature of waste discharges, the character of outflows, and

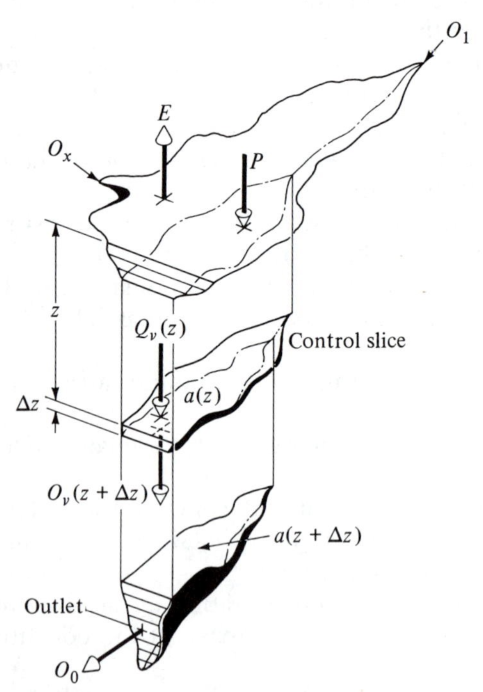


***Fig. 13-4.*** *One-dimensional idealization of a reservoir. (After Gerald T. Orlob, and Lawrence G. Selna, "Temperature Variations in Deep Reservoirs,"* Proc. ASCE, J. Hyd. Div., *96, No. HY2, Proc. Paper 7063 (Feb. 1970): 393.)*

meteorologic conditions. Computations can be performed over time periods ranging from hours to 1 day. Table 13-1 gives input requirements of the model and shows the water quality variables which can be simulated.

For computational facility, each lake layer is considered as a discrete continuously stirred tank reactor.[26,28] Hydrodynamic calculations determine the inflow and outflow from each hydraulic element. Changes in concentration of water quality constituents or biota are related to conditions existing in the assumed reactor. Mass balance equations describing the physical, chemical, and biological processes affecting the ecosystem variables are used to determine the rates of change. Constituent concentrations as a function of time are computed using numerical techniques.

## 13-4 Lombardo and Franz Model

Lombardo and Franz have developed a model (QUALITY) which simulates the water quality dynamics of rivers and impoundments.[27,28] It is used jointly with the Hydrocomp Hydrologic Simulation Program (HSP) for calculating the hydrologic response of a watershed.[34,35] By using both models, the hydrologic and water quality interactions of a watershed can be simulated.

The Hydrocomp model is based on the subdivision of a watershed into land segments and stream reaches. As noted in Chapter 10, the HSP system includes three models, LIBRARY, LANDS, and CHANNEL. LIBRARY is the data handling component; LANDS calculates the channel inflow volumes from the input rainfall and evapotranspiration; and CHANNEL routes the computed channel inflows for each reach progressively downstream using the kinematic wave routing method. Point sources and diversions may be specified along the stream system. The model is not designed to account for tidal effects.

The water quality model (QUALITY) is a separate module used with LIBRARY and LANDS and CHANNELS if desired. Quality can simulate water quality changes in channel flows and, if used in conjunction with LANDS, the surface runoff quality as well. Streams are represented by a series of reaches in the model and lakes by a system of three layers; each reach and lake layer is modeled as a continuously stirred tank reactor. The partial differential equations that describe the water quality dynamics are solved using a multiple-step explicit solution. Dispersion is assumed negligible. The model can accommodate a watershed with any number of lakes.

To operate the model, the stream reach and system parameters are input. Water quality indices that can be simulated are temperature; biochemical oxygen demand; coliforms (total, fecal, fecal

**Table 13-1** *Variables of a Lake and Estuary Ecologic Model*

| | Model Variables | | | |
|---|---|---|---|---|
| | *Estuarial System* | | *Lake System* | |
| Parameters | *Hydraulic* | *Ecologic* | *Hydraulic* | *Ecologic* |
| Climate (Zones) | | | | |
| Latitude | *I* | — | *I* | — |
| Longitude | *I* | — | *I* | — |
| Atmospheric pressure | *I* | — | *I* | — |
| Cloud cover | *I* | — | *I* | — |
| Wind speed | *I* | — | *I* | — |
| Dry bulb temperature | *I* | — | *I* | — |
| Wet bulb temperature | *I* | — | *I* | — |
| Wind direction | *I* | — | *I* | — |
| Evaporation | *O* | *C* | *O* | *C* |
| Short-wave radiation | *O* | *C* | *O* | *C* |
| Net solar radiation | *O* | *C* | *O* | *C* |
| Geometry | | | | |
| Function | | | | |
| Surface area | *I* | *C* | *I* | *C* |
| Side slope | *I* | *I* | *I* | *I* |
| Elevation | *I* | — | *I* | *I* |
| Volume | *C* | *C* | *C* | *C* |
| Channel | | | | |
| Length | *I* | *I* | — | — |
| Width | *I* | *I* | — | — |
| Friction factor | *I* | — | — | — |
| Hydrology | | | | |
| External flow | | | | |
| Rivers | *I* | *I* | *I* | *I* |
| Tide (1 point) | *I* | — | — | — |
| Waste discharges | *I* | *I* | — | — |
| Outflow | *I* | *I* | *I* | *I* |
| Internal flow | | | | |
| Channel flows | *O* | *C* | — | — |
| Surface overflow | *O* | *C* | *O* | *C* |
| Quality Constituent[a] | | | | |
| Temperature | — | *O* | *O* | *O* |
| Toxicity | — | *O* | — | — |
| Total dissolved solid | — | *O* | — | *O* |
| Coliform | — | *O* | — | *O* |
| BOD | — | *O* | — | *O* |
| Oxygen | — | *O* | — | *O* |

**Table 13-1** *(Cont.)*

| Parameters | Model Variables | | | |
|---|---|---|---|---|
| | *Estuarial System* | | *Lake System* | |
| | *Hydraulic* | *Ecologic* | *Hydraulic* | *Ecologic* |
| Total carbon (inorganic) | — | *C* | — | *C* |
| $PO_4$—P | — | *O* | — | *O* |
| Alkalinity | — | *O* | — | *O* |
| $NH_3$—N | — | *O* | — | *O* |
| $NO_2$—N | — | *O* | — | *O* |
| $NO_3$—N | — | *O* | — | *O* |
| Algae (2 groups) | — | *O* | — | *O* |
| Zooplankton | — | *O* | — | *O* |
| Fish (2 groups) | — | *O* | — | *O* |
| Benthic animal | — | *O* | — | *O* |
| Detritus | — | *O* | — | *O* |
| $CO_2$ | — | *O* | — | *O* |
| pH | — | *O* | — | *O* |
| System Coefficients | | | | |
| Light extinction | | | | |
| Background | — | *I* | — | *I* |
| Algal suspension | — | *I* | — | *I* |
| Reaeration | | | | |
| Oxygen | — | *I* | — | *I* |
| $CO_2$—C | — | *I* | — | *I* |
| Decay rate | | | | |
| BOD | — | *I* | — | *I* |
| Detritus | — | *I* | — | *I* |
| Coliform | — | *I* | — | *I* |
| $NH_3$—N | — | *I* | — | *I* |
| $NO_2$—N | — | *I* | — | *I* |
| Temperature coefficient | — | *I* | — | *I* |
| Algae | | | | |
| Respiration | — | *I* | — | *I* |
| Settling velocity | — | *I* | — | *I* |
| Oxygenation factor | — | *I* | — | *I* |
| Chemical composition (C, N, P) | — | *I* | — | *I* |
| Maximum specific growth | — | *I* | — | *I* |
| Half-saturation constants (light, carbon, nitrogen, phosphorus) | — | *I* | — | *I* |
| Zooplankton | | | | |
| Respiration | — | *I* | — | *I* |
| Mortality coefficients | — | *I* | — | *I* |

***Table 13-1*** *(Cont.)*

| Parameters | Model Variables | | | |
|---|---|---|---|---|
| | *Estuarial System* | | *Lake System* | |
| | *Hydraulic* | *Ecologic* | *Hydraulic* | *Ecologic* |
| Digestive efficiency | — | *I* | — | *I* |
| Chemical composition (C, N, P) | — | *I* | — | *I* |
| Preference for algae | — | *I* | — | *I* |
| Maximum specific growth | — | *I* | — | *I* |
| Half-saturation constants (algae, detritus) | — | *I* | — | *I* |
| Fish (2 groups) | | | | |
| Respiration | — | *I* | — | *I* |
| Mortality coefficients | — | *I* | — | *I* |
| Digestive efficiency | — | *I* | — | *I* |
| Chemical composition (C, N, P) | — | *I* | — | *I* |
| Maximum specific growth | — | *I* | — | *I* |
| Half-saturation constants (zooplankton, benthic animal) | — | *I* | — | *I* |
| Benthic Animal | | | | |
| Respiration | — | *I* | — | *I* |
| Mortality coefficients | — | *I* | — | *I* |
| Digestive efficiency | — | *I* | — | *I* |
| Chemical composition (C, N, P) | — | *I* | — | *I* |
| Maximum specific growth | — | *I* | — | *I* |
| Half-saturation constant (detritus) | — | *I* | — | *I* |
| Others | | | | |
| Chemical composition (C, N, P) | | | | |
| Zooplankton pellet | — | *C* | — | *C* |
| Fish pellet | — | *C* | — | *C* |
| Benthic pellet | — | *C* | — | *C* |
| Detritus | — | *I* | — | *I* |

*I* = Input.
*C* = Calculated.
*O* = Output.
— Not applicable.
[a] Requires initial conditions.

*Source:* C. W. Chen and G. T. Orlob, "Ecologic Simulation for Aquatic Environments, Final Report" OWRR Project No. C-2044, Department of the Interior, Office of Water Resources Research, Dec. 1972.

streptococci); algae—chlorophyll *a*; zooplankton; sediment; organic nitrogen; dissolved oxygen; total dissolved solids; nitrite-N; ortho-phosphate; ammonia-N; conservative constituents. A complete discussion of the model is given in Refs. 27, 28, and 35.

## 13-5 QUAL-I and DOSAG-I Simulation Models

QUAL-I and DOSAG-I were developed by the Texas Water Development Board for use in stream quality simulation studies.[32,36] QUAL-I is designed to simulate spatial and temporal variations of temperature, biochemical oxygen demand, dissolved oxygen (DO), and conservative minerals in streams and canals. DOSAG-I is used to simulate the spatial variation of BOD/DO in a system of streams and canals. It is not as accurate as QUAL-I and cannot be used for routing temperature and conservative minerals. QUAL-I is best suited for detailed studies of a stream under a prescribed condition, whereas DOSAG-I is better designed for use over a wide range of conditions when less detail is required. QUAL-I will be described in some detail and its use illustrated by an application of it to the Salt Creek Watershed in Nebraska. The program for QUAL-I was written by William A. White.[32] The following discussion is similar to that of the Texas Water Development Board given in Ref. 32.

QUAL-I is a set of interrelated quality routing models, the operation of which produces a history and spatial distribution of the temperature, biochemical oxygen demand/dissolved oxygen (BOD/DO), and up to three conservative minerals. The framework for its use is a one-dimensional, fully mixed, branching stream or canal system with multiple waste inputs and withdrawals. The models can be operated independently or simultaneously, with one model providing input to another if it is necessary. There are seven options that may be selected:

1. Route temperature, BOD/DO, and conservative minerals.
2. Route temperature and BOD/DO.
3. Route BOD/DO.
4. Route conservative minerals and temperature.
5. Route temperature.
6. Route BOD/DO and conservative minerals.
7. Route conservative minerals.

Flow augmentation requirements based on preselected minimum allowable dissolved oxygen concentrations can be set by the user. Restrictions on the use of the program are as follows:

1. Maximum number of reaches = 25.
2. Maximum number of waste inputs = 25.
3. Maximum number of headwaters = 5.
4. Maximum number of junctions = 5.

5. Maximum number of computational elements = 500.

In QUAL-I, the routing calculations are initiated at the most upstream locations (headwaters) of the stream or canal system. Incremental inflows and waste inputs or withdrawals are entered into the calculations as they occur. A set of simultaneous equations equal in number to the number of computational elements in the system is thus generated at the downstream point. These are solved, advancing the solution forward in time. An iterative process is continued until steady state conditions are reached. This is the approximate flow time from the uppermost point in the system to its terminus.

To use the program, it is first necessary to determine which segments of the stream or canal system are to be simulated. A schematic diagram of the stream system similar to the one shown in Fig. 13-5 facilitates this. The next step is to determine the degree of detail needed. This should be based on the availability and reliability

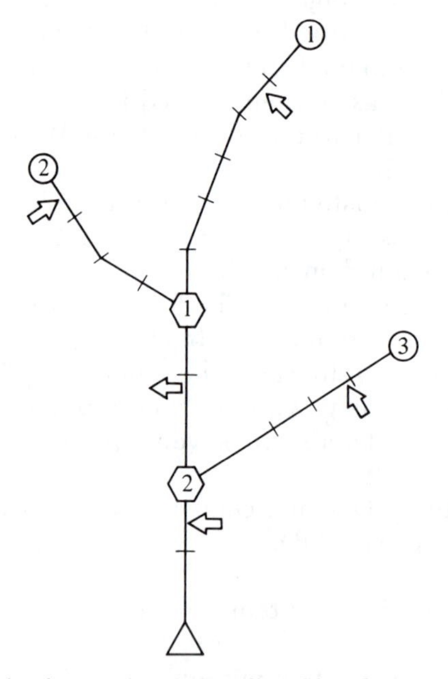


○ , 3 headwaters (maximum allowable = 5)
⇨| , 4 waste discharges (maximum allowable = 25)
⇦| , 1 withdrawal (maximum allowable = 25)
⬡ , 2 junctions (maximum allowable = 5)
|—| , 18 reaches (maximum allowable = 25).
Each reach may be divided into a maximum of 20 computational elements with a maximum of 500 of these per system

***Fig. 13-5.*** *A schematic diagram of a hypothetical stream system.*

of streamflow and water quality data, stream geometry, and the number and location of waste inputs or withdrawals. The stream is then subdivided into reaches having nearly uniform characteristics: reaches, headwaters, waste discharges or withdrawals, and junctions are ordered sequentially from the uppermost point in the system.

The final step is to determine the degree of resolution required. This decision should reflect some knowledge of the prototype behavior. For example, if the dissolved oxygen concentration changes from saturated to critical and returns to saturated through an interval of about 5 river mi, a degree of resolution of less than 1 mi is appropriate.[32] This decision permits each reach to be divided into computational elements or control volumes as shown in Fig. 13-5. Complete program details and procedures for use are given in Ref. 32.

### Application of QUAL-I to the Salt Creek Watershed

An example of the application of QUAL-I will illustrate the nature of the model output and its usefulness. The information presented follows the discussion of use of the model by the Nebraska Natural Resources Commission.[30]

The Nebraska Natural Resources Commission in cooperation with the Environmental Protection Agency developed the Salt Creek water quality model for determining the assimilative capacity of Salt Creek.[30] The QUAL-I modeling technique was selected. Water quality parameters examined in the Salt Creek system were biochemical oxygen demand (BOD) and dissolved oxygen (DO). Sulfate, chloride, and conductivity were also considered but are not discussed herein.

QUAL-I routes these parameters through the stream system on an hourly basis. It assumes that the major transport mechanisms, advection and dispersion, are only along the principal direction of flow of the stream. It allows for multiple waste discharges, withdrawals, tributary flows, and incremental runoff and has the capability to compute required dilution flows for flow augmentation to meet specified dissolved oxygen levels.

The model was applied to the lower Platte River network (including Salt Creek) as shown in Fig. 13-6.

The dissolved oxygen and biochemical oxygen demand portions of the model are based on the Streeter and Phelps mass action equilibrium equation for determining the oxygen sag curve.[40] This curve represents the deficit in dissolved oxygen downstream from a wastewater source.

The Salt Creek Model was developed principally for use with low flows and did not include agricultural pollutants or urban runoff. It was verified using flows and sampling data from 1972. Constants and flow rates were adjusted for use with the 1970 seven-day, 10-year low flows and the 1985 seven-day, 10-year low flows. A seven-day,

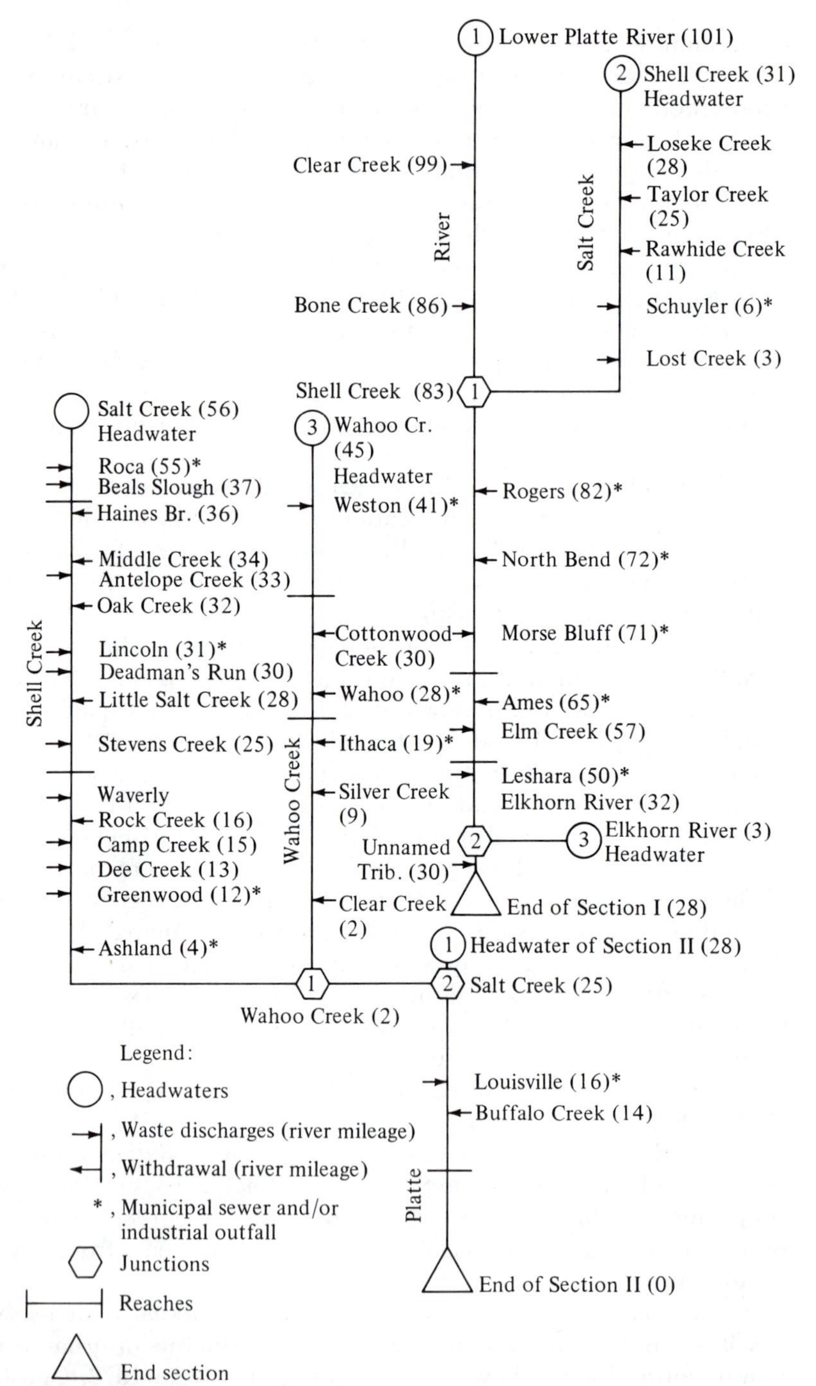


**Fig. 13-6.** *Lower Platte River basin schematic diagram QUAL-I. (Nebraska Natural Resources Commission.)*

10-year low flow represents the average low flow for seven consecutive days which would occur with a frequency of once in 10 years.

The results of two simulation runs are given in Figs. 13-7 and 13-8. The first illustrates the impact of limited wastewater treatment at Lincoln, Nebraska, during a drought. The second shows the effects of eliminating Lincoln's waste flows from Salt Creek. As shown in Fig. 13-8, with no waste flows from the city, the dissolved oxygen concentration in Salt Creek is significantly increased. Although the model is not an exact replica of Salt Creek, it gives an interpretation of how the stream would act under various conditions of stress. Such information is of great value in the planning and management of local and regional wastewater treatment facilities.

## 13-6 EPA Storm Water Management Model

In Chapter 11 it was noted that the EPA Storm Water Management Model was designed to simulate the quantity and quality of storm water flows from urbanized areas.[38] The program consists of a control

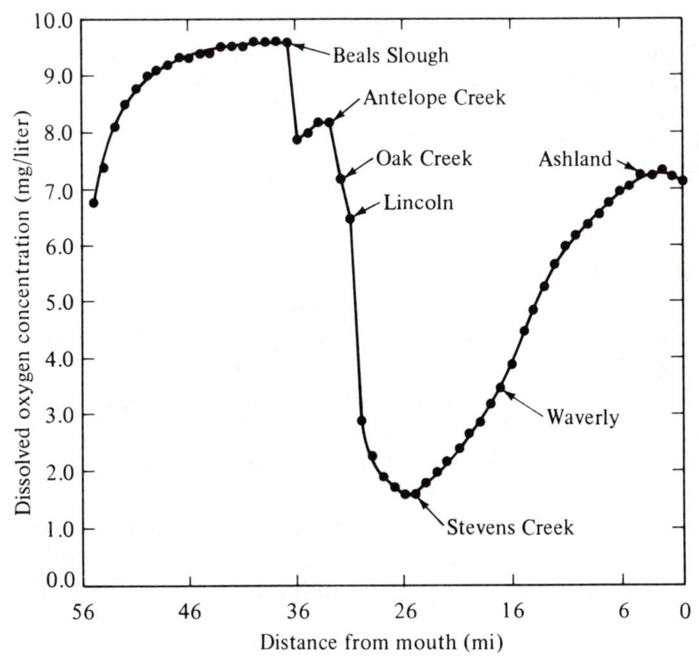


***Fig. 13-7.*** *Results of Simulation 1. Dissolved oxygen concentration on Salt Creek (Nebraska Natural Resources Commission). Note: 1985 projected low stream flow for Lincoln. One Sewage Treatment Plant. Effluent BOD 102 mg/1. Oak Creek DO 3.5 mg/1. Beals Slough DO 5.3 mg/1. 1985 projected actual wastewater flow.*

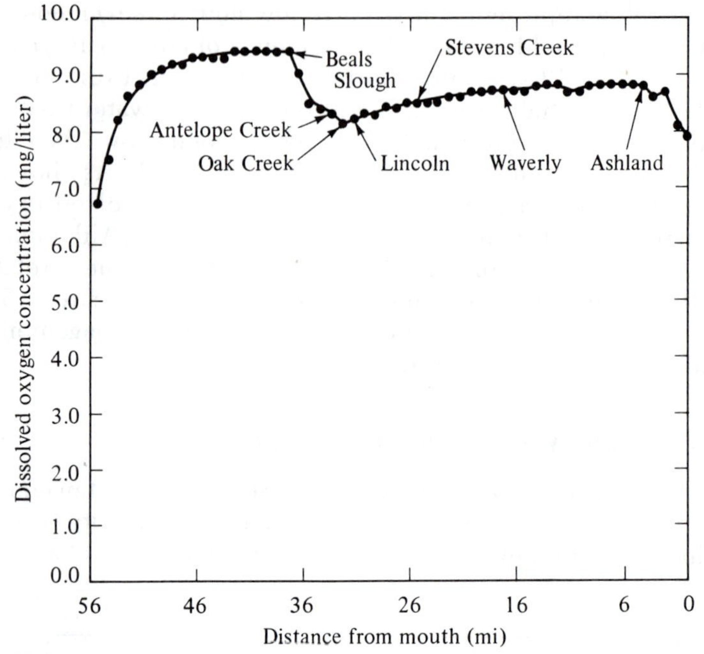


***Fig. 13-8.*** *Results of Simulation 2. Dissolved oxygen concentration on Salt Creek (Nebraska Natural Resources Commission). Note: Waste source loadings, 45 mg/l $BOD_5$. Lincoln flow = 0.1 cfs. 1970 low flow in stream. No Lincoln Sewage Treatment Plants. Oak Creek DO 8.0 mg/l, 1970 wastewater flow.*

segment and five computational blocks: (1) executive, (2) runoff, (3) transport, (4) storage, and (5) receiving water.

Input to the runoff water quality model consists of flow hydrographs developed in the quantity model. Output takes the form of pollutographs for each pollutant modeled. Pollutographs and hydrographs are introduced into the transport block where they are summed and modified by the addition of dry weather flow and infiltration to produce the final outfall characteristics.

It is assumed that the amount of pollutant which can be removed during a rainfall is dependent on the storm duration and initial quantity of the pollutant. This can be modeled by a first order differential equation of the form

$$-\frac{dP}{dt} = kP \tag{13-11}$$

or

$$P_0 - P = P_0(1 - e^{-kt}) \tag{13-12}$$

where

$P_0$ = the pollutant originally on the ground (lb)
$P$ = the pollutant after time $t$ (lb)
$k$ = a constant

The value of $k$ is assumed to be directly proportional to the rate of the runoff. Therefore, $k = br$, where $b$ is a constant and $r$ is the runoff intensity. Based upon available data, a value of 4.6 is assigned $b$. For each time step, the runoff rate is determined from the hydrograph and a value of $P$, which becomes the new value of $P_0$ for the next step, is calculated.

## 13-7 University of Cincinnati Urban Runoff Model

The University of Cincinnati Urban Runoff Model is similar to the EPA model (Chapter 11). Calculation of pollutant concentration in runoff is based on the assumption that the rate of pollutant removal depends on the amount of pollutant initially in the drainage area and the rainfall intensity.[39] The equation used is

$$-\frac{dP}{dt} = KqP \tag{13-13}$$

or

$$\frac{-dP}{P} = Kq\,dt \tag{13-14}$$

After integration, this becomes

$$P = P_0 e^{-KV_t} \tag{13-15}$$

where

$P$ = the amount of pollutant remaining on the ground at time $t$ (lb.)
$P_0$ = the initial amount of pollutant
$K$ = a constant
$V_t$ = volume of runoff up to time $t$ (ft$^3$)
$q$ = discharge (in./hr) per unit area

During a storm, the amount of pollutant washed into the sewer system in a time interval $t$ is

$$P_1 - P_2 = P_0(e^{-KV_1} - e^{-KV_2}) \tag{13-16}$$

The amount of solids washed into a storm sewer during a rainstorm is assumed proportional to the square of the runoff intensity. This equation is as follows:

$$r = \lambda q^2$$

where

$r$ = the fraction of solids carried off
$\lambda$ = the proportionality factor
$q$ = the runoff intensity (in./hr)

The amount of solids brought into the system in a time interval may be expressed as:

$$P_1 - P_2 = P_0 \lambda \bar{q}^2 (e^{-KV_1} - e^{-KV_2}) \quad \textbf{(13-17)}$$

where

$\bar{q}$ = the mean runoff intensity
$V_1, V_2$ = the runoff volumes at times $t_1$, $t_2$ ($ft^3$)

Soluble pollutants are routed downstream at the same velocity as the flow and are summed in the same manner as flows to determine final values. Provision is made for sediment transport.

## 13-8 Groundwater Quality Modeling

The complexity of modeling groundwater flow systems was pointed out in Chapters 8 and 10. Unfortunately, the degree of difficulty is even greater for groundwater quality models and thus the state of the art of these models is much less advanced than for surface water quality models. The importance of groundwater quality in water resource management and planning must be taken into account, however, and reliable models are needed. Gelhar, Wilson, and others have addressed this question.[37]

## Problems

13-1. Select two water quality models for conservative constituents and compare their characteristics.

13-2. Select two water quality models for nonconservative constituents and compare their characteristics.

13-3. Obtain the computer program for QUAL-I and make several simulation runs for a watershed of your choice.

13-4. Discuss how water quality models can complement traditional approaches to water quality planning.

## References

1. Weibel, S. R., R. B. Weidner, J. M. Cohen, and A. G. Christianson, "Pesticides and Other Contaminants in Rainfall and Runoff as a Factor in Stream Pollution," *J. Amer. Water Works Association,* Aug. 1966.
2. Wolman, M. G., and A. P. Schick, "Effects of Construction on Fluvial Sediment, Urban and Suburban Areas of Maryland," Water Resources Research, Vol. 3, No. 2, Second Quarter 1967.

3. Sawyer, C. N., "Fertilization of Lakes by Agricultural and Urban Drainage," *J. New England Water Works Association*, 61, 1947.
4. Sylvester, R. O., and G. C. Anderson, "A Lake's Response to Its Environment," *Proc. ASCE, J. Sanitary Engineering Div.*, 90, No. SA1, Feb. 1964.
5. Smith, R. L., W. J. O'Brien, A. R. LeFeuvre, and E. C. Pogge, "Development and Evaluation of a Mathematical Model of the Lower Reaches of the Kansas River Drainage System," Civil Engineering Department, University of Kansas, Lawrence, Jan. 1967.
6. Thayer, R. P., and R. G. Krutchkoff, "A Stochastic Model for Pollution and Dissolved Oxygen in Stream," Water Resources Research Center Publication, Virginia Polytechnic Institute, Blacksburg, Va., Aug. 1966.
7. Loucks, D. P., et al., "Linear Programming Models for Water Pollution Control," *Management Science*, 14, No. 4, Dec. 1967.
8. Steele, T. D., "Seasonal Variations in Chemical Quality of Surface Water in the Pescadero Creek Watershed, San Mateo County, California," Ph.D. Dissertation, Stanford University, Stanford, Calif., May 1968.
9. Ledbetter, J. O., and E. F. Gloyna, "Predictive Techniques for Water Quality Inorganics," *Proc. ASCE, J. Sanitary Engineering Div.*, 90, No. SA1, Feb. 1964, Part 1.
10. Wischmeier, W. H., "Cropping-Management Factor Evaluations for a Universal Soil-Loss Equation," Proc. Soil Science Society, 1960.
11. Woods, P. C., "Management of Hydrologic Systems for Water Quality Control," Contribution No. 121, Water Resources Center, University of California, Berkeley, Calif., June 1967.
12. Weibel, S. R., R. J. Anderson, and R. L. Woodward, "Urban Land Runoff as a Factor in Stream Pollution," *J. Water Pollution Control Federation*, 36, No. 7, July 1964.
13. Zumberge, J. H., and J. C. Ayers, "Hydrology of Lakes and Swamps," in V. T. Chow (ed.), *Handbook of Applied Hydrology* (New York: McGraw-Hill Book Company, 1964).
14. Viessman, Warren, Jr., "Estimation of Lake Flushing Rates for Water Quality Control Planning and Management," Proceedings of Conference on the Reclamation of Maine's Dying Lakes, University of Maine, Orono, Me., Mar. 1971.
15. Hutchinson, G. E., "A Treatise on Limnology—Volume I" (New York: John Wiley & Sons, Inc., 1957).
16. Liggett, J. A., "Circulation in Shallow Homogeneous Lakes," *Proc. ASCE, J. Hyd. Div.*, No. HY2, Mar. 1969.
17. Liggett, J. A., "Unsteady Circulation in Shallow Homogeneous Lakes," *Proc. ASCE, J. Hyd. Div.*, No. HY4, July 1969.
18. Liggett, J. A., "Cell Method in Computing Lake Circulation," *Proc. ASCE, J. Hyd. Div.*, No. HY3, Mar. 1970.
19. Liggett, J. A., and K. K. Lee, "Properties of Circulation in Stratified Lakes," *Proc. ASCE, J. Hyd. Div.*, No. HY1, Jan. 1971.
20. Csanady, G. T., "Wind-Driven Summer Circulation in the Great Lakes," *J. Geophys. Research*, 73, No. 8, Apr. 1968.
21. Harleman, D. R. F., et al., "The Feasibility of a Dynamic Model of Lake Michigan," Publication No. 9, Great Lakes Research Divisions, University of Michigan, Ann Arbor, 1962.

22. Henson, E. B., A. S. Bradshaw, and D. C. Chandler, "The Physical Limnology of Cayuga Lake, New York," Memoir 378, Agriculture Experiment Station, New York State College of Agriculture, Cornell University, Ithaca, N.Y., Aug. 1961.
23. Lee, K. K., and J. A. Liggett, "Computation of Wind Driven Circulation in Stratified Lakes," *Proc. ASCE, J. Hyd. Div.*, 96, No. HY10, Proc. Paper 7634, Oct. 1970.
24. Verber, J. L., "Current Profiles to Depth in Lake Michigan," Publication No. 13, Great Lakes Research Division, University of Michigan, Ann Arbor, 1965.
25. Chen, C. W., and G. T. Orlob, "Ecologic Simulation for Aquatic Environments, Annual Report," OWRR Project No. C-2044, U.S. Department of the Interior, Office of Water Resources Research, Aug. 1971.
26. Chen, C. W., and G. T. Orlob, "Ecologic Simulation for Aquatic Environments, Final Report," OWRR Project No. C-2044, U.S. Department of the Interior, Office of Water Resources Research, Dec. 1972.
27. Lombardo, P. S., and D. D. Franz, "Mathematical Model of Water Quality in Rivers and Impoundments," Hydrocomp, Inc., Palo Alto, Calif., Dec. 1972.
28. Lombardo, P. S., "Critical Review of Currently Available Water Quality Models," Hydrocomp, Inc., Palo Alto, Calif., July 1973.
29. Dysart, B. C., III, "Water Quality Planning in the Presence of Interacting Pollutants," Forty-second Annual Conference, Water Pollution Control Federation, Dallas, Tex., Oct. 1969.
30. Nebraska Natural Resources Commission, "Lower Platte River Basin Water Quality Management Plan," Lincoln, Neb., June 1974.
31. Viessman, Warren, Jr., "Assessing the Quality of Urban Drainage," Public Works, Oct. 1969.
32. Texas Water Development Board, "QUAL-I—Simulation of Water Quality in Streams and Lands," Program Documentation and Users' Manual, National Technical Information Service, PB 202973, Springfield, Va., 1970.
33. Orlob, G. T., and L. G. Selna, "Temperature Variations in Deep Reservoirs," *Proc. ASCE, J. Hyd. Div*, Vol. 96, No. HY2, Feb. 1970.
34. Crawford, N. H., and R. K. Linsley, "The Synthesis of Continuous Streamflow Hydrographs on a Digital Computer," Tech. Rept. 12, Department of Civil Engineering, Stanford University, Palo Alto, Calif., 1966.
35. Hydrocomp, Inc., *The Hydrocomp Simulation Network Operations Manual*, Palo Alto, Calif., 1969.
36. Texas Water Development Board, "DOSAG-I Simulation of Water Quality in Streams and Canals, Program Documentation and Users' Manual," National Technical Information Service, Springfield, Va., 1970.
37. Gelhar, L. W., and J. L. Wilson, "Ground Water Quality Modeling," Proceedings of Second National Ground Water Quality Symposium, U.S. Environmental Protection Agency, Washington, D.C., 1974.
38. Metcalf and Eddy, Inc., University of Florida, Water Resources Engineers, Inc., *Storm Water Management Model*, Environmental Protection Agency, Vol. 1, 1971.

39. Division of Water Resources, Department of Civil Engineering, University of Cincinnati, "Urban Runoff Characteristics," Water Pollution Control Research Series, EPA, 1970.
40. Streeter, N. W., and E. B. Phelps, U.S. Public Health Service, Bulletin 146, 1925.
41. Clark, J. W., Warren Viessman, Jr., and M. J. Hammer, *Water Supply and Pollution Control*, 3rd ed. (New York: Dun-Donnelly Publishing Corporation, Inc., 1977).

# Appendix

## Matrix Definition

A matrix is a set of quantities arranged in a rectangular array of $m$ rows and $n$ columns. A matrix can be expressed in the following manner:

$$P = [P_{ij}] = \begin{bmatrix} P_{11} & P_{12} & \cdots & P_{1n} \\ P_{21} & P_{22} & \cdots & P_{2n} \\ \cdot & \cdot & & \cdot \\ \cdot & \cdot & & \cdot \\ \cdot & \cdot & & \cdot \\ P_{m1} & P_{m2} & \cdots & P_{mn} \end{bmatrix}$$

$P$ stands for the entire array of quantities whereas the elements of the matrix are denoted by $P_{ij}$. A rectangular matrix of $m$ rows and $n$ columns is said to be of order $m \times n$, and if $m = n$, the matrix is square.

### FUNDAMENTAL OPERATIONS

## Transpose of a Matrix

The transpose $P^T$ of a matrix $P$ is found by simply interchanging rows for columns in the matrix $P$.

$$P = \begin{bmatrix} P_{11} & P_{12} & P_{13} \\ P_{21} & P_{22} & P_{23} \end{bmatrix} \qquad P^T = \begin{bmatrix} P_{11} & P_{21} \\ P_{12} & P_{22} \\ P_{13} & P_{23} \end{bmatrix}$$

## Determinant of a Matrix

The determinant of the elements of a square matrix $P$ is called the determinant of the matrix $P$ and is represented by $|P|$.

## Cofactor

The cofactor of an element in the $i$th row and $j$th column of a square matrix $P$ is the determinant formed by considering the matrix as a determinant, suppressing the $i$th row and the $j$th column, and multiplying by $(-1)^{i+j}$.

## The Adjoint Matrix

The adjoint matrix of a square matrix $P$ is formed by replacing each element by its cofactor and transposing. It is denoted by adj $P$.

## Equality of Matrices

Two matrices $P$ and $Z$ are equal if and only if they are the same order and all their corresponding elements are equal; that is, $P = Z$, provided that $P$ and $Z$ have the same order and $P_{ij} = Z_{ij}$.

## Matrix Multiplication

Matrix multiplication has been defined in such a manner that operations involving linear transformations can be concisely expressed. The definition of multiplication states that two matrices can be multiplied together only when the number of columns of the first is equal to the number of rows of the second. Matrices satisfying this condition are termed conformable matrices. If two matrices $P$ and $U$ are conformable, their product is defined by the equation $PU = Q$, where

$$Q_{ij} = \sum_{k=1}^{p} P_{ik} U_{kj}$$

and the orders of $P$, $U$, and $Q$ are $m \times p$, $p \times n$, and $m \times n$, respectively. As an example of the definition, consider the matrices

$$P = \begin{bmatrix} p_{11} & p_{12} & p_{13} \\ p_{21} & p_{22} & p_{23} \\ p_{31} & p_{32} & p_{33} \end{bmatrix} \qquad U = \begin{bmatrix} u_{11} & u_{12} \\ u_{21} & u_{22} \\ u_{31} & u_{32} \end{bmatrix}$$

Here $P$ is a matrix of order $3 \times 3$ and $U$ of order $3 \times 2$; hence these matrices are conformable. The rule above gives the following product matrix, of order $3 \times 2$.

$$Q = PU = \begin{bmatrix} (p_{11}u_{11} + p_{12}u_{21} + p_{13}u_{31}) & (p_{11}u_{12} + p_{12}u_{22} + p_{13}u_{32}) \\ (p_{21}u_{11} + p_{22}u_{21} + p_{23}u_{31}) & (p_{21}u_{12} + p_{22}u_{22} + p_{23}u_{32}) \\ (p_{31}u_{11} + p_{32}u_{21} + p_{33}u_{31}) & (p_{31}u_{12} + p_{32}u_{22} + p_{33}u_{32}) \end{bmatrix}$$

In general, if a matrix of order $m \times p$ is multiplied by one of order $p \times n$, the product is a matrix of order $m \times n$.

## The Inverse Matrix

If the determinant $P$ of a square matrix $P$ does not vanish, $P$ is said to be *nonsingular;* it then possesses an inverse matrix $P^{-1}$ such that $PP^{-1} = I = P^{-1}P$, where $I$ is the unit matrix of the same order as $P$. The inverse matrix $P^{-1}$ of the matrix $P$ is given by

$$P^{-1} = \frac{\text{adj}\, P}{|P|}$$

## Solution of a System of Linear Equations

The Gauss-Jordan elimination method is the best general procedure available for solving a system of linear equations $PU = Q$, where $P$ is an arbitrary $n \times n$ matrix, $U$ an arbitrary $n \times 1$ matrix, and $Q$ the $n \times 1$ solution vector. The method consists of performing elementary transformations (adding or subtracting a multiple of a row to some other row, interchanging two rows, or multiplying a row by a constant) on the $n \times n + 1$ augmented matrix $P$, $U$ until $P$ has been reduced to a unit matrix, when $U$ will become $P^{-1}U$. If matrix $P$ is ill-conditioned (elements along the principal diagonal are less than the other matrix elements), it is highly desirable to choose the so-called *pivotal elements* to be the largest in magnitude columnwise in order to avoid large roundoff errors.

The technique is best explained by the following example.

$$u_1 - 2u_2 + 3u_3 = 5$$

$$2u_1 + u_2 + u_3 = 0$$

$$3u_1 + u_2 + u_3 = 0$$

Steps involved in finding the solution are as follows.

$$\begin{bmatrix} 1 & -2 & 3 & 5 \\ 2 & 1 & 1 & 0 \\ 3 & 1 & 1 & 0 \end{bmatrix}$$

Original matrix.

First pivot chosen.

$$\begin{bmatrix} 3 & 1 & 1 & 0 \\ 2 & 1 & 1 & 0 \\ 1 & -2 & 3 & 5 \end{bmatrix}$$ Rows 1 and 3 interchanged.

$$\begin{bmatrix} 1 & \frac{1}{3} & \frac{1}{3} & 0 \\ 0 & \frac{1}{3} & \frac{1}{3} & 0 \\ 0 & -\frac{7}{3} & \frac{8}{3} & 5 \end{bmatrix}$$ Row 1 divided by first pivot. Rows 2 and 3 modified by row 1. Second pivot chosen.

$$\begin{bmatrix} 1 & \frac{1}{3} & \frac{1}{3} & 0 \\ 0 & -\frac{7}{3} & \frac{8}{3} & 5 \\ 0 & \frac{1}{3} & \frac{1}{3} & 0 \end{bmatrix}$$ Rows 2 and 3 are interchanged.

$$\begin{bmatrix} 1 & 0 & \frac{5}{7} & \frac{5}{7} \\ 0 & 1 & -\frac{8}{7} & -\frac{15}{7} \\ 0 & 0 & \frac{5}{7} & \frac{5}{7} \end{bmatrix}$$ Row 2 divided by second pivot. Rows 1 and 3 modified by row 2. Third pivot chosen.

$$\begin{bmatrix} 1 & 0 & 0 & 0 \\ 0 & 1 & 0 & -1 \\ 0 & 0 & 1 & 1 \end{bmatrix}$$ Row 3 divided by third pivot. Rows 1 and 2 modified by row 3.

The solution vector after this final step is simply the last column on the right:

$$u_1 = 0, \qquad u_2 = -1, \qquad u_3 = 1$$

## Error

An estimate of the error in computing $P^{-1}$ can be found by computing the inverse of $P^{-1}$ and expressing it as follows:

$$(P^{-1})^{-1} = P + 2E$$

where $E = e_{ij}$ is the error matrix for $P^{-1}$.

# B

# Appendix

***Table B-1*** *Water Properties, Constants, and Conversion Factors*

Gas Constants ($R$)

$R = 0.0821$ (atm)(liter)/(g-mol)(°K)
$R = 1.987$ g-cal/(g-mol)(°K)
$R = 1.987$ Btu/(lb-mol)(°R)

Heat of Vaporization of Water at 1.0 atm

540 cal/g = 970 Btu/lb

Acceleration of Gravity (Standard)

$g = 32.17$ ft/sec$^2$ = 980.6 cm/sec$^2$

Specific Heat of Air

$C_P = 0.238$ cal/(g)(°C)

Heat of Fusion of Water

79.7 cal/g = 144 Btu/lb

Density of Dry Air @ 0°C and 760 mm

0.001293 g/cm$^3$

Conversion Factors

1 second-foot-day per square mile = 0.03719 inch
1 inch of runoff per square mile = 26.9 second-foot-days
= 53.3 acre-feet
= 2,323,200 cubic feet
1 cubic foot per second = 0.9917 acre-inch per hour
= 1 sec-ft = 1 cusec
1 horsepower = 0.746 kilowatt
= 550 foot-pounds per second
$e = 2.71828$
$\log_{10} e = 0.43429$
$\log_e 10 = 2.30259$

Metric Equivalents

| | |
|---|---|
| 1 foot | = 0.3048 meter |
| 1 mile | = 1.609 kilometers |
| 1 acre | = 0.4047 hectare |
| | = 4047 square meters |
| 1 square mile ($mi^2$) | = 259 hectares |
| | = 2.59 square kilometers ($km^2$) |
| 1 acre foot (acre-ft) | = 1233 cubic meters |
| 1 million cubic feet (mcf) | = 28,320 cubic meters |
| 1 cubic foot per second (cfs) | = 0.02832 cubic meters per second |
| | = 1.699 cubic meters per minute |
| 1 second-foot-day (cfsd) | = 2447 cubic meters |
| 1 million gallons (mg) | = 3785 cubic meters |
| | = 3.785 million liters |
| 1 million gallons per day (mgd) | = 694.4 gallons per minute (gpm) |
| | = 2.629 cubic meters per minute |
| | = 3785 cubic meters per day |

***Table B-2*** *Properties of Water*

| Temp., °F | Specific Gravity | Unit Weight ($lb/ft^3$) | Heat of Vaporization (Btu/lb) | Kinematic Viscosity ($ft^2/sec$) | Vapor Pressure | | |
|---|---|---|---|---|---|---|---|
| | | | | | *Mb* | *Psi* | *in. Hg* |
| 32 | 0.99987 | 62.416 | 1073 | $1.93 \times 10^{-5}$ | 6.11 | 0.09 | 0.18 |
| 40 | 0.99999 | 62.423 | 1066 | $1.67 \times 10^{-5}$ | 8.36 | 0.12 | 0.25 |
| 50 | 0.99975 | 62.408 | 1059 | $1.41 \times 10^{-5}$ | 12.19 | 0.18 | 0.36 |
| 60 | 0.99907 | 62.366 | 1054 | $1.21 \times 10^{-5}$ | 17.51 | 0.26 | 0.52 |
| 70 | 0.99802 | 62.300 | 1049 | $1.06 \times 10^{-5}$ | 24.79 | 0.36 | 0.74 |
| 80 | 0.99669 | 62.217 | 1044 | $0.929 \times 10^{-5}$ | 34.61 | 0.51 | 1.03 |
| 90 | 0.99510 | 62.118 | 1039 | $0.828 \times 10^{-5}$ | 47.68 | 0.70 | 1.42 |
| 100 | 0.99318 | 61.998 | 1033 | $0.741 \times 10^{-5}$ | 64.88 | 0.95 | 1.94 |

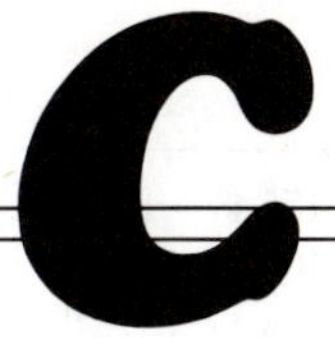

# Appendix

**Table C-1** *Areas under the Normal Curve*

$$F(z) = \int_0^z \frac{1}{\sqrt{2\pi}} e^{-z^2/2}\, dz$$

| z | .00 | .01 | .02 | .03 | .04 | .05 | .06 | .07 | .08 | .09 |
|---|---|---|---|---|---|---|---|---|---|---|
| .0 | .0000 | .0040 | .0080 | .0120 | .0159 | .0199 | .0239 | .0279 | .0319 | .0359 |
| .1 | .0398 | .0438 | .0478 | .0517 | .0557 | .0596 | .0636 | .0675 | .0714 | .0753 |
| .2 | .0793 | .0832 | .0871 | .0910 | .0948 | .0987 | .1026 | .1064 | .1103 | .1141 |
| .3 | .1179 | .1217 | .1255 | .1293 | .1331 | .1368 | .1406 | .1443 | .1480 | .1517 |
| .4 | .1554 | .1591 | .1628 | .1664 | .1700 | .1736 | .1772 | .1808 | .1844 | .1879 |
| .5 | .1915 | .1950 | .1985 | .2019 | .2054 | .2088 | .2123 | .2157 | .2190 | .2224 |
| .6 | .2257 | .2291 | .2324 | .2357 | .2389 | .2422 | .2454 | .2486 | .2518 | .2549 |
| .7 | .2580 | .2611 | .2642 | .2673 | .2704 | .2734 | .2764 | .2794 | .2823 | .2852 |
| .8 | .2881 | .2910 | .2939 | .2967 | .2995 | .3023 | .3051 | .3078 | .3106 | .3133 |
| .9 | .3159 | .3186 | .3212 | .3238 | .3264 | .3289 | .3315 | .3340 | .3365 | .3389 |
| 1.0 | .3413 | .3438 | .3461 | .3485 | .3508 | .3531 | .3554 | .3577 | .3599 | .3621 |
| 1.1 | .3643 | .3665 | .3686 | .3708 | .3729 | .3749 | .3770 | .3790 | .3810 | .3830 |
| 1.2 | .3849 | .3869 | .3888 | .3907 | .3925 | .3944 | .3962 | .3980 | .3997 | .4015 |
| 1.3 | .4032 | .4049 | .4066 | .4082 | .4099 | .4115 | .4131 | .4147 | .4162 | .4177 |
| 1.4 | .4192 | .4207 | .4222 | .4236 | .4251 | .4265 | .4279 | .4292 | .4306 | .4319 |
| 1.5 | .4332 | .4345 | .4357 | .4370 | .4382 | .4394 | .4406 | .4418 | .4430 | .4441 |
| 1.6 | .4452 | .4463 | .4474 | .4485 | .4495 | .4505 | .4515 | .4525 | .4535 | .4545 |

*(Continued on next page)*

| $z$ | .00 | .01 | .02 | .03 | .04 | .05 | .06 | .07 | .08 | .09 |
|---|---|---|---|---|---|---|---|---|---|---|
| 1.7 | .4554 | .4564 | .4573 | .4582 | .4591 | .4599 | .4608 | .4616 | .4625 | .4633 |
| 1.8 | .4641 | .4649 | .4656 | .4664 | .4671 | .4678 | .4686 | .4693 | .4699 | .4706 |
| 1.9 | .4713 | .4719 | .4726 | .4732 | .4738 | .4744 | .4750 | .4756 | .4762 | .4767 |
| 2.0 | .4772 | .4778 | .4783 | .4788 | .4793 | .4798 | .4803 | .4808 | .4812 | .4817 |
| 2.1 | .4821 | .4826 | .4830 | .4834 | .4838 | .4842 | .4846 | .4850 | .4854 | .4857 |
| 2.2 | .4861 | .4865 | .4868 | .4871 | .4875 | .4878 | .4881 | .4884 | .4887 | .4890 |
| 2.3 | .4893 | .4896 | .4898 | .4901 | .4904 | .4906 | .4909 | .4911 | .4913 | .4916 |
| 2.4 | .4918 | .4920 | .4922 | .4925 | .4927 | .4929 | .4931 | .4932 | .4934 | .4936 |
| 2.5 | .4938 | .4940 | .4941 | .4943 | .4945 | .4946 | .4948 | .4949 | .4951 | .4952 |
| 2.6 | .4953 | .4955 | .4956 | .4957 | .4959 | .4960 | .4961 | .4962 | .4963 | .4964 |
| 2.7 | .4965 | .4966 | .4967 | .4968 | .4969 | .4970 | .4971 | .4972 | .4973 | .4974 |
| 2.8 | .4974 | .4975 | .4976 | .4977 | .4977 | .4978 | .4979 | .4980 | .4980 | .4981 |
| 2.9 | .4981 | .4982 | .4983 | .4983 | .4984 | .4984 | .4985 | .4985 | .4986 | .4986 |
| 3.0 | .4987 | .4987 | .4987 | .4988 | .4988 | .4989 | .4989 | .4989 | .4990 | .4990 |
| 3.1 | .4990 | .4991 | .4991 | .4991 | .4992 | .4992 | .4992 | .4992 | .4993 | .4993 |
| 3.2 | .4993 | .4993 | .4994 | .4994 | .4994 | .4994 | .4994 | .4995 | .4995 | .4995 |
| 3.3 | .4995 | .4995 | .4996 | .4996 | .4996 | .4996 | .4996 | .4996 | .4996 | .4997 |
| 3.4 | .4997 | .4997 | .4997 | .4997 | .4997 | .4997 | .4997 | .4997 | .4998 | .4998 |
| . | . | | | | | | | | | |
| . | . | | | | | | | | | |
| . | . | | | | | | | | | |
| 4.0 | .499968 | | | | | | | | | |

*Source:* After C. E. Weatherburn, *Mathematical Statistics* (London: Cambridge University Press, 1957) (for $z = 0$ to $z = 3.1$); C. H. Richardson, *An Introduction to Statistical Analysis* (New York: Harcourt, Brace & World, Inc., 1944) (for $z = 3.2$ to $z = 3.4$); A. H. Bowker and G. J. Lieberman, *Engineering Statistics* (Englewood Cliffs, N. J.: Prentice-Hall, Inc., 1959) (for $z = 4.0$ and 5.0).

**Table C-2** *K Values for Pearson Type-III Distribution*

| Skew Coeff., | Recurrence Interval in Years | | | | | | | | | | |
|---|---|---|---|---|---|---|---|---|---|---|---|
| | *1.0101* | *1.0526* | *1.1111* | *1.2500* | *2* | *5* | *10* | *25* | *50* | *100* | *200* |
| | Percent Chance | | | | | | | | | | |
| $C_s$ | *99* | *95* | *90* | *80* | *50* | *20* | *10* | *4* | *2* | *1* | *0.5* |
| | | | | | Positive Skew | | | | | | |
| 3.0 | −0.667 | −0.665 | −0.660 | −0.636 | −0.396 | 0.420 | 1.180 | 2.278 | 3.152 | 4.051 | 4.970 |
| 2.9 | −0.690 | −0.688 | −0.681 | −0.651 | −0.390 | 0.440 | 1.195 | 2.277 | 3.134 | 4.013 | 4.909 |
| 2.8 | −0.714 | −0.711 | −0.702 | −0.666 | −0.384 | 0.460 | 1.210 | 2.275 | 3.114 | 3.973 | 4.847 |
| 2.7 | −0.740 | −0.736 | −0.724 | −0.681 | −0.376 | 0.479 | 1.224 | 2.272 | 3.093 | 3.932 | 4.783 |
| 2.6 | −0.769 | −0.762 | −0.747 | −0.696 | −0.368 | 0.499 | 1.238 | 2.267 | 3.071 | 3.889 | 4.718 |
| 2.5 | −0.799 | −0.790 | −0.771 | −0.711 | −0.360 | 0.518 | 1.250 | 2.262 | 3.048 | 3.845 | 4.652 |
| 2.4 | −0.832 | −0.819 | −0.795 | −0.725 | −0.351 | 0.537 | 1.262 | 2.256 | 3.023 | 3.800 | 4.584 |
| 2.3 | −0.867 | −0.850 | −0.819 | −0.739 | −0.341 | 0.555 | 1.274 | 2.248 | 2.997 | 3.753 | 4.515 |
| 2.2 | −0.905 | −0.882 | −0.844 | −0.752 | −0.330 | 0.574 | 1.284 | 2.240 | 2.970 | 3.705 | 4.444 |
| 2.1 | −0.946 | −0.914 | −0.869 | −0.765 | −0.319 | 0.592 | 1.294 | 2.230 | 2.942 | 3.656 | 4.372 |
| 2.0 | −0.990 | −0.949 | −0.895 | −0.777 | −0.307 | 0.609 | 1.302 | 2.219 | 2.912 | 3.605 | 4.398 |
| 1.9 | −1.037 | −0.984 | −0.920 | −0.788 | −0.294 | 0.627 | 1.310 | 2.207 | 2.881 | 3.553 | 4.223 |

*(Continued on next page)*

***Table C-2*** *(Cont.)*

| Skew Coeff., $C_s$ | Recurrence Interval in Years *1.0101* | *1.0526* | *1.1111* | *1.2500* | *2* | *5* | *10* | *25* | *50* | *100* | *200* |
|---|---|---|---|---|---|---|---|---|---|---|---|
| | Percent Chance *99* | *95* | *90* | *80* | *50* | *20* | *10* | *4* | *2* | *1* | *0.5* |
| | | | | | Positive Skew | | | | | | |
| 1.8 | −1.087 | −1.020 | −0.945 | −0.799 | −0.282 | 0.643 | 1.318 | 2.193 | 2.848 | 3.499 | 4.147 |
| 1.7 | −1.140 | −1.056 | −0.970 | −0.808 | −0.268 | 0.660 | 1.324 | 2.179 | 2.815 | 3.444 | 4.069 |
| 1.6 | −1.197 | −1.093 | −0.994 | −0.817 | −0.254 | 0.675 | 1.329 | 2.163 | 2.780 | 3.388 | 3.990 |
| 1.5 | −1.256 | −1.131 | −1.018 | −0.825 | −0.240 | 0.690 | 1.333 | 2.146 | 2.743 | 3.330 | 3.910 |
| 1.4 | −1.318 | −1.168 | −1.041 | −0.832 | −0.225 | 0.705 | 1.337 | 2.128 | 2.706 | 3.271 | 3.828 |
| 1.3 | −1.383 | −1.206 | −1.064 | −0.838 | −0.210 | 0.719 | 1.339 | 2.108 | 2.666 | 3.211 | 3.745 |
| 1.2 | −1.449 | −1.243 | −1.086 | −0.844 | −0.195 | 0.732 | 1.340 | 2.087 | 2.626 | 3.149 | 3.661 |
| 1.1 | −1.518 | −1.280 | −1.107 | −0.848 | −0.180 | 0.745 | 1.341 | 2.066 | 2.585 | 3.087 | 3.575 |
| 1.0 | −1.588 | −1.317 | −1.128 | −0.852 | −0.164 | 0.758 | 1.340 | 2.043 | 2.542 | 3.022 | 3.489 |
| .9 | −1.660 | −1.353 | −1.147 | −0.854 | −0.148 | 0.769 | 1.339 | 2.018 | 2.498 | 2.957 | 3.401 |
| .8 | −1.733 | −1.388 | −1.166 | −0.856 | −0.132 | 0.780 | 1.336 | 1.993 | 2.453 | 2.891 | 3.312 |
| .7 | −1.806 | −1.423 | −1.183 | −0.857 | −0.116 | 0.790 | 1.333 | 1.967 | 2.407 | 2.824 | 3.223 |
| .6 | −1.880 | −1.458 | −1.200 | −0.857 | −0.099 | 0.800 | 1.328 | 1.939 | 2.359 | 2.755 | 3.132 |
| .5 | −1.955 | −1.491 | −1.216 | −0.856 | −0.083 | 0.808 | 1.323 | 1.910 | 2.311 | 2.686 | 3.041 |
| .3 | −2.029 | −1.524 | −1.231 | −0.855 | −0.066 | 0.816 | 1.317 | 1.880 | 2.261 | 2.615 | 2.949 |
| .3 | −2.104 | −1.555 | −1.245 | −0.853 | −0.050 | 0.824 | 1.309 | 1.849 | 2.211 | 2.544 | 2.856 |
| .2 | −2.178 | −1.586 | −1.258 | −0.850 | −0.033 | 0.830 | 1.301 | 1.818 | 2.159 | 2.472 | 2.763 |
| .1 | −2.252 | −1.616 | −1.270 | −0.846 | −0.017 | 0.836 | 1.292 | 1.785 | 2.107 | 2.400 | 2.670 |
| .0 | −2.326 | −1.645 | −1.282 | −0.842 | 0 | 0.842 | 1.282 | 1.751 | 2.054 | 2.326 | 2.576 |
| | | | | | Negative Skew | | | | | | |
| − .1 | −2.400 | −1.673 | −1.292 | −0.836 | 0.017 | 0.846 | 1.270 | 1.716 | 2.000 | 2.252 | 2.482 |
| − .2 | −2.472 | −1.700 | −1.301 | −0.830 | 0.033 | 0.850 | 1.258 | 1.680 | 1.945 | 2.178 | 2.388 |
| − .3 | −2.544 | −1.726 | −1.309 | −0.824 | 0.050 | 0.853 | 1.245 | 1.643 | 1.890 | 2.104 | 2.294 |
| − .4 | −2.615 | −1.750 | −1.317 | −0.816 | 0.066 | 0.855 | 1.231 | 1.606 | 1.834 | 2.029 | 2.201 |
| − .5 | −2.686 | −1.774 | −1.323 | −0.808 | 0.083 | 0.856 | 1.216 | 1.567 | 1.777 | 1.955 | 2.108 |
| − .6 | −2.755 | −1.797 | −1.328 | −0.800 | 0.099 | 0.857 | 1.200 | 1.528 | 1.720 | 1.880 | 2.016 |
| − .7 | −2.824 | −1.819 | −1.333 | −0.790 | 0.116 | 0.857 | 1.183 | 1.488 | 1.663 | 1.806 | 1.926 |
| − .8 | −2.891 | −1.839 | −1.336 | −0.780 | 0.132 | 0.856 | 1.166 | 1.448 | 1.606 | 1.733 | 1.837 |
| − .9 | −2.957 | −1.858 | −1.339 | −0.769 | 0.148 | 0.854 | 1.147 | 1.407 | 1.549 | 1.660 | 1.749 |
| −1.0 | −3.022 | −1.877 | −1.340 | −0.758 | 0.164 | 0.852 | 1.128 | 1.366 | 1.492 | 1.588 | 1.664 |
| −1.1 | −3.087 | −1.894 | −1.341 | −0.745 | 0.180 | 0.848 | 1.107 | 1.324 | 1.435 | 1.518 | 1.581 |
| −1.2 | −3.149 | −1.910 | −1.340 | −0.732 | 0.195 | 0.844 | 1.086 | 1.282 | 1.379 | 1.449 | 1.501 |
| −1.3 | −3.211 | −1.925 | −1.339 | −0.719 | 0.210 | 0.838 | 1.064 | 1.240 | 1.324 | 1.383 | 1.424 |
| −1.4 | −3.271 | −1.938 | −1.337 | −0.705 | 0.225 | 0.832 | 1.041 | 1.198 | 1.270 | 1.318 | 1.351 |
| −1.5 | −3.330 | −1.951 | −1.333 | −0.690 | 0.240 | 0.825 | 1.018 | 1.157 | 1.217 | 1.256 | 1.282 |
| −1.6 | −3.388 | −1.962 | −1.329 | −0.675 | 0.254 | 0.817 | 0.994 | 1.116 | 1.166 | 1.197 | 1.216 |
| −1.7 | −3.444 | −1.972 | −1.324 | −0.660 | 0.268 | 0.808 | 0.970 | 1.075 | 1.116 | 1.140 | 1.155 |
| −1.8 | −3.499 | −1.981 | −1.318 | −0.643 | 0.282 | 0.799 | 0.945 | 1.035 | 1.069 | 1.087 | 1.097 |
| −1.9 | −3.553 | −1.989 | −1.310 | −0.627 | 0.294 | 0.788 | 0.920 | 0.996 | 1.023 | 1.037 | 1.044 |
| −2.0 | −3.605 | −1.996 | −1.302 | −0.609 | 0.307 | 0.777 | 0.895 | 0.959 | 0.980 | 0.990 | 0.995 |
| −2.1 | −3.656 | −2.001 | −1.294 | −0.592 | 0.319 | 0.765 | 0.869 | 0.923 | 0.939 | 0.946 | 0.949 |
| −2.2 | −3.705 | −2.006 | −1.284 | −0.574 | 0.330 | 0.752 | 0.844 | 0.888 | 0.900 | 0.905 | 0.907 |
| −2.3 | −3.753 | −2.009 | −1.274 | −0.555 | 0.341 | 0.739 | 0.819 | 0.855 | 0.864 | 0.867 | 0.869 |
| −2.4 | −3.800 | −2.011 | −1.262 | −0.537 | 0.351 | 0.725 | 0.795 | 0.823 | 0.830 | 0.832 | 0.833 |
| −2.5 | −3.845 | −2.012 | −1.250 | −0.518 | 0.360 | 0.711 | 0.771 | 0.793 | 0.798 | 0.799 | 0.800 |
| −2.6 | −3.889 | −2.013 | −1.238 | −0.499 | 0.368 | 0.696 | 0.747 | 0.764 | 0.768 | 0.769 | 0.769 |
| −2.7 | −3.932 | −2.012 | −1.224 | −0.479 | 0.376 | 0.681 | 0.724 | 0.738 | 0.740 | 0.740 | 0.741 |
| −2.8 | −3.973 | −2.010 | −1.210 | −0.460 | 0.384 | 0.666 | 0.702 | 0.712 | 0.714 | 0.714 | 0.714 |
| −2.9 | −4.013 | −2.007 | −1.195 | −0.440 | 0.390 | 0.651 | 0.681 | 0.683 | 0.689 | 0.690 | 0.690 |
| −3.0 | −4.051 | −2.003 | −1.180 | −0.420 | 0.396 | 0.636 | 0.660 | 0.666 | 0.666 | 0.667 | 0.667 |

*Source:* After ***Water Resources Council, Bulletin No. 15***, December 1967.

**Table C-3** *Uniformly Distributed Random Numbers*

| | | | | |
|---|---|---|---|---|
| 53 74 23 99 67 | 61 32 28 69 84 | 94 62 67 86 24 | 98 33 41 19 95 | 47 53 53 38 09 |
| 63 38 06 86 54 | 99 00 65 26 94 | 02 82 90 23 07 | 79 62 67 80 60 | 75 91 12 81 19 |
| 30 30 58 21 46 | 06 72 17 10 94 | 25 21 31 75 96 | 49 28 24 00 49 | 55 65 79 78 07 |
| 63 43 36 82 69 | 65 51 18 37 88 | 61 38 44 12 45 | 32 92 85 88 65 | 54 34 81 85 35 |
| 98 25 37 55 26 | 01 91 82 81 46 | 74 71 12 94 97 | 24 02 71 37 07 | 03 92 18 66 75 |
| 02 63 21 17 69 | 71 50 80 89 56 | 38 15 70 11 48 | 43 40 45 86 98 | 00 83 26 91 03 |
| 64 55 22 21 82 | 48 22 28 06 00 | 61 54 13 43 91 | 82 78 12 23 29 | 06 66 24 12 27 |
| 85 07 26 13 89 | 01 10 07 82 04 | 59 63 69 36 03 | 69 11 15 83 80 | 13 29 54 19 28 |
| 58 54 16 24 15 | 51 54 44 82 00 | 62 61 65 04 69 | 38 18 65 18 97 | 85 72 13 49 21 |
| 34 85 27 84 87 | 61 48 64 56 26 | 90 18 48 13 26 | 37 70 15 42 57 | 65 65 80 39 07 |
| 03 92 18 27 46 | 57 99 16 96 56 | 30 33 72 85 22 | 84 64 38 56 98 | 99 01 30 98 64 |
| 62 95 30 27 59 | 37 75 41 66 48 | 86 97 80 61 45 | 23 53 04 01 63 | 45 76 08 64 27 |
| 08 45 93 15 22 | 60 21 54 46 91 | 98 77 27 85 42 | 28 88 61 08 94 | 69 62 03 42 73 |
| 07 08 55 18 40 | 45 44 75 13 90 | 24 94 96 61 02 | 57 55 66 83 15 | 73 42 37 11 61 |
| 01 85 89 95 66 | 51 10 19 34 88 | 15 84 97 19 75 | 12 76 39 43 78 | 64 63 91 08 25 |
| 72 84 71 14 35 | 19 11 58 49 26 | 50 11 17 17 76 | 86 31 57 20 18 | 95 60 78 46 75 |
| 88 78 28 16 84 | 13 52 53 94 53 | 75 45 69 30 96 | 73 89 65 70 31 | 99 17 43 48 76 |
| 45 17 75 65 57 | 28 40 19 72 12 | 25 12 74 75 67 | 60 40 60 81 19 | 24 62 01 61 16 |
| 96 76 28 12 54 | 22 01 11 94 25 | 71 96 16 16 88 | 68 64 36 74 45 | 19 59 50 88 92 |
| 43 31 67 72 30 | 24 02 94 08 63 | 38 32 36 66 02 | 69 36 38 25 39 | 48 03 45 15 22 |
| 50 44 66 44 21 | 66 06 58 05 62 | 68 15 54 35 02 | 42 35 48 96 32 | 14 52 41 52 48 |
| 22 66 22 15 86 | 26 63 75 41 99 | 58 42 36 72 24 | 58 37 52 18 51 | 03 37 18 39 11 |
| 96 24 40 14 51 | 23 22 30 88 57 | 95 67 47 29 83 | 94 69 40 06 07 | 18 16 36 78 86 |
| 31 73 91 61 19 | 60 20 72 93 48 | 98 57 07 23 69 | 65 95 39 69 58 | 56 80 30 19 44 |
| 78 60 73 99 84 | 43 89 94 36 45 | 56 69 47 07 41 | 90 22 91 07 12 | 78 35 34 08 72 |
| 84 37 90 61 56 | 70 10 23 98 05 | 85 11 34 76 60 | 76 48 45 34 60 | 01 64 18 39 96 |
| 36 67 10 08 23 | 98 93 35 08 86 | 99 29 76 29 81 | 33 34 91 58 93 | 63 14 52 32 52 |
| 07 28 59 07 48 | 89 64 58 89 75 | 83 85 62 27 89 | 30 14 78 56 27 | 86 63 59 80 02 |
| 10 15 83 87 60 | 79 24 31 66 56 | 21 48 24 06 93 | 91 98 94 05 49 | 01 47 59 38 00 |
| 55 19 68 97 65 | 03 73 52 16 56 | 00 53 55 90 27 | 33 42 29 38 87 | 22 13 88 83 34 |
| 53 81 29 13 39 | 35 01 20 71 34 | 62 33 74 82 14 | 53 73 19 09 03 | 56 54 29 56 93 |
| 51 86 32 68 92 | 33 98 74 66 99 | 40 14 71 94 58 | 45 94 19 38 81 | 14 44 99 81 07 |
| 35 91 70 29 13 | 80 03 54 07 27 | 96 94 78 32 66 | 50 95 52 74 33 | 13 80 55 62 54 |
| 37 71 67 95 13 | 20 02 44 95 94 | 64 85 04 05 72 | 01 32 90 76 14 | 53 89 74 60 41 |
| 93 66 13 83 27 | 92 79 64 64 72 | 28 54 96 53 84 | 48 14 52 98 94 | 56 07 93 89 30 |

*Source:* After L. R. Beard, *Statistical Methods in Hydrology*, U.S. Army Corps of Engineers, 1962.

# Glossary

$A$ drainage area
$A$ area
$A$ cross-section area
$A$ exchange coefficient
$A_C$ area coefficient
$A_c$ cross-sectional area
ADT average daily traffic
$A_i$ area between isochrones
$A_L$ flow area at left node
AMC antecedent moisture condition
$A_R$ flow area at right node
$A_s$ water surface area
$A_t$ partial derivative of A with respect to time
$\bar{A}_u$ average area of basins of order u
$AWC$ porosity drainable only by evapotranspiration
$a$ limiting value
$a$ best estimate of skewness
$a$ acceleration
$a$ reference level
$a$ regression constant
$a$ stage-discharge coefficient
$a$ positive integer
$a$ antilog adder
$B$ Bowen ratio
$B$ summation of monthly consumptive use factors
$B$ water surface width
$B$ watershed parameter
BOD biochemical oxygen demand
$b$ limiting value
$b$ saturated depth of aquifer
$b$ reference level
$b$ constant
$b$ regression coefficient
$b$ stage-storage coefficient
$b$ soil moisture coefficient
$b$ positive integer
$b_j$ standardized correlation coefficient
$C$ empirical coefficient
$C$ constant shape factor
$C$ constituent concentration
$C$ weir discharge coefficient
$C$ pipe discharge coefficient
$C$ Chézy's resistance coefficient
$C$ loss coefficient
$C$ runoff coefficient
$C$ regression coefficient
$C$ regional flood coefficient
$C_{IC}$ concentration of a constituent stored within a system
$C_{II}$ constituent input to a system per unit time
$C_{IO}$ constituent output from a system per unit time

$C_M$ Manning slope-roughness coefficient
CN runoff curve number
$C_o$ Muskingum storage coefficient
$C_p$ coefficient accounting for flood wave and storage conditions
$C_p$ specific heat of water
$C_p$ specific heat of air
$C_p$ peak flow coefficient
$C_s$ coefficient of skewness
$C_t$ coefficient representing variations of watershed slopes and storage
*CUML* accumulative storm losses
$C_v$ coefficient of variation
$C_1$ Muskingum storage coefficient
$C_2$ Muskingum storage coefficient
*c* turbulent exchange coefficient
*c* Courant condition coefficient
*c* wave celerity
*c* positive integer
*c* soil moisture coefficient
*D* original duration of one unit hydrograph
*D* effective snowmelt
*D* coefficient of determination
*D* snowmelt
*D* depth of snowpack
*D* 6-hr snowmelt
*D* rainstorm duration
*D* detention storage volume
*D* overland flow distance
*D* percent error in topographic index
*D* surface detention
$D_d$ overland detention requirement depth
$D_e$ volume of surface detention
*DH* ruggedness number
DO dissolved oxygen
*DR* average height of direct runoff ordinate
$D_r$ overland flow depth
DRH direct runoff hydrograph
*DT* time increment
$D_t$ water depth at time t
*d* angle
$d^R$ grain diameter
$D^I$ the possible durations of unit hydrograph
E evaporation
E event
*E* voltage
*E* evapotranspiration depth or rate
*E* recession exponent
$E_p$ potential evapotranspiration rate
$E_s$ reservoir evaporation
ET evapotranspiration
*e* partial pressure of water vapor
*e* vapor pressure
*e* vapor pressure of air
*e* random error
$e_a$ the vapor pressure of the air
$e_o$ saturation vapor pressure
exp symbol for exponential of *e*
*F* accumulated mass infiltration
*F* force
*F* fraction of watershed area in forest
*F* frequency factor
*F* accumulated infiltration depth
*F* flood event
*FD* abbreviation for downstream force
$F_f$ friction force
$F_g$ force of gravity
$F_i$ hydrostatic force
*FU* abbreviation for upstream force
$F(x)$ cumulative distribution function
$F(x,y)$ cumulative distribution function
*f* infiltration rate
*f* function symbol
*f* pipe friction factor
$f_c$ equilibrium infiltration rate
$f_c$ ultimate infiltration rate
$f_o$ initial infiltration capacity
$f_o$ initial infiltration rate
$f_p$ percent deficiency in liquid water of snowpack
$f(\tau)$ the ordinate of the IUH at time $\tau$
$f(x)$ probability density function
$f(x/y)$ conditional probability
$f'$ first derivative of function
G groundwater flow
G moisture freely drained by gravity
G daily net allwave radiation absorbed by snow in open
$G_h$ hydraulic gradient
GI growth index
$G(Q)$ annual exceedance probability of flow Q
$G_s$ seepage
$G'$ a complex function
g acceleration of growth
g acceleration of gravity
g acceleration constant of gravity
g skew coefficient
*H* relative humidity
*H* depth of water on soil surface
*H* reservoir depth above spillway
*H* head
*H* weighted sum of flow rates
*H* flow depth

$H_c$ heat transferred from air by convection
$H_e$ latent heat of vaporization derived from condensation
$H_g$ heat conduction from ground
$H_m$ heat equivalent of snowmelt
$H_p$ rainfall heat content
$H_q$ internal energy change in snowpack
$H_{rs}$ absorbed solar radiation
$H_{r1}$ net long-wave radiation exchange between snowpack and surroundings
$H_1$ abbreviation for net flow
$H_2$ abbreviation for net flow
$h$ contour interval
$h$ head
$h$ depth of unconfined aquifer
$I$ inflow
$I$ infiltration
$I$ inflow rate
$I$ rainfall intensity
$I_a$ initial rainfall abstraction
$I_c$ channel inflow
$I_g$ groundwater inflow
$I_i$ ith inflow rate
$I_p$ peak inflow rate
$I_t$ infiltration rate at time t
IUH instantaneous unit hydrograph
$I'$ ratio of inflow rate to time
**$I\sigma$ overland inflow**
$i$ rainfall intensity
$i$ number of periods of rainfall excess
$i$ current
$i$ square root of $-1$
$i$ lateral inflow rate
$i(t-\tau)$ rainfall intensity at time $(t-\tau)$
$i(x,t)$ lateral inflow rate
$j$ the number of unit hydrograph ordinates
$K$ a recession constant
$K$ frequency factors
$K$ hydraulic conductivity
$K$ thermal conductivity of soil
$K$ a constant
$K$ Muskingum storage coefficient
$K$ baseflow recession constant
$K$ permeability
$K$ storage coefficient
$K$ infiltration decay coefficient
$K_f$ field coefficient of permeability
$K_i$ Muskingum crest routing coefficient
$K_j$ frequency factor for ith flow
$K_o$ initial loss rate coefficient
$K_s$ standard coefficient of permeability
$K$ loss rate coefficient
$k'$ constant
$k$ saturated permeability
$k$ specific or intrinsic permeability
$k$ proportionality constant
$k$ recession constant
$k$ Pearson Type III random deviate
$k_o$ von Kármán's coefficient
$k_s$ seasonal consumptive use coefficient
$L$ the latent heat of vaporization
$L$ stream length
$L$ length of main stream channel from the outlet to the divide
$L$ length
$L$ lag time
$L$ length of overland flow
$L$ channel length
$L$ length of gutter
$L_{ea}$ length along the main channel to a point nearest the watershed centroid
$L_i$ volume of water intercepted
$L_i$ interception loss
$L_i$ loss rate during ith increment
$L_t$ instantaneous loss rate
$\bar{L}_u$ mean length of stream segment
LW net long-wave radiation
$\bar{L}$ mean travel distance which is equal to the length of that portion of the sewer which flows full
$l$ total length of grid line segments (horizontal and vertical)
$l$ length to divide in feet
$ln_e$ base $e$ logarithm
$M$ mass
$M$ melt
$M$ drainage area
$M_c$ convection melt
$M_c$ convective snowmelt depth
$M_{ce}$ combined convection-condensation melt
$M_e$ condensation melt
M.F. minimal freeboard
$M_p$ snowmelt due to rainfall
$M_p$ rainfall snowmelt depth
$M_r$ radiation melt
$M_r$ hourly radiation snowmelt depth
$m$ molecular weight
$m$ average moisture content
$m$ mass
$m$ aquifer thickness
$m$ degree of cloudiness
$m$ rate of melt
$m$ stage-storage coefficient
$m$ mass of fluid element
$m$ regression coefficient

$m$ cross-sectional area
$m_i$ drawdown at well i
$m_o$ initial moisture content
$m_t$ total drawdown
$N$ coefficient
$N$ time in days
$N$ number of future trials
$N$ annual precipitation
N.F. normal freeboard
$N_R$ Reynald's number
$N_u$ number of stream segments of order u
$n$ number of gram molecules
$n$ total number of contour intersections by the horizontal and vertical grid lines
$n$ number of storm hydrograph ordinates
$n$ number of years or observations
$n$ porosity
$n$ number of spaces
$n$ regression parameter
$n$ stage-discharge coefficient
$n$ Manning roughness coefficient
$n$ shape parameter
$n$ an empirical constant
$n$ weighted Manning coefficient for the main sewer
$n$ integer number
$O$ outflow
$O$ outflow rate
$O_c$ channel outflow
$O_i$ ith outflow rate
$O_p$ peak outflow rate
$O_s$ net seepage
$P$ precipitation
$P$ gross precipitation
$P$ capillary potential
$P$ probability
$P$ amount of pollutant remaining on ground at time t
$P$ precipitation depth
$P$ accumulated rainfall depth
$P_a$ the antecedent precipitation index
$P_b$ a base precipitation value below which Q is zero
$P_d$ daily rainfall
$P_e$ excess precipitation
$P_g$ percent infiltration that reaches groundwater storage
$P_n$ net precipitation
$P_o$ initial amount of pollutant
$P_p$ piezometric potential
$P_R$ the period of rise
$P_r$ rainfall intensity
$P_r$ rainfall
$P_r$ percent detention retained by upper zone
$P_s$ snowfall
$P_t$ rain intensity
$P_u$ unsaturated pore volume
$PX$ rainfall depth
$P(x/y)$ conditional probability
$\bar{P}$ mean of 24-hr annual maximum rain depths
$P^*$ rainfall prior to runoff
$P_{100}$ 100-year rainfall depth
PMP probable maximum precipitation
$PMP_{24}$ 24-hr probable maximum precipitation depth
$p$ pressure
$p$ atmospheric pressure
$p$ monthly daytime hours
$p$ pore pressure
$p_a$ total pressure of moist air
$p_i$ ith random number
$Q$ outflow
$Q$ streamflow
$Q$ flood magnitudes
$Q$ groundwater flow
$Q$ time rate of flow of a specified air property
$Q$ daily snowmelt runoff
$Q$ discharge
$Q$ equivalent depth of runoff
$Q$ flow from gutter
$Q_a$ incoming long-wave radiation from the atmosphere
$Q_{ar}$ reflected long-wave radiation
$Q_{bs}$ long-wave radiation emitted by the water
$Q_e$ energy used in evaporation
$Q_e$ moisture transfer to snow surface per unit time
$Q_g$ ground and channel storage
$Q_g$ flow rate in gutter
$Q_h$ energy conducted from water mass as sensible heat
$Q_i$ ith flow in sequence
$Q_j$ ith annual flow
$Q_m$ inflow rate to gutter
$Q_o$ increase in stored energy by the water
$Q_o$ upstream gutter inflow
$Q_P$ peak discharge
$Q_P$ peak runoff rate
$Q_p$ peak runoff rate
$Q_r$ reflected solar radiation
$Q_rK$ Horton number

$Q_s$ solar radiation incident at the water surface
$Q_s$ hydrograph ordinate
$Q_{sm}$ change in available soil moisture
$Q_t$ discharge at time t
$Q_t$ snow quality
$Q(t)$ surface runoff rate at time t
$Q_{t/P_R}$ **% flow/0.25 $P_R$ at any given $t/P_R$**
$Q_u$ unit hydrograph ordinate
$Q_v$ net energy advected (net energy content of incoming and outgoing water) into the water body
$Q_w$ energy advected by evaporated water
$Q_w$ overland flow discharge rate
$Q_w$ flow rate from subcatchment
$Q_{10}$ ten-year peak flow rate
$\hat{Q}_{10(ATP)}$ trial value of 10-year flood rate
$Q_{2.33}$ mean annual flood rate
$Q_{50}$ 50-year flood rate
$q$ specific humidity
$q$ specific discharge
$q$ discharge per unit area
$q$ runoff intensity
$q$ shape factor
$q$ flow rate
$q$ runoff rate per unit of area
$q$ overland flow discharge rate
$q_i$ direct runoff hydrograph ordinates
$q_L$ overland flow
$q_o$ initial discharge
$q_o$ flow rate at inflection of hydrograph
$q_p$ peak flow rate
$q_t$ the discharge at any time t after flow $q_o$
$\bar{q}$ mean runoff intensity
$R$ surface runoff
$R$ universal gas constant
$R$ risk
$R$ multiple correlation coefficient
$R$ resistance
$R$ recharge intensity
$R$ return flow
$R$ hydraulic radius
$R$ rainfall intensity
$R$ runoff depth
$R$ baseflow recession constant
$R$ rainfall factor
$R_a$ total radiation
$R_i$ excess rainfall intensity
$R_L$ hydraulic radius at left node
R.O. daily generated snowmelt runoff over snow-covered area
$R_R$ hydraulic radius at right node
$R_t$ rainfall intensity at time t
$\bar{R}$ weighted hydraulic radius of the main sewer flowing full
$r$ radius of circular lake
$r$ radius
$r$ coefficient of correlation
$r$ fraction of solids removed by runoff
$r$ evapotranspiration opportunity
$r$ hydraulic radius
$r_j$ correlation coefficient
$r_x$ correlation coefficient of monthly flows
$r_y$ correlation coefficient of logarithms of monthly flows
S storage
S total porosity
S storage potential of soil
S weighted physical slope of the main sewer
S potential maximum retention
S energy gradient
S overland slope
S groundwater aquifer storage coefficient
S basin retention depth
S lake storage at time t
SA available porosity
$S_c$ storage coefficient
$S_d$ maximum depression storage
$S_d$ total depression storage
$S_f$ amount of water stored in snowpack
$S_f$ friction slope
$S_{fL}$ friction slope at node on the left
$S_{fR}$ friction slope at node on the right
$S_i$ interception storage
SLP slope of channel
$s_n$ standard deviation of annual maximum rain depths
$S_p$ total storage potential
SPS standard project storm
$SR_i$ excess rain rate during ith increment
$S_t$ tranitory storage
$S_y$ specific yield
$S_y$ standard deviation of observed runoff
$S_{yx}$ standard error of estimated runoff
$\bar{S}$ weighted physical slope of the main sewer
$s$ slope of the line (P/Q)
$s$ drawdown
$s$ surface retention rate
$s$ standard deviation
$s^2$ **best estimate of $\sigma^2$**
$T$ transpiration
$T$ absolute temperature

$T$ temperature
$T$ return period
$T$ coefficient of transmissivity
$T$ air temperature
$T$ wind stress
$T$ water surface width
$T$ travel time overland
$T$ peak flow rate
$T$ topographic index
$T_a$ temperature of saturated air
$T_a$ air temperature
$T_b$ time base of the hydrograph
$T_b$ temperature of an arbitrary datum usually taken as 0°C
$T_e$ temperature of evaporated water (°C)
$T_{max}$ daily maximum temperature
$T_o$ duration of excess rain
$T_o$ water surface temperature (°C)
$T_P$ total porosity
$T_P$ time of peak outflow rate
$T_P$ spillway width
$T_R$ top width of reservoir
$T_r$ absolute temperature
$T_r$ average recurrence interval
$T_s$ mean snowpack temperature below zero degrees centigrade
$TTL$ length of reservoir
$T_w$ wet bulb temperature
$\hat{T}_{AP}$ estimate of the topographic index
$t$ temperature
$t$ time
$t$ base time of a synthetic unit hydrograph
$t$ random normal variate
$t_c$ time of concentration
$t_c$ time to raise snowpack temperature to zero degrees
$t_e$ time to equilibrium of runoff rate
$t_f$ time needed to fill storage $S_f$
$t_i$ random normal variate
$t_i$ time of concentration of ith subarea
$t_L$ time coordinate of node on the left
$t_l$ lag time
$t_o$ time of inflection of hydrograph
$t_p$ time from the beginning of rainfall to peak discharge
$t_p$ time to peak
$t_p$ time coordinate of point P
$t_p$ time of peak flow
$t_p$ time to peak runoff rate
$t_p$ time of peak inflow rate
$t_R$ desired unit hydrograph duration
$t_R$ time coordinate of node on the right
$t_r$ original unit hydrograph duration
$t_s$ duration of runoff-producing rain
$t_s$ mean snowpack temperature below zero degrees centigrade
$t_t$ delay time
$t_1$ lag time
$t_{1R}$ adjusted lag time
$U$ consumptive use of water
$U$ wind velocity
$u$ average wind velocity
$u$ velocity in x direction
$u$ dummy variable
$V$ volume
$V$ wind velocity
$V$ volume of water stored
$V$ volume of runoff
$V$ rate of transmission of melt through snowpack
$V$ runoff volume
$V$ aquifer volume
$V$ average velocity
$V_L$ velocity at the node on the left
$V_m$ velocity at the middle node
$V_P$ velocity at point P
$V_R$ velocity at the node on the right
$V_t$ volume of runoff up to time t
$v$ rate of change of depression storage with respect to time
$v$ velocity in y direction
$v$ wind velocity
$W$ precipitable water
$W$ wind velocity
$W$ water equivalent of snowpack
$W$ withdrawal
$W$ width of watershed
$W$ width of overland flow
$W_c$ cold constant
$W_o$ initial water equivalent of snowpack
$W(u)$ well function of u
$w$ velocity in z direction
$X$ random variable throughout its range
$X_r$ random variate
$x$ random variable
$x$ weighting factor
$x$ exponent in weir equation
$x$ increment of DT
$x_L$ space coordinate of node on the left
$x_P$ space coordinate of point P
$x_R$ space coordinate of node on the right
$\bar{x}$ moisture supply rate
$\bar{x}$ first moment about the origin
$Y$ average watershed slope in percent
$Y$ snowmelt runoff
$Y$ length of spillway crest
$Y$ pipe cross-sectional area

$Y$ flow depth
$Y_i$ flow in month i
$Y_L$ depth of flow at the node on the left
$Y_M$ depth of flow at the middle node
$Y_P$ depth of flow at point P
$Y_{P_k}$ kth estimate of peak flow depth
$Y_R$ depth of flow at the node on the right
$Y_t$ water depth in gutter at time t
$y$ depth
$y$ depth of cross section
$y$ random variable
$\bar{y}$ centroidal distance
$Z1L$ berm side slope on left
$Z2L$ channel side slope on left
$Z1R$ berm side slope on right
$Z2R$ channel side slope on right
$z$ elevation
$z$ deviate from the mean
$z_o$ roughness parameter
$\alpha$ angle
$\alpha$ regression constant
$\alpha$ skewness
$\beta$ regression coefficient
$\beta$ fluid compressibility
$\Gamma$ gamma function
$\Gamma(n)$ gamma function of n
$\gamma$ scale parameter
$\zeta_z$ vorticity
$\theta$ angle
$\theta$ angle of flow
$\kappa$ von Kármán's constant
$\kappa$ a coefficient dependent upon wind velocity, atmospheric pressure, and other factors
$\lambda$ proportionality factor
$\lambda$ multiplier
$\mu$ mean value
$\mu$ dynamic viscosity
$\mu'_r$ rth moment about the origin
$\mu'_{r,s}$ joint moment of order r and s
$\mu_x$ mean of monthly flows
$\mu_y$ mean of logarithms of monthly flows
$\rho$ density of water
$\rho$ mass density of air
$\rho$ fluid density
$\rho_s$ density of snowpack
$\rho_w$ vapor density
$\sigma$ overland flow supply rate
$\sigma$ standard deviation
$\sigma$ Stefan's constant
$\sigma_j$ standard deviation
$\sigma_x$ standard deviation of monthly flows
$\sigma_{x,y}$ covariance
$\sigma_y$ standard deviation of logarithms of monthly flows
$\sigma^2$ variance
$\phi$ velocity potential
$\psi$ stream function
$\nabla$ vector operator "del"

# Author Index

Numbers in parentheses indicate note numbers. Italic numbers indicate pages on which complete references occur.

# Subject Index